Klaus Buchholz, Volker Kasche,
and Uwe T. Bornscheuer

Biocatalysts and Enzyme Technology

Second, Completely Revised, and Enlarged Edition

WILEY-BLACKWELL

The Authors

Prof. Dr. Klaus Buchholz
Institut für Technische Chemie
Technologie d. Kohlenhydrate
Hans-Sommer-Strasse 10
38106 Braunschweig
Germany

Prof. Dr. Volker Kasche
Riensberger Str. 104
28359 Bremen
Germany

Prof. Dr. Uwe T. Bornscheuer
University of Greifswald
Institute of Biochemistry
Felix-Hausdorff-Str. 4
17487 Greifswald
Germany

Cover
Penicillin Acylase (Source: PDB, pdb code: 1JX9)
Background: 250 L Setup,
© Rentschler Biotechnologie GmbH

Library of Congress Card No.: applied for

British Library Cataloguing-in-Publication Data
A catalogue record for this book is available from the British Library.

Bibliographic information published by the Deutsche Nationalbibliothek
The Deutsche Nationalbibliothek lists this publication in the Deutsche Nationalbibliografie; detailed bibliographic data are available on the Internet at http://dnb.d-nb.de.

Wiley-Blackwell is an imprint of John Wiley & Sons, formed by the merger of Wiley's global Scientific, Technical, and Medical business with Blackwell Publishing.

Print ISBN: 978-3-527-32989-2
ePDF ISBN: 978-3-527-63292-3
oBook ISBN: 978-3-527-63291-6

Cover Design Grafik-Design Schulz, Fußgönheim

Typesetting Thomson Digital, Noida, India
Printing and Binding Markono Print Media Pte Ltd, Singapore

Printed on acid-free paper

For
Diana, Helene, Melanie, and Peter
Karin, Maria, Anna, Andreas, Magdalena, Johann, and Richard
Tanja and Annika

Contents

Preface to the Second Edition

We have been very pleased by the success of the first English edition of our book. We are especially grateful that it serves as primary source for teaching courses in *Biocatalysis and Enzyme Technology* at many universities around the world. We would also like to thank the readers who pointed out corrections and to those who made useful suggestions for this second edition.

More than 7 years have passed since the first English edition was published and we have observed substantial and exciting developments in all areas of biocatalysis. Hence, we did not simply update the first edition with new references and add singular sentences, but substantially expanded and reorganized the book. For instance, the importance of enzyme discovery and protein engineering is now treated in a separate new section (Chapter 3). Although biocatalysis primarily refers to the use of isolated (and immobilized) enzymes, we decided to cover also the use of designed whole cells for biotransformations in the new Chapter 5 to allow the reader to get a glimpse on the emerging field of metabolic engineering, where the understanding and biochemical characterization of enzymes is of course an important aspect. Furthermore, we have included several new case studies in Chapter 12, to exemplify how biocatalysis can be performed on the large scale and which criteria are important to establish a novel process. These include starch processing and glucose isomerization, biofuels from biomass, and the production of 7-ACA by direct hydrolysis of cephalosporin C as examples for current as well as potential industrial processes. In addition, enzymatic routes for the synthesis of advanced pharmaceutical intermediates for the drug Lipitor are covered.

Furthermore, we have added a few new sections on topics of current interest: process sustainability and ecological considerations, process integration, biofilm catalysis, microbial fuel cells, and regulations that influence the production and use of enzymes.

We have also expanded and updated the exercises and decided that also the solutions be directly given in the book – in Appendix B – so that the students can first try to answer the questions by themselves and then look up the solutions.

Finally, we would like to thank Ulrich Behrendt, Sonja Berensmeier, Matthias Höhne, Zoya Ignatova, Hans-Joachim Jördening, Burghard König, Sven Pedersen,

Ralf Pörtner, Klaus Sauber, and Antje Spiess for valuable discussions, revisions, and suggestions while preparing this book.

June 2012

Klaus Buchholz/Braunschweig
Volker Kasche/Bremen
Uwe T. Bornscheuer/Greifswald

Preface

To the First German Edition

Biotechnology is the technical application of biological systems or parts thereof to provide products and services to meet human needs. It can, besides other techniques, contribute towards doing this in a sustainable manner. Since, in the majority of cases, renewable raw materials and biological systems are used in biotechnological processes, these processes can – and should – be performed practically without waste, as all of the byproducts can be recycled.

The development of natural and engineering science fundamentals for the design of such processes remains a challenge to biotechnology – a field that originated from the overlapping areas of biology, chemistry, and process engineering.

The requisite education for a career in biotechnology consists, in addition to a basic knowledge of each of these fields, of further biotechnological aspects which must provide an overview over the entire field and a deeper insight into different areas of biotechnology. The biotechnological production of various materials is performed either in fermenters using living cells (technical microbiology), or with enzymes – either in an isolated form or contained in cells – as biocatalysts. Indeed, the latter aspect has developed during recent years to form that area of biotechnology known as enzyme technology, or applied biocatalysis.

The aim of the present textbook is to provide a deeper insight into the fundamentals of enzyme technology and applied biocatalysis. It especially stresses the following inter-relationships: A thorough understanding of enzymes as biocatalysts and the integration of knowledge of the natural sciences of biology (especially biochemistry), cell and molecular biology; physico-chemical aspects of catalysis and molecular interactions in solutions; heterogeneous systems and interphase boundaries; and the physics of mass transfer processes. The same applies to the inter-relations between enzyme technology and chemical and process engineering, which are based on the above natural sciences.

In less than a century since the start of industrial enzyme production, enzyme technology and its products have steadily gained increasing importance. In the industrial production of materials to meet the demands of everyday life, enzymes play an important role – and one which is often barely recognized. Their application ranges from the production of processed foods such as bread, cheese, juice and beer, to pharmaceuticals and fine chemicals, to the processing of leather and textiles, as process aids in detergents, and also in environmental engineering.

Meeting the demand for these new products – which increasingly include newly developed and/or sterically pure pharmaceuticals and fine chemicals – has become an important incentive for the further development of biocatalysts and enzyme technology. Of similar importance is the development of new sustainable production processes for existing products, and this is detailed in Chapter 1, which forms an Introduction.

Enzymes as catalysts are of key importance in biotechnology, similar to the role of nucleic acids as carriers of genetic information. Their application as isolated catalysts justifies detailed examination of the fundamentals of enzymes as biocatalysts, and this topic is covered in Chapter 2. Enzymes can also be analyzed on a molecular level, and their kinetics described mathematically. This is essential for an analytical description and the rational design of enzyme processes. Enzymes can also catalyze a reaction in both directions – a property which may be applied in enzyme technology to achieve a reaction end-point both rapidly and with a high product yield. The thermodynamics of the catalyzed reaction must also be considered, as well as the properties of the enzyme. The amount of enzyme required for a given conversion of substrate per unit time must be calculated in order to estimate enzyme costs, and in turn the economic feasibility of a process. Thus, the quantitative treatment of biocatalysis is also highlighted in Chapter 2.

When the enzyme costs are too high, they can be reduced by improving the production of enzymes, and this subject is reviewed in Chapter 3 (*Chapter 4 in the present book*).

In Chapter 4 (*here Chapter* 5), applied biocatalysis with free enzymes is described, together with examples of relevant enzyme processes. When single enzyme use is economically unfavorable, the enzymes can be either reused or used for continuous processes in membrane reactors (Chapter 4; *here Chapter* 5) or by immobilization (Chapters 5 and 6; *here Chapters* 6 *and* 7). The immobilization of isolated enzymes is described in detail in Chapter 5, while the immobilization of microorganisms and cells, with special reference to environmental technology, is detailed in Chapter 6.

In order to describe analytically the processes associated with immobilized biocatalysts that are required for rational process design, the coupling of reaction and diffusion in these systems must be considered. To characterize immobilized biocatalysts, methods which were developed previously for analogous biological and process engineering (heterogeneous catalysis) systems can be used (Chapter 7; *here Chapter* 8).

Details of reactors and process engineering techniques in enzyme technology are provided in Chapter 8 (*here Chapter* 9), while the analytical applications of free and immobilized enzymes is treated in Chapter 10 (*not covered in the present book*).

Within each chapter an introductory survey is provided, together with exercises and references to more general literature and original papers citing or relating the content of that chapter.

This textbook is designed to address both advanced and graduate students in biology, chemistry and biochemical, chemical and process engineering, as well as scientists in industry, research institutes and universities. It should provide a solid foundation that covers all relevant aspects of research and development in applied biocatalysis/enzyme technology. It should be remembered that these topics are not of equal importance in all cases, and therefore selective use of the book – depending on the individual reader's requirements – might be the best approach to its use.

In addition to a balanced methodological basis, we have also tried to present extensive data and examples of new processes, in order to stress the relevance of these in industrial practice.

From our point of view it is also important to stress the interactions, which exist beyond the scientific and engineering context within our society and environment. The importance and necessity of these interactions for a sustainable development has been realized during the past two decades, and this has resulted in new economic and political boundary conditions for scientific and engineering development. Problems such as allergic responses to enzymes in detergents and, more recently, to enzymes produced in recombinant organisms, have direct influences on enzyme technology/applied biocatalysis. Therefore, an integrated process design must also consider its environmental impact, from the supply and efficient use of the raw materials to the minimization and recycling of the byproducts and waste. Political boundary conditions derived from the concept of sustainability, when expressed in laws and other regulations, necessitates due consideration in research and development. The design of sustainable processes is therefore an important challenge for applied biocatalysis/enzyme technology. Ethical aspects must also be considered when gene technology is applied, and this is an increasing consideration in the production of technical and pharmaceutical enzymes. The many interactions between research and development and economic and political boundary conditions must be considered for all applications of natural and engineering sciences. Most importantly, this must be appreciated during the early phases of any development, with subsequent evaluation and selection of the best alternative production processes to meet a variety of human needs, as is illustrated in the following scheme:

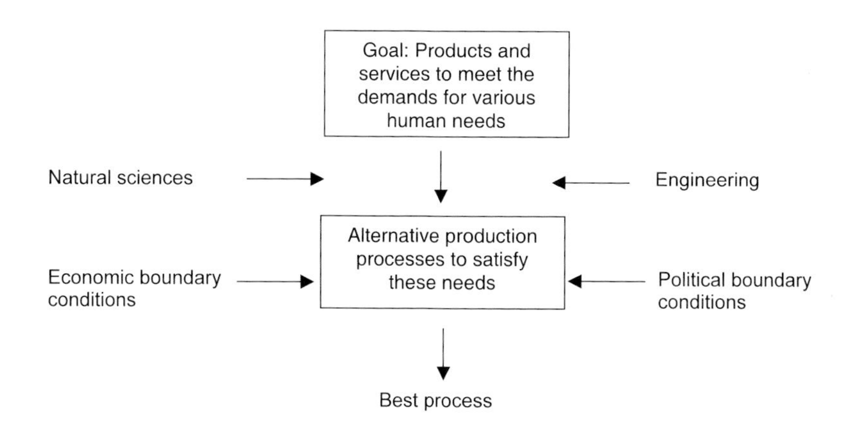

This book has been developed from our lecture notes and materials, and we also thank all those who provided valuable help and recommendations for the book's production. In particular, we thank Dipl.-Ing. Klaus Gollembiewsky, Dr. Lieker, Dr. Noll-Borchers, and Dipl. Chem. André Rieks.

Klaus Buchholz
Volker Kasche

To the First English Edition

The basic philosophy of the previous German edition is retained, but the contents have been revised and updated to account for the considerable development in enzyme technology/applied biocatalysis since the German edition was prepared some 10 years ago. Hence, a new chapter (Chapter 3) has been added to account for the increasing importance of enzymes as biocatalysts in organic chemistry. Recent progress in protein design (by rational means and directed evolution) has been considerably expanded in Chapter 2. The final chapter has been amended with more detailed case studies to illustrate the problems that must be solved in the design of enzyme processes. An appendix on information retrieval using library and internet resources has also been added, and we thank Thomas Hapke (Subject Librarian for Chemical Engineering at the Library of the Technical University Hamburg-Harburg) for help in the preparation of this material. The chapter on enzymes for analytical purposes has been removed in this English edition as it now is beyond the scope of this textbook.

We thank Prof. Dr. L. Jaenicke and Prof. Dr. J.K.P. Weder for their very constructive suggestions for corrections and improvements of the German edition.

The authors of this edition thank Prof. Dr. Andreas Bommarius, Dr. Aurelio Hidalgo, Dr. Janne Kerovuo, Dr. Tanja Kummer, Dr. Dieter Krämer, Dr. Brian Morgan, Sven Pedersen, Poul Poulsen, Prof. Dr. Peter Reilly, Dr. Klaus Sauber, Dr. Wilhelm Tischer, and Dr. David Weiner for valuable discussions, revisions and suggestions while preparing this book.

January 2005

Klaus Buchholz
Volker Kasche
Uwe T. Bornscheuer

1
Introduction to Enzyme Technology

1.1
Introduction

Biotechnology offers an increasing potential for the production of goods to meet various human needs. In enzyme technology – a subfield of biotechnology – new processes have been and are being developed to manufacture both bulk and high added-value products utilizing enzymes as biocatalysts, in order to meet needs such as food (e.g., bread, cheese, beer, vinegar), fine chemicals (e.g., amino acids, vitamins), and pharmaceuticals. Enzymes are also used to provide services, as in washing and environmental processes, or for analytical and diagnostic purposes. The driving force in the development of enzyme technology, both in academia and in industry, has been and will continue to be

- the development of new and better products, processes, and services to meet these needs, and/or
- the improvement of processes to produce existing products from new raw materials such as biomass.

The goal of these approaches is to design innovative products and processes that not only are competitive but also meet criteria of sustainability. The concept of sustainability was introduced by the World Commission on Environment and Development (WCED, 1987) with the aim to promote a necessary ". . . development that meets the needs of the present without compromising the ability of future generations to meet their own needs." This definition is now part of the *Cartagena Protocol on Biosafety to the Convention on Biological Diversity*, an international treaty governing the movements of living modified organisms (LMOs) resulting from modern biotechnology from one country to another. It was adopted on January 29, 2000 as a supplementary agreement to the Convention on Biological Diversity and entered into force on September 11, 2003 (http://bch.cbd.int/protocol/text/). It has now been ratified by 160 states. To determine the sustainability of a process, criteria that evaluate its economic, environmental, and social impact must be used (Gram *et al.*, 2001; Raven, 2002; Clark and Dickson, 2003). A positive effect in all these three fields is required for a sustainable process. Criteria for the quantitative evaluation

Biocatalysts and Enzyme Technology, Second Edition. Klaus Buchholz, Volker Kasche, and Uwe T. Bornscheuer

of the economic and environmental impact are in contrast with the criteria for the social impact, easy to formulate. In order to be economically and environmentally more sustainable than an existing process, a new process must be designed not only to reduce the consumption of resources (e.g., raw materials, energy, air, water), waste production, and environmental impact, but also to increase the recycling of waste per kilogram of product (Heinzle, Biwer, and Cooney, 2006).

1.1.1
What are Biocatalysts?

Biocatalysts either are proteins (*enzymes*) or, in a few cases, may be nucleic acids (*ribozymes*; some RNA molecules can catalyze the hydrolysis of RNA). These ribozymes were detected in the 1980s and will not be dealt with here (Cech, 1993). Today, we know that enzymes are necessary in all living systems, to catalyze all chemical reactions required for their survival and reproduction – rapidly, selectively, and efficiently. Isolated enzymes can also catalyze these reactions. In the case of enzymes, however, the question whether they can also act as catalysts outside living systems had been a point of controversy among biochemists in the beginning of the twentieth century. It was shown at an early stage, however, that enzymes could indeed be used as catalysts outside living cells, and several processes in which they were applied as biocatalysts have been patented (see Section 1.3).

These excellent properties of enzymes are utilized in enzyme technology. For example, they can be used as biocatalysts, either as isolated enzymes or as enzyme systems in living cells, to catalyze chemical reactions on an industrial scale in a sustainable manner. Their application covers the production of desired products for all human material needs (e.g., food, animal feed, pharmaceuticals, bulk and fine chemicals, detergents, fibers for clothing, hygiene, and environmental technology), as well as for a wide range of analytical purposes, especially in diagnostics. In fact, during the past 50 years the rapid increase in our knowledge of enzymes – as well as their biosynthesis and molecular biology – now allows their rational use as biocatalysts in many processes, and in addition their modification and optimization for new synthetic schemes and the solution of analytical problems.

This introductory chapter outlines the technical and economic potential of enzyme technology as part of biotechnology. Briefly, it describes the historical background of enzymes, as well as their advantages and disadvantages, and compares these to alternative production processes. In addition, the current and potential importance and the problems to consider in the rational design of enzyme processes are also outlined.

1.1.2
Bio- and Chemocatalysts – Similarities and Differences

Berzelius, in 1835, conceived the pioneering concept of catalysis, including both chemo- and biocatalysis, by inorganic acids, metals such as platinum, and enzymes

(Berzelius, 1835). It was based on experimental studies on both bio- and chemocatalytic reactions. The biocatalytic system he studied was starch hydrolysis by diastase (a mixture of amylases). In both systems, the catalyst accelerates the reaction, but is not consumed. Thus, bio- and chemocatalysis have phenomenological similarities. The main differences are the sources and characteristics of these catalysts. Chemocatalysts are designed and synthesized by chemists, and are in general low molecular weight substances, metal catalysts, complexes of metals with low molecular weight organic ligands, such as Ziegler-Natta and metallocene catalysts, and organocatalysts (Fonseca and List, 2004). In contrast, biocatalysts are selected by evolution and synthesized in living systems. Furthermore, enzymes (including ribonucleic acid-based biocatalysts) are macromolecules, their highly sophisticated structure being essential for their function, and notably for their regio-, chemo-, and enantioselectivity.

Due to development of gene and recombinant technologies in the past 40 years, enzymes that previously only could be obtained in limited amounts from microorganisms and tissues can now be synthesized in nearly unlimited quantities in suitable microorganisms. Further, based on the development in biochemistry, bioinformatics, and micro- and molecular biology, new tools have been developed to improve the properties of enzymes for their use in biocatalytic processes. They are rational protein design and *in vitro* evolution in combination with high-throughput screening tools. Very recently, also the *de novo* computational design of enzymes was described, but so far these show little activity in the same range as catalytic antibodies (Jiang *et al.*, 2008; Röthlisberger *et al.*, 2008).

Until the first oil crisis of 1973, the development and application of bio- and chemocatalysis occurred in – at that time – nonoverlapping fields. Biocatalysis was mainly studied by biochemists, biochemical engineers, microbiologists, physiologists, and some physical organic chemists (Jencks, 1969). It was mainly applied in the food, fine chemical, and pharmaceutical industries and medicine (see Section 1.3). Chemocatalysis was mainly studied by chemical engineers and chemists. It was applied in the production of bulk chemicals such as acids and bases, and products derived from coal and oil (fuel, plastics, etc.). This resulted for a long time in a small exchange of fundamental results between those who studied and applied bio- and chemocatalysis. The analytical description of heterogeneous catalysis, where the catalyst is located only in a part of the system, was first developed and verified experimentally for living systems in the 1920s by biochemists. Contrary to homogeneous catalysis, this description involves the coupling of the reaction with mass transfer. This applies also for heterogeneous chemocatalysis. The same description as for living systems was derived independently by chemical engineers in the end of the 1930s (see Chapter 10).

The detailed mechanism of the catalyzed reactions has now been determined for many bio- and chemocatalysts. This knowledge that is continuously increasing yields information that can be used to design improved bio- and chemocatalysts. This, however, requires a closer cooperation of those working with these catalysts. Fortunately, due to the increasing use of enzymes by organic chemists in the past decades, this cooperation has increased markedly.

1.2 Goals and Potential of Biotechnological Production Processes

Biomass – that is, renewable raw materials – has been and will continue to be a sustainable resource that is required to meet a variety of human material needs. In developed countries such as Germany, biomass covers ≈30% of the raw material need – equivalent to ~7000 kg per person per year. The consumption of biomass for different human demands is shown schematically in Figure 1.1. This distribution of the consumption is representative for a developed country in the regions that have a high energy consumption during the winter. However, the consumption of energy (expressed as tons of oil equivalent per capita in 2007) showed a wide range, from 8 in the United States to 4 in Germany and the United Kingdom, 1.5 in China, and 0.5 in India (IEA, 2010).

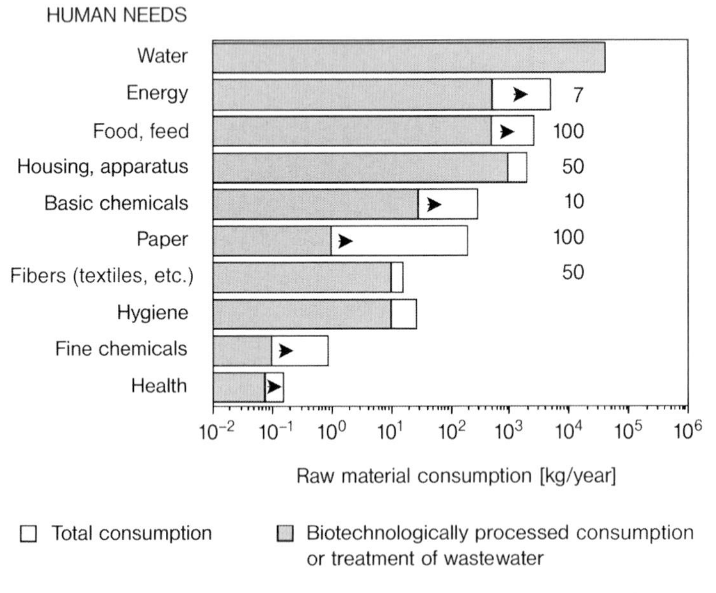

Figure 1.1 Consumption of raw materials for various human needs per person and year in Germany 1992. The water consumption is only for household use. These numbers are still valid. The energy consumption per capita has hardly changed since then. However, now (2010) 11% is derived from renewable resources (biomass, solar, water, wind) (AGEB, 2010). The arrowheads indicate the current increase in biotechnological processing of the products for different demands. For food and animal feed, only renewable raw materials (biomass) can be used; the figures to the right give the percentage for biomass of the raw materials currently used for the production. They can, especially for energy, only increase when they do not interfere with the biomass demand for food and feed. Due to the low material demands for hygiene, fine chemicals, and health products, 0–100% of the raw materials can be biomass, depending on the product. After the use of the products, the unavoidable waste must be recycled in a sustainable manner. Besides wastewater, this results in about 1000 kg of solid waste per year (soil, building materials, plastics, sludge, etc.). Energy is measured in coal equivalents.

This is mainly due to differences in energy use for housing, transport, and the production of other material needs. In less-developed countries, although the fraction of biomass as raw material to meet human demands is higher than that in the developed countries, the total consumption is smaller.

Biomass – in contrast to nonrenewable raw materials such as metals, coal, and oil – is renewable in a sustainable manner when the following criteria are fulfilled:

- the C, N, O, and salt cycles in the biosphere are conserved, and
- the conditions for a sustainable biomass production through photosynthesis and biological turnover of biomass in soil and aqueous systems are conserved (Beringer, Lucht, and Schaphoff, 2011).

Currently, these criteria are not fulfilled on a global level, one example being the imbalance between the CO_2 production to meet energy requirements and its consumption by photosynthesis in the presently decreasing areas of rain forests. This leads to global warming and other consequences that further violate these criteria. International treaties – for example, the Kyoto Convention and the Convention on Biological Diversity – have been introduced in an attempt to counteract these developments and to reach a goal that fulfills the above criteria (see Section 1.1).

Only when the above sustainability criteria are fulfilled, biomass can be used as raw material to meet the human demands illustrated in Figure 1.1. The needs for human food and animal feed must be met completely by biomass, though when these needs of highest priority are met, biomass can be used to fulfill the other demands shown in Figure 1.1. This applies especially to those areas with lower total raw material consumption than for food. From this point, it also follows that a large consumption of biomass to meet energy demands is only possible in countries with a low population density and a high biomass production.

By definition, biotechnological processes are especially suited to the production of compounds from biomass as the raw material (Figure 1.2). The amount produced in, and economic importance of, such processes is detailed in Table 1.1.

This also involves the development of suitable concepts, methods, and equipment to obtain more sustainable processes. From the information provided in

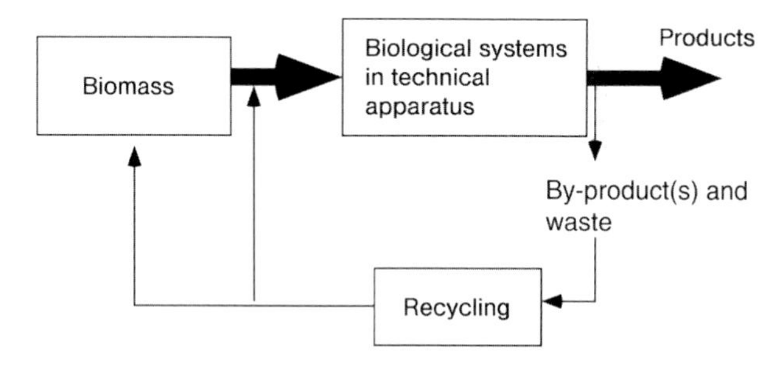

Figure 1.2 Schematic view of an ideal sustainable biotechnological production process. Biomass as a regenerable resource is converted into desired products with minimal waste and by-product production. The waste and by-products must be completely recycled.

Table 1.1 Yearly production and value of biotechnologically produced products to meet human needs.

Human need	Product (year)	World production (t $year^{-1}$)	Value ($\times 10^9$ euros $year^{-1}$)	Production method	
				Biotechnological	Chemical
Food and feed	Beer/wine (2009)	195 000 000 (a)	≈300	F, E	
	Cheese (2009)	19 400 000 (a)	≈100	F, E	
	Baker's yeast (1992)	1 800 000	?	F	
	Vegetable oils (partly used for biodiesel) (2009)	107 000 000 (a)	?	E	+
	Vinegar (10% acetic acid)	>1 500 000*		F	
Fine chemicals including feed and food additives	Amino acids (2006)	3 000 000 (c)	3 (c)	F, E	+
	Glucose–fructose syrup	12 000 000 (g)	5	E	
	Vitamin C (2006)	80 000 (c)	1 (c)	F, E	+
	Aspartame (dipeptide) (2006)	15 000 (c)	?	F, E	+
	Citric acid (2006)	1 500 000 (c)	1.2 (c)	F	
	Herbicides, insecticides	>2 200 000*	?	E	+
	Enzymes (2010)	>10 000	2.6 (d)	F	
Basic chemicals	Products from biomass				
	Biodiesel from vegetable oils (2007)	9 000 000 m^3 (e)			
	Bioethanol (2007)	50 000 000 m^3 (e)		E, F	
	Acrylamide	>150 000*	?	E	+
	1,3-Propanediol	>100 000*		F	
	Polylactic acid	140 000 (g)		F	+
	Acetic acid (2003)	3 400 000 (b)			+

Fibers for textiles	Cotton (2008)	25 000 000 (a)	30 (a)	E	+
	Wool (2008)	2 000 000 (a)	3.5 (a)	E	+
	Linen (2008)	600 000 (a)	0.3 (a)	F, E	+
Paper	All forms (2009)	370 000 000 (a) (50% recycled)		E	+
Hygienics/detergents	Biotensides	?	?	E, F	+
	Washing powder (2003)	24 000 000 (b)		F	+
Therapeuticals	Antibiotics	>60 000*	≈35 (h)	F, E	+
	Insulin (2010)	>10	10 (f)	F, E	
	Recombinant proteins (factor VIII, interferons, tPA, hormones, growth factors, etc.) (2008)	?	50 (g)	F, E	
	Monoclonal antibodies (2008)	?	20 (g)	F	
Diagnostics (estimated figures)	Monoclonal antibodies	?	>1	F, E	+
	DNA/protein chips	?	?	F, E	+
	Enzyme-based	<1	>1	F, E	+
Environment	Clean water (only Germany)	13 000 000 000	13	F	+
	Clean air/soil (soil remediation)	?	?	F	+
Comparison					
Chemical industry	All products (2009)	>1 000 000 000*	≈1900 (i)		
	Chemical catalysts (2009)	?	12 (j)		

All production data are from 2003 to 2010; values are only given where sources give the present data or when they can be estimated, based on prices in the European Union (EU). F = fermentation; E = enzyme technology.

Sources: (a) FAOSTAT (http://faostat.fao.org/site/339/default.aspx); (b) UN (2003); (c) Soetaert and Vandamme (2010); (d) Novozymes (2011); (e) FAO (2008); (f) Novo Nordisk (2011); (g) Buchholz and Collins (2010); (h) Hamad (2010); (i) CEFIC (2010); (j) Bryant (2010).
* Estimated as newer data are not available for open access.

Figure 1.1, it also follows that biotechnology has a major potential in the development of sustainable processes to meet all human needs.

Enzyme technology is a part of biotechnology that is defined in the internationally accepted Cartagena Convention (see Section 1.1) as follows:

"Biotechnology means any technological application that uses biological systems, living organisms, or derivatives thereof, to make or modify products or processes for specific use."

with the following amendment:

"The use must be sustainable, this means the use of components of biological diversity in a way and at a rate that does not lead to long-term decline of biological diversity, thereby maintaining its potential to meet the needs and aspirations of present and future generations."

This requires that traditional classical – as well as new biotechnological – processes must be improved and/or developed in order to be sustainable (Figure 1.2). The fundamentals needed for the development of such processes in the interdisciplinary field of biotechnology require the close cooperation of biologists, chemists, and biochemical and chemical engineers.

1.3 Historical Highlights of Enzyme Technology/Applied Biocatalysis

1.3.1 Early Developments

Applied biocatalysis has its roots in the ancient manufacture and preservation of food and alcoholic drinks, as can be seen in old Egyptian pictures. Cheese making has always involved the use of enzymes, and as far back as about 400 BC, Homer's Iliad mentions the use of a kid's stomach for making cheese.

With the development of modern natural science during the eighteenth and nineteenth centuries, applied biocatalysis began to develop a more scientific basis. In 1833, Payen and Persoz investigated the action of extracts of germinating barley in the hydrolysis of starch to yield dextrin and sugar, and went on to formulate some basic principles of enzyme action (Payen and Persoz, 1833):

- small amounts of the preparation were able to liquefy large amounts of starch,
- the material was thermolabile, and
- the active substance could be precipitated from aqueous solution by alcohol, and thus be concentrated and purified. This active substance was called *diastase* (a mixture of amylases).

In 1835, the hydrolysis of starch by diastase was acknowledged as a catalytic reaction by Berzelius. In 1839, he also interpreted fermentation as being caused by a catalytic force, and postulated that a body – by its mere presence – could, by affinity to the fermentable substance, cause its rearrangement to the products (Hoffmann-Ostenhof, 1954).

The application of diastase was a major issue from the 1830s onwards, and the enzyme was used to produce dextrin that was used mainly in France in bakeries, and also in the production of beer and wines from fruits. The process was described in more detail, including its applications and economic calculations, by Payen (1874) (Figure 1.3). Indeed, it was demonstrated that the use of the enzymes in malt (amylases, amyloglucosidases) in this hydrolytic process was more economic than that of sulfuric acid.

Lab preparations were also used to produce cheese (Knapp, 1847), and Berzelius later reported that 1 part of lab ferment preparation coagulated 1800 parts of milk, and that only 0.06 parts of the ferment was lost. This provided further evidence for Berzelius' hypothesis that ferments were indeed catalysts.

About two decades later, the distinction of organized and unorganized ferments was proposed (Wagner, 1857), and further developed by Payen (1874). These investigators noted that fermentation appeared to be a contact (catalytic) process of a degradation or addition process (with water), and could be carried out by two substances or bodies:

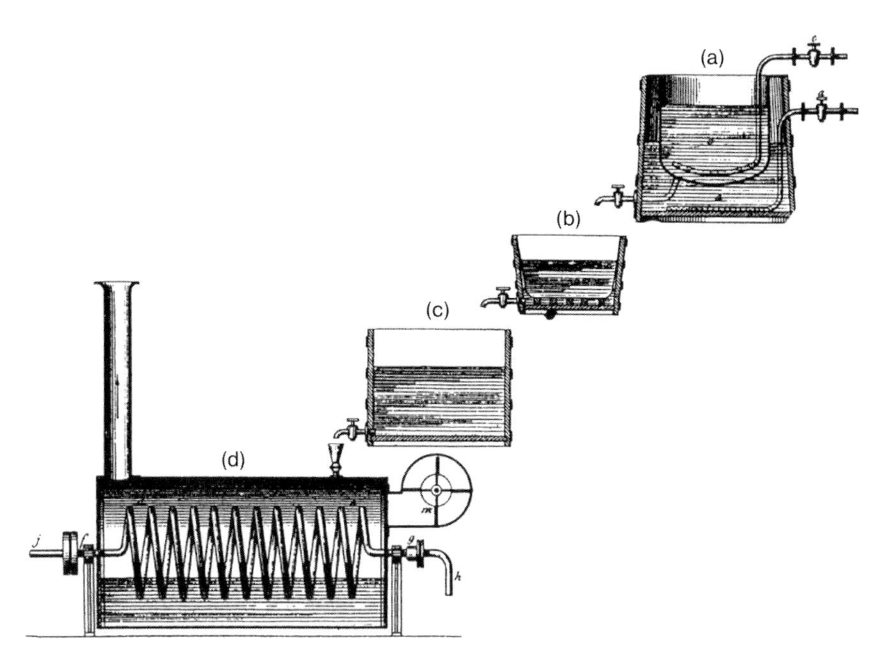

Figure 1.3 Process for dextrin production, with reaction vessel (a), filter (b), reservoir (c), and concentration unit (d).

- A nitrogen-containing organic (unorganized) substance, such as protein material undergoing degradation.
- A living (organized) body, a lower class plant, or an "infusorium," an example being the production of alcohol by fermentation.

It is likely that the effect is the same, insofar as the ferment of the organized class produces a body of the unorganized class – and perhaps a large number of singular ferments. Consequently, in 1878, Kühne named the latter class of substances *enzymes*.

Progress in the knowledge of soluble ferments (enzymes) remained slow until the 1890s, mainly due to a scientific discussion where leading scientists such as Pasteur denied the existence of "unorganized soluble ferments" that had no chemical identity. Consequently, the subject of enzymatic catalysis remained obscure, and was considered only to be associated with processes in living systems. In the theory of fermentation, a degree of mystery still played a role: Some *vital factor, "le principe vital,"* which differed from chemical forces, was considered to be an important principle in the chemical processes associated with the synthesis of materials isolated from living matter. But Liebig and his school took an opposite view, and considered fermentation simply to be a decay process.

In 1874, in Copenhagen, Denmark, Christian Hansen started the first company (Christian Hansen's Laboratory) for the marketing of standardized enzyme preparations, namely, rennet for cheese making (Buchholz and Poulson, 2000).

1.3.2
Scientific Progress Since 1890: The Biochemical Paradigm; Growing Success in Application

From about 1894 onward, Emil Fischer elaborated the essential aspects of enzyme catalysis. The first aspect was *specificity*, and in a series of experiments Fischer investigated the action of different enzymes using several glycosides and oligosaccharides. For this investigation, he compared invertin and emulsin. He extracted invertin from yeast – a normal procedure – and showed that it hydrolyzed the α-, but not the β-methyl-D-glucoside. In contrast, emulsin – a commercial preparation from Merck – hydrolyzed the β-, but not the α-methyl-D-glucoside. Fischer therefore deduced the famous picture of a "lock and key," which he considered a precondition for the potential of an enzyme to have a chemical effect on the substrate. In this way, he assumed that the "geometrical form of the (enzyme) molecule concerning its asymmetry corresponds to that of the natural hexoses" (sugars) (Fischer, 1909).

The second aspect referred to the protein nature of enzymes. In 1894, Fischer stated that among the agents that serve the living cell, the proteins are the most important. He was convinced that enzymes were proteins, but it took more than 20 years until the chemical nature of enzymes was acknowledged. Indeed, Willstätter, as late as 1927, still denied that enzymes were proteins (Fruton, 1976). A few years after Fischer's initial investigations, Eduard Buchner published a series of papers (Buchner, 1897, 1898) that signaled a breakthrough in fermentation and enzymology. In his first paper on alcoholic fermentation without yeast cells, he stated, in a remarkably short and precise manner, that "... a separation of the (alcoholic) fermentation from the living yeast cells was not successful up to now." In subsequent reports, he described a process that solved this problem (Buchner, 1897), and provided experimental details for the preparation of a cell-free pressed juice from yeast cells that transformed sugar into alcohol and carbon dioxide. Buchner presented the proof that (alcoholic) fermentation did not require the presence of "... such a complex apparatus as is the yeast cell." The agent was in fact a soluble substance – without doubt a protein body – that he called *zymase* (Buchner, 1897). In referring to the deep controversy on his findings and theory, and in contradiction to the ideas of Pasteur (see above), Buchner insisted that his new experimental findings could not be disproved by older theories.

After a prolonged initial period of about a hundred years, during which time a number of alternative and mysterious theories were proposed, Buchner's elaborate results brought about a new biochemical paradigm. It stated – in strict contrast to the theories of Pasteur – that enzyme catalysis, including complex phenomena such as alcoholic fermentation, was a chemical process not necessarily linked to the presence and action of living cells, nor requiring a vital force – a *vis vitalis*. With this, the technical development of enzymatic processes was provided with a new, scientific basis on which to proceed in a rational manner.

The activity in scientific research on enzymes increased significantly due to this new guidance, and was reflected in a pronounced increase in the number of papers published on the subject of soluble ferments from the mid-1880s onward (Buchholz and Poulson, 2000). Further important findings followed within a somewhat short time. In 1898, Croft-Hill performed the first enzymatic synthesis – of isomaltose – by allowing a yeast extract (α-glycosidase) to act on a 40% glucose solution (Sumner and Somers, 1953). In 1900, Kastle and Loevenhart showed that the hydrolysis of fat and other esters by lipases was a reversible reaction, and that enzymatic synthesis could occur in a dilute mixture of alcohol and acid. Subsequently, this principle was utilized in the synthesis of numerous glycosides by Fischer and coworkers in 1902, and by Bourquelot and coworkers in 1913 (Wallenfels and Diekmann, 1966). In 1897, Bertrand observed that certain enzymes required dialyzable substances to exert catalytic activity, and these he termed *coenzymes*.

The final proof that enzymes were in fact proteins was the crystallization of urease by Sumner in 1926, and of further enzymes (e.g., trypsin) by Northrup and Kunitz in 1930–1931. In all known cases, the pure enzyme crystals turned out to be proteins (Sumner, 1933).

Despite these advances, the number of new applications of enzymes remained very small. In the United States, J. Takamine began isolating bacterial amylases in the 1890s, in what was later to become known as Miles Laboratories. In 1895, Boidin discovered a new process for the manufacture of alcohol, termed the "amyloprocess." This comprised cooking of the cereals, inoculation with a mold that formed saccharifying enzymes, and subsequent fermentation with yeast (Uhlig, 1998). The early applications and patents on enzyme applications in the food industry (which numbered about 10 until 1911) have been reviewed by Neidleman (1991).

At the beginning of the twentieth century, plant lipases were produced and utilized for the production of fatty acids from oils and fats, typically in the scale of 10 tons per week (Ullmann, 1914). Likewise, in the chill proofing of beer, proteolytic enzymes have been used successfully since 1911 in the United States (Tauber, 1949). Lintner, as early as 1890, noted that wheat diastase interacts in dough making, and studied the effect extensively. As a result, the addition of malt extract came into practice, and in 1922 American bakers used 30 million pounds (13.5×10^6 kg) of malt extract valued at US$ 2.5 million (Tauber, 1949).

The use of isolated enzymes in the manufacture of leather played a major role in their industrial-scale production. For the preparation of hides and skins for tanning, the early tanners kept the dehaired skins in a warm suspension of the dungs of dogs and birds. In 1898, Wood was the first to show that the bating action of the dung was caused by the enzymes (pepsin, trypsin, lipase) that it contained. In the context of Wood's investigations, the first commercial bate, called Erodine, was prepared from cultures of *Bacillus erodiens*, based on a German patent granted to Popp and Becker in 1896. In order to produce Erodine, bacterial cultures were adsorbed onto wood meal and mixed with ammonium chloride (Tauber, 1949).

In 1907, Röhm patented the application of a mixture of pancreatic extract and ammonium salts as a bating agent (Tauber, 1949). Röhm's motivation as a chemist was to find an alternative to the unpleasant bating practices using dungs. Although the first tests with solutions of only ammonia failed, Röhm was aware of Buchner's studies on enzymes. He came to assume that enzymes might be the active principle in the dung, and so began to seek sources of enzymes that were technically feasible. His tests with pancreatic extract were successful, and on this basis in 1907 he founded his company, which successfully entered the market and expanded rapidly. In 1908, the company sold 10 tons of a product with the trade name Oropon, followed by 53 and 150 tons in the subsequent years. In 1913, the company (Figure 1.4) was employing 22 chemists, 30 other employees, and 48 workers (Trommsdorf, 1976). The US-based company – later the Rohm and Haas Company (now

Figure 1.4 The factory of the Rohm and Haas Company, Darmstadt, Germany, 1911.

subsidiary of Dow) – was founded in 1911. The example of Röhm's company (now a part of Evonik, Germany) illustrates that although the market for this new product was an important factor, knowledge of the principles of enzyme action was equally important in providing an economically and technically feasible solution.

This success of the enzymatic bating process was followed by new applications of pancreatic proteases, including substitution therapy in maldigestion, desizing of textile fibers, wound treatment, and the removal of protein clots in large-scale washing procedures. During the 1920s, however, when insulin was discovered in pancreas, the pancreatic tissue became used as a source of insulin in the treatment of diabetes. Consequently, other enzyme sources were needed in order to provide existing enzymatic processes with the necessary biocatalysts, and the search successfully turned to microorganisms such as bacteria, fungi, and yeasts.

1.3.3
Developments Since 1950

Between 1950 and 1970, a combination of new scientific and technical knowledge, market demands for enzymes for use in washing processes, starch processing, and cheaper raw materials for sweeteners and optically pure amino acids stimulated the further development of enzyme technology. As a result, an increasing number of enzymes that could be used for enzyme processes were found, purified, and characterized. Among these were penicillin amidase (or acylase), used for the hydrolysis of penicillin and first identified in 1950, followed some years later by glucose isomerase, which is used to isomerize glucose to the sweeter molecule, fructose. With the new techniques of enzyme immobilization, enzymes could be reused and their costs in enzyme processes reduced. Although sucrose obtained from sugarcane or sugar beet was the main sweetener used, an alternative raw material was starch, which is produced in large quantities from corn (mainly in the United States). Starch can be hydrolyzed to glucose, but on a weight basis glucose is less sweet than sucrose or its hydrolysis products, glucose and fructose. As glucose isomerase can isomerize glucose to fructose, starch became an alternative sweetener source. The process was patented in 1960, but it lasted almost 15 years until the enzyme process to convert starch to glucose–fructose syrups became industrialized. This was in part due to an increase in sucrose prices and due to the introduction of immobilized glucose isomerase as a biocatalyst. It is interesting to note that scientists working in this field discussed the political and social consequences for the main sugar-producing countries, before this process was introduced on an industrial scale (Wingard, 1974). Although European engineers and scientists had contributed strongly to the development of this process, it is applied only minimally in Europe (~1% of world production) due to the protection of sucrose production from sugar beet.

Enzyme immobilization was first introduced to enable the reuse of costly enzymes. Some of the initial attempts to do this were described during the early parts of the past century (Hedin, 1915), but the enzymes when adsorbed to charcoal proved to be very unstable. Around the time of 1950, several groups began to immobilize enzymes on other supports (Michel and Evers, 1947; Grubhofer and Schleith, 1954; Manecke 1955, cited in Silman and Katchalski, 1966). Georg Manecke was one of the first to succeed in making relatively stable immobilized systems of proteins on polymer carriers, and although he was granted a patent on his method he could not convince industry of the importance to further develop this invention. Rather, it was a group of chemists working with Ephraim Katchalski-Katzir in Israel who opened the eyes of industry to the world of immobilized enzymes (among Katchalski-Katzir's coworkers were Klaus Mosbach and Malcolm Lilly who later made important contributions to establish enzyme technology). The first industrial applications of immobilized enzymes, besides the isomerization of glucose to fructose to produce high-fructose corn syrups (HFCS), were in the production of optically pure amino acids (Tosa *et al.*, 1969) and the hydrolysis of penicillin G (Carleysmith and Lilly, 1979, together with Beecham Pharmaceuticals (now Glaxo SmithKline, UK), and G. Schmidt-Kastner (Bayer, Germany), in cooperation with the penicillin producer Gist Brocades (now DSM, The Netherlands)).

Even today, the largest immobilized enzyme product in terms of volume is immobilized glucose isomerase. As these products were introduced, they became more efficient, and stable biocatalysts were developed that were cheaper and easy to use. As a result, the productivity of commercial immobilized glucose isomerase increased from ~500 kg HFCS kg^{-1} immobilized enzyme product (in 1975) to ~15 000 kg HFCS kg^{-1} immobilized enzyme product (in 1997) (Buchholz and Poulson, 2000).

During the 1960s, enzyme production gained speed only in modest proportions, as reflected by the growing sales of bacterial amylases and proteases. Indeed, the annual turnover of the enzyme division of Novo Industri (now Novozymes), the leading enzyme manufacturer at the time, did not exceed $1 million until 1965. However, with the appearance of the detergent proteases, the use of enzymes increased dramatically, and during the late 1960s, everybody wanted Biotex, the protease-containing detergent. At the same time, an acid/enzyme process to produce dextrose using glucoamylase was used increasingly in starch processing. As a consequence, by 1969 – within only a 4-year period – Novo's enzyme turnover exceeded US$ 50 million annually, and in 2009 Novozymes' turnover was approximately US$ 1400 million. The present global market is estimated to be around 2.5 billion € (Novozymes, 2011) (Figure 1.5a), and this has been reflected in the increased employment within the enzyme-producing industry (Figure 1.5b). The main industrial enzyme processes with free or immobilized enzymes as biocatalysts are listed in Table 1.2.

The introduction of gene technology during the 1970s provided a strong impetus for both improved and cheaper biocatalysts, and also widened the scope of application. Productivity by recombinant microorganisms was dramatically improved, as was enzyme stability, and this led to a considerable lowering of prices and improvements in the economics of enzyme applications. Today, most of the enzymes used as biocatalysts in enzyme processes – except for food processing – are recombinant. The recent development of techniques of site-directed mutagenesis, gene shuffling, and directed evolution has opened the perspective of modifying the selectivity and specificity of enzymes (see Section 2.11 and Chapter 3).

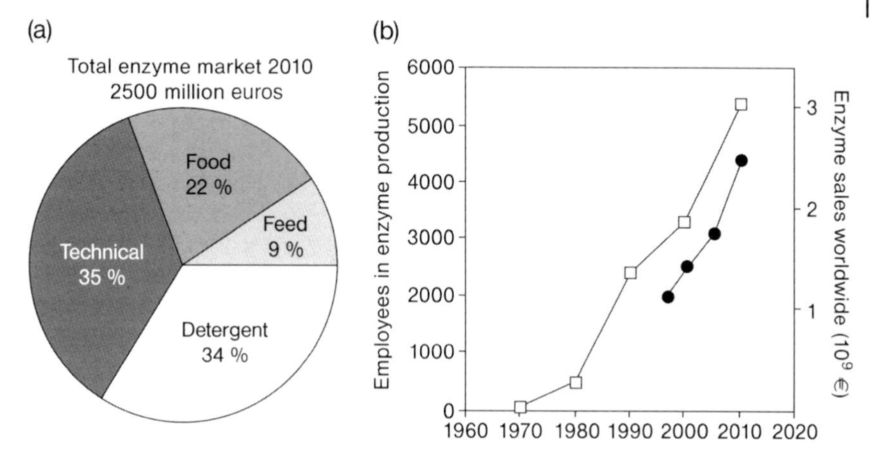

Figure 1.5 (a) Market for enzymes used as biocatalysts for different purposes (2010), and (b) the increase in the application of enzymes reflected in the number of employees in the industry producing enzymes for biocatalytic purposes and their worldwide sales since 1970. Number of Novozymes employees that has about 50% of the world market for such enzymes (squares), and value of their worldwide sales (filled circles) (Novozymes yearly reports, last one from 2010). The value of the world production of technical enzymes is much larger than that shown in (a), as many companies that use enzymes as biocatalysts produce them in-house in order to have a safe and stable enzyme supply and/or protect their proprietary knowledge.

More detailed accounts on the scientific and technological development can be found in articles by Sumner and Somers (1953), Ullmann (1914), Tauber (1949), Neidleman (1991), Roberts *et al.* (1995), Buchholz and Poulson (2000), and Buchholz and Collins (2010). A profound analysis of the background of biotechnology and "Zymotechnica" has been presented by Bud (1992, 1993).

1.4 Biotechnological Processes: The Use of Isolated or Intracellular Enzymes as Biocatalysts

Biotechnological processes use one or more enzymes with or without cofactors or cosubstrates as biocatalysts (Figure 1.6). When more enzymes and cosubstrate regeneration (ATP, NADH) are required, fermentation processes with living cells are more effective than processes with isolated enzymes. In environmental biotechnology, mixed living cultures are mainly used as biocatalysts (see Chapter 9). Now cells can be designed where enzymes have been deleted that catalyze side reactions, avoiding or reducing by-product formation, and/or enzymes that catalyze consecutive reactions that reduce the product yield (Figure 1.6, I) (Keasling, 2010). Uses of such designed cells to produce desired products in fermentations will be covered in Chapter 5. For enzyme processes, which utilize few enzymes ($\leq$3) without any cosubstrate regeneration, those with isolated enzymes or enzymes in

Table 1.2 Products produced in quantities larger than 1000 t $year^{-1}$ by different companies with enzymes or cells as biocatalysts.

Product	Enzyme/designed cells	Free or immobilized isolated enzyme or cell	Companies
>10 000 000 t a^{-1}			
HFCS	Amylase	Free	Several
	Glucoamylase	Free	
	Glucose isomerase	Immobilized	
Ethanol (gasoline additive)	Amylase	Free	Several
	Glucoamylase	Free	
	Hemicellulases	Free	
>100 000 t a^{-1}			
Acrylamide	Nitrilase in resting cells	Immobilized resting cells	Nitto, DSM
Cocoa butter	Lipase	Immobilized	Fuji Oil, Unilever
Isomaltulose	Sucrose mutase in dead cells	Immobilized dead cells	Südzucker, Cerestar
Polylactic acid	Designed cells		Cargill-Dow Polymers LLC
1,3-Propanediol	Designed cells		DuPont
Transesterification of fats and oils	Lipase	Immobilized	Several
>10 000 t a^{-1}			
6-Aminopenicillanic acid	Penicillin amidase	Immobilized	Several
Lactose-free milk or whey	β-Galactosidase lactase	Free or immobilized	Several
>1000 t a^{-1}			
Acarbose	Selected high-yield cell strain after mutagenesis	Immobilized living cells	Bayer
7-Aminocephalos-poranic acid (7-ACA)	Engineered cephalosporin C amidase	Immobilized	Novartis
7-Aminodesace-toxycephalos-poranic acid (7-ADCA)	Engineered glutaryl amidase	Immobilized	DSM
(*S*)-Aspartic acid	Aspartase	Immobilized (?)	Tanabe
Aspartame	Thermolysin	Immobilized	Toso, DSM
(*S*)-Methoxyiso-propylamine	Lipase	Immobilized	BASF
(*R*)-Pantothenic acid	Aldolactonase		Fuji Chemical Industry Co., Ltd
(*R*)-Phenylglycine	Hydantoinase, carbamoylase	Immobilized	Several
(*S*)-Amino acids	Aminoacylase	Free	Degussa, Tanabe

1000 to >10 t a^{-1}			
Amoxicillin	Penicillin amidase	Immobilized	DSM
Cephalexin	Penicillin amidase	Immobilized	DSM
(*S*)-DOPA	β-Tyrosinase	Immobilized	Ajinomoto
Insulin	Carboxypeptidase B	Free	Novo Nordisk
	Lysyl endopeptidase	Free	Sanofi
	Trypsin	Free	Eli Lilly
Sterically pure alcohols and amines	Lipase	Immobilized	BASF
(*R*)-Mandelic acid	Nitrilase	Immobilized	BASF

Some other products produced in the range 10–1000 t year^{-1} in recently developed enzyme processes are also included.

(either) dead (or living) cells have the following advantages compared with fermentations:

- Higher space–time yields can be obtained than with living cells; smaller reactors can then be used, reducing processing costs.
- The risk that a desired product is converted by other enzymes in the cells can be reduced.
- The increased stability and reuse of immobilized biocatalysts allows continuous processing for up to several months.

For such enzyme processes:

1) The required intra- or extracellular enzyme must be produced in sufficient quantities and purity (free from other disturbing enzymes and other compounds).
2) Cells without intracellular enzymes that may disturb the enzyme process must be selected or designed.
3) The enzyme costs must be less than 5–10% of the total product value.

Tables 1.1 and 1.2 provide information about important enzyme products and the processes in which they are produced. In many cases, enzyme and chemical processes are combined to obtain these products. This will also be the case in the future when both processes fulfill the economic and sustainability criteria listed above. When this is not the case, new processes must be developed that better fulfill these criteria. This is illustrated by the first two large-scale enzyme processes, namely, the hydrolysis of penicillin (Figure 1.7) and the production of glucose–fructose syrup from starch (Figure 1.8). Both processes were developed some 30 years ago, and the first replaced a purely chemical process that was economically unfavorable, and less sustainable, compared with the enzyme process (Tischer, 1990; Heinzle, Biwer, and Cooney, 2006).

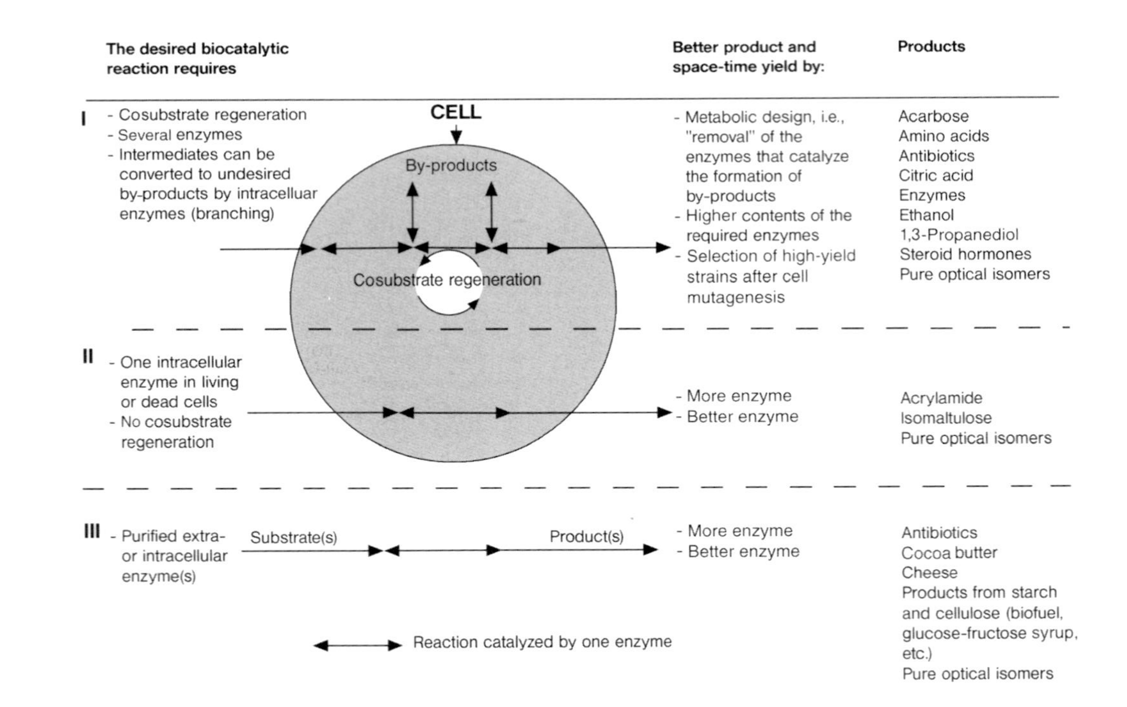

Figure 1.6 Classification of biocatalytic processes with enzymes as biocatalysts. I must be performed with enzymes in living cells, II can be performed with enzymes in living or dead cells, and III with isolated enzymes. For I now mainly designed cells are used, and for II high-yield cells for one-step reactions that do not require cosubstrate regeneration are used. Processes I–III will be covered in this book.

1.5
Advantages and Disadvantages of Enzyme-Based Production Processes

In Figures 1.7 and 1.8, the enzyme processes for the hydrolysis of penicillin and the production of glucose–fructose syrup are compared with previously used procedures that had the same aims. In the case of penicillin hydrolysis, the chemical process uses environmentally problematic solvents and toxic compounds, leading to toxic wastes that are difficult to recycle. The process is, therefore, not sustainable. The enzyme process is more sustainable than the previous process, and leads to a

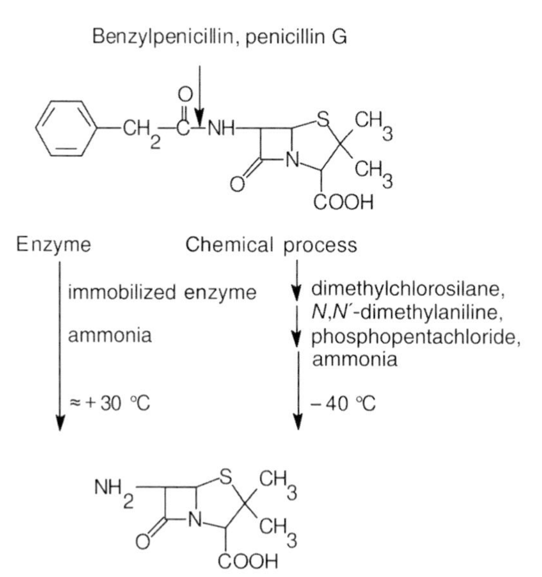

Required for the production of 500 t 6-APA

For hydrolysis:

Enzyme process		Chemical process	
1000 t	penicillin G	1000 t	penicillin G
45 t	ammonia	300 t	dimethylchlorosilane
≈ 1 t	immobilized enzyme	800 t	*N,N*-dimethylaniline
10000 m³	water	600 t	phosphopentachloride
		160 t	ammonia
		4200 m³	dichloromethane
		4200 m³	*n*-butanol

For downstream processing:

Enzyme process	Chemical process
acetone	hydrochloric acid
ammonium bicarbonate	butyl acetate
	acetone

Figure 1.7 Comparison of the old (chemical) and the new (enzyme) process for the hydrolysis of penicillin G. The product, 6-aminopenicillanic acid (6-APA), is used for the synthesis of semisynthetic penicillins with side chains other than phenylacetic acid. In the enzyme process, the by-product phenylacetic acid can be recycled in the production of penicillin by fermentation (from Tischer, 1990).

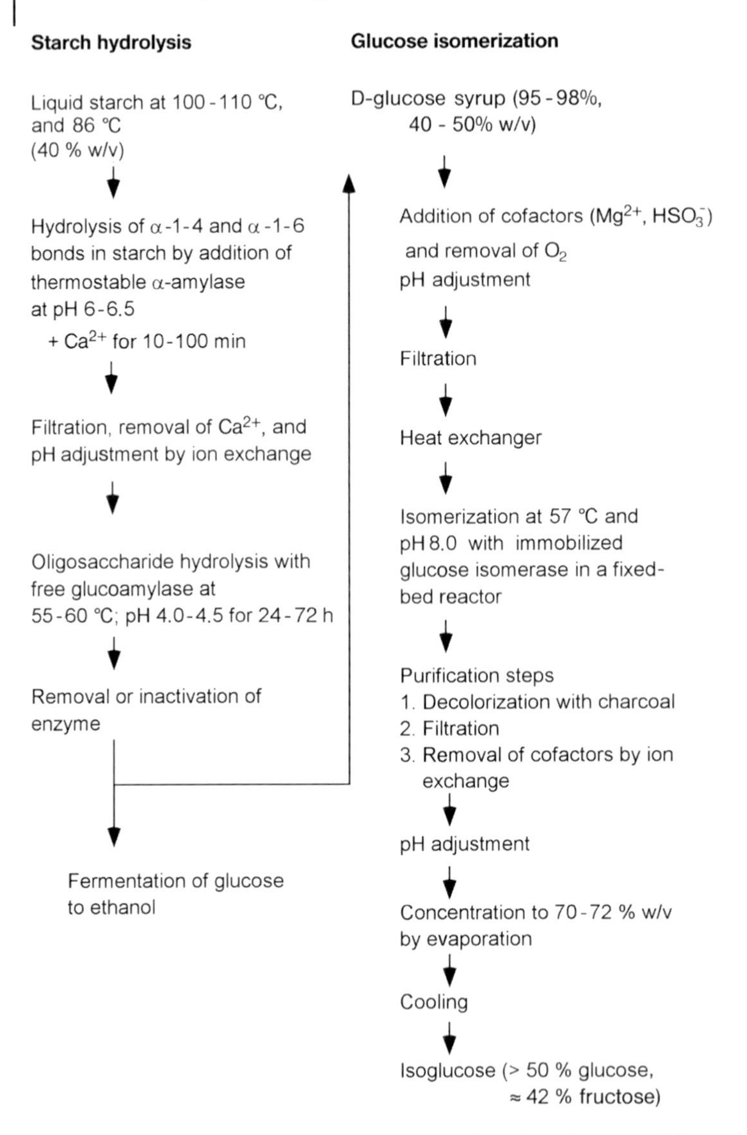

Figure 1.8 The enzyme process for the hydrolysis of starch to glucose, and the isomerization of glucose to fructose (w/v = weight per volume) (see Chapters 7, 8, and 12).

considerable reduction in waste that in turn reduces the processing costs (Tischer, 1990; Heinzle, Biwer, and Cooney, 2006). In this process, the product yield could also be increased to >95%. The hydrolysis of starch and isomerization of glucose cannot be performed chemically at reasonable cost, as each would result in lower yields, unwanted by-products, and the considerable production of waste acids. These processes illustrate some of the advantages of enzyme processes compared with alternative processes (Box 1.1). However, it must be remembered that the use

Box 1.1: Advantages and disadvantages of cells and enzymes as biocatalysts in comparison with chemical catalysts.

Advantages	• Stereo- and regioselective • Low temperatures (0–110 °C) required • Low energy consumption • Active at pH 2–12 • Less by-products • Nontoxic when correctly used • Can be reused (immobilized) • Can be degraded biologically • Can be produced in unlimited quantities
Disadvantages	• Cells and enzymes are – unstable at high temperatures – unstable at extreme pH values – unstable in aggressive solvents – inhibited by some metal ions – hydrolyzed by peptidases • Some enzymes – are still very expensive – require expensive cosubstrates • When inhaled or ingested, enzymes are, as all foreign proteins, potential allergens

of enzymes as biocatalysts may be limited by their biological and chemical properties (see Chapter 2).

Enzymes are proteins that are essential for living systems and, in the right place, they catalyze all chemical conversions required for the system's survival and reproduction. However, in the wrong place, they can be harmful to an organism. Peptidases from the pancreas are normally transported into the intestine where they are necessary for the digestion of proteins to amino acids. The amino acids are transported into blood vessels and distributed to different cells, where they are used for the synthesis of new proteins. Under shock situations or pancreatic insufficiencies, these peptidases may be transported from the pancreas directly into the bloodstream where they may cause harmful blood clotting. To prevent this from occurring, the blood contains inhibitors for pancreatic peptidases.

Enzymes are normal constituents of food and, as with all orally ingested proteins, they are hydrolyzed in the stomach and intestine. However, if enzymes or other proteins are inhaled as small particles or aerosols in the lungs, they can be transferred directly into the bloodstream. There, they are recognized as foreign proteins and induce an immune reaction – that is, the production of antibodies against them. This may also lead to enzyme or protein allergies. These risks must be considered in the production and use of enzymes and other proteins, and simple measures can be taken to minimize them. The enzyme-producing companies neglected this, when they introduced enzymes in detergents for household use around 1970. The number of allergy cases increased rapidly among the employees and users, due to the inhalation of the small enzyme particles. Fortunately, the companies reacted rapidly and covered the enzymes used in washing powders in a drying process with a wax layer, yielding particles so large (>100 μm) that they cannot be inhaled into the lung. This reduced the number of new allergy cases rapidly. Likewise, when enzymes are used in the liquid phase, aerosol formation must be prevented.

Some proteins (enzymes) can also be transferred from the digestive tract into the bloodstream and cause allergies. This applies to proteins that are digested very slowly in the stomach and intestine (Fuchs and Astwood, 1996; Jank and Haslberger, 2003). The slow digestion has been correlated with a high thermal stability, and enzymes used on an industrial scale as biocatalysts should therefore be rapidly hydrolyzed by peptidases in the digestive system in order to minimize the allergy risk. This applies especially to enzymes that cannot easily be used in closed systems, and particularly those used in food processing. Regulations that control the use of enzymes as biocatalysts are given in Chapter 6.

1.6 Goals and Essential System Properties for New or Improved Enzyme Processes

1.6.1 Goals

The advantages detailed in Box 1.1 are not sufficient alone for the industrial use of enzyme processes. Sustainability goals, derived from the criteria outlined in Section 1.1, must also be considered (Table 1.3).

Enzyme processes have become competitive and have been introduced into industry when they attain these goals better than alternative processes. This, however, also requires that these goals be quantified such that the amount of product and by-products (or waste) produced with a given amount of enzyme in a given time must be determined. For this aim, enzyme processes – as with all catalyzed chemical processes – can be divided into two categories (Figure 1.9):

1) **Equilibrium-controlled processes:** the desired product concentration or property has a maximum at the end point of the process (B in Figure 1.9); the chemical equilibrium is independent of the properties of the catalyst (enzyme), but is dependent on pH and temperature.
2) **Kinetically controlled processes:** the desired product concentration or property (such as fiber length or smoothness in textiles or paper) reaches a maximum (A in Figure 1.9), the concentration or properties of which depend on the properties of the catalyst (enzyme, see Chapter 2), pH, and temperature. The process must be stopped when the maximum is reached.

Table 1.3 Economic and environmental sustainability goals that can be realized in enzyme processes (modified from Uhlig, 1998).

Goals	Means to achieve the goals	Products/processes
Cost reduction	Yield increase	Penicillin and cephalosporin C hydrolysis
	Biocatalyst reuse and increased productivity by immobilization	Glucose isomerization
	Better utilization of the raw material	Isomaltulose production
		Juice and wine production
	Reduction of process costs for	
	• Filtration	Sterile filtration of plant extracts
	• Energy	Low-temperature washing powder
	• Desizing of fibers	Desizing with enzymes
	• Cheese ripening	Increase rate of process with enzymes
	• Malting in beer production	
	Reduction of residence time in starch processing	
Improvement of biological properties and quality	Produce only isomers with the desired biological property	Racemate resolution
	Improved preservation of foods	Juice concentrates
	Improvement of technical properties	Protein modification, flour for baking, transesterification of vegetable oils, biodiesel
	Improved taste (sweetness)	Glucose isomerization to glucose–fructose syrup
Utilization of new regenerable sources of raw materials	Utilization of wastes from food and paper production (such as whey, filter cakes from vegetable oil production, waste water)	Drinks from whey Animal feed Biofuels (Biodiesel, Biogas Ethanol)
Reduction of environmental impact	Reduction of nonrecyclable waste	Penicillin and cephalosporin C hydrolysis, leather production, paper bleaching
	Waste recycling	Utilization of whey

In both cases, the time to reach the maximum product concentration or property depends on the properties and amount of enzyme used, and on the catalyzed process (endo- or exothermal, pH and temperature dependence of equilibrium constants, solubility and stability of substrates, products, etc.). This must be considered in the rational design of enzyme processes. Another difference to consider is that in these processes the enzymes are used at substrate concentrations

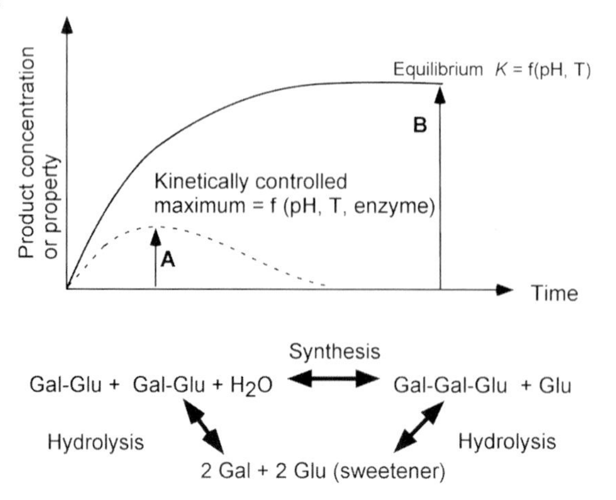

Figure 1.9 Time dependence (progress curves) of equilibrium- (solid line) and kinetically (broken line) controlled processes catalyzed by enzymes. The suitable end points of these processes are those where the maximum product concentration or property is achieved – that is, A for the kinetically and B for the equilibrium-controlled process. Such processes are illustrated with the hydrolysis of lactose in milk or whey. Whey is an inevitable by-product in cheese production, where mainly protein and fats are precipitated in the milk by addition of a "coagulating" enzyme (chymosin or rennin, a carboxyl acid peptidase, EC 3.4.23.4). The remaining liquid phase (whey) contains ~5% sugars (mainly lactose, a disaccharide galactosylglucose), 1% protein, 1% amino acids, and 1% ions (Ca^{2+}, Na^{+}, phosphate ions, etc.). Previously, with mainly small dairies, whey was used as a feed, or condensed to various sweet local products. Today, in large cheese-producing dairies, up to 10^7 tons of whey is formed each year that cannot be used as before. The main content lactose cannot be used as a sweetener as a part of the population cannot tolerate lactose (this fraction is higher in parts of Asia and Africa). When it is hydrolyzed, its sweetness is increased. The enzyme β-glucosidase that catalyzes the hydrolysis of lactose also catalyzes the kinetically controlled synthesis of tri- and tetrasaccharides. When lactose is consumed, these oligosaccharides are hydrolyzed. This also illustrates the formation of undesired by-products in enzyme-catalyzed processes. The oligosaccharides are by-products in the equilibrium-controlled hydrolysis of lactose. On the other hand, in the kinetically controlled synthesis of the oligosaccharide that can be used as prebiotics, by-product formation is due to the hydrolysis to monosaccharides (Illanes, Wilson, and Raiman, 1999; Bruins *et al.*, 2003). The formation of the by-products must be minimized by selecting suitable process conditions and biocatalysts.

(up to 1 M) that are much higher than those in living systems (≤0.01 M). At the substrate and product concentrations used in enzyme technology, the formation of undesired by-products in catalyzed and uncatalyzed bimolecular reactions cannot be neglected (see the next section and Case Study 3 in Chapter 12).

1.6.2
Essential System Properties for Rational Design of an Enzyme Process

The steps to be considered for the design of enzyme processes that are within the scope of this book may be illustrated based on the equilibrium-controlled hydrolysis of the substrate lactose to the products glucose and galactose, as shown in Figure 1.9. This process was developed to reduce the lactose content in milk products so that those who suffer from lactose intolerance can consume them. It can also be used to increase the sweetness of products derived from whey (Illanes, Wilson, and Raiman, 1999). The steps are summarized in Figure 1.10.

A high substrate content is favorable in order to reduce downstream processing costs. In milk, the lactose content cannot be changed, but in whey it can be increased by nanofiltration. The upper limit is given by the solubility (150–200 g l^{-1}), which is lower than that for other disaccharides such as sucrose. As both substrates and products have no basic or acidic functional groups, the equilibrium constant should not depend on pH, but on the temperature. This dependence must be known in order to select a suitable process temperature (T), though the selection also depends on the properties of the biocatalyst. Its selectivity (ratio of hydrolysis to synthesis rates) must be high in order to minimize the formation of by-products (oligosaccharides) in a kinetically controlled process. In addition, its catalytic properties and stability as a function of pH and temperature must also be known in order to calculate the amount of biocatalyst required to reach the end point of the process within a given time.

When other constraints have been identified, a process window in a pH–T-plane can be found where it can be carried out with optimal yield and minimal biocatalyst costs. The maximal yield of an equilibrium-controlled process as a function of pH and T is only defined by the catalyzed reaction. When this maximum is outside the optimal process window of the enzyme, the process can be improved as follows: either by screening for a better biocatalyst or by changing the properties of the enzyme by protein engineering, so that the process can be carried out at pH and T-values where this maximum can be reached (see Section 2.11 and Chapter 3).

In order to reduce the enzyme costs, the enzyme production can be improved (see Chapter 6) or the enzymes used in a reusable form. This can be achieved by their immobilization to porous particles that can easily be filtered off at the end of the process (see Chapters 8 and 11). In these systems, the kinetics differ from those of systems with free enzymes, as the mass transfer inside and to and from the particles with the biocatalyst causes the formation of concentration and pH gradients that influence rates and yields (this topic is dealt with in Chapter 10).

Once the process conditions and its end point have been chosen in the process window, the enzyme costs per kilogram of product are influenced by the type of reactor (batch, continuous stirred tank, or fixed-bed reactor) selected to carry out the process (see Chapter 11). The procedure to design an enzyme process (summarized in Figure 1.10) will be illustrated in more detail as case studies for the design of classical (HFCS production) and newer (7-ACA production by direct hydrolysis of cephalosporin C biofuel production from biomass, and lipitor side chain synthesis) enzyme processes (see Chapter 12).

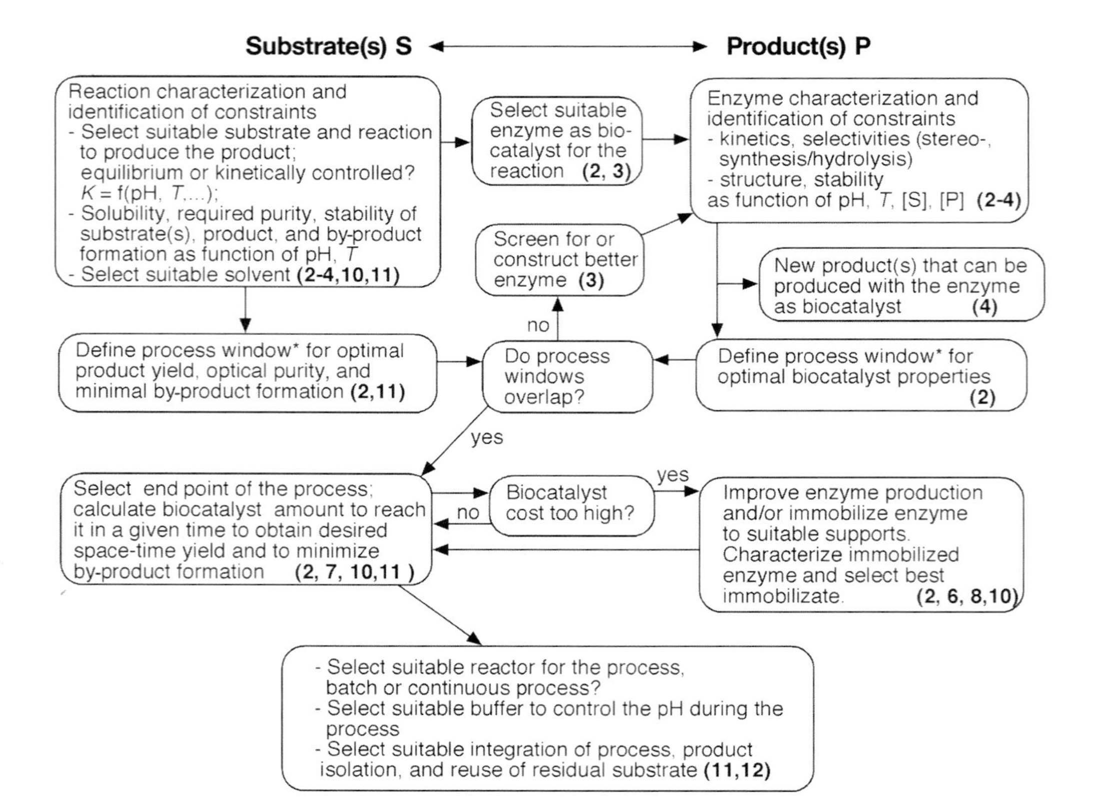

Figure 1.10 Steps to be considered in the design of an enzyme process with isolated enzymes or an enzyme in dead cells (processes II and III in Figure 1.6) to produce existing or new products (bold numbers refer to chapters in this book). *Process window = the range in a pH–*T*- (or pH–[S]-, pH–[P]-, *T*–[S]-, *T*–[P]-) plane where the reaction can be carried out with a given yield or optical purity, and where the properties (activity, selectivity, stability) of the biocatalyst are optimal. For multienzyme processes with living cells (process I in Figure 1.6), the optimal process design requires metabolic engineering, and other boundary conditions than those above must be considered. With optimal metabolic engineering, almost quantitative conversion of the substrate(s) to the desired product can be obtained. Due to the many reactions involved, the time course of the reactions cannot be described as quantitatively as the processes above. The design of these processes will be covered in Chapter 5.

1.6.3
Current Use and Potential of Enzyme Technology

Current amounts of products obtained with industrial biotechnical and enzyme processes are detailed in Tables 1.1 and 1.2, respectively. The values of the enzymes used as biocatalysts for different applications are shown in Figure 1.5. Besides industrial applications, many enzymes are used for analytical purposes, mainly in diagnostics, though on a weight basis less than 1‰ of all produced enzymes are used for these applications. Some enzymes are produced in increasing amounts for therapeutic purposes; this applies especially to recombinant enzymes such as factor VIII, tPA, and urokinase that cannot be produced in sufficient amounts from natural sources (blood serum or urine) (Buchholz and Collins, 2010, Section 7.5.6). Another advantage of the recombinant production of these enzymes is that possible contamination with pathogenic human viruses (HIV, herpes) can be avoided.

A large number of new enzyme processes (>100) introduced during the past 30 years have recently been reviewed in detail (Liese, Seelbach, and Wandrey, 2006). The type of process used, the compounds produced, and the enzymes used for these processes have been analyzed statistically (Straathof, Panke, and Schmid, 2002). These data show that hydrolases, lyases, and oxidoreductases are used in two-thirds of all processes, while only about 1% of the about 4000 known enzymes are used in larger amounts for enzyme technological and therapeutic purposes. During the past 30 years, the three-dimensional structures and detailed mechanisms of the reactions that they catalyze have been determined for many of the enzymes considered to be important in enzyme technology. This information allows a more rational improvement of their properties that is essential for their application. Based on the above discussion and on the information shown in Figure 1.5b, the number of new enzyme processes is expected to increase further during the next few decades. The rational and sustainable design of these processes – and the improvement of existing processes – requires the interdisciplinary cooperation of (bio)chemists, micro- and molecular biologists, and (bio)chemical engineers. The (bio)chemist must determine the mechanism and properties of the catalyzed process, the kinetics of the enzyme-catalyzed process, and other relevant properties of substrate, product, and free and immobilized enzymes (stability, solubility, pH and temperature dependence of equilibrium constants, selectivities), and select the suitable support for the immobilization together with the engineer. This also provides information about the properties of the enzyme that should be improved (specificity, selectivity, pH optimum, stability, metal ion requirement, fermentation yield), and this is a task for the micro- or molecular biologist. The method by which this problem may be solved is either to screen for better enzymes in nature or to promote molecular *in vitro* evolution (see Section 2.11 and Chapter 3). Finally, the engineer must use this knowledge to scale up the process to the production scale. In improving the latter procedure, however, the engineer will also identify problems that must be solved by the (bio)chemists and micro- and molecular biologists, and this is illustrated in Figure 1.6. The number of processing steps can be reduced when the enzymes used can be applied at the same pH value and have the same requirements for metal ions, but this problem has not yet been sufficiently solved. For an exception, see glucose isomerization in Section 8.4.1.

Fields where large amounts of enzymes will be required in order to realize more sustainable new enzyme processes to meet human needs include the following:

- The production of optically pure fine chemicals. It is expected that, in future, only the isomer with desirable biological activity will be approved for use by regulatory authorities. Many pharmaceuticals and fine chemicals are still provided only as racemates, the resolution of which for any process has a maximal yield of 50%. For a sustainable process, the other 50% must be racemized, and to solve this problem the rational integration of chemical and enzyme processes is required in the development of dynamic kinetic resolution processes or asymmetric synthesis (Breuer *et al.*, 2004; Bornscheuer *et al.*, 2012).
- The synthesis of antibiotics (Bruggink, 2001).
- The synthesis of pharmaceutical intermediates (building blocks) (Pollard and Woodley, 2006).
- Paper production or recycling to reduce waste and energy consumption (Bajpai, 1999; Schäfer *et al.*, 2007; Aehle, 2007, Section 5.2.6).
- The regio- and stereoselective synthesis of oligosaccharides for food and pharmaceutical purposes (Seibel and Buchholz, 2010).
- The selective glycosylation of peptides, proteins, and other drugs (Pratt and Bertozzi, 2005; Wong, 2005).
- In the modification of lipids, fats, and oils (*trans*-fatty acid-free fats/oils, diglyceride oils, phospholipid reduction) (Biermann *et al.*, 2011).
- Environmental biotechnology (Jördening and Winter, 2005).
- The synthesis of biofuels from biomass (see Case Study 2 in Chapter 12).
- For the production of bulk products from biomass in biorefineries (Busch, Hirth, and Liese, 2006; Kamm and Kamm, 2007).

The latter two will become of increasing importance as a shift from limited fossil to renewable bio-based raw materials is required in the future.

Exercises

1.1 How was it shown that enzymes can act as catalysts outside living cells? In which enzyme process was this knowledge first applied?

1.2 How can the process in Figure 1.8 be improved by a reduction in the number of processing steps? What must be done to achieve this? (*Hint*: See Case Study 1 in Chapter 12.)

1.3 Explain the relevance of Figure 1.10 for enzyme technology. Which system properties must be known in addition to the properties of the biocatalyst to improve the yields of these processes?

1.4 Which properties of the enzyme and the catalyzed process must be known to minimize by-product formation in the production of oligosaccharides from lactose, as shown in Figure 1.9? (*Hint*: Use Figure 1.10 to answer this question.)
1.5 Test whether Figure 1.1 is in agreement with your consumption pattern.
1.6 How can the allergic and toxic risks due to enzymes be avoided in enzyme technology?

Literature

Overview of Enzyme Technology

The following books give an overview of enzyme technology from the point of view of the biotechnological and chemical industry (enzyme producers and users). Besides the established and new applications of free and immobilized enzymes, some also cover health, legal, and economic aspects of enzyme technology.

Aehle, W. (ed.) (2007) *Enzymes in Industry*, 3rd edn, Wiley-VCH Verlag GmbH, Weinheim.
Atkinson, B. and Mavituna, F. (1991) *Biochemical Engineering and Biotechnology Handbook*, 2nd edn, Stockton Press, New York.
Bommarius, A.S. and Riebel, B. (2004) *Biocatalysis*, Wiley-VCH Verlag GmbH, Weinheim.
Kirst, H.A., Yeh, W-.K., and Zmijewski, M.J., Jr. (eds) (2001) *Enzyme Technologies for Pharmaceutical and Biotechnological Applications*, Marcel Dekker, New York.
Liese, A., Seelbach, K., and Wandrey, C. (eds) (2006) *Industrial Biotransformations*, 2nd edn, Wiley-VCH Verlag GmbH, Weinheim.
Soetaert, W. and Vandamme, E. (eds) (2010) *Industrial Biotechnology*, Wiley-VCH Verlag GmbH, Weinheim.
Uhlig, H. (1998) *Industrial Enzymes and Their Applications*, John Wiley & Sons, Inc., New York.
Whitaker, J.R., Voragen, A.G.J., and Wong, D.W.S. (eds) (2003) *Handbook of Food Enzymology*, Marcel Dekker, New York.

General Books and Encyclopedias on Bio- and Chemocatalysis

Anonymous (2009) *Ullmann's Encyclopedia of Industrial Chemistry*, Wiley-VCH Verlag GmbH, Weinheim.
Jencks, W.P. (1969) *Catalysis in Chemistry and Biochemistry*, McGraw-Hill, New York.

Historical Development

These articles and books cover the historical development of biotechnology and enzyme technology.

Buchholz, K. and Collins, J. (2010) Chapters 1–4, in *Concepts in Biotechnology – History, Science and Business*, Wiley-VCH Verlag GmbH, Weinheim.
Bud, R. (1992) The zymotechnic roots of biotechnology. *Br. J. Hist. Sci.*, **25**, 127–144.
Bud, R. (1993) *The Uses of Life: A History of Biotechnology*, Cambridge University Press.
McLaren, A.D. and Packer, L. (1970) Some aspects of enzyme reactions in heterogeneous systems. *Adv. Enzymol.*, **33**, 245–303.
Mosbach, K. (ed.) (1976) *Immobilized Enzymes, Methods in Enzymology*, vol. **44**, Academic Press, New York.
Mosbach, K. (ed.) (1987) *Immobilized Enzymes and Cells, Methods in Enzymology*, vols. **135–137**, Academic Press, New York.
Silman, I.M. and Katchalski, E. (1966) Water insoluble derivatives of enzymes, antigens and antibodies. *Annu. Rev. Biochem.*, **35**, 873–908.

International Treaties that Influence the Application of Biotechnology and Enzyme Technology

Cartagena Protocol on Biosafety, http://bch.cbd.int/protocol/text/.

References

AGEB (2010) Energieverbrauch in Deutschland, Daten für das 1. Quartal, www.ag-energiebilanzen.de.

Bajpai, P. (1999) Applications of enzymes in the pulp and paper industry. *Biotechnol. Prog.*, **15**, 147–157.

Beringer, T., Lucht, W., and Schaphoff, S. (2011) Bioenergy production potential of global biomass plantations under environmental and agricultural constraints. *GBC Bioenergy.* doi: 101111/j.1757-1707.2010.01088x.

Berzelius, J.J. (1835) Einige Ideen über eine bei der Bildung organischer Verbindungen in der lebenden Natur wirksame aber bisher nicht bemerkte Kraft. *Jahresber. Fortschr. Chem.*, **15**, 237–245.

Biermann, U., Bornscheuer, U.T., Meier, M.A.R., Metzger, J.O., and Schäfer, H.J. (2011) New developments for the chemical utilization of oils and fats as renewable raw materials. *Angew. Chem., Int. Ed.*, **50**, 3854–3871.

Bornscheuer, U.T., Huisman, G.W, Kazlauskas, R.J., Lutz, S., Moore J.C., and Robins K. (2012) Engineering the third wave of biocatalysis. *Nature*, **485**, 185–194.

Breuer, M., Ditrich, K., Habicher, T., Hauer, B., Keßeler, M., Stürmer, R., and Zelinski, T. (2004) Industrial methods for the production of optically active intermediates. *Angew. Chem., Int. Ed.*, **43**, 788–824.

Bruggink, A. (ed.) (2001) *Synthesis of β-Lactam Antibiotics*, Kluwer Academic Publishers, Dordrecht.

Bruins, M.E., Strubel, M., van Lieshout, J.F.T., Janssen, A.E.M., and Boom, R.M. (2003) Oligosaccharide synthesis by the hyperthermostable β-glucosidase from *Pyrococcus furiosus*: kinetics and modelling. *Enzyme Microb. Technol.*, **33**, 3–11.

Bryant, R. (2010) Europe's place within the global fine chemical industry. *Chim. Oggi*, **28**, 14–16.

Buchholz, K. and Poulson, P.B. (2000) Overview of history of applied biocatalysis, in *Applied Biocatalysis* (eds A.J.J. Straathof and P. Adlercreutz), Harwood Academic Publishers, Amsterdam, pp. 1–15.

Buchner, E. (1897) Alkoholische Gährung ohne Hefezellen. *Ber. Dtsch. Chem. Ges.*, **30**, 117–124.

Buchner, E. (1898) Über zellfreie Gährung. *Ber. Dtsch. Chem. Ges.*, **31**, 568–574.

Busch, R., Hirth, T., and Liese, A. (2006) The utilization of renewable resources in German industrial production. *Biotechnol. J.*, **1**, 770–776.

Carleysmith, S.W. and Lilly, M.D. (1979) Deacylation of benzylpenicillin by immobilised penicillin acylase in a continuous four-stage stirred-tank reactor. *Biotechnol. Bioeng.*, **21**, 1057–1073.

Cech, T.R. (1993) Catalytic RNA: structure and mechanism. *Biochem. Soc. Trans.*, **21**, 229–234.

CEFIC (European Chemical Industry Council) (2010) *The European Chemical Industry in a Worldwide Perspective*, CEFIC, Brussels.

Clark, W.C. and Dickson, N.M. (2003) Sustainable science: the emerging research program. *Proc. Natl. Acad. Sci. USA*, **100**, 8059–8061.

FAO (2008) *Biofuels: Prospects, Risks and Opportunities*, FAO, Rome.

Fischer, E. (1909) *Untersuchungen über Kohlenhydrate und Fermente*, Springer, Berlin.

Fonseca, M.H. and List, B. (2004) Combinatorial chemistry and high-throughput screening for the discovery of organocatalysts. *Curr. Opin. Chem. Biol.*, **8**, 319–326.

Fruton, J.S. (1976) The emergence of biochemistry. *Science*, **192**, 327–334.

Fuchs, R.L. and Astwood, J. (1996) Allergenicity assessment of foods derived from genetically modified plants. *Food Technol.*, 83–88.

Gram, A., Treffenfeldt, W., Lange, U., McIntyre, T., and Wolf, O. (2001) *The Application of Biotechnology to Industrial Sustainability*, OECD Publications Service, Paris.

Hamad, B. (2010) The antibiotics market. *Nat. Rev. Drug Discov.*, **9**, 675–676.

Hedin, S.G. (1915) CT Kap. 4, in *Grundzüge der Physikalischen Chemie in ihrer Beziehung zur Biologie*, J.F. Bergmann Verlag, Wiesbaden.

Heinzle, E., Biwer, A., and Cooney, C. (2006) *Development of Sustainable Bioprocesses*, John Wiley & Sons, Inc., New York, pp. 105–107.

Hoffmann-Ostenhof, O. (1954) *Enzymologie*, Springer, Wien.

IEA (International Energy Agency) (2010) *Key World Energy Statistics 2009*, IEA, Paris, www.iea.org.

Illanes, A., Wilson, L., and Raiman, L. (1999) Design of immobilized enzyme reactors for the continuous production of fructose syrup from whey permeate. *Bioprocess Eng.*, **21**, 509–551.

Jank, B. and Haslberger, A.G. (2003) Improved evaluation of potential allergens in GM food. *Trends Biotechnol.*, **21**, 249–250.

Jiang, L., Althoff, E.A., Clemente, F.R., Doyle, L., Röthlisberger, D., Zanghellini, A., Gallaher, J.L., Betker, J.L., Tanaka, F., Barbas, C.F., 3rd, Hilvert, D., Houk, K.N., Stoddard, B.L., and Baker, D. (2008) *De novo* computational design of retro-aldol enzymes. *Science*, **319**, 1387–1391.

Jördening, H.-J. and Winter, J. (2005) *Environmental Biotechnology*, Wiley-VCH Verlag GmbH, Weinheim.

Kamm, B. and Kamm, M. (2007) Biorefineries – multi product processes. *Adv. Biochem. Eng. Biotechnol.*, **105**, 175–204.

Keasling, J.D. (2010) Manufacturing molecules through metabolic engineering. *Science*, **330**, 1355–1358.

Knapp, F. (1847) *Lehrbuch der chemischen Technologie*, F. Vieweg und Sohn, Braunschweig.

Neidleman, S.L. (1991) Enzymes in the food industry: a backward glance. *Food Technol.*, **45**, 88–91.

Novo Nordisk (2011) Annual Report for 2010, Novo Nordisk, Copenhagen, www.novonordisk.com.

Novozymes (2011) Annual Report for 2010, Novozymes, Copenhagen, www.novozymes.com.

Payen, A. (1874) *Handbuch der technischen Chemie*, vol. **II** (eds F. Stohmann and C. Engler), E. Schweizerbartsche Verlagsbuchhandlung, Stuttgart, p. 127.

Payen, A. and Persoz, J.F. (1833) Mémoire sur la diastase, les principaux produits de ses réactions, et leurs applications aux arts industriels. *Ann. Chim. Phys., 2me Sér.*, **53**, 73–92.

Pollard, D.J. and Woodley, J.M. (2006) Biocatalysis for pharmaceutical intermediates: the future is now. *Trends Biotechnol.*, **26**, 66–73.

Pratt, M.R. and Bertozzi, C.R. (2005) Synthetic glycopeptides and glycoproteins as tools for biology. *Chem. Soc. Rev.*, **34**, 58.

Raven, P.H. (2002) Science, sustainability, and the human prospect. *Science*, **297**, 954–958.

Roberts, S.M., Turner, N.J., Willets, A.J., and Turner, M.K. (1995) *Biocatalysis*, Cambridge University Press, Cambridge, p. 1.

Röthlisberger, D., Khersonsky, O., Wollacott, A.M., Jiang, L., DeChancie, J., Betker, J., Gallaher, J.L., Althoff, E.A., Zanghellini, A., Dym, O., Albeck, S., Houk, K.N., Tawfik, D.S., and Baker, D. (2008) Kemp elimination catalysts by computational enzyme design. *Nature*, **453**, 190–195.

Schäfer, T., Borchert, T.W., Nielsen, V.S., and Skagerlind, P. (2007) Industrial enzymes. *Adv. Biochem. Eng. Biotechnol.*, **105**, 59–131.

Seibel, J. and Buchholz, K. (2010) Tools in oligosaccharide synthesis: current research and application. *Adv. Carbohydr. Chem. Biochem.*, **63**, 101–138.

Straathof, A., Panke, S., and Schmid, A. (2002) The production of fine chemicals by biotransformations. *Curr. Opin. Biotechnol.*, **13**, 548–556.

Sumner, J.B. (1933) The chemical nature of enzymes. *Science*, **78**, 335.

Sumner, J.B. and Somers, G.F. (1953) *Chemistry and Methods of Enzymes*, Academic Press, New York, pp. XIII–XVI.

Tauber, H. (1949) *The Chemistry and Technology of Enzymes*, John Wiley & Sons, Inc., New York.

Tischer, W. (1990) Umweltschutz durch technische Biokatalysatoren, in *Symposium Umweltschutz durch Biotechnik*, Boehringer Mannheim GmbH.

Tosa, T., Mori, T., Fuse, N., and Chibata, I. (1969) Studies on continuous enzyme reactions. 6. Enzymatic properties of DEAE-Sepharose aminoacylase complex. *Agric. Biol. Chem.*, **33**, 1047–1056.

Trommsdorf, E. (1976) *Dr. Otto Röhm – Chemiker und Unternehmer*, Econ, Düsseldorf.

Ullmann, F. (1914) *Enzyklopädie der technischen Chemie*, vol. **5**, Urban und Schwarzenberg, Berlin, p. **445**.

UN (United Nations) (2003) *Industrial Commodity Statistics Yearbook 2001*, United Nations, New York.

Wagner, R. (1857) *Die chemische Technologie*, O. Wiegand, Leipzig.

Wallenfels, K. and Diekmann, H. (1966) Glykosidasen, in *Hoppe-Seyler/Thierfelder Handbuch der physiologisch- und pathologisch-chemischen Analyse*, vol. **6**, Springer, Berlin, pp. 1156–1210.

WCED (World Commission on Environment and Development) (1987) *Our Common Future*, Oxford University Press, Oxford.

Wingard, L. (1974) Enzyme engineering a global approach, *New Scientist*, **64**, 565–566.

Wong, C.H. (2005) Protein glycosylation: new challenges and opportunities. *J. Org. Chem.*, **70**, 4219–4225.

Internet Resources for Enzyme Technology

See Appendix A.

2
Basics of Enzymes as Biocatalysts

For the development of a new enzyme process, the following questions must be answered:	To do this, the following must be known or performed:
How can a suitable enzyme be selected?	Enzyme classification (Section 2.2)
What structural properties of enzymes are important for their application in enzyme technology?	Structure (primary, secondary, tertiary, quaternary); amino acid residues with functional groups on the enzyme surface or clefts in it (Section 2.3)
How can the biological function of enzymes as biocatalysts be described and applied for equilibrium- and kinetically controlled processes? Mechanism? Quantitative measures for the biological function of enzymes? Variations in these for enzymes with the same function from different sources?	The enzyme function can be described by • Substrate binding, characterized by a dissociation constant K_m • A monomolecular catalytic reaction between the enzyme and the part of the bound substrate that is changed in the reaction, characterized by a first-order rate constant (turnover number or k_{cat}) • Finally, the dissociation of the product(s) from the enzyme. When they are bound, they cause product inhibition, characterized by a dissociation constant K_i This also determines the substrate and stereospecificity of enzymes (Sections 2.4–2.6)
How can the substrate and stereospecificity be determined from enzyme kinetic properties? How do the enzyme kinetic properties	The determination of k_{cat}, K_m, and K_i from initial rate measurements $v = f(k_{cat}, K_m, K_i[\text{substrate}][\text{product}])$

Biocatalysts and Enzyme Technology, Second Edition. Klaus Buchholz, Volker Kasche, and Uwe T. Bornscheuer

depend on pH, *T*, ionic strength, inhibitors, and primary structure?	as a function of pH, *T*, ionic strength, [inhibitor] (Section 2.7)
Which system properties determine the end point, with maximal product yield, of enzyme processes? How much enzyme is required to reach this end point in a given time?	How pH, *T*, ionic strength, inhibitors, activators, k_{cat}, K_m, K_i, and stereoselectivity influence the equilibrium and steric purity at the end points, and the time to reach these in equilibrium-controlled and kinetically controlled processes (Section 2.8)
How can enzyme processes be carried out at substrate concentrations up to ≈1 M, even with substrates that are slightly soluble in aqueous solutions?	Solubilities of substrates and products should be known. Carry out the enzyme process either in aqueous suspensions or emulsions or in organic solvents, ionic liquids, or supercritical gases (Section 2.9)
Which factors influence the stability of enzymes in enzyme processes?	Factors that influence enzyme de- and renaturation (Section 2.10)
Which properties of an enzyme must be improved for their use as biocatalysts in equilibrium- and kinetically controlled processes?	See table in Section 2.11. There the range within these properties differ in natural enzyme variants are discussed. For protein engineered enzymes this is covered in Chapter 3

2.1 Introduction

The biochemical basis of enzymology is fundamental for the successful development and application of enzyme processes. The subject is described in detail in several excellent textbooks (see the literature list), but the following sections include a summary of some general aspects and principles such as the classification, structure, function (binding of substrate(s) and catalysis), and substrate, stereo-, and regiospecificity of enzymes, as well as the mechanisms and kinetics of enzyme-catalyzed reactions.

Some aspects of enzymes and enzyme-catalyzed processes that are important for enzyme technology are, however, often not treated in the biochemically oriented literature. These include the following:

- enzymes can be used as biocatalysts for equilibrium-controlled processes in both directions, or to obtain nonequilibrium concentrations of products in kinetically controlled synthesis;
- enzymes are not strictly stereospecific;

- the end points of enzyme processes and the amount of enzyme required to reach these in a given time;
- the physicochemical properties of the catalyzed process and its process window for optimal yields;
- enzymes can be used to catalyze reactions with nonnatural substrates also in aqueous suspensions and organic solvents, supercritical gases, and ionic liquids;
- stability of enzymes, substrates, and products.

These topics will be treated in more detail in this chapter. Finally, the properties of natural enzymes that must be changed to obtain a better enzyme for a specific enzyme process are summarized in Section 2.11. How this can be achieved by

- screening natural enzyme variants, and
- protein engineering (site-directed mutagenesis, random or directed evolution)

will be covered in Chapter 3.

2.2
Enzyme Classification

The system that has long been used by the Enzyme Commission to classify known enzymes is shown in Table 2.1. Every enzyme is given four numbers after the abbreviation EC (Webb, 1992; www.chem.qmul.ac.uk/iubmb/enzyme). The first number indicates one of the six possible reaction types that the enzyme can catalyze; the second number defines the chemical structures that are changed in this process; the third defines the properties of the enzyme involved in the catalytic reaction or further characteristics of the catalyzed reaction; and the fourth number is a running number. This classification system now covers more than 4000 enzymes. It emphasizes the function of the enzyme as a catalyst of a process in one direction, as is mostly the case in living systems. As a catalyst, the enzyme can, however, also catalyze the reverse reaction, and this property is often used in enzyme technology. This point has already been made in the list of the enzymes that are currently applied in enzyme processes (see Table 1.2), where enzymes with the same function but from different sources have the same EC number. The quantitative properties with which the enzyme carries out its function (kinetics) vary with the source (organism) of the enzyme. This is due to the fact that the primary structure of the enzyme with the same EC number can differ much from source to source (see Sections 2.3 and 2.4). Thus, in addition to the EC number, the source of the enzyme must always be stated.

The above-described EC classification system is based on the biochemical function of enzymes in living systems. This database contains no quantitative data on the properties of the enzymes. Newer enzyme databases (a list with description and evaluation of Internet databases of importance for enzyme technology is given in Appendix A) provide information on

- new classifications of enzymes as superfamilies based on their amino acid sequence, three-dimensional (3D) structure, and function: for all enzymes

Table 2.1 The enzyme classification system developed by the Enzyme Commission (EC), a commission of IUPAC (International Union of Pure and Applied Chemistry).

Enzyme classes and subclasses (functions)	Remarks
1. Oxidoreductases (oxidation–reduction reactions)	Cosubstrate required Two-substrate reactions:
1.1	At –CH–OH
1.2	At –C=O
1.3	At –C=C–
2. Transferases (group transfer reactions)	Two-substrate reactions, one substrate must be activated
2.1. C1 groups	
2.2. Aldehyde or keto groups	
2.3. Acyl groups	
2.4. Glycosyl groups	
3. Hydrolases (strictly transferases that transfer groups to H_2O, that is, hydrolysis reactions)	Two-substrate reactions, one of these is H_2O
3.1. Ester bonds	
3.2. Glycoside bonds	
3.3. Ether bonds	
3.4. Peptide bonds	
3.5. Amide bonds	
4. Lyases (nonhydrolytic bond-breaking reactions)	One-substrate reactions → bond breaking Two-substrate reactions ← bond formation
4.1. C–C	
4.2. C–O	
5. Isomerases (isomerization reactions)	One-substrate reactions
5.1. Racemizations	
5.2. *cis–trans*-Isomerizations	
5.3. Intramolecular oxidoreductases	
6. Ligases (bond formation reactions)	Require ATP as cosubstrate Two-substrate reactions
6.1. C—O	
6.2. C—S	
6.3. C—N	
6.4. C—C	

The classification based on the second number is incomplete here; the third (enzyme property) and fourth (running number) EC numbers are not covered.

based on domain structures, CATH Protein Structure Classification (www.cathdb.info/); Carbohydrate Active Enzyme database covering these enzymes from EC classes 2, 3, and 4 (www.cazy.org; Cantarel *et al.*, 2009); Lipase Database (www.led.uni-stuttgart.de); Peptidase Database (merops.sanger.ac.uk; Rawlings, Barrett, and Bateman, 2010);

- their properties (www.brenda-enzymes.org); this database gives extensive quantitative data for enzymes (enzyme kinetic constants for substrates, binding constants for activators and inhibitors, pH and temperature stability, and other information of importance for enzyme technology) for the same enzyme from different sources;

- the enzyme(s) that can catalyze a specific reaction (www.genome.jp/kegg/ligand.html).

These databases are based on literature data, cite these, and are regularly updated. With the further development of bioinformatics, it will be possible to combine these databases in order to select the enzymes that can catalyze a given reaction.

Once an enzyme has been identified as a suitable biocatalyst for a process, its development also involves the screening of other sources (microorganisms) for the same enzyme with properties that are better suited to this purpose than the original enzyme. Besides the catalytic properties of the enzyme, the end points of the enzyme-catalyzed process, the selectivity and stability of the enzyme under process conditions, and the properties of the source microorganism (enzyme yield, safety class) must be considered in selecting the optimal biocatalyst. Recently, this method to find better enzymes has been amended by protein engineering of already available enzymes (see Chapter 3).

2.3 Enzyme Synthesis and Structure

The *primary structure* – that is, the amino acid sequence – is defined by the base sequence in the structural gene (DNA) and the genetic code (Figure 2.1). Only the information in one strand of the double-stranded DNA, the sense strand, is used here. It is first transcribed into mRNA (*transcription*), the information from which is then translated in the synthesis of the polypeptide with the correct primary structure (*translation*). The synthesis direction is always $NH_3^+ \rightarrow COO^-$ – that is, the polypeptide chain starts with an amino group. This process also involves tRNAs, having a complementary code for the amino acids that can be bound to the corresponding code on the mRNA. Before this, the tRNA must be acylated by the correct amino acid in a reaction requiring energy (ATP), yielding an aminoacylated tRNA (Figure 2.1). The genetic code is redundant; 64 codons code for 20 amino acids, start and stop of protein synthesis. For some amino acids (Arg, Ser, Leu), there are six codons; for Met and Trp, only one. From this follows that there must be more than 20 different tRNAs in a living cell. The number and concentrations of these differ from cell to cell. The codon usage, that is, which of the redundant codes is used for an amino acid, also differs from cell to cell. It seems also to depend on where the corresponding amino acid is incorporated in the protein structure (Saunders and Deane, 2010). Thus, the rate of protein synthesis is determined not only by the mRNA content, but also by the contents of the different tRNAs, the amino acids, and the codon usage in the cell (http://www.kazusa.or.jp/codon). This is important to consider when the gene for a desired enzyme is transferred, transcribed, and translated in a recombinant organism (Sections 6.2.2 and 6.3).

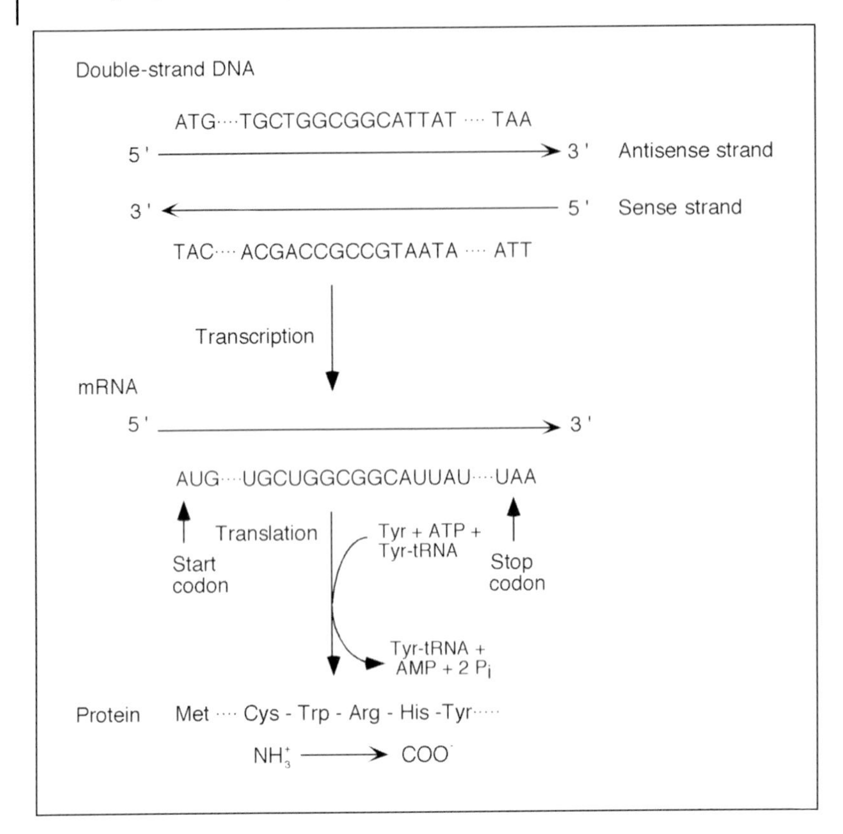

Figure 2.1 During transcription (DNA → mRNA), only one strand of DNA (sense strand) is transcribed, that is, used for the synthesis of complementary mRNA. During ribosomal protein synthesis, the latter is translated into an amino acid sequence or the primary protein structure. The rate of translation is also influenced by the concentrations of the amino acid-specific tRNAs. This and the codon usage, the frequency in the use of the different codons for an amino acid, differ from cell to cell. For Tyr as shown in the figure, there are two codons, TAC and TAT.

For intracellular proteins, the main *secondary structure* elements, the α-helix and the β-sheet, are formed spontaneously either co- or posttranslationally, and are stabilized by hydrogen bonds or hydrophobic interactions between amino acid residues, respectively. These secondary structures are folded spontaneously into *domains*, stabilized by hydrophobic, charge–charge, charge–dipole, and dipole–dipole interactions. The *tertiary* or *three-dimensional structure* of an enzyme often consists of more than one domain, and is stabilized by such interactions between amino acid residues far away from each other along the primary structure. The domains are separated by linker polypeptides without a secondary structure; this gives the tertiary structure a flexibility that is essential for the function of the enzyme. The spontaneous folding of the translation product competes with bimolecular aggregation reactions between partly folded translation products. The latter reaction cannot be neglected at the high protein content in the cell, and leads to inactive enzyme products. To prevent this loss, living systems have developed a special class of proteins (chaperones) that can act as folding catalysts and prevent the aggregation reaction of intracellular enzymes.

The spontaneous folding of some extracellular enzymes must be inhibited in the cell by other chaperones in order to allow translocation of the unfolded polypeptide through the cell membrane. The folding then occurs spontaneously, but extracellularly.

Many enzymes that are used in enzyme technology consist of oligomers with identical subunits (such as glucose isomerase, β-galactosidase, alcohol dehydrogenase, some aldolases, some lipases, urease). This *quaternary structure* is formed spontaneously by noncovalent interactions between the subunits. Information on this and the following posttranslational modifications for enzymes can be found in the BRENDA database.

The folded translation product of a large number of enzymes is often not catalytically active. This applies generally for extracellular and periplasmic enzymes. Different posttranslational modifications are required to activate the enzyme precursors (pro-enzymes):

- Noncovalent binding of organic (pyridoxal-5′-phosphate (PLP), biotin, etc.) or inorganic (metal ions) cofactors. More than 30% of all enzymes require the tight binding of metal ions for their activation or biological activity. To these belong metal ions that are directly involved in the catalytic reactions either as free ions or complexed to organic cofactors (Frausto Da Silva and Williams, 2001):
  - Co^{+}, Co^{2+} in oxidoreductases requiring the cofactor vitamin B_{12} containing Co^{+}; glucose isomerase from *Streptomyces* strains requires Co^{2+}, and can, due to the toxicity of this ion, not be used for the isomerization of glucose. For this aim, the enzyme that binds Mg^{2+} is used (see below).
  - Cu^{+}, Cu^{2+} in many oxidoreductases (cytochrome oxidase, laccase, lysine oxidase, etc.).
  - Fe^{2+}, Fe^{3+} in many oxidoreductases (catalases, cytochrome c and P450, lipoxygenase, etc.).
  - Mg^{2+} in kinases, enolase, ATPases, and recombinant glucose isomerase (wild-type enzyme from *Bacillus coagulans* does not bind Mg^{2+} tightly, and the ion is partly lost during the purification of the enzyme; see Figure 1.8 and Section 8.4.1).
  - Mn^{2+} in transferases and hydrolases (galactosyl transferase, arginase, etc.), or Mn^{2+}, Mn^{3+} in oxidoreductases (Mn catalase, Mn peroxidase, etc.).
  - Ni^{2+} in urease.
  - Zn^{2+} in metallopeptidases (carboxypeptidase A and B, thermolysin), other hydrolases (alkaline phosphatases, β-lactamase, collagenase, hydantoinase, etc.), carbonic anhydrase, and oxidoreductases (alcohol dehydrogenase, etc.).

  Not directly involved in the catalytic reaction, but required for correct folding and maturation of many extracellular hydrolases is
  - Ca^{2+}; the tightly bound ion was first observed in 3D structures and has been found to be essential for the correct folding, posttranslational processing, and membrane transport of extracellular enzymes such as peptidases (proteinase K,

subtilisin, thermitase), α-amylases, penicillin amidases, and some lipases (Kasche *et al.*, 1999, 2005). More weakly bound Ca^{2+} has been found to be essential for the proteolytic processing of the pro-enzymes trypsinogen and chymotrypsinogen to their biological active forms. This ion has also been found to stabilize many enzymes.

- Covalent binding of cofactors and other molecules (hemes, mono- and oligosaccharides (only for extracellular enzymes in eukaryotic cells), phosphate groups, etc.).
- Proteolytic processing of the polypeptide chain. For extracellular enzymes, this applies first for the N-terminal signal peptide required for the binding of the pre-pro-enzyme (pre stands for the signal peptide) to the cytoplasmic part of the membrane protein translocation system. The signal peptide is hydrolyzed by peptidases that belong to this system and are localized on the outer membrane side (Dalbey and Robinson, 1999). The formed pro-enzyme is then activated by proteolytic processing reactions (hydrolysis of one or more peptide bonds in the pro-enzyme) in or outside the membrane. The first step has for some amidases and peptidases (pro-calpain, pro-penicillin amidase, pepsinogen, the first peptidases in processing chains of physiological importance as apoptosis, blood clotting and its reversal, activation of digestive enzymes, etc.) been shown to be an intramolecular autoproteolytic reaction (Kasche *et al.*, 1999). Many hydrolases that are used in large amounts in enzyme technology (amylases, galactosidases, lipases, glutaryl and penicillin amidases, peptidases as subtilisins, trypsins, etc.) are proteolytically processed extracellular enzymes. Crude preparations of these enzymes generally contain several proteolysis products with the same enzyme function, but different activities. This must be considered when they are applied to enzyme processes where consistent properties of the enzyme used are required. Enzyme kinetic constants must be determined for pure enzyme preparations, and this is especially important when the properties of the same enzyme from different sources or mutant enzymes are compared.

For all posttranslational modifications, the surface of the folded translation product is important. The functional groups involved in the covalent and noncovalent binding of cofactors, and the peptide bonds hydrolyzed in proteolytic processing, must be localized on this surface (Table 2.2). About 40–50% of the amino acid residues on the surface of an enzyme are hydrophobic. This is important for the oligomerization of subunit enzymes, and is applied in the purification of enzymes with hydrophobic or mixed-mode adsorbents (see Chapter 6).

The binding of substrates, inorganic ions, or organic molecules to enzymes may lead to large changes in the tertiary and quaternary structure (Figure 2.2), when the enzyme's properties may be markedly changed. Hexokinase without bound glucose is a weak hydrolase (ATPase), but with glucose it becomes a transferase with a markedly reduced hydrolase function. This illustrates that a flexible tertiary and quaternary structure is important for the function of an enzyme. When the enzyme is immobilized (see Chapters 8 and 10), this flexibility must be retained.

Table 2.2 Average content of amino acids with functional groups, and their frequency as active-site residues in enzymes.

Amino acid	Content (%)	Functional group	Frequency in active site (%)
Ser	7.8	—OH	4
Lys	7	$—NH_2$	9
Thr	6.5	—OH	3
Asp	4.8	—COOH	15
Glu	4.8	—COOH	11
Arg	3.8	$—NH_2$	11
Tyr	3.4	—OH	5
Cys	3.4	—SH	6
His	2.2	—N—	18
Met	1.6	—S—	<1
Trp	1.2	—N—	1

A large fraction of these are localized on or in clefts of the protein surface and may interact with molecules in solution (Creighton, 1993; Bartlett *et al.*, 2002).

2.4
Enzyme Function and Its General Mechanism

Enzyme function is defined by the reactions in which that enzyme acts as a catalyst. This generally involves three steps, each of which can be subdivided into different reactions:

1) Noncovalent binding of one or two substrates to functional groups in the *binding subsites* S_i before and S_i' behind the chemical bond that is changed in the enzyme-catalyzed process, on the enzyme surface. The numbering i starts from 1 in both directions from this bond. This convention applied for peptides is shown in Figure 2.3, but it can also be applied for other substrates.

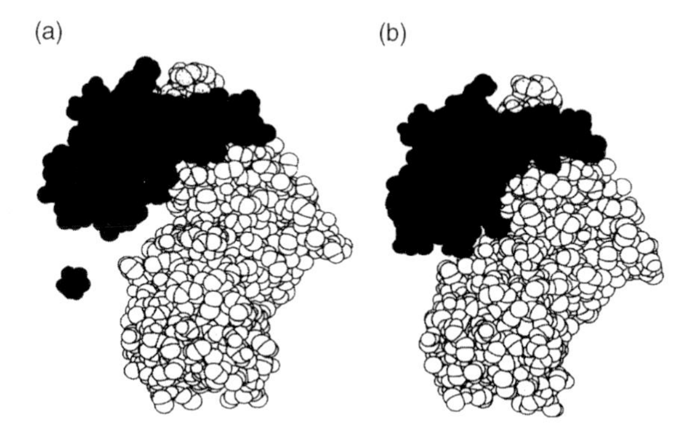

Figure 2.2 Change in the tertiary structure of an enzyme due to substrate binding, illustrated for the binding of glucose to hexokinase: (a) without glucose; (b) after binding of glucose (from Darnell, Lodish, and Baltimore, 1986; copyright 1986 Scientific American Books, with permission by W.H. Freeman and Co.).

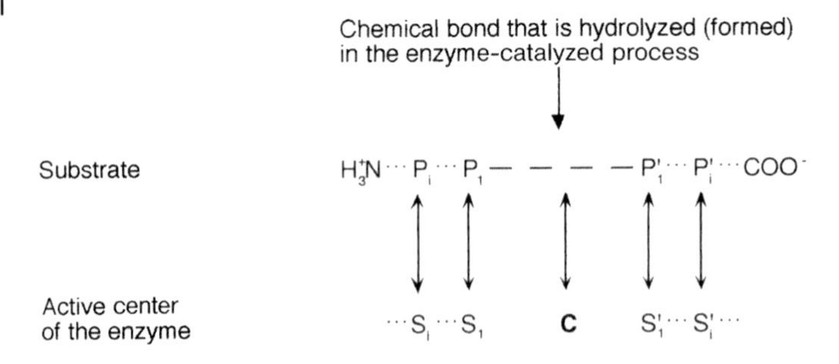

Figure 2.3 Conventions used to describe the substrate specificity of enzymes and the binding (S-, S′-) and catalytically active (C) subsites of the active site of the enzyme, illustrated for peptidases (Schechter and Berger, 1967). This can be applied to all enzymes that catalyze the formation or breaking of a covalent bond. The number of subsites involved depends on the size of the molecules involved. The substrate sequence specificity $P_i \ldots P_1 - P_1' \ldots P_i'$ for the hydrolysis by a peptidase, determined by the binding sites $S_i \ldots S_1 S_1' \ldots S_i'$ can be found in the MEROPS database (Rawlings, Barrett, and Bateman, 2010).

2) When the substrate is correctly bound to the binding subsites, the functional groups in the *catalytically active amino acid residues* C in clefts on the enzyme surface can interact with the part of the substrate leading either to the formation of covalent bonds (such as for the peptidases that form acyl-enzyme intermediates) or noncovalent interactions or to its chemical transformation. In the first case, this involves two reactions: first the formation of a covalent acyl-enzyme intermediate and then the deacylation of the acyl-enzyme.
3) Dissociation of the products formed in step 2 from the binding subsites.

The *active site* of the enzyme is the sum of the binding subsites and the catalytically active amino acid residues (catalytic subsite).

The active sites of many industrially important enzymes, such as α- and β-amylase, carboxypeptidase A, α-chymotrypsin, glycoamylase, glucose isomerase, glutaryl amidase (or cephalosporin amidase), lipase, penicillin amidase, and trypsin, have been mapped. This has mainly been done by the determination of their 3D structure with and without substrates or products. This gives direct information on the enzyme mechanism and the molecular interactions (such as ion–ion, hydrogen bonds, hydrophobic bonds) between the active site and the substrate that directly influence the rate of the enzyme-catalyzed reactions. This information can be used for a rational design by site-directed mutagenesis to improve the enzyme for a specific application as a biocatalyst (see Chapter 3).

The detailed mechanisms with which some hydrolases hydrolyze linear biopolymers (proteins, the linear starch amylose) into monomers are shown in Figures 2.4

and 2.5. They illustrate general enzyme mechanisms, and the regio-, stereo-, and substrate specificity of enzymes in the EC classes 2–6 (Table 2.1). For oxidoreductases (EC class 1) that also involve oxidations and reductions, some mechanisms involving such reactions are given in Chapter 4.

The hydrolysis of biopolymers generally requires two enzyme types:

- Endohydrolases that hydrolyze bonds between monomers inside the polymer chain.
- Exohydrolases that hydrolyze the bond between the last two monomers.

For instance, amylose and amylopectin (a branched starch molecule) consist of uncharged identical monomers, whereas proteins have amino acids with charged and uncharged nonidentical side chains. The hydrolysis mechanisms for the exopeptidase carboxypeptidase A (EC 3.4.17.1, peptide family M 14 (MEROPS), a metallopeptidase from pancreas) and the exoglycosidase glucoamylase (EC 3.2.1.3, glycosidase family 15, from *Aspergillus niger*) are given in Figure 2.4. In Figure 2.5, the corresponding mechanisms are given for the bovine endopeptidase α-chymotrypsin (EC 3.4.21.1, peptidase family S1, a serine peptidase) and the endoamylase α-amylase (EC 3.2.1.1, glycosidase family 13, from *A. niger*).

In the binding step, both charged and uncharged functional groups in the binding subsites are involved. The amino acid residues in these sites are moved, when they bind the substrate, to allow for a better interaction with the substrate (induced fit). This has been verified by 3D structure determinations of the enzyme without and with bound substrate that demonstrates the importance of a flexible tertiary structure as emphasized above (Figure 2.2). In the binding step, the enzyme can discriminate between charged and uncharged substrates. Due to the positively charged Arg_{145} in the S'_1 subsite of carboxypeptidase A, it can bind and hydrolyze the negatively charged C-terminal amino acid in a polypeptide. Hence, it is an exopeptidase. The endopeptidase α-chymotrypsin has a negative charge in the S'_1 subsite and therefore cannot hydrolyze such peptides; rather, it can only hydrolyze peptide bonds in a polypeptide in front of neutral or positively charged amino acid residues. In this step and in subsequent catalytic reactions, the enzymes can also discriminate between stereoisomers. This requires at least two binding interactions with groups bound to asymmetric C atoms in the substrate. This is illustrated by the mechanism in Figure 2.4a. When the C-terminal R′ and carboxyl group of an (*S*)-amino acid[1)] are bound correctly to functional groups that bind R′ and Arg_{145} in the S'_1 subsites of carboxypeptidase A, the N–H group in the amide bond that is hydrolyzed is correctly oriented with respect to the catalytically active amino acid residues C. For a C-terminal (*R*)-amino acid, only two of these three interactions can occur simultaneously. Carboxypeptidase is thus stereospecific for C-terminal (*S*)-amino acid residues; that is, they are hydrolyzed much faster than C-terminal (*R*)-amino acids. The interactions between the functional groups in the binding

1) In this book, stereoisomers will be written using (*R*)- or (*S*)-, instead of the older D- or L-, designation. The latter will, however, be used for saccharides, as it is still mainly used for these compounds.

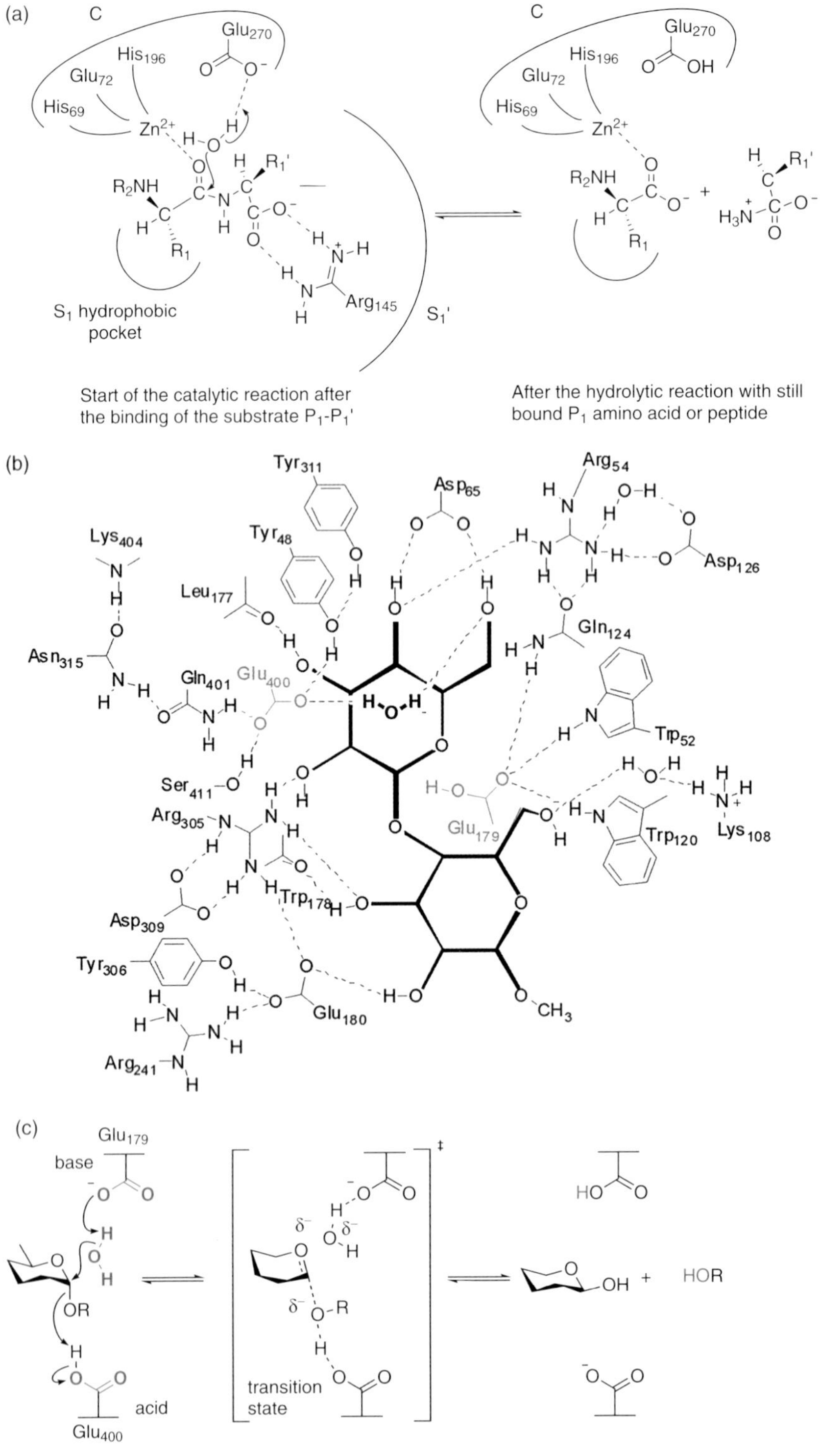

(a)

C
Glu270
His196
Glu72
His69
Zn^{2+}
O^-
H-O
H
O
H
R_1'
R_2NH
C
N
C
H
C
O^-
H
O
R_1
H
N^+-H
H
N
Arg145
H
S_1 hydrophobic pocket
S_1'

Start of the catalytic reaction after the binding of the substrate P_1-P_1'

C
Glu270
His196
Glu72
OH
His69
Zn^{2+}
O
R_2NH
C
O^-
H
C
R_1
+
H
R_1'
C
H_3N^+
C
O^-
O

After the hydrolytic reaction with still bound P_1 amino acid or peptide

(b)

Tyr311
Asp65
Arg54
O-H
Lys404
Tyr48
Asp126
Leu177
Gln124
Asn315
Gln401
Glu400
H-O-H
Trp52
Ser411
Glu179
Arg305
Trp120
Lys108
Trp178
Asp309
Tyr306
Glu180
CH_3
Arg241

(c)

Glu179
base
O
O
H
O
H
OR
H
O
O
acid
Glu400

‡
δ-
O
H
O
δ-
O
H
H
δ-
O-R
H
O
O
transition state

HO
O
OH
+
HOR
$^-$O
O

sites and the P_i and P'_1 amino acid side chains can discriminate between different amino acid residues in this position, and demonstrate the substrate specificity of enzymes. Detailed information on the sequence specificity of peptidases (Figure 2.3) has been compiled in the MEROPS database.

Glucoamylase (known industrially as amyloglucosidase) is an exo-acting inverting glycoside hydrolase (α-(1→4)-D-glucan glucohydrolase) that catalyzes the release of β-D-glucose. It is exo-acting since it hydrolyzes bonds only from the non-reducing end of starch, illustrating the regioselectivity of enzymes. It is inverting since β-glucose is formed from an α-glycosidic bond. However, in solution the initial product undergoes spontaneous mutarotation to the equilibrium mixture of α- and β-D-glucose.

Figure 2.4 Mechanism of the enzyme-catalyzed hydrolysis of peptides and di- and oligosaccharides bound in the binding subsites of the active site by the exoenzymes carboxypeptidase A and glucoamylase.
(a) Hydrolysis of C-terminal amino acids by pancreatic carboxypeptidase A. It has a P_1 specificity for Gly or Phe, and a P'_1 specificity for amino acids with aliphatic or aromatic side chains. The S'_1 binding site consists of Arg_{145} that forms hydrogen and ion–ion bonds with the P'_1 carboxyl group, and a hydrophobic pocket (Try_{198}, Tyr_{248}, and Phe_{279}) that binds the hydrophobic P'_1 side chains. The S_1 binding site is a hydrophobic pocket that has not yet been clearly identified. The catalytically active amino acid residues C are Glu_{270} that interacts with a bound water molecule, and His_{69}, Glu_{72}, and His_{196} that bind the Zn^{2+} ion required for the activity. This ion also interacts with O of the peptide bond. The catalytic reaction starts with a proton transfer from the bound H_2O to Glu_{270}. Then the formed OH^- attacks C in the peptide bond carbonyl group, yielding a gem-diol intermediate. In the last catalytic reaction step, this proton is transferred to NH of the peptide bond, leading to its cleavage. This is a general acid–general base reaction mechanism (Xu and Guo, 2009; Kilshtain and Warshel, 2009). R_1 is the P_1 amino acid side chain; R_2 is H, amino acid, or peptide acyl group; R'_1 is the P'_1 amino acid side chain.
(b) A two-dimensional representation of the interactions of the amino acid residues in the active site and neighboring residues of glucoamylase from *A. niger* with bound methyl α-maltoside (bold face). Note that this distorts the distance between neighboring amino acid residues. The catalytically active residues and the bound water molecule are in red and bold face, respectively. The active site is in the shape of a well where only the nonreducing end of a di- or oligosaccharide can be bound at the bottom of the well (lower part of the substrate). In this well, the catalytically active residues Glu_{179} and Glu_{400} (numbering for *A. niger* glucoamylase) are located so that they can act on the first α-(1–4) bond from the nonreducing end, yielding the exo-activity of this enzyme. The tight binding of the substrate by multiple hydrogen bonds is obvious, and is responsible for high specificity. In most fungal glycoamylases, the active site well continues into a *O*-glycosylated linker between the catalytic domain and a starch binding domain, which allows the catalytic domain to attack a larger area of a starch granule while tethered to its surface (a mechanism similar to that of cellulases) (see Section 12.2, Figure 12.5) (adapted from Coutinho and Reilly, 1994; Aleshin *et al.*, 2003).
(c) Simplified scheme of the mechanism for the hydrolysis of a di- or oligosaccharide catalyzed by glucoamylase (the acid is Glu_{400} and the base is Glu_{179}, numbering for *A. niger* glucoamylase). The general acid–general base reaction mechanism starts with a proton transfer from the bound water molecule to Glu_{400}. Then OH^- acts as a nucleophile on C-1, causing the hydrolysis of the terminal glucose. Finally, the proton of Glu_{179} is transferred to the leaving mono- or oligosaccharide. In the final step, rapid proton exchange from Glu_{400} to water, and from water to Glu_{179}, occurs (not shown).

The acid catalyst Glu_{179} must be uncharged, whereas the base catalyst Glu_{400} must be charged (Figure 2.4c). The enzyme is usually used as an industrial biocatalyst at pH 4–5. Thus, the acid catalyst must have a pK value above 5 that is higher than the normal value for carboxyl groups. This high value may probably be due to the charged carboxyl group Glu_{180} near Glu_{179}. This explains the pH dependence of the activity of this enzyme. It reduces at pH values above the pK value of the carboxyl group of this residue. The enzyme can also hydrolyze and synthesize α-(1–6)-glycosidic bonds. This explains the formation of the by-products isomalto-oligosaccharides by glucoamylase in the reaction shown in Figure 2.4c. The transition state glucosyl residue can undergo nucleophilic attack by another glucose or oligosaccharide molecule instead of by water, notably at high sugar concentrations (as in technical processes), with formation of an α-(1–6)-glycosidic bond, leading to isomalto-oligosaccharides. At the end of the technical hydrolysis of hydrolyzed starch with this enzyme to produce glucose for glucose isomerization (see Figure 1.8), the main by-product is the disaccharide isomaltose. It amounts to about 1–2% of the hydrolyzed starch or up to around 200 000 tons per year (see Table 1.1). Thus, it is important to reduce the formation of this by-product (see Exercise 2.20).

A number of starch-converting enzymes belong to a single family: the α-amylases or family 13 glycosyl hydrolases. This group of enzymes shares a number of common characteristics such as a $(\beta/\alpha)_8$ barrel structure, the hydrolysis or formation of α-glycosidic bonds, and a number of conserved amino acid residues in the active site. The three-dimensional structures, mechanistic principles deduced from structure–function relationships, and properties such as kinetics, selectivity, and stability of these enzymes have been investigated, reported, and summarized in the databases CAZy and BRENDA that are regularly updated.

Amylolytic enzymes bind substrate glucosyl residues at an array of consecutive subsites that extends throughout the active-site cleft. In typical endo-acting enzymes such as α-amylases, the substrate binding region comprises from 5 to 11 subsites (Sauer *et al.*, 2000). The structure of the binding region determines substrate and product selectivity. The established catalytic mechanism of the α-amylase family, with α-retaining activity, involves two catalytic residues in the active site: a glutamate residue as general acid/base catalyst and an aspartate residue as nucleophile (Glu_{230} and Asp_{206} in *A. niger* numbering) (Figure 2.5a). As for glycoamylase, the acid catalyst Glu residue must have a high pK value, as it must be uncharged at the pH values (≈6) where the enzyme is used to hydrolyze starch.

The catalytic step, hydrolysis in Figure 2.4 and acylation and deacylation in Figure 2.5, starts with a proton transfer from or to the catalytic amino acid residues. The withdrawal of a proton from water by Glu_{270} in carboxypeptidase A leads to a nucleophilic attack of OH^- on the peptide bond that is hydrolyzed without the formation of a covalent acyl-enzyme intermediate. The same applies for glycoamylase where the proton transfer from water to the base Glu_{400} leads to a nucleophilic attack of OH^- on the glycoside bond that is hydrolyzed without the formation of a covalent glycosyl-enzyme intermediate. For α-amylase in Figure 2.5a, the proton transfer is from the acid/base catalyst Asp_{206} to O in the glycoside bond that leads to a nucleophilic attack by the charged O in Asp_{206}, resulting in the glycosylation of this residue and

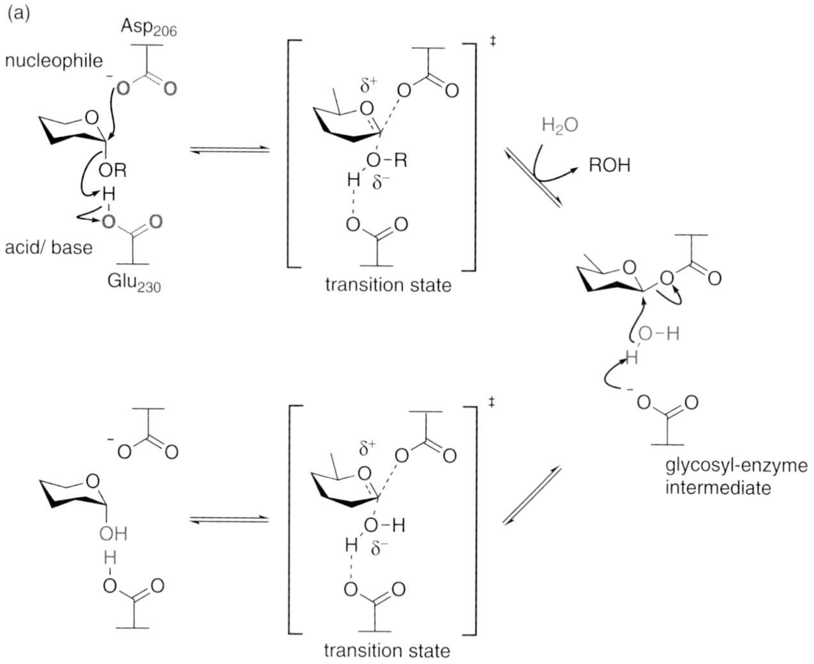

(a) Asp206 nucleophile acid/ base Glu230 transition state H2O ROH glycosyl-enzyme intermediate transition state

Figure 2.5 Mechanism of the enzyme-catalyzed hydrolysis of oligosaccharides and peptides bound in the binding subsites of the active site by the endohydrolases α-amylase from *A. niger* and bovine α-chymotrypsin involving the formation of covalent acyl- or glycosyl-enzyme intermediates.
(a) Hydrolysis of amylose by α-amylase. Hydrolysis with net retention of configuration is most commonly achieved via a two-step, double displacement mechanism involving a covalent glycosyl-enzyme intermediate, as is shown in the figure. Each step passes through an oxocarbenium ion-like transition state. Reaction occurs with acid/base and nucleophilic assistance provided by two amino acid side chains, Glu_{230} and Asp_{206}, located 5.5 Å apart. In the first step (often called the glycosylation step), Asp_{206} residue plays the role of a nucleophile, attacking the anomeric center between two substrate glucosyl units to form a covalent glucosyl-enzyme intermediate. At the same time, Glu_{270} functions as an acid catalyst and protonates the glycosidic oxygen and the bond is cleaved. The protonated glucosyl unit at subsite +1 leaves the active site while a water molecule moves in. In the second step (known as the deglycosylation step), the glucosyl-enzyme is hydrolyzed by water, with the other residue now acting as a base catalyst deprotonating the water molecule as it attacks. A third conserved residue, a second aspartate (Asp_{297}), binds to the OH-2 and OH-3 groups of the substrate through hydrogen bonds and plays an important role in substrate distortion (van der Maarel *et al.*, 2002). When another glucose molecule, instead of water, enters the active site, a new glycosidic bond, preferentially with an α-(1–6) bond, is formed (Kelly, Dijkhuisen, and Leemhuis, 2009). The p*K* value of the acid/base group cycles between high and low values during catalysis to optimize it for its role at each step of catalysis. For α-chymotrypsin 2.5b) see next two pages.

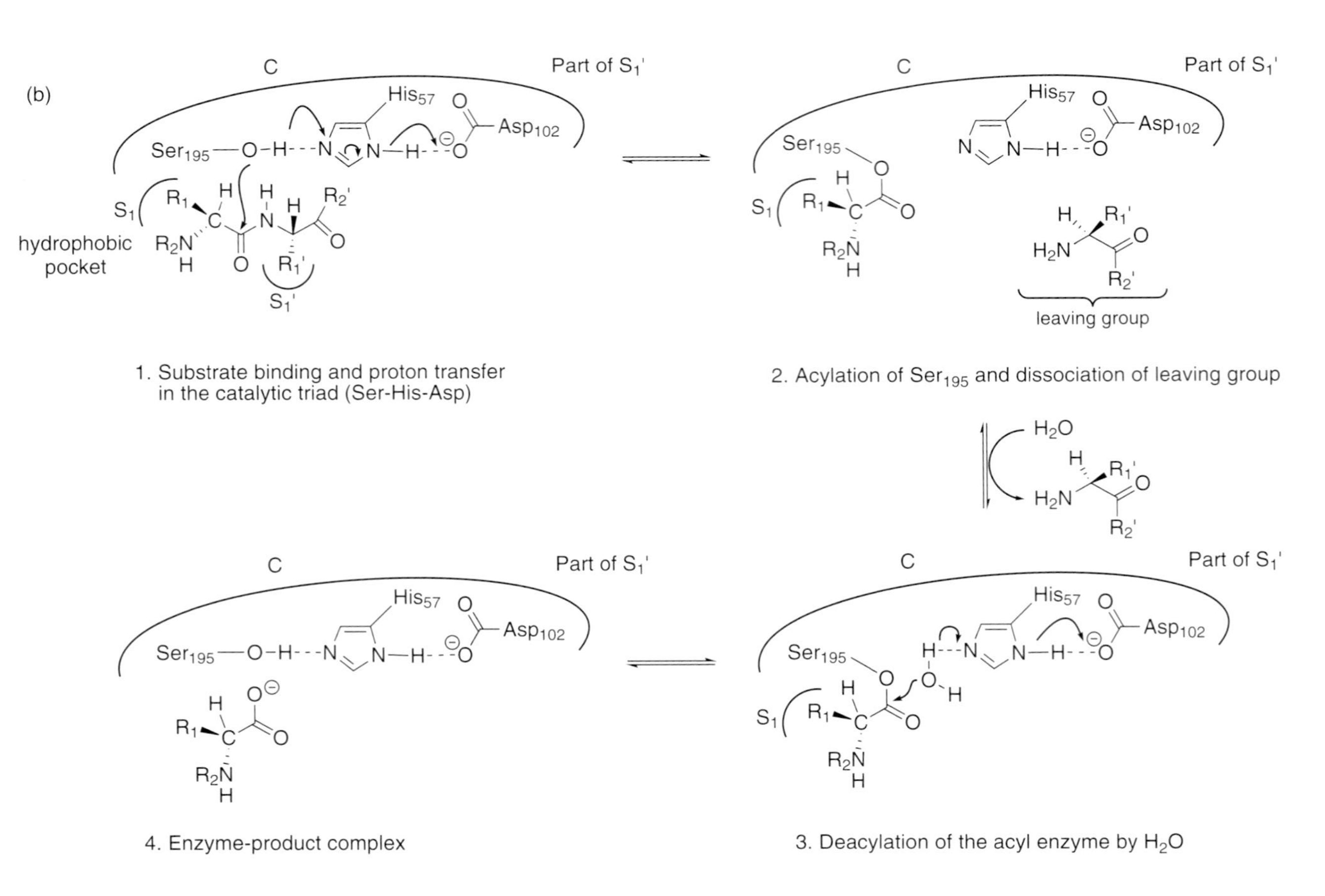

(b)
C
Part of S_1'
His_{57}
Asp_{102}
Ser_{195}
S_1
hydrophobic pocket
S_1'
leaving group
H_2O
1. Substrate binding and proton transfer in the catalytic triad (Ser-His-Asp)
2. Acylation of Ser_{195} and dissociation of leaving group
3. Deacylation of the acyl enzyme by H_2O
4. Enzyme-product complex

formation of a covalent glycosyl-enzyme intermediate. The latter is then hydrolyzed by a nucleophilic attack by water assisted by the acid/base catalyst Glu_{230}. In Figure 2.5b, the charge and proton transport in the catalytic triad Ser_{195}–His_{57}–Asp_{102} in α-chymotrypsin activates the O in Ser for a nucleophilic attack on C in the peptide bond. A mechanism similar to Figure 2.5b applies for lipases (EC 3.1.1.3) that also have a catalytic triad (see Figure 4.26). Such proton transfer mechanisms start the catalytic step in many enzyme-catalyzed reactions, not only for hydrolytic enzymes (Dodson and Wlowader, 1998; Brannigan *et al.*, 1995). For peptidases, this leads to an acylation of Ser (or Cys) – that is, the formation of a covalent acyl-enzyme – and the formation of a C-terminal peptide or amino acid. In both cases, the involved His and Ser residues must be uncharged to allow for the proton transport. This shows how the rates of these enzyme-catalyzed processes depend on pH. Below the pK values of this His residue, the rate of the catalytic step will decrease. Following acylation, the acyl-enzyme is deacylated by a nucleophilic attack of H_2O.

Finally, the product(s) must dissociate from the binding subsites. As they are part of the substrate, their binding to the active site is weaker. The binding of the product in the active site inhibits the binding of a substrate. This leads to *product inhibition* that cannot be avoided in enzyme processes, which is important to consider in enzyme processes, since the product concentration increases during the reaction toward the desired end point. In these processes, much larger substrate concentrations are also used (up to 3 M, as in the isomerization of glucose) compared to living systems ($\leq$10 mM).

As catalysts, enzymes can catalyze a reaction in both directions. This applies also for the reactions shown in Figures 2.4 and 2.5. Carboxypeptidase A can therefore also catalyze the condensation of an amino acid to the C-terminal end of a peptide

Figure 2.5 (b) Hydrolysis of peptide bond with the serine endopeptidase α-chymotrypsin. This enzyme catalyzes the hydrolysis of peptide bonds in the polypeptide chain where the S_1 binding subsite preferably binds aromatic amino acid residues (Tyr, Trp, Phe). It consists of a hydrophobic pocket (Ser_{189}, Ser_{190}, Met_{192}, Trp_{215}, Gly_{216}, Ser_{217}) that can bind the hydrophobic side chains of the P_1 aromatic amino acids. The S_1' binding site includes the negatively charged Asp_{102}, and can therefore not bind unprotected C-terminal amino acids; thus, this enzyme acts as an endohydrolase. It is less specific than the S_1 binding subsite; the highest specificity is for Leu, Ser, and due to the negative charge it is specific for Lys and Arg. The serine endopeptidases have the catalytic triad Ser_{195}–His_{57}–Asp_{102} in the catalytic subsite. The catalytic reactions start with a proton transfer from Ser to His, and from His to Asp in this triad that results in a nucleophilic attack by the serine oxygen on C in the peptide bond carbonyl group of the substrate. This leads to the formation of an acyl-enzyme intermediate and a leaving peptide whose amino group finally binds the proton that was transferred from Ser to His. The acyl-enzyme is then hydrolyzed due a nucleophilic attack by water activated by the catalytic triad, releasing the other part of the original peptide (Berg *et al.*, 2002). Other nucleophiles can also attack the acyl-enzyme; thus, other products can be formed in this kinetically controlled reaction (see Section 2.6). These condensation products, however, are later hydrolyzed until the equilibrium state has been reached. R_1 and R_1' are amino acid side chains; R_2 and R_2' are protecting groups or other amino acids of a peptide.

or an amino acid. Glycoamylase also catalyzes the hydrolysis and synthesis of (1–6)-glycoside bonds. When this enzyme hydrolyzes maltose or oligosaccharides, it can also catalyze the synthesis of isomaltose or isomalto-oligosaccharides that have α-(1–6)-glycoside bonds. These are unwanted by-products in the hydrolysis of starch, which are favored at the high concentrations used in technical processes. As an enzyme cannot change the equilibrium constant of a reaction, the maximum product and by-product concentrations of the reactions catalyzed by carboxypeptidase A and glycoamylase are the equilibrium concentrations that depend on pH only for peptide synthesis (Figure 2.20) that increase with increasing substrate concentrations. For the reactions in Figure 2.5 catalyzed by α-amylase and α-chymotrypsin that involve the formation of a covalent glycosyl- or acyl-enzyme intermediate, another process than the equilibrium-controlled synthesis catalyzed by carboxypeptidase A and glycoamylase is possible. Other nucleophiles than water can deacylate the acyl-enzyme. For α-chymotrypsin they can be the amino group at the N-end of amino acid amides (the enzyme is an endoenzyme!) or peptides leading to the formation of products with a new C-terminal amino acid or peptide. These products can, however, be hydrolyzed by the enzyme. When the rate of the synthetic reaction is larger than that for the hydrolytic reaction, product concentrations much larger than the equilibrium concentrations can be obtained (Figure 1.9). Such kinetically controlled processes can only be catalyzed by hydrolases that form acyl- or glycosyl-enzyme intermediates (see Section 2.6). Then they act as transferases. As for transferases these reactions require activated substrates (Table 2.1). There are also exopeptidases (serine carboxypeptidases, EC 3.4.16, peptidase family S10) and exoglycosidases (see Figure 1.9, lactase, EC 3.2.1.23, glycosyl hydrolase family 1) that form acyl- or glycosyl-enzyme intermediates. They can be used for kinetically controlled synthesis of desired condensation products (peptides or oligosaccharides). For oligosaccharides, however, glucosyl transferases are more efficient than exoglycosidases for this purpose (see Section 8.4.1).

As shown in Figures 2.4 and 2.5, the amino acid residues in the active center are located far from each other in the primary structure, but must be near each other in the tertiary structure. The primary structure of enzymes with the same function and EC number from different sources can differ by up to 70%. As the function is conserved, it can be assumed that the structure of the active site and its sterical orientation is conserved in these natural enzyme variants. This has been verified for some enzymes studied in detail. By the occurrence of spontaneous *mutations* (changes in the bases in the structural gene in the DNA), the cytochrome P450 gene from *Paracoccus denitrificans* has been changed during evolution so that the same gene from tuna fish encodes for a cytochrome with 70% change in amino acids in the primary structure (Salemme, 1977). The same has been shown to apply for enzymes of technological interest, such as glucoamylase (Sauer *et al.*, 2000; Aleshin *et al.*, 2003), glucose isomerase (Hartley *et al.*, 2000), lipases (Kazlauskas, 1994), and penicillin amidases (Brannigan *et al.*, 1995; McDonough, Klei, and Kelly, 1999; Kasche, Galunsky, and Ignatova, 2003) (Figure 2.6). The spontaneous mutation frequency in living systems is 10^{-10} to 10^{-9} wrongly incorporated nucleotides per base pair and cell division. For an enzyme with a molecular weight of 10^5 Da, this implies that one

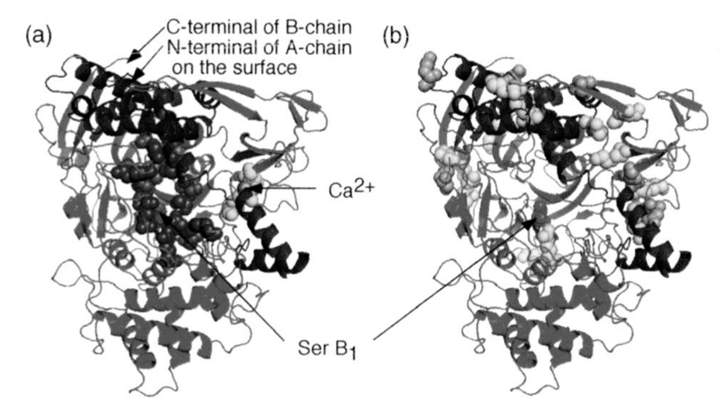

Figure 2.6 Location of the strictly conserved, that is, identical amino acid, residues (as balls), with known functions in six penicillin amidases (from *E. coli*, *P. rettgeri*, *K. cytrophila*, *A. viscosus*, *A. faecalis*, *B. megaterium*), in the 3D ribbon structure of PA from *E. coli*. The 3D structures have been determined for *E. coli* (PDB 1PNK without and 1GK9 with bound substrate), *P. rettgeri* (PDB 1CP9), and *A. faecalis* (PDB 3ML0). The 146 strictly conserved residues were determined by a multiple amino acid alignment based on the primary sequence of the pro-penicillin amidases; the numbering starts from the N-terminal of the A-chain (blue). The homology (% of strictly conserved residues compared with the *E. coli* enzyme) varies from ≈30% (*A. viscosus*) to almost 90% (*K. cytrophila*). The strictly conserved catalytically active Ser_{264} ((a) red and (b) green), the N-terminal of the B-chain (green) is formed by proteolytic processing that removes a linker peptide between the A- and B-chains (see Section 6.4.1). (a) Thirteen strictly conserved residues (brown) are, as expected, located in and around the active site, and two (yellow) around the tightly bound Ca^{2+} ion (gray) that holds the A- and B-chains together, and also is required for the correct folding, membrane transport, and processing of the pro-enzyme (Brannigan *et al.*, 1995; McDonough, Klei, and Kelly, 1999; Kasche, Galunsky, and Ignatova, 2003, 2005). (b) Strictly conserved eight positively (yellow) and eight negatively (orange) charged amino acid residues are mainly located far from the active site. They stabilize the enzyme at neutral pH by salt bridges, and determine the stability of the enzyme as a function of pH (see Section 2.10, Figure 2.28). The other strictly conserved residues are located far from the active site or the Ca^{2+} binding site. Their importance for the enzyme function and its activity are still unknown.

enzyme mutant (with one amino acid exchange) is formed after about 10^4–10^5 cell divisions (Lewin, 1997). Only those cells in which the enzyme function is conserved after so-called neutral mutations survive. Mutations that lead to changes in the active site and cause loss of enzyme function are lethal for the organism, and enzymes with these properties disappear during evolution. These results also show that most spontaneous mutations in enzyme genes are neutral (Kimura and Ohta, 1973). They cause the large differences observed in the primary structure of the same enzyme from different sources, and result in the natural enzyme variants with different properties of interest for their application. For a specific enzyme process, the optimal enzyme must be found among the natural variants formed by neutral mutations during evolution. The differences in enzyme properties in these are generally due to changes in less conserved amino acids that can be found by multiple alignment of the primary structures of these variants (see Figure 3.1). These less conserved amino acids are candidates to improve an enzyme by site-directed mutagenesis. Thus, protein engineering, metagenome resources, and search in databases are now alternatives for the screening of enzyme variants in living systems. The gene from a natural enzyme can be mutated *in vitro* in a test tube, where much larger mutation frequencies can be obtained than in living systems. The mutated enzymes with desired properties for the specific application are then selected by suitable screening methods (see Chapter 3).

2.5 Free Energy Changes and the Specificity of Enzyme-Catalyzed Reactions

Once an enzyme has bound substrate(s) in the active site, the catalytically active functional groups in the amino acid residues and cofactors can act as nucleophiles or electrophiles on the groups on the substrate(s) that are transformed in the catalyzed reaction. The binding interactions are both electrostatic (such as ion–ion, ion–dipole, and hydrogen bonds) and hydrophobic. In the enzyme–substrate(s) complex, both the structure of the substrate(s) and the enzyme active center are changed so that the functional groups are spatially oriented near the groups on the substrate(s) with which they interact with a precision within 10^{-2} nm. This leads to a reduction in activation energies in comparison with the uncatalyzed reaction that increases the rate of the catalyzed reaction by factors of up to more than 10^{10}. This is shown schematically in the free energy diagrams (Figures 2.7 and 2.8). In

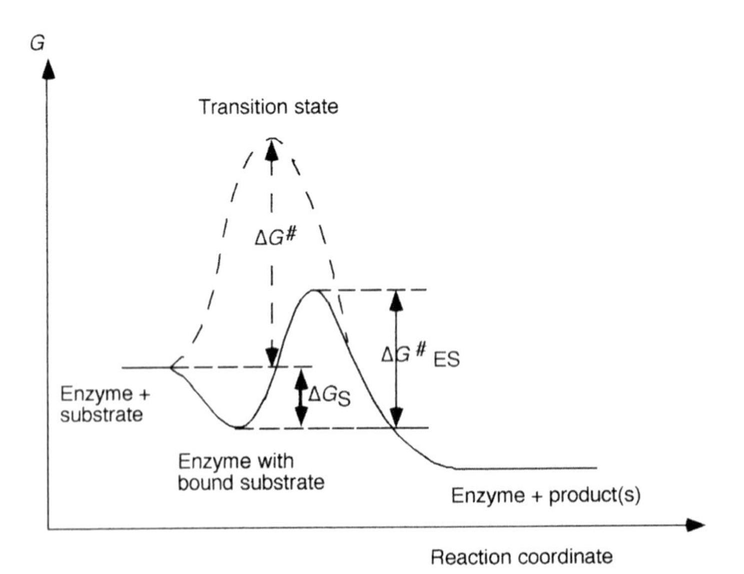

Figure 2.7 Free energy diagram for uncatalyzed (dashed line) and enzyme-catalyzed (solid line) reactions involving one binding step and one chemical reaction (see Figure 2.4). The apparent activation energy of the latter ($\Delta G^{\#}_{ES} - \Delta G_S$) is much smaller than that for the uncatalyzed reaction ($\Delta G^{\#}$); ΔG_S is free energy change due to substrate binding.

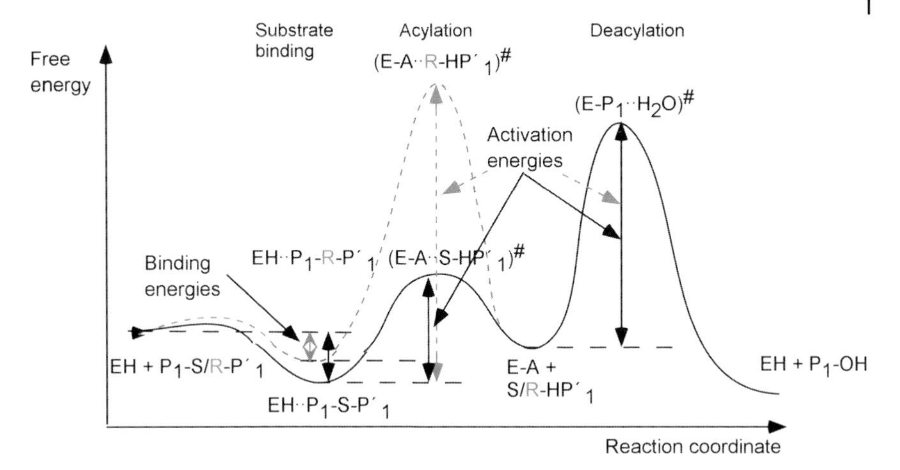

Figure 2.8 Free energy diagram for an enzyme-catalyzed reaction involving the formation of an acyl-enzyme intermediate, consisting of one binding process and two chemical reactions as for the peptide or amide bond hydrolysis shown in Figure 2.5 with either a *S*- (solid line) or *R*-amino acid (dashed line) in the P'_1 position, respectively (see Figure 2.3). The acyl-enzyme E-A is formed by acylation by the amino acid in the P_1 position.

comparison with the uncatalyzed reaction, the catalyzed reaction requires at least two reaction steps – binding, followed by the catalyzed chemical reaction.

Knowledge of the detailed reaction mechanism is required to analyze all factors that may influence the kinetics of an enzyme-catalyzed process. The free energy diagram in Figure 2.7 can be used for a thermodynamic discussion of the peptide bond hydrolysis catalyzed by carboxypeptidase A shown in Figure 2.4. Where this hydrolysis involves the formation of covalent acyl-enzyme intermediates (see Figure 2.5), this diagram is insufficient. In this case, the diagram in Figure 2.8 must be used, where both the acylation and deacylation reactions, but not product binding, are included.

The free energy changes for substrate binding and the activation energies for the chemical reactions in Figures 2.7 and 2.8 can be related to dissociation and rate constants for the substrate binding and catalytic steps of an enzyme-catalyzed reaction. The free energy change ΔG_S defines a dissociation constant K_S from which the commonly used Michaelis–Menten constant K_m can be derived, and is used as a quantitative measure for the substrate binding (see Section 2.7.1). On a molecular level, it is the sum of the free energies for the binding interactions between the substrate and functional groups in the active site of the enzyme. The activation energy $\Delta G^{\#}_{ES}$ defines the first-order rate constant k_{cat} that is a quantitative measure of the catalytic steps(s) in an enzyme-catalyzed reaction. It is also called the *turnover number.* When the chemical step involves two reactions (Figure 2.8), the turnover number depends on the first-order rate constants for both reactions, as will be derived in Section 2.7.1.

One enzyme can catalyze the conversion of different substrates. The interactions with the functional groups in the active site lead to different free energy changes

(Figures 2.7 and 2.8). Therefore, K_m and k_{cat} vary from substrate to substrate. This is the thermodynamic basis for the *substrate specificity* of enzymes. Changes in K_m and k_{cat} are also expected for reactions with the same substrate and enzymes with the same function from different sources.

The main cause for the acceleration of an enzyme-catalyzed versus an uncatalyzed reaction of up to more than a factor 10^{10} was until recently considered to be mainly due to the decrease in the activation energy. The enzyme structure in the transition states in Figures 2.7 and 2.8 was considered to be complementary to the transition state of the substrate. This causes a stabilization of this state that leads to a reduction in the activation energy. The structure of the enzyme is, however, already changed when it binds the substrate (induced fit, as shown in Figure 2.2) (Done *et al.*, 1998). For the evolution of enzymes, it is favorable when the binding energy ΔG_S also contributes to the reduction of the activation energy. Therefore, both the substrate binding and catalytic reaction contribute to the substrate specificity of an enzyme and the acceleration of the rate compared with the uncatalyzed reaction (Menger, 1992; Schowen, 2003; Garcia-Viloca *et al.*, 2004).

In Figure 2.8, the free energy diagram for the hydrolysis of a dipeptide or a N-acylated amino acid where the C-terminal amino acid is either a (*R*)- or (*S*)-enantiomer is shown for an enzyme-catalyzed reaction involving acyl-enzyme intermediates. The corresponding reaction mechanisms are given in Figure 2.5. Due to a different orientation of the residues of the (*R*)- and (*S*)-substrate relative to the functional groups in the active site of the enzyme, both the binding and activation energies differ for the two substrates. In the case shown, this applies only for the binding and acylation step for a (*S*)-specific enzyme. For a N-terminal (*R*,*S*)-amino acid in a dipeptide, differences in the activation energy are also expected for the deacylation step. This shows that enzymes are *stereospecific*, but not always strictly, so both substrates can be hydrolyzed by the enzyme. Both substrate binding and catalysis contribute to the stereospecificity. From the above discussion, it can also be concluded that the stereospecificity should differ for enzymes with the same function from different sources. This discussion on the stereospecificity should also apply for the *regiospecificity* of enzymes.

Although the above analysis on the thermodynamic basis for the substrate and stereospecificity of enzymes was derived for hydrolases, it is equally applicable to all types of enzymes. For a more quantitative analysis of all the factors that influence the kinetics and yields of enzyme-catalyzed reactions, the kinetic constants K_m and k_{cat} must be determined (see Section 2.7).

2.6
Equilibrium- and Kinetically Controlled Reactions Catalyzed by Enzymes

When enzymes catalyze equilibrium-controlled reactions (see Figures 2.4 and 2.5), they accelerate the rate to reach the equilibrium, but do not influence the equilibrium constant or the end point of the reaction (Figure 2.9). In enzyme technology, high product yields are essential in order to obtain a process that is competitive.

The equilibrium (thermodynamic) yield is only determined by the initial substrate concentration, the solubilities of substrate and product, and the equilibrium constant that can be influenced by the system properties such as pH, *T*, *P*, and so on. From this, a process window in the latter properties can be derived where a given yield (such as >90%) can be obtained. Among the enzymes that can catalyze this process, the best enzyme is the one with the best properties (activity, stability) inside this process window. This demonstrates that in the rational design of an enzyme process one must consider the properties of both the enzyme and the catalyzed process.

For an equilibrium-controlled, enzyme-catalyzed resolution of racemates, the equilibrium yield of the desired enantiomer is not influenced by the properties of the enzyme. In this case, however, the steric purity of the desired enantiomer must also be considered. This is influenced by the stereospecificity of the enzyme (Figures 2.8 and 2.9a). Thus, for these processes this property must also be considered in the selection of the optimal enzyme.

A detailed analysis of the reaction mechanisms in Figure 2.5 shows that H_2O acts as a nucleophile in the deacylation step. However, might other molecules (R–OH or R–NH_2) also act as nucleophiles here? When this is the case, the hydrolases could also act as transferases, transferring acyl or glycosyl groups to R–OH or R–NH_2. This has been shown to apply for hydrolases that form covalent intermediates, such as acyl-enzymes (Figure 2.5), glycosyl-enzymes, or intermediates such as cyclic phosphates observed in RNase-catalyzed reactions (Kasche, 1986). These reactions require the use of an activated substrate (such as an ester or amide). The general reaction scheme of such enzyme-catalyzed, *kinetically controlled reactions* and experimental data for the synthesis of benzylpenicillin are shown in Figure 2.9. In these processes, the nucleophiles H_2O and NH (6-aminopenicillanic acid (6-APA) in Figure 2.9b) compete in the deacylation reaction of the acyl-enzyme (see Figure 2.5b). In the former reaction, the enzyme acts as a hydrolase with an apparent hydrolase rate constant k_H. In the latter reaction, where the enzyme acts as a transferase with an apparent transferase rate constant k_T, a condensation product is formed that can also be hydrolyzed by the enzyme. The concentration of the latter increases until its synthesis rate equals the hydrolysis rate, where the maximal product concentration is observed. The final condensation product concentration is given by the equilibrium constant for the reaction:

$$\mathrm{AN} + \mathrm{H_2O} \rightleftharpoons \mathrm{AOH} + \mathrm{NH}$$

Thus, in kinetically controlled reactions much larger concentrations of the product AN than for an equilibrium-controlled reaction can be obtained. The maximum product concentration depends on the selectivity $(k_T/k_H)_{app}$ and the rate with which AN is hydrolyzed by the enzyme. Thus, in contrast to the equilibrium-controlled process the maximal product yield in the kinetically controlled process depends on the properties of the enzyme. The relationship between $(k_T/k_H)_{app}$ and intrinsic rate constants of the enzyme can be derived from Scheme (2.1).

The ratio $(k_T/k_H)_{app}$ is determined from the initial rates of the formation of the condensation product AN (v_T) and hydrolysis product AOH (v_H). From Figure 2.9

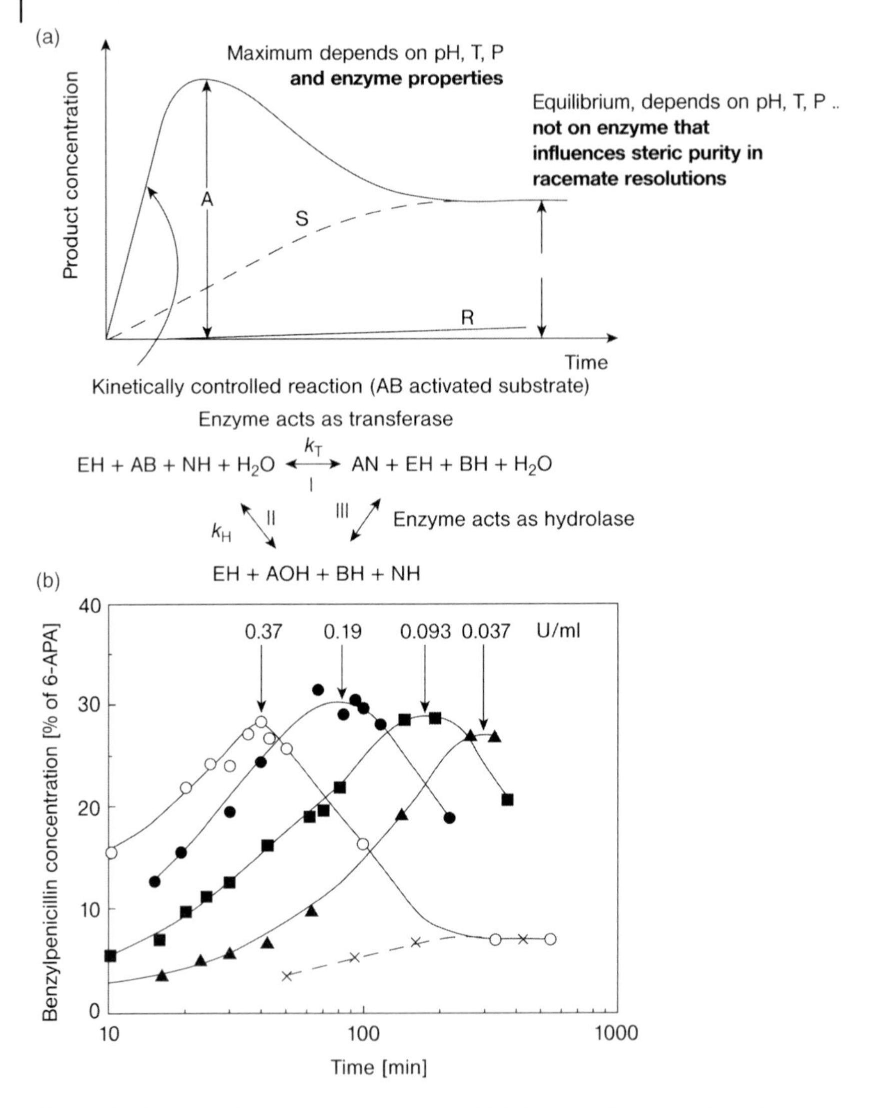

Figure 2.9 Schematic time dependence (a) and experimental results (b) for kinetically and equilibrium-controlled reactions catalyzed by hydrolases with acyl-enzyme and other covalent intermediates.
(a) This also shows that the steric purity of the product in equilibrium-controlled racemate resolutions depends on the stereoselectivity of the used enzyme.
(b) Kinetically (solid line) and equilibrium-controlled (dashed line) synthesis of penicillin G (AN) from equal concentrations of phenylacetyl glycine (AB) or phenylacetic acid (AOH) and 6-aminopenicillanic acid (NH) at different enzyme (penicillin amidase from *E. coli*) concentrations given in U ml^{-1} (pH 6.0, 25 °C). The kinetically controlled maximum is (within experimental error) independent of the concentration of the enzyme, and is much larger than the equilibrium concentration.

and Eq. (2.1), the following relationship is derived:

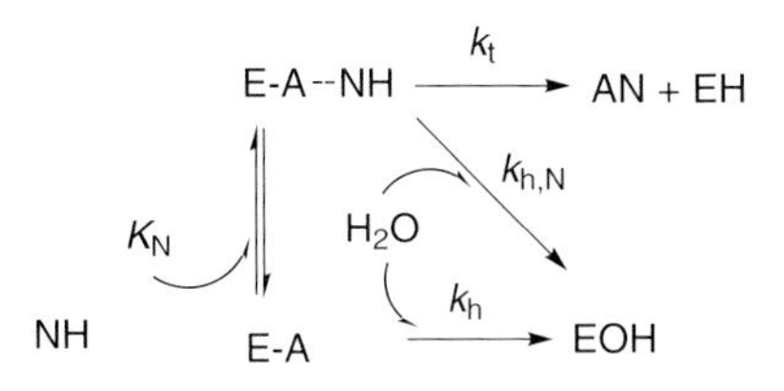

Scheme 2.1

$$(\nu_T/\nu_H) = (k_T/k_H)_{app}[NH]/[H_2O]$$
$$= k_t[E-A\cdots NH]/(k_h[H_2O][E-A] + k_{h,N}[H_2O][E-A\cdots NH]), \quad (2.1)$$

assuming that equilibrium is established in the nucleophile binding, that is,

$$K_N = [E-A][NH]/[E-A\cdots NH], \quad (2.2)$$

then

$$(k_T/k_H)_{app} = k_t/(k_h K_N + k_{h,N}[NH]), \quad (2.3)$$

that is, $(k_T/k_H)_{app}$ depends on both intrinsic enzyme properties and [NH].

Such kinetically controlled processes can be applied in enzyme technology for the production of condensation products catalyzed by hydrolases that can also act as transferases (Kasche, 1986, 2001). Some processes have already been realized on an industrial scale (production of insulin and other peptides such as the sweetener aspartame; synthesis of semisynthetic cephalosporins and penicillins; numerous kinetic resolutions of racemates, synthesis of chiral products from prochiral precursors, etc.) (Bornscheuer and Kazlauskas, 2005; Bruggink, 2001; Kasche, 2001; Liese *et al.*, 2006; Zmiejewski *et al.*, 1991). Hydrolases that can be used as biocatalysts for such processes and their apparent transferase to hydrolase ratio $(k_T/k_H)_{app}$ derived from initial rate measurements are listed in Table 2.3. The ratio $(k_T/k_H)_{app}$ is also a measure for the P_1' specificity of peptidases and amidases (see Figure 2.3) and can be used to map these (Kasche, 2001; Galunsky and Kasche, 2002).

The concentrations of condensation products (biopolymers such as nucleic acids, polysaccharides, and proteins) in living cells are much higher than those that exist at thermodynamic equilibrium in these systems. The synthesis of these compounds at concentrations much higher than those that can exist at equilibrium occurs in kinetically controlled processes from activated substrates (trinucleotides such as ATP, nucleotide-activated monosaccharides, sucrose, adenylated amino acids). The transferases that catalyze these reactions have been optimized during evolution so that their transferase activity is much larger than the hydrolase activity (Table 2.3). These enzymes and their activated substrates are still too expensive for use in enzyme technology, although exceptions here are dextran- and levansucrases that can synthesize polysaccharides from the activated substrate sucrose (Buchholz and Monsan, 2003).

Table 2.3 Transferases and hydrolases that can be used as transferases in kinetically controlled synthesis of condensation products and their selectivities $(k_T/k_H)_{app}$ for different nucleophiles (Kasche, 1986).

Enzyme	Nucleophile	$(k_T/k_H)_{app}$
2. Transferases		
DNA polymerase	DNA	10^7
Hexokinase	Glucose	10^6
Dextransucrase	Glucose	
3. Hydrolases		
Lipase[a)]	Acids	
	Alcohols	10
	Amines	
Alkaline phosphatase	Tris	10^2–10^3
RNase I	Alcohols	10
	Nucleoside	10^2–10^3
Glycosidases	Alcohols	10^2
β-Galactosidase	Lactose	10^2–10^3
Serine-, thiol-	(*S*)-Amino acids	10–10^2
Peptidases	(*S*)-Amino acid esters	10^2–10^4
	(*S*)-Amino acid amides	10^2–10^5
	(*R*)-Amino acid amides	10–10^3
	Alcohols	–10^2
	Tris	–10^2
Amidases		
Penicillin amidase	6-Aminopenicillanic acid	10^3–10^4
	(*S*)-Amino acids	–10^4
	Alcohols	1

a) For reactions with lipases (esterifications are often carried out in almost water-free systems), the above data apply for aqueous systems.

It can be seen from the data in Table 2.3 that many hydrolases may be used as biocatalysts for the kinetically controlled synthesis of condensation products. However, in order to obtain large condensation product yields, much higher nucleophile concentrations (up to 1 M) must be used compared to those in living systems ($<10^{-2}$ M). These concentrations must be chosen so that $(k_T/k_H)_{app})([NH]/[H_2O]) \gg 1$ (see Table 2.3 and Eq. (2.1)).

2.7 Kinetics of Enzyme-Catalyzed Reactions

Enzymes are used as catalysts to decrease the time required to reach the end point of equilibrium- and kinetically controlled processes (see Figure 2.9). This time is a function of the enzyme concentration and enzyme properties (binding of substrate and product, activation energy of the catalytic reactions), as discussed in Section 2.5. In order to calculate this time, these properties must be determined based on a kinetic analysis of the enzyme-catalyzed reaction (see Section 2.7.1). The end point of an enzyme-catalyzed process must be selected to give the highest possible product yield. When the end point is the equilibrium state of the reaction, the yield cannot be influenced by the properties of the enzyme. The latter, however, influence the yield at the product maximum in a kinetically controlled process. The product yield depends also on initial substrate concentrations, pH, temperature, solvent, and ionic strength. This implies that the selection of the process conditions for an enzyme process depends on the pH, ionic strength, solvent, and temperature dependence of the catalyzed process (see Section 2.8). These can be unfavorable for the enzyme used as biocatalyst, when it is unstable at the conditions optimal for the process. From this it follows that the pH, temperature, and ionic strength dependence of the properties of the enzyme (binding, catalysis, stability) must also be known in order to determine the best process conditions (see Section 2.7.2).

2.7.1 Quantitative Relations for Kinetic Characteristics and Selectivities of Enzyme-Catalyzed Reactions

2.7.1.1 Turnover Number (k_{cat}) and Michaelis–Menten Constant (K_m)

The enzyme-catalyzed reactions discussed in Sections 2.4–2.6, as well as for all other enzymes, can be described by one of the following kinetic schemes. *Note*: The enzyme in this section, in contrast to the usual abbreviation E, is written EH; this is done to stress that in many enzyme-catalyzed processes proton transfer steps are directly involved (see Figure 2.5; Benkovic and Hammes-Schiffer, 2003) and to indicate the pH dependence of these processes. Later, the common E will be used:

1) One-substrate (AB) reactions (only for lyases in one direction and isomerases in both directions).

$$\mathrm{EH + AB} \underset{k_{-1}}{\overset{k_1}{\rightleftharpoons}} \mathrm{(EH{\cdot\cdot}AB)} \begin{cases} \overset{k_2}{\rightleftharpoons} \mathrm{EH + AB'} \quad \text{(isomerases)} \\ \underset{k_{-2}}{\rightleftharpoons} \mathrm{EH + A + B} \ \text{(lyases)} \end{cases} \tag{2.4}$$

2) Two-substrate (AB and NH) reactions (for all other enzymes).
3) Both substrates cannot be bound simultaneously in the active center.

$$\underbrace{\mathrm{EH + AB} \underset{k_{-1}}{\overset{k_1}{\rightleftharpoons}} \mathrm{(EH{\cdot\cdot}AB)}}_{\text{substrate binding}} \underbrace{\underset{k_{-2}}{\overset{k_2}{\rightleftharpoons}} \mathrm{(E{\cdot\cdot}A) + BH} \underset{k_{-3}}{\overset{k_3}{\rightleftharpoons}}}_{\text{catalytic reactions}} \mathrm{(NH{\cdot\cdot}EA)} \underset{k_{-4}}{\overset{k_4}{\rightleftharpoons}} \mathrm{EH + AN} \tag{2.5}$$

(NH enters and BH leaves at the k_3 step.)

4) Both substrates can be bound simultaneously in the active center.

$$\begin{array}{ccc} & (EH\cdot\cdot AB) + NH & \\ EH + AB + NH & & (NH\cdot\cdot EH\cdot\cdot AB) \underset{k_{-2}}{\overset{k_2}{\rightleftharpoons}} (NH\cdot\cdot EA) + BH \underset{k_{-3}}{\overset{k_3}{\rightleftharpoons}} EH + AN + BH \\ & (NH\cdot\cdot EH) + AB & \end{array}$$

substrate binding — catalytic reactions (2.6)

Most enzyme-catalyzed reactions can be described by one of these three schemes. In the case where more than one substrate NH participate as nucleophiles, as in kinetically controlled processes, the reaction schemes (2.5) and (2.6) will include more branching reactions. From these schemes, relationships that provide quantitative measures for the enzyme properties, substrate binding, and rate of the enzyme-catalyzed reaction can be derived. They are derived assuming that the following boundary conditions are fulfilled:

1) Substrate concentration ≫ enzyme concentration:

$$[NH]_0, [AB]_0 \gg [EH]_0 \quad (\text{subscript 0 for } t = 0).$$

2) Only the initial rate is measured. Then $[NH]_0$ and $[AB]_0$ are practically constant, and the reverse reactions after the binding of the substrate(s) can be neglected.
3) Equilibrium in the binding of the substrates is obtained. This is an assumption that is problematic.
4) Steady state in the concentration of the intermediates EH ··· AB, H ··· EH ··· AB, NH ··· EA, and so on ("···" denotes noncovalent interactions; EA = covalent enzyme–substrate intermediate (such as an acyl- or glycosyl-enzyme)).

With conditions 1–4 and mass conservation relationships, a set of linear equations can be derived, from which the reaction rate can be expressed in the rate constants of the reaction schemes (2.4)–(2.6) and the known concentrations. Most enzyme-catalyzed reactions, also in enzyme technology, are two-substrate reactions. They will therefore be considered here.[2] The rate of product formation (or sub-

2) In biochemistry textbooks, generally only one-substrate reactions (Eq. 2.4) are described. The rate of product formation is then $v = k_2[EH \cdots AB]$. The mass conservation relationship for the enzyme is

$$[EH]_0 = [EH] + [EH \cdots AB]$$

and the steady-state condition for $[EH \cdots AB]$ gives

$$k_1[EH][AB]_0 = (k_{-1} + k_2)[EH \cdots AB].$$

From these two equations, we obtain

$$[EH \cdots AB] = [EH]_0[AB]_0/((k_{-1}+k_2)/k_1 + [AB]_0).$$

The rate of product formation is then

$$v = k_2[EH]_0[AB]_0/(K_m + [AB]_0),$$

where the turnover number k_2 and the Michaelis–Menten constant $K_m = (k_{-1} + k_2)/k_1$ are the quantitative measures for the catalytic and substrate binding properties of the enzyme.

strate consumption) for the reaction scheme 2.5 is

$$v = \mathrm{d}[\mathrm{AN}]/\mathrm{d}t = k_4[\mathrm{EA}\cdots\mathrm{NH}]. \tag{2.7}$$

From the linear set of equations, an expression for $[\mathrm{EH}\cdots\mathrm{AN}]$ in the known concentrations $[\mathrm{EH}]_0$, $[\mathrm{AB}]_0$, and $[\mathrm{NH}]_0$ and the rate constants can be derived. After insertion in Eq. (2.7), the following expression in the form of a Michaelis–Menten equation of a one-substrate reaction is obtained:

$$v = \frac{k_{\mathrm{cat},AB}[\mathrm{AB}]_0[\mathrm{EH}]_0}{K_{\mathrm{m},AB}[\mathrm{AB}]_0}, \tag{2.8}$$

with the turnover number

$$k_{\mathrm{cat},AB} = \frac{k_2 k_3 k_4 [\mathrm{NH}]_0}{k_3(k_4 + k_2)[\mathrm{NH}]_0 + k_2 k_4} \tag{2.9}$$

and the Michaelis–Menten constant

$$K_{\mathrm{m},AB} = \frac{(k_{-1} + k_2)k_3 k_4 [\mathrm{NH}]_0}{(k_3(k_4 + k_2)[\mathrm{NH}]_0 + k_2 k_4)k_1}. \tag{2.10}$$

For large $[\mathrm{NH}]_0$ (as in hydrolysis reactions with $[\mathrm{NH}]_0 = \mathrm{H_2O}$), Eqs. (2.9) and (2.10) become

$$k_{\mathrm{cat},AB} = \frac{k_2 k_4}{k_2 + k_4} \tag{2.11}$$

and

$$K_{\mathrm{m},AB} = \frac{(k_{-1} + k_2)k_4}{(k_2 + k_4)k_1}, \tag{2.12}$$

that is, for the same enzyme these quantities are characteristic for the substrate AB only under these conditions. For smaller $[\mathrm{NH}]_0$, they also depend on this concentration. Equation (2.8) was first derived in 1913 by Michaelis and Menten. It is a general expression for the rate of enzyme-catalyzed reactions, and plots of the rate as a function of the substrate concentration are called Michaelis–Menten plots (Figure 2.10).

The turnover number k_{cat} has the dimension s^{-1}, and is the number of molecules converted per second by one enzyme molecule under substrate saturation conditions. It is only influenced by the rates of the catalytic reactions (k_2, k_4), and is a measure for the activation energy $G^{\#}_{\mathrm{ES}}$ in Figures 2.7 and 2.8. The Michaelis–Menten constant K_{m} has the dimension $\mathrm{mol\ l^{-1}}$ (M), and is the substrate concentration at which 50% of the maximal velocity V_{max} ($=k_{\mathrm{cat}}[\mathrm{EH}]_0$) has been reached. It is an apparent equilibrium constant and a measure for the substrate binding energy ΔG_{S} in Figures 2.7 and 2.8. It is determined by the equilibrium constant for

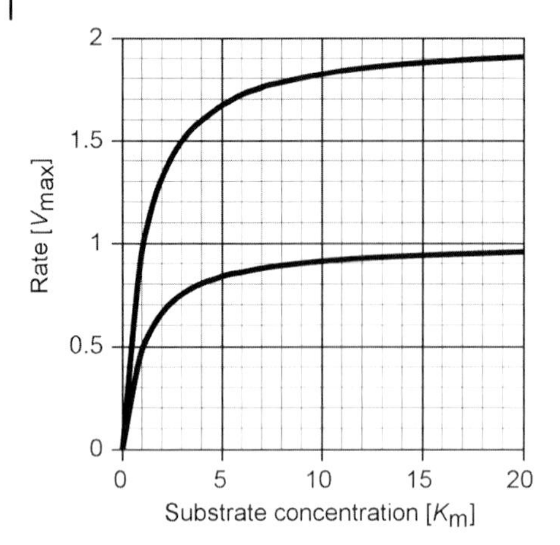

Figure 2.10 Michaelis–Menten plot – that is, rate ν (Eq. (2.8)) in units of V_{max} – for an enzyme-catalyzed reaction as a function of the substrate concentration (in K_m units) at enzyme concentrations $[EH]_0$ (lower) and $2[EH]_0$ (upper curve). Note that $\nu = 0.9V_{max}$ requires a substrate content of $\approx 10K_m$.

the substrate binding and rate constants for the catalytic reactions. The smaller the K_m value, the better the binding of the substrate to the enzyme and the efficiency of substrate conversion. The turnover number k_{cat} and the Michaelis–Menten constant K_m are used to characterize the kinetic properties of an enzyme. They are influenced by both the enzyme and the substrate. The ratio k_{cat}/K_m (dimension: $M^{-1}\,s^{-1}$), an apparent second-order rate constant, is called *specificity constant*, and is a quantitative measure for the substrate specificity of an enzyme. The specificity constant cannot be larger than the rate of a diffusion-controlled bimolecular reaction ($<10^8$–$10^9\,M^{-1}\,s^{-1}$) (Fersht, 1999). For the determination of the substrate specificity of one enzyme and different substrates or of different enzymes for one substrate, k_{cat} and K_m must be determined. When the enzyme concentration is unknown, as in homogenates, $V_{max} = k_{cat}[EH]_0$ is determined. The turnover number can only be determined for enzymes for which the active site can be titrated – that is, when the active enzyme concentration $[EH]_0$ can be determined. This is possible for enzymes such as serine peptidases, penicillin amidases, and lipases that can be practically irreversibly covalently acylated in the active site (Fersht, 1999). V_{max} is a measure for the amount of enzyme expressed in activity units. For this, the enzyme unit (U) is still used, although it is not a Système Internationale (SI) unit. One U is the amount of enzyme that catalyzes the conversion of 1 μmol substrate per minute under defined conditions (pH, *T*). One SI unit for enzyme activity (katal, abbreviated kat) is the amount of enzyme that converts 1 mol of substrate per second, but this is hardly used in enzymological literature. One kat is equivalent to 60×10^6 U.

Different methods have been developed to determine k_{cat} and K_m from the determinations of ν at different substrate concentrations. One suitable method was developed by Eadie and Hofstee. Equation (2.8) can be linearized as

$$\frac{v}{[\mathrm{AB}]_0[\mathrm{EH}]_0} = \frac{k_{\mathrm{cat}}}{K_{\mathrm{m}}} - \frac{v}{[\mathrm{EH}]_0 K_{\mathrm{m}}}. \tag{2.13}$$

$v/[\mathrm{AB}]_0[\mathrm{EH}]_0$ as a function of $v/[\mathrm{EH}]_0$ provides a linear plot (the Eadie–Hofstee plot), from which k_{cat} and K_m can be determined from the x- and y-intercepts. Frequently, these quantities are determined by another linear plot ($1/v$ is plotted against $1/[\mathrm{AB}]_0$; Lineweaver–Burk plot). It has been shown that the experimental error in the determination of k_{cat} and K_m is larger for the Lineweaver–Burk plot than for the Eadie–Hofstee plot (Deranleau, 1969), but even for the latter plot the errors are not negligible ($\pm 10\%$) when the error in rate determination is $\pm 2\%$.

Values of k_{cat} and K_m for different substrates (same enzyme) and different enzymes (same substrate) are listed in Tables 2.4 and 2.5. These tables provide information about the substrate specificity such as the binding of different P_1' and P_1 residues, hydrolysis of ester or amide bonds – that is, activation of these bonds by the catalytic subsite C in Figure 2.3, and the interactions between the active site and the substrate. It is clear from the data in Table 2.4 that the substrate specificity for one enzyme can vary by several orders of magnitude for different substrates, and that the specificity constants for some enzymes approach the rate of a diffusion-controlled reaction $k_{cat}/K_m \approx 10^8\,\mathrm{M}^{-1}\,\mathrm{s}^{-1}$. Trypsin and α-chymotrypsin are better esterases than amidases, and have different P_1 specificities. The difference in esterase and amidase activity is mainly caused by differences in the substrate binding (trypsin) or the catalytic reactions (α-chymotrypsin). These endopeptidases cannot hydrolyze a C-terminal unprotected amino acid as the carboxyl group carries a negative charge. The substrate is repelled by the negatively charged active site of these enzymes (see Figure 2.5). The data in Table 2.4 for α-chymotrypsin and penicillin amidase also show that the specificity constants for the same substrate vary considerably with the source of the enzyme.

2.7.1.2 Stereoselectivities for Equilibrium- and Kinetically Controlled Reactions

It is clear from the data in Table 2.5 that the same substrate can be hydrolyzed by different enzymes. It is also apparent that the stereoselectivity of enzymes differs, and from this it follows that hydrolases, as well as other enzymes, have no strict stereospecificity. (The stereospecificity gives qualitative information on whether the enzyme prefers (*S*)- or (*R*)-substrates.) Enzymes that act preferably on (*S*)-enantiomers can also convert (*R*)-enantiomers, and vice versa. The results in Table 2.5 show that the stereoselectivity of the studied enzymes in all cases is due to changes in both k_{cat} and K_m for the enantiomeric substrates. The binding of the less specific substrate results in a larger distance between the bond that is changed in the catalytic reaction and the catalytic subsite C than for the preferred substrate (Figure 2.11). This results in different activation energies for the transformation of the (*S*)- and (*R*)-enantiomers (see Figure 2.8). The data for penicillin amidase also show that the stereospecificity of an enzyme can be different in the S_1 and S_1' binding subsites. The stereoselectivity is increased and the stereospecificity reversed when an additional binding site is

Table 2.4 Turnover number k_{cat} and Michaelis–Menten constant K_m for different hydrolases from different sources.

Enzyme	Substrate ($P_i \cdots P_1 \downarrow P_1' \cdots P_1$)	Temperature (°C)	pH	k_{cat} (s^{-1})	K_m (mM)	k_{cat}/K_m ($M^{-1}\ s^{-1}$)
Trypsin (bovine)	Benzoyl-Arg↓NH_2	25.5	7.8	27	2.1	13 000
	Benzoyl-Arg↓OEt	25.0	8.0	19	0.02	1 000 000
	Z-Lys↓OMe	30	8.2	101	0.23	440 000
	Z-Lys↓Ala	40	8.2	0		≈0
	Z-Lys↓Ala–Ala–Ala	40	8.2	4.6	3.7	1200
α-Chymotrypsin (bovine)	Acetyl-Tyr↓OEt	25.0	7.0	160	3.7	42 000
	Acetyl-Tyr↓NH_2	25.0	7.8	0.28	7	40
	Benzoyl-Tyr↓OEt	25.0	7.8	78	4	20 000
	Benzoyl-Phe↓OEt	25.0	7.8	37	6	6300
	Benzoyl-Met↓OEt	25.0	7.8	0.77	0.8	1000
	Ac-ProAlaProPhe↓Ala	37	8.0	0		≈0
	Ac-ProAlaProPhe↓AlaAlaNH_2	37	8.0	37	0.83	44 000
Mouse NZB	Acetyl-Tyr↓OEt	25	7.0	250	2.1	120 000
Mouse A/sn	Acetyl-Tyr↓OEt	25	7.0	210	2.1	100 000
Carboxypeptidase A	Carbobenzoyl-Gly↓Try	25	7.5	89	2	17 000
	Carbobenzoyl-Gly↓Leu	25	7.5	10	28	390
Penicillin amidase						
E. coli		25	7.8	48	0.01	4 800 000
A. faecalis	Penicillin G	25	7.8	80	0.008	10 000 000
K. citrophila		25	7.8	60	0.02	3 000 000
Adenosine triphosphatase	ATP	25	7.0	104	0.012	8 300 000
Urease	Urea	20.8	7.1	20 000	4	5 000 000

NH in Eqs. (2.5) and (2.6) = H_2O; ↓ = bond that is hydrolyzed (from Laidler, 1958 and the laboratory of Kasche).

Table 2.5 Stereoselectivity in the hydrolysis of the same substrate with different hydrolases (pH 7.5, 25 °C, $I = 0.2$ M) (Michaelis, 1991; Lummer *et al.*, 1999; Galunsky and Kasche, 2002).

Substrate P_1-↓-P_1'	K_m (mM)	k_{cat} (s^{-1})	k_{cat}/K_m ($M^{-1}\ s^{-1}$)	Stereoselectivity E_{eq} $(k_{cat}/K_m)_S/(k_{cat}/K_m)_R$
Hydrolysis with penicillin amidase (EC 3.5.1.11 from *E. coli*)				
(*S*)-Phg-↓-O-Me	20	11	550	0.5
(*R*)-Phg-↓-O-Me	32	35	1100	
N-Acetyl-(*S*)-Phg-↓-O-Me			1.3	>13
N-Acetyl-(*R*)-Phg-↓-O-Me			<0.1	
(*S*)-Phg-↓-NH_2	11	6.4	580	0.27
(*R*)-Phg-↓-NH_2	17	36	2100	
Hydrolysis with α-chymotrypsin (EC 3.4.21.1, bovine)				
(*S*)-Phg-↓-O-Me	50	0.46	9.2	14
(*R*)-Phg-↓-O-Me	140	0.08	0.57	
N-Acetyl-(*S*)-Phg-↓-O-Me	4	0.8	200	285
N-Acetyl-(*R*)-Phg-↓-O-Me	7	0.005	0.7	
Hydrolysis with proteinase K (EC 3.4.21.14 from *Tritirachium album*)				
(*S*)-Phg-↓-O-Me			0.6	2
(*R*)-Phg-↓-O-Me			0.3	

↓ indicates the bond that is hydrolyzed. Phg: phenylglycyl.

involved in binding of the substrate (Figure 2.11). This follows from the results for *N*-acetylated substrates in Table 2.5, where the acetyl group is bound in the S_2 subsite.

The stereoselectivity of enzymes is increasingly applied to produce pure enantiomers from prochiral (prosterogenic) compounds (asymmetric synthesis) or racemic mixtures (kinetic resolution). An α-keto acid is a prochiral compound that in living cells is transformed to a (*S*)-amino acid in reactions catalyzed by transaminases (transferases) or dehydrogenases (oxidoreductases). One of the first industrial enzyme processes was the production of (*S*)-amino acids from a racemic mixture of *N*-acetylated (*R,S*)-amino acids using a (*S*)-specific aminoacylase (a hydrolase) (Tosa *et al.*, 1969). The deacylated (*S*)-amino acid was separated from the *N*-acetyl-(*R*)-amino acid by crystallization or ion-exchange chromatography. The isolated *N*-acetyl-(*R*)-amino acid was racemized, after which almost all of the racemic mixture could be transformed into the (*S*)-amino acid (see Chapter 4).

The enzyme reactions that can be used for the production of pure enantiomers are shown in Figure 2.12. In the kinetic resolution of racemates II–IV, ≤50% of the racemic mixture can be transformed to the desired enantiomer. To increase the yield, the latter must be separated from the other enantiomer that can be racemized in either chemical or enzymatic, catalyzed by racemases, processes. This can also be achieved when the kinetic resolution of racemates and the racemization can be carried out simultaneously, and this is referred to as "dynamic kinetic resolution." The application of enzymes for racemate resolutions or enantiomer production

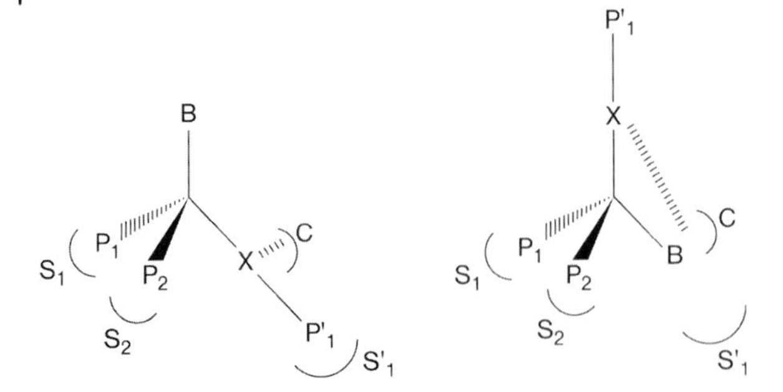

Figure 2.11 Binding interactions that determine the stereospecificity of an enzyme. They are illustrated for an enzyme that is specific for an (*S*)-enantiomer (with groups P_1, P_2, B, and X-P'_1 bound to the asymmetric C atom) that has binding interactions with the S_1, S_2, and S'_1 binding subsites. The correct binding of this enantiomer is shown to the left. The catalytic subsite (C) is then near the bond changed in the catalytic reaction. For the (*R*)-enantiomer, this and the other three binding interactions cannot occur simultaneously. Only two of the three possible binding interactions are possible. In the case shown to the right, one of these cases leads to a large distance between the catalytic site and the bond to be changed. Thus, either the binding of the (*R*)-enantiomer is weaker than that for the (*S*)-enantiomer or the catalytic reaction for the (*R*)-enantiomer is slower as the activation energy will increase (see Figure 2.8).

from prochiral compounds is covered in detail in Chapter 4. The stereoselectivities of the enzymes used in these processes must be known, but can be determined as follows for equilibrium- and kinetically controlled reactions.

The stereoselectivity E of an enzyme-catalyzed reaction is defined as the ratio of the rates of the (*S*)- and (*R*)-enantiomer consumption (Chen *et al.*, 1982):

$$E = v_S / v_R, \tag{2.14}$$

For equilibrium-controlled reactions such as hydrolysis reactions, the stereo- or enantioselectivity has been shown to be

$$E_{eq} = (k_{cat}/K_m)_S / (k_{cat}/K_m)_R, \tag{2.15}$$

that is, the ratio of the specificity constants of the two enantiomers. The intrinsic enzyme property E_{eq} can be calculated from the determination of these constants, based on initial rate measurements, for the isolated enantiomers. Kinetically controlled reactions are only used for the resolution of racemic nucleophile mixtures (see Section 2.6, Eq. (2.1), and Figure 2.12). The kinetically controlled resolution of a racemic mixture of the activated substrate (*R,S*)-A-B with a nucleophile NH is not a suitable racemate resolution procedure as it gives two products with the desired enantiomer (*S*)-A-OH and (*S*)-A-N)). Thus, for kinetically controlled racemate resolutions, the relationship for E for only the nucleophile binding subsite (S'_1 stereoselectivity) must be derived using the definition given in Eq. (2.14). The intrinsic (concentration-independent) S'_1 stereoselectivity can only be determined from initial

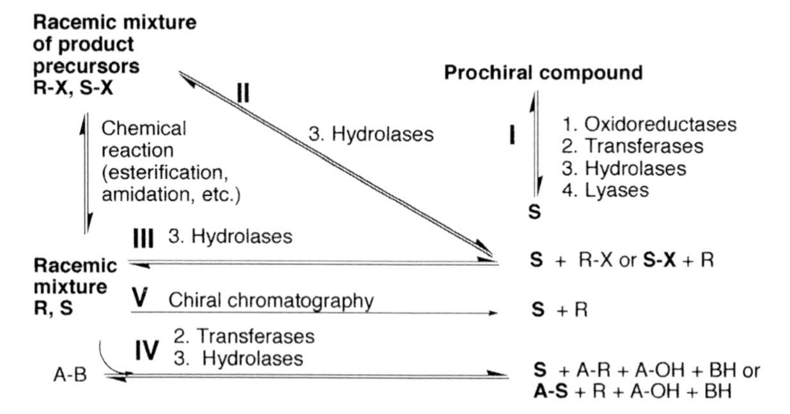

Figure 2.12 Different enzyme and chromatographic separation processes for the production of pure (*S*)-enantiomers from racemic mixtures by kinetic resolution or asymmetric synthesis from prochiral compounds. The class of enzymes that can catalyze the different reactions is given based on the classification given in Section 2.2. The enzyme processes I–IV are equilibrium controlled, with the exception of process IV, which is catalyzed by hydrolases in aqueous systems that are kinetically controlled reactions. In process I, a 100% product yield can be obtained, whereas in the other processes (II–V) only 50% of the racemic mixture can be converted to the desired enantiomer. To obtain 100% yield, the unwanted enantiomer must be racemized. A similar scheme can be designed for the production of the (*R*)-enantiomer.

rates of the formation of the (*S*)- and (*R*)-enantiomeric products using a racemic nucleophile mixture (Galunsky and Kasche, 2002).[3] Under the assumption that equilibrium is obtained in nucleophile binding, it is

$$E_{\mathrm{kin}} = \nu_{\text{A-}(S)\text{-N}}/\nu_{\text{A-}(R)\text{-N}} = (k_t/K_N)_S/(k_t/K_N)_R. \tag{2.16}$$

The relationships for E determined by Eqs. (2.15) and (2.16) can be used for a thermodynamic analysis of the pH, T, and ionic strength dependence of E. This requires a large number of measurements, but allows the determination of E in the range from $\approx 10^{-4}$ to $\approx 10^4$.

A simpler and more rapid, but less accurate, estimation of E is based on the determination of the enantiomer excess of the product ee_P ($ee_P = |[P_S] - [P_R]|/([P_S] + [P_R])$) or substrate ee_S ($ee_S = |[S_S] - [S_R]|/([S_S] + [S_R])$) and the extent of the reaction when the enantiomeric excess is measured:

$$c = 1 - ([S_S] + [S_R])/([S_S]_0 + [S_R]_0),$$

3) Previously, the ratio $(k_T/k_H)_S/(k_T/k_H)_R$, determined from separate measurements with the two enantiomers, was used as a measure for the stereoselectivity in kinetically controlled reactions. This ratio, however, depends on [NH] – that is, it is concentration dependent. It should therefore not be used as a measure for the stereoselectivity of the enzyme.

where the subscript 0 denotes the initial concentration, for an equilibrium- or kinetically controlled resolution of a racemic mixture from the relationships

$$E = \frac{\ln[1 - c(1 + ee_P)]}{\ln[1 - c(1 - ee_P)]} \tag{2.17}$$

or

$$E = \frac{\ln[(1 - c)(1 - ee_S)]}{\ln[(1 - c)(1 + ee_S)]}. \tag{2.18}$$

The E-values calculated from Eq. (2.17) or (2.18) equal the values determined from the initial rates of Eq. (2.15) or (2.16) only when the reverse reaction in equilibrium-controlled processes, product inhibition, and product hydrolysis in kinetically controlled synthesis (Figure 2.9a) can be neglected. It may be assumed that the different product enantiomers inhibit the enzyme differently. When this is the case, the steric purity of the desired product enantiomer is also influenced by product inhibition (Straathof and Jongejan, 1997).

The experimental error in determining E from Eqs. (2.17) and (2.18) is much larger than that for the E-values determined from Eqs. (2.15) and (2.16), especially at ee_P or ee_S values close to 1. Therefore, Eqs. (2.17) and (2.18) can only be used to determine E values in the range from $\approx 10^{-2}$ to 10^2.

Figure 2.13 provides a summary of the range of published stereoselectivities for enzymes in equilibrium- and kinetically controlled reactions. For applications in enzyme technological kinetic resolution of racemates, the stereoselectivity E should be >100 or <0.01 for (S)- and (R)-specific enzymes, respectively, in order to obtain products with high steric purity.

2.7.2 Dependence of k_{cat}, K_m, and Selectivities on pH, Temperature, Inhibitors, Activators, and Ionic Strength in Aqueous Solutions

In Section 2.7.1, quantitative relationships for "undisturbed" enzyme kinetics and stereoselectivity were derived. These were based on a discussion of the reactions occurring when the substrate is bound in the active site. Other compounds bound to the enzyme and changes in pH and temperature can influence these reactions. This can result in either an increase (activation) or a decrease (inhibition) in the rate of the enzyme-catalyzed process. Quantitative studies on these changes can provide valuable information about the groups that control the enzyme function (which amino acid residues are in the active center, how enzymes can be selectively inhibited, etc.). This information in turn can be applied to the search for better enzymes by screening natural enzyme sources or creating improved new enzymes by protein engineering (see Chapter 3). The quantitative relationship describing the influence of these factors can be derived from Eqs. (2.8),(2.11),(2.12) and (2.14)–(2.18). How these properties are changed in aqueous suspensions or nonconventional solvents (organic solvents, supercritical gases, ionic liquids) will be discussed in Section 2.9.

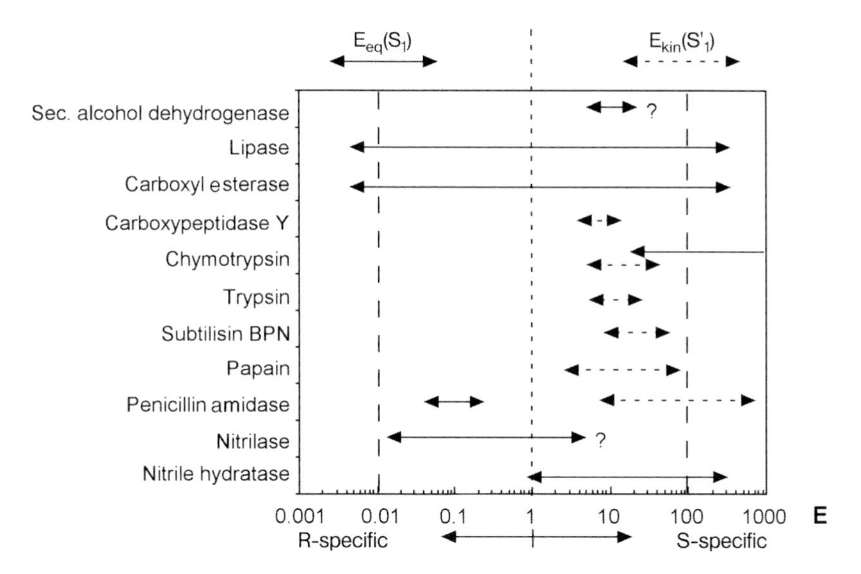

Figure 2.13 Range of published stereoselectivities E_{eq} (Eq. (2.15)) for the S_1 binding subsite (acyl binding in Figure 2.3) and E_{kin} (Eq. (2.16)) for the S'_1 binding subsite (leaving group in equilibrium-controlled hydrolysis or nucleophile binding subsite in kinetically controlled synthesis) for enzymes (mainly hydrolases). Note that penicillin amidase has different stereospecificities in both binding subsites. The same applies for lipases and esterases, where the binding subsites are the acid and alcohol binding sites.

2.7.2.1 pH Dependence

k_{cat} and K_m Acidic and basic groups occur on the surface of enzymes, but they can also form part of the active site. Hence, their ionization state can influence the k_{cat} and K_m values of the enzyme. In Section 2.7.1, the pH dependence of k_{cat} and K_m was not considered, and the values were considered to be apparent (pH dependent). Here, this situation will be illustrated for α-chymotrypsin (see Figure 2.5). When the histidine in the active site of α-chymotrypsin is protonated (EH), k_{cat} is 0 and the enzyme function is lost. Only that fraction of the enzyme where the histidine group is uncharged (f_) contributes to the activity of the enzyme, at which time the apparent k_{cat} value is

$$k_{cat} = f_{-} k^0_{cat} = \frac{k^0_{cat}}{1 + [H^+]/K_1}, \tag{2.19}$$

that is, a function of pH where pK_1 is the pK value for the histidine in the active site.

The same can apply for K_m when another group with $pK = pK_2$ influences the binding of the substrate. We assume that both the charged and uncharged groups with this pK can bind the substrate with different K_m values. Then, the apparent K_m value as a function of pH is

$$K_m = \frac{K'_m}{1 + K_2/[H^+]} + \frac{K''_m}{1 + [H^+]/K_2}. \tag{2.20}$$

When Eqs. (2.19) and (2.20) are inserted into Eq. (2.8), we obtain the pH dependence of the rate:

$$v = \frac{k^0_{cat}[EH]_0[AB]_0/(1 + [H^+]/K_1)}{[AB]_0 + K'_m/(1 + K_2/[H^+]) + K''_m/(1 + [H^+]/K_2)}. \tag{2.21}$$

When $[H^+] > K_1$ (or $pH < pK_1$) and $K_1 > K_2$, v will increase with pH ($[H^+]$ decreases). When $pH > pK_1$ and pK_2, $[H^+]/K_1$ is ≈ 0 and the denominator will increase with pH, and v will decrease. From this, it follows that v will have a maximum between pK_1 and pK_2 (pH optimum). This is generally observed in enzyme kinetics (Fersht, 1999). From Figure 2.14, pK_1 and pK_2 can be determined. When $k_{cat} = (1/2)k^0_{cat}$, $[H^+]$ equals K_1, or $pH = pK_1$. The amino acid residues in enzymes have similar pK values as the isolated amino acids. Thus, curves such as those in Figure 2.14 can provide qualitative information on the type of amino acids that give rise to the pH dependence of k_{cat} and K_m. For α-chymotrypsin, such kinetic evidence has been supported by structural studies on the active site (see Figure 2.5a). The amino acid residue with the pK_1 value ≈ 7 is His_{57}. The K_m value is influenced by a residue with $pK_2 \approx 9$ that must be an amino group. Equation (2.21) shows that

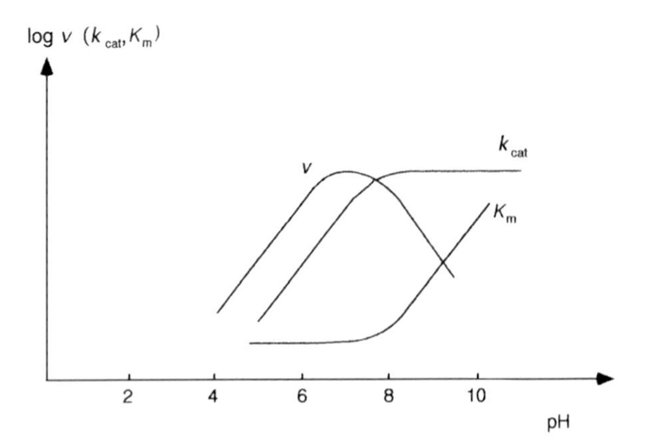

Figure 2.14 pH dependence of v, k_{cat}, and K_m for α-chymotrypsin-catalyzed hydrolysis of an uncharged substrate (from Bender *et al.*, 1964).

a pH optimum can only occur when two different amino acid residues control the pH dependence of k_{cat} and K_m. This applies for most enzymes (exception: pepsin in the stomach with pH 1–2). Quantitative studies to determine k_{cat} and K_m must thus be carried out in buffered systems. Especially when acids or bases are formed or consumed in an enzyme-catalyzed reaction, the buffers used must have a sufficient buffering capacity at the pH where the measurements are carried out. It is also important that the buffers do not participate in the enzyme-catalyzed reaction, so that the use of buffers with amino groups that may deacylate acyl-enzymes are excluded. Therefore, Tris buffers must not be used as a buffer in enzyme kinetic studies (Kasche and Zöllner, 1982). Charged substrates may also influence the pH dependence of k_{cat} and K_m, either directly or indirectly, as they influence the ionic strength (Lummer *et al.*, 1999).

Selectivities In kinetically controlled reactions, the reactive group of the nucleophile is often an amino group, and it must be uncharged in order to act as a nucleophile. In the equation for the selectivity of such reactions (Eq. (2.3)), k_t depends on the pK value of this amino group. It decreases with decreasing pH below this value. Kinetically controlled synthesis must therefore be carried out at pH values at least one pH unit above the pK for the amino group of the nucleophile.

The stereoselectivity E_{eq} has been found to depend on pH for alcohol dehydrogenase and penicillin amidase (Lummer *et al.*, 1999). This can occur when the pK of a group that influences k_{cat} is changed differently by the bound (*S*)- or (*R*)-substrate. The observed changes were less than one order of magnitude.

2.7.2.2 Temperature Dependence

k_{cat} and K_m In Section 2.7.1, it was shown that k_{cat} is a rate constant for a monomolecular reaction and K_m a dissociation constant. From the temperature dependence of these based on free energy changes (Figures 2.7 and 2.8), the temperature dependence of v (Eq. (2.8)) can be written as

$$v = \frac{A[\mathrm{EH}]_0[\mathrm{AB}]_0\, \mathrm{e}^{-\Delta G^{\#}_{\mathrm{ES}}/RT}}{[\mathrm{AB}]_0 + B\, e^{-\Delta G_{\mathrm{S}}/RT}}, \tag{2.22}$$

where A and B are constants. The activation energy $\Delta G^{\#}_{\mathrm{ES}}$ is always >0, and ΔG_{S} can be <0 or >0. This implies that k_{cat} always increases with temperature, whereas K_m can increase or decrease with temperature. $\Delta G^{\#}_{\mathrm{ES}}$ and ΔG_{S} can be determined from the temperature dependence of k_{cat} and K_m using Arrhenius plots. As $|\Delta G^{\#}_{\mathrm{ES}}| > |\Delta G_{\mathrm{S}}|$, v will generally increase with temperature up to a “temperature optimum” (Figure 2.15), above which v will decrease. Enzymes have a flexible structure that, at higher temperatures, leads to unfolding (denaturation) as well as structural changes in the active center that reduce v. The temperature optimum for enzymes in aqueous solutions varies with their source from about 40–50 °C for enzymes from mesophilic organisms up to 100 °C for enzymes from thermophilic

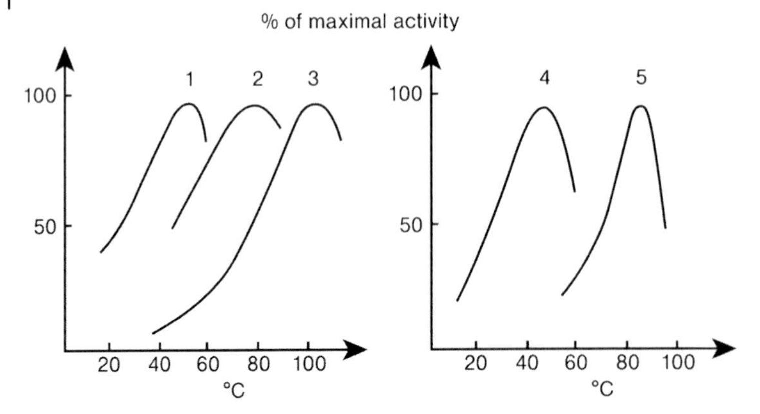

Figure 2.15 Temperature dependence of the activity of different enzymes. Amylase from pancreas (1), *Bacillus subtilis* (2), and *Bacillus licheniformis* (3); peptidase from pancreas (4) and *B. subtilis* (5) (from Godfrey, 1996).

organisms. As a rule, enzyme processes should be carried out about 10–20 °C below the "temperature optimum" in order to avoid rapid denaturation of the enzyme. (The word "optimum" is here a misleading designation.)

Selectivities The stereoselectivity E_{eq} for equilibrium-controlled processes (Eq. (2.15)) is important when enzymes are used for the equilibrium-controlled kinetic resolution of racemates. From the thermodynamic quantities in Figures 2.7 and 2.8, the temperature dependence of k_{cat} and K_m, the following equation for ln $E_{eq}(T)$ can be derived:

$$\ln E_{eq} = -\frac{(\Delta(G^{\#}_{ES})_S - \Delta(\Delta G^{\#}_{ES})_R) - (\Delta(G_S)_R - \Delta(G_S)_S)}{RT}, \qquad (2.23)$$

that is, the enantioselectivity is, as expected, temperature dependent – an effect that was first shown experimentally in 1989 (Phillips, 1996). In general, the differences in activation energies ($\Delta G^{\#}$) are larger than the differences in the binding energies (ΔG). Then, E_{eq} should decrease with temperature. This has also been observed in most studies on the temperature dependence of E_{eq} (Phillips, 1996; Galunsky, Ignatova, and Kasche, 1997; Sakai *et al.*, 1997). Figure 2.16 shows that E_{eq} can be increased by up to a factor of 10, or even reversed, by a reduction in temperature.

For kinetically controlled racemate resolutions, similar expressions as Eq. (2.23) for the temperature dependence of E_{kin} can be derived. Where its temperature dependence has been studied, it has been found that E_{kin} in most cases decreases with increasing temperature (Kasche *et al.*, 1996; Galunsky, Ignatova, and Kasche, 1997). This, and the temperature dependence of E_{kin}, has been used to increase the yield and steric purity of the product in a kinetically controlled racemate resolution in the production of β-lactam antibiotics that is carried out at 5 °C (Zmiejewski *et al.*, 1991).

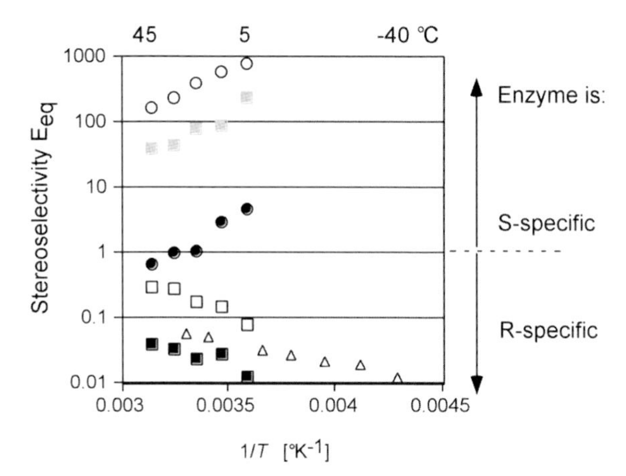

Figure 2.16 Temperature dependence of the stereoselectivity E_{eq} for different binding subsites of α-chymotrypsin and penicillin amidase in aqueous solution, and lipase in organic solvent. Bovine α-chymotrypsin: (S_1 selectivity, see Figure 2.3) hydrolysis of *N*-acetyl-(*R,S*)-phenylglycine methyl ester (closed circles) and (*R,S*)-phenylglycine methyl ester (open circles) at pH 7.5 (Galunsky, Ignatova, and Kasche, 1997). Penicillin amidase from *E. coli*: (S_1 selectivity) hydrolysis of (*R,S*)-phenylglycine amide (open squares), (*R,S*)-hydroxyphenylglycine ester (filled squares), and *N*-acetyl-(*R,S*)-phenylglycine (gray squares, S_1' selectivity) at pH 7.5 (Kasche *et al.*, 1996). Lipase from *P. cepacia* (triangles): (nucleophile binding site) esterification of racemic azirine with vinyl acetate in diethyl ether (Sakai *et al.*, 1997). Note that stereospecificity of penicillin amidase differs in the S_1 and S_1' subsites.

2.7.2.3 Binding of Activator and Inhibitor Molecules

Every compound that directly or indirectly changes the structure of the active center changes the rate v (Eq. (2.8)) of an enzyme-catalyzed reaction. When v increases, the compound is an activator, whereas inhibitors lead to a decrease in v. An enzyme can have binding sites for these also outside the active center (Figure 2.17).

The enzyme may bind a competitive inhibitor in the active site:

$$\mathrm{EH} + \mathrm{I_c} \underset{k_d}{\overset{k_a}{\longleftrightarrow}} \mathrm{EH} \cdots \mathrm{I_c}, \quad K_i = k_d/k_a. \tag{2.24}$$

When such inhibitors are bound, the enzyme cannot perform its function – that is, it is competitively inhibited. This type of inhibitors can be

1) Products (product inhibition) such as the H^+-ion for hydrolases when this ion is formed during the reaction, and can protonate His_{57} in α-chymotrypsin that reduces the rate of acylation (see Figure 2.5, 2.14, and Eq. (2.19)).
2) Substrates that can acylate enzymes almost irreversibly – that is, they are deacylated very slowly. To this subgroup belong neurotoxins or penicillins in Gram-positive bacteria.
3) Similar effects are caused by chelating compounds such as EDTA that can bind metal ions essential for the activity of the enzyme (see Section 2.3).

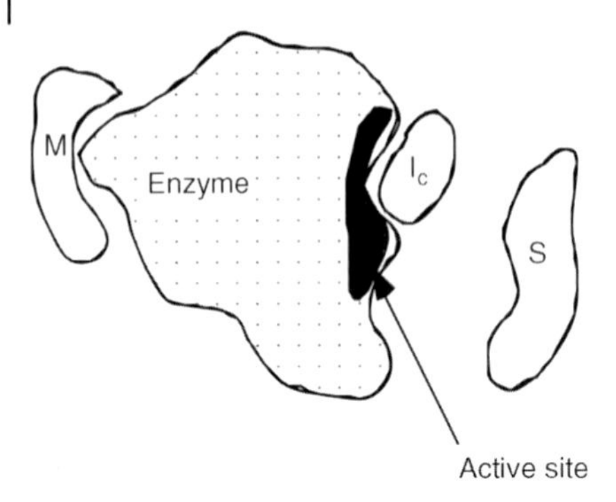

Figure 2.17 Interactions of enzymes with inhibitors (I_c is a competitive inhibitor), modulators M (activators or noncompetitive inhibitors), and substrates S.

When the inhibitor or activator is bound outside the active site, the equilibrium $M + EH \leftrightarrow M \cdots EH$ must be considered besides the enzyme reactions (Figure 2.17 and Eqs. (2.4)–(2.6)). The conformation (tertiary structure) of $M \cdots EH$ differs from EH. The active center is not involved, and the enzyme-catalyzed reaction can occur with $M \cdots EH$. As the structure of the active center is changed in $M \cdots EH$, k_{cat} and K_m can differ from those for EH. When v increases with the concentration of M, the latter is an *activator* for the enzyme. This frequently applies for metal ions such as Ca^{2+} for trypsin or Mg^{2+} for phosphodiesterase. The opposite case where v decreases with the concentration of M is *noncompetitive inhibition,* when M is a noncompetitive inhibitor. The decrease in v can be either due to an increase in K_m in $M \cdots EH$ or when k_{cat} for $M \cdots EH$ is smaller than that for EH. Both cases can occur simultaneously, at which point we have a *mixed inhibition.* Expressions for k_{cat} and K_m in the presence of competitive and noncompetitive inhibitors (where $K_{m.EH} = K_{m.EH \cdots M}$ and $k_{cat.EH \cdots M} = 0$) are given in Table 2.6. The type of inhibition can be determined by measuring $v = f([S])$ at constant inhibitor concentration and plotting the data using Eq. (2.13) (Figure 2.18).

Substrates can also be inhibitors, for example, when they are bound incorrectly in the active site or bound outside the active site. An example where this applies is for hydrolases that hydrolyze oligosaccharides or the two-substrate reactions (Eqs. (2.5) and (2.6)) where v is measured keeping the concentration of one substrate constant while the content of the other substrate is changed. This is frequently observed at high substrate concentrations that are of interest for enzyme technology. In these cases, maxima in v as a function of the substrate concentration can be observed. For noncompetitive substrate inhibition with $[S] = [I] = [AB]_0$ (S for substrate, $[NH] = 0$), Table 2.6 provides the following expression for Eq. (2.8) for this case:

$$v = \frac{k_{cat}[EH][S]}{K_m + [S](1 + K_m/K_i) + [S]^2/K_i}. \tag{2.25}$$

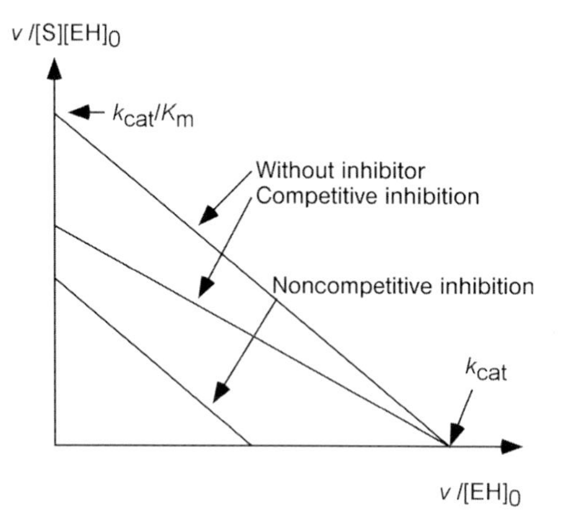

Figure 2.18 Graphical determination of the type of inhibition from Eadie–Hofstee plots with and without a constant concentration inhibitor. The type of substrate inhibition (competitive or noncompetitive) cannot be determined from such curves.

Table 2.6 Influence of competitive and noncompetitive inhibitors on k_{cat} and K_m.

	Competitive inhibitor	Noncompetitive inhibitor
k_{cat}	Unchanged	$k_{cat}/(1+[I]/K_I)$
K_m	$K_m(1+[I]/K_I)$	Unchanged

[I] = free inhibitor concentration.

2.7.2.4 **Influence of Ionic Strength**

Enzymes are polyelectrolytes, and the active site may consist of charged amino acid residues. Several substrates of interest in enzyme technology, such as amino acids, peptides, organic acids, and nucleotides, contain acidic or basic functional groups. In enzyme technology, these substrates are used at concentrations of up to ≈1 M, at which the ionic strength I is given by

$$I = 0.5\sum c_i z_i^2, \tag{2.26}$$

where c_i is the ion concentration and z_i is the ion charge of the ion i.

The ionic strength (I) influences the activity of the ions, the rates, and the equilibrium constants. The activity of a charged substrate S is, at ionic strengths up to ≈0.1 M, given by

$$\{S\} = [S]\exp\left(-\frac{z_S^2 A\sqrt{I}}{1+B\sqrt{I}}\right) = [S]\gamma, \tag{2.27}$$

where γ is the activity coefficient, z_S is the charge, and A and B are constants.

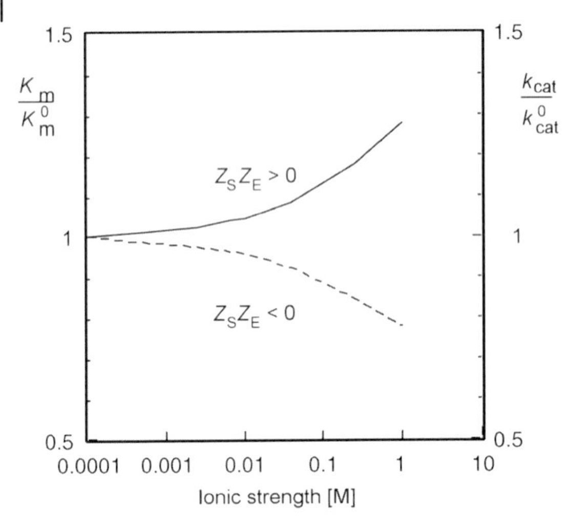

Figure 2.19 Ionic strength dependence of k_{cat} and K_m of an enzyme-catalyzed reaction for different charges of the substrate and the active site $|Z_E| = |Z_S| = 1$.

The influence of the ionic strength and ion–ion interactions on the rate (k_{cat}) and equilibrium constant (K_m) can be expressed as follows, with $A \approx 0.5$ and $B \approx 1$ in Eq. (2.27) (Martinek, Yatsimirski, and Berezin, 1971; Dale and White, 1982):

$$k_{cat} = k_{cat}^{o} \exp(z_S z_E \sqrt{I}/(1 + \sqrt{I})) \tag{2.28}$$

and

$$K_m = K_m^{o} \exp - (z_S z_E \sqrt{I}(1 + \sqrt{I})), \tag{2.29}$$

where the superscript "o" applies to the ionic strength-independent constants.

This ionic strength dependence is shown graphically in Figure 2.19, and shows that ionic strength effects cannot be neglected at concentrations >0.1 M. As I also influences the pK groups that control the pH dependence of the enzyme, it may also influence the pH optimum. Few data are available on the influence of I on k_{cat}, K_m, and selectivities, though some are provided in the references to this section. The selectivity and yield in the kinetically controlled synthesis of semisynthetic β-lactam antibiotics (ampicillin, cephalexin) have been found to decrease with ionic strength that reduces the binding of the nucleophile to the charged binding site on the enzyme (Kasche, 1986, and unpublished data).

2.8
End Points of Enzyme Processes and Amount of Enzyme Required to Reach the End Point in a Given Time

The time dependence and possible end points of enzyme-catalyzed reactions are shown in Figures 1.9 and 2.9. At the end point, a maximal product yield or optimal product quality is desired. In equilibrium-controlled processes, the enzyme cannot influence the yield, but it can influence the time to reach the end point. In kinetically controlled synthesis (Figures 1.9 and 2.9), or in processes where fibers are treated by enzymes either to reach an optimal fiber length or to polish the fibers (treatment of recycled paper or to improve cotton or wool fibers for textile production), the product concentration or product quality has a maximum at which the process must be stopped by removing the enzyme, or by appropriate adjustment of the residence time in continuous processes (see Chapter 11). This maximum is (in contrast to the equilibrium-controlled process) influenced by the properties of the enzyme. In kinetic racemate resolutions and asymmetric synthesis, it is not only the yield that is important in the selection of a suitable end point, but rather the steric purity of the product. The selection of end points for such processes will be discussed at the end of this section.

In a technological process, the enzyme costs should be less than 5–10% of the total process costs. Hence, in order to estimate costs, the amount of enzyme required to reach the end point within a given time must be determined. For equilibrium-controlled processes, this can be done as follows.

From the given time t and the desired change in substrate concentration, from the initial $[S]_0$ to the final concentration at the selected end point $[S]_t$, the required space–time yield (STY) is

$$STY = \frac{[S]_0 - [S]_t}{t}. \tag{2.30}$$

The STY has the dimensions of (mol substrate per unit time and reactor volume), and is an important quantity in enzyme technology as it determines the reactor volume. It is a given quantity that is determined by previous processing steps. When all penicillin G produced per day by fermentation (see Chapter 1) must be hydrolyzed to 6-APA, this gives the minimal STY required in the enzyme process. The equations derived in Section 2.7.1 are insufficient to calculate the enzyme amount required to obtain the necessary STY as they are based on initial rates where only a fraction of the substrate is converted to product. The reverse reactions and product inhibition must also be considered. Therefore, the analytical description of initial rate enzyme kinetics, which may be found in biochemically oriented textbooks on enzymology, must be amended for enzyme technological purposes. The enzyme amount required to reach the end point in a given time can be derived as follows. The rate of the change in the substrate content is

$$\frac{d[S]}{dt} = -\frac{V_{max}(t)}{f([S],[P])}, \tag{2.31}$$

where $V_{max}(t)$ is given by the following relationship:

$$V_{max}(t) = V_{max,0}\, e^{-k_i t}, \tag{2.32}$$

assuming that the enzyme is inactivated in a first-order reaction with the rate constant k_i, and that $V_{max,0}$ is the initial enzyme concentration. The differential equation (Eq. (2.31)) can be solved after variable separation, and provides the following relationship for $V_{max,0}$:

$$V_{max,0} = \frac{\int_{[S]_t}^{[S]_0} f([S],[P])d[S]}{(1-e^{-k_i t})/k_i}, \tag{2.33}$$

where the denominator tends to t when k_i approaches 0 – that is, enzyme inactivation can be neglected. The function $f([S],[P])$ for a process where the reverse reaction can be neglected can be derived from Eq. (2.8) and the relationships in Table 2.6. It is

$$\frac{K_m + [S]}{[S]} \tag{2.34}$$

when substrate and product inhibition can be neglected,

$$\frac{K_m + (1 + K_m/K_i)[S] + [S]^2/K_i}{[S]} \tag{2.35}$$

for noncompetitive substrate inhibition, and

$$\frac{K_m(1 + [S]_0/K_i) + (1 - K_m/K_i)[S]}{[S]} \tag{2.36}$$

for competitive product inhibition.

In many important equilibrium-controlled enzyme processes, the theoretical yield is below 100% substrate conversion. This applies for

- the isomerization of glucose, with about 50% substrate conversion,
- the hydrolysis of penicillin and cephalosporin C, with about 90–95% substrate conversion,
- the hydrolysis of maltose, with ≈95% substrate conversion.

For these processes, 100% substrate conversion can only be obtained at pH values and temperatures far outside the range where the enzymes used as biocatalysts are stable. For these processes, the back reaction (see Section 2.7.1) cannot be neglected, and Eqs. (2.33)–(2.36) can only be used for rough estimates, as Eqs. (2.8) and (2.33) do not include the reverse reaction. For one-substrate reactions, such as the isomerization of glucose (Glu) to fructose (Fru), the following Michaelis–Menten equation can be derived:

$$v = \frac{(k_{cat}/K_m)[Glu] - (k'_{cat}/K'_m)[Fru]}{1 + ([Glu]/K_m) + ([Fru]/K'_m)}[EH]_0, \tag{2.37}$$

where the prime applies for the reverse reaction. Corresponding equations for the hydrolysis of penicillin can be found in the literature (Spiess *et al.*, 1999).

In the kinetically controlled synthesis, the suitable end point is the maximum product concentration $[AN]_{max}$ (see Figure 2.9). It is

$$[AN]_{max} = \frac{[AB]_0[NH]_0}{(k_H/k_T)_{app}[H_2O] + [NH]_0} \tag{2.38}$$

when the rate of hydrolysis of AN can be neglected, or

$$[\mathrm{AN}]_{\max} = \left(\frac{(k_{\mathrm{T}}/k_{\mathrm{H}})_{\mathrm{app}}}{(k_{\mathrm{T}}/k_{\mathrm{H}})_{\mathrm{app}}[\mathrm{NH}] + \mathrm{H_2O}} \right) \frac{(k_{\mathrm{cat}}/K_{\mathrm{m}})_{\mathrm{AB.NH}}}{(k_{\mathrm{cat}}/K_{\mathrm{m}})_{\mathrm{AN.NH}}} [\mathrm{NH}][\mathrm{AB}] \qquad (2.39)$$

when the rate of hydrolysis of AN cannot be neglected (Kasche, 1986, 2001). From these relationships, it follows that $[\mathrm{AN}]_{\max}$ depends on the properties of the enzyme, but not on the enzyme content. This has been confirmed in other studies (Youshko *et al.*, 2002). Analytical equations from which the amount of enzyme required to reach these concentrations in a given time can be determined have not yet been derived. For both equilibrium- and kinetically controlled reactions, short reaction times to reach the end point are required when the product is unstable. One example where this applies is the hydrolysis and synthesis of β-lactam antibiotics (see Chapter 12, Case Study 3).

Optimal yields at the end points for equilibrium- and kinetically controlled processes are generally obtained at pH and temperature values outside the range where the enzymes used as biocatalysts have their optimal properties. From this, it follows that *in the design of enzyme processes, the properties of both the enzyme and the enzyme-catalyzed process must be considered.* This will be illustrated for the influence of temperature and pH.

2.8.1
Temperature Dependence of the Product Yield

For equilibrium-controlled processes, thermodynamic data are required to analyze the temperature dependence of the product yield, but unfortunately these are still missing for many important enzyme processes. Some data are provided in Table 2.7.

For kinetically controlled processes, it has been observed that the yield decreases with temperature (Kasche, 1986). Few data exist relating to investigations of whether the reactions are either endo- or exothermal (Michaelis, 1991).

2.8.2
pH Dependence of the Yield at the End Point

For equilibrium-controlled processes, where acid or bases are produced or consumed, the equilibrium constant and the product yield are a function of pH. This applies for the hydrolysis and synthesis of peptides, peptide antibiotics (such as cephalosporins and penicillins), lipids, and esters. Calculation of the equilibrium as a function of pH can be performed when the pK values of the substrates and products are known. The apparent association constant K_{app} for an equilibrium condensation of a base and an acid is, when all activity coefficients (Eq. (2.27)) are 1,

Table 2.7 Thermodynamic data for enzyme-catalyzed reactions (end point = equilibrium) (Goldberg, Tewari, and Bhat, 2004; http://xpdb.nist.gov/enzyme_thermodynamics/).

Process	Enthalpy change (kJ mol^{-1})[a]	Process is	Change in yield with increasing temperature
Reduction of formate	−15	Exothermal	Smaller
Hydrolysis of			
Penicillin G	20	Endothermal	Larger
Penicillin V	<0	Exothermal	Smaller
Sucrose	−15	Exothermal	Smaller
Lactose	0.5	?	?
Maltose	−4	Exothermal	Smaller
Isomaltose	6	Endothermal	Larger
Cellulose	<0	Exothermal	Smaller
Starch	<0	Exothermal	Smaller
Lipids	?	?	?
Peptide bond	−5[b]	Exothermal	Smaller
Formation of Asp from fumarate and NH_3	−24	Exothermal	Smaller
Isomerization of glucose	3	Endothermal	Larger

a) At pH 7–8, 25–40 °C.
b) Between protected amino acids (also in polypeptides).

$$K_{app} = \frac{(\sum[\text{condensation product}])[H_2O]}{(\sum[\text{acid}])(\sum[\text{base}])} = K_{ref}\frac{(\sum[\text{condensation product}])/[\text{reference condensation product}]}{[(\sum[\text{acid}])/[\text{reference acid}]][(\sum[\text{base}])/[\text{reference base}]]}, \tag{2.40}$$

where the sums are over all proteolytic states of the substrates and the product K_{ref} is a reference association constant applying to the condensation of only one of the different protonated states of the acid and base. K_{ref} applies to a reference pH value and is thus independent of pH. The ratios in the nominator and denominator in Eq. (2.40) can be determined from dissociation schemes, as shown in Eq. (2.41).

$$H_3{}^+N - R_i - COOH \overset{K_1}{\longleftrightarrow} H_3{}^+N - R_i - COO^- + H^+ \overset{K_2}{\longleftrightarrow} H_2N - R_i - COO^- + 2H^+ \tag{2.41}$$

In Figure 2.20, K_{app}/K_{ref} has been calculated as a function of pH for some condensation reactions. The yields increase with K_{app}. For the hydrolysis of the condensation products, the yields increase with decreasing K_{app}. For the synthesis of peptides, the highest yields are obtained at neutral pH values, where the serine peptidases that can be used as biocatalysts have their pH optimum (see Figure 2.14). For the synthesis of penicillins with penicillin amidase and esters with serine peptidases, the highest yields are obtained at pH values much below the pH optima of these enzymes. Penicillin amidases also have a low pH stability at pH 5

(Figure 2.28), and the association constants are too low to obtain high yields. Thus, these enzymes are not suitable as biocatalysts for the equilibrium-controlled synthesis of these condensation products. The opposite applies when they are used as biocatalysts for the hydrolysis of the condensation products. From Figure 2.20, it follows that the hydrolysis of penicillin with a high product yield should be carried out at pH >9. At this pH, however, the possible biocatalyst – penicillin amidase from *E. coli* – has a low pH stability (Figure 2.28). The process with this enzyme must therefore be carried out under suboptimal conditions. This again demonstrates that the properties of both the process and the enzyme must be considered to select the optimal condition for an enzyme process.

For kinetically controlled processes, the pH dependence of the maximum product concentration can be derived from Eqs. (2.38) and (2.39). This is, however, difficult to perform as the pH dependence of the rate constants in these equations cannot easily be determined. These processes must be carried out at pH values where the nucleophile is uncharged – that is, above the pK values of these. Only the uncharged nucleophiles can deacylate the acyl-enzyme. Therefore, penicillin

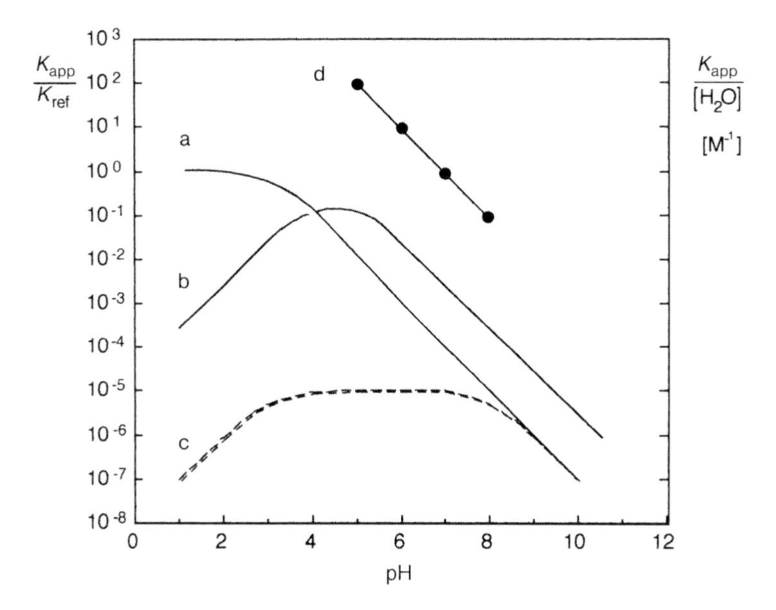

Figure 2.20 Calculated K_{app}/K_{ref} from Eq. (2.40) and experimental values ($K_{app}/[H_2O]$) of pH dependence of the association constants for the formation of esters (a), peptide antibiotics (b, d), and peptides (c). (a) Ester synthesis from a carboxylic acid (p$K_1' = 3$); reference states: the uncharged compounds. (b) Synthesis of penicillin G (p$K_1''' = 2.9$) from phenylacetic acid (p$K_1' = 4.9$) and 6-APA (p$K_1'' = 2.9$, p$K_2'' = 4.6$); reference states: uncharged product and neutral acid and positively charged 6-APA. (c) Synthesis of a dipeptide from an amino-protected (p$K_1' = 3.0$) and carboxyl-protected (p$K_2'' = 7.8$) amino acid; reference states: the uncharged compounds. (d) Experimental data for the synthesis of penicillin G at 25 °C ($I = 0.2$ M). Single, double, and triple primes denote the dissociation constants for the acid, base, and condensation product, respectively; subscript 1 denotes the acid dissociation constant and subscript 2 denotes the base dissociation constant.

synthesis is carried out at pH values >5 and peptide synthesis at pH >9 (Kasche, 1986).

2.8.3
End Points for Kinetic Resolutions of Racemates

In these processes, not only high product yields but also a high steric purity of the product is essential. The steric purity is given by the enantiomeric excess of the product (ee_P) or substrate (ee_S); the yield of the desired enantiomer (product or remaining substrate) is less than 50% of the initial racemate concentration (see Figure 2.12 and Chapter 4). These experimental conditions must also be chosen so that more than 90% of the desired enantiomer is formed in the equilibrium- or kinetically controlled process. For equilibrium-controlled processes, this requires that the equilibrium constant and its pH and *T*-dependence is known (see above). For kinetically controlled processes, this can be achieved when the concentration ratio $[AB]_0/[NH]_0$ and the selectivity ν_T/ν_H in Eq. (2.1) are larger than 1 (Kasche, 1986).

Selection of the end point based on steric purity can be made based on plots of ee_P and ee_S as a function of the extent of the reaction (Figure 2.21). From this figure, it follows that for $ee_P \geq 0.95$ and product yields ≥45% of the initial racemate concentration, (*S*)-specific enzymes with $E \geq 100$ or (*R*)-specific enzymes with $E \leq 0.01$ must be used. The amount of enzyme required to reach this end point in a given time for equilibrium-controlled processes can be calculated from Eq. (2.33) when product inhibition and the competitive inhibition of the undesired enantiomer are considered.

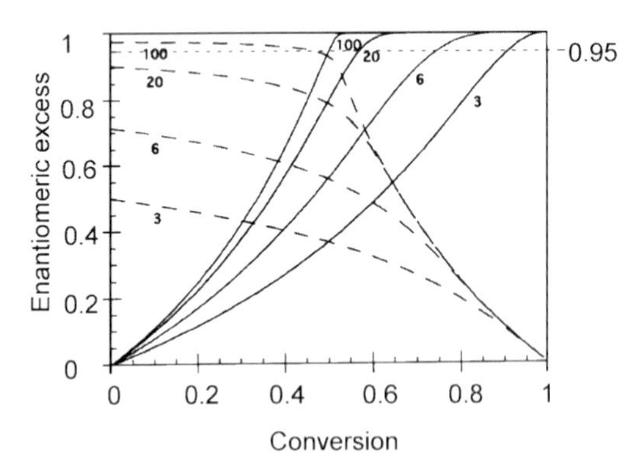

Figure 2.21 The enantiomeric excess ee_P of the product (dashed line) and ee_S for the remaining substrate (solid line) as a function of the extent of the reaction for kinetic racemate resolutions catalyzed by (*S*)- or (*R*)-specific enzymes with different *E*-values (the *E*-values for the (*R*)-specific enzymes are the inverse of the given *E*-values). The horizontal dotted line shows that a product yield of ≥45% with ee ≥0.95 requires a (*S*)-specific enzyme with $E \geq 100$.

2.9
Enzyme-Catalyzed Processes with Slightly Soluble Products and Substrates

Enzymes have been evolved over a billion years to secure the survival of living systems where their biological function is carried out in aqueous systems. Enzymes catalyze reactions in these systems where the substrate concentrations are comparatively small (<10 mM). In enzyme processes, at least one order of magnitude higher concentrations are desired to reduce processing costs. This may be above the solubility (in water) of many hydrophobic natural compounds or organic molecules that are substrates or products in enzyme processes. This section covers methods to solve this problem:

1) The enzyme reactions can be carried out in aqueous suspensions (or emulsions) with precipitated substrate and/or product (see Section 2.9.1). This is possible when the rate of mass transfer (solid to liquid, or emulsion to liquid) is not rate limiting (Figure 2.22). This requires that the size of the solid particles or droplets in the emulsion be small (Eq. (2.42)), at which point high *STY* can be achieved (Figure 2.23).
2) When method 1 is not possible, the reaction can be carried out in different one- or two-phase systems with solvents in which the substrates and products are soluble (Table 2.8). Besides traditional organic solvents, supercritical gases such as CO_2 or ionic liquids can be used here (Figure 2.24). The organic solvents and ionic liquids used must be nontoxic and easy to recycle. In such systems, equilibria that are unfavorable in water can be shifted in the desired direction (Eq. (2.43) and Figure 2.25).

In many processes, one of the substrates (such as alcohols) can be used as a solvent. Such systems are especially used in enzyme processes where lipase (EC 3.1.1.3) is used as a biocatalyst. This enzyme catalyzes the conversion of slightly soluble hydrophobic substrates at organic–aqueous interphases also in living systems. In these "dry" systems, water is required for optimal enzyme activity (Figure 2.26), and a minimal number of water molecules must be bound to the enzyme to provide it with the flexibility required to carry out its function.

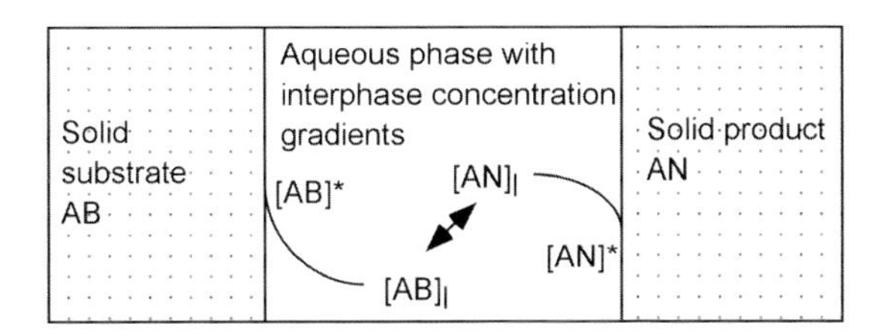

Figure 2.22 The concentration gradients at the phase boundaries in enzyme-catalyzed processes in aqueous suspensions involving insoluble substrates or products. The asterisk denotes the concentration at the phase boundary at equilibrium between the solid and aqueous phases.

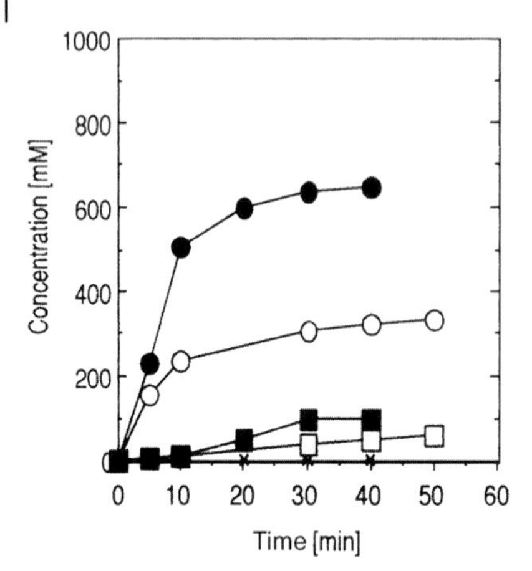

Figure 2.23 Kinetically controlled synthesis of soluble *N*-acetyl-(*S*)-Tyr-(*S*)-Arg-NH_2 (ATAA, circles), a precursor of the dipeptide (Tyr-Arg) kyotorphin, from *N*-acetyl-(*S*)-Tyr-OEt (ATEE) as a suspension and soluble (*S*)-Arg-NH_2 with the peptidase α-chymotrypsin (CT). *Conditions*: enzyme content 10 μg ml^{-1}, 25 °C, pH 9.0 (carbonate buffer, $I = 0.2$ M); the pH is kept constant during the reaction. *Squares*: the soluble hydrolysis product *N*-acetyl-(*S*)-Tyr. *Closed symbols*: starting concentrations are 800 mM (*S*)-Arg-NH_2 and 750 mM (total content) ATEE as a suspension. *Open symbols*: starting concentrations are 400 mM (*S*)-Arg-NH_2 and 400 mM (total content) ATEE as a suspension. The dissolved ATEE concentration in the aqueous system during the reaction was <1 mM (Kasche and Galunsky, 1995).

2.9.1
Enzyme-Catalyzed Processes in Aqueous Suspensions

In an aqueous suspension with a slightly soluble substrate that is converted by an enzyme in the aqueous phase, a concentration gradient is formed in the interphase between the solvent and the solid-phase substrate (Figure 2.22). This causes a mass transfer of the substrate from the solid to liquid phase. The rate of dissolution of the substrate is

$$k_{L,AB}a_{AB}([AB]^* - [AB]) = \frac{Sh\, D_{AB} 3\alpha}{2(r_{AB})^2}([AB]^* - [AB]_1), \tag{2.42}$$

where $k_{L,AB}$ and a_{AB} are the mass transfer coefficient and surface area per volume unit, respectively, *Sh* is the Sherwood number, D_{AB} is the diffusion coefficient of the substrate, r_{AB} is the average radius of the solid substrate particles, and α is the volume fraction of the solid particles in the suspension. This rate can be estimated for a 10% suspension ($\alpha = 0.1$) of a substrate with MW = 500 Da ($D_{AB} \approx 5 \times 10^{-6}$ $cm^2\,s^{-1}$), $Sh = 10$, $r_{AB} = 10^{-2}$ cm, and a concentration gradient of 1 mM. This gives a rate of dissolution of (Eq. (2.42))

$$75 \times 10^{-6}\ M\,s^{-1} = 4500\ U\,l^{-1}.$$

With a stationary substrate concentration in the aqueous phase $>K_m$, and an enzyme content of 10 μM, the rate of the enzymatic reaction equals the rate of dissolution when $k_{cat} \geq 8.3\,s^{-1}$. From this and Table 2.4, it follows that many enzymes have turnover numbers above this value. Therefore, it should be possible to carry out enzyme processes in suspensions (emulsions) of slightly soluble substrates with total concentrations of more than 1 M successfully and with high space–time yield. This has been demonstrated in some studies (Figure 2.23) (Kasche and Galunsky, 1995; Cao *et al.*, 1997; Vulfson *et al.*, 2001).

One reason why such processes have not yet been studied in more detail or applied in enzyme technology might be that slightly soluble products can be formed. This may be favorable in equilibrium-controlled processes with slightly soluble products, where the equilibrium is shifted in the direction of the product. However, in enzyme processes carried out in suspensions, the enzyme must be removed once the end point has been reached. This is difficult with soluble enzymes, but it is possible when the enzyme is immobilized (see Chapters 8–10) in nonporous or porous supports that can be separated from the insoluble substrate or product. This separation is especially easy with magnetic supports (Bozhinova *et al.*, 2004).

For an insoluble product, precipitation of the product in the pores of a porous support must be avoided to allow reuse of the immobilized enzyme. The solubility is inversely proportional to r_{AB} (Freundlich, 1909). Thus, precipitation can be avoided when supports with either small or no pores are used (Kasche and Galunsky, 1995; Bozhinova *et al.*, 2004; see also Section 10.8).

2.9.1.1 Changes in Rates, k_{cat}, K_m, and Selectivities in These Systems Compared with Homogeneous Aqueous Solutions

The enzyme-catalyzed reaction occurs in the aqueous phase. Thus, k_{cat}, K_m, and the stereoselectivities are not changed. This also applies for the rate and the selectivity $(k_T/k_H)_{app}$, when the rate of dissolution is not limiting (Eq. (2.42)), for the same bulk concentration. When the water content is low in these systems, it may be difficult to control the pH for reactions where acids or bases are either consumed or formed. This may reduce the rates and change $(k_T/k_H)_{app}$ for kinetically controlled processes (Vulfson *et al.*, 2001).

2.9.2 Enzyme-Catalyzed Processes in Nonconventional Solvents Where Products and Substrates Are Dissolved (and the Enzyme Suspended)

Biological systems are generally aqueous solutions with up to 20% soluble and some proportion of slightly soluble components such as membranes, fats, waxes, structural proteins (hair, wool), and cellulose – that is, they are heterogeneous systems. The slightly soluble fraction consists of hydrophobic, low molecular weight molecules (lipids) or biopolymers. Many biologically active lipids, vitamins, and polycyclic or peptide hormones are slightly soluble in water, and in biological systems the concentrations at which they are present and biologically active are below their solubility limit ($\leq$1 μM).

Initially, the enzymatic technological production of hydrophobic compounds was considered to be difficult due to their low solubility. That this does not apply for aqueous suspension systems was outlined in Section 2.9.1. The possibility of using one- or two-phase systems with solvents to increase the solubility of hydrophobic compounds in enzyme processes was demonstrated more than 100 years ago, and has also been extensively studied during the past 20 years (Kasche *et al.*, 1988; Overmeyer *et al.*, 1999; Halling, 1994, 2000; Mesiano *et al.*, 1999; Vulfson *et al.*, 2001; Lamare, Legoy, and Graber, 2004; Trivedi *et al.*, 2006; Cantone *et al.*, 2007). Such solvents, where the enzyme-catalyzed reactions occur in the aqueous phase or in the phase with suspended enzyme, include (Table 2.8)

- water-miscible solvents (methanol, ethanol, acetone, DMSO, dimethylformamide, etc.);
- solvents that are not miscible with water (liquid alkanes, higher alcohols, ethyl acetate, ethers, chlorinated hydrocarbons, etc.);
- supercritical solvents such as CO_2 (with the limitation that many substances exhibit a low solubility in these);
  similar to these are
- inert gases (such as N_2) as solvents, enriched or saturated with volatile substrates; and
- ionic liquids (salts that are liquid at room temperature).

The use of organic solvents in industrial processes is problematic due to their dangerous properties (toxicity, flammability, etc.), and this has led to strict regulations for their use and containment. In many countries, regulations derived from laws for the protection of the environment require the use of less dangerous materials in the processing industry when this is possible. The search for environmentally more friendly compounds must be documented. This applies to solvents such as halogenated hydrocarbons that destroy the ozone layer in the atmosphere. They have therefore been replaced by other solvents, and must not be used for enzyme

Table 2.8 Systems in which enzyme-catalyzed processes can be carried out in the presence of organic or supercritical solvents, ionic liquids, and gases.

Two-phase system			One-phase system	
I. Solvent not miscible with water[a)]	II. Enzyme with water suspended in organic solvents or ionic liquids	III. Immobilized enzyme suspended in organic or supercritical solvents, ionic liquids, or gases	IV. Buffer saturated with organic solvent	V. Buffer with water-miscible organic solvent

a) The aqueous phase is usually small ($\leq$1%).

technological purposes. Other organic solvents must be recycled in the production plant, as their emission is (or must be) limited by environmental regulations. Supercritical solvents such as CO_2 are also environment-friendly alternatives, but whether this also applies to ionic liquids – which have been used only relatively recently in enzyme technology – requires further study, notably with regard to their influences on biological systems (Poliakoff *et al.*, 2002; Park and Kazlauskas, 2003). Ionic liquids (ILs) are low melting point ($<100\,^{\circ}C$) salts that represent a new class of nonaqueous solvents. In contrast to conventional organic solvents, they do not exhibit a vapor pressure, so their reuse should be facilitated. As they do not evaporate, they are also considered as "green solvents" (Seddon, 2003). Newer data question this. It has been found that low concentrations of certain ILs can substantially reduce or stop the growth of microorganisms (Ganske and Bornscheuer, 2006). Data on their toxicity and biodegradability have recently been reviewed (Ranke *et al.*, 2007).

The most often used ILs in biocatalysis are 1-alkyl-3-methylimidazolium salts (Figure 2.24). The solvent properties of these can be modulated considerably by changing the cation or anion. Thus, ionic liquids of different polarity are available, which makes them either water-miscible (e.g., BMIM-BF_4) or immiscible (e.g., BMIM-PF_6). Recent studies have also revealed that not only ILs replace organic solvents, but that the performance of an enzyme is also influenced in ILs. Ionic liquids have in some cases been shown to increase regio- and enantioselectivity of enzymatic reactions, activity, and stability (Kragl, Eckstein, and Kraftzik, 2002; Park and Kazlauskas, 2003; van Rantwijk, Madeira Lau, and Sheldon, 2003). Proteases, lipases, esterases, glycosidases, and oxidoreductases were each found to be active in ILs (Moniruzzaman *et al.*, 2010). As yet, the widespread use of these solvents – especially in industrial processes – has been hampered by their very high price (typically 800-fold more than conventional solvents), along with a need for more efficient methods for their reuse and/or product isolation. Another problem here is the lack of comparative studies of the same enzyme process using these different solvents and aqueous systems. Such data are needed to evaluate the possible use and potential of these solvents in industrial enzyme processes. Until now only a few industrial processes exist with these nonconventional solvents.

Enzyme processes with these solvents can be carried out in the one- or two-phase systems, as shown in Table 2.8. The additional solvents can influence the enzyme or the enzyme-catalyzed reaction either *directly* or *indirectly*.

The direct influence is caused by the binding of solvent molecules to the hydrophobic parts of the enzyme surface, which results in structural changes in the

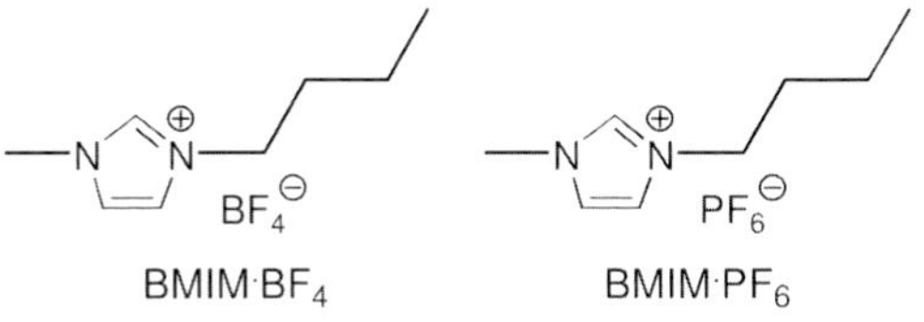

BMIM·BF_4 BMIM·PF_6

Figure 2.24 Structures of the two most often used ionic liquids 1-butyl-3-methylimidazolium tetrafluoroborate (BMIM-BF_4) or its hexafluorophosphate (BMIM-PF_6).

active site. The bound solvent changes the kinetic properties of the enzyme. This can be studied in system IV in Table 2.8, where indirect effects are excluded. Existing data show that bound organic solvents (hexane, octane, hexanol, ethyl acetate) influence enzymes in different ways. For example, the secondary and tertiary structures of α-chymotrypsin, penicillin amidase, and proteinase K were changed, as shown by their fluorescence and circular dichroism spectra in buffers saturated with the organic solvents – that is, by the binding of organic molecules to the enzymes (Michaelis, 1991). This caused different changes in the S, S′ binding and catalytic sites. For α-chymotrypsin and penicillin amidase, the specificity constants were hardly influenced. However, the selectivity (stereo- or synthesis/hydrolysis ratio in kinetically controlled synthesis reactions) in the S_1' binding site was changed by almost an order of magnitude for α-chymotrypsin, but unchanged for penicillin amidase in the presence of bound organic molecules. For proteinase K, the specificity constant for the hydrolysis of *N*-acetyl-L-tyrosine ethyl ester increased by almost an order of magnitude in the presence of bound organic solvent (Kasche, Galunsky, and Michaelis, 1991).

The indirect effect of the solvent is caused by its influence on the end point due to the partition of substrates and products in the two-phase systems (I–III) in Table 2.8. The solvents change the equilibrium in equilibrium-controlled processes (this is shown schematically in Figure 2.25). The apparent equilibrium constant for the two-phase system K_{biph} is given by the following relationship:

$$K_{\text{biph}} = K_{\text{aq}} \frac{(1+\alpha p_{\text{C}})(1+\alpha)}{(1+\alpha p_{\text{A}})(1+\alpha p_{\text{B}})}, \tag{2.43}$$

where α is the volume ratio $V_{\text{org}}/V_{\text{aq}}$. From this equation, it follows that higher yields in C are only obtained when $p_{\text{C}} > p_{\text{A}} \approx p_{\text{B}}$ – that is, when the product is much more hydrophobic than the substrates.

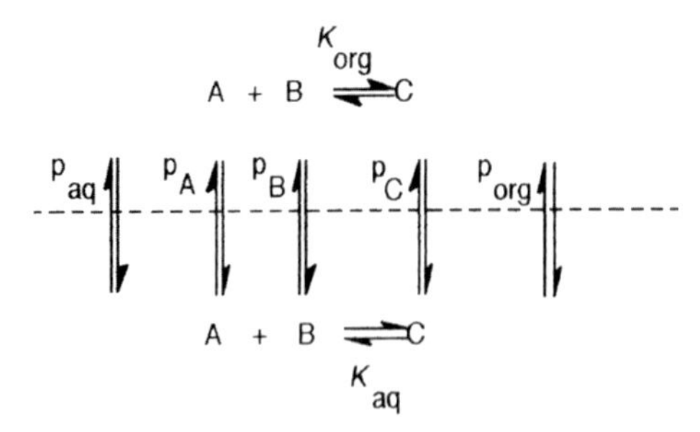

Figure 2.25 Enzyme-catalyzed process in a two-phase system. The enzyme-catalyzed process can only occur in the aqueous phase or at the interphase. K_{aq} and K_{org} are the equilibrium constants in the aqueous and organic phases, respectively; p_{A}, p_{B}, and p_{C} ($=c_{\text{org}}/c_{\text{aq}}$) are partition coefficients; p_{aq} and p_{org} are the partition coefficients for the solvents.

The situation is different when the organic solvent itself is one of the substrates such as a liquid ester, alcohol, lipid, and so on, in which the other substrates can be dissolved. This is the case for many biotransformations that can be catalyzed by lipase, an enzyme that has evolved for the bioconversion of hydrophobic compounds. This enzyme has a wide substrate specificity, and is used in enzyme processes such as the transesterification of lipids and racemate resolution of hydrophobic compounds used as fine chemicals or pharmaceuticals (see Chapter 4).

2.9.2.1 Changes in Rates, k_{cat}, K_m, and Selectivities in These Systems Compared with Homogeneous Aqueous Solutions

k_{cat}, K_m, and Rates In the two-phase system I, and the one-phase systems IV and V in Table 2.8, k_{cat} and K_m are changed due to bound organic molecules. Whether this also applies for ionic liquids that are not miscible with water has not yet been studied. The rates are reduced as the concentration of the substrate is lower than its solubility in water.

In the other two-phase systems (II and III), the dependence of the rate on the water content is shown in Figure 2.26. The studies on biocatalysis in these solvents have also raised the question: How much water is required to maintain the catalytic function of an enzyme? Enzymes have been evolved for catalysis in the presence of water. Therefore, enzymes should not function as biocatalysts in the absence of water. This has been confirmed in studies on the systems I, II, and III in Table 2.8. The results can be summarized as shown in Figure 2.26. Observations from the direct effects of bound organic solvents show that the kinetic properties of the enzymes are changed by less than an order of magnitude. So, how can the maximum in Figure 2.26 be explained? This must be caused by at least two different effects: At low water content, the solvent and the enzyme compete for the water

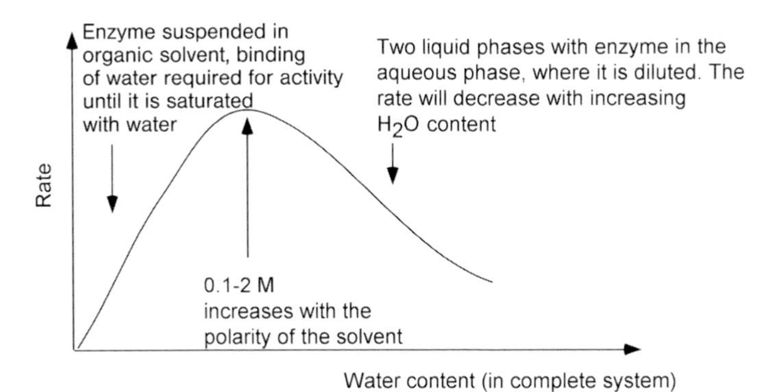

Figure 2.26 The rate of an enzyme-catalyzed reaction with low (systems II and III in Table 2.8) and high (system I in Table 2.8) water content as a function of total water content. The total enzyme amount in the system is kept constant.

molecules dissolved in the organic solvent. The enzyme activity increases until it has bound the water molecules that are required for its optimal function. When the water amount is increased further two phases are formed (to system I), and the enzyme that is located in the aqueous phase is diluted. This causes the decrease in the rate, as shown in Figure 2.26.

Also, the substrate content in the aqueous phase is reduced. At the phase boundary, it will always be smaller than its solubility in water – that is, lower than that at the phase boundary shown in Figure 2.22. Thus, the mass transfer between the phases and the reduced flexibility of the enzyme (the conformation is not changed in nonaqueous solvents) cause a reduction in the rates in such two-phase systems up to a factor of 10^{-3}, compared to aqueous solutions or suspensions with dissolved or suspended substrates (Halling, 1994; Kasche and Galunsky, 1995; Schmitke, Wescott, and Klibanov, 1996). From Figure 2.25 it follows that the enzyme processes should be carried out at the water content with the maximum rate in system II or III in Table 2.8. However, in the system II the suspended free enzymes tend to aggregate, leading to an insufficient usage of the enzymes' active sites. In order to avoid any reduction in rate caused by this effect, the enzymes should be used in a disaggregated form, and this can be achieved either by immobilization or by using covalent modifications to reduce the formation of enzyme aggregates.

Stereoselectivity Most racemate resolutions with organic solvents or ionic liquids are carried out in the two-phase systems II or III in Table 2.8. In these, both direct and indirect effects cause the observed changes in stereoselectivity. Most published data show that the changes are larger than those in aqueous solution (Carrea, Ottolina, and Riva, 1995; Halling, 2000). Although no conclusive explanation for this has yet been given, it has been shown that it is strongly influenced by how the enzyme was incubated (pH) and lyophilized before being added to the organic solvent, as well as by the presence of metal ions that also influence the rates (Halling, 2000). As for aqueous systems, the stereoselectivity of immobilized lipase in supercritical CO_2 has been found to decrease with increasing temperature in the temperature range 50–120 °C (Overmeyer *et al.*, 1999).

Selectivity in Kinetically Controlled Processes Due to the low water content, hydrolysis of products is not expected for kinetically controlled processes, such as transesterifications, in systems II and III. These reactions can therefore be treated as equilibrium-controlled processes.

The conclusion here is that enzyme processes involving slightly soluble substrates or products can be carried out in aqueous suspensions or emulsions with higher space–time yields than in the systems with other solvents shown in Table 2.8. The use of organic or supercritical solvents to solubilize substrates for such processes is favorable in cases where the solvent is nontoxic, can be safely contained, transported, and recycled, or when one of the substrates can be used as the solvent. This is often the case in lipase-catalyzed processes.

2.10
Stability, Denaturation, and Renaturation of Enzymes

In enzyme processes, it is desired to use the biocatalyst for long periods in order to reduce enzyme costs. In living systems, enzymes have not evolved for long-term use as their biological half-life is generally in the range of minutes to several days. Optimal use of the enzyme requires that factors influencing its stability are known, with stability being reduced as little as possible under the process conditions.

The biologically active tertiary and quaternary structure of pro-enzymes or unprocessed enzymes is formed spontaneously – that is, their native structure is thermodynamically stable (Creighton, 1993). For enzymes formed from pro-enzymes or pro-enzymes that require tightly bound metal ions for their correct folding, and biological activity of the formed enzyme, the folded enzyme may not be the thermodynamically most stable tertiary structure. This has consequences for the stability and renaturation of an enzyme that has been denatured (unfolded) without covalent changes in the primary structure. Enzymes, whose native structure is the thermodynamically most stable conformation, can be reversibly denatured (Figure 2.27). The unfolded enzyme will refold to the native structure in dilute solution, where bimolecular reactions, yielding inactive enzyme aggregates, are suppressed. This is applied in the renaturation of inclusion bodies (see Chapter 6). For native enzymes, formed from pro-enzymes or with tightly bound metal ions, no renaturation of the unfolded enzyme or pro-enzyme to active enzyme or pro-enzyme was observed in the absence of the pro-peptide or metal ion. In the presence of the latter, 50% of the activity of the enzyme and 100% conversion of the renatured pro-enzyme to active enzyme have been observed (Ignatova *et al.*, 2005). Studies on the reversibility of enzyme denaturation are of basic interest, but still of minor practical importance for enzymes used in enzyme processes. For these the factors that influence their denaturation and stabilize or destabilize the active enzyme structure must be considered. The stability of native enzymes can be reduced when interactions that stabilize the native structure (e.g., hydrophobic bonds, hydrogen bonds, binding ion–ion interactions such as $NH^+\cdots COO^-$) are weakened by chaotropic solvents (urea) and increased temperature or pH values that lead to unfolding, loss of function, and, ultimately, to denaturation of the enzyme (Figure 2.27).

The stability of an enzyme is also influenced when substrates and organic solvent molecules are bound to the enzyme (see Section 2.9). It has been shown that substrates and organic solvents immiscible with water stabilize enzymes, whereas solvents that are miscible with water destabilize enzymes. This knowledge obtained during the last century has recently been confirmed in a review over organic solvent-tolerant enzymes (Doukyu and Ogino, 2010). In supercritical CO_2, the stability of enzymes is reduced when the water content in the supercritical gas increases (Kasche *et al.*, 1988). Enzymes can be stabilized when destabilizing interactions such as repulsive ion–ion interactions are weakened at high ionic strengths, or in the presence of doubly charged "cross-linking" ions such as Ca^{2+} or HPO_4^{2-}. The stability of an enzyme is also a function of its primary structure (Figure 2.28). For

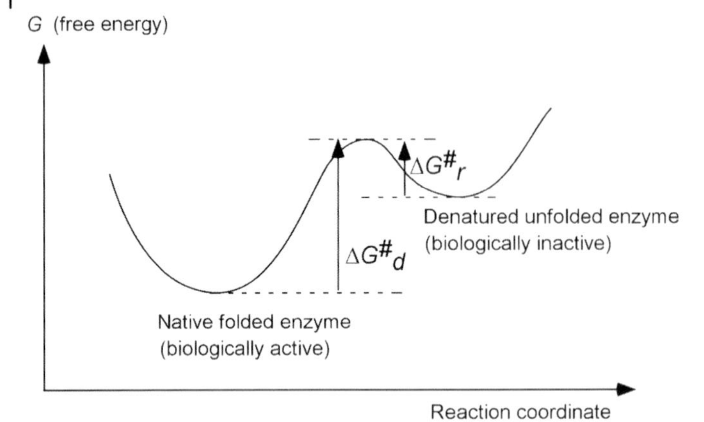

Figure 2.27 Free energy diagram for the reversible denaturation of an enzyme with the activation energies for denaturation ($\Delta G_d^{\#}$) and renaturation ($\Delta G_r^{\#}$). The native structure is formed spontaneously at normal temperatures. At higher temperatures or extreme pH values, the equilibrium is shifted to the unfolded (denatured) enzyme. The denatured enzyme can be renatured by slowly changing the temperature or pH to the original value. A rapid change "freezes" the denatured state, due to the slow rate of refolding. The renaturation must also be carried out at low enzyme concentrations to avoid aggregation of the denatured enzymes. A comparison with Figure 2.7 shows that enzymes are stabilized when they bind substrates.

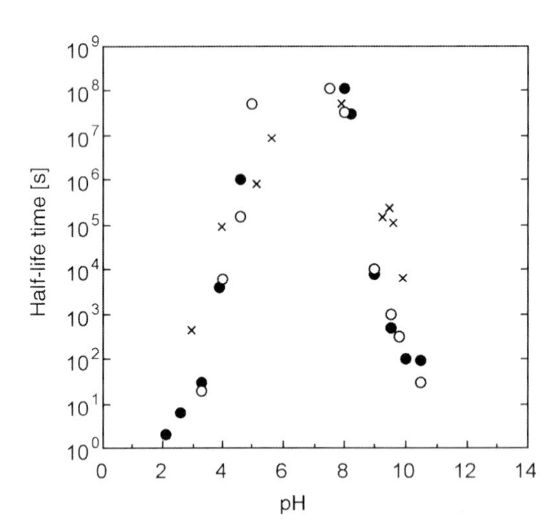

Figure 2.28 The pH stability of different penicillin amidase forms from *E. coli* (closed circles: $PA_{7.0}$; open circles: $PA_{6.7}$; subscripts indicate their isoelectric point) and *A. faecalis* (cross symbols) at 25 °C. The half-life time of the monomolecular denaturation was measured in buffer with ionic strength 1 M by activity determinations (Wiesemann, 1991; A. Rieks and V. Kasche, unpublished data).

penicillin amidase, it has also been shown that pH stability can be increased by covalent immobilization in porous particles (see Chapters 8 and 10).

Different factors that may cause reversible or irreversible (leading to covalent changes in the primary structure) enzyme denaturation during enzyme-catalyzed processes are listed in Table 2.9. Intramolecular autoproteolysis is a newly discovered process in which a peptide bond hydrolysis within a polypeptide chain is catalyzed by amino acid residues, in the same polypeptide chain, near this bond (Kasche *et al.*, 1999). These processes can also cause irreversible enzyme denaturation.

The half-lives of the enzymes range from a few days (lipase, β-glucosidase) to >100 days (penicillin amidase, more stable lipases) under optimal process

Table 2.9 Denaturation of an enzyme by factors that influence rates and yields in enzyme processes, and how this can be minimized.

Factor/purpose	Denaturation by	Reduce denaturation by
Temperature increase/higher rates and yields	Unfolding (reversible); chemical modification (irreversible)	Use temperatures ≥20 °C below temperature optimum; use enzyme with higher temperature optimum
Temperature decrease/higher yields	Dissociation of oligomeric enzymes (reversible)	Cross-link the oligomeric enzyme
Shear forces/increase mass transfer	Unfolding (reversible)	Cross-link or immobilize the enzyme
Increase or decrease pH/higher rates and yields	Unfolding (reversible) when pH >> or << pI	Cross-link or immobilize enzyme, or use enzyme with better pH stability
O_2/increase rate with oxidases; H_2O_2/bleaching agent in washing powder	Oxidation of –SH or methionine (irreversible)	Find or construct more stable enzymes
High substrate content/increase yields and reduce product purification costs	Chemical modification (irreversible); glucose can react with amino groups of lysine	Find or construct more stable enzymes
Organic solvents	Unfolding (reversible)	Immobilize or cross-link enzyme
Peptidase in enzyme preparation	Hydrolysis of peptide bond (irreversible)	Use a more pure enzyme or immobilize the enzyme
Intramolecular autoproteolysis	Hydrolysis of peptide bond (irreversible)	Find or construct a more stable enzyme
Heavy metal ions as impurities in water, buffer compounds	Irreversible inhibition of –SH	Use water and buffer with minimal heavy metal ion content
UV or ionizing radiation/sterilization	Chemical modification (irreversible)	Immobilize enzyme

Table 2.10 Rate constant for the irreversible denaturation and chemical changes in lysozyme at 100 °C (Ahern and Klibanov, 1988).

Process	First-order rate constant (h^{-1}) at pH		
	4	6	8
All processes	0.49	4.1	50
Deamidation of Asn	0.45	4.1	18
Hydrolysis of Asp–X peptide bonds	0.12	0	0
Destruction of Cys	0	0	6
Formation of "wrong" structures (reversible process (?))	0	0	32

conditions, though the reason for these widely ranging stabilities of enzymes remains incompletely understood. All enzymes are stabilized in the presence of their substrates or products. Mainly due to that the binding results in a more compact enzyme structure. This is important for enzyme processes with high substrate concentrations (>1 M), as for glycosidases (see Section 7.1.2). Rates for irreversible thermal denaturation are listed in Table 2.10. One possible way to avoid denaturation caused by factors listed in Table 2.9 and to reduce enzyme costs would be to screen for more stable enzymes in living systems, or to develop and construct better enzymes by protein engineering.

2.11 Better Enzymes by Natural Evolution, *In Vitro* Evolution, or Rational Enzyme Engineering

One of the main problems in enzyme technology is to identify or design an optimal enzyme for a specific enzyme process. These processes can only be realized on an industrial scale when the enzyme costs can be reduced to less than 5% of the costs to transform the substrate(s) to the desired product(s). As shown in Section 2.8, a process window for optimal yields in a pH–T-plane for an equilibrium-controlled process can be constructed based on thermodynamic data for the process. From Eq. (2.33), it follows that the amount of enzyme required to reach this yield in a given time depends on

1) the enzyme kinetic constants (k_{cat}, K_m, and K_i);
2) the stability (chemical, pH, or temperature) of the enzyme under process conditions, given by the half-life, determined from the apparent first-order rate constant k_i (Eq. (2.32)).

   For equilibrium-controlled racemate resolutions, the steric purity of the product(s) also depends on

3) the enantioselectivity E_{eq} of the enzyme (Eq. (2.15)) determined by the enzyme kinetic constants in property 1. This also applies for kinetically controlled

racemate resolutions. In this case, however, the enantioselectivity E_{kin} (Eq. (2.16)) depends also on other rate and equilibrium constants than for equilibrium-controlled processes.

For other kinetically controlled processes, the maximal product yield is additionally influenced by

4) the selectivity for the synthesis and hydrolysis reactions $(k_T/k_H)_{app}$ (Eqs. (2.3), (2.38),(2.39)).

For a given enzyme process, the biocatalyst costs can be reduced when a better enzyme, with respect to the properties 1–4 above, can be found among natural variants or constructed by protein engineering. How these properties must change, compared to an enzyme used for a given process, is shown in Table 2.11. From this it follows that an increase in k_{cat} or k_{cat}/K_m alone does not necessarily lead to an improved enzyme for an equilibrium-controlled process. Here also changes in K_i must be considered (Fox and Clay, 2008). That an increase in K_i also contributes to improve an enzyme for such a process has recently been shown (Shin *et al.*, 2009). By protein engineering an improved cephalosporinase, where all these properties were increased, was constructed with which cephalosporin C could be hydrolyzed in one step. This replaced the previously used two-step procedure (see Chapter 12, Case Study 3).

That neutral (or silent) mutations cause variations in the primary structure and the properties 1–4 of natural enzyme variants of the same enzyme were outlined in Section 2.4. These mutations are very important for the evolution of living systems as they can result in a better adaptation of organisms in changing environments (higher or lower pH, temperature, substrate or salt content, new substrates, etc.). On the enzyme level, this leads to the large variation in the primary structure

Table 2.11 How must the enzyme properties 1–4 be changed to obtain a better enzyme, with a higher space–time yield per mole of enzyme, than the one currently used for a given process?.

Process enzyme property	Equilibrium-controlled process; enzyme cannot influence maximal yield (at equilibrium)	Equilibrium-controlled racemate resolution	Kinetically controlled reaction; enzyme can influence maximal yield	Kinetically controlled racemate resolution
k_{cat} (turnover number)	Increase			
K_m (substrate binding)	Decrease			
K_i (product inhibition)	Increase			
k_i (enzyme stability)	Decrease	Decrease	Decrease	Decrease
E_{eq}		Increase		
$(k_T/k_H)_{app}$			Increase	
E_{kin}				Increase

observed for the same enzyme from different organisms that differ in properties 1–4. For enzyme technology, this implies that better enzymes for a process can be found by screening for the enzyme in different living organisms or for the enzyme gene in DNA from environmental samples ("metagenomes" (see Chapter 3); Lorenz *et al.*, 2002; Steele *et al.*, 2009). This allows screening for enzymes in the more than 99% of all microorganisms that have not yet been – or cannot be – cultivated. The metagenome is a mixture of DNA from living and dead cells found in environmental samples that contain genes for enzymes; these genes can be expressed in suitable host organisms. Suitable environments to screen for an enzyme or its gene are those with pH and temperature values in the optimal process window, and with high contents of the substrates or analogous compounds that are used in the process.

In this context, it is of interest to analyze the ranges within which properties 1–4 can differ for the same enzyme due to natural evolution, and how much they can be changed by protein engineering. The latter is covered in detail in the next chapter.

2.11.1 Changes in Enzyme Properties by Natural Evolution

2.11.1.1 k_{cat} and K_m

The protein (including enzymes) content in a living cell cannot be increased above the value of $\approx$20% observed in prokaryotic and eukaryotic cells (Atkinson, 1969). New metabolic reactions that require additional enzymes and their genes can therefore only be obtained when enzymes and their genes required for other metabolic reactions are

- lost during the evolution; and/or
- improved during the evolution so that less enzyme is required to catalyze these metabolic reactions.

Both cases are observed during the evolution of living systems. Eukaryotic cells such as mammalian cells have lost genes for the enzymes required to synthesize certain amino acids, to allow the incorporation of new genes that code for enzymes that catalyze new metabolic reactions. The synthesis of enzymes requires much energy (substrate). An organism that synthesizes an enzyme with improved k_{cat} value requires less enzyme and energy to catalyze the metabolic reaction than organisms with a less active enzyme. For organisms living at the same ambient temperature, the one with the higher k_{cat} has an evolutionary advantage, and this leads to variations in turnover number at a constant temperature for the same enzyme from different organisms. The changes in k_{cat} during evolution are, however, limited by a maximum value of the specificity constant (k_{cat}/K_m) of $\sim 10^8\,M^{-1}\,s^{-1}$ (this is discussed in Section 2.7.1) and the substrate concentrations in living systems. It is not advantageous for an organism to synthesize enzyme molecules that cost energy, when only a fraction of those molecules will catalyze metabolic reactions. This occurs when $[S] \ll K_m$. Therefore, enzymes in living systems have been selected during evolution with K_m values in the range of the substrate concentrations in order to minimize energy costs for their synthesis (Fersht, 1999; Hochachka and

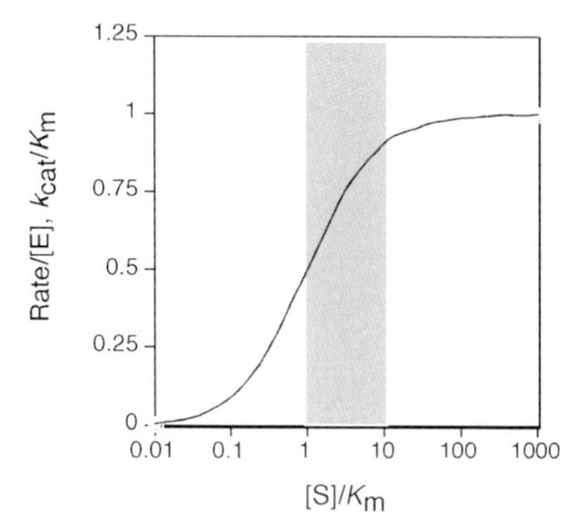

Figure 2.29 The specific rate of an enzyme calculated from Eq. (2.8) and expressed in k_{cat}/K_m units as a function of the substrate content [S] in the living cell or its environment expressed in K_m units. It is an evolutionary advantage to use as much as possible of the synthesized enzyme for catalytic conversion of the substrate. This occurs when [S] is 1–10K_m (shaded area). At higher ratios, the specific rate is independent of [S]; this is unfavorable for metabolic regulation.

Somero, 2002). This situation is shown graphically in Figure 2.29. From the boundary condition for the substrate specificity, it follows that the upper limit of k_{cat} obtainable by natural evolution is $\approx 10^8$[S] s^{-1} – that is, it is limited by the substrate content in living systems and their environment.

For organisms that can survive over a large temperature range (e.g., prokaryotes, fishes), the temperature dependence of k_{cat} and K_m is also of evolutionary importance. For organisms living at low temperatures (psychrophiles), the enzyme that catalyzes a metabolic reaction must be evolved so that its k_{cat} value is higher at this temperature than for an enzyme that catalyzes the same reaction in an organism living at a higher temperature (meso- or thermophiles). When this is not the case, the psychrophilic organism must produce more of the enzyme, and this is an evolutionary disadvantage. The variation in the k_{cat} values observed at one temperature for lactate dehydrogenase from psychro-, meso-, and thermophilic organisms has been found to differ by one order of magnitude (Fields and Somero, 1998). As these organisms can live at different temperatures, the rates of the reaction must be regulated to become less dependent on temperature. This requires enzymes for which the K_m values increase with temperature. This applies to lactate dehydrogenases, as discussed above, with K_m values at one temperature having also been found to vary by up to one order of magnitude (Fields and Somero, 1998).

For nonnatural substrates, the k_{cat}, K_m, and specificity constants have been found to differ by up to one order of magnitude for the same enzyme from different organisms (Table 2.12). These changes did not differ markedly for specific and less specific substrates (see also Table 2.4).

In enzyme technology, much higher substrate contents (up to 1 M) are employed than in living systems, and this is generally much higher than the K_m values of the

Table 2.12 (a) Kinetic constants and selectivities of penicillin amidases (PA) from different sources and two genetically constructed hybrid enzymes with the A- and B-chains from different sources (hybrid I $A_{Kc}B_{Ec}$ and hybrid II $A_{Ec}B_{Kc}$; Ec = *E. coli*, Kc = *K. citrophila*): k_{cat}, K_m, and specificity constant for the hydrolysis of benzylpenicillin (BP), 2-nitro-5-phenylacetamidobenzoic acid (NIPAB), phenylglycine amide ($PhgNH_2$), and phenylglycine methyl ester (PhgOMe) at pH 7.5 ($I = 0.2$) and 25 °C; (b) stereoselectivity E_{eq} for the hydrolysis of (*R*,*S*)-$PhgNH_2$, (*R*,*S*)-pOH$PhgNH_2$ (S_1 stereoselectivity) and PAA-(*R*,*S*)-Phe (S_1' stereoselectivity) at pH 7.5 ($I = 0.2$) and 25 °C; selectivity $(k_T/k_H)_{app}$ in kinetically controlled synthesis of amoxicillin from 20 mM (*R*)-pOH$PhgNH_2$ and 20 mM 6-aminopenicillanic acid at pH 6.5 ($I = 0.2$) and 25 °C, and cephalexin from 200 mM (*R*)-$PhgNH_2$ and 200 mM 7-aminodesacetoxycephalosporanic acid at pH 7.5 ($I = 0.2$) and 25 °C.

(a)

PA from	BP			NIPAB			(*R*)-$PhgNH_2$			(*R*)-PhgOMe		
	k_{cat} (s^{-1})	K_m (μM)	k_{cat}/K_m ($M^{-1}s^{-1}$)	k_{cat} (s^{-1})	K_m (μM)	k_{cat}/K_m ($M^{-1}s^{-1}$)	k_{cat} (s^{-1})	K_m (mM)	k_{cat}/K_m ($M^{-1}s^{-1}$)	k_{cat} (s^{-1})	K_m (mM)	k_{cat}/K_m ($M^{-1}s^{-1}$)
E. coli	50	10	5.0×10^6	20	20	1.0×10^6	16	16	1.0×10^3	16	15	1.1×10^3
A. faecalis	80	8	1.0×10^7	82	12	6.8×10^6	7	6	1.2×10^3	43	6	7.2×10^3
K. citrophila	60	18	3.3×10^6	30	20	1.5×10^6	15	20	7.5×10^2	13	13	1.0×10^3
A. viscosus	140	20	7.0×10^6	5	130	3.8×10^4	45	480	9.4×10^1	42	13	3.2×10^3
Hybrid I	100	12	8.3×10^6	40	20	2.0×10^6	15	21	7.1×10^2	18	21	8.6×10^2
Hybrid II	115	18	6.4×10^6	6	20	3.0×10^5	14	25	5.6×10^2	—	—	—

(b)

PA from	(*R*,*S*)-$PhgNH_2$ hydrolysis	(*R*,*S*)-pOH$PhgNH_2$ hydrolysis	PAA-(*R*,*S*)-Phe hydrolysis	Amoxicillin synthesis	Cephalexin synthesis
	E_{eq}	E_{eq}	E_{eq}	$(k_T/k_H)_{app}$	$(k_T/k_H)_{app}$
E. coli	0.5	0.03	1000	3200	3300
A. faecalis	0.4	n.d.	250	2500	800
K. citrophila	0.3	0.2	n.d	1900	n.d.
A. viscosus	0.7	n.d	n.d	n.d.	n.d.
Hybrid I	0.4	0.2	n.d	1600	n.d.
Hybrid II	n.d.	n.d.	n.d	2500	n.d.

n.d. = not determined.

enzymes used. In these cases, the enzyme activity under process conditions can only be increased by using enzyme variants with higher k_{cat} values. From the above analysis, it follows that similar changes in k_{cat} as observed for natural variants should be expected for enzymes improved by *in vitro* evolution or rational enzyme engineering.

2.11.1.2 **Enzyme Stability**

During evolution, organisms have been selected that can live in environments of different pH (range 1–11) and temperature (range 0 to >100 °C). This implies that enzymes in organisms have evolved to achieve a better biological activity at pH and temperature values in the environment of these organisms compared to the same enzymes in organisms that survive better under other conditions (pH, *T*). Organisms produce both intra- and extracellular enzymes, but due to intracellular proteolysis (protein turnover) the intracellular enzymes have biological half-lives that are not limited by their pH and temperature stability (see Section 6.3). From this, it follows that intracellular enzymes have not been thoroughly selected for pH and temperature stability during evolution.

Extracellular enzymes are produced to hydrolyze biopolymers and to transform organic compounds into products that can be taken up and metabolized in the cells. They can also be used to detoxify compounds outside the cells that are toxic to the cells. For the cells, it is an evolutionary advantage when the pH and temperature stability of these enzymes is increased as they then have to produce less enzyme to transform extracellular compounds into molecules required for their survival. One enzyme process where pH stability of the enzyme used is essential is in the hydrolysis of penicillin G (see Sections 2.7 and 2.10, and Exercise 2.14). The enzyme first used in this process was the extracellular (periplasmic) penicillin amidase from *E. coli*. Due to its limited pH stability, it could only be used up to pH ≈8, but at this pH <95% of penicillin is hydrolyzed in the equilibrium-controlled process. By introducing the same enzyme from *A. faecalis* that could be used at pH ≈ 9 due to its better pH stability (Figure 2.28), the hydrolysis yield was increased to ≈99% (Verhaert *et al.*, 1997; Ignatova *et al.*, 1998).

From this it follows that the pH and temperature stability for intracellular enzymes could be improved by *in vitro* evolution or by rational protein design more than for those extracellular enzymes selected for better pH and temperature stability.

The chemical stability of an enzyme is a measure of its chemical modification due to reactions with metabolites and other organic molecules in the environment. Hydroxyl groups of sugars can react with amino groups on enzymes, while O_2 or H_2O_2 can oxidize Met or Cys, and so on. In natural systems, the concentration of these compounds is much less than that used in enzyme processes, and therefore the rate of enzyme modification by such bimolecular reactions is much higher in the latter situation. In this case, chemical stability is a measure of the covalent modification of the enzyme by the substrates and products of the enzyme process. Thus, it could be assumed that during evolution enzymes have not been selected for optimal chemical stability, and that the same enzyme from different organisms can vary widely with regard to its chemical stability. Although this point has not yet been studied in detail, it follows that this property of enzymes could be improved by *in vitro* evolution or by rational protein design.

2.11.1.3 Stereoselectivity

With the exception of amino acids, hydroxy acids, carbohydrates, and some other compounds, optically active molecules are only present as one enantiomer in living systems. This is also reflected in the list of enzymes that catalyze racemization reactions (EC 5.1). The enzymes involved in the biosynthesis of sterically pure compounds from racemic mixtures of the above compounds must therefore have been selected for high stereoselectivity during evolution. This applies to the synthesis of oligomers and polymers from monomers. Ribosomal peptide (protein) synthesis involves only (*S*)-amino acids, whereas peptide antibiotics that are synthesized nonribosomally contain both (*S*)- and (*R*)-amino acids (β-lactam antibiotics as penicillin, gramicidin S, cyclosporin, tyrocidine, etc.) (Kleinkauf and von Döhren, 1990). The latter are mainly produced in prokaryotes. Some ribosomal antibacterial peptides from eukaryotes (e.g., dermorphin from frog skin) contain (*R*)-amino acids that are formed by racemization reactions after the peptide synthesis from (*S*)-amino acids. In natural systems, these oligomers and polymers are hydrolyzed by hydrolases. As the monomers can be racemized in prokaryotes, the hydrolases in these are not selected for high stereoselectivities as the transferases that catalyze the polymerization reactions. Therefore, many peptidases can be used for the kinetically controlled synthesis of peptide bonds between (*S*)- and (*R*)-amino acids (Kasche, 2001; see also Figure 2.12).

In living systems, sterically pure compounds are synthesized from prochiral compounds containing double bonds or are keto acids. The enzymes catalyzing these reactions, such as dehydratases (EC 4.2.1) and oxygenases (EC 1.13) for additions to double bonds, or transaminases (EC 2.6.1) and dehydrogenases (EC 1), must have been selected for high stereoselectivity during evolution.

Enzymes that in natural systems mainly catalyze reactions involving optically pure substrates or substrates that are optically inactive are not expected to be selected for high stereoselectivity during evolution. Lipases have a wide range of stereoselectivity among the natural enzyme variants (see Figure 2.13). The stereoselectivity of penicillin amidases, which still have unknown biological functions, has both (*S*)- and (*R*)-specific binding subsites, and the stereoselectivity is known to vary by up to one order of magnitude for natural enzyme variants (see Table 2.11).

As with the stability of enzymes, it is possible to increase the stereoselectivity of enzymes by *in vitro* evolution or by rational enzyme engineering (see Chapter 3).

2.11.1.4 Selectivity in Kinetically Controlled Synthesis of Condensation Products

In natural systems, processes are catalyzed by transferases that have been selected for high $(k_T/k_H)_{app}$ values during evolution (see Table 2.3), whereas in enzyme processes the hydrolases are mainly used as biocatalysts. This property varies up to a factor of 4 for the same hydrolase from different sources for the same reaction (see Table 2.12). These enzymes have not been selected with regard to this property during evolution.

Exercises

2.1 For the ionization of Tris (for formula, see Merck Index) and phosphoric acid, the heat of ionization is 40 and $\approx 0\,kJ\ mol^{-1}$, respectively. In which buffer would you study the pH and temperature dependence of enzyme-catalyzed reactions? What other reasons can you give for the advice to avoid the use of Tris as buffer when you study enzyme kinetics? ($pK_{Tris} \approx 8.5$) (see Table 2.3).

2.2 Why do oxidoreductases require cosubstrates (and which)?

2.3 Explain Figures 2.4 and 2.5. Why does carboxypeptidase A hydrolyze at the C-end and α-chymotrypsin not? (*Hint*: The latter enzyme has a negative charge in the S_1' subsite). What influence has the ionic strength here? In what steps can the enzyme function be divided? Can product inhibition be avoided?

2.4 What other similarities, besides the conserved active site structure, do you find in Figure 2.6?

2.5 Derive the Michaelis–Menten equation.

2.6 A pyrophosphatase from potatoes catalyzes the hydrolysis of inorganic pyrophosphate at pH 5.3. V_{max} was determined at different temperatures:

Temperature (°C)	15	25	35	40
V_{max} (μM min^{-1})	6.53	10.47	16.79	20.65

Calculate the activation energy for the enzyme-catalyzed process and compare it with that for the uncatalyzed reaction ($=121\,kJ\ mol^{-1}$). What activation energy is determined here? (*Hint*: See Figure 2.7).

2.7 Define k_{cat}, K_m, and k_{cat}/K_m. What information do they give? Why does the last quantity have an upper limit?

2.8 What information do Tables 2.4 and 2.5 provide about the P_1, P_1' substrate specificity and stereoselectivity of the studied enzymes? What is the esterase-/amidase activity ratio of the peptidases?

2.9 Glycolipids are biodegradable detergents. Which enzymes could catalyze the synthesis and hydrolysis of such compounds?

2.10 The below figure gives the Arrhenius plots for the thermal denaturation at pH 7.5 (closed circles, left scale) of penicillin amidase from *E. coli* and k_{cat} (right scale) for the hydrolysis of (*R*)-phenylglycine amide (broken line) and (*S*)-phenylglycine amide (thick line) by the same enzyme. Calculate and compare the activation energies for the different processes. What activation energy is determined in Figure 2.8?

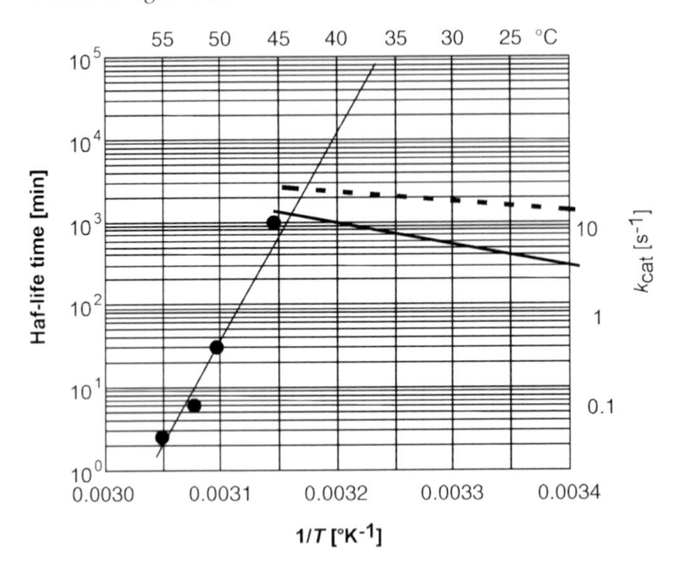

2.11 Transglutaminase (EC 2.3.2.13) catalyzes the formation of a peptide bond between polypeptide chains between the carboxyl group of glutamine (acyl donor) and the amino group of lysine (acyl acceptor) (Hamada, 1994). In this reaction, ammonia is formed. Try to formulate a reaction mechanism and discuss where such a peptide can be formed in or on a protein. Can this enzyme be used to catalyze reactions where the enzyme is labeled with fluorescent dyes or other ligands, the immobilization of proteins on surfaces? What reactive groups must these ligands or surfaces have? For what purposes can such labeled proteins be used? Is this an equilibrium- or kinetically controlled process? What is the optimal pH for this reaction? What other applications can you propose for this enzyme in enzyme technology?

2.12 Why is substrate and product inhibition important in enzyme technology?

2.13 What properties should organic solvents used in enzyme technology have?

2.14 The below figure gives the pH and temperature dependence of K_{app} (as dissociation constant = [6-aminopenicillanic acid][phenylacetic acid]|[penicillin G]) for the hydrolysis of penicillin G. Determine the process window in a pH–*T*-plane for 95 and 99% yields for the hydrolysis of 300 mM penicillin G. Which enzyme in Figure 2.28 should be used as the biocatalyst? What other factors can limit the size of this process window? (See also Exercise 2.10.) What buffers are suitable to control the pH during the hydrolysis? (*Hint*: A buffer with a high buffer capacity at the optimal pH for the process, and that does not increase wastewater treatment costs.)

2.15 How much does the temperature change during the hydrolysis of 300 mM penicillin G or 1 M saccharose in an adiabatic reactor? (*Hint*: See Table 2.7.) What are the consequences for enzyme technology?

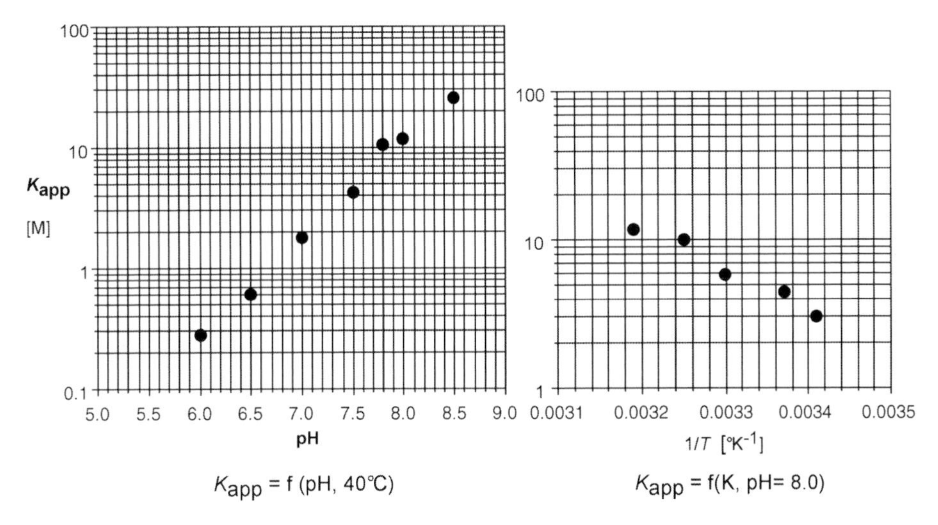

K_{app} = f (pH, 40°C) K_{app} = f(K, pH= 8.0)

2.16 Enzyme processes with slightly soluble substrates can be carried out in aqueous suspensions with up to ≥90% solid substrates. When H^+ is formed or consumed in the process, the pH will change during the reaction. Discuss the possibility of keeping the pH constant (and the product yield high) by adding base or acid at different solid contents. At what solid contents would you carry out such a reaction?

2.17 The below graph provides the ionic strength dependence of the selectivity (open symbols) and the maximal yield (closed symbols) in the kinetically controlled synthesis of the β-lactam antibiotic ampicillin from 200 mM (*R*)-phenylglycine amide and 6-APA) at pH 7.5 and 25 °C. The hydrolysis of the amide does not depend on *I*. Try to explain the cause for this ionic strength dependence. (*Hints*: p*K* values in Figure 2.20, enzyme binding site charge, and how does 6-APA inhibit the hydrolysis of the amide.)

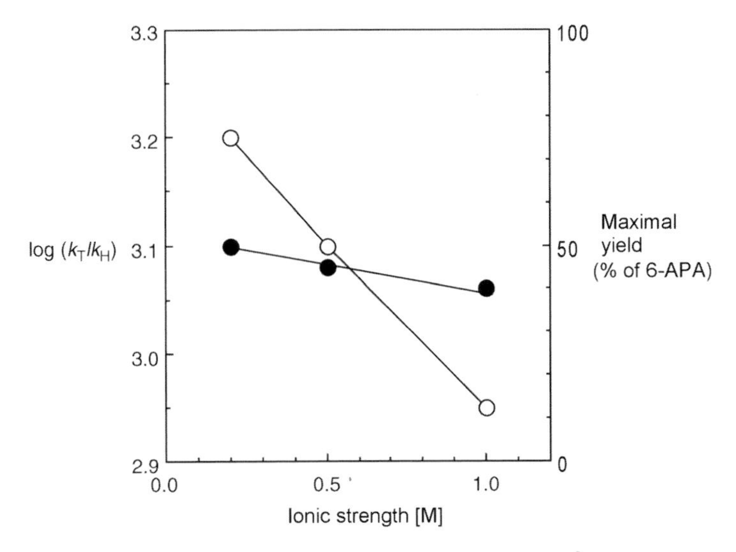

2.18 Derive and integrate Eq. (2.33) for competitive product inhibition.

2.19 How can a better enzyme be found or constructed for a given enzyme process? Discuss this for the following processes:

a. The hydrolysis of penicillin G based on the figure in Exercise 2.14, Figure 2.28, and Tables 2.7 and 2.11.
b. The isomerization of glucose based on Figure 1.8 and Tables 2.7 and 2.11.

2.20 The equilibrium constants (=[glucose]^2/[disaccharide]) for the hydrolysis of maltose and isomaltose at 298 K are >500 and 15 M, respectively; the enthalpy changes for the reactions are given in Table 2.7 (Tewari and Goldberg, 1989).

a. Calculate the equilibrium concentrations of glucose and isomaltose after the hydrolysis of a starch solution with 1 M glucose.
b. How can the amount of the by-product isomaltose be reduced in this process? Include both system and process (kinetics) and enzyme properties.

2.21 Pyruvate decarboxylase can catalyze the carboxylation of acetaldehyde to pyruvate. At what pH and in which solvents would you carry out this reaction to obtain a high product yield? (*Hints*: What is the pK value for pyruvic acid, CO_2 is used for carboxylation) (Miyazaki *et al.*, 2001).

2.22 Peptide hormones or neuropeptides are synthesized as parts of ribosomically produced polypeptides. Some of the biologically active peptides are amidated at the C-terminal end (Kleinkauf and von Döhren, 1990). To these belong the following peptide hormones: oxytocin (stimulates contraction of smooth muscles; Cys-Tyr-Ile-Gln-Asn-Cys-Pro-Leu-Gly-NH_2) and vasopressin (antidiuretic; Cys-Tyr-Phe-Gln-Asn-Cys-Pro-Arg-Gly-NH_2), where the two Cys form an S–S bond.

a. Discuss how this amidated peptide can be produced in an enzyme process from the non-amidated peptide. What properties must the enzyme used as biocatalyst have? (*Hints*: Figures 2.3–2.5 and Table 2.4.)
b. Find out more about the biosynthesis of these peptide using Internet resources, and suggest processes for their biotechnological production.

2.23 That insulin isolated from animal pancreas can be used to treat diabetes was first observed by Romanian and Canadian scientists around 1920. Until 1980 only insulin isolated from animal pancreas was used to treat diabetes. The primary structure of animal insulin was determined in 1951 by Sanger. It was the first primary structure of a protein that was determined. Based on this, the chemical synthesis of separate A- and B-chains (a) was achieved by groups in China, Germany, and the United States in the 1960s. The yield of insulin after mixing the chains under conditions where the S–S could be formed (lines in (a)) was very low. Based on this, it was postulated that insulin *in vivo* is synthesized as proinsulin with a pro-peptide that allowed the formation of the correct sulfur bridges before it is removed by hydrolysis with peptidases. This was experimentally confirmed around 1967 (Steiner *et al.*, 1967). The 3D structure of insulin was determined by Hodgkin in 1969. She found that the

C-terminal of the B-chain is near the N-terminal of the A-chain. Based on these observations and the development of gene technology, almost all insulin is now produced as a recombinant pro-protein from which human insulin is derived by enzyme technology, as a correctly folded extracellular mini-pro-insulin (b) produced in yeast cells with a signal peptide and exported out of the cells without the signal peptide. To obtain high yields, the C-terminal Thr in the B-chain is missing in the mini-proinsulin. The other method is to produce proinsulin as inclusion bodies in *E. coli*. After their solubilization, it is purified and refolded to biologically active proinsulin. For more information about the fascinating history of insulin, see Wikipedia.

A Human insulin

B Mini-proinsulin

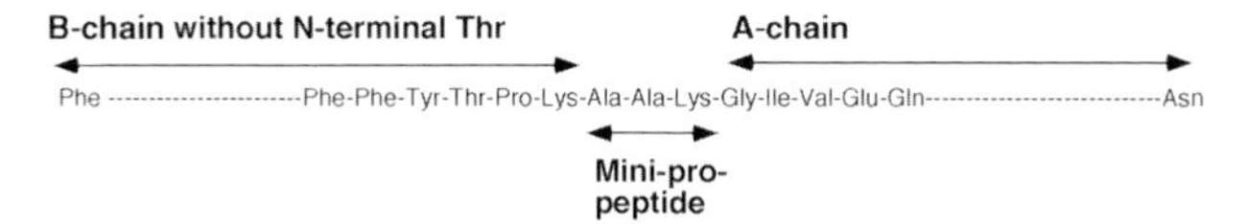

C Proinsulin

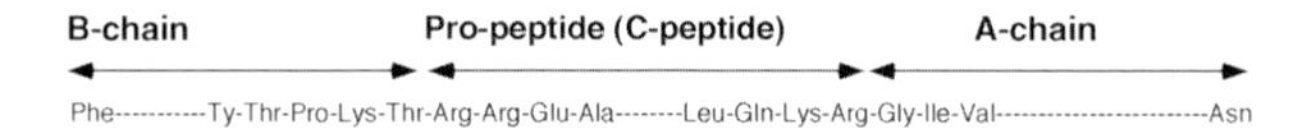

Use the MEROPS database to select the peptidases that can be used to produce human insulin (a) from folded mini-proinsulin (b) and proinsulin (c). Reduce the search to two amino acids on both sites of the bond that must be hydrolyzed. Note that a Thr must be added to the B-chain of mini-proinsulin. Discuss whether the necessary hydrolytic reactions are equilibrium or kinetically controlled. Can the peptidases used hydrolyze other bonds in human insulin?

Literature

Textbooks

These textbooks cover biosynthesis of enzymes, the biological, organic, and physical–chemical basics, kinetics, and structure–function analysis of enzyme-catalyzed reactions.

Berg, J.M., Tymoczko, J.L., and Stryer, L. (2002) *Biochemistry*, 5th edn, W.H. Freeman, New York.

Bisswanger, H. (2002) *Enzyme Kinetics*, 2nd edn, Wiley-VCH Verlag GmbH, Weinheim.

Creighton, T.E. (1993) *Proteins: Structures and Molecular Properties*, W.H. Freeman, New York.

Fersht, A.R. (1999) *Structure and Mechanism in Protein Science: A Guide to Enzyme Catalysis and Protein Folding*, W.H. Freeman, New York.

Frausto Da Silva, J.J.R.F. and Williams, R.J.P. (2001) *The Biological Chemistry of the Elements: The Inorganic Chemistry of Life*, 2nd edn, Oxford University Press, Oxford.
Laidler, K.J. (1958) *The Chemical Kinetics of Enzyme Action*, Clarendon Press, Oxford.

Handbooks and/or Internet Resources

Schomburg, D. and Salzmann, I. (2001) *Springer Handbook of Enzymes*, 2nd rev. edn, Springer, Berlin, www.brenda-enzymes.org.
Webb, E.C. (1992) *Enzyme Nomenclature*, Academic Press, San Diego, CA, www.chem.qmul.ac.uk/iubmb/enzyme.

References

Ahern, T.J. and Klibanov, A.M. (1988) Analysis of processes causing thermal inactivation of enzymes. *Methods Biochem. Anal.*, **33**, 91–127.
Aleshin, A.F., Feng, P.-H., Honzatko, R.B., and Reilly, P.J. (2003) Crystal structure and evolution of a prokaryotic glucoamylase. *J. Mol. Biol.*, **327**, 61–73.
Atkinson, D.E. (1969) Limitation of metabolite concentrations and the conservation of solvent capacity in the living cell. *Curr. Top. Cell Regul.*, **1**, 29–43.
Bartlett, G.J., Porter, C.T., Borkakoti, N., and Thornton, J.M. (2002) Analysis of catalytic residues in enzyme active sites. *J. Mol. Biol.*, **324**, 105–121.
Bender, M.L., Clement, G.E., Kézdy, F.J., and D'A. Heck, H. (1964) The correlation of the pH (pD) dependence and the stepwise mechanism of α-chymotrypsin-catalyzed reactions. *J. Am. Chem. Soc.*, **86**, 3680–3689.
Benkovic, S.J. and Hammes-Schiffer, S. (2003) A perspective on enzyme catalysis. *Science*, **301**, 1196–1202.
Bornscheuer, U.T. and Kazlauskas, R.J. (2005) *Hydrolases in Organic Synthesis – Regio- and Stereoselective Biotransformations*, 2nd edn, Wiley-VCH Verlag GmbH, Weinheim.
Bozhinova, D., Galunsky, B., Yueping, G., Franzreb, M., Köster, R., and Kasche, V. (2004) Evaluation of magnetic polymer micro-beads as carriers for immobilized biocatalysts for selective and stereoselective transformations. *Biotechnol. Lett.*, **26**, 343–350.
Brannigan, J.A., Dodson, G., Duggleby, H.J., Moody, P.C.E., Smith, J.L., Tomchik, D.R., and Murzin, A.G. (1995) A protein catalytic framework with an N-terminal nucleophile is capable of self-activation. *Nature*, **378**, 416–419.
Bruggink, A. (ed.) (2001) *Synthesis of β-Lactam Antibiotics*, Kluwer Academic Publishers, Dordrecht.
Bryan, P.N. (2000) Protein engineering of subtilisin. *Biochim. Biophys. Acta*, **1543**, 203–222.
Buchholz, K. and Monsan, P. (2003) Dextransucrase, in *Handbook of Food Enzymology* (eds J.R. Whitaker, A.G.J. Voragen, and D.W.S. Wong), Marcel Dekker, New York.
Cantarel, B.L., Coutino, P.M., Rancurel, C., Bernard, T., Lombard, V., and Henrissat, B. (2009) The Carbohydrate-Active EnZymes database (CAZy): an expert resource for glycogenomics. *Nucleic Acids Res.*, **37**, D233–D238.
Cantone, S., Hanefeld, U., and Basso, A. (2007) Biocatalysis in non-conventional media – ionic liquids, supercritical fluids and the gas phase. *Green Chem.*, **9**, 954–971.
Cao, L., Fischer, A., Bornscheuer, U.W., and Schmid, R.D. (1997) Lipase-catalyzed solid phase synthesis of sugar fatty acid esters. *Biocatal. Biotransform.*, **14**, 269–283.
Carrea, G., Ottolina, G., and Riva, S. (1995) Role of solvents in the control of enzyme selectivity in organic media. *Trends Biotechnol.*, **13**, 63–70.
Chen, C.S., Fujimoto, Y., Girdaukas, G., and Shih, C.J. (1982) Quantitative analysis of biochemical kinetic resolution of enantiomers. *J. Am. Chem. Soc.*, **104**, 7294–7299.
Coutinho, P.M. and Reilly, P.J. (1994) Structure–function relationships in the catalytic and starch binding domains of glucoamylase. *Protein Eng.*, **7**, 393–400.
Dalbey, R. and Robinson, C. (1999) Protein translocation into and across the bacterial plasma membrane and the plant thylakoid membrane. *Trends Biochem. Sci.*, **24**, 17–22.
Dale, B.E. and White, D.H. (1982) Ionic strength: a neglected variable in enzyme

technology. *Enzyme Microb. Technol.*, **5**, 227–229.
Darnell, J., Lodish, H., and Baltimore, D. (1986) *Molecular Cell Biology*, Scientific American Books, New York, p. 72.
Deranleau, D.A. (1969) Theory of the measurement of weak molecular complexes. I. General considerations. *J. Am. Chem. Soc.*, **91**, 4044–4049.
Dodson, G. and Wlowader, A. (1998) Catalytic triads and their relatives. *Trends Biochem.*, **23**, 347–352.
Done, S.H., Brannigan, J.A., Moody, P.C.E., and Hubbard, R.E. (1998) Ligand-induced conformational change in penicillin acylase. *J. Mol. Biol.*, **284**, 463–475.
Doukyu, N. and Ogino, H. (2010) Organic solvent-tolerant enzymes. *Biochem. Eng. J.*, **48**, 270–282.
Duggleby, H.J., Tolley, S.P., Hill, C.P., Dodson, E.J., Dodson, G.G., and Moody, P. C. (1995) Penicillin acylase has a single amino acid catalytic centre. *Nature*, **373**, 264–268.
Fields, P.A. and Somero, G.N. (1998) Hot spots in cold adaptation: localized increases in conformational flexibility in lactate dehydrogenase A_4 orthologs of Antarctic notothenioid fishes. *Proc. Natl. Acad. Sci. USA*, **95**, 11476–11481.
Fox, R.J. and Clay, M.D. (2008) Catalytic effectiveness, a measure of enzyme proficiency for industrial applications. *Trends Biotechnol.*, **27**, 137–140.
Freundlich, H. (1909) *Kapillarchemie*, Akad. Verlagsgesellschaft, Leipzig.
Galunsky, B., Ignatova, Z., and Kasche, V. (1997) Temperature effects on S_1- and S_1'-enantioselectivity of α-chymotrypsin. *Biochim. Biophys. Acta*, **1343**, 130–138.
Galunsky, B. and Kasche, V. (2002) Determination of the enantioselectivity for kinetically controlled condensations catalysed by amidases and peptidases. *Adv. Synth. Catal.*, **344**, 1115–1119.
Ganske, F. and Bornscheuer, U.T. (2006) Growth of *Escherichia coli*, *Pichia pastoris* and *Bacillus cereus* in the presence of the ionic liquids [BMIM][BF_4] and [BMIM][PF_6] and organic solvents. *Biotechnol. Lett.*, **28**, 465–469.
Garcia-Viloca, M., Gao, J., Karplus, M., and Truhlar, D.G. (2004) How enzymes work: analysis by modern rate theory and computer simulations. *Science*, **303**, 186–195.
Godfrey, T. (ed.) (1996) *Industrial Enzymology*, 2nd edn (eds T. Godfrey and J. Reichelt), Stockton Press, New York, pp. 435–482.
Goldberg, R.N., Tewari, Y.B., and Bhat, T.N. (2004) Thermodynamics of enzyme-catalyzed reactions – a database for quantitative biochemistry. *Bioinformatics*, **20**, 2874–2877.
Halling, P.J. (1994) Thermodynamic predictions for biocatalysis in nonconventional media: theory, tests, and recommendations for experimental design and analysis. *Enzyme Microb. Technol.*, **16**, 178–205.
Halling, P.J. (2000) Biocatalysis in low-water media: understanding effects of reaction condition. *Curr. Opin. Chem. Biol.*, **4**, 74–80.
Hamada, J.S. (1994) Deamidation of food proteins to improve functionality. *Crit. Rev. Food Sci. Nutr.*, **34**, 283–292.
Hartley, B.S., Hanlon, N., Jackson, R.J., and Rangarajan, M. (2000) Glucose isomerase: insight into protein engineering for increased stability. *Biochim. Biophys. Acta*, **1543**, 294–335.
Hochachka, P.W. and Somero, G.N. (2002) *Biochemical Adaptation: Mechanism and Process in Physiological Evolution*, Oxford University Press, Oxford.
Ignatova, Z., Stoeva, S., Galunsky, B., Hörnle, C., Nurk, A., Piotraschke, E., Voelter, W., and Kasche, V. (1998) Proteolytic processing of penicillin amidase from *A. faecalis* in *E. coli* yields several active forms. *Biotechnol. Lett.*, **20**, 977–982.
Ignatova, Z., Wischnewski, F., Notbohm, H., and Kasche, V. (2005) Pro-sequence and Ca^{2+}-binding: implications for folding and maturation of Ntn-hydrolase penicillin amidase from *E. coli*. *J. Mol. Biol.*, **348**, 999–1014.
Kasche, V. (1986) Mechanism and yields in enzyme catalyzed equilibrium and kinetically controlled synthesis of β-lactam antibiotics, peptides and other condensation products. *Enzyme Microb. Technol.*, **8**, 4–16.
Kasche, V. (2001) Proteases in peptide synthesis, in *Proteolytic Enzymes. A Practical Approach* (eds M. Beynon and J. Bond),

Oxford University Press, Oxford, pp. 265–292.

Kasche, V. and Galunsky, B. (1995) Enzyme catalyzed biotransformations in aqueous two-phase systems with precipitated substrate and/or product. *Biotechnol. Bioeng.*, **45**, 261–267.

Kasche, V., Galunsky, B., and Ignatova, Z. (2003) Fragments of pro-peptide activate mature penicillin amidase of *Alcaligenes faecalis*. *Eur. J. Biochem.*, **270**, 4721–4728.

Kasche, V., Galunsky, B., and Michaelis, G. (1991) Binding of organic solvent molecules influences the P_1'–P_2' stereo- and sequence specificity of α-chymotrypsin in kinetically controlled peptide synthesis. *Biotechnol. Lett.*, **13**, 75–80.

Kasche, V., Galunsky, B., Nurk, A., Piotraschke, E., and Rieks, A. (1996) The dependency of the stereoselectivity of penicillin amidases – enzymes with *R*-specific S_1- and *S*-specific S_1'-subsites – on temperature and primary structure. *Biotechnol. Lett.*, **18**, 455–460.

Kasche, V., Ignatova, Z., Märkl, H., Plate, W., Punckt, N., Schmidt, D., Wiegandt, K., and Ernst, B. (2005) Ca^{2+} is a cofactor required for membrane transport and maturation and is a yield-determining factor in high cell density penicillin amidase production. *Biotechnol. Prog.*, **21**, 432–438.

Kasche, V., Lummer, K., Nurk, A., Piotraschke, E., Riecks, A., Stoeva, S., and Voelter, W. (1999) Intramolecular autoproteolysis initiates the maturation of penicillin amidase from *E. coli*. *Biochim. Biophys. Acta*, **1433**, 76–86.

Kasche, V., Schlothauer, R., and Brunner, G. (1988) Enzyme denaturation in supercritical CO_2: stabilizing effect of S—S bonds during the depressurization step. *Biotechnol. Lett.*, **10**, 569–574.

Kasche, V. and Zöllner, R. (1982) Tris (hydroxymethyl)methylamine is acylated when it reacts with acyl-chymotrypsin. *Hoppe Seyler's Z. Phys. Biol. Chem.*, **363**, 531–534.

Kazlauskas, R.J. (1994) Elucidating structure–mechanism relationships in lipases: prospects for predicting and engineering catalytic properties. *Trends Biotechnol.*, **12**, 464–472.

Kelly, R.M., Dijkhuisen, L., and Leemhuis, H. (2009) Starch and α-glucan acting enzymes, modulating their properties by directed evolution. *J. Biotechnol.*, **140**, 184–193.

Kilshtain, A.V. and Warshel, A. (2009) On the origin of the catalytic power of carboxypeptidase A and other metalloenzymes. *Proteins*, **77**, 536–550.

Kimura, M. and Ohta, T. (1973) Mutation and evolution at the molecular level. *Genetics*, **72**, 19–35.

Kleinkauf, H. and von Döhren, H. (1990) Bioactive peptide analogues: *in vivo* and *in vitro* production. *Prog. Drug Res.*, **34**, 287–317.

Kragl, U., Eckstein, M., and Kraftzik, N. (2002) Enzyme catalysis in ionic liquids. *Curr. Opin. Biotechnol.*, **13**, 565–571.

Lamare, S., Legoy, M.-D., and Graber, M. (2004) Solid/gas bioreactors: powerful tools for fundamental research and efficient technology for industrial applications. *Green Chem.*, **6**, 445–458.

Lewin, B. (1997) *Genes*, Oxford University Press, Oxford, p. 94.

Liese, A., Seelbach, K., and Wandrey, C. (eds) (2006) *Industrial Biotransformations*, 2nd Ed., Wiley-VCH, Weinheim.

Lorenz, P., Liebeton, K., Niehaus, F., and Eck, J. (2002) Screening for novel enzymes for biocatalytic processes: accessing the metagenome as a resource of novel functional sequence space. *Curr. Opin. Biotechnol.*, **13**, 572–577.

Lummer, K., Riecks, A., Galunsky, B., and Kasche, V. (1999) pH-dependence of penicillin amidase enantioselectivity for charged substrates. *Biochim. Biophys. Acta*, **1433**, 327–334.

Martinek, K., Yatsimirski, A.K., and Berezin, I.V. (1971) Effect of ionic strength on the steady state kinetics of α-chymotrypsin catalyzed reactions. *Mol. Biol. (USSR)*, **5**, 96–109.

McDonough, M.A., Klei, H.E., and Kelly, J.A. (1999) Crystal structure of penicillin G acylase from the Bro1 mutant strain of *Providencia rettgeri*. *Protein Sci.*, **8**, 1971–1981.

Menger, F.M. (1992) Analysis of ground-state and transition-state effects in enzyme catalysis. *Biochemistry*, **31**, 5368–5373.

Mesiano, A.J., Beckman, E.J., and Russel, A.J. (1999) Supercritical biocatalysis. *Chem. Rev.*, **99**, 623–634.

Michaelis, G. (1991) *Peptidase-katalysierte Peptidsynthese*. Thesis, TU Hamburg-Harburg.

Miyazaki, M., Shibue, M., Ogino, K., Nakamura, H., and Maeda, H. (2001) Enzymatic synthesis of pyruvic acid from acetaldehyde and carbon dioxide. *Chem. Commun.*, **21**, 1800–1801.

Moniruzzaman, M., Nakashima, K., Kamiya, N., and Goto, M. (2010) Recent advances of enzymatic reactions in ionic liquids. *Biochem. Eng. J.*, **48**, 295–314.

Overmeyer, A., Schrader-Lippert, S., Kasche, V., and Brunner, G. (1999) Lipase-catalyzed kinetic resolution of racemates at temperatures from 40 to 160 °C in supercritical CO_2. *Biotechnol. Lett.*, **21**, 65–69.

Park, S. and Kazlauskas, R.J. (2003) Biocatalysis in ionic liquids – advantages beyond green technology. *Curr. Opin. Biotechnol.*, **14**, 432–437.

Phillips, R.S. (1996) Temperature modulation of the stereochemistry of enzymatic catalysis: prospects for exploitation. *Trends Biotechnol.*, **14**, 13–16.

Poliakoff, M., Fitzpatrick, J.M., Farren, T.R., and Anastas, P.T. (2002) Green chemistry: science and politics of change. *Science*, **297**, 807–810.

Ranke, J., Stolte, S., Störmann, R., Arning, J., and Jastorff, B. (2007) Design of sustainable chemical products – the example of ionic liquids. *Chem. Rev.*, **107**, 2183–2206.

Rawlings, N.D., Barrett, A.J., and Bateman, A. (2010) MEROPS: the peptidase database. *Nucleic Acids Res.*, **38**, D227–D233.

Sakai, T., Kawabata, I., Kishimoto, T., Ema, T., and Utaka, M. (1997) Enhancement of the enantioselectivity in lipase-catalyzed kinetic resolution of phenyl-2*H*-azirine-2-mehanol by lowering the temperature. *J. Org. Chem.*, **62**, 4906–4907.

Salemme, F.R. (1977) Structure and function of cytochrome c. *Annu. Rev. Biochem.*, **46**, 299–329.

Sauer, J., Sigurskjold, B.W., Christensen, U., Frandsen, T.P., Mirgodskaya, E., Harrison, M., Roepstorff, P., and Svensson, B. (2000) Glucoamylase: structure/function relationships, and protein engineering. *Biochim. Biophys. Acta*, **1543**, 275–293.

Saunders, R. and Deane, C.M. (2010) Synonymous codon usage influences the local protein structure observed. *Nucleic Acids Res.*, **38**, 6719–6728.

Schechter, I. and Berger, A. (1967) On size of active site in protease. I. Papain. *Biochem. Biophys. Res. Commun.*, **27**, 157–162.

Schmitke, J.L., Wescott, C.R., and Klibanov, A.M. (1996) The mechanistic dissection of the plunge in enzymatic activity upon transition from water to anhydrous solvents. *J. Am. Chem. Soc.*, **118**, 3360–L 3365.

Schowen, R.L. (2003) How an enzyme surmounts the activation energy barrier. *Proc. Natl. Acad. Sci. USA*, **100**, 11931–11932.

Seddon, K.R. (2003) Ionic liquids. A taste of the future. *Nat. Mater.*, **2**, 1–2.

Shin, Y.C., Jeon, J.Y.J., Jung, K.H., Park, M.R., and Kim., Y. (2009) Cephalosporin C acylase mutant and method for preparing 7-ACA using same. US Patent 7,592,168.

Spiess, A., Schlothauer, R., Hinrichs, J., Scheidat, B., and Kasche, V. (1999) pH gradients in heterogeneous biocatalysts and their influence on rates and yields of the catalysed processes. *Biotechnol. Bioeng.*, **62**, 267–277.

Steele, H.L., Jäger, K.E., Daniel, R., and Streit, W.R. (2009) Advances in recovery of novel biocatalysts from metagenomes. *J. Mol. Microbiol. Biotechnol.*, **16**, 25–37.

Steiner, D.F., Cunningham, D., Spigelman, L., and Aten, B. (1967) Insulin biosynthesis: evidence for a precursor. *Science*, **157**, 697–700.

Straathof, A.J.J. and Jongejan, J.A. (1997) The enantiomeric ratio: origin, determination and prediction. *Enzyme Microb. Technol.*, **21**, 559–671.

Tewari, Y.B. and Goldberg, R.N. (1989) Thermodynamics of hydrolysis of disaccharides. *J. Biol. Chem.*, **264**, 3966–3971.

Tosa, T., Mori, T., Fuse, N., and Chibata, I. (1969) Studies on continuous enzyme reactions. 6. Enzymatic properties of DEAE-Sepharose aminoacylase complex. *Agric. Biol. Chem.*, **33**, 1047–1056.

Trivedi, A.H., Spiess, A., Daussmann, T., and Büchs, J. (2006) Study on mesophilic and thermophilic alcohol dehydrogenases in gas-phase reaction. *Biotechnol. Prog.*, **22**, 454–458.

van der Maarel, M.J., van der Veen, B., Uitdehag, J.C., Leehuis, H., and Dijkhuisen, L. (2002) Properties and application of starch-converting enzymes of the α-amylase family. *J. Biotechnol.*, **94**, 137–155.

van Rantwijk, F., Madeira Lau, R., and Sheldon, R.A. (2003) Biocatalytic transformations in ionic liquids. *Trends Biotechnol.*, **21**, 131–138.

Verhaert, R.M.D., Riemens, A.M., van der Laan, J.M., van Duin, J., and Quax, W.J. (1997) Molecular cloning and analysis of the gene encoding the thermostable penicillin G acylase from *Alcaligenes faecalis*. *Appl. Environ. Microbiol.*, **63**, 3412–3418.

Vulfson, E.N., Halling, P.J., and Holland, H. L. (eds) (2001) *Enzymes in Nonaqueous Solvents: Methods and Protocols*, Humana Press, Totowa, NJ.

Wiesemann, T. (1991) *Enzymmodifikation für analytische und präparative Zwecke: natürliche und künstliche Penicillinamidase-Varianten*. Dissertation, TU Hamburg-Harburg.

Youshko, M.I., Chilov, G.G., Shcherbakova, T. A., and Svedas, V.K. (2002) Quantitative characterization of the nucleophile reactivity in penicillin acylase-catalyzed acyl transfer reactions. *Biochim. Biophys. Acta*, **1599**, 134–140.

Xu, D. and Guo, H. (2009) Quantum mechanical/molecular mechanical and density functional theory studies of a prototypical zinc peptidase (carboxypeptidase A) suggest a general acid–general base mechanism. *J. Am. Chem. Soc.*, **131**, 9780–9788.

Zmiejewski, M.J., Jr., Briggs, B.S., Thompson, A.R., and Wright, I.G. (1991) Enantioselective acylation of a beta-lactam intermediate in the synthesis of Loracarbef using penicillin G amidase. *Tetrahedron Lett.*, **32**, 1621–1622.

3
Enzyme Discovery and Protein Engineering

Why enzyme discovery?	Constant need for new enzymes: • Different substrate specificity, stereoselectivity, and so on • Different properties (pH or temperature optimum)
Why protein engineering?	Alteration of protein properties to meet process conditions • Improved stability (pH, temperature, solvent) • Better expression, better folding • Altered selectivities (substrate range, stereoselectivity, etc.) • Lowered substrate or product inhibition
How to perform protein engineering?	Rational protein design Directed (molecular) evolution Focused directed evolution Computational design

3.1
Enzyme Discovery

The traditional method to identify new enzymes is based on screening, for example, of soil samples, unusual habitats, industrial sites, or strain collections by enrichment cultures or from plant or animal tissues. This requires an assay method to clearly identify the desired activity of the enzyme searched for. Once a source such as a microorganism has been identified, standard protein purification methods are used to purify the enzyme to homogeneity and to perform detailed biochemical characterization, that is, substrate scope, determination of pH and temperature profile, dependency of cofactors such as NAD(P)H, FAD, FMN, and metal ions, molecular weight, quaternary structure, and the kinetic constants V_{max},

Biocatalysts and Enzyme Technology, Second Edition. Klaus Buchholz, Volker Kasche, and Uwe T. Bornscheuer

K_{cat}, and K_m (see also Chapter 6 and Section 2.11.1). If it is intended to clone and recombinantly express the protein of interest, various approaches can be used such as identification of the encoding gene from the DNA of the source organism using degenerate primers designed from the sequences found for homologous enzymes in databases (Figure 3.1), from N-terminal protein sequences, or by shotgun cloning and expression of fragments.

These standard methods have been substantially facilitated by the development of recombinant DNA techniques, which simplified the access to enzymes and now allow the production of biocatalysts in recombinant form in quantities sufficient for laboratory and industrial use and also at constant quality. This was shown, for instance, for the hydroxynitrile lyase from *Hevea brasiliensis* (Griengl, Schwab, and Fechter, 2000; Johnson, Zabelinskaja-Mackova, and Griengl, 2000) or pure isoenzymes of pig liver esterase (Hummel *et al.*, 2007; Musidlowska, Lange, and Bornscheuer, 2001).

However, only a small number of microbes can successfully be cultivated in the laboratory. Depending on the habitat, it was estimated that <1% of the microbes present in an environmental sample can be grown under standard laboratory

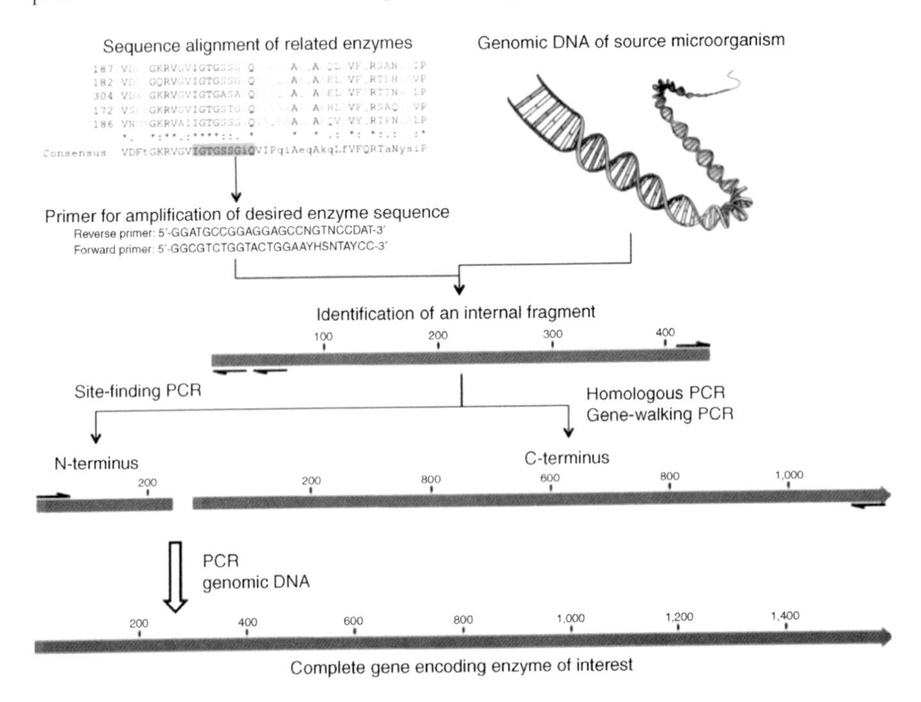

Figure 3.1 Example for the identification of an enzyme-encoding gene. First, degenerate primers are designed based on an alignment of homologous proteins (using BLAST and ClustalW, see sections below). These primers are then used to amplify a sequence fragment from genomic DNA of the microorganism. In order to get access to the full-length gene, various methods such as gene walking, homologous PCR, inverse PCR, or site-finding PCR can be used. Once the entire gene has been identified, it can be subcloned, transformed, and expressed in the recombinant host organism.

conditions (Amann, Ludwig, and Schleifer, 1995; Hugenholtz, Goebel, and Pace, 1998; Pace, 1997; Staley and Konopka, 1985). Reasons are that proper cultivation conditions are unknown, growth rates are too slow, or the growth of a microbe depends on other species providing nutrients. The metagenome approach overcomes these limitations by circumventing the cultivation step. For this, the complete genomic DNA is extracted from an environmental sample, fragmented, and cloned yielding the corresponding metagenome libraries (Streit and Daniel, 2010). These libraries can then be either screened for the desired activity (with a function-based assay) or subjected to sequence-based approach in which genes are identified based on homology to already described enzymes while the entire DNA is sequenced. Both approaches have their advantages and disadvantages. The hit rate in the activity-based approach is usually lower because the gene needs to be functionally expressed in the host microorganism, which requires correct positioning of the metagenome sequence; promoter and codon usage must fit to the host and correct posttranslational modifications of the enzyme must be ensured. If enzyme activity depends on several subunits or requires additional proteins, for example, for electron transfer, then the functional expression becomes even more challenging. On the other hand, enzymes discovered by this activity-based approach are already cloned and accessible in a functional way. As the costs and time required for DNA sequencing substantially dropped by orders of magnitude in the past few years, entire metagenome libraries can now be sequenced at reasonable budgets and timelines. This has the advantage that novel enzymes and even complete pathways can be quickly discovered using bioinformatic tools. Together with sequence alignments from databases, this approach facilitates identification of novel enzymes and the choice of an appropriate expression system. Current reviews nicely summarize recently discovered enzymes using both approaches (Fernández-Arrojo *et al.*, 2010; Uchiyama and Miyazaki, 2009).

In the activity-based approach, the major limitation is the availability of reliable high-throughput assays. This is the reason why still most of the enzymes isolated from metagenome libraries are hydrolases (Bayer *et al.*, 2010; Chen, Guo, and Dang, 2011; Okamura *et al.*, 2010; Park, Shin, and Kim, 2011; Yu *et al.*, 2011). Though esterases, lipases, and proteases form the largest group, an increasing number of other enzyme classes have been accessed from metagenome libraries, among them several industrially interesting polysaccharide-degrading enzymes such as xylanases (Mo *et al.*, 2010; Wang *et al.*, 2011a), glucanases (Liu *et al.*, 2011), or cellulases (Kwon *et al.*, 2010), which are of interest for the production of bulk and fine chemicals from renewable resources or β-galactosidases for the food industry (Wang *et al.*, 2010).

Besides the classical enrichment cultivation and the metagenome approach (Lorenz and Eck, 2005), a third way to access new enzymes is *in silico* screening. Protein sequences obtained from sequencing of single enzymes, entire genomes, or microbial consortia (i.e., from the Sargasso Sea (Venter *et al.*, 2004)) are deposited in public databases and currently >12 million protein sequences are available (http://www.ncbi.nlm.nih.gov/RefSeq/). However, only a tiny fraction of all these sequences corresponds to enzymes, which have been produced in the laboratory and their biochemical properties and function have been experimentally confirmed. For the vast majority, a possible function is postulated by sequence comparison, but this annotation can often be wrong or at least misleading, as shown in two recent examples for a monooxygenase (Kirschner, Altenbuchner, and Bornscheuer, 2007; Kirschner and Bornscheuer, 2006) and certain transaminases (Höhne *et al.*, 2010). The more information about the desired enzyme (class) exists, the easier it is to discover novel biocatalysts in public databases. If several enzymes with a distinct activity are already described in the literature, an alignment of their protein sequences together with experimentally confirmed activities or specificities can reveal conserved regions within the sequences and a simple BLAST search can yield new and useful biocatalysts. There are numerous examples for the discovery of new enzymes *in silico* including nitrilases (Kaplan *et al.*, 2011), a ligase (Kino *et al.*, 2010), an epoxide hydrolase (van Loo *et al.*, 2006), and a reductase for the highly selective reduction yielding ethyl (*S*)-4-chloro-3-hydroxybutanoate with excellent optical purity (99% ee) at high substrate concentration (600 g l^{-1}) (Wang *et al.*, 2011b). By curtailing the sequences to be searched through, one is even able to increase the chances that the new catalysts fulfill additional criteria. Scanning the genomes of (hyper-)thermophilic organisms, for example, will increase the chances to find a thermostable enzyme. Thus, Machielsen *et al.* (2006) identified a thermostable ADH in the genome of *Pyrococcus furiosus* with a half-life of 130 min at 100 °C and Fraaije *et al.* (2005) discovered a thermostable Baeyer–Villiger monooxygenase with a promiscuous sulfur oxidizing activity. An *in silico* search strategy enabled to identify 20 novel (*R*)-selective amine transaminases from 5000 sequences available in public databases. The algorithm was based on detailed analysis of reaction mechanisms and residues involved in substrate binding in the pyridoxal-5′-phosphate-dependent enzyme superfamily of a certain fold class. This analysis could predict both substrate specificity (ketones/-amines, not α-keto acids/α-amino acids) and enantiopreference ((*R*)- not (*S*)-selectivity) successfully. From the 20 sequences, 17 proteins could be functionally expressed using synthetic genes in *E. coli* and for the 7 most active amine transaminases application in asymmetric synthesis could be demonstrated (Schätzle *et al.*, 2011).

Many more examples can be found in recent reviews. Stewart (2006), for example, illustrated the potential of genome mining strategies exemplified for yeast dehydrogenases, while Furuya and Kino (2010) focused on aspects of the genome mining approach that are relevant for the discovery of novel P450 biocatalysts. The identification of biosynthetic gene clusters of indole alkaloids by this approach was summarized recently by Li (2011).

A comparison between the power of *in vivo* versus *in silico* screening of mutants has recently been made by Barakat, Barakat, and Love (2010) as exemplified for the rational redesign of an unstable variant of streptococcal protein G.

Still, not all enzymes found in nature are suitable for a certain synthetic problem – for example, the enzyme's activity, stability, substrate specificity, and enantioselectivity are not always satisfactory. These limitations can be overcome by protein engineering using various methods as outlined in the next sections.

3.2 Strategies for Protein Engineering

Strategies for protein engineering can be divided into two major concepts to adjust enzyme properties toward the desired application: rational design or directed (molecular) evolution (Kazlauskas and Bornscheuer, 2009; Lutz and Bornscheuer, 2009; Bornscheuer *et al.*, 2012).

Rational design uses structural and mechanistic information and molecular modeling for the prediction of changes in the protein structure in order to alter or induce the desired properties. The advances in computer technology have helped in creating better protein models to improve predictions for rational design, but structure–activity relationships are still not trivial. Once amino acid residues are identified from structural analysis to be crucial for the protein properties, these are introduced into the protein by site-directed mutagenesis and after production of the enzyme the variant can be analyzed for the altered function. In directed evolution, very large mutant libraries are created by random changes, screened for the desired property, and the variants showing promising results are subjected to further rounds of evolution. Both strategies have their advantages and limitations: rational protein design depends on the availability of the protein 3D structure and sufficient knowledge of the contribution of residues to enzyme properties, whereas directed evolution strongly depends on suitable mutagenesis methods and high-throughput screening or selection methods to identify the desired variants in a fast and reliable manner. Nowadays, more researchers use combined methods of these two strategies, dubbed focused directed evolution or semi-rational design. Here, information available from related protein structures, families, and mutants already identified is combined and then used for targeted randomization of certain areas of the protein (Figure 3.2) (Behrens *et al.*, 2011).

Typical goals of protein engineering are to stabilize the protein, to improve its reactivity with target substrates, to extend the substrate range (Section 2.11), or to expand the reaction mechanism to new reactions such as catalytic promiscuity (Bornscheuer and Kazlauskas, 2004) as summarized in Box 3.1 (Bornscheuer and Kazlauskas, 2011). Proteins used as biocatalysts often must tolerate high temperatures or unusual reaction conditions. In the first example of protein engineering of an industrial enzyme, Genencor engineered a protease to tolerate the bleaching agent H_2O_2 in laundry detergents (Estell, Graycar, and Wells, 1985). Other examples dealt with expanding the substrate range of an enzyme (Reetz *et al.*, 2005) or improving or inverting stereoselectivity (Bartsch, Kourist, and Bornscheuer, 2008; Magnusson *et al.*, 2005; Terao *et al.*, 2007; Zha *et al.*, 2001). Inversion of stereoselectivity was also reported for an aldolase (Figure 3.3).

Protein engineering is difficult because the biocatalysts must meet multiple criteria to be useful for an application. For example, if the goal of protein engineering is

Box 3.1: Strategies for protein engineering (adapted from Bornscheuer and Kazlauskas, 2011).

Choosing locations for changes – rational design strategies

- Mutations throughout the protein
- Mutations at selected regions
  - -Near active site
  - -At flexible sites
  - -Sites identified by structure-based modeling
  - -Sites identified by sequence comparison
- Less or more conserved residues

Creating variants – mutagenesis strategies

- Stabilize protein first so that a larger fraction of variants are stable?
- Single amino acid substitutions
  - -Error-prone PCR
  - -Saturation mutagenesis
  - -Limited amino acid replacements
  - -Genetic drift libraries
- Multiple substitutions
  - -Gene synthesis to add specific changes
  - -Stepwise accumulation of single amino acid substitutions
  - -Simultaneous mutagenesis of multiple sites
  - -Simultaneous mutagenesis of a few sites
- Recombination methods
  - -Homology-dependent methods such as DNA-shuffling
  - -Homology-independent methods
- Circular permutations (rare)
- Insertions and deletions

Finding solutions – screening strategies

- Select for improved variant
- Screen for improved variants
  - -True substrate or more convenient substrate analogue
  - -True application conditions or more convenient simplified conditions
  - -Exhaustive versus partial screening of library
- Test all variants to find contributions of each substitutions to desired properties

to improve the selectivity, the changes introduced should not affect protein expression or stability. For instance, directed evolution of an esterase identified a more enantioselective variant with three amino acid substitutions (Schmidt *et al.*, 2006). Unfortunately, while bacteria produced high amounts of the soluble starting

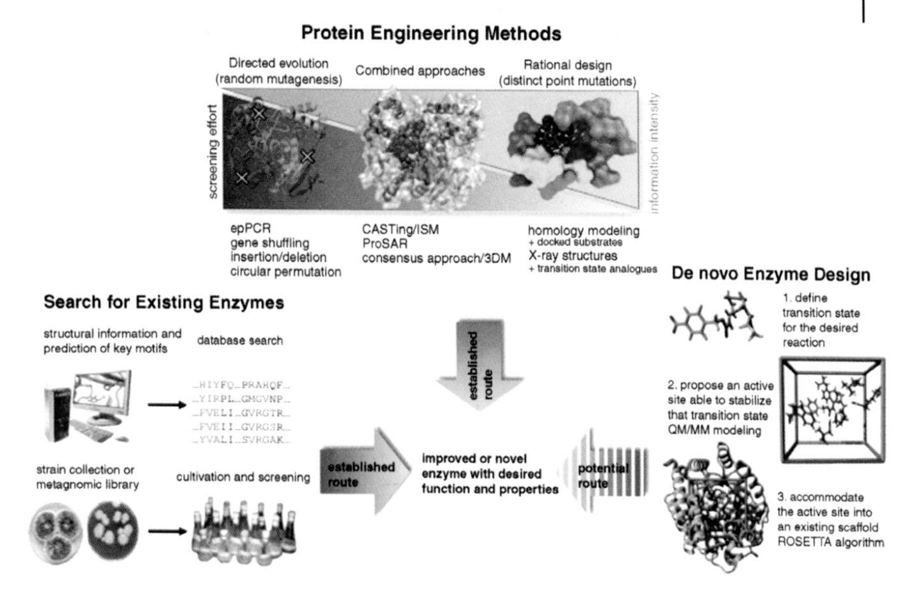

Figure 3.2 Overview of approaches for protein discovery and engineering by rational, evolutionary, or combined methods (taken from Behrens *et al.*, 2011).

esterase, they produced only small amounts of the variant, and most of it in an unfolded insoluble form. Back mutations identified a double mutant that was both highly enantioselective and efficiently produced in soluble form.

3.2.1
Rational Protein Design

The best starting point for rational protein design is the high-resolution 3D structure of the enzyme, usually determined by X-ray crystallography, which is deposited

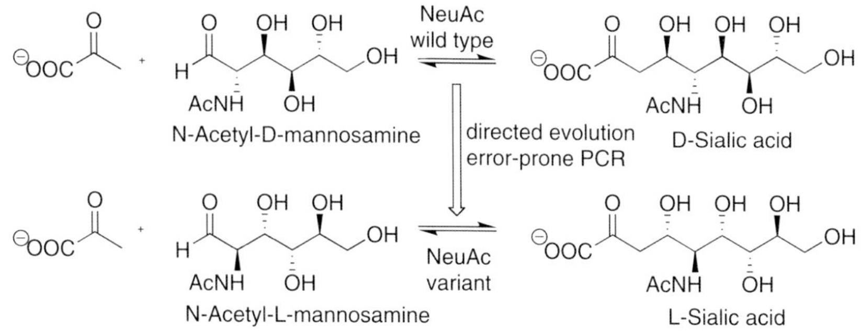

NeuAc wild type
N-Acetyl-D-mannosamine
D-Sialic acid
directed evolution error-prone PCR
NeuAc variant
N-Acetyl-L-mannosamine
L-Sialic acid

Figure 3.3 Inversion of the stereospecificity of *N*-acetylneuraminic acid aldolase (NeuAc) toward sialic acid by directed evolution (Wada *et al.*, 2003).

at the Brookhaven Protein Database (www.pdb.org). Preferentially, the structure also contains a substrate (or substrate analogue) bound to the active site as this provides detailed information about the residues involved in substrate binding and pockets binding substrate substituents. In the absence of a 3D structure, a model of the protein structure can be generated automatically based on a known homologous structure (e.g., by Phyre (Kelley and Sternberg, 2009), SwissModell (Britton, Brand, and Markovetz, 1974; Padhi, Bougioukou, and Stewart, 2009), or Robetta (Bougioukou *et al.*, 2009)). Software for molecular modeling now runs on standard personal computers and is rather easy to use (e.g., Yasara (Hulley *et al.*, 2010)); tools for sequence comparison (e.g., BLAST:[1] http://blast.ncbi.nlm.nih.gov/Blast.cgi (Woo and Silverman, 1995)) and alignment (e.g., ClustalW: http://www.ebi.ac.uk/Tools/msa/clustalw2/ (Larkin *et al.*, 2007)) enable quick identification of novel sequences and conserved motifs (Figure 3.4).

Rational design uses structural and mechanistic information and molecular modeling for the prediction of changes in the protein structure in order to change its properties. The advances in computer technology have helped in creating better protein models to improve predictions for rational design, but structure–activity relationships are still not trivial.

3.2.2
Directed (Molecular) Evolution

An alternative that emerged during the mid-1990s was the strategy of directed evolution (also called molecular or *in vitro* evolution). This "evolution in the test tube" comprises essentially two steps: (1) random mutagenesis of the gene encoding the enzyme; and (2) identification of desired biocatalyst variants within these mutant libraries by screening or selection (Figure 3.5); commonly, best mutations are finally combined as often synergistic effects are observed.

Prerequisites for directed evolution are the availability of the gene(s) encoding the enzyme(s) of interest, a suitable (usually microbial) expression system, an effective method to create mutant libraries, and a suitable screening or selection system. Many detailed protocols for this are available, including several books (2003a, 2003b; Brakmann and Johnsson, 2002; Brakmann and Schwienhorst, 2004) and reviews (Böttcher and Bornscheuer, 2010; Neylon, 2004; Reetz, 2011; Turner, 2003).

3.2.2.1 Methods to Create Mutant Libraries

A broad range of methods has been developed to create mutant libraries. These can be divided into two approaches: (1) a nonrecombining mutagenesis, in which one parent gene is subjected to random mutagenesis leading to variants with point mutations; or (2) recombining methods, in which several parental genes (usually showing high sequence homology) are randomized. This results in a library of chimeras rather than an accumulation of point mutations.

1) BLAST, basic local alignment search tool, is a software that allows comparison of nucleotide or protein sequences within a database.

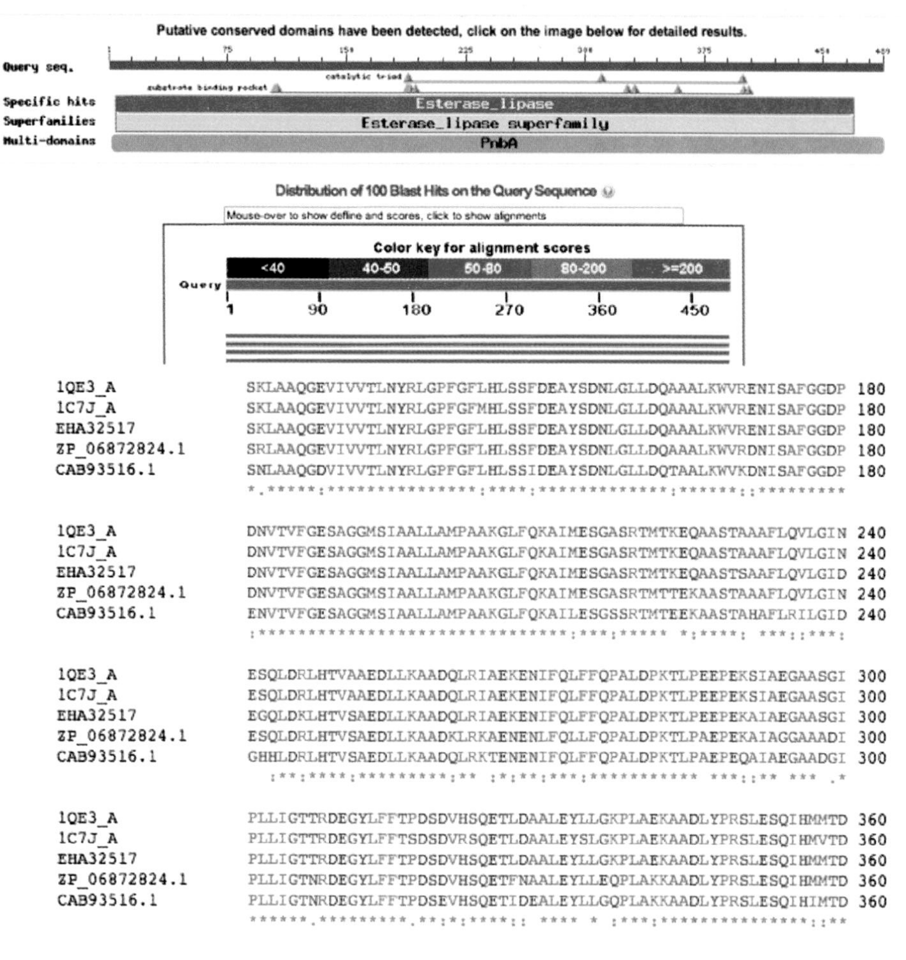

Figure 3.4 Starting from a given protein sequence, the software tool BLAST enables identification of related proteins (top) as exemplified here for esterases. These can then be compared in detail (bottom) using alignment software such as ClustalW to identify identical (*), homologous (. or :), and differing amino acids (blank) as well as conserved motifs.

One challenge in directed evolution experiments is the coverage of a sufficiently large sequence space – that is, the creation and analysis of as many variants as possible. When considering a protein (enzyme) consisting of 200 amino acids, the number of possible variants of a protein by introduction of M substitutions in N amino acids can be calculated from the formula $19^M[N!/(N-M)!M!]$ (from the 20 proteinogenic amino acids, one is already present in the wild-type enzyme; therefore, the value 19). Thus, for two random mutations already more than 7 million

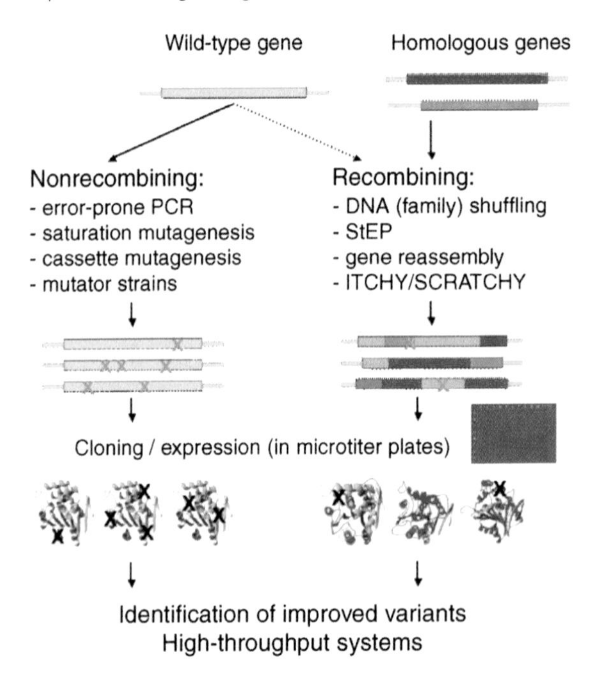

Figure 3.5 Principle of directed evolution. The gene(s) encoding the wild-type or homologous enzyme(s) are subjected to random mutagenesis using nonrecombining or recombining methods. The resulting mutant libraries are then cloned and expressed (often in microtiter plates). The desired improved variants are identified by high-throughput screening systems, usually using microtiter plate-based assays or selection (e.g., using agar plate assays; not shown).

variants are possible; with three or more substitutions, the creation and screening of a library becomes very challenging (Table 3.1).

The still most often used method for the creation of libraries is the error-prone polymerase chain reaction (epPCR) in which conditions are used that lead to the introduction of approximately 1–10 nucleotide mutations per 1000 base pairs (Cadwell and Joyce, 1992). This is achieved by changing the reaction conditions – that is, to use Mn^{2+} salts instead of Mg^{2+} salts (the polymerase is

Table 3.1 Sequence space of possible variants for a protein consisting of 200 amino acids at a given number of substitutions.

Substitutions (*M*)	Number of variants (sequence length *N* = 200)
1	3800
2	7 183 900
3	9 008 610 600
4	8 429 807 368 950
5	6 278 520 528 393 960

magnesium dependent), the use of *Taq* polymerase from *Thermus aquaticus*, and variations in the concentrations of the four deoxynucleotides (dNTPs). It should be noted that due to the biased exchange of nucleotides and the degenerate genetic code only an average of 6 of all possible 19 amino acid substitutions are accessible by this method. This can be overcome by the use of an optimized polymerase such as Mutazyme from Stratagene. Another approach utilizes mutator strains, for example, the *E. coli* derivative *Epicurian coli* XL1-Red, which lacks DNA repair mechanisms (Bornscheuer, Altenbuchner, and Meyer, 1998). The introduction of a plasmid bearing the gene encoding the protein of interest leads to mutations during replication. Both methods introduce point mutations, and several iterative rounds of mutation followed by identification of best variants are usually required to obtain a biocatalyst with the desired properties.

Alternatively, methods of recombination (also referred to as sexual mutagenesis) can be used. The first example was the DNA (or gene) shuffling developed by Stemmer (1994a, 1994b), in which DNase degrades the gene followed by recombination of the fragments using PCR with and without primers. This process mimics natural recombination and has been proven in various examples as a very effective tool to create desired enzymes. Later, this method was further refined and termed "DNA family shuffling" or "molecular breeding," enabling the creation of chimeric libraries from a family of genes.

The Arnold laboratory developed two variations of DNA shuffling: the staggered extension process (StEP) is based on a modified PCR protocol using a set of primers and short reaction times for annealing and polymerization. Truncated oligomers dissociate from the template and anneal randomly to different templates leading to recombination. Several repetitions allow the formation of full-length genes (Zhao *et al.*, 1998). Many other methods are covered in the reviews and books cited above.

3.2.2.2 **Assay Systems**

The major challenge in directed evolution is the identification of desired variants within the mutant libraries. Suitable assay methods should enable a fast, very accurate, and targeted identification of desired biocatalysts out of libraries comprising usually 10^3–10^6 mutants. In principle, two different approaches can be applied, namely, screening or selection.

Selection Selection-based systems have been used traditionally to enrich certain microorganisms. For *in vitro* evolution, selection methods are less frequently used as they usually can only be applied to enzymatic reactions that occur in the metabolism of the host strain. On the other hand, selection-based systems allow a considerably higher throughput compared to screening systems (see below). Often, selection is performed as a complementation – that is, an essential metabolite is produced only by a mutated enzyme variant. For instance, a growth assay was used to identify monomeric chorismate mutases. Libraries were screened using media lacking L-tyrosine and L-phenylalanine (MacBeath, Kast, and Hilvert, 1998). In a similar manner, complementation of biochemical pathways has also been used to identify mutants of an enzyme involved in tryptophan biosynthesis. HisA and TrpF (isomerases involved in the biosynthesis of histidine and tryptophan, respectively) have a similar tertiary structure, and the aminoaldose substrates (ProFAR and PRA) used are very similar except for a different residue at the amino functionality. Using random mutagenesis and selection by complementation on media lacking tryptophan, several HisA variants that catalyze the TrpF reaction both *in vivo* and *in vitro* were identified (Juergens *et al.*, 2000). One of these variants also retained significant HisA activity.

Stemmer's group subjected four genes of cephalosporinases – enzyme causing antibiotic resistance by hydrolysis of cephalosporins – from *Enterobacter, Yersinia, Citrobacter,* and *Klebsiella* species to error-prone PCR or DNA shuffling. Libraries from four generations (a total of 50 000 colonies) were assayed by selection on agar plates with increasing concentrations of moxalactam (a β-lactam antibiotic). Only those clones could survive that were able to hydrolyze the β-lactam antibiotic. The best variants from epPCR gave only an 8-fold increased activity, but the best chimeras from multiple gene shuffling showed 270–540-fold resistance to moxalactam (Crameri *et al.*, 1998). Sequencing of a mutant revealed low homology compared to the parental genes (Figure 3.6) and a total of 33 amino acid substitutions and 7 crossovers were found. These changes would have been rather impossible to achieve using epPCR and single-gene shuffling only; thus, these investigations demonstrate the power of DNA shuffling.

Mutants of an esterase from *Pseudomonas fluorescens* (PFE) produced by directed evolution using the mutator strain *Epicurian coli* XL1-Red were assayed for altered substrate specificity using a selection procedure (Bornscheuer, Altenbuchner, and Meyer, 1998). The key to the identification of improved variants acting on a sterically hindered 3-hydroxy ester – which was not hydrolyzed by the wild-type esterase – was an agar plate assay system based on pH indicators, thus leading to a change in color upon hydrolysis of the ethyl ester. Parallel assaying of replica-plated

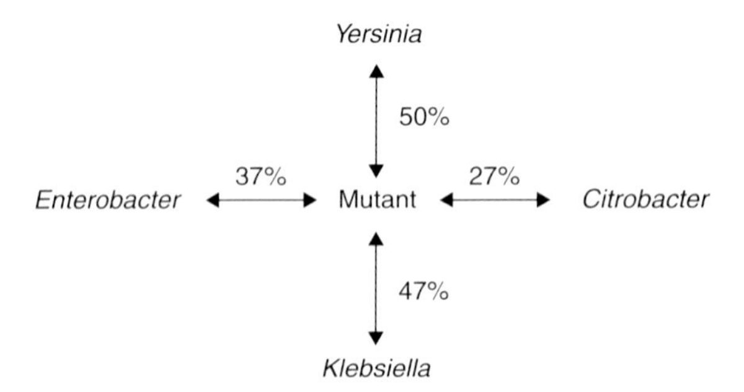

Figure 3.6 Homology between mutants of a cephalosporinase gene obtained by DNA shuffling of parental genes from four different species (Crameri *et al.*, 1998).

colonies on agar plates supplemented with the glycerol derivative of the 3-hydroxy ester was used to refine the identification, because only *E. coli* colonies producing active esterases had access to the carbon source glycerol, thus leading to enhanced growth and in turn larger colonies. By this strategy, a double mutant was identified, which efficiently catalyzed hydrolysis.

Screening Screening-based systems (not to be confused with the use of the term "screening" for the identification of microorganisms) are much more frequently used. Due to the very high number of variants generated by directed evolution, common analytical tools such as gas chromatography and HPLC are less useful, as they are usually too time-consuming. High-throughput GC–MS or NMR techniques have also been described, but these require the availability of rather expensive equipment and, in the case of screening for enantioselective biocatalysts, the use of deuterated substrates. Phage display, ribosome display, and fluorescence-activated cell sorting (FACS) have also been used to screen within mutant libraries containing on the order of $>10^6$ variants (Aharoni, Griffiths, and Tawfik, 2005; Becker *et al.*, 2008; Fernández-Àlvaro *et al.*, 2011), but they are not generally applicable.

The most frequently used methods are based on photometric and fluorimetric assays performed in microtiter plate (MTP)-based formats in combination with high-throughput robot assistance. These allow a rather accurate screening of several tens of thousands of variants within a reasonable time, and provide information about the enzymes investigated, notably their activity by determining initial rates or end points and their stereoselectivity by using both enantiomers of the compound of interest. One versatile example is the use of umbelliferone derivatives (Figure 3.7). Esters or amides of umbelliferone are rather unstable, especially at extreme pH and at elevated temperatures, but these ether derivatives are very stable as the fluorophore is linked to the substrate via an ether bond. Only after the enzymatic reaction and treatment with sodium periodate and bovine serum albumin (BSA) is the fluorophore released

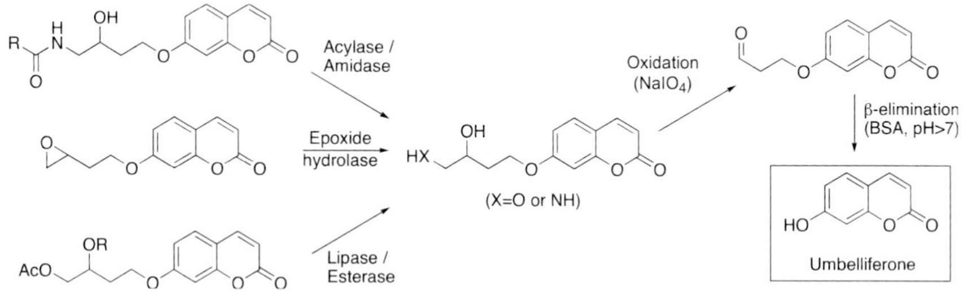

Figure 3.7 Fluorogenic assay based on umbelliferone derivatives. Enzyme activity yields a product, which upon oxidation with sodium periodate and treatment with bovine serum albumin (BSA) yields umbelliferone (Reymond and Wahler, 2002).

(Reymond and Wahler, 2002). Other techniques are outlined in the following sections. Three recent extensive reviews cover the use of protein engineering methods to obtain better enzymes for organic synthesis (Behrens *et al.*, 2011; Reetz, 2011; Strohmeier *et al.*, 2011).

3.2.2.3 Examples

Protein engineering is now a mature technology and numerous successful applications can be found in the literature, especially after the methods for

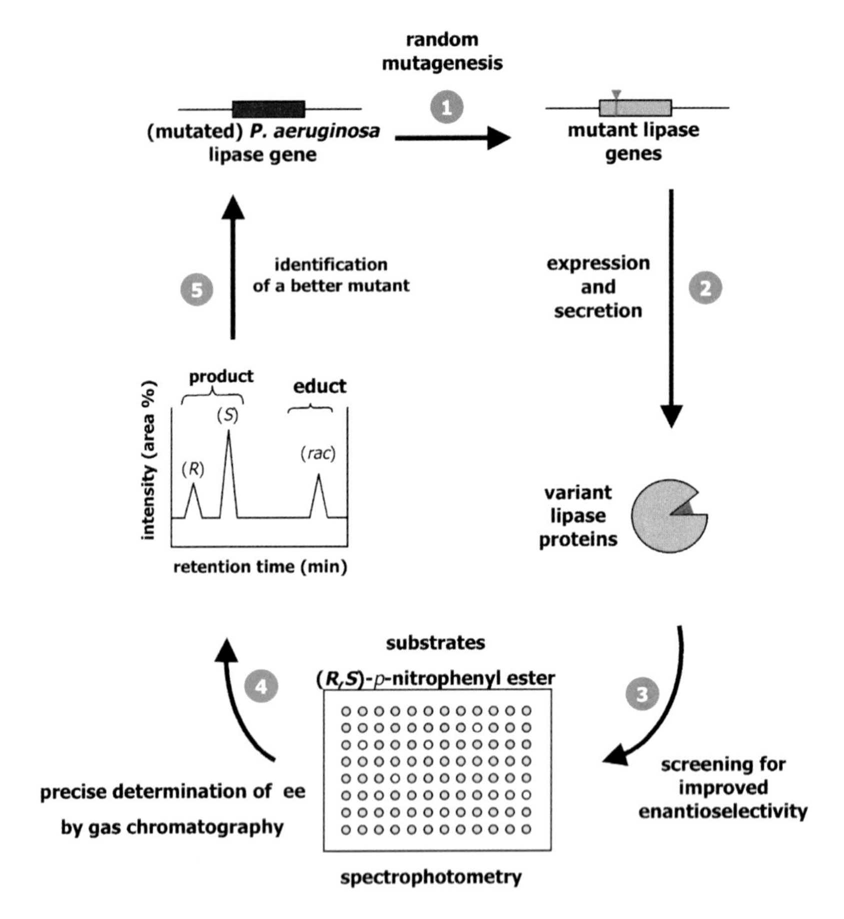

Figure 3.8 Overview of the directed evolution of a lipase from *P. aeruginosa* for the enantioselective resolution of 2-methyl decanoate. In the first step (1), the lipase gene was subjected to random mutagenesis; next, the mutated genes were expressed and secreted (2). Screening for improved enantioselectivity was based on a spectrophotometric assay using optically pure (*R*)- or (*S*)-*p*-nitrophenyl esters of the substrate (3). Hit mutants with improved enantioselectivity were then verified by gas chromatography (4). The cycle was repeated several times to identify best mutants (5) (Reetz *et al.*, 2001).

directed evolution were invented and applied in the early 1990s. Thus, only a few selected examples are given in more detail here and readers are strongly encouraged to read the reviews and book cited and to consult the recent literature.

In the first example addressing the stereoselectivity of an enzyme, Reetz and coworkers turned a nonenantioselective (2% ee, $E = 1.1$) lipase from *Pseudomonas aeruginosa* PAO1 into a variant with very good selectivity ($E > 51$, >95% ee) in the kinetic resolution of 2-methyl decanoate. The identification of variants was based on optically pure (*R*)- and (*S*)-*p*-nitrophenyl esters of 2-methyl decanoate in a spectrophotometric screening. In the first step, the wild-type lipase gene was subjected to several rounds of random mutagenesis by epPCR, leading to a variant with an $E = 11$ (81% ee) followed by saturation mutagenesis ($E = 25$). Key to further doubling of enantioselectivity was a combination of DNA shuffling, combinatorial cassette mutagenesis, and saturation mutagenesis, which led to a maximal recombination of best variants. The best mutant ($E > 51$) contained six amino acid substitutions, and a total of approximately 40 000 variants were screened (Reetz, 2001). The overall strategy is illustrated in Figure 3.8, and the overall changes in enantioselectivity using the combination of different approaches for random mutagenesis are summarized in Figure 3.9.

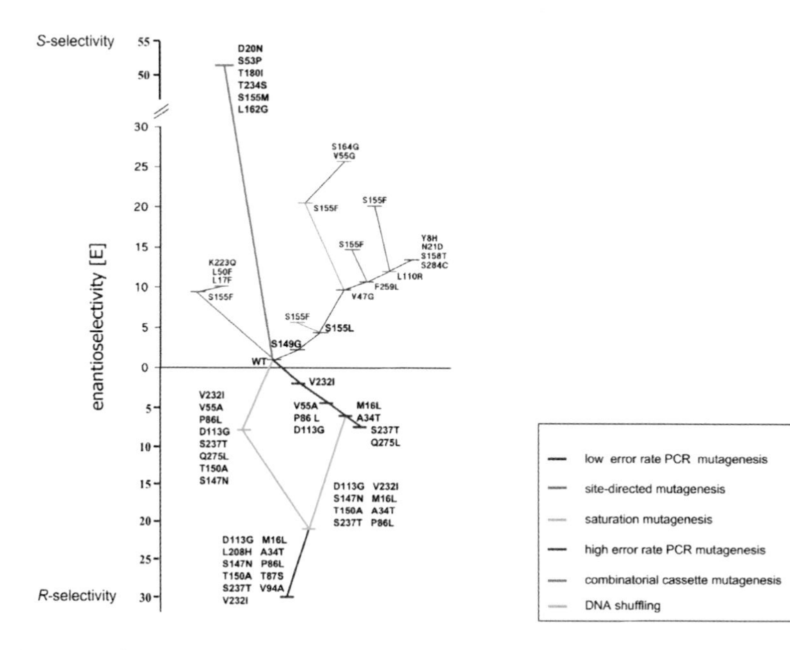

Figure 3.9 Changes in enantioselectivity of a lipase from *P. aeruginosa* using methods of directed evolution. Starting from the nonselective wild type (WT, $E = 1.1$), the combination of various genetic tools led to the creation and identification of variants with high (*S*)-selectivity ($E = 51$) and with good (*R*)-selectivity ($E = 30$) (Reetz *et al.*, 2001).

The Arnold group reported the inversion of enantioselectivity of a hydantoinase from D-selectivity (40% ee) to moderate L-preference (20% ee at 30% conversion) by a combination of epPCR and saturation mutagenesis. Only one amino acid substitution was sufficient to invert enantioselectivity. Thus, production of L-methionine from (*R,S*)-5-(2-methylthioethyl)hydantoin in a whole-cell system of recombinant *E. coli* containing also a L-carbamoylase and a racemase at high conversion became feasible (May, Nguyen, and Arnold, 2000).

Even if a biocatalyst with proper substrate specificity (and stereoselectivity) has already been identified, the requirements for a cost-effective process are not always fulfilled. Enzyme properties such as pH, temperature, and solvent stability are very difficult to improve using "classical" methods such as immobilization techniques or site-directed mutagenesis. Again, directed evolution has been shown to be a versatile tool to meet this challenge.

The first example dealt with an esterase from *Bacillus subtilis* to hydrolyze the *p*-nitrobenzyl ester of Loracarbef, a cephalosporin antibiotic. Unfortunately, the wild-type enzyme was only weakly active in the presence of dimethylformamide (DMF), which must be added to dissolve the substrate. A combination of epPCR and DNA shuffling led to the generation of a variant with 100 times higher activity compared to the wild type in 15% DMF (Moore and Arnold, 1996). Later, the thermostability of this esterase could also be increased by ~14 °C by directed evolution. In a similar manner, the performance of subtilisin E was improved 470-fold for tolerance of the solvent DMF.

Researchers at Novozymes (Denmark) subjected a heme peroxidase from *Coprinus cinereus* (CiP) to multiple rounds of directed evolution in an effort to produce a mutant suitable for use in laundry detergents (Cherry and Fidantsef, 2003). Mutants obtained by epPCR and site-directed mutagenesis were screened for improved stability by measuring residual activity after incubation under conditions mimicking those in a washing machine (e.g., pH 10.5, 50 °C, 5–10 mM peroxide). Subsequent *in vivo* shuffling led to dramatic improvements in oxidative stability, yielding a mutant with 174 times higher thermal stability and 100 times improved oxidative stability of wild-type CiP.

It is often assumed that improving a biocatalyst in one direction is countered with a loss of other enzyme characteristics. However, it was also shown that it is possible to increase the thermostability of a cold-adapted protease to 60 °C while maintaining high activity at 10 °C (Miyazaki *et al.*, 2000). The best psychrophilic subtilisin S41 variant contained only seven amino acid substitutions resembling only a tiny fraction of the usual 30–80% sequence difference found between psychrophilic enzymes and mesophilic counterparts.

In another example, researchers at Maxygen (USA) and Novozymes (Denmark) simultaneously screened for four properties in a library of family-shuffled subtilisins: activity at 23 °C, thermostability, organic solvent tolerance, and pH profile, and reported variants with considerably improved characteristics for all parameters (Ness *et al.*, 1999). Further examples are summarized in Table 3.2.

Table 3.2 Survey of recent examples for enzymes optimized by protein engineering (adapted from Behrens *et al.*, 2011).

Enzyme/property	Methods	Result and comments	Reference
Thermostability			
Lactate dehydrogenase	RD and SSM	Broadening of cofactor acceptance from NAD(H) to NADP(H)	Richter, Zienert, and Hummel (2011)
Xylanase	RD	Half-life at 50 °C and catalytic efficiency increased 15 and 1.3 times, respectively	Joo *et al.* (2010)
Pyranose-2-oxidase	SRD	Melting temperature (T_M) increased by 14 °C and improved catalytic properties	Spadiut *et al.* (2009)
β-Amylase	RD	Half-denaturation time at 60 °C increased from 6 to 35 min	Yamashiro *et al.* (2010)
Endoglucanase	epPCR	Triple mutant had 92% longer half-life at 60 °C on carboxymethyl cellulose	Liu *et al.* (2010)
Asparaginase	StEP, SSM	T_M increased by 10 °C and half-life at 50 °C increased from 3 to 160 h	Kotzia and Labrou (2009)
Lipase	B-FIT	Increased thermolability with T_{50} value reduced from 72 to 36 °C	Reetz, Soni, and Fernandez (2009)
Substrate specificity			
Monooxygenase	RD	190-fold increased initial oxidation rate for 2-phenylethanol	Brouk, Nov, and Fishman (2010)
Choline oxidase	RD and DE	Activity toward MTEA increased fivefold and k_{cat}/K_M(MTEA)/k_{cat}/K_M(choline) changed from 0.01 to 0.2	Ribitsch *et al.* (2010)
Styrene monooxygenase	RD	Exchange of one amino acid led to acceptance of bulkier α-ethylstyrene	Qaed *et al.* (2011)
Transaldolase	SSM	Improved acceptance of nonphosphorylated substrates dihydroxyacetone and glyceraldehyde	Schneider *et al.* (2010)
Xylose reductase	SSM and epPCR	Narrowing substrate acceptance to reduce unwanted formation of arabitol from arabinose while retaining xylose reductase activity	Nair and Zhao (2008)
Toluene monooxygenase		Enantioselective oxidation of aromatic sulfides by point mutation yielding methyl phenyl sulfoxide with >98% ee (pro-*S*)	Feingersch *et al.* (2008)
Galactose oxidase	epPCR	Catalyzes oxidation of secondary alcohols	Escalettes and Turner (2008)
Limonene epoxide hydrolase	ISSM	Catalyzes the asymmetrization of cyclopentene oxide stereoselectively to form the (*R*,*R*)- or the (*S*,*S*)-diol	Zheng and Reetz (2010)
Lipase	CAST	Hydrolysis of α-substituted *p*-nitrophenyl esters (95–99% ee)	Engstroem *et al.* (2010)

(*continued*)

Table 3.2 (Continued)

Enzyme/property	Methods	Result and comments	Reference
Cofactor specificity			
Alcohol dehydrogenase	RD	Improved preference for NADP(H) over NAD(H)	Campbell, Wheeldon, and Banta (2010)
Alcohol dehydrogenase	RD	Single amino acid exchange broadened coenzyme acceptance from pure NAD(H) to both NAD(H) and NADP(H)	Stiti *et al.* (2011)
Enantioselectivity/enantiopreference			
Lipase	RD	*E*-value for bulky substrate increased from 5 (WT) to >200	Ema *et al.* (2010)
Lipase	CAST	$E = 111$ for kinetic resolution of an axially chiral allene, *p*-nitrophenyl-4-cyclohexyl-2-methylbuta-2,3-dienoate	Carballeira *et al.* (2007)
P450 monooxygenase	ISSM	Changed preference from 43% ee (*S*) for WT to 83% ee (*R*)	Tang, Li, and Zhao (2010)
Alcohol dehydrogenase	RD	Reduction of benzylic and heteroarylic ketones with anti-Prelog configuration by a single mutant	Musa *et al.* (2009)
Ammonia lyase	RD	Stereoselectivity of 3-methylaspartate ammonia lyase enhanced by active-site mutation	Raj *et al.* (2009)
Increased activity			
Glucose-6-oxidase	RD	400-fold increased activity toward glucose	Lippow *et al.* (2010)
Dehalogenase	RD and SSM	32-fold higher activity toward 1,2,3-trichloropropane	Pavlova *et al.* (2009)
Cocaine hydrolase	RD	Computational design of mutant with 2000-fold improved catalytic efficiency	Zheng *et al.* (2008)
Aldolase	epPCR, SSM, and shuffling	60-fold higher k_{cat}/K_M for catalyzing the addition of pyruvate to D-erythrose 4-phosphate to form DHAP	Ran and Frost (2007)
Phenylalanine ammonia lyase	DE	15-fold increased reaction rate	Bartsch and Bornscheuer (2010)
Galactose oxidase	SSM	Mutations to optimize codon usage enable drastic enhancement of expression (>100 000-fold) and SSM in active site increased activity against galactose by 60%	Deacon and McPherson (2011)
Nitrogenase	RD	Variants show increased production of H_2 as side product during N_2 fixation	Masukawa *et al.* (2011)

Abbreviations: (S)RD: (semi-)rational design; (I)SSM: (iterative) site-directed saturation mutagenesis; epPCR: error-prone PCR; DE: directed evolution; CAST: combinatorial active-site saturation test; RM: random mutagenesis; WT: wild type; B-FIT: B-factor iterative test; T_M: melting temperature.

3.2.3
Focused Directed Evolution

Both strategies mentioned above have drawbacks that limit their practical use. Rational design can be restricted because quite often the information needed is missing or incomplete and extensive computation might be necessary. On the other hand, directed evolution is limited by the size of the library to be screened versus the availability of fast and reliable high-throughput screening methods (Figure 3.2, top).

A combination is often an alternative to overcome the limitations of both methods. Focusing, for example, on certain enzyme areas that are believed to be linked to the desired property can drastically reduce the screening effort. Combinatorial active-site saturation test (CAST) is a method pursuing this strategy by saturation mutagenesis – the replacement of the amino acid at a given position by all other 19 proteinogenic amino acids. For this, a small set of amino acids in the vicinity of the active site are chosen and mutated randomly followed by screening for best hits. One single CAST approach can result in an impressive enhancement of enzyme features as demonstrated for an epoxide hydrolase (Reetz and Sanchis, 2008). ISM (iterative saturation mutagenesis) introduces this evolution factor into the CASTing strategy (Reetz *et al.*, 2010). The best variants from a first CASTing round serve as template for the next round. Bearing in mind that mutations can have additive or cooperative effects, the order of positions to be mutated and screened might be crucial, leading to different pathways. Following all pathways is work intensive, as each step in a pathway includes a whole library that needs to be screened (Reetz *et al.*, 2005; Reetz, Wang, and Bocola, 2006). Weinreich *et al.* (2006) tested all 120 paths to an improved variant of a β-lactamase as a model enzyme and found that only 18 paths improved resistance to β-lactams at each stage. Similarly, only 8 paths increased the enantioselectivity of an epoxide hydrolase at each step (55 paths if one includes neutral mutations) (Reetz and Sanchis, 2008).

Introduction of statistical methods can also greatly improve mutagenesis strategies. A computer program analyzes sets of sequence–activity data, preferentially gained from different mutagenesis experiments, and extracts information about the influence of individual mutations. The ProSAR (protein sequence–activity relationships) strategy developed by Fox *et al.* (2007) was used to guide the development of a halohydrin dehalogenase by the data obtained from different protein engineering approaches (rational design, random mutagenesis, gene shuffling, saturation mutagenesis), sorting the single mutations into beneficial, neutral, and deleterious ones and keeping only the beneficial mutations for further rounds of mutagenesis. The authors highlight this feature, as the problem with random directed evolution might be that certain mutations present in positive mutants will undoubtedly be taken along into further mutation rounds, although the mutation itself might not be beneficial. ProSAR detects and eliminates these. The final variant in the halohydrin dehalogenase example contained 35 amino acid substitutions leading to a 4000-fold increased volumetric productivity in the synthesis of a Lipitor® side chain (Fox *et al.*, 2007).

To increase enzyme stability against high temperature or organic solvents, the "consensus approach" can be the key to success (Lehmann *et al.*, 2000). It is assumed that the most abundant amino acids at each position in a set of homologous enzymes contribute more than average to protein stability. Comparison of sequences within large enzyme families can identify conserved and differing amino acids, which then guides planning mutations to be introduced into the starting protein. This strategy was used to increase thermostability and organic solvent tolerance of a glucose-1-dehydrogenase raising the apparent melting point in the best variant to >80 °C and the half-life in organic solvents by >2500-fold (Vazquez-Figueroa *et al.*, 2008). In another approach, a database (3DM) with 1751 structurally related proteins from the α/β-hydrolase enzyme superfamily served as basis to create "small, but smart" libraries (Kourist *et al.*, 2010; Kuipers *et al.*, 2010). This 3DM analysis-based library was designed to cover only amino acids frequently occurring at a given position in this superfamily. This concept enabled to improve the enantioselectivity ($E_{mutant} = 80$, $E_{WT} = 3.2$) of an esterase from *P. fluorescens* (PFE) by mutating four residues near the active site and also improved the activity up to 240-fold (Jochens and Bornscheuer, 2010). This method was also used to increase thermoactivity of PFE by 8 °C (Jochens, Aerts, and Bornscheuer, 2010). In another example, identification and mutagenesis of flexible residues using the B-FIT approach[2] in combination with ISM gave drastically improved thermoactivity (from 45 to 93 °C) for a lipase from *B. subtilis* (Reetz *et al.*, 2010).

Another concept to take advantage of synergistic effects of simultaneous mutations is to reduce the library size by eliminating duplicate codons. Most saturation mutagenesis uses the NNK codon (N = A, T, G or C; K = G or T), which yields 32 possible codons to encode 20 amino acids and thus 12 codons encode for duplicate amino acids. This has considerable consequences on the size of the library to be screened (Table 3.3). For instance, when three amino acids are varied simultaneously, the NNK approach creates 32 768 codons, but only 8000 (= 20 × 20 × 20) different proteins. This requires screening 98 163 colonies to ensure by a threefold oversampling a ~95% coverage of all possible variants. An alternative is the NDT

Table 3.3 Influence of NNK or NDT codon on the size of the library needed to be screened in simultaneous saturation mutagenesis experiments.

NNK codon			NDT codon		
AA sites	**Codons**	**Colonies**	**AA sites**	**Codons**	**Colonies**
1	32	94	1	12	34
2	1024	3066	2	144	420
3	32 768	98 163	3	1728	5175
4	1 048 576	3 141 251	4	20 736	62 118

2) B-FIT, B-factor iterative test. The B-factor refers to flexible residues in a protein structure and can be deduced from the X-ray crystallography data.

codon (D = G, A, or T), which consists of 12 codons that code for 12 different amino acids (note that acidic, basic, neutral, and aromatic amino acids are still included). Here, three simultaneous substitutions yield 1728 possible protein or codon variants and only 5175 variants need to be screened for ~95% coverage. Hence, use of the NDT codon substantially reduces screening efforts (Reetz, Kahakeaw, and Lohmer, 2008), but on the other hand this approach omits eight possible amino acids. In one example where the enantiopreference of an esterase was inverted, a tryptophan substitution was found crucial for success (Bartsch, Kourist, and Bornscheuer, 2008), but this amino acid is not covered by the NDT codon and consequently would have been missed by this directed evolution concept.

3.3 Computational Design of Enzymes

Although numerous enzymes have been identified, some important organic reactions seem to have no natural enzymatic counterpart. In recent years, significant advances have been made toward the design of tailor-made enzymes from scratch. Although the relationship between an enzyme's structure and its activity is still poorly understood, examples have been published using extensive computational power to design enzymes that catalyze reactions not observed in nature. However, these proteins still do not reach the high activity of natural biocatalysts.

The principal strategy in creating a new enzyme is based on the fact that an enzyme can catalyze a chemical transformation, because, among other reasons, it stabilizes the transition state of the reaction. First of all, the basic mechanism of the reaction must be known in detail, and then, focusing on the transition state, amino acid residues that can stabilize or even interact with the substrate are positioned in the relevant order, thus forming the putative active site. This step already requires sophisticated computing using quantum mechanical (QM) calculations to find the right amino acids and position them as exactly as possible. The next step consists of finding a natural scaffold that can be used to accommodate these residues in their proper positioning. The group of Baker was able to develop and enhance algorithms to perform this task (Leaver-Fay *et al.*, 2011).

Using these methods, it was possible to design a retro-aldolase, a Diels-Alderase, and a Kemp eliminase. The retro-aldolase reaction is catalyzed by a lysine residue via a Schiff base or an imine intermediate. Since the proposed mechanism consisted of several steps, the above-mentioned general method needed to be extended for the design of active sites that are compatible with multiple transition states. In total, 32 designs, belonging to four different motifs, were found to show retro-aldolase activity with rate enhancements of up to 10^4 over the noncatalyzed reaction, but still with slow turnover rates of about 9×10^{-3} min^{-1} (Jiang *et al.*, 2008).

The Kemp eliminase was proposed to use a general base, being either aspartate or glutamate or a histidine–aspartate/glutamate dyad. Of the 59 selected variants, 39 were aspartate or glutamate dependent and 20 had a dyad in the active site and 8 designs showed activity. A promising variant relied on a His–Asp dyad and showed a k_{cat} of $2 \times 10^{-2}\ s^{-1}$. By random mutagenesis, this could be further enhanced to $1.3\ s^{-1}$ (Röthlisberger *et al.*, 2008). Another variant was further optimized by various techniques ending up with mutants showing a turnover number of $5–8\ s^{-1}$ and catalytic efficiencies of $5 \times 10^4\ M^{-1}\ s^{-1}$ (Khersonsky *et al.*, 2011).

Designing an enzyme capable of catalyzing the Diels–Alder reaction was supposed to be quite challenging, since bimolecular bond formation is involved. A carbonyl oxygen of a glutamine or asparagine residue was used to form a hydrogen bond to the diene intermediate, whereas the dienophile was hydrogen bonded by the hydroxyl group of a serine, threonine, or tyrosine. With these assumptions, 10^{19} possible active-site variants were computed and about 10^6 could be matched into protein scaffolds. These variants were further evaluated and in the end 50 variants were obtained as soluble proteins. Only two designs had Diels-Alderase activity, and after further optimization by mutagenesis, a variant gave a rate of $k_{cat}/k_{uncat} = 89$, which is an effective turnover rate of about $2\ s^{-1}$. Interestingly, the most active variant was strictly stereospecific, as intended by the design, and yielded >97% of the desired product (Siegel *et al.*, 2010).

Exercises

3.1 Explain the term "directed evolution."

3.2 Explain the difference between recombining and nonrecombining methods (asexual versus sexual methods).

3.3 Describe the error-prone PCR. Which experimental conditions are varied?

3.4 Explain with an example a selection method. What are the advantages and disadvantages of this approach?

3.5 Explain the iterative saturation mutagenesis? How many combinations are possible if four positions (A, B, C, D) are mutated?

References

Aharoni, A., Griffiths, A.D., and Tawfik, D.S. (2005) High-throughput screens and selections of enzyme-encoding genes. *Curr. Opin. Chem. Biol.*, **9**, 210–216.

Amann, R.I., Ludwig, W., and Schleifer, K.-H. (1995) Phylogenetic identification and *in situ* detection of individual microbial cells without cultivation. *Microbiol. Rev.*, **59**, 143–169.

Arnold, F.H. and Georgiou, G. (eds) (2003a) *Directed Enzyme Evolution: Screening and Selection Methods*, vol. **230**, Humana Press, Totowa, NJ.

Arnold, F.H. and Georgiou, G. (eds) (2003b) *Directed Evolution Library Creation: Methods and Protocols*, vol. **231**, Humana Press, Totowa, NJ.

Barakat, N.H., Barakat, N.H., and Love, J.J. (2010) Combined use of experimental and computational screens to characterize protein stability. *Protein Eng. Des. Sel.*, **23**, 799–807.

Bartsch, S. and Bornscheuer, U.T. (2010) Mutational analysis of phenylalanine ammonia lyase to improve reactions rates for various substrates. *Protein Eng. Des. Sel.*, **23**, 929–933.

Bartsch, S., Kourist, R., and Bornscheuer, U.T. (2008) Complete inversion of enantioselectivity towards acetylated tertiary alcohols by a double mutant of a *Bacillus subtilis* esterase. *Angew. Chem., Int. Ed.*, **47**, 1508–1511.

Bayer, S., Kunert, A., Ballschmiter, M., and Greiner-Stoeffele, T. (2010) Indication for a new lipolytic enzyme family: isolation and characterization of two esterases from a metagenomic library. *J. Mol. Microbiol. Biotechnol.*, **18**, 181–187.

Becker, S., Hobenreich, H., Vogel, A., Knorr, J., Wilhelm, S., Rosenau, F., Jaeger, K.E., Reetz, M.T., and Kolmar, H. (2008) Single-cell high-throughput screening to identify enantioselective hydrolytic enzymes. *Angew. Chem., Int. Ed.*, **47**, 5085–5088.

Behrens, G.A., Hummel, A., Padhi, S.K., Schätzle, S., and Bornscheuer, U.T. (2011) Discovery and protein engineering of biocatalysts for organic synthesis. *Adv. Synth. Catal.*, **353**, 2191–2215.

Bornscheuer, U.T., Altenbuchner, J., and Meyer, H.H. (1998) Directed evolution of an esterase for the stereoselective resolution of a key intermediate in the synthesis of epothilones. *Biotechnol. Bioeng.*, **58**, 554–559.

Bornscheuer, U.T., Huisman, G., Kazlauskas, R.J., Lutz, S., Moore, J., and Robins, K. (2012) Engineering the third wave in biocatalysis, *Nature*, **485**, 185–194

Bornscheuer, U.T. and Kazlauskas, R.J. (2004) Catalytic promiscuity in biocatalysis: using old enzymes to form new bonds and follow new pathways. *Angew. Chem., Int. Ed.*, **43**, 6032–6040.

Bornscheuer, U.T. and Kazlauskas, R.J. (2011) Survey of protein engineering strategies. *Curr. Protoc. Prot. Sci.*, **66**, 26.7.1–26.7.14.

Böttcher, D. and Bornscheuer, U.T. (2010) Protein engineering of microbial enzymes. *Curr. Opin. Microbiol.*, **13**, 274–282.

Bougioukou, D.J., Kille, S., Taglieber, A., and Reetz, M.T. (2009) Directed evolution of an enantioselective enoate-reductase: testing the utility of iterative saturation mutagenesis. *Adv. Synth. Catal.*, **351**, 3287–3305.

Brakmann, S. and Johnsson, K. (eds) (2002) *Directed Molecular Evolution of Proteins*, Wiley-VCH Verlag GmbH, Weinheim.

Brakmann, S. and Schwienhorst, A. (eds) (2004) *Evolutionary Methods in Biotechnology: Clever Tricks for Directed Evolution*, Wiley-VCH Verlag GmbH, Weinheim.

Britton, L.N., Brand, J.M., and Markovetz, A.J. (1974) Source of oxygen in the conversion of 2-tridecanone to undecyl acetate by *Pseudomonas cepacia* and *Nocardia* sp. *Biochim. Biophys. Acta*, **369**, 45–49.

Brouk, M., Nov, Y., and Fishman, A. (2010) Improving biocatalyst performance by integrating statistical methods into protein engineering. *Appl. Environ. Microbiol.*, **76**, 6397–6403.

Cadwell, R.C. and Joyce, G.F. (1992) Randomization of genes by PCR mutagenesis. *PCR Methods Appl.*, **2**, 28–33.

Campbell, E., Wheeldon, I.R., and Banta, S. (2010) Broadening the cofactor specificity of a thermostable alcohol dehydrogenase using rational protein design introduces novel kinetic transient behavior. *Biotechnol. Bioeng.*, **107**, 763–774.

Carballeira, J.D., Krumlinde, P., Bocola, M., Vogel, A., Reetz, M.T., and Bäckvall, J.-E. (2007) Directed evolution and axial chirality: optimization of the enantioselectivity of *Pseudomonas aeruginosa* lipase towards the kinetic resolution of a racemic allene. *Chem. Commun.*, 1913–1915.

Chen, R.P., Guo, L.Z., and Dang, H.Y. (2011) Gene cloning, expression and characterization of a cold-adapted lipase from a psychrophilic deep-sea bacterium *Psychrobacter* sp. C18. *World J. Microbiol. Biotechnol.*, **27**, 431–441.

Cherry, J.R. and Fidantsef, A.L. (2003) Directed evolution of industrial enzymes: an update. *Curr. Opin. Biotechnol.*, **14**, 438–443.

Crameri, A., Raillard, S.A., Bermudez, E., and Stemmer, W.P. (1998) DNA shuffling of a family of genes from diverse species accelerates directed evolution. *Nature*, **391**, 288–291.

Deacon, S.E. and McPherson, M.J. (2011) Enhanced expression and purification of fungal galactose oxidase in *Escherichia coli* and use for analysis of a saturation mutagenesis library. *ChemBioChem*, **12**, 593–601.

Ema, T., Kamata, S., Takeda, M., Nakano, Y., and Sakai, T. (2010) Rational creation of mutant enzyme showing remarkable enhancement of catalytic activity and enantioselectivity toward poor substrates. *Chem. Commun.*, **46**, 5440–5442.

Engström, K., Nyhlen, J., Sandstroem, A.G., and Bäckvall, J.-E. (2010) Directed evolution of an enantioselective lipase with broad substrate scope for hydrolysis of α-substituted esters. *J. Am. Chem. Soc.*, **132**, 7038–7042.

Escalettes, F. and Turner, N.J. (2008) Directed evolution of galactose oxidase: generation of enantioselective secondary alcohol oxidases. *ChemBioChem*, **9**, 857–860.

Estell, D.A., Graycar, T.P., and Wells, J.A. (1985) Engineering an enzyme by site-directed mutagenesis to be resistant to chemical oxidation. *J. Biol. Chem.*, **260**, 6518–6521.

Feingersch, R., Shainsky, J., Wood, T.K., and Fishman, A. (2008) Protein engineering of toluene monooxygenases for synthesis of chiral sulfoxides. *Appl. Environ. Microbiol.*, **74**, 1555–1566.

Fernández-Àlvaro, E., Snajdrova, R., Jochens, H., Davids, T., Böttcher, D., and Bornscheuer, U.T. (2011) A combination of *in vivo* selection and cell sorting for the identification of enantioselective biocatalysts. *Angew. Chem., Int. Ed.*, **50**, 8584–8587.

Fernández-Arrojo, L., Guazzaroni, M.-E., López-Cortés, N., Beloqui, A., and Ferrer, M. (2010) Metagenomic era for biocatalyst identification. *Curr. Opin. Biotechnol.*, **21**, 725–733.

Fox, R.J., Davis, S.C., Mundorff, E.C., Newman, L.M., Gavrilovic, V., Ma, S.K., Chung, L.M., Ching, C., Tam, S., Muley, S., Grate, J., Gruber, J., Whitman, J.C., Sheldon, R.A., and Huisman, G.W. (2007) Improving catalytic function by ProSAR-driven enzyme evolution. *Nat. Biotechnol.*, **25**, 338–344.

Fraaije, M.W., Wu, J., Heuts, D.P.H.M., van Hellemond, E.W., Spelberg, J.H.L., and Janssen, D.B. (2005) Discovery of a thermostable Baeyer–Villiger monooxygenase by genome mining. *Appl. Microbiol. Biotechnol.*, **66**, 393–400.

Furuya, T. and Kino, K. (2010) Genome mining approach for the discovery of novel cytochrome P450 biocatalysts. *Appl. Microbiol. Biotechnol.*, **86**, 991–1002.

Griengl, H., Schwab, H., and Fechter, M. (2000) The synthesis of chiral cyanohydrins by oxynitrilases. *Trends Biotechnol.*, **18**, 252–256.

Hartmans, S. and de Bont, J.A.M. (1986) Acetol monooxygenase from *Mycobacterium* Py1 cleaves acetol into acetate and formaldehyde. *FEMS Microbiol. Lett.*, **36**, 155–158.

Höhne, M., Schätzle, S., Jochens, H., Robins, K., and Bornscheuer, U.T. (2010) Rational assignment of key motifs for function guides *in silico* enzyme identification. *Nat. Chem. Biol.*, **6**, 807–813.

Hugenholtz, P., Goebel, B.M., and Pace, N.R. (1998) Impact of culture-independent studies on the emerging phylogenetic view of bacterial diversity. *J. Bacteriol.*, **180**, 4765–4774.

Hulley, M.E., Toogood, H.S., Fryszkowska, A., Mansell, D., Stephens, G.M., Gardiner, J.M., and Scrutton, N.S. (2010) Focused directed evolution of pentaerythritol tetranitrate reductase by using automated anaerobic kinetic screening of site-saturated libraries. *ChemBioChem*, **11**, 2433–2447.

Hummel, A., Brüsehaber, E., Böttcher, D., Trauthwein, H., Doderer, K., and Bornscheuer, U.T. (2007) Isoenzymes of pig-liver esterase reveal striking differences in enantioselectivities. *Angew. Chem., Int. Ed.*, **46**, 8492–8494.

Jiang, L., Althoff, E.A., Clemente, F.R., Doyle, L., Rothlisberger, D., Zanghellini, A., Gallaher, J.L., Betker, J.L., Tanaka, F., Barbas, C.F., 3rd, Hilvert, D., Houk, K.N., Stoddard, B.L., and Baker, D. (2008) *De novo* computational design of retro-aldol enzymes. *Science*, **319**, 1387–1391.

Jochens, H., Aerts, D., and Bornscheuer, U.T. (2010) Thermostabilization of an esterase by alignment-guided focussed directed evolution. *Protein Eng. Des. Sel.*, **23**, 903–909.

Jochens, H. and Bornscheuer, U.T. (2010) Natural diversity to guide focused directed evolution. *ChemBioChem*, **11**, 1861–1866.

Johnson, D.V., Zabelinskaja-Mackova, A.A., and Griengl, H. (2000) Oxynitrilases for asymmetric C—C bond formation. *Curr. Opin. Chem. Biol.*, **4**, 103–109.

Joo, J.C., Pohkrel, S., Pack, S.P., and Yoo, Y.J. (2010) Thermostabilization of *Bacillus*

circulans xylanase via computational design of a flexible surface cavity. *J. Biotechnol.*, **146**, 31–39.
Juergens, C., Strom, A., Wegener, D., Hettwer, S., Wilmanns, M., and Sterner, R. (2000) Directed evolution of a (beta-alpha)$_8$-barrel enzyme to catalyze related reactions in two different metabolic pathways. *Proc. Natl. Acad. Sci. USA*, **97**, 9925–9930.
Kaplan, O., Bezouška, K., Malandra, A., Veselá, A., Petříčková, A., Felsberg, J., Rinágelová, A., Křen, V., and Martínková, L. (2011) Genome mining for the discovery of new nitrilases in filamentous fungi. *Biotechnol. Lett.*, **33**, 309–312.
Kazlauskas, R.J. and Bornscheuer, U.T. (2009) Finding better protein engineering strategies. *Nat. Chem. Biol.*, **5**, 526–529.
Kelley, L.A. and Sternberg, M.J. (2009) Protein structure prediction on the Web: a case study using the Phyre server. *Nat. Protoc.*, **4**, 363–371.
Khersonsky, O., Rothlisberger, D., Wollacott, A.M., Murphy, P., Dym, O., Albeck, S., Kiss, G., Houk, K.N., Baker, D., and Tawfik, D.S. (2011) Optimization of the *in-silico*-designed Kemp eliminase KE70 by computational design and directed evolution. *J. Mol. Biol.*, **407**, 391–412.
Kino, K., Noguchi, A., Arai, T., and Yagasaki, M. (2010) Identification and characterization of a novel L-amino acid ligase from *Photorhabdus luminescens* subsp. *laumondii* TT01. *J. Biosci. Bioeng.*, **110**, 39–41.
Kirschner, A., Altenbuchner, J., and Bornscheuer, U.T. (2007) Cloning, expression, and characterization of a Baeyer–Villiger monooxygenase from *Pseudomonas fluorescens* DSM 50106 in *E. coli*. *Appl. Microbiol. Biotechnol.*, **73**, 1065–1072.
Kirschner, A. and Bornscheuer, U.T. (2006) Kinetic resolution of 4-hydroxy-2-ketones catalyzed by a Baeyer–Villiger monooxygenase. *Angew. Chem., Int. Ed.*, **45**, 7004–7006.
Kotzia, G.A. and Labrou, N.E. (2009) Engineering thermal stability of L-asparaginase by *in vitro* directed evolution. *FEBS J.*, **276**, 1750–1761.
Kourist, R., Jochens, H., Bartsch, S., Kuipers, R., Padhi, S.K., Gall, M., Böttcher, D., Joosten, H.J., and Bornscheuer, U.T. (2010) The alpha/beta-hydrolase fold 3DM database (ABHDB) as a tool for protein engineering. *ChemBioChem*, **11**, 1635–1643.
Kuipers, R.K., Joosten, H.J., van Berkel, W.J., Leferink, N.G., Rooijen, E., Ittmann, E., van Zimmeren, F., Jochens, H., Bornscheuer, U., Vriend, G., dos Santos, V. A., and Schaap, P.J. (2010) 3DM: systematic analysis of heterogeneous superfamily data to discover protein functionalities. *Proteins*, **78**, 2101–2113.
Kwon, E.J., Jeong, Y.S., Kim, Y.H., Kim, S.K., Na, H.B., Kim, J., Yun, H.D., and Kim, H. (2010) Construction of a metagenomic library from compost and screening of cellulase- and xylanase-positive clones. *J. Korean Soc. Appl. Biol. Chem.*, **53**, 702–708.
Larkin, M.A., Blackshields, G., Brown, N.P., Chenna, R., McGettigan, P.A., McWilliam, H., Valentin, F., Wallace, I.M., Wilm, A., Lopez, R., Thompson, J.D., Gibson, T.J., and Higginsn D.G., (2007) Bioinformatics. **23**, 2947–2948.
Leaver-Fay, A., Tyka, M., Lewis, S.M., Lange, O.F., Thompson, J., Jacak, R., Kaufman, K., Renfrew, P.D., Smith, C.A., Sheffler, W., Davis, I.W., Cooper, S., Treuille, A., Mandell, D.J., Richter, F., Ban, Y.E.A., Fleishman, S.J., Corn, J.E., Kim, D.E., Lyskov, S., Berrondo, M., Mentzer, S., Popovic, Z., Havranek, J.J., Karanicolas, J., Das, R., Meiler, J., Kortemme, T., Gray, J.J., Kuhlman, B., Baker, D., and Bradley, P. (2011) Rosetta3: an object-oriented software suite for the simulation and design of macromolecules. *Methods Enzymol.*, **487**, 545–574.
Lehmann, M., Pasamontes, L., Lassen, S.F., and Wyss, M. (2000) The consensus concept for thermostability engineering of proteins. *Biochim. Biophys. Acta*, **1543**, 408–415.
Li, S.M. (2011) Genome mining and biosynthesis of fumitremorgin-type alkaloids in ascomycetes. *J. Antibiot.*, **64**, 45–49.
Lippow, S.M., Moon, T.S., Basu, S., Yoon, S. H., Li, X., Chapman, B.A., Robison, K., Lipovsek, D., and Prather, K.L.J. (2010) Engineering enzyme specificity using computational design of a defined-

sequence library. *Chem. Biol.*, **17**, 1306–1315.

Liu, J.A., Liu, W.D., Zhao, X.L., Shen, W.J., Cao, H., and Cui, Z.L. (2011) Cloning and functional characterization of a novel endo-beta-1,4-glucanase gene from a soil-derived metagenomic library. *Appl. Microbiol. Biotechnol.*, **89**, 1083–1092.

Liu, W.J., Zhang, X.Z., Zhang, Z.M., and Zhang, Y.H.P. (2010) Engineering of *Clostridium phytofermentans* endoglucanase Cel5A for improved thermostability. *Appl. Environ. Microbiol.*, **76**, 4914–4917.

Lorenz, P. and Eck, J. (2005) Metagenomics and industrial applications. *Nat. Rev. Microbiol.*, **3**, 510–516.

Lutz, S. and Bornscheuer, U.T. (eds) (2009) *Protein Engineering Handbook*, Wiley-VCH Verlag GmbH, Weinheim.

MacBeath, G., Kast, P., and Hilvert, D. (1998) Redesigning enzyme topology by directed evolution. *Science*, **279**, 1958–1961.

Machielsen, R., Uria, A.R., Kengen, S.W.M., and van der Oost, J. (2006) Production and characterization of a thermostable alcohol dehydrogenase that belongs to the aldo-keto reductase superfamily. *Appl. Environ. Microbiol.*, **72**, 233–238.

Magnusson, A.O., Takwa, M., Hamberg, A., and Hult, K. (2005) An *S*-selective lipase was created by rational redesign and the enantioselectivity increased with temperature. *Angew. Chem., Int. Ed.*, **44**, 4582–4585.

Masukawa, H., Inoue, K., Sakurai, H., Wolk, C.P., and Hausinger, R.P. (2011) Site-directed mutagenesis of the *Anabaena* sp. strain PCC 7120 nitrogenase active site to increase photobiological hydrogen production. *Appl. Environ. Microbiol.*, **76**, 6741–6750.

May, O., Nguyen, P.T., and Arnold, F.H. (2000) Inverting enantioselectivity by directed evolution of hydantoinase for improved production of L-methionine. *Nat. Biotechnol.*, **18**, 317–320.

Miyazaki, K., Wintrode, P.L., Grayling, R.A., Rubingh, D.N., and Arnold, F.H. (2000) Directed evolution study of temperature adaptation in a psychrophilic enzyme. *J. Mol. Biol.*, **297**, 1015–1026.

Mo, X.C., Chen, C.L., Pang, H., Feng, Y., and Feng, J.X. (2010) Identification and characterization of a novel xylanase derived from a rice straw degrading enrichment culture. *Appl. Microbiol. Biotechnol.*, **87**, 2137–2146.

Moore, J.C. and Arnold, F.H. (1996) Directed evolution of a *para*-nitrobenzyl esterase for aqueous–organic solvents. *Nat. Biotechnol.*, **14**, 458–467.

Musa, M.M., Lott, N., Laivenieks, M., Watanabe, L., Vieille, C., and Phillips, R.S. (2009) A single point mutation reverses the enantiopreference of *Thermoanaerobacter ethanolicus* secondary alcohol dehydrogenase. *ChemCatChem*, **1**, 89–93.

Musidlowska, A., Lange, S., and Bornscheuer, U.T. (2001) By overexpression in the yeast *Pichia pastoris* to enhanced enantioselectivity: new aspects in the application of pig liver esterase. *Angew. Chem., Int. Ed.*, **40**, 2851–2853.

Nair, N.U. and Zhao, H.M. (2008) Evolution in reverse: engineering a D-xylose-specific xylose reductase. *ChemBioChem*, **9**, 1213–1215.

Ness, J.E., Welch, M., Giver, L., Bueno, M., Cherry, J.R., Borchert, T.V., Stemmer, W.P., and Minshull, J. (1999) DNA shuffling of subgenomic sequences of subtilisin. *Nat. Biotechnol.*, **17**, 893–896.

Neylon, C. (2004) Chemical and biochemical strategies for the randomization of protein encoding DNA sequences: library construction methods for directed evolution. *Nucleic Acids Res.*, **32**, 1448–1459.

Okamura, Y., Kimura, T., Yokouchi, H., Meneses-Osorio, M., Katoh, M., Matsunaga, T., and Takeyama, H. (2010) Isolation and characterization of a GDSL esterase from the metagenome of a marine sponge-associated bacteria. *Mar. Biotechnol.*, **12**, 395–402.

Pace, N.R. (1997) A molecular view of microbial diversity and the biosphere. *Science*, **276**, 734–740.

Padhi, S.K., Bougioukou, D.J., and Stewart, J.D. (2009) Site-saturation mutagenesis of tryptophan 116 of *Saccharomyces pastorianus* old yellow enzyme uncovers stereocomplementary variants. *J. Am. Chem. Soc.*, **131**, 3271–3280.

Park, S.Y., Shin, H.J., and Kim, G.J. (2011) Screening and identification of a novel esterase EstPE from a metagenomic DNA library. *J. Microbiol.*, **49**, 7–14.

Pavlova, M., Klvana, M., Prokop, Z., Chaloupkova, R., Banas, P., Otyepka, M., Wade, R.C., Tsuda, M., Nagata, Y., and Damborsky, J. (2009) Redesigning dehalogenase access tunnels as a strategy for degrading an anthropogenic substrate. *Nat. Chem. Biol.*, **5**, 727–733.

Qaed, A.A., Lin, H., Tang, D.F., and Wu, Z.L. (2011) Rational design of styrene monooxygenase mutants with altered substrate preference. *Biotechnol. Lett.*, **33**, 611–616.

Raj, H., Weiner, B., Veetil, V.P., Reis, C.R., Quax, W.J., Janssen, D.B., Feringa, B.L., and Poelarends, G.J. (2009) Alteration of the diastereoselectivity of 3-methylaspartate ammonia lyase by using structure-based mutagenesis. *ChemBioChem*, **10**, 2236–2245.

Ran, N. and Frost, J.W. (2007) Directed evolution of 2-keto-3-deoxy-6-phosphogalactonate aldolase to replace 3-deoxy-D-arabino-heptulosonic acid 7-phosphate synthase. *J. Am. Chem. Soc.*, **129**, 6130–6139.

Reetz, M.T. (2001) Combinatorial and evolution-based methods in the creation of enantioselective catalysts. *Angew. Chem., Int. Ed.*, **40**, 284–310.

Reetz, M.T. (2011) Laboratory evolution of stereoselective enzymes: a prolific source of catalysts for asymmetric reactions. *Angew. Chem., Int. Ed.*, **50**, 138–174.

Reetz, M.T., Bocola, M., Carballeira, J.D., Zha, D.X., and Vogel, A. (2005) Expanding the range of substrate acceptance of enzymes: combinatorial active-site saturation test. *Angew. Chem., Int. Ed.*, **44**, 4192–4196.

Reetz, M.T., Kahakeaw, D., and Lohmer, R. (2008) Addressing the numbers problem in directed evolution. *ChemBioChem*, **9**, 1797–1804.

Reetz, M.T. and Sanchis, J. (2008) Constructing and analyzing the fitness landscape of an experimental evolutionary process. *ChemBioChem*, **9**, 2260–2267.

Reetz, M.T., Soni, P., and Fernandez, L. (2009) Knowledge-guided laboratory evolution of protein thermolability. *Biotechnol. Bioeng.*, **102**, 1712–1717.

Reetz, M.T., Soni, P., Fernandez, L., Gumulya, Y., and Carballeira, J.D. (2010) Increasing the stability of an enzyme toward hostile organic solvents by directed evolution based on iterative saturation mutagenesis using the B-FIT method. *Chem. Commun.*, **46**, 8657–8658.

Reetz, M.T., Wang, L.W., and Bocola, M. (2006) Directed evolution of enantioselective enzymes: iterative cycles of CASTing for probing protein-sequence space. *Angew. Chem., Int. Ed.*, **45**, 1236–1241.

Reetz, M.T., Wilensek, S., Zha, D., and Jaeger, K.-E. (2001) Directed evolution of an enantioselective enzyme through combinatorial multiple-cassette mutagenesis. *Angew. Chem., Int. Ed.*, **40**, 3589–3591.

Reymond, J.L. and Wahler, D. (2002) Substrate arrays as enzyme fingerprinting tools. *ChemBioChem*, **3**, 701–708.

Ribitsch, D., Winkler, S., Gruber, K., Karl, W., Wehrschutz-Sigl, E., Eiteljorg, I., Schratl, P., Remler, P., Stehr, R., Bessler, C., Mussmann, N., Sauter, K., Maurer, K.H., and Schwab, H. (2010) Engineering of choline oxidase from *Arthrobacter nicotianae* for potential use as biological bleach in detergents. *Appl. Microbiol. Biotechnol.*, **87**, 1743–1752.

Richter, N., Zienert, A., and Hummel, W. (2011) A single-point mutation enables lactate dehydrogenase from *Bacillus subtilis* to utilize NAD^+ and $NADP^+$ as cofactor. *Eng. Life Sci.*, **11**, 26–36.

Röthlisberger, D., Khersonsky, O., Wollacott, A.M., Jiang, L., DeChancie, J., Betker, J., Gallaher, J.L., Althoff, E.A., Zanghellini, A., Dym, O., Albeck, S., Houk, K.N., Tawfik, D.S., and Baker, D. (2008) Kemp elimination catalysts by computational enzyme design. *Nature*, **453**, 190–195.

Schätzle, S., Steffen-Munsberg, F., Thontowi, A., Höhne, M., Robins, K., and Bornscheuer, U.T. (2011) Enzymatic asymmetric synthesis of enantiomerically pure aliphatic, aromatic and arylaliphatic amines with (*R*)-selective amine transaminases. *Adv. Synth. Catal.*, **353**, 2439–2445.

Schmidt, M., Hasenpusch, D., Kähler, M., Kirchner, U., Wiggenhorn, K., Langel, W., and Bornscheuer, U.T. (2006) Directed evolution of an esterase from *Pseudomonas fluorescens* yields a mutant with excellent enantioselectivity and activity for the kinetic resolution of a chiral building block. *ChemBioChem*, **7**, 805–809.

Schneider, S., Gutierrez, M., Sandalova, T., Schneider, G., Clapes, P., Sprenger, G. A., and Samland, A.K. (2010) Redesigning the active site of transaldolase TalB from *Escherichia coli*: new variants with improved affinity towards nonphosphorylated substrates. *ChemBioChem*, **11**, 681–690.

Siegel, J.B., Zanghellini, A., Lovick, H.M., Kiss, G., Lambert, A.R., St Clair, J.L., Gallaher, J.L., Hilvert, D., Gelb, M.H., Stoddard, B.L., Houk, K.N., Michael, F.E., and Baker, D. (2010) Computational design of an enzyme catalyst for a stereoselective bimolecular Diels–Alder reaction. *Science*, **329**, 309–313.

Spadiut, O., Leitner, C., Salaheddin, C., Varga, B., Vertessy, B.G., Tan, T.C., Divne, C., and Haltrich, D. (2009) Improving thermostability and catalytic activity of pyranose 2-oxidase from *Trametes multicolor* by rational and semi-rational design. *FEBS J.*, **276**, 776–792.

Staley, J.T. and Konopka, A. (1985) Measurement of *in situ* activities of nonphotosynthetic microorganisms in aquatic and terrestrial habitats. *Annu. Rev. Microbiol.*, **39**, 321–346.

Stemmer, W.P.C. (1994a) Rapid evolution of a protein *in vitro* by DNA shuffling. *Nature*, **370**, 389–391.

Stemmer, W.P.C. (1994b) DNA shuffling by random fragmentation and reassembly: *in vitro* recombination for molecular evolution. *Proc. Natl. Acad. Sci. USA*, **91**, 10747–10751.

Stewart, J.D. (2006) *Advances in Applied Microbiology*, vol. 59 (eds S.S. Allen, I. Laskin, and M.G. Geoffrey), Academic Press, pp. 31–52.

Stiti, N., Adewale, I.O., Petersen, J., Bartels, D., and Kirch, H.H. (2011) Engineering the nucleotide coenzyme specificity and sulfhydryl redox sensitivity of two stress-responsive aldehyde dehydrogenase isoenzymes of *Arabidopsis thaliana*. *Biochem. J.*, **434**, 459–471.

Streit, W.R. and Daniel, R. (2010) *Metagenomics: Methods and Protocols*, Springer.

Strohmeier, G.A., Pichler, H., May, O., and Gruber-Khadjawi, M. (2011) Application of designed enzymes in organic synthesis. *Chem. Rev.*, **111**, 4141–4164.

Tang, W.L., Li, Z., and Zhao, H. (2010) Inverting the enantioselectivity of P450pyr monooxygenase by directed evolution. *Chem. Commun.*, **46**, 5461–5463.

Terao, Y., Ijima, Y., Miyamoto, K., and Ohta, H. (2007) Inversion of enantioselectivity of arylmalonate decarboxylase via site-directed mutation based on the proposed reaction mechanism. *J. Mol. Catal. B*, **45**, 15–20.

Turner, N.J. (2003) Directed evolution of enzymes for applied biocatalysis. *Trends Biotechnol.*, **21**, 474–478.

Uchiyama, T. and Miyazaki, K. (2009) Functional metagenomics for enzyme discovery: challenges to efficient screening. *Curr. Opin. Biotechnol.*, **20**, 616–622.

van Loo, B., Kingma, J., Arand, M., Wubbolts, M.G., and Janssen, D.B. (2006) Diversity and biocatalytic potential of epoxide hydrolases identified by genome analysis. *Appl. Environ. Microbiol.*, **72**, 2905–2917.

Vazquez-Figueroa, E., Yeh, V., Broering, J.M., Chaparro-Riggers, J.F., and Bommarius, A.S. (2008) Thermostable variants constructed via the structure-guided consensus method also show increased stability in salts solutions and homogeneous aqueous–organic media. *Protein Eng. Des. Sel.*, **21**, 673–680.

Venter, J.C., Remington, K., Heidelberg, J.F., Halpern, A.L., Rusch, D., Eisen, J.A., Wu, D., Paulsen, I., Nelson, K.E., Nelson, W., Fouts, D.E., Levy, S., Knap, A.H., Lomas, M. W., Nealson, K., White, O., Peterson, J., Hoffman, J., Parsons, R., Baden-Tillson, H., Pfannkoch, C., Rogers, Y.-H., and Smith, H.O. (2004) Environmental genome shotgun sequencing of the Sargasso Sea. *Science*, **304**, 66–74.

Wada, M., Hsu, C.-C., Franke, D., Mitchell, M., Heine, A., Wilson, I., and Wong, C.-H. (2003) Directed evolution of *N*-acetylneuraminic acid aldolase to catalyze enantiomeric aldol reactions. *Bioorg. Med. Chem.*, **11**, 2091–2098.

Wang, K., Li, G., Yu, S.Q., Zhang, C.T., and Liu, Y.H.A. (2010) A novel metagenome-derived beta-galactosidase: gene cloning, overexpression, purification and characterization. *Appl. Microbiol. Biotechnol.*, **88**, 155–165.

Wang, G.Z., Luo, H.Y., Wang, Y.R., Huang, H.Q., Shi, P.J., Yang, P.L., Meng, K., Bai, Y. G., and Yao, B. (2011a) A novel cold-active xylanase gene from the environmental DNA of goat rumen contents: direct cloning, expression and enzyme characterization. *Bioresour. Technol.*, **102**, 3330–3336.

Wang, L.-J., Li, C.-X., Ni, Y., Zhang, J., Liu, X., and Xu, J.-H. (2011b) Highly efficient synthesis of chiral alcohols with a novel NADH-dependent reductase from *Streptomyces coelicolor. Bioresour. Technol.*, **102**, 7023–7028.

Weinreich, D.M., Delaney, N.F., Depristo, M.A., and Hartl, D.L. (2006) Darwinian evolution can follow only very few mutational paths to fitter proteins. *Science*, **312**, 111–114.

Woo, J.C.G. and Silverman, R.B. (1995) Monoamine oxidase B catalysis in low aqueous medium. Direct evidence for an imine product. *J. Am. Chem. Soc.*, **117**, 1663–1664.

Yamashiro, K., Yokobori, S., Koikeda, S., and Yamagishi, A. (2010) Improvement of *Bacillus circulans* beta-amylase activity attained using the ancestral mutation method. *Protein Eng. Des. Sel.*, **23**, 519–528.

Yu, E.Y., Kwon, M.A., Lee, M., Oh, J.Y., Choi, J.E., Lee, J.Y., Song, B.K., Hahm, D.H., and Song, J.K. (2011) Isolation and characterization of cold-active family VIII esterases from an arctic soil metagenome. *Appl. Microbiol. Biotechnol.*, **90**, 573–581.

Zha, D.X., Wilensek, S., Hermes, M., Jaeger, K.E., and Reetz, M.T. (2001) Complete reversal of enantioselectivity of an enzyme-catalyzed reaction by directed evolution. *Chem. Commun.*, 2664–2665.

Zhao, H., Giver, L., Shao, Z., Affholter, J.A., and Arnold, F.H. (1998) Molecular evolution by staggered extension process (StEP) *in vitro* recombination. *Nat. Biotechnol.*, **16**, 258–261.

Zheng, F., Yang, W.C., Ko, M.C., Liu, J.J., Cho, H., Gao, D.Q., Tong, M., Tai, H.H., Woods, J.H., and Zhan, C.G. (2008) Most efficient cocaine hydrolase designed by virtual screening of transition states. *J. Am. Chem. Soc.*, **130**, 12148–12155.

Zheng, H.B. and Reetz, M.T. (2010) Manipulating the stereoselectivity of limonene epoxide hydrolase by directed evolution based on iterative saturation mutagenesis. *J. Am. Chem. Soc.*, **132**, 15744–15751.

4
Enzymes in Organic Chemistry

Learning Objectives

Why enzymes in organic synthesis?	Alternative to chemical methods: • High regio- and stereoselectivity • Milder reaction conditions • Environmentally more friendly
Which enzyme(s)?	Oxidoreductases (Section 4.2.1) Transferases (Section 4.2.2) Hydrolases (Section 4.2.3) Lyases (Section 4.2.4) Isomerases/racemases (Section 4.2.5)
Which reaction system? (Section 2.9)	Aqueous Aqueous and water-miscible organic solvent Aqueous and water-immiscible organic solvent Pure organic solvent Other solvents (supercritical fluids, ionic liquids)
Which "chemistry"?	(Dynamic) kinetic resolution versus asymmetric synthesis Cofactor-free or cofactor-dependent enzymes

4.1
Introduction

The first applications of enzymes in organic chemistry date back more than a century. As early as 1908, Rosenthaler used a hydroxynitrile lyase-containing extract for the preparation of (*R*)-mandelonitrile from benzaldehyde and hydrogen cyanide (HCN) (Rosenthaler, 1908). Since then, an increasing number of enzymes have been identified, and their use in organic chemistry has steadily increased in parallel. In particular, since the mid-1970s the number of reports of enzyme utilization as well as the number of industrialized enzyme-related processes has increased substantially.

Biocatalysts and Enzyme Technology, Second Edition. Klaus Buchholz, Volker Kasche, and Uwe T. Bornscheuer

Several reasons can be identified for this development, including the following:

- More organic chemists accept the use of biocatalysts.
- Biocatalysis may save additional reaction steps compared to organic synthesis.
- Enzymes are often highly chemo-, regio-, and stereospecific.
- Biocatalysis is usually a safer and "greener" technology.
- A substantially increased demand for optically pure compounds, especially for pharmaceutical applications.
- Easier production of biocatalysts due to recombinant expression systems.
- Modern protein engineering methods allow the straightforward tailor design of desired biocatalysts for a given process.
- Many enzymes are commercially available.

The most important application of enzymes in organic chemistry is in the synthesis of optically active compounds. This is due to the excellent stereospecificity shown by many enzymes, which makes them attractive alternatives to asymmetric organic syntheses or reactions starting from the chiral pool.

Often, the targets for organic synthesis are identical when comparing an enzymatic approach with the use of transition metal catalysis; for instance, a chiral alcohol can be made by stereoselective hydrogenation using a Pd catalyst or by using a ketoreductase in the presence of a cofactor such as NAD(P)H. The decision between biocatalysis or chemocatalysis has to be made case-by-case and major determinants are the optical purity of the product, the efficiency of each process, the possibility to scale it up, and the overall process costs, which must include, for instance, the need to remove transition metal traces from the product and solvent recycling.

In addition, enzymes are used for the synthesis of chemicals lacking a chiral center where mild reaction conditions and cleaner reactions are the key advantages to use biocatalysts. Prominent examples for this are the production of acrylamide at >150 000 tons per annum scale and of nicotinamide (both using a nitrile hydratase), and the synthesis of esters for use in cosmetic applications (see also Table 1.3).

In the following examples, the applications of biocatalysts in organic chemistry are organized based on the division suggested by the Enzyme Commission. Thus, a brief introduction, the reaction principle, and selected examples are provided for each enzyme class. Note that with a few exceptions only those reactions using isolated enzymes are covered.

During the past two decades, the application of enzymes in organic synthesis has emerged as an extremely broad field and a wide variety of types of enzymes and examples of their use in chemoenzymatic syntheses have been described. Within this chapter, it would be impossible to provide sufficient coverage of all developments. Hence, only selected enzymes are covered here, the intention being to provide a basic introduction to the subject. Thus, readers are encouraged to consult the broad range of excellent books and reviews cited in the literature list, as many of these provide a much more in-depth coverage of the uses of enzymes in organic chemistry.

4.1.1 Kinetic Resolution or Asymmetric Synthesis

Enzymatic syntheses of optically active compounds can start from kinetic resolution of racemic mixtures or can be performed via asymmetric synthesis (Figure 4.1). Kinetic resolution will only lead to a maximum yield of 50% unless the unwanted enantiomer is racemized. This can be achieved by using a racemase or by chemical racemization. If kinetic resolution and racemization are performed simultaneously, the process is named dynamic kinetic resolution (DKR) and this is exemplified in Section 4.2.3.1. In a few cases, kinetic mixtures can be resolved to 100% enantiomerically pure product by enantioconvergent reactions using epoxide hydrolases (EHs) (Section 4.2.3.4) or by the formation of homochiral mixtures as shown for alkyl sulfatases (Gadler and Faber, 2007).

In contrast, asymmetric synthesis allows in principle the production of one enantiomer at 100% yield. Examples include the desymmetrization of prostereogenic compounds using, for example, an alcohol dehydrogenase (ADH) in the reduction of a ketone (Section 4.2.1.1) or the formation of a chiral compound by, for example, C—C bond formation using a hydroxynitrile lyase (HNL) (Section 4.2.4.1).

The performance of an enzyme in a kinetic resolution can easily be judged by the *E*-value (enantioselectivity, enantiomeric ratio). This value is the ratio of the reaction rates (V_{max}/K_m) of the enzyme for one enantiomer over the other. Chen and coworkers developed simple equations to calculate the *E*-value from the optical purity (% ee, enantiomeric excess) of the substrate or product and the conversion (Chen *et al.*, 1982, 1987). The theoretical background and equations to calculate this value are outlined in Section 2.7.1. A nonselective enzyme has an *E*-value of 1, preparative useful *E*-values are >50, and desired *E*-values are >100 as only then

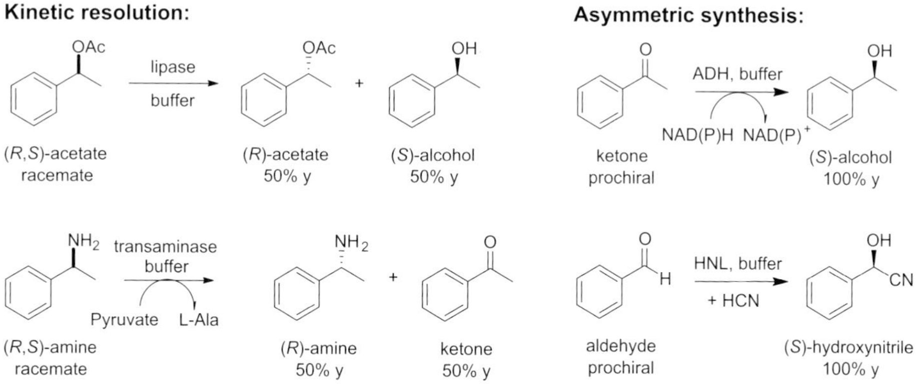

Kinetic resolution:

Asymmetric synthesis:

Figure 4.1 Enzymatic reactions can be performed as kinetic resolutions yielding at maximum 50% product as shown for a lipase-catalyzed or a transaminase-catalyzed kinetic resolution. In contrast, asymmetric synthesis gives in principle 100% of the desired optically pure product exemplified for a reduction using an alcohol dehydrogenase (ADH/ketoreductase) and the cofactor NADH or by a C—C coupling reaction using a hydroxynitrile lyase.

optically pure product and substrate can be isolated at 50% conversion. Note that in asymmetric synthesis, the enantiomeric purity is independent of the conversion.

The (*R*,*S*)-nomenclature is recommended for the assignment of absolute configurations. In the scientific literature on amino acids and sugars, the D,L-nomenclature is still common practice and is therefore used in the following chapters, where the examples for these compounds are treated in greater detail.

4.2
Examples

4.2.1
Oxidoreductases (EC 1)

Although it is estimated that about 25% of all presently known enzymes are oxidoreductases, the most useful enzymes for preparative applications are still dehydrogenases or reductases. Mono- and dioxygenases, oxidases, peroxidases, and enoate reductases also belong to this class. Most of these enzymes require NADH or NADPH as a cofactor and, due to this cofactor dependency, recycling is necessary in order to conduct cost-effective processes, unless the reaction is performed in a whole-cell system. Currently, most oxidoreductases are not used as isolated biocatalysts and are therefore not covered extensively in the following sections.

4.2.1.1 Dehydrogenases (EC 1.1.1.-, EC 1.2.1.-, EC 1.4.1.-)

Synthesis of Alcohols The most important application of oxidoreductases in organic chemistry is in the reduction mode, as this yields chiral compounds such as alcohols, hydroxy acids, or amino acids. In the case of alcohol dehydrogenases, a hydrogen and two electrons are transferred from the reduced nicotinamide moiety to an acceptor molecule such as a ketone or an α-keto acid. In many cases, this reaction is highly stereoselective and the hydride is delivered by the dehydrogenase either from the *re*- or from the *si*-face of the carbonyl, yielding the corresponding (*R*)- or (*S*)-products (Figure 4.2). Many ADHs were found to obey the Prelog

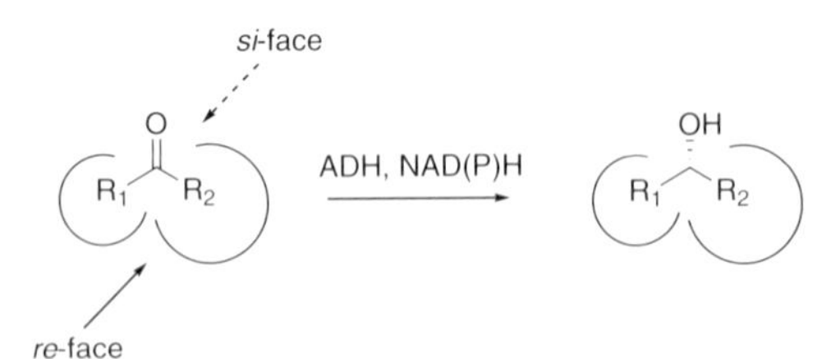

Figure 4.2 According to the Prelog rule, the size of the substituents R_1 and R_2 (here, $R_1 < R_2$) determines whether the carbonyl of a ketone is attacked by the hydride either from the *re*- or from the *si*-face. In the example shown above, the (*S*)-product is formed, if a sequence rule of $R_2 > R_1$ is assumed.

Table 4.1 Alcohol dehydrogenases from various sources and their selectivities.

ADH	Attack from	Configuration
Yeast	*Re*	Prelog
Horse liver	*Re*	Prelog
Thermoanaerobium brockii	*Re*	Prelog
Curvularia falcata	*Re*	Prelog
Pseudomonas sp.	*Si*	Anti-Prelog
Lactobacillus kefir	*Si*	Anti-Prelog
Mucor javanicus	*Si*	Anti-Prelog

rule – which is based on the size of substituents and allows to predict which enantiomer will be produced – although exceptions have also been described (anti-Prelog). An overview of different ADHs and their selectivities is provided in Table 4.1.

In the literature, older examples for the use of ADHs deal with whole-cell systems, such as baker's yeast. However, the recent progress in molecular biology and enzyme discovery (Chapter 3) allowed to identify hundreds of new ADHs in the past decade and hence many new ADHs are now available from commercial suppliers. In addition, these enzymes are now often used as isolated enzymes coupled with efficient cofactor regeneration systems, which do not require a separate biocatalyst. Instead, NAD(P)H is directly recycled by the ADH used to produce the optically pure alcohol in the presence of isopropanol. The key to success was the high stability of the ADH at high concentrations of isopropanol and acetone (Figures 4.3 and 4.4) (Stampfer *et al.*, 2002).

Synthesis of Amino Acids Amino acids (α-amino carboxylic acids) are widely used in nutrition, medical applications, and organic synthesis. They can be produced by one of the following routes:

- Cultivation of microorganisms, especially overproducers of certain amino acids.
- Extraction from protein hydrolysates.
- Enzymatic synthesis from prochiral precursors or via kinetic resolution of racemates.
- Chemical synthesis.

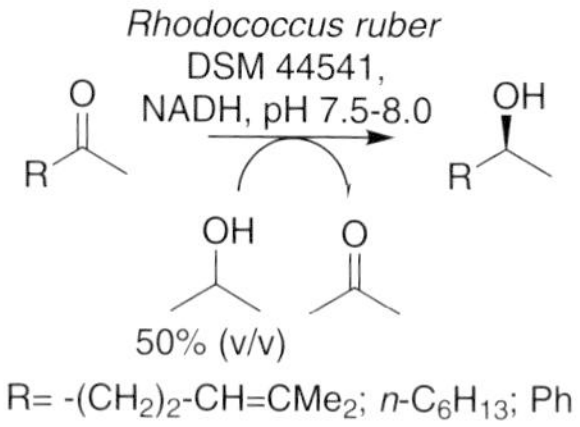

Rhodococcus ruber
DSM 44541,
NADH, pH 7.5-8.0
O
R
OH
R
OH
O
50% (v/v)
R= $-(CH_2)_2-CH{=}CMe_2$; n-C_6H_{13}; Ph

Figure 4.3 Example for the synthesis of optically active alcohols using an alcohol dehydrogenase. The cofactor NADH is regenerated directly from isopropanol.

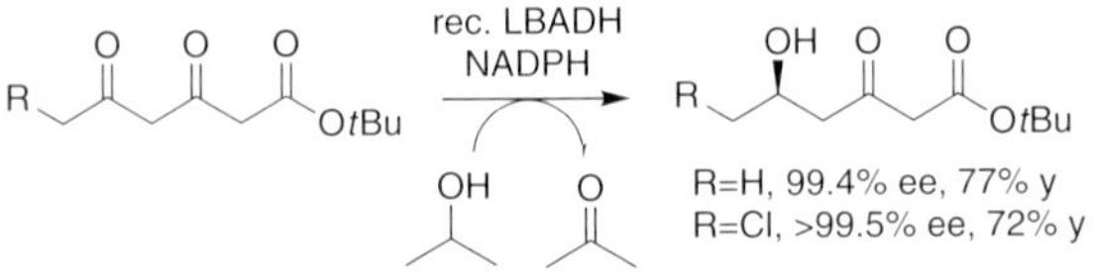

Figure 4.4 Example for the synthesis of optically active 3-oxo-5-hydroxycarboxylic acids using a recombinant alcohol dehydrogenase from *Lactobacillus brevis* (LBADH) exemplifying the regio- and stereoselectivity of this ADH. The cofactor NADPH is regenerated directly from isopropanol.

The first two methods provide access only to natural L-amino acids, whereas the last two approaches allow for the synthesis of nonnatural amino acids and the D-enantiomers. L-Glutamate (>1 500 000 tons) and L-lysine (>1 000 000 tons) are amino acids obtained by cultivation with *Corynebacterium glutamicum* as the main producer for which metabolic engineering (see Chapter 5) has extensively been studied (Wendisch, Bott, and Eikmanns, 2006). *C. glutamicum* could also be engineered to produce L-serine (Peters-Wendisch *et al.*, 2005). Protein hydrolysates are used for the isolation of, for instance, L-leucine, L-asparagine, L-arginine, and L-tyrosine.

The enzymatic synthesis can start from racemic amino acids obtained by chemical synthesis (i.e., the Strecker method) followed by kinetic resolution using esterases (i.e., enantioselective hydrolysis of amino acid carboxylic acid esters; see Section 4.2.3.2), acylases/amidases (i.e., enantioselective hydrolysis of amides; see Section 4.2.3.3), or hydantoinases (see Section 4.2.3.7). Alternatively, prochiral precursors can be subjected to reductive amination using amino acid dehydrogenases (AADHs). With a few exceptions, only L-amino acids are accessible. Although a variety of AADHs have been described in the literature (Table 4.2) (Bommarius, 2002), only a limited number are currently used. A broad variety of sequences of AADHs have been identified, and several 3D crystal structures have been solved. Although sequences vary considerably between AADHs, the residues involved in catalysis and nicotinamide cofactor binding are highly conserved.

The catalytic mechanism of AADHs has been proposed to follow via the formation of an imine. As shown for leucine dehydrogenase (LeuDH) (Figure 4.5), the α-keto acid and ammonia react in the amination reaction with stabilization via hydrogen bonding between the carboxylic group and the oxyanion by two lysine

Table 4.2 Amino acid dehydrogenases from various sources and their cofactor dependencies.

AADH	Coenzyme	Source
Alanine-DH	NADH	Bacteria
Glutamate-DH	NAD(P)H	Bacteria, yeast, fungi, plants
Serine-DH	NADH	Plant
Valine-DH	NAD(P)H	Bacteria, plant
Lysine-DH	NADH	Bacteria, human
Phenylalanine-DH	NADH	Bacteria

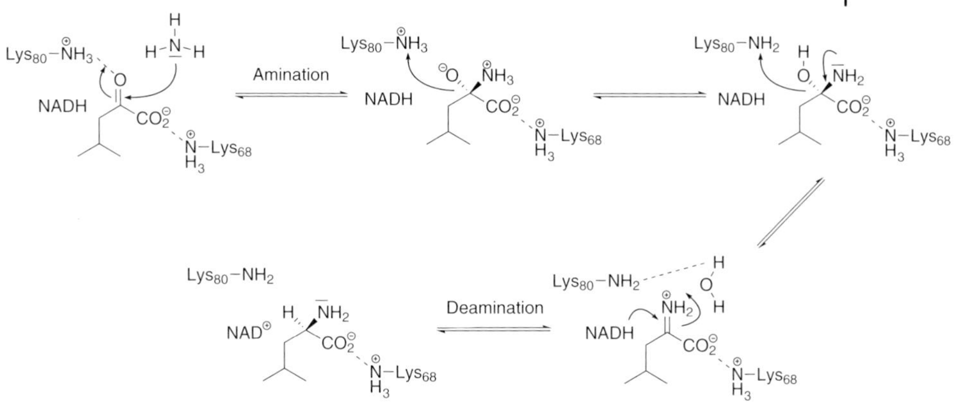

Figure 4.5 Proposed reaction mechanism for LeuDH.

residues (Lys_{68} and Lys_{80}). After formation of the imine, a water molecule hydrogen bonded to Lys_{80} and hydride transfer from NADH yield the α-amino acid and NAD^+ (Ohshima and Soda, 2000).

Leucine and phenylalanine dehydrogenases (LeuDH, PheDH) are the most important AADHs, especially as they are not restricted to their natural substrates, accept a broad range of other precursors, and efficiently allow the synthesis of non-proteinogenic α-amino acids from the corresponding prostereogenic α-keto acids (Figure 4.6).

As pointed out above, cofactor regeneration is required for the cost-effective application of dehydrogenases. An elegant solution for this was developed by Kula and Wandrey, based on the use of formate dehydrogenase (FDH), for example, from the yeast *Candida boidinii* using ammonium formate as cosubstrate (Figure 4.6). Carbon dioxide formed as by-product is highly volatile, and this leads to a favorable shift of the equilibrium. This allows for total turnover numbers (moles of product per mole of cofactor) of up to 600 000, as demonstrated for the continuous synthesis of phenylalanine (Hummel *et al.*, 1987). The overall reaction is best performed in an enzyme membrane reactor, in which the AADH and the FDH are retained in the reactor compartment using ultrafiltration membranes (see also Section 7.4). Covalent coupling of the cofactor NADH to

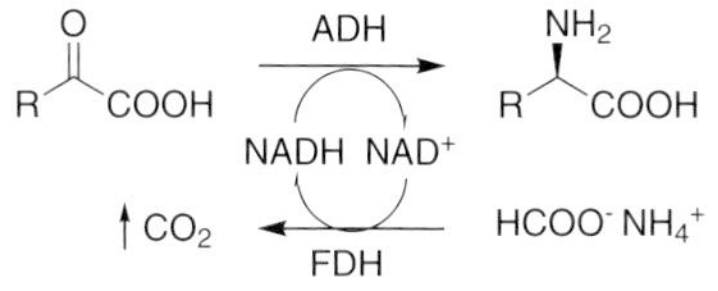

Figure 4.6 Example for the synthesis of optically active L-amino acids using an amino acid dehydrogenase. The cofactor NADH is regenerated using a formate dehydrogenase from *C. boidinii*. The equilibrium is shifted toward NADH by using ammonium formate yielding carbon dioxide as by-product.

Table 4.3 Examples for the synthesis of L-amino acids using the principle shown in Figure 4.6 (adapted from Bommarius, 2002).

ADH[a)]	Product	Product concentration. (mmol l^{-1})	Conversion (%)	STY[b)] (g l^{-1} day^{-1})	Enzyme consumption[c)] (U kg^{-1})
Leucine-DH	L-Leu	80	80	250	300 (300)
Leucine-DH	L-Leu	70	70	72	730 (350)
Leucine-DH	L-Met	240	60	143	n.d.
Leucine-DH	L-^{t}Leu	425	85	640	1000 (2000)
Alanine-DH	L-Ala	184	46	134	4700 (2600)
Phenylalanine-DH	L-Phe	114	95	456	1500 (150)

n.d., not determined.
a) AADH, amino acid dehydrogenase.
b) STY, space–time yield.
c) Values in parentheses refer to the consumption of the cofactor regenerating enzyme.

polyethylene glycol (PEG) can be used to avoid leakage through the membrane. An overview of amino acids obtained by this approach is provided in Table 4.3, while further examples and more details can be found in reviews (Bommarius, 2002; Ohshima and Soda, 2000).

From the examples shown in Table 4.3, L-*tert*-leucine (L-^{t}Leu) is the only nonproteinogenic amino acid. L-^{t}Leu is an important building block for a range of pharmaceuticals, as peptide bonds involving this amino acid are only slowly hydrolyzed by peptidases. (*Note*: According to the Enzyme Commission, proteases should now be named “peptidases,” and this term is used throughout the book.) As biocatalytic production using acylase, amidase, or hydantoinase/carbamoylase routes failed, reductive amination using LeuDH in combination with the FDH for cofactor recycling was the method of choice for large-scale production (Bommarius *et al.*, 1992). Other nonnatural amino acids accessible by this route include L-neopentylglycine, L-β-hydroxyvaline, and 6-hydroxy-L-norleucine.

4.2.1.2 **Oxygenases**

Oxygenases are enzymes that introduce either one (monooxygenases) or two (dioxygenases) oxygen atoms into their substrates. Typically, NADH or NADPH serves as reduction equivalents via electron transfer proteins such as reductases. The major interest in these enzymes for organic synthesis is due to their high regio- and stereoselectivity. Moreover, many of these reactions are difficult to perform by chemical methods, especially if nonactivated hydrocarbon moieties need to be transformed. Despite the fact that numerous oxygenases are known, their application in organic synthesis is still limited due to a number of problems. These include limited availability of sufficient amounts of enzyme, insufficient stability and often very low specific activity, requirement of costly cofactors, and the presence of a reductase. Many enzymes are also membrane bound, which further hampers their application. Some of these problems were overcome by the use of whole-cell systems, preferentially with overexpression of the oxygenase. Dioxygenases have been only used in whole-cell systems, and are not covered here.

P450 Monooxygenases (EC 1.14.13.-) P450 monooxygenases (also named cytochrome P450 enzymes) are widely distributed in nature, and play a key role in various steps of primary and secondary metabolism, as well as in the detoxification of xenobiotic compounds. A range of reactions is catalyzed by these enzymes (Figure 4.7), which all include the transfer of molecular oxygen to nonactivated aliphatic or aromatic X–H bonds (X = –C, –N, –S) (Goldstein and Faletto, 1993). Furthermore, a remarkable number of P450 enzymes are able to epoxidize –C=C– double bonds (Lewis, 1996). For these oxygenation reactions, P450 enzymes require cofactors such as NADPH or NADH as reduction equivalents. P450 enzymes show characteristic spectral properties, notably the maximum absorption at 450 nm in the differential spectrum with carbon monoxide, which gave them their name (Omura and Sato, 1964).

The P450 superfamily is one of the largest and oldest gene families. In 2007, the number of P450 encoding sequences was estimated at over 800 (http://www.icgeb.trieste.it/~p450srv/new/p450.html). The cytochrome P450 engineering database provides access to numerous P450 families, sequences, and structures and enables homology searches (http://www.cyped.uni-stuttgart.de/). The genome project of the plant *Arabidopsis thaliana* has led to the identification of more than 270 putative P450 genes (Nelson, 1999). The classification of P450 genes is based on primary sequence homologies. P450 genes are identified by the abbreviation CYP (cytochrome P), followed by a number denoting the family, a letter designating the subfamily (when two or more exist), and a numeral representing the individual gene within the subfamily. For example, *CYP*4A1 represents the *first* gene in the P450 subfamily *A* of the P450 gene family *4*.

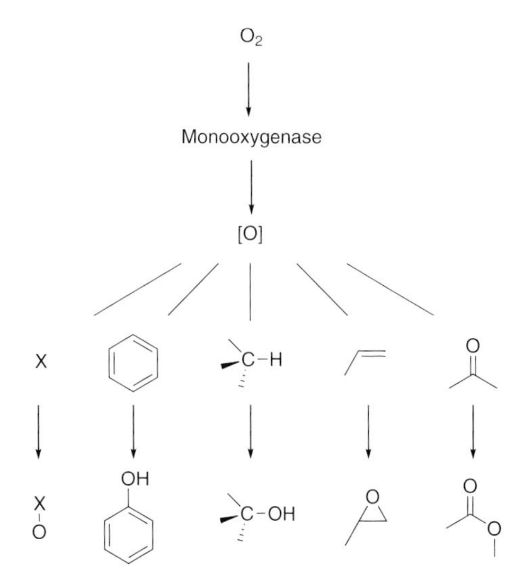

Figure 4.7 Overview of reactions catalyzed by monooxygenases.

Depending on the mechanism for the electron transfer system (reductase system), P450 enzymes are divided into four classes:

- **Class I:** These mainly occur in mitochondrial systems and most bacteria. The electron transfer systems consist of a FAD domain as reductase and a further iron–sulfur protein.
- **Class II:** These are often located in the endoplasmic reticulum and require only a single protein for the electron transfer, a FAD/FMN reductase.
- **Class III:** These do not require reduction equivalents. The P450 enzymes directly convert peroxygenated substrates, which have already "incorporated" the oxygen.
- **Class IV:** Only a few enzymes are known, which receives its electrons directly from NADH.

Postulated Reaction Cycle The postulated reaction cycle of P450 enzymes is shown in Figure 4.8 (see also Figure 4.11). This reaction typifies how enzymatic

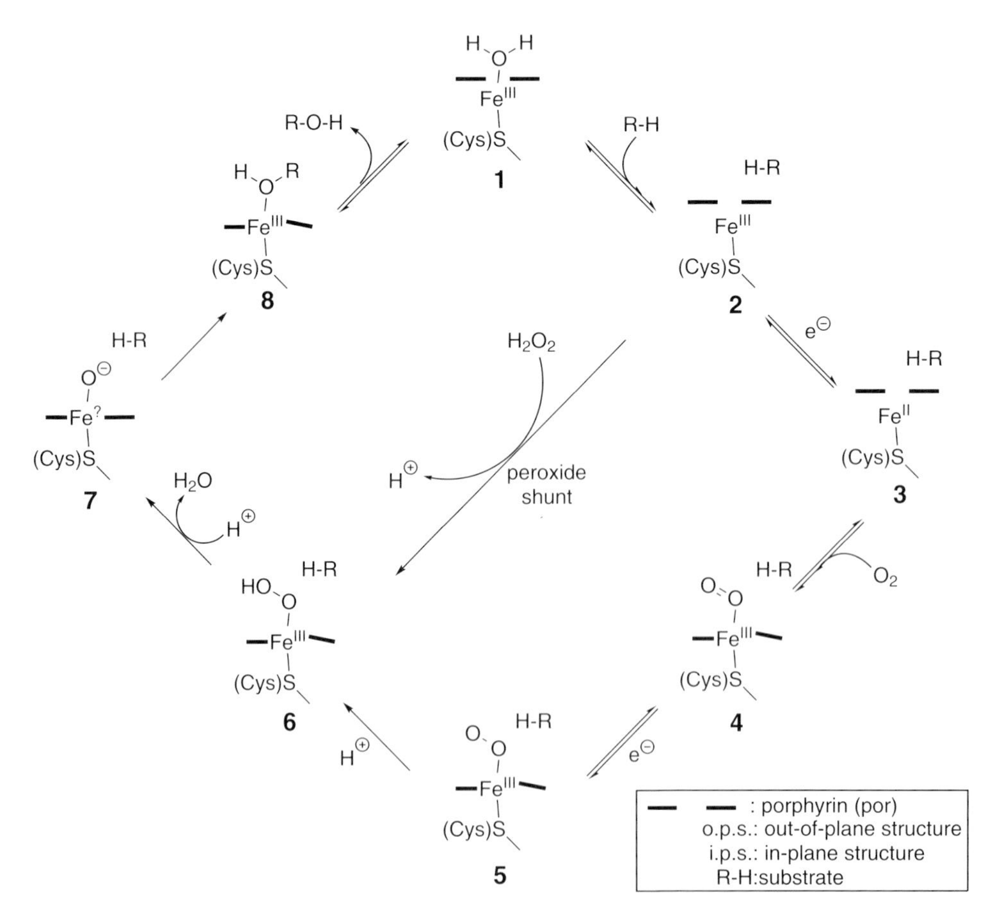

Figure 4.8 Reaction cycle of a hydroxylation catalyzed by a P450 monooxygenase.

oxygenation is carried out under physiological conditions using molecular oxygen as oxidant (Urlacher, Lutz-Wahl, and Schmid, 2004). Activation of oxygen takes place at the iron–protoporphyrin (heme), which is sixfold coordinated. It has a conserved thiolate residue as the fifth ligand and, in the inactive ferric form, a water molecule as its sixth ligand (stage 1 in Figure 4.8). First, the substrate is bound and the water molecule is displaced (stage 2), after which the ferric enzyme is reduced to a ferrous state by one-electron transfer (stage 3). Molecular oxygen is bound, resulting in a ferrous dioxy species (stage 4). A second reduction followed by a proton transfer leads to an iron–hydroperoxo intermediate (stage 5), which upon cleavage of the O—O bond releases water and an activated iron–oxo ferryl species (stage 6). This iron–oxo ferryl oxidizes the substrate, and the product is subsequently released. During this catalytic cycle, several highly reactive oxygen species are formed, which are believed to cause the rather rapid inactivation of P450 enzymes. Those P450s, which are able to perform hydroxylation using the "peroxide shunt" pathway, utilizing hydrogen peroxide as electron and oxygen donor, are very useful in biocatalysis.

Several enzyme structures of CYP have been resolved that, despite their low sequence homology, share a close structural similarity. As the activated intermediates are not covalently bound, fewer structures in complex with their natural substrates (for instance, for CYP101, $P450_{cam}$, and CYP102) have been determined.

Further basic information on P450s can be found in a number of books and reviews (Cirino and Arnold, 2002; Schmid and Urlacher, 2007; Urlacher and Eiben, 2006).

Synthesis of Hydroxylated Carboxylic Acids Hydroxylated carboxylic acids have various (potential) applications as polymer building blocks or as intermediates in antibiotic synthesis. In lactonized form, they can serve as perfume ingredients.

A considerable number of P450s are known to catalyze the hydroxylation of fatty acids, though in most cases only medium- to long-chain fatty acids (C_{12}–C_{18}) are accepted as substrates. In addition, hydroxylation occurs at several positions, namely, terminal (ω-position) and subterminal (ω–1- and ω–2-positions) to yield a product mixture (Schwaneberg and Bornscheuer, 2000). The substrate specificity of a P450 enzyme from *Bacillus megaterium* could be efficiently altered by means of rational protein design, and especially by directed evolution (see also Section 3.2.2). The P450 produced by the strain *B. megaterium* (P450 BM3) is able to catalyze the hydroxylation of fatty acids with the highest turnover numbers yet reported for P450 monooxygenases (in the range of $>1000\,\mathrm{Eq\,min^{-1}}$). The encoding gene was cloned and functionally expressed in *Escherichia coli* (Narhi and Fulco, 1986; Ruettinger, Wen, and Fulco, 1989). P450 BM3 (CYP102) is especially suitable for biocatalysis, as it is a water-soluble natural fusion protein containing the P450 and reductase part on one polypeptide chain. In addition, its crystal structure has been determined. The replacement of Arg_{47} with Glu resulted in an ability of this P450 BM3 mutant to hydroxylate *N*-alkyltrimethylammonium compounds, which was explained by an inversion of the substrate binding conditions (Oliver *et al.*, 1997). P450 BM3, heterologously expressed in *E. coli*, has been used *in vivo* to produce

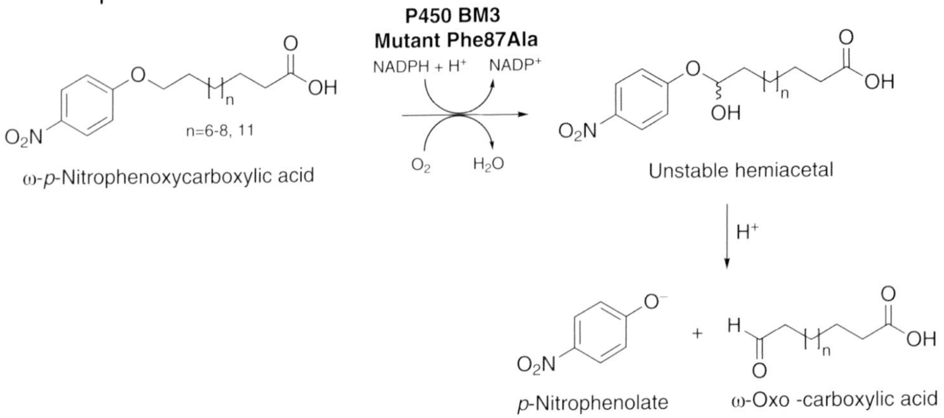

Figure 4.9 Principle of the colorimetric pNCA assay allowing the determination of the fatty acid hydroxylating activity of P450 BM3 mutant F87A.

mixtures of chiral 12-, 13-, and 14-hydroxypentadecanoic acids on a preparative scale at high optical purity (Schneider *et al.*, 1998). Furthermore, P450 BM3 and its mutant Phe87Ala can be expressed in gram scale and efficiently purified in a single step for further enzyme-based biotransformation reactions on a preparative scale (Schwaneberg *et al.*, 1999b).

The key to the successful engineering of P450 BM3 was the development of an elegant chromophoric assay (Figure 4.9), which allows determination of the P450 BM3 wild-type and mutant activity without background reaction (Schwaneberg *et al.*, 1999a). As shown in Figure 4.9, hydroxylation of the terminal position of the substrate yields an unstable hemiacetal, which dissociates spontaneously into the ω-oxo-carboxylic acid and the chromophore *p*-nitrophenolate. The latter can be easily quantified at 410 nm using a spectrophotometer. This pNCA assay created the basis for the directed evolution of P450 BM3 as it allows screening of variants in a high-throughput format.

The development of P450 BM3 mutants with improved hydroxylation activity began with the identification of eight mutation sites from the X-ray crystallographic structure of P450 BM3, which were then subjected to random mutagenesis. After screening with the pNCA assay, the best variants were combined, yielding a biocatalyst with five mutations, which efficiently hydroxylated C_8-pNCA while maintaining its activity for C_{10} and C_{12} fatty acids (Li *et al.*, 2001). Arnold and coworkers discovered that P450 BM3 also hydroxylates short-chain alkanes such as octane, yielding a mixture of 4-octanol, 3-octanol, 2-octanol, 4-octanone, and 3-octanone. Also, the hydroxylation of propane to yield 2-propanol was reported (Glieder, Farinas, and Arnold, 2002). This can be regarded as a major breakthrough, as upon further improvement commodity chemicals might become available by biotransformation routes. Later, other variants were created by directed evolution, which also exhibited enantioselectivity resulting in the formation of (*S*)-2-octanol (40% ee) or (*R*)-2-hexanol (40–55% ee) (Peters *et al.*, 2003).

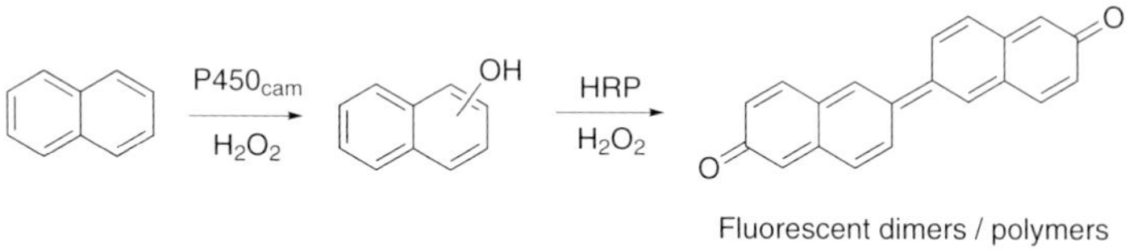

Figure 4.10 Principle of the assay used to identify $P450_{cam}$ variants hydroxylating naphthalene. The resulting naphthols are oxidatively coupled with horseradish peroxidase (HRP) to fluorescent polymers. All reactions occur intracellularly in the recombinant *E. coli* cell.

Directed evolution was also applied to alter the substrate specificity of $P450_{cam}$ from *Pseudomonas putida* to convert naphthalene more efficiently into hydroxylation products. Mutants were identified by coexpressing them with horseradish peroxidase (HRP), which converts the products of the P450 reaction into fluorescent compounds amenable to digital imaging screening (Joo, Lin, and Arnold, 1999) (Figure 4.10).

The examples mentioned above are summarized in Figure 4.11.

Baeyer–Villiger Monooxygenases (EC 1.14.14.16, EC 1.14.13.22) Baeyer–Villiger monooxygenases (BVMOs) catalyze the biocatalytic counterpart of the chemical Baeyer–Villiger oxidation (Walsh and Chen, 1988), which was first described by Adolf Baeyer and Victor Villiger more than a century ago (Baeyer and Villiger, 1899). The chemical oxidation usually uses peracids. The mechanism is generally accepted to proceed by a two-step process, which was initially proposed by Criegee (1948). It was also shown that certain BMVOs catalyze the synthesis of chiral sulfoxides (Secundo *et al.*, 1993).

The first examples of enzymatic Baeyer–Villiger oxidation (BVO) date back to 1948 for a fermentation of cholestanone with *Proactinomyces erythropolis*

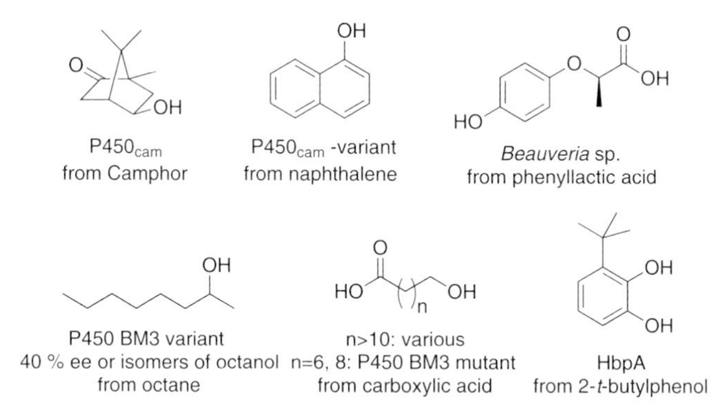

Figure 4.11 Selected examples of products obtained from monooxygenase-catalyzed hydroxylations. Note that most examples represent the use of engineered P450s.

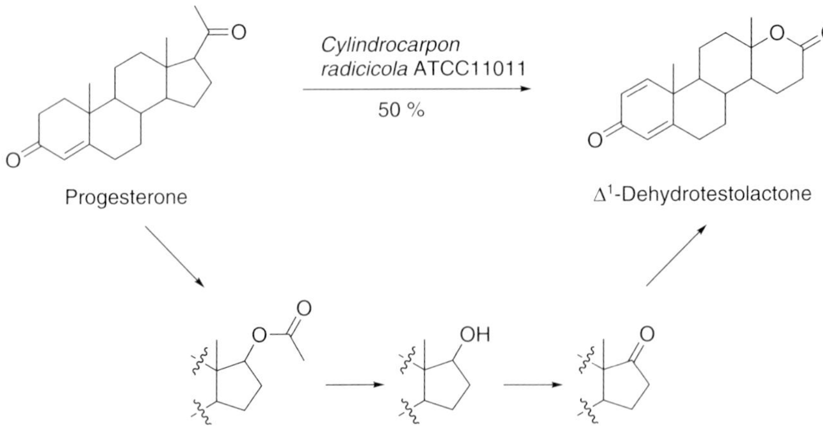

Figure 4.12 Early example of an enzyme-catalyzed Baeyer–Villiger oxidation.

(Turfitt, 1948). A few years later, a double BVO was suggested for the side-chain degradation of progesterone (Figure 4.12) (Fried, Thoma, and Klingsberg, 1953).

During the early 1980s, a range of publications about biocatalysis with newly discovered BVMOs appeared. One enzyme originates from *P. putida* (Grogan *et al.*, 1993; Taylor and Trudgill, 1986), and the other is produced by the S2 organism *Acinetobacter calcoaceticus* (Gagnon *et al.*, 1994). For both enzymes, it could be demonstrated that they accept a relatively broad range of mono- and bicyclic ketones. In the case of racemic substrates, the oxidation often proceeded with good to high enantioselectivity. During the 1990s, the BVMO from *Acinetobacter* sp. was cloned and functionally expressed in yeast (Kayser, Chen, and Stewart, 1999; Stewart, Reed, and Kayser, 1996) and later also in *E. coli* (Mihovilovic *et al.*, 2001). Moreover, it could be shown that a whole-cell biotransformation is possible, although side reactions – such as a reduction of the ketone substrate to an alcohol by an endogenous alcohol dehydrogenase present in the whole-cell system – can occur. While only few BMVOs were available for a long time, the number of enzymes increased immensely by the identification of a conserved sequence motif and the following isolation of genes from sequenced genomes. The phenylacetone monooxygenase (PAMO) from *Thermobifida fusca*, active on aromatic substrates, could be isolated this way (Fraaije *et al.*, 2005). The group of Grogan cloned 29 new BVMOs by the structure-based approach from the genomes of the actinomycetes *Mycobacterium tuberculosis* (Bonsor *et al.*, 2006) and *Rhodococcus jostii* (Szolkowy *et al.*, 2009). The discovery of the first BVMO structure of the thermostable PAMO (Malito *et al.*, 2004) was a major breakthrough to enable rational protein engineering. Five years later, a second BVMO structure of an enzyme from a *Rhodococcus* strain was resolved, the structure of a novel CHMO in a complex with NADP (Mirza *et al.*, 2009). Based on the PAMO structure, the enzyme could be evolved in several studies yielding a variant that now converts 2-phenylcyclohexanone enantioselectively and to accept substrates commonly converted by CHMO-type BVMOs (Reetz and Wu, 2008, 2009).

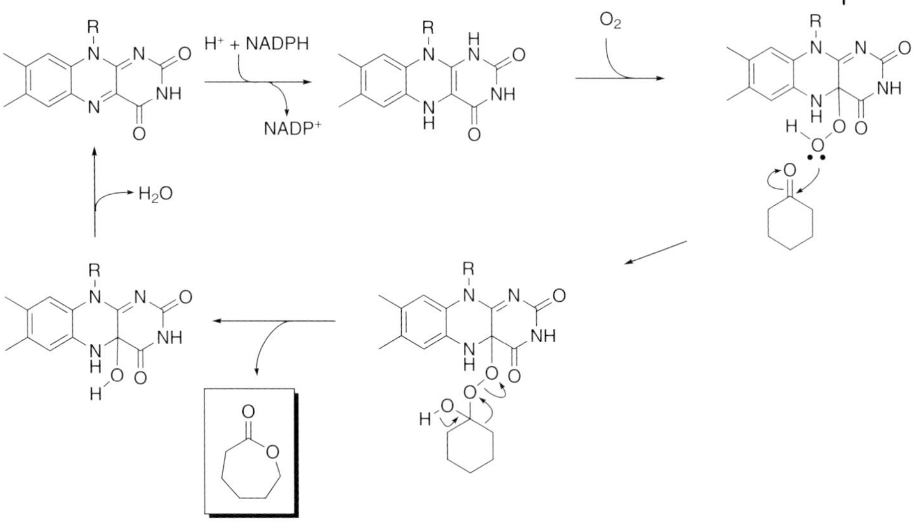

Figure 4.13 Proposed mechanism of an enzyme-catalyzed Baeyer–Villiger oxidation.

Mechanism During the enzymatic Baeyer–Villiger oxidation, one oxygen atom is introduced into a ketone precursor molecule, yielding either the corresponding lactone or an ester (depending on whether a cyclic or a noncyclic ketone is used). The second oxygen atom is converted into water.

For their catalytic action, BVMOs require a cofactor (NADH or NADPH) and they are usually flavin dependent (FMN or FAD). Figure 4.13 shows the mechanism for the oxidation of cyclohexanone to ε-caprolactone by BVMO from *A. calcoaceticus* NCIMB 9871 (Donoghue, Norris, and Trudgill, 1976).

Recently, a fusion protein between a BVMO covalently linked to a soluble $NADP^+$-dependent phosphite dehydrogenase was created (Torres Pazmino *et al.*, 2008), enabling the usage of phosphite as cheap electron donor for efficient cofactor recycling. Interestingly, kinetic parameters, substrate specificity, and stereoselectivity were not negatively influenced as the fusion enzymes displayed a very similar biocatalytic behavior as compared to wild-type enzymes.

By far the best studied BVMO is the enzyme produced by *A. calcoaceticus* NCIMB 9871, which is also available as recombinant enzyme as mentioned above. Examples for the application of this biocatalyst are the conversion of mono- and bicyclic ketones (Figure 4.14). A quite similar substrate spectrum was reported for a BVMO from camphor-induced *P. putida*, which produces three different Baeyer–Villiger monooxygenases: two almost identical enzymes require NADH (each type is generated by induction with either (−)- or (+)-camphor), and the third BVMO utilizes NADPH. All three enzymes have recently been cloned and expressed in *E. coli* (Kadow *et al.*, 2011; Kadow, *et al.*, 2012).

Also, acetophenone-converting BVMOs were described originating from, for example, *Arthrobacter* (Cripps, 1975), *Nocardia* (Cripps, Trudgill, and Whateley,

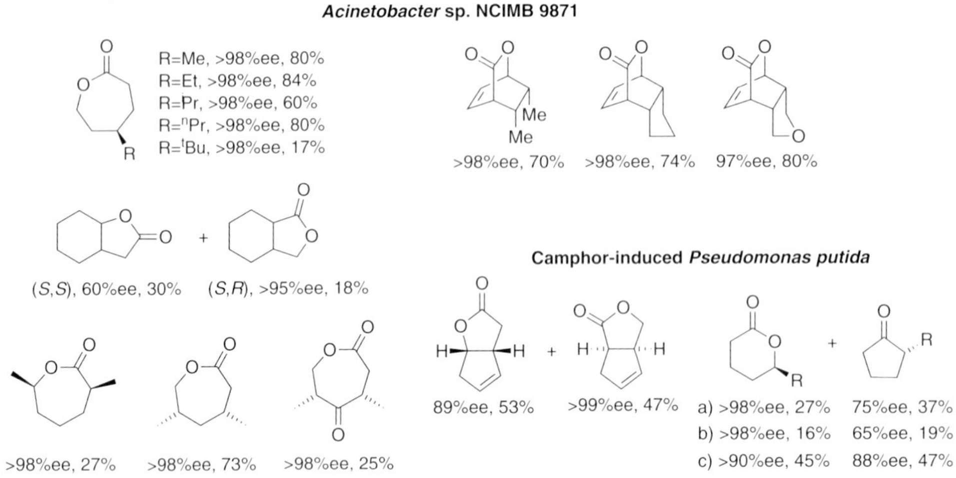

Acinetobacter sp. NCIMB 9871

R=Me, >98%ee, 80%
R=Et, >98%ee, 84%
R=Pr, >98%ee, 60%
R=^{n}Pr, >98%ee, 80%
R=^{t}Bu, >98%ee, 17%

>98%ee, 70% >98%ee, 74% 97%ee, 80%

(*S*,*S*), 60%ee, 30% (*S*,*R*), >95%ee, 18%

>98%ee, 27% >98%ee, 73% >98%ee, 25%

Camphor-induced *Pseudomonas putida*

89%ee, 53% >99%ee, 47%

a) >98%ee, 27% 75%ee, 37%
b) >98%ee, 16% 65%ee, 19%
c) >90%ee, 45% 88%ee, 47%

Figure 4.14 Selected examples of BVMO-catalyzed conversion of mono- and bicyclic ketones using BVMO from *Acinetobacter* sp. or camphor-induced *P. putida*. (a) R = CH_2CO_2Et; (b) R = CH_2OAc; (c) R = $CH_2CH_2O(CH_2)_2OMe$.

1978), and *Pseudomonas* species (Kamerbeek *et al.*, 2001; Tanner and Hopper, 2000). These enzymes accept a broad range of ring-substituted acetophenones, which were converted into the corresponding achiral acetate esters.

Conversion of aliphatic straight-chain ketones was described for monooxygenases from *Mycobacterium* sp. (Hartmans and deBont, 1986), *Pseudomonas* (Forney, Markovetz, and Kallio, 1967), and *Nocardia* sp. (Britton, Brand, and Markovetz, 1974). Also the formation of optically pure products and remaining substrates could be shown for the BVMO-catalyzed kinetic resolution of β-hydroxyketones (Kirschner and Bornscheuer, 2006) (Figure 4.15). Interestingly, not only N-protected β-aminoketones were converted to the expected amino alcohols, but also the formation of β-amino acids was observed (Rehdorf, Mihovilovic, and Bornscheuer, 2010). Thus, the oxygen is inserted by certain BVMOs at either side of the keto functionality allowing also regioselective oxidations.

An extensive coverage of the state-of-the-art research regarding BVMOs can be found in recent reviews (Gonzalo, Mihovilovic, and Fraaije, 2010; Leisch, Morley, and Lau, 2011; Rehdorf and Bornscheuer, 2009, 2010; Balke *et al.*, 2012).

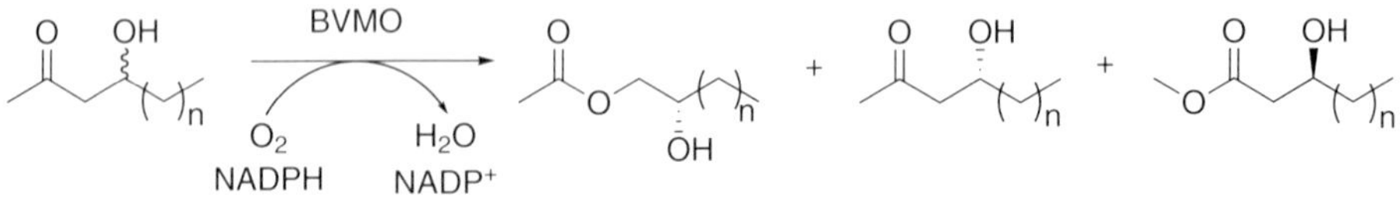

BVMO
O_2 NADPH
H_2O $NADP^+$

Figure 4.15 A BVMO from *P. fluorescens* converts β-hydroxyketones into the monoacetate of a 1,2-diol as major product and the β-hydroxyacid methyl ester as minor by-product.

4.2.1.3 **Peroxidases (EC 1.11.1.10)**

Heme iron-containing peroxidases are capable of oxidizing organic substrates using a peroxide, usually hydrogen peroxide, as oxidant. The best studied enzyme for organic synthesis is a chloroperoxidase from *Caldariomyces fumago* that has been shown to catalyze stereoselective oxidation of aromatic compounds, yielding the corresponding alcohols, and the epoxidized alkenes (Figure 4.16). Also, (*Z*)-3-heptene was converted into its isomeric alcohols (Zaks and Dodds, 1995).

4.2.1.4 **Enoate Reductases (EC 1.4.1.31)**

Enoate reductases (EREDs, also called ene reductases) catalyze the asymmetric reduction of activated C=C bonds (Stürmer *et al.*, 2007; Toogood, Gardiner, and Scrutton, 2010). The activating group is usually an electron-withdrawing group (EWG), which includes aldehyde, ketone, carboxylic acid, carboxylic ester, anhydride, lactone, imide, or nitro groups (Figure 4.17). The importance of this biocatalytic asymmetric reduction is that it can produce two adjacent sp^3 chiral centers from a prostereogenic activated alkene in a single step. Furthermore, the reduction occurs by the net addition of hydrogen (H_2) in a *trans*-stereospecific manner, which provides a potential synthetic route for enzymatic asymmetric *trans*-hydrogenation of activated C=C bond (Figure 4.17).

The biocatalytic asymmetric reduction is catalyzed by a number of flavin-dependent enoate reductases from different microbial sources and also from the so-called old yellow enzyme (OYE) family. Enzymes from the OYE family of flavin mononucleotide (FMN)-containing oxidoreductases were long known for their stereoselective C=C bond reduction. Their catalytic mechanism proceeds with (i) reduction of the noncovalently bound FMN cofactor at N_5 by initial transfer of the *pro*-(*R*)-hydride of NAD(P)H, known as reductive half-reaction followed by (ii) delivery of the hydride from the FMN_{red} at N_5 to the β-carbon of the activated alkene, known as the oxidative half-reaction. FMN_{red} delivers the hydride to the substrate from the side facing toward it, while there is a concomitant proton transfer at the α-carbon of the activated alkene from the tyrosine of the enzyme, which completes the stereospecific net *trans* H_2 addition (Figure 4.17).

The synthetic application of this reaction has steered to explore a number of enoate reductases from various yeasts and microbial sources, with the first enzyme

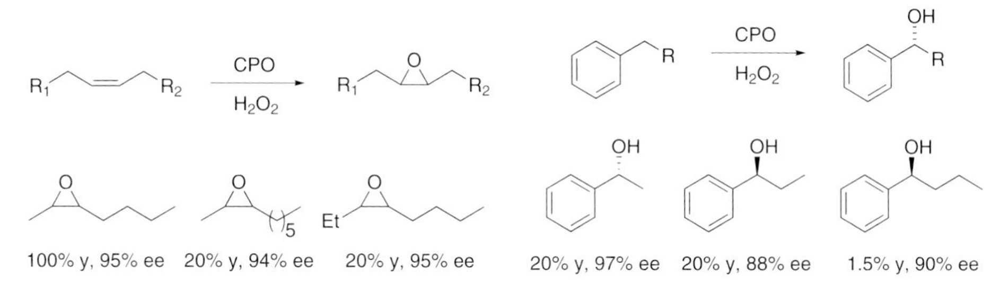

Figure 4.16 Examples of epoxidations and benzylic hydroxylations catalyzed by chloroperoxidase (CPO) from *C. fumago*.

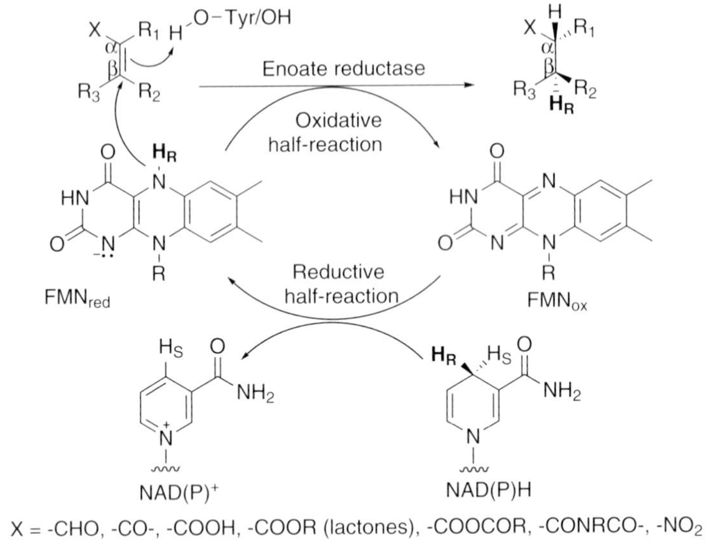

Figure 4.17 Catalytic mechanism of stereoselective *trans*-hydrogenation of C=C bond by flavin mononucleotide (FMN) containing old yellow enzyme. The H_R of NAD(P)H is transferred to the β-carbon via FMN_{red} while a proton from the tyrosine or water is transferred to the α-carbon of the activated alkene, giving a net *trans*-addition product.

purified from *Saccharomyces pastorianus* (OYE1). Toogood, Gardiner, and, Scrutton (2010) and Stürmer *et al.* (2007) described various other sources of enoate reductase in their recent reviews. Many of the enoate reductases were characterized and cloned to express the heterologous proteins, which were subsequently used for biocatalysis. The substrate profile of the various OYE-like family members include the C=C bond reduction of substituted and nonsubstituted α,β-unsaturated cyclic and acyclic aldehydes, ketones, imides, nitroalkenes, carboxylic acids, and esters, maleimides, cyclic and acyclic enones, terpenoids, cyanides, nitrate esters, and also the nitro group reduction of nitroaromatics (TNT), glycerol trinitrate (GTN), and pentaerythritol tetranitrate (PETN). Several examples are shown in Figure 4.18.

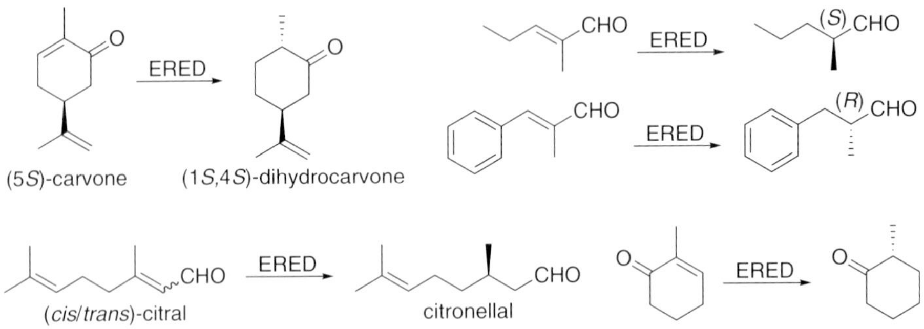

Figure 4.18 Examples for enoate reductase-catalyzed reductions to yield optically pure products.

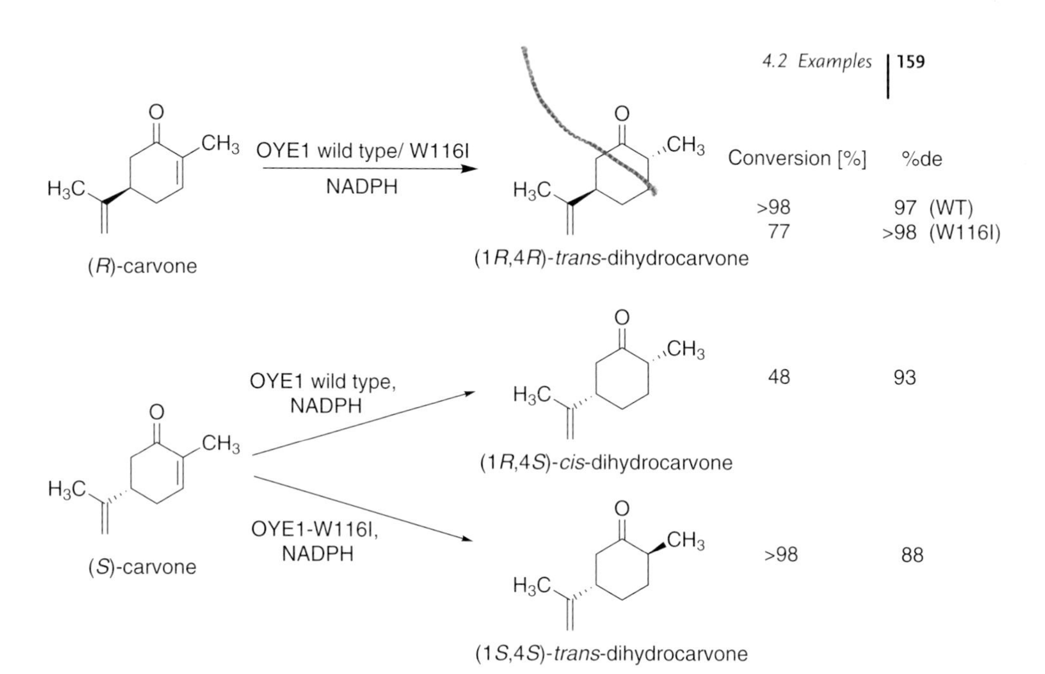

Figure 4.19 OYE1 wild type and W116I form products with opposite stereochemistry from (*S*)-carvone. With (*R*)-carvone, both enzymes produced the same diastereomer.

Despite broad substrate selectivity of OYE1 (*S. pastorianus*), it lacked higher activity toward the reduction of bulky 3-alkyl-substituted 2-cyclohexenones. Aiming to take over the issue, Padhi, Bougioukou, and Stewart (2009) carried out site saturation mutagenesis at position 116 of the OYE1 and discovered two new variants W116I and W116F, which showed opposite stereopreference for certain substrates. The OYE1-W116I reduced (*R*)- and (*S*)-carvone to enantiomeric products compared to their diastereomers made by wild type (Figure 4.19). The inverted stereoselectivity was due to the substrate binding in a flipped orientation in the active site followed by net *trans* H_2 addition. OYE1-W116I also showed opposite stereopreference for the C=C reduction of (*R*)-perillaldehyde to produce *trans*-dihydroperillaldehyde compared to the *cis*-form made by wild-type OYE1. However, another variant OYE1-W116F reduced neral to (*R*)-citronellal compared to (*S*)-citronellal formed by the wild type.

In the first report on directed evolution of the enoate reductase YqjM from *Bacillus subtilis*, a homologue of the old yellow enzyme (Hulley *et al.*, 2010), a library of variants was generated, which showed increased activity and enhanced and inverted enantioselectivity in the reduction of a number of cyclic enones (Bougioukou *et al.*, 2009).

4.2.1.5 **Monoamine Oxidases**

Monoamine oxidases (EC 1.4.3.4) catalyze the oxidative deamination of monoamines leading to the formation of the corresponding aldehydes. Mechanistic

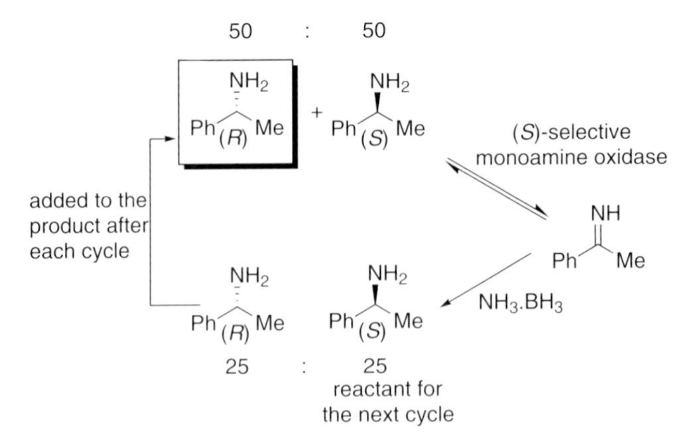

Figure 4.20 Deracemization of α-MBA using a (*S*)-selective monoamine oxidase coupled with ammonia–borane reduction of the intermediate imine to form racemic amine. In the first round, 50% (*S*)-amine is oxidized to the imine yielding after chemical reduction 25% (*R*)-amine and 25% (*S*)-amine, the latter being oxidized again. After seven to eight rounds, the racemic starting material is completely converted into optically pure (*R*)-α-MBA.

investigation of this reaction, however, revealed that the reaction proceeds through the formation of the imine, which could also be isolated (Woo and Silverman, 1995). Turner *et al.* exploited this enzymatic transformation in the deracemization of racemic amines to synthesize the corresponding enantiopure products. The deracemization process involves stereoinversion of one enantiomer to the other by repeated cycles of monoamine oxidase-catalyzed stereoselective oxidation to the imine combined with a nonselective chemical reduction of the imine to the amine that notably works in a one-pot reaction (Figure 4.20).

Using directed evolution, Alexeeva *et al.* improved the selectivity of the monoamine oxidase from *Aspergillus niger* (MAO-N) toward the oxidation of α-methylbenzylamine (α-MBA). The MAO-N mutant Asn336Ser showed 47-fold higher activity toward (*S*)-α-MBA and 5.8-fold higher selectivity toward (*S*)-α-MBA over (*R*)-α-MBA. Combining this engineered protein with ammonia–borane as a reducing agent, (*R*)-α-MBA could be synthesized in 77% yield and 93% ee from its racemic substrate (Alexeeva *et al.*, 2002). The mutant also showed broad substrate selectivity and high enantioselectivity toward a range of primary amines (Carr *et al.*, 2003). Directed evolution of MAO-N produced another variant (Ile246Met/Arg259Lys/Arg260Lys/Asn336Ser/Met348Lys), which catalyzed the deracemization of a structurally diverse range of secondary amines (Carr *et al.*, 2005). Similarly, the MAO-N-5 variant Ile246Met/Asn336Ser/Met348Lys/Thr384Asn/Asp385Ser catalyzed (*S*)-selective oxidation of tertiary amines and hence could be used in the deracemization process for the synthesis of enantiopure cyclic tertiary amines (Dunsmore *et al.*, 2006). An engineered monoamine oxidase (MAO-N-5) was used in the chemoenzymatic synthesis of the alkaloid (+)-crispine and its derivatives (Figure 4.21) (Bailey *et al.*, 2007).

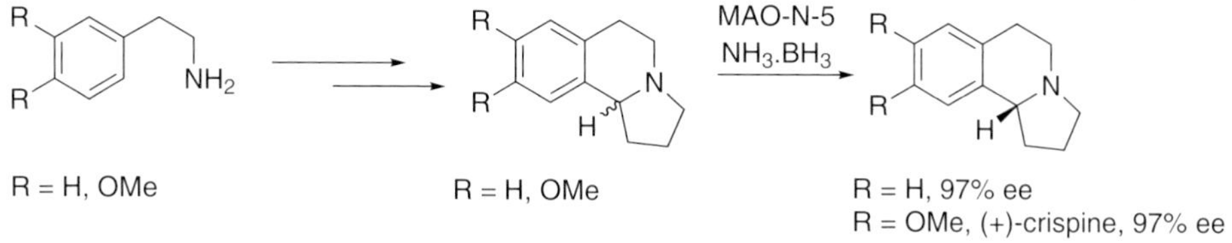

R = H, OMe

R = H, OMe

MAO-N-5
$NH_3.BH_3$

R = H, 97% ee
R = OMe, (+)-crispine, 97% ee

Figure 4.21 Chemoenzymatic synthesis of (+)-crispine and its framework using a deracemization catalyzed by a mutant of MAO-N.

4.2.2
Transaminases

Transaminases (TAs, EC 2.6.1.x) belong to the enzyme class of transferases (see also glycosyltransferases, Chapter 8) and catalyze the transfer of an amino group from an amino donor (amine or α-amino acid) to an amino acceptor (ketone or α-keto acid) utilizing the cofactor pyridoxal-5′-phosphate (PLP) (Figure 4.22). They are usually named α-transaminase if α-amino acids are formed and ω-transaminase if amines are the product; alternatively, the latter are referred to as amine transaminase (ATA). The transamination can be performed either as kinetic resolution of a racemic amine (or α-amino acid) (Figure 4.23a) or as an asymmetric synthesis starting from a prostereogenic ketone (or α-keto acid) (Figure 4.23b) (Höhne and Bornscheuer, 2009). By means of a combination of both strategies, deracemization is also possible: the ketone generated in a kinetic resolution can subsequently be used as substrate in an asymmetric synthesis employing a TA of opposite enantiopreference (Koszelewski *et al.*, 2009).

Whereas α-transaminases require the presence of a carboxylic acid group in the α-position to the keto or amine functionality and hence only allow the formation of α-amino acids, amine transaminases are much more versatile as they accept a very broad range of ketones or amines. The first groundbreaking work in this field was done in the late 1980s by the US-based company, Celgene (Matcham and Bowen, 1996). In the last decade, ω-transaminases have been extensively studied and they have been identified in a mere dozen organisms, were biochemically characterized, and also overexpressed in microbial hosts such as *E. coli* with the enzyme from *Vibrio fluvialis* as probably the most intensively studied ω-transaminase.

A major limitation in the asymmetric synthesis starting from prostereogenic ketones is the unfavorable equilibrium. Shin and Kim reported that α-methylbenzylamine is formed from acetophenone in only 0.5% yield, even if a 10-fold excess of alanine was used as amine donor (Shin and Kim, 1999). Hence, powerful methods were established to shift the reaction equilibrium (Höhne *et al.*, 2008; Koszelewski *et al.*, 2008a, 2008b; Matcham *et al.*, 1999). The most often used system is based on lactate dehydrogenase to generate lactic acid from the pyruvate. This reaction requires the cofactor NADH and hence a cofactor recycling system or whole cells need to be used. An efficient equilibrium shift was found by combination with a glucose dehydrogenase for cofactor recycling (Koszelewski *et al.*, 2008b).

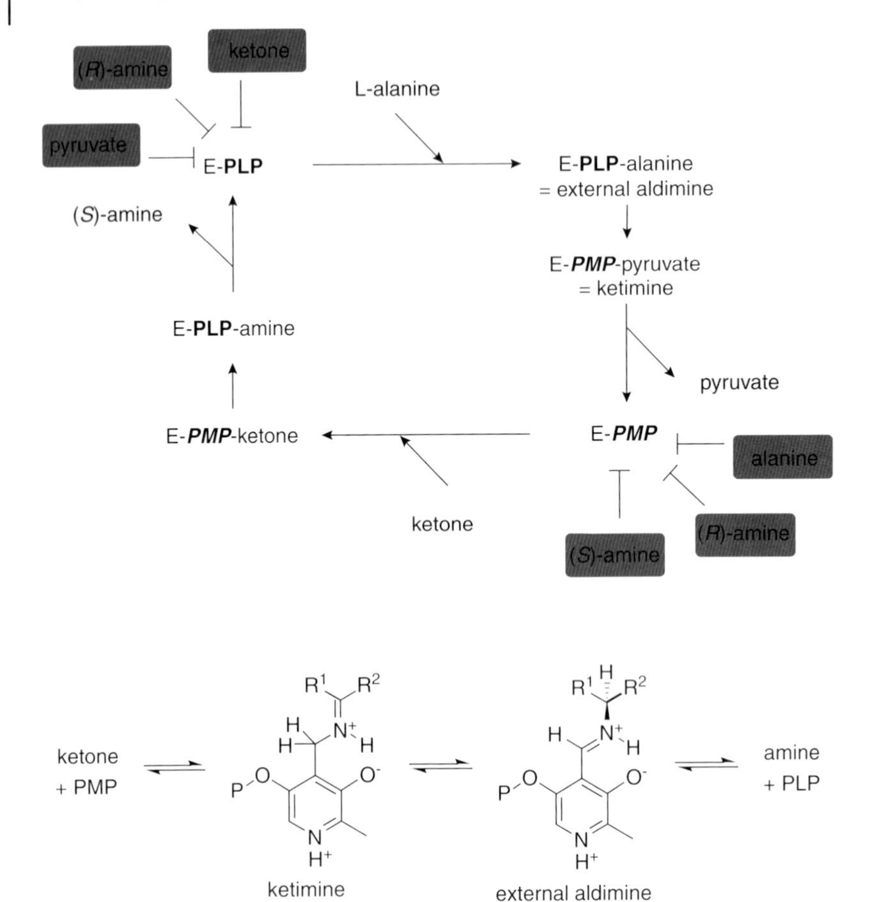

Figure 4.22 Reaction cycle for asymmetric synthesis of amines with ω-TA. Although all reactions are fully reversible, only simple reaction arrows are shown to indicate the direction of the desired asymmetric synthesis and the chronological order in which the substrates and products (shaded in light gray) have to be bound or released from the enzyme. During the reaction cycle, two forms of the free enzyme exist (E-PLP and E-PMP). Inhibition may be caused by binding of substrates (shaded in dark gray) to the "wrong" free enzyme forming abortive complexes, which results in inhibition of the enzyme.

A range of ketones (50 mM) were efficiently converted to the respective amines at high to quantitative conversions and with excellent enantiomeric purities (>98% ee). A very elegant way is the recycling of pyruvate with ammonia, NADH, and AADH to alanine (Koszelewski *et al.*, 2008a). The formed NAD^+ can be regenerated with the well-established formate dehydrogenase cofactor recycling system. In the overall reaction, a ketone is converted with ammonium formate yielding optically pure amine, water, and carbon dioxide, which essentially resembles an asymmetric reductive amination catalyzed by an amine dehydrogenase. In most cases, high yields of 90–99% were obtained; only α-MBA gave unsatisfactory 6%

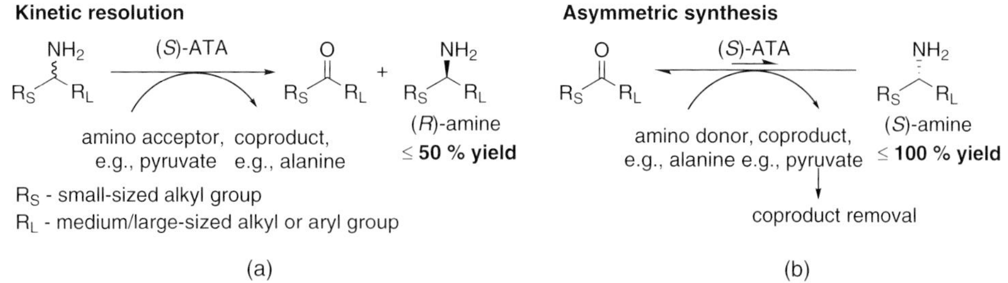

Figure 4.23 Strategies for the synthesis of optically active amines using amine transaminases. In a kinetic resolution (a), the amine transaminase converts in the ideal case only one of the amine enantiomers to the corresponding ketone. The remaining enantiomer can be isolated in high optical purity and at a (theoretical) maximum yield of 50%. In an asymmetric synthesis (b), a prostereogenic ketone is aminated enantioselectively, yielding directly the optically active amine at a theoretical yield of 100%. Common cosubstrates for amine transaminases are pyruvate and alanine. As the equilibrium favors ketone formation, high yields in asymmetric synthesis can only be achieved by shifting the equilibrium, for example, by enzymatic removal of the formed coproduct pyruvate. The same principles also apply to α-transaminases useful for the production of α-amino acids.

conversion. Another alternative is the use of pyruvate decarboxylase (PDC) to yield acetaldehyde and CO_2 (Höhne *et al.*, 2008). This has the advantage that no additional enzyme and no cofactor recycling of NADH are needed and the reaction equilibrium is irreversibly shifted due to carbon dioxide formation. Furthermore, PDCs are commercially available or a crude extract from *Zymomonas mobilis* can be used. This recently published principle was applied in the asymmetric synthesis of 1-*N*-Boc-protected 3-aminopyrrolidine and -piperidine, which are interesting building blocks for organic synthesis. It must be noted that with the ω-transaminase from *V. fluvialis* acetaldehyde formed was also aminated to yield ethanamine as undesired by-product.

For enzymes with broad substrate specificity and high solvent tolerance, isopropylamine (Matcham *et al.*, 1999) can be used as alternative amino donor, yielding acetone as by-product. This method was used in the asymmetric synthesis of (*S*)-methoxyisopropylamine serving as building block for (*S*)-dimethenamide, the active ingredient of the herbicide Outlook® (Breuer *et al.*, 2004) (Figure 4.24), and in the synthesis of the antidiabetic drug Sitagliptin® (see below).

Product inhibition in this example could be overcome by directed evolution (Matcham *et al.*, 1999) yielding a variant with enhanced thermal and chemical stability so that the removal of acetone (generated from the isopropylamine) was possible at 50 °C under reduced pressure. The most effective mutant afforded >99% optically pure (*S*)-product from 2 M methoxyacetone and 2.5 M isopropylamine at 93% yield after 7 h.

Until recently, almost all ω-transaminases were (*S*)-selective. Hence, the (*R*)-enantiomer was not accessible by asymmetric synthesis. The company Codexis together with Merck & Co. (USA) recently engineered a (*R*)-selective enzyme

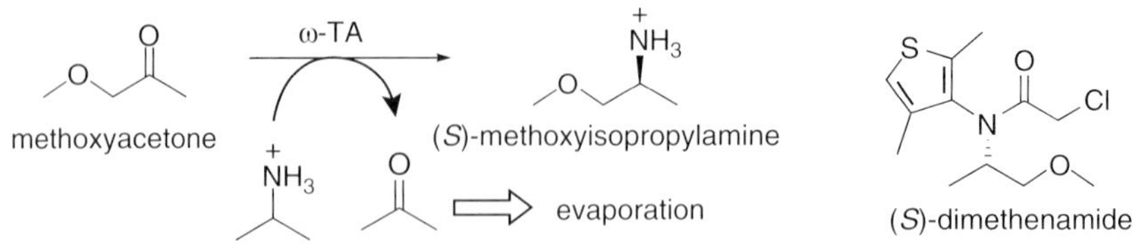

Figure 4.24 Asymmetric synthesis of (S)-methoxyisopropylamine.

(ATA-117). Starting from the wild type with virtually no activity (0.2% conversion!) toward the target substrate, the final variant converted 200 g l^{-1} prositagliptin ketone to Sitagliptin® with 99.95% ee at 92% yield (Figure 4.25) (Savile *et al.*, 2010). This variant contained 27 mutations and was obtained by a combination of homology model-based rational design followed by 10 rounds of directed evolution including random and site saturation mutagenesis as well as DNA shuffling. The biocatalytic process not only reduced the total waste and eliminated all heavy metals, but even increased the overall yield by 10% and the productivity (kg l^{-1} day^{-1}) by 53% compared to the rhodium-catalyzed process. Both routes have recently been compared (Desai, 2011).

A range of so far unknown (R)-selective amine transaminases was identified by an *in silico* screening approach (Höhne *et al.*, 2010). Identification and careful analysis of conserved motifs in (R)- and (S)-selective amino acid transaminases with 3D structures belonging to the PLP-dependent enzymes of fold class IV enabled to design an algorithm to filter 5000 sequences from public databases. This resulted in 20 protein sequences and the corresponding enzymes were then expressed recombinantly in *E. coli* using synthetic genes. Seventeen enzymes could be functionally expressed and all were confirmed to have the desired activity and (R)-enantiopreference. Seven of them were recently applied successfully in the asymmetric synthesis of various (R)-amines having >99% ee at >99% yield (Schätzle *et al.*, 2011). This example demonstrates how useful it is to explore the sequences available from public databases.

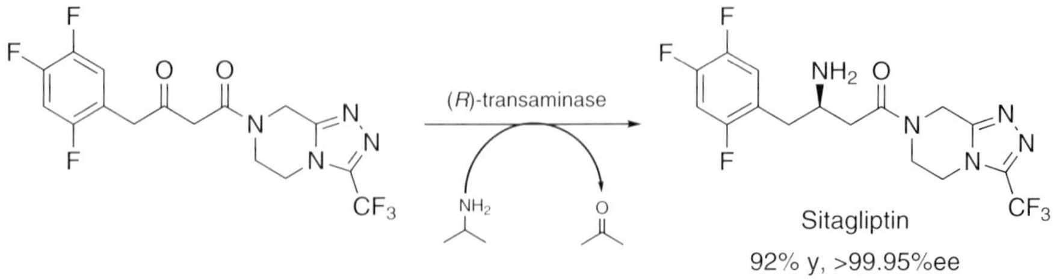

Figure 4.25 An engineered (R)-selective amine transaminase is used in a large-scale process for the production of the drug Sitagliptin®.

4.2.3
Hydrolases (EC 3.1)

4.2.3.1 Lipases (EC 3.1.1.3)

Lipases (triacylglycerol hydrolases) are probably the most frequently used hydrolases in organic synthesis (Bornscheuer and Kazlauskas, 2005; Schmid and Verger, 1998). They are widely found in nature (animals, human, bacteria, yeast, fungi, plants), and a considerable number of enzymes are commercially available. Their natural function is the hydrolysis and re-esterification of triglycerides – that is, natural fats and oils. The reaction is catalyzed by a catalytic triad composed of Ser, His, and Asp (sometimes Glu) similar to serine peptidases (see Figure 2.5a) and carboxyl esterases (EC 3.1.1.1). The mechanism for ester hydrolysis or formation is essentially the same for lipases and esterases, and is composed of four steps:

1) The substrate reacts with the active-site serine, yielding a tetrahedral intermediate stabilized by the catalytic His and Asp residues.
2) The alcohol is released and a covalent acyl-enzyme complex is formed.
3) Attack of a nucleophile (water in hydrolysis, alcohol in (trans-)esterification) forms again a tetrahedral intermediate.
4) The intermediate collapses to yield the product (an acid or an ester) and free enzyme (Figure 4.26).

Besides their use in organic synthesis for the production of optically active compounds (which are covered below in detail), lipases are used in a variety of other areas. Applications include laundry detergents, cheese making, modification of

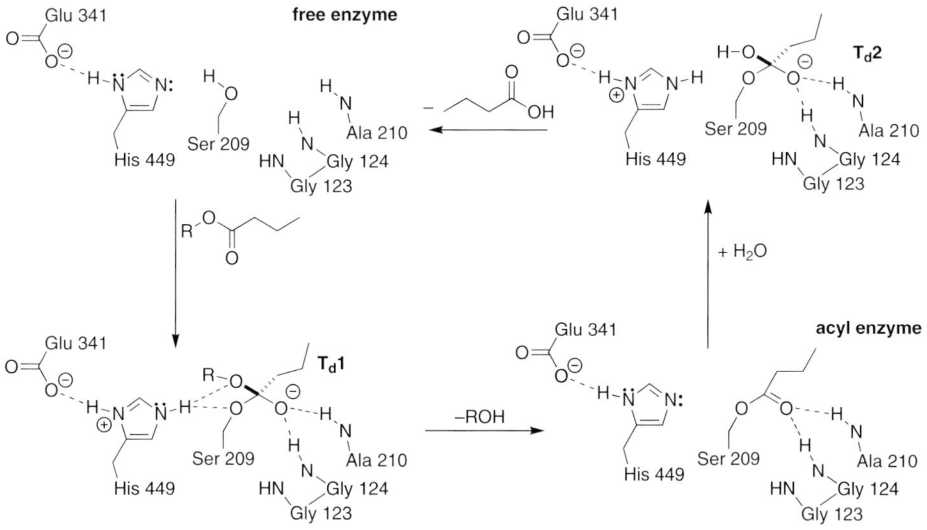

Figure 4.26 Mechanism of lipase-catalyzed ester hydrolysis of a butyrate ester. Numbering of amino acid residues is for lipase from *C. rugosa* (CRL).

natural fats and oils (e.g., the synthesis of cocoa butter equivalents, monoglycerides, incorporation of polyunsaturated fatty acids for human nutrition, fatty acid methyl esters to serve as biodiesel), and synthesis of sugar esters and simple ester used in personal care (e.g., myristyl myristate, decyl cocoate; Biermann *et al.*, 2011).

Lipases are distinguished from esterases (see below) by their substrate specificity: lipases accept long-chain fatty acids (in triglycerides) as substrates, whereas esterases prefer short-chain fatty acids. More generally, it can be stated that lipases readily accept water-insoluble substrates, while esterases prefer water-soluble compounds. A further difference was found in the 3D structures of these enzymes: lipases contain a hydrophobic oligopeptide (often called a *lid* or *flap*) – which is not present in esterases – covering the entrance to the active site. Lipases preferentially act at a water–organic solvent (or oil) interface, which presumably accounts for a movement of the *lid* making the active site accessible for the substrate. This phenomenon is referred to as "interfacial activation." It could be shown that after removal of the lid by genetic engineering, the activity of a lipase was improved in solution, mainly for applications in the laundry/detergent area. Further characteristic structural features of lipases are the α,β-hydrolase fold (Jochens *et al.*, 2011; Ollis *et al.*, 1992) and a consensus sequence around the active-site serine (Gly-X-Ser-X-Gly, where X denotes any amino acid).

The further classification of lipases can be based on substrate specificity toward triglycerides: nonspecific lipases hydrolyze all three fatty acids of a triglyceride, while *sn*1,3-regiospecific enzymes hydrolyze only the primary ester bonds of triglycerides, yielding *sn*2-monoglycerides. Some lipases also show distinct fatty acid chain length and saturation selectivity. For instance, lipase B from *Geotrichum candidum* shows high selectivity toward *cis*Δ9 fatty acids, such as oleic acid (Baillargeon and McCarthy, 1991). Recently, lipase A from *Candida antarctica* (CAL-A) could be engineered to high selectivity for *trans* fatty acids (Brundiek et al., 2012). Other classifications are based on their molecular weight, amino acid sequences, or sequence motifs (Pleiss *et al.*, 2000).

The main interest in the application of lipases in organic chemistry is due to the following reasons:

- Lipases are highly active in a broad range of nonaqueous solvents.
- They often exhibit excellent stereoselectivity.
- They accept a broad range of esters other than triglycerides.
- They accept nucleophiles other than water (i.e., alcohols, amines).

For the synthesis of optically pure compounds, a kinetic resolution is usually the method of choice yielding (under optimum conditions) 50% of either enantiomer in optically pure form. A kinetic resolution can be performed by hydrolysis of a racemic substrate (i.e., an acetate; Figure 4.27) or by acylation in an organic solvent (Figure 4.28). Note that if the enzyme is (*R*)-selective, the (*R*)-alcohol is formed in the hydrolysis reaction, but the (*S*)-alcohol is left unreacted in the acylation reaction.

As lipases are very stable in a range of organic solvents, acylation is often the preferred reaction. In order to drive the kinetic resolution to completion, either one substrate is used in excess or activated acyl donors are used (Figure 4.29). Using enol

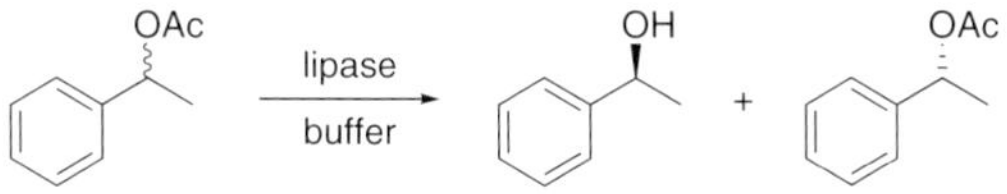

Figure 4.27 Example for a lipase-catalyzed kinetic resolution by hydrolysis. A preparative separation of an ester/alcohol mixture is straightforward by distillation or chromatography.

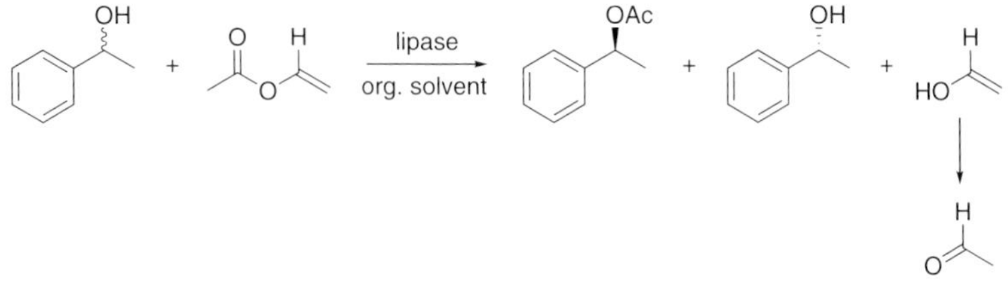

Figure 4.28 Example for a lipase-catalyzed kinetic resolution by acylation. The use of the enol ester vinyl acetate ensures an irreversible reaction as the by-product acetaldehyde is generated via a keto–enol tautomerization from the vinyl alcohol.

esters such as vinyl acetate or isopropenyl acetate renders the reaction practically irreversible as the vinyl alcohol generated undergoes keto–enol tautomerization to carbonyl compounds (i.e., acetaldehyde from vinyl esters, acetone from isopropenyl esters) (Figure 4.28). However, acetaldehyde might inactivate the enzyme via formation of a Schiff base with lysine residues. Trifluoroethyl esters are more expensive, and the thiols generated from *S*-ethyl thioesters have an undesirable flavor. Anhydrides such as acetic acid anhydride are cheap, but the free acid generated causes a drop in pH and might also participate in a slower background acylation.

Currently, more than 1000 examples (!) for the synthesis of optically active compounds with lipases can be found in the literature (Bornscheuer and Kazlauskas, 2005). In general, reactions proceed with good to excellent enantioselectivity for secondary alcohols, but only moderate selectivity is found for primary alcohols. Researchers have tried to predict the outcome of a lipase-catalyzed resolution based on the substrate structure and the type of lipase. Kazlauskas and coworkers developed empirical rules, which allow the prediction of the fast reacting enantiomer of a primary or a secondary alcohol (Kazlauskas *et al.*, 1991; Weissfloch and Kazlauskas, 1995). According to this rule, high enantioselectivities can be achieved for substrates bearing a medium-sized (e.g., methyl-) and large-sized (e.g., phenyl-) substituent (Figure 4.30). In contrast to secondary and primary alcohols, fewer

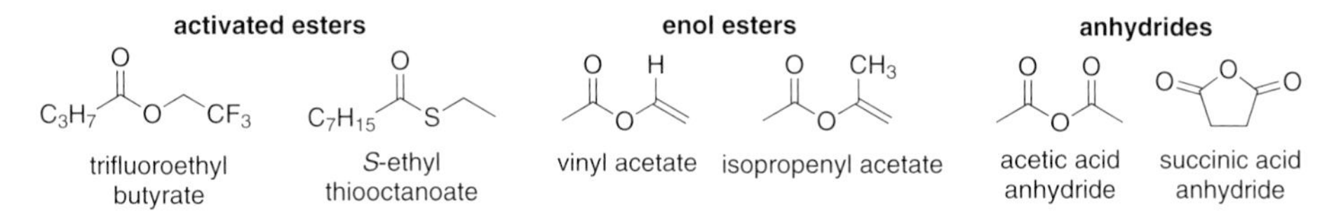

Figure 4.29 Some acyl donors used in lipase-catalyzed kinetic resolutions.

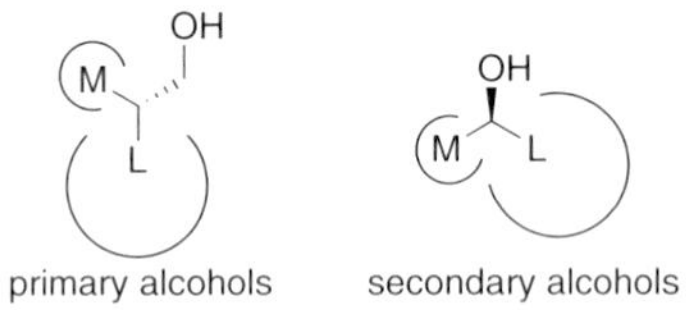

primary alcohols secondary alcohols

Figure 4.30 An empirical rule ("Kazlauskas rule") developed for lipase from *B. cepacia* (BCL) summarizing the enantiopreference for primary and secondary alcohols. The scheme shows the favored enantiomer. For primary alcohols, this rule is only reliable if no oxygen is attached to the stereocenter. Note that BCL shows an opposite enantiopreference for primary alcohols.

examples for the application of lipases can be found for the kinetic resolution of carboxylic acids (Figure 4.31). Interestingly, lipases, which show high selectivity for alcohols (e.g., PCL, CAL-B), are much less selective in the resolution of carboxylic acids, and vice versa (e.g., CRL).

From the considerable number of lipases available, only a few have been shown to be broadly applicable, to exhibit high enantioselectivity, and to have sufficient stability. These are porcine pancreatic lipase (PPL), lipase B from *Candida antarctica* (CAL-B, trade name: Novozyme NZ435), lipases from *Burkholderia cepacia* (BCL, former names: *Pseudomonas cepacia, Pseudomonas fluorescens*, Amano PS), *Candida rugosa* (CRL, former name: *Candida cylindracea*, Amano AY), *Rhizomucor miehei* (RML, trade name: Lipozyme RMIM), and *Thermomyces lanuginosus* (TLL, trade name: Lipozyme TLIM); the latter two are mostly used in lipid modification. For this area, an overview for the use of enzymes and especially lipases can be found in a review (Biermann *et al.*, 2011).

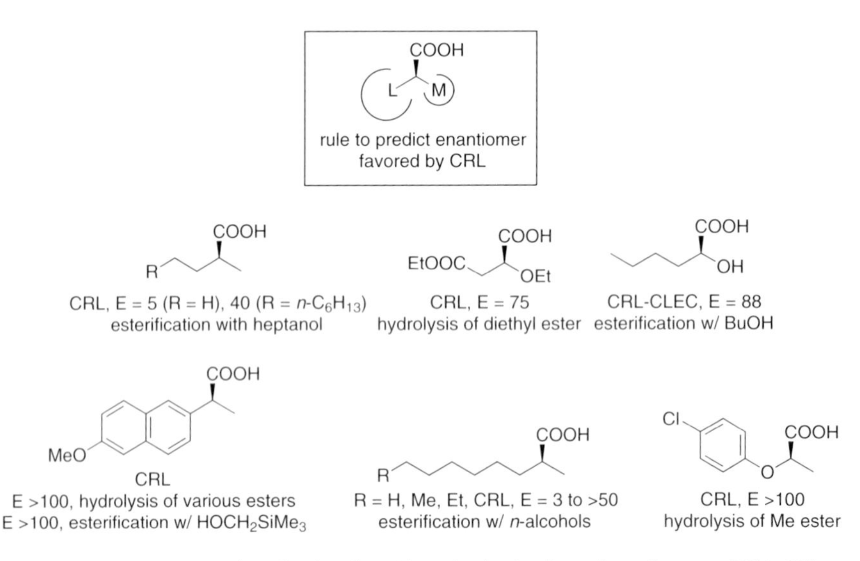

Figure 4.31 Some examples of carboxylic acid resolved using lipase from *C. rugosa* (CRL). CRL-CLEC is a cross-linked enzyme crystal preparation of CRL. Higher *E*-values might be due to the removal of interfering isoenzymes during crystallization.

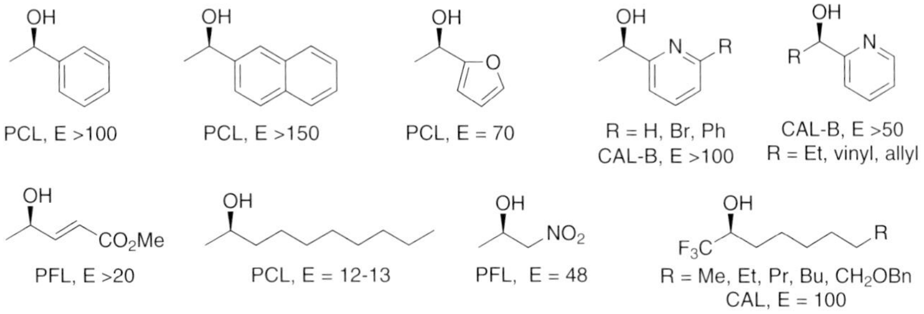

OH
PCL, E >100
PCL, E >150
O
PCL, E = 70
N
R
R = H, Br, Ph
CAL-B, E >100
CAL-B, E >50
R = Et, vinyl, allyl
CO2Me
PFL, E >20
PCL, E = 12-13
NO2
PFL, E = 48
F3C
R = Me, Et, Pr, Bu, CH2OBn
CAL, E = 100

Figure 4.32 Some examples of secondary alcohols resolved using lipases. In all cases, vinyl acetate served as acyl donor. Note the changes in enantioselectivity with varying substrate structure.

Figures 4.32 and 4.33 show a few selected examples of substrates resolved using these lipases.

Conversion and Resolution of Tertiary Alcohols Until a decade ago, only a few lipases (e.g., CRL and lipase A from *C. antarctica*) were known to accept tertiary alcohols. Due to their bulky structure, it was believed that these compounds do not fit into the active site of the enzyme. It was then discovered that a certain amino acid motif (GlyGlyGly(Ala)X motif, where X denotes any amino acid) located in the oxyanion binding pocket of lipases and esterases determines activity toward tertiary alcohols (Henke *et al.*, 2002). All enzymes bearing this motif (e.g., pig liver esterase, several acetylcholine esterases, an esterase from *B. subtilis*, BS2) were active toward several acetates of tertiary alcohols, while enzymes bearing the more common GlyX motif did not hydrolyze the model compounds. However, the enantioselectivity of these enzymes is usually rather low. This could be overcome by rational protein design and site-directed mutagenesis yielding variants of BS2 (Gly105Ala or Gln188Asp) with significantly higher selectivity (Heinze *et al.*, 2007; Henke *et al.*, 2003). More recently, the enantioselectivity could also be inverted by two mutations E188D/M193C, which were discovered by high-throughput screening of saturation mutagenesis libraries (Bartsch *et al.*, 2008) (Figure 4.34). The use of hydrolases in the kinetic resolution of tertiary alcohols has recently been reviewed (Kourist and Bornscheuer, 2011; Kourist *et al.*, 2008).

HO
AcO
R
PCL, E >100, R = i-Pr, t-Bu, Ph
E = 4 - 7, R = Et, n-Bu, n-decyl, benzyl
OAc
PCL, E ~33
O
PCL, E = 15
N
Ph
PCL, E >50 (-40°C)

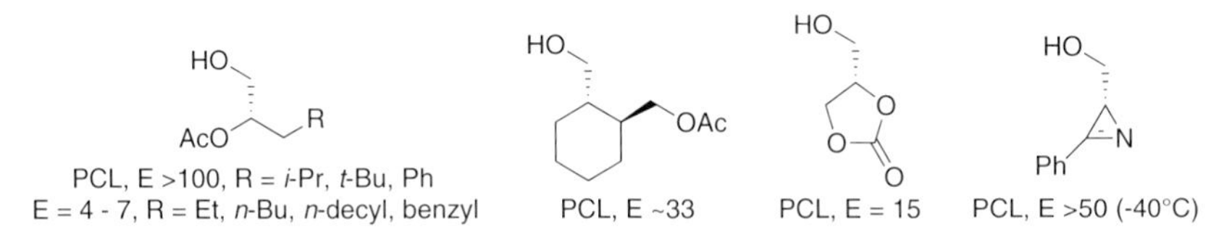

Figure 4.33 Some examples of primary alcohols resolved using lipase from *P. cepacia* (PCL). In all cases, vinyl acetate served as acyl donor. Note the changes in enantioselectivity with varying substrate structure and the lower *E*-values compared to secondary alcohols.

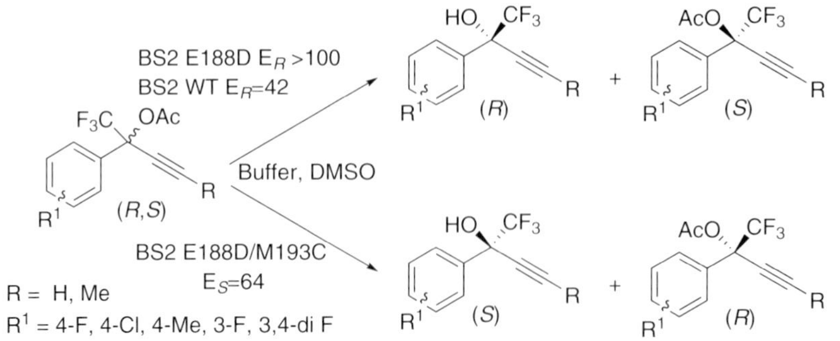

Figure 4.34 Kinetic resolution of aryl aliphatic tertiary alcohol acetates using a double mutant of esterase BS2 (E188D/M193C) resulted in higher or inverted enantioselectivity.

Industrial Applications Optically pure amines are versatile synthons for a wide variety of products. Interestingly, lipases were found most efficient for the synthesis of these compounds by kinetic resolution. An example is the highly enantioselective acylation of (*R*,*S*)-1-phenethylamine in a process established at BASF (Figure 4.35). The (*R*)-amide is separated from the (*S*)-amine by distillation or extraction, and the free (*R*)-amine is released through basic hydrolysis. As the lipase from *Burkholderia* sp. shows broad substrate specificity, a wide variety of aryl alkyl amines, alkyl amines, and amino alcohols could be resolved, in some cases on the multi-ton scale. Undesired enantiomers can be racemized and the acylating agent (such as methoxyacetic acid ethyl ester) can be recovered. BASF produces more than 2500 tons annually with this process (Balkenhohl *et al.*, 1997).

Researchers at DSM developed a lipase-catalyzed process for the production of a captopril intermediate, (*R*)-3-chloro-2-methyl propionate. All lipases preferentially hydrolyzed the (*S*)-enantiomer, and up to 98% ee at 32% conversion was observed with a lipase from *C. cylindracea* (Figure 4.36).

In addition, further industrial processes based on lipase catalysis have been established in the past few years (Breuer *et al.*, 2004; Liese *et al.*, 2006; Strohmeier *et al.*, 2011; Tao *et al.*, 2009).

Dynamic Kinetic Resolutions As already outlined in Section 4.1.1, a kinetic resolution of a racemate can only yield at maximum 50% product. In order to achieve a complete conversion of both enantiomers, a dynamic kinetic resolution can be used. Such a strategy can also make the synthesis of optically pure compounds

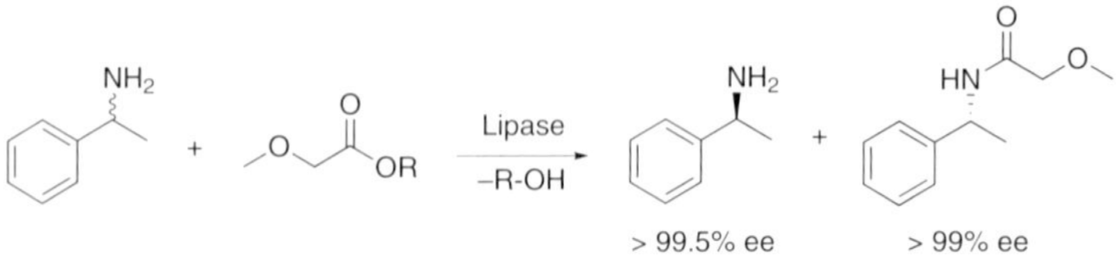

Figure 4.35 Lipase-catalyzed kinetic resolution of amines in the BASF process.

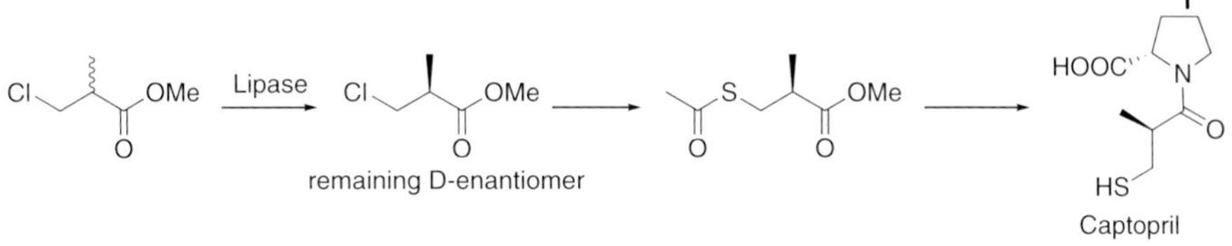

Figure 4.36 Lipase-catalyzed kinetic resolution of a building block for the synthesis of captopril.

more competitive to an asymmetric synthesis using, for example, alcohol dehydrogenases and a prochiral substrate.

The requirements for a DKR are as follows: (1) the substrate must racemize faster than the subsequent enzymatic reaction proceeds; (2) the product must not racemize; and (3) as in any asymmetric synthesis, the enzymatic reaction must be highly stereoselective. The principle for this is exemplified in Figure 4.37, and many examples have been described in recent reviews (Martin-Matute and Bäckvall, 2007; Pàmies and Bäckvall, 2004).

The earliest example of a DKR was the synthesis of optically pure α-amino acids from hydantoins. Racemization of the hydantoin occurs at alkaline pH or with the aid of a racemase (see Section 4.2.3.7). Later, dynamic kinetic resolutions have been described for desymmetrizations of chemically labile secondary alcohols, thiols, and amines (i.e., cyanohydrins, hemiacetals, hemithioacetals). Furthermore, *in situ* deracemization via nucleophilic displacement has been demonstrated for 2-chloropropionate (92% yield, 86% ee) using lipase from *C. cylindracea* in an aminolysis supported by triphenylphosphonium chloride (Bdjìc *et al.*, 2001).

Other approaches are combinations of enzymatic resolution with metal-catalyzed racemization. These reactions usually proceed either via hydrogen transfer or via π-allyl complex formation. Bäckvall and coworkers developed a hydrogen transfer system based on a ruthenium catalyst with *p*-chloroethyl acetate as acyl donor. (It should be noted that the use of metals, and especially ruthenium, can lead to problems in recycling/regeneration, and there may be environmental problems associated with the overall process.) Enol esters – with the exception of isopropenyl acetate – cannot be used due to side reactions. On the other hand, no additions of

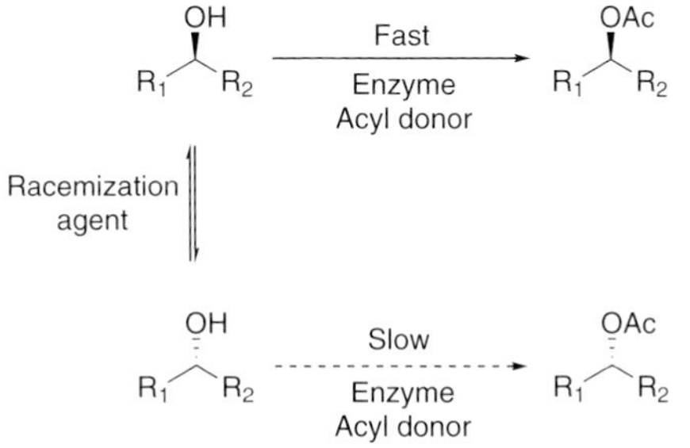

Figure 4.37 Principle of dynamic kinetic resolutions.

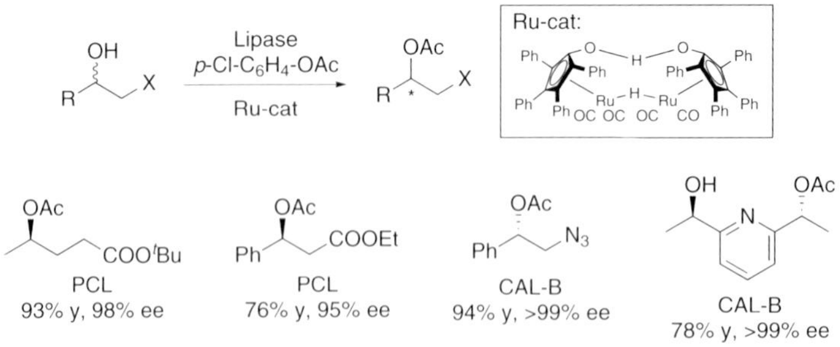

Figure 4.38 Examples of the dynamic kinetic resolution of secondary alcohols using a ruthenium catalyst.

ketones or external bases are required, which often affect the reaction performance. Selected examples are shown in Figure 4.38.

Kim and coworkers improved the DKR of allylic acetates using Pd(0) catalysts in tetrahydrofuran. 2-Propanol serves as acyl acceptor, and the unreactive enantiomer is racemized by $Pd(PPh)_3$ with added diphosphine at room temperature (Figure 4.39). A series of linear allylic acetates were deracemized in high enantiomeric excess (97–99% ee) and with moderate to good yields (61–78%).

4.2.3.2 **Esterases (EC 3.1.1.1)**

Although a considerable number of carboxyl esterases are known and have been overexpressed in suitable hosts (Table 4.4), only a few of them are used commercially for the synthesis of optically pure compounds as usually lipases show higher stability in organic solvents and broader substrate ranges. However, the number of commercial esterases has considerably increased in the past few years and they are available from, for example, Fluka, Amano, Codexis, and Enzymicals. The application of microbial esterases has been reviewed (Bornscheuer, 2002).

One useful enzyme is the carboxyl esterase NP (NP from naproxen, a non-steroidal anti-inflammatory drug) originating from *B. subtilis* (Quax and Broekhuizen, 1994). Besides naproxen, various other 2-arylpropionic acids were resolved with high enantioselectivity (Figure 4.40) (Azzolina *et al.*, 1995). Carboxyl esterase NP has a molecular weight of 32 kDa, a pH optimum between pH 8.5 and 10.5, and a temperature optimum between 35 and 55 °C. Carboxyl esterase NP is produced as an intracellular protein, for which the structure is unknown. In a pilot-scale process, (*R*,*S*)-naproxen methyl ester was hydrolyzed in the presence of Tween 80 to increase substrate solubility at pH 9.0. The (*S*)-acid was separated from the

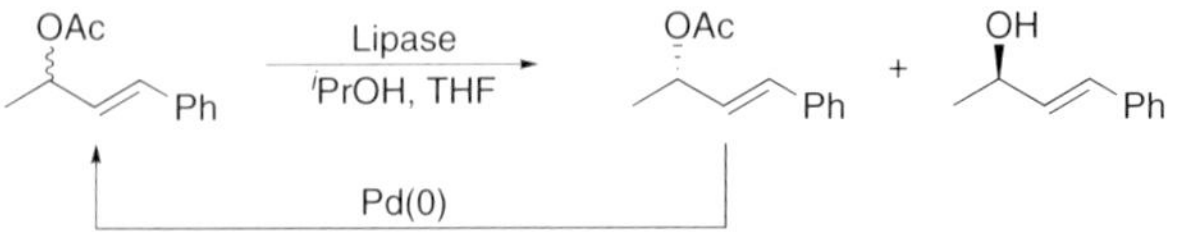

Figure 4.39 Example of the dynamic kinetic resolution of an allylic alcohol using Pd(0).

Table 4.4 Comparison of (recombinant) microbial esterases.

Origin[a]	Biochemical properties	Substrate selectivity	Remarks
B. gladioli ATCC 10248 (EstB)	392 aa, 42 kDa	Triglycerides (>C_6), deacylates cephalosporins	S-x-x-K motif, β-lactamase-like
B. gladioli ATCC 10248 (EstC)	298 aa, 32 kDa	—	G-x-S-x-G motif, similar to plant hydroxynitrile lyases
P fluorescens DSM 50106	36 kDa, $T_{opt.}$ 43 °C	Lactones, ethyl caprylate, moderate enantioselectivity	G-x-S-x-G motif, similar to a haloperoxidase
P. fluorescens SIKW1	27 kDa, homodimer	High *E*-value for α-phenyl ethanol	Altered substrate specificity and improved enantioselectivity by directed evolution
P. putida MR2068	29 kDa, homodimer, $T_{opt.}$ 70 °C	Alkyl dicarboxylic acid methyl esters, high selectivity ($E > 100$)	—
Bacillus acidocaldarius	34 kDa, $T_{opt.}$ 70 °C	Moderate selectivity ($E \sim 18$)	Similar to hormone-sensitive lipase
B. subtilis NRRL B8079	489 aa, 54 kDa, $T_{opt.}$ 52 °C (66.5 °C for best mutant)	*p*-Nitrobenzyl ester of Loracarbef	Evolved by directed evolution for increased stability in DMF and thermostability
B. stearothermophilus	—	Moderate enantioselectivity	Thermostable mutants
B. subtilis (Thai18)[b]	32 kDa, $T_{opt.}$ 35–55 °C	High *E* for 2-arylpropionic acids	Structure known, stable mutants by SDM[c]
Pyrococcus furiosus DSM 3638	$T_{opt.}$ 100 °C, $t_{1/2}$ 50 min at 126 °C	MU-Ace[c]	—
Lactococcus lactis[d]	258 aa, 30 kDa	Tributyrin, C_6-phospholipids	G-x-S-x-G motif, function unclear
Rh. ruber DSM 43338[e]	—	Linalool acetate ($E > 100$)	Two esterases with opposite enantiopreference
Rhodococcus sp. H1	34 kDa, tetramer	Heroin	G-x-S-x-G motif, conserved His_{86}
Rhodococcus sp. MB1	574 aa, 65 kDa, monomer	Cocaine	Similar to aminopeptidases
S. diastatochromogenes	326 aa, 31 kDa	Moderate enantioselectivity	—
Saccharomyces cerevisiae IFO 2347	28 kDa, homodimer, $T_{opt.}$ 25 °C	Isoamyl acetate, isobutyl acetate	—

a) Overexpressed in *E. coli*, if not stated otherwise.
b) Overexpressed in *B. subtilis*.
c) MU-Ace, 4-methylumbelliferyl acetate; SDM, site-directed mutagenesis.
d) Overexpressed in *L. lactis*.
e) Nonrecombinant purified enzymes.

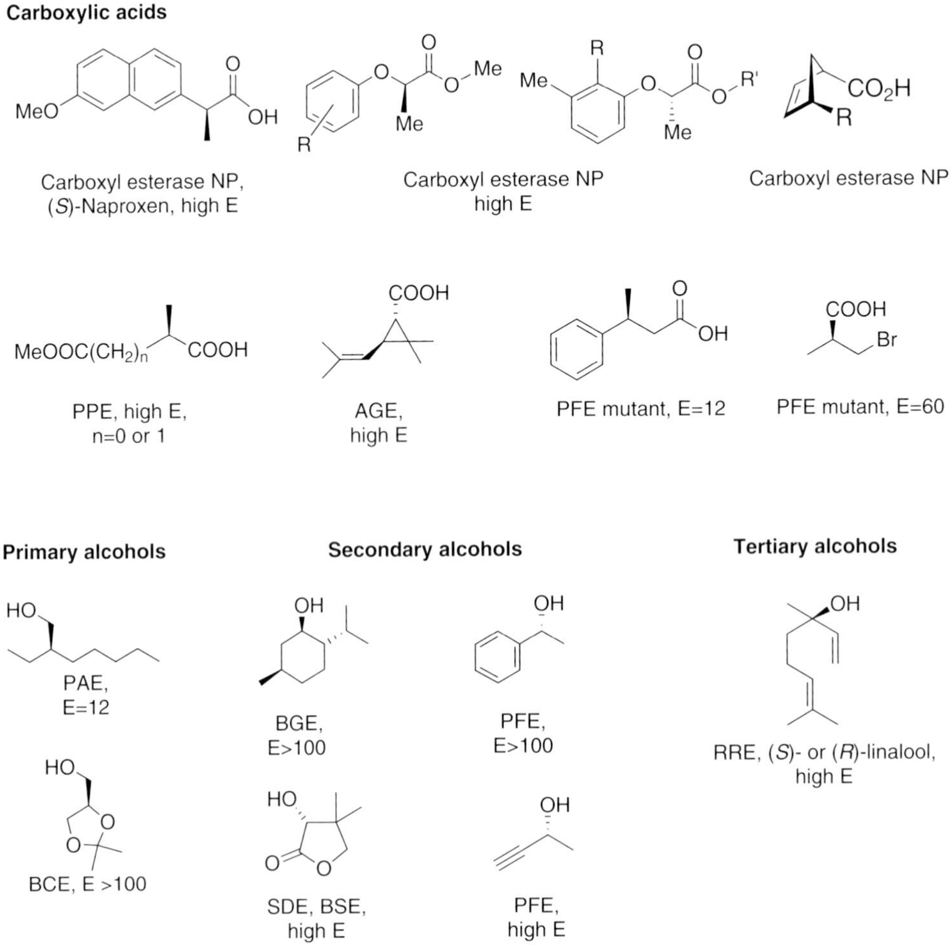

Carboxylic acids

Carboxyl esterase NP, (*S*)-Naproxen, high E

Carboxyl esterase NP high E

Carboxyl esterase NP

PPE, high E, n=0 or 1

AGE, high E

PFE mutant, E=12

PFE mutant, E=60

Primary alcohols

PAE, E=12

BCE, E >100

Secondary alcohols

BGE, E>100

PFE, E>100

SDE, BSE, high E

PFE, high E

Tertiary alcohols

RRE, (*S*)- or (*R*)-linalool, high E

Figure 4.40 Examples of esterase-catalyzed resolutions. AGE, *A. globiformis* esterase; BCE, *Bacillus coagulans* esterase; BGE, *Burkholderia gladioli* esterase; BSE, *B. stearothermophilus* esterase; PAE, *Pseudomonas aeruginosa* esterase; PFE, *P. fluorescens* esterase; PPE, *P. putida* esterase; RRE, *Rh. ruber* esterase; SDE, *Streptomyces diastatochromogenes* esterase.

remaining (*R*)-methyl ester and the latter racemized using an organic base. This reaction yields (*S*)-naproxen with excellent optical purity (99% ee) at an overall yield of 95%.

Another efficient kinetic resolution was achieved in the synthesis of (+)-*trans*-(1*R*,3*R*)-chrysanthemic acid, which is an important precursor of pyrethrin insecticides (Figure 4.40). Here, an esterase from *Arthrobacter globiformis* catalyzed the sole formation of the desired enantiomer (>99% ee, at 38% conversion). The enzyme was purified and the gene cloned in *E. coli* (Nishizawa *et al.*, 1995). In a 160 g scale process, hydrolysis was performed at pH 9.5 at 50 °C. The acid produced was separated through a hollow-fiber membrane module and the esterase proved to be very stable over four cycles, each of 48 h.

Further selected examples of the application of microbial carboxyl esterases in the synthesis of optically pure compounds are summarized in Figure 4.40.

Pig Liver Esterase Pig liver esterase (PLE) has been widely used in organic synthesis, as it accepts a broad range of substrates, which are often converted with excellent stereoselectivity. Besides the kinetic resolution of various racemates, PLE was shown to be efficient in the desymmetrization of prostereogenic and *meso*-compounds (Jones, 1990; Jones *et al.*, 1985; Lam *et al.*, 1986).

PLE was originally isolated from pig liver by extraction, and consists of several isoenzymes with the α-, β-, and γ-subunits as the most dominant ones. In the literature, it was debated whether these isoenzymes differ (Öhrner *et al.*, 1990) or not (Lam *et al.*, 1988) in their enantioselectivity. The first functional overexpression of recombinant PLE (the γ-isoenzyme) in the yeast *Pichia pastoris* (Lange *et al.*, 2001) revealed that the recombinant enzyme shows substantially higher *E*-values in the resolution of acetates of secondary alcohols (Figure 4.41). For instance, the resolution of (*R*,*S*)-1-phenyl-2-butyl acetate proceeded with $E = 1$–4 using commercial PLE, but with $E > 100$ using the recombinant enzyme, which contains only a single isoenzyme (Musidlowska *et al.*, 2001). Later, five further isoenzymes were identified,

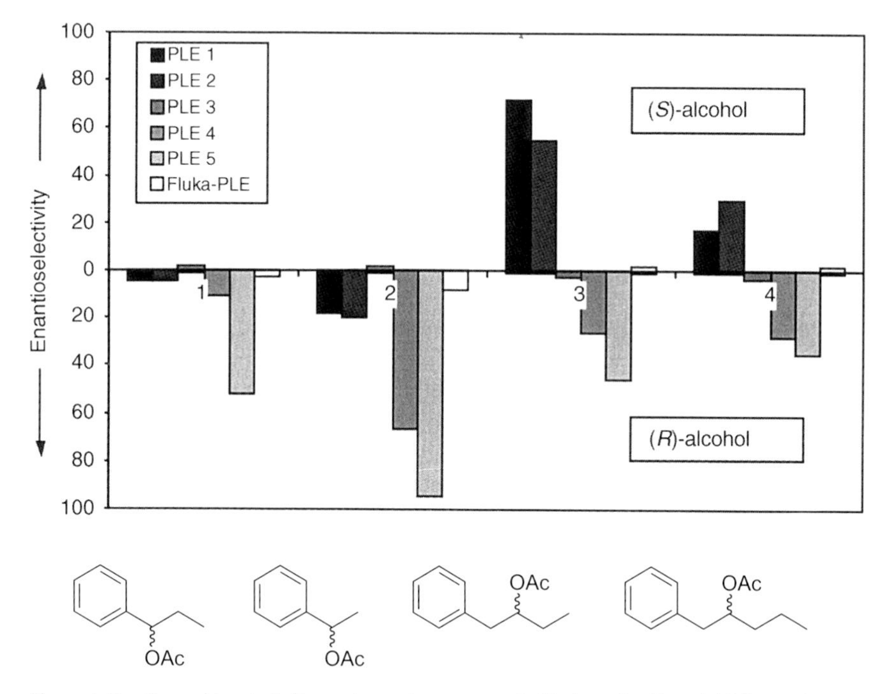

Figure 4.41 Recombinant pig liver esterase isoenzymes (1–5) show significantly higher and even inverted enantioselectivity in the kinetic resolution of four acetates of secondary alcohols compared to a crude preparation from Fluka.

cloned, and expressed in *E. coli* or *P. pastoris*, which all show distinct chemo- and enantioselectivity (Fig. 4.41) and hence allow to obtain either the (*R*)- or (*S*)-enantiomer of a chiral alcohol (Hummel *et al.*, 2007). One isoenzyme (APLE) is used in large-scale kinetic resolutions of methyl-(4*E*)-5-chloro-2-isopropyl-4-pentenoate by the company DSM (Hermann *et al.*, 2008).

4.2.3.3 **Peptidases, Acylases, and Amidases**

Peptidases (EC 3.4.-.-) and amidases (EC 3.5.-.-) catalyze both the formation and hydrolysis of amide links. Although their natural role is hydrolysis, they are also used to form amide bonds. Two different strategies have been applied for this, namely, a thermodynamic or a kinetic control (Figure 4.42; see also Sections 2.6 and 2.8).

In thermodynamically controlled syntheses, the reaction conditions are changed to shift the equilibrium toward synthesis instead of hydrolysis. The hydrolysis of peptides is favored by $-5\ \mathrm{kJ\ mol^{-1}}$, and is driven mainly by the favorable solvation of the carboxylate and ammonium ions; it also depends on the pH (see Figure 2.20). One common way to shift the equilibrium toward synthesis is to replace the water with an organic solvent. The organic solvent suppresses the ionization of the starting materials, and also reduces the concentration of water. Other common ways of shifting the equilibrium are to increase the concentrations of the starting materials, or to choose protective groups that promote precipitation of the product.

In kinetically controlled syntheses, an activated carboxyl component (usually an ester or amide) is used; this reacts with the enzyme to form an acyl-enzyme intermediate, which then reacts either with an amine to form the desired amide or with water to form a carboxylic acid. Because the starting material is an activated

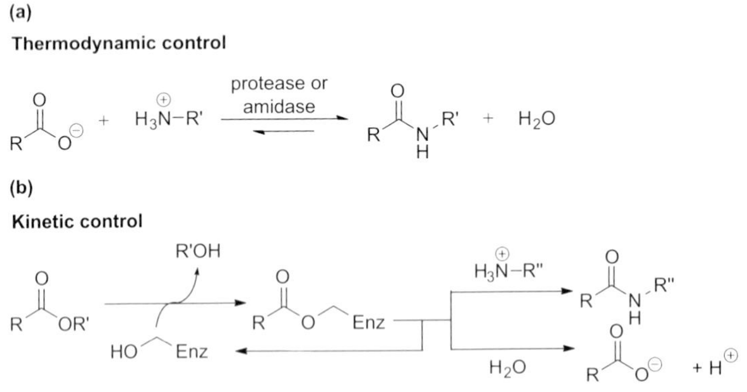

(a)

Thermodynamic control

$$RCOO^{\ominus} + H_3N^{\oplus}{-}R' \underset{}{\overset{\text{protease or amidase}}{\rightleftharpoons}} RC(O)NHR' + H_2O$$

(b)

Kinetic control

$$RC(O)OR' \xrightarrow{-R'OH,\ +HO{-}Enz} RC(O)O{-}Enz \begin{cases} \xrightarrow{H_3N^{\oplus}{-}R''} RC(O)NHR'' \\ \xrightarrow{H_2O} RCOO^{\ominus} + H^{\oplus} \end{cases} \quad (+\ HO{-}Enz)$$

Figure 4.42 Synthesis of amide bonds using proteases and amidases. (a) Thermodynamic control shifts the equilibrium toward synthesis by changing the reaction conditions. For example, organic solvents are added to reduce the concentration of water and to suppress ionization of the starting materials. (b) Kinetic control starts with an activated carboxyl component (e.g., an ester or an amide) and forms an acyl-enzyme intermediate. The acyl-enzyme intermediate then reacts with an amine to form the amide. In a competing side reaction, water may react with the acyl-enzyme intermediate.

carboxyl component, reactions are faster in the kinetically controlled approach than in the thermodynamically controlled approach (see Figure 2.9). Because the kinetically controlled approach requires an acyl-enzyme intermediate, only cysteine and serine hydrolases (e.g., the peptidase subtilisin, the amidase penicillin G amidase) are suitable. Metallopeptidases such as thermolysin function only in thermodynamically controlled syntheses.

Although peptidase-catalyzed peptide synthesis was first reported in 1901 (Savjalov, 1901), peptidases have been used much more frequently in peptide synthesis since the late 1970s (for reviews, see Bordusa, 2002; Kasche, 2001; Schellenberger and Jakubke, 1991). The advantages of an enzyme-catalyzed peptide synthesis are mild conditions, no racemization, a minimal need for protective groups, and high regio- and enantioselectivity.

The largest scale application (producing hundreds to thousands of tons) of peptidase-catalyzed peptide synthesis is the thermolysin-catalyzed synthesis of aspartame, a low-calorie sweetener (Figure 4.43) (Oyama, 1992). Precipitation of the product drives this thermodynamically controlled synthesis, and the high regioselectivity of thermolysin ensures that only the α-carboxyl group in aspartate reacts. Thus, there is no need to protect the β-carboxylate. The high enantioselectivity allows the use of racemic amino acids as only the L-enantiomer reacts. In addition, peptidases are also important in the production of insulin.

Subtilisin accepts a broader range of substrates than other peptidases, so it is also used for amide couplings involving unnatural substrates (Moree *et al.*, 1997). When coupling a D-amino acid, it is best to use it as the nucleophile rather than as the carboxyl donor because subtilisin is more tolerant of changes in the nucleophile than in the carboxyl group.

Acylases and amidases also catalyze the formation of peptide bonds. An important application is the hydrolysis and synthesis of the β-lactam peptide antibiotics (penicillins and cephalosporins). About 40 000 tons of penicillin are hydrolyzed each year in an equilibrium-controlled process catalyzed by penicillin amidases to produce 6-aminopenicillanic acid (6-APA) (see Figure 1.7 and Exercise 2.14; see also Section 12.3). Processes for the synthesis of a range of semisynthetic β-lactam antibiotics starting from 6-APA or 7-aminocephalosporanic acid (7-ACA) have been

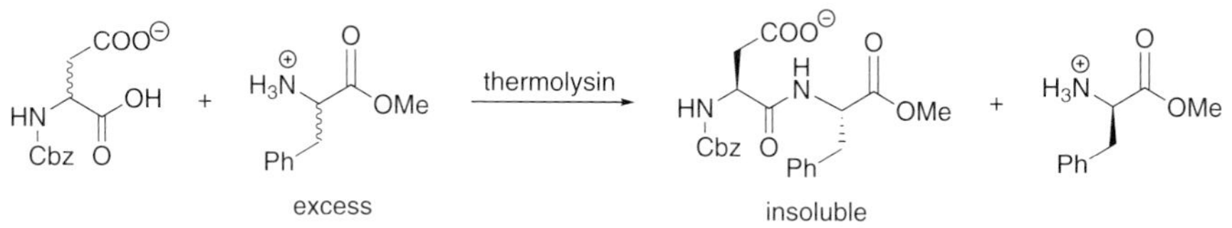

Figure 4.43 Commercial process for the production of aspartame (α-L-aspartyl-L-phenylalanine methyl ester). Thermolysin catalyzes the coupling of an *N*-Cbz-protected aspartic acid with phenylalanine methyl ester. The product forms an insoluble salt with excess of phenylalanine methyl ester. This precipitation drives this thermodynamically controlled peptide synthesis. The high regioselectivity of thermolysin for the α-carboxylate allows the β-carboxylate in aspartate to be left unprotected. The enantioselectivity of thermolysin allows the use of racemic starting materials.

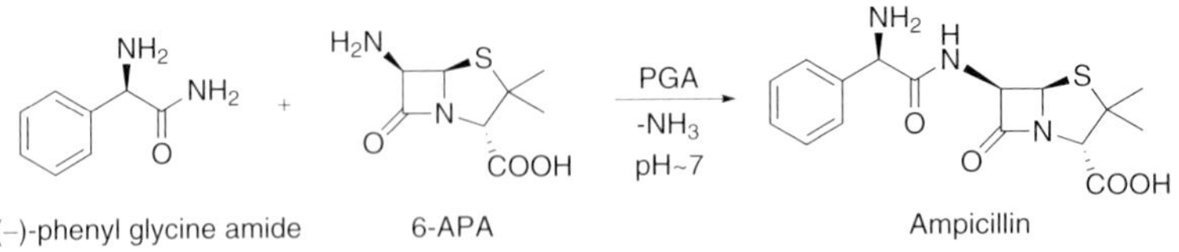

Figure 4.44 The penicillin acylase (PGA)-catalyzed coupling of 6-aminopenicillanic acid with (*R*)-(−)-phenylglycine to produce ampicillin, a β-lactam antibiotic.

developed (see Section 12.3; Liese *et al.*, 2006). For the synthesis, penicillin amidase is used in a kinetically controlled approach, and the antibiotic can be obtained in yields >95% (of 6-APA or 7-ACA) in aqueous solution when the synthesis is performed with a more than twofold excess of activated substrate (Figure 4.44). The production of 7-ACA from cephalosporin C is described in detail in Section 12.3.

Biochemical properties and the mechanism of peptidases and amidases are covered in more detail in Figure 2.5 (see also Sections 2.7 and 2.10). More information can also be found in a variety of books (e.g., Liese *et al.*, 2006; Patel, 2006; Tao *et al.*, 2009).

4.2.3.4 **Epoxide Hydrolases (EC 3.3.2.3)**

Optically pure epoxides are versatile building blocks in organic synthesis. Several chemical methods for the preparation of optically pure epoxides have been developed, as described by Sharpless or Jacobsen and Katsuki. The Sharpless epoxidation (Katsuki and Martin, 1996) is limited to allylic alcohols, but the Jacobsen–Katsuki reaction (Hosoya *et al.*, 1994; Linker, 1997) has been improved to become a versatile method for obtaining optically pure epoxides from alkenes. These and related chemical methods are used on an industrial scale (Breuer *et al.*, 2004).

Alternatively, epoxides can be resolved by using epoxide hydrolases, which catalyze the hydrolysis of an epoxide to furnish the corresponding vicinal diol. The reaction proceeds via an S_N2-specific opening of the epoxide, leading to the formation of the corresponding *trans*-configurated 1,2-diols.

Epoxide hydrolases do not require cofactors and have been found in a variety of sources, including mammals, plants, insects, yeasts, filamentous fungi, and bacteria. These enzymes are catalytically active in the presence of organic solvents, and often show high regio- and enantioselectivity. In general, EHs are used as washed whole cells or as lyophilized cell-free extracts, because isolated enzymes show less stability. Although they have been intensively studied with respect to their metabolic function, mammalian EHs are not applied in biocatalysis, mainly due to their limited availability.

The first reports on microbial EHs were published in 1991 and 1993, but since then a broad range of EHs from various microorganisms have been identified and extensively characterized. The structures of several EHs from mammalian, bacterial, and fungal origin have been determined and belong to the typical α/β-hydrolase fold.

The proposed mechanism of epoxide ring opening involves the attack of a nucleophilic carboxylate residue at one end of the epoxide that has been activated by

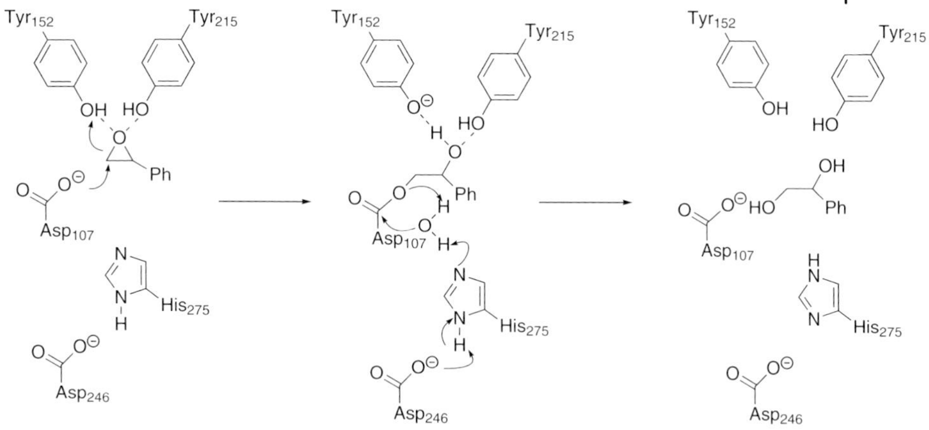

Figure 4.45 Mechanism proposed for epoxide hydrolase from *A. radiobacter.*

protonation. This leads to an α-hydroxyester intermediate covalently bound to the active site of the enzyme. This intermediate is hydrolyzed by the nucleophilic attack of a water molecule, which is activated by a histidine, followed by the release of the diol product and regeneration of the enzyme. From the crystal structure, it was concluded that the activation of the epoxide moiety is guided by two tyrosines (Figure 4.45) (Nardini *et al.*, 1999).

The enzymatic hydrolysis of terminal epoxides may proceed via attacking either the less-hindered oxirane carbon, leading to retention of configuration (most common), or at the stereogenic center, which results in an inversion of configuration (Faber *et al.*, 1996) (Figure 4.46).

During the past decades, a wide range of microbial EHs were identified and examples for the resolution of epoxides using enzymes of bacterial origin are shown in Figure 4.47. Aliphatic and aryl aliphatic substrates were converted with high enantioselectivity, though in general higher selectivities are observed for monosubstituted aryl aliphatic epoxides.

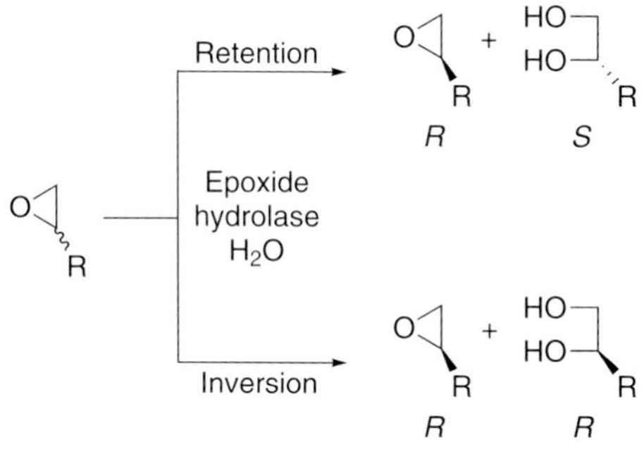

Figure 4.46 Hydrolysis of epoxides can proceed with retention or inversion of configuration.

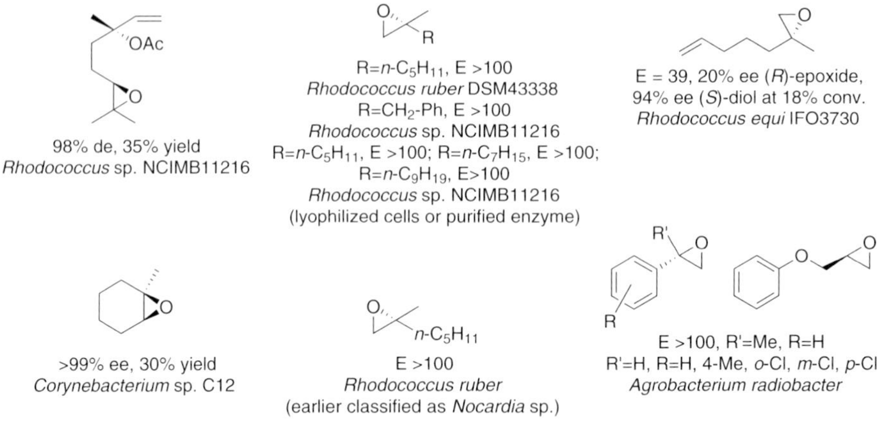

OAc
O
98% de, 35% yield
Rhodococcus sp. NCIMB11216

O
R
R=n-C_5H_{11}, E >100
Rhodococcus ruber DSM43338
R=CH_2-Ph, E >100
Rhodococcus sp. NCIMB11216
R=n-C_5H_{11}, E >100; R=n-C_7H_{15}, E >100;
R=n-C_9H_{19}, E>100
Rhodococcus sp. NCIMB11216
(lyophilized cells or purified enzyme)

O
E = 39, 20% ee (R)-epoxide,
94% ee (S)-diol at 18% conv.
Rhodococcus equi IFO3730

O
>99% ee, 30% yield
Corynebacterium sp. C12

O
n-C_5H_{11}
E >100
Rhodococcus ruber
(earlier classified as *Nocardia* sp.)

R'
O
R
O
O
E >100, R'=Me, R=H
R'=H, R=H, 4-Me, o-Cl, m-Cl, p-Cl
Agrobacterium radiobacter

Figure 4.47 Examples of the resolution of epoxides using bacterial epoxide hydrolases.

Several strains from *Nocardia* sp. (H8, EH1, TB1, later designated as *Rhodococcus ruber*) have been identified as EH producers. In particular, *Nocardia* sp. EH1 shows high enantioselectivity at 50% conversion in the resolution of 2-methyl-1,2-epoxyheptane ($E > 100$). However, the introduction of a phenyl group into the side chain decreases enantioselectivity ($E = 5.6$). The enzyme was purified to homogeneity via a four-step procedure. It is a monomer with a molecular weight of 34 kDa, a pH optimum of 8–9, and a temperature optimum of 35–40 °C. The pure enzyme is much less stable than a whole-cell preparation, but it is stabilized by the addition of Tween 80 or Triton X-100. Immobilization on DEAE-cellulose doubled the specific activity and allowed five repeated batch reactions, though enantioselectivity was slightly lowered (Kroutil *et al.*, 1998a, 1998b). Using *Nocardia* sp. EH1, the synthesis of naturally occurring (R)-(−)-mevalonolactone was achieved by deracemization of 10 g 2-benzyl-2-methyloxirane. The enzymatic reaction gave the corresponding (S)-diol, while the addition of catalytic amounts of sulfuric acid hydrolyzed the remaining (R)-epoxide with inversion of configuration, thus allowing the isolation of (S)-diol in an overall yield of 94% at 94% ee. Subsequent chemical steps afforded (R)-(−)-mevalonolactone in a total yield of 55% (Orru *et al.*, 1997, 1998). Hydrolysis of (±)-*cis*-2,3-epoxyheptane with rehydrated lyophilized cells of *Nocardia* sp. EH1 proceeded in an enantioconvergent fashion, and only (2R,3R)-heptane-2,3-diol was obtained as the sole product (Kroutil *et al.*, 1996).

The recombinant microbial EH cloned from *Agrobacterium radiobacter* AD1 and overexpressed in *E. coli* (Rink *et al.*, 1997) accepts a broad range of styrene oxide derivatives and phenyl glycidyl ether, which are converted with excellent enantioselectivity (Spelberg *et al.*, 1998). The enzyme has a molecular weight of 34 kDa, and the catalytic triad was proposed to consist of Asp_{107}, His_{275}, and Asp_{246}.

The first preparative-scale epoxide hydrolysis was reported by the group of Furstoss, who discovered that the fungus *A. niger* enantioselectively converts epoxy geraniol N-phenyl carbamate to yield the (S)-epoxide in high optical purity (96% ee) (Zhang *et al.*, 1991). The resolution of p-nitrostyrene oxide proceeded with

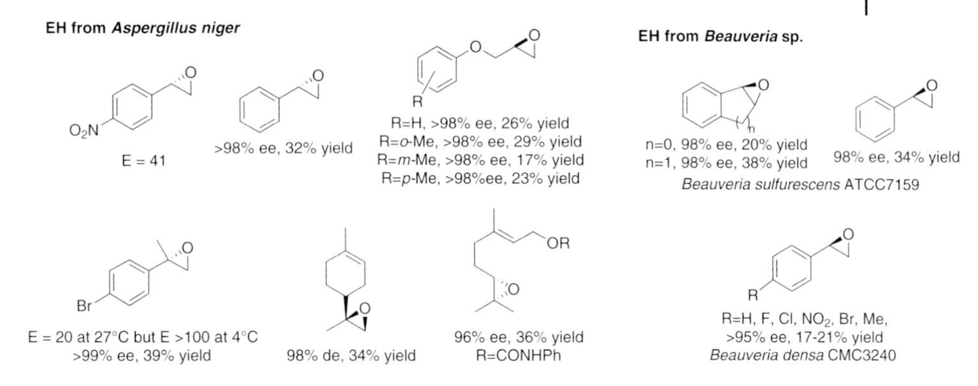

Figure 4.48 Examples of the resolution of epoxides using yeast and fungal epoxide hydrolases.

acceptable enantioselectivity ($E = 41$), and up to 20% DMSO could be added without any significant loss of activity. EH activity was also discovered in *Beauveria sulfurescens* ATCC 7159, which converts styrene oxide as well as indene and tetrahydronaphthalene oxides with high enantioselectivity (Pedragosa-Moreau *et al.*, 1996) but opposite enantiopreference. These and further examples are summarized in Figure 4.48.

Interestingly, *A. niger* and *B. sulfurescens* produced the (*R*)-diol in the hydrolysis of styrene oxide. Thus, the reaction catalyzed by *A. niger* proceeded with retention of configuration (via attack at C-2), whereas the hydrolysis with *B. sulfurescens* occurred with inversion of configuration (via attack at C-1, benzylic position). Employing a mixture of both organisms permitted the enantioconvergent synthesis of (*R*)-1-phenyl-1,2-dihydroxyethane in 92% yield and 89% ee (Figure 4.49). In the search for an enantioselective EH capable of resolving indene oxide, a precursor to the side chain of HIV peptidase inhibitor MK639, researchers at Merck found that out of 80 fungal strains investigated, *Diplodia gossypina* ATCC 16391 and

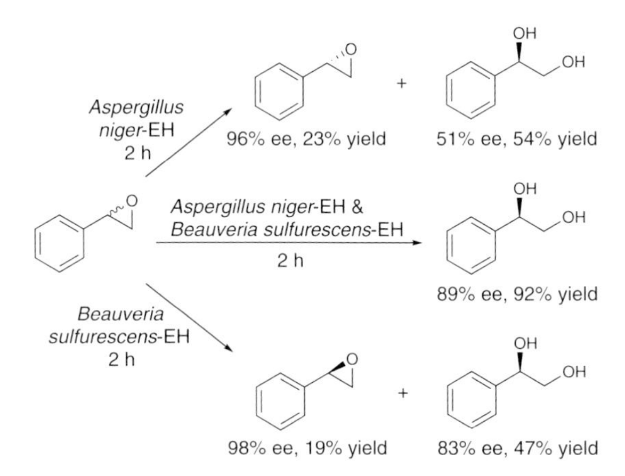

Figure 4.49 Resolution of styrene oxide using fungal epoxide hydrolases from *A. niger* or *B. sulfurescens* or a mixture of both for an enantioconvergent synthesis.

Lasiodiplodia theobromae MF5215 showed excellent enantioselectivity, yielding exclusively the desired (1*S*,2*R*)-enantiomer. Two other strains from *Gilmaniella humicola* MF5363 and from *Alternaria tenuis* MF4352 showed opposite enantiopreference. Preparative biotransformation using whole cells of *D. gossypina* ATCC 16391 allowed isolation of optically pure (1*S*,2*R*)-indene oxide in 14% yield after a 4 h reaction time (Zhang *et al.*, 1995).

More examples of the characterization and application of EHs can be found in reviews (Archelas and Furstoss, 2001; Smit, 2004; Steinreiber and Faber, 2001).

4.2.3.5 **Dehalogenases (EC 3.8.1.5)**

One interesting alternative to epoxide hydrolases are dehalogenases (EC 3.8.1.5, official name: haloalkane dehalogenase (HAD), also named halohydrin dehalogenases or halohydrin hydrogen-halide lyases). They catalyze the reversible dehalogenation of vicinal haloalcohols by an intramolecular displacement of a halogen to yield an epoxide and a halide (de Vries and Janssen, 2003; Janssen, 2004). The mechanism proposed for a HAD from *A. radiobacter* – based on the sequence similarity with the short-chain dehydrogenase/reductase protein family and the 3D structure of the enzyme from *A. radiobacter* – suggests activation of the epoxide by hydrogen bonding to a tyrosine (Figure 4.50).

The most useful feature of these enzymes is that they also accept nucleophiles other than water and thus not only 1,2-diols can be formed. It could be shown that not only halides (Cl^-, Br^-, I^-), but also CN^-, NO_2^-, and N_3^- are accepted by these enzymes. The HAD from *A. radiobacter* AD1 also had excellent regio- and enantioselectivity in the kinetic resolution of *p*-nitrostyrene oxide (Spelberg *et al.*, 2002, 2001) (Figure 4.51).

4.2.3.6 **Nitrilases (EC 3.5.5.1) and Nitrile Hydratases (EC 4.2.1.84)**

Nitriles are important precursors for the synthesis of carboxylic acid amides and carboxylic acids. The chemical hydrolysis of nitriles requires strong acid or base at high temperatures. Nitrile-hydrolyzing enzymes have the advantage that they react under mild conditions and do not produce large amounts of by-products. In

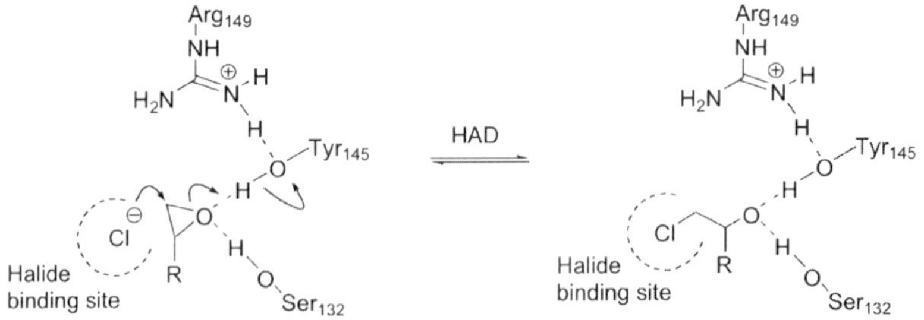

Figure 4.50 Proposed mechanism for the reversible epoxide ring opening catalyzed by *A. radiobacter* AD1 haloalcohol dehalogenase. The Arg–Tyr pair is involved in leaving group protonation.

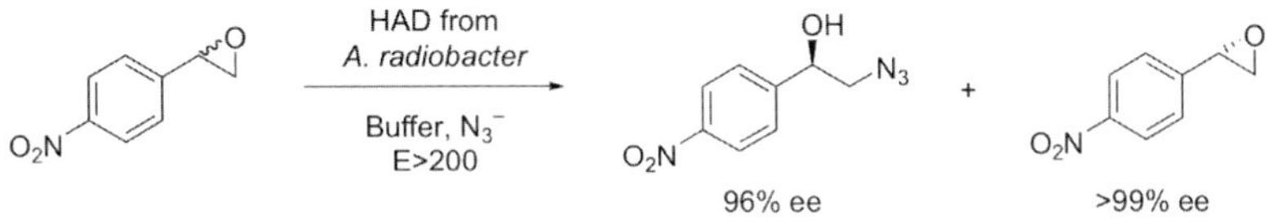

Figure 4.51 Haloalcohol dehalogenase-catalyzed ring opening of *p*-nitrostyrene oxide in the presence of azide proceeds with excellent regio- and enantioselectivity and yields an azido alcohol as product.

addition, the enzyme can be regio- and stereoselective. Two different enzymatic pathways can be used to hydrolyze nitriles (Figure 4.52). Nitrilases (EC 3.5.5.1) directly catalyze the conversion of a nitrile into the corresponding acid plus ammonia. In the other pathway, a nitrile hydratase (NHase; EC 4.2.1.84; a lyase) catalyzes the hydration of a nitrile to the carboxyl amide, which may be converted to the carboxylic acid and ammonia by an amidase (EC 3.5.1.4).

Purified nitrilases and nitrile hydratases are usually unstable, and thus the biocatalyst is often used as a whole-cell preparation. The nitrile-hydrolyzing activity must be induced first and common inducers are benzonitrile, isovaleronitrile, crotononitrile, and acetonitrile, though the inexpensive inducer urea also works. In addition, inducing with ibuprofen or ketoprofen nitriles can yield enantioselective enzymes (Layh *et al.*, 1997). After induction, preparative conversion is usually performed by adding the nitriles either during cultivation or by employing resting cells. The most commonly used strains are from *Rhodococcus* sp., and the most important ones are subspecies of *Rhodococcus rhodochrous*.

Nitrilases are cysteine hydrolases, which act via an enzyme-bound imine intermediate using a Glu–Lys–Cys catalytic triad. All nitrilases are inactivated by thiol reagents (e.g., 5,5′-dithiobis(2-nitrobenzoic acid)), indicating that they are sulfhydryl enzymes.

Two different groups of nitrile hydratases have been described, which require Fe(III) or Co(III) ions (Nagasawa, Takeuchi, and Yamada, 1991). The *Rh. rhodochrous* J1 strain produces two kinds of NHases that differ in their molecular weight (520 and 130 kDa). The high molecular weight NHase acts preferentially on aliphatic nitriles, whereas the smaller enzyme also has high affinity toward aromatic nitriles. Most nitrile hydratases accept aliphatic nitriles only (e.g., from *Arthrobacter* sp. J-1, *Brevibacterium* R312, *Pseudomonas chlororaphis* B23); however, strains have also been described that can act on arylalkylnitriles, arylacetonitriles, and heterocyclic nitriles.

Yamada's group showed by the use of electron spin resonance (ESR) studies that nitrile hydratases are nonheme ferric iron-containing enzymes. Further spectroscopic studies suggested that the enzyme also binds the cofactor pyrroloquinoline

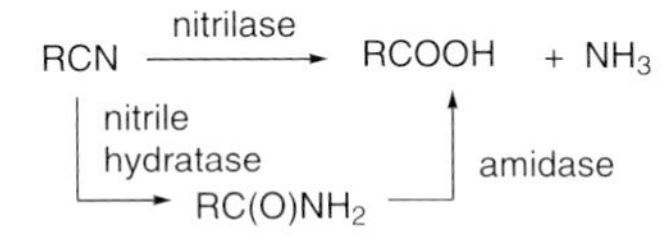

Figure 4.52 Hydrolysis of nitriles follows two different pathways.

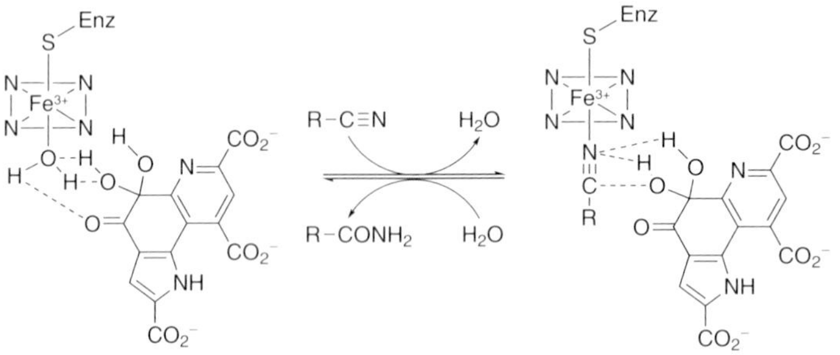

Figure 4.53 Proposed mechanism of nitrile hydration catalyzed by a nitrile hydratase involving Fe(III) and PQQ.

quinone (PQQ), leading to the proposed mechanism shown in Figure 4.53 (Sugiura *et al.*, 1987).

The crystal structure of nitrile hydratase from *Rhodococcus* sp. R312 suggested that the enzyme consists of two subunits (α, β), which contain one iron atom per α, β unit. The α-subunit consists of a long N-terminal and a C-terminal domain that forms a novel fold, which can be described as a four-layered α–β–β–α structure. The two subunits form a tight heterodimer that is the functional unit of the enzyme. The active site is located in a cavity at a subunit–subunit interface. The iron center is formed by residues from the α-subunit only – three cysteine thiolates and two main chain amide nitrogen atoms are ligands – although the iron center is located between the α- and β-subunits. Three possible catalytic roles for the metal ion in NHases were proposed, in which the metal ion always acts as a Lewis acid activating the nitrile for hydration. Further information on structure, regulation, and application of metallo nitrile hydratases can be found in a review (Kobayashi and Shimizu, 1998).

Various NHase genes have been cloned and characterized, and the amidase gene was found to be closely located to the NHase gene; this supported the theory that both enzymes are involved in the two-step degradation of nitriles to carboxylic acids.

Two applications based on nitrile hydratase from *Rh. rhodochrous* (Nagasawa and Yamada, 1995) have been commercialized (Figure 4.54). The large-scale production

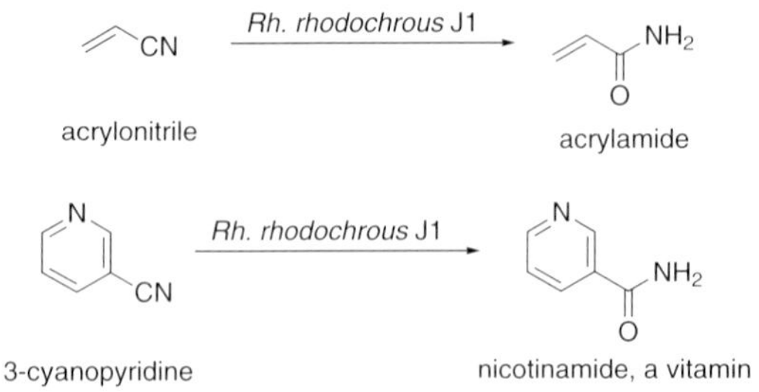

Figure 4.54 Commercial production of acrylamide and nicotinamide using resting cells of *Rh. rhodochrous* J1.

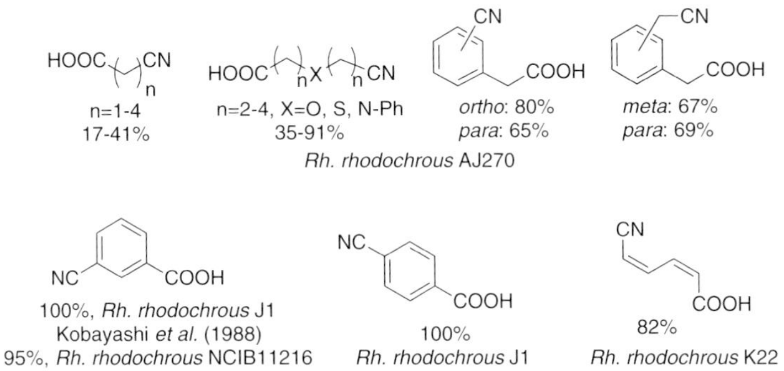

HOOC–(CH2)n–CN, n=1-4, 17-41%
HOOC–X–CN, n=2-4, X=O, S, N-Ph, 35-91%
ortho: 80%, para: 65%
meta: 67%, para: 69%
Rh. rhodochrous AJ270
100%, Rh. rhodochrous J1
Kobayashi et al. (1988)
95%, Rh. rhodochrous NCIB11216
100%
Rh. rhodochrous J1
82%
Rh. rhodochrous K22

Figure 4.55 Examples of the regioselective hydrolyses using nitrile hydratases and nitrilases.

of the commodity chemical acrylamide is performed by Nitto Chemical (Yokohama, Japan) on a >150 000 tons per year scale. Initially, strains from *Rhodococcus* sp. N-774 or *P. chlororaphis* B23 were used, but the current process uses the 10-fold more productive strain *Rh. rhodochrous* J1. The productivity is >7000 g acrylamide per gram of cells at a conversion of acrylonitrile of 99.97%. The formation of acrylic acid is barely detectable at the reaction temperature of 2–4 °C. In laboratory-scale experiments with resting cells, up to 656 g acrylamide per liter reaction mixture was achieved. Besides acrylamide, a wide range of other amides can also be produced, including acetamide ($150\,g\,l^{-1}$), isobutyramide ($100\,g\,l^{-1}$), methacrylamide ($200\,g\,l^{-1}$), propionamide ($560\,g\,l^{-1}$), and crotonamide ($200\,g\,l^{-1}$).

Furthermore, *Rh. rhodochrous* J1 also accepts aromatic and aryl aliphatic nitriles as substrates. For example, the conversion of 3-cyanopyridine to nicotinamide (a vitamin in animal feed supplementation) is catalyzed, and the process was industrialized by Lonza (Switzerland) on a >3000 tons per year scale.

A wide variety of dinitriles can be hydrolyzed with moderate to excellent regioselectivity to the corresponding monocarboxylic acids (Figure 4.55).

In the NHase/amidase system, it is usually only the amidase that shows stereoselectivity, while the nitrile hydratase is nonselective. Examples of kinetic resolutions include precursors of nonsteroidal anti-inflammatory drugs (e.g., ketoprofen, ibuprofen, naproxen) (Figure 4.56).

The strain *Alcaligenes faecalis* ATCC 8750 can be used for the production of mandelic acid (Yamamoto *et al.*, 1991). *A. faecalis* contains a nitrilase and an amidase,

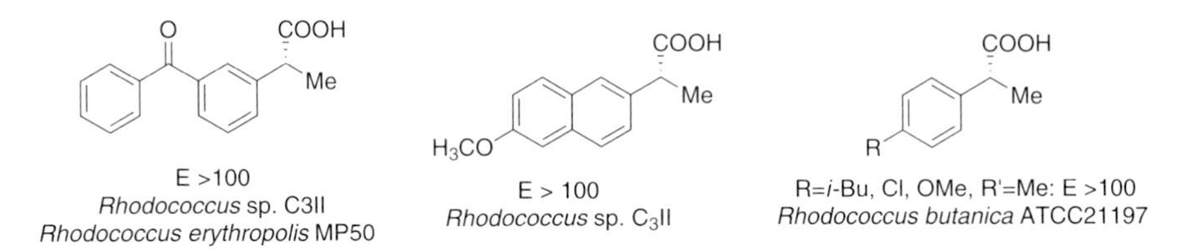

Figure 4.56 Examples of the synthesis of nonsteroidal anti-inflammatory drugs by hydrolysis of nitriles.

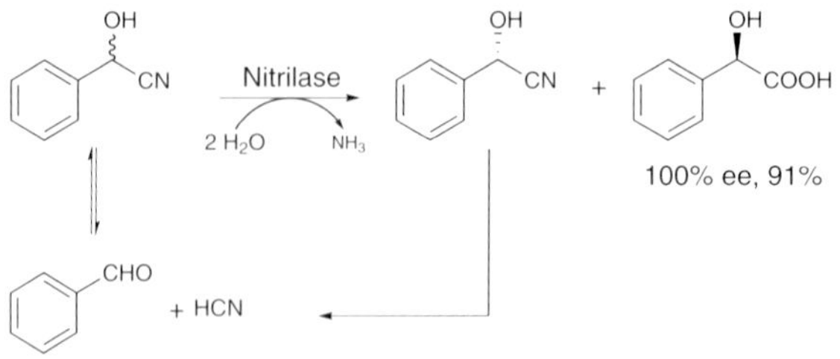

Figure 4.57 Synthesis of (*R*)-(−)-mandelic acid by dynamic kinetic resolution using a nitrilase involves *in situ* recycling of (*S*)-mandelonitrile by disproportion into benzaldehyde and HCN followed by formation of (*R*,*S*)-mandelonitrile.

but no NHase. Using resting cells, (*R*)-(−)-mandelic acid was formed in excellent optical purity (100% ee). Moreover, the remaining (*S*)-mandelonitrile racemized, resulting in an overall yield of 91% mandelonitrile after a 6 h reaction time. It was suggested that the rapid racemization was due to an equilibrium between mandelonitrile and benzaldehyde/HCN, because mandelic acid could also be obtained when only benzaldehyde and HCN were used as substrates (Figure 4.57).

By classic enrichment culture and screening, only approximately 15 nitrilases had been described. Researchers at Diversa (San Diego, USA; now Verenium Inc.) discovered more than 200 unique nitrilases from genomic libraries obtained from DNA extracted from environmental samples (DeSantis *et al.*, 2002). Twenty-seven enzymes afforded mandelic acid in >90% ee under conditions of dynamic kinetic resolution (Figure 4.57). Out of these biocatalysts, one nitrilase afforded (*R*)-mandelic acid in 86% yield and 98% ee. Further studies showed that this nitrilase also converted a broad range of mandelic acid derivatives and analogues with high activity and similar stereoselectivity. Another nitrilase exhibited high activity and broad substrate tolerance toward aryllactic acid derivatives, which were also converted in a dynamic kinetic resolution (Table 4.5).

A few groups have already investigated the hydrolysis of prochiral dinitriles. The hydrolysis of 3-hydroxyglutaronitriles revealed that Bn- or Bz-protecting groups

Table 4.5 Preparation of optically active aryllactic acid derivatives using a nitrilase under dynamic kinetic resolution conditions (modified from DeSantis *et al.*, 2002).

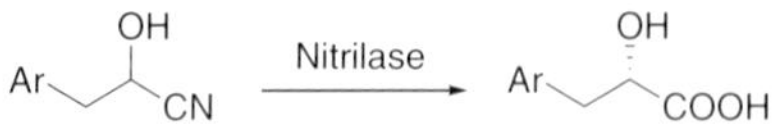

Ar	Relative activity (%)	ee (%)
C_6H_5	25	96
2-Me-C_6H_5	160	95
2-Br-C_6H_5	121	95
3-F-C_6H_5	22	99
1-Naphthyl	64	96
2-Pyridyl	10.5	99

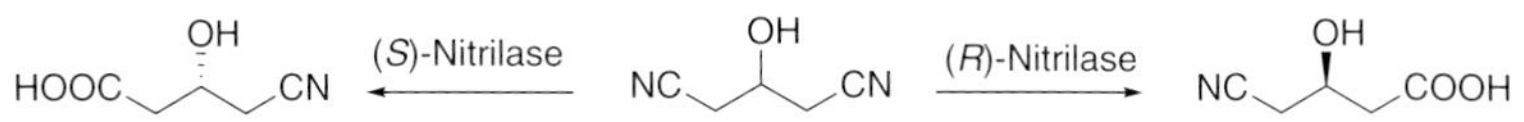

Figure 4.58 Desymmetrization of prochiral hydroxyglutaronitrile using engineered nitrilases.

were required to achieve acceptable enantiomeric excess for the monocarboxylic acid (Crosby, Parratt, and Turner, 1992). For the nitrilases discovered by Diversa, it could be shown that a range of enzymes show high conversion (>95%) and selectivity (>90% ee). The best enzyme gave 98% yield and 95% ee for the (*R*)-product (DeSantis *et al.*, 2002). In addition, 22 enzymes that afford the opposite enantiomer with 90–98% ee were discovered (Figure 4.58). In a later study, the most effective (*R*)-nitrilase was optimized by directed evolution (see Section 3.2.2) to withstand high substrate concentrations while maintaining high enantioselectivity. The best variant obtained by a "gene site saturation mutagenesis" technique contained a single mutation (Ala190His) and allowed the production of the (*R*)-acid at 3 M substrate concentration with 96% yield at 98.5% ee (DeSantis *et al.*, 2003).

More examples on the characterization and application of nitrilases and nitrile hydratases/amidases can be found in a review (Wieser and Nagasawa, 1999).

4.2.3.7 **Hydantoinases (EC 3.5.2.-)**

Hydantoinases are valuable enzymes for the production of optically pure D- and L-amino acids. (Hydantoinase is the commonly used name; more specifically, they are cyclic amidases. An alternative name is dihydropyrimidinase.) These enzymes catalyze the reversible hydrolytic cleavage of hydantoins and 5′-monosubstituted hydantoins. In combination with carbamoylases (EC 3.5.1.-), the reaction yields L- or D-amino acids, depending on the stereoselectivity of the enzymes (Figure 4.59). To shift the yield above 50%, a dynamic kinetic resolution starting from racemic hydantoins is also possible by either working at slightly alkaline pH values (pH >8) or using hydantoin-specific racemases. Since racemic hydantoins are readily available by chemical synthesis (e.g., Strecker synthesis), the synthesis of nonnatural amino acids is also feasible, if the enzyme has an appropriate substrate specificity.

It has been known since the 1940s that some microorganisms can grow on D,L-5-monosubstituted hydantoins, if these were used as sole carbon and nitrogen source. Especially since the 1970s, a broad range of hydantoinases and carbamoylases have been discovered by screening, owing to their importance in the synthesis of optically pure amino acids. Nowadays, hydantoinases from, for example, *Arthrobacter* sp., *Nocardia* sp., *Bacillus* sp., and *Pseudomonas* sp. have been described and a range of D-amino acids are accessible by using them. For a hydantoinase from *Arthrobacter aurescens* DSM 3745, it was shown that about 2.5 moles of Zn^{2+} is required per mole of subunit, and that these metal ions have both a catalytic and a structural function. Crystal structures have been elucidated for hydantoinases from *A. aurescens* and *Thermus* sp. and the D-carbamoylase from *Agrobacterium*. For this D-carbamoylase, the catalytic center was identified to consist of a glutamine, a lysine, and a cysteine residue (Altenbuchner, Siemann-Herzberg, and Syldatk, 2001; Syldatk *et al.*, 1999).

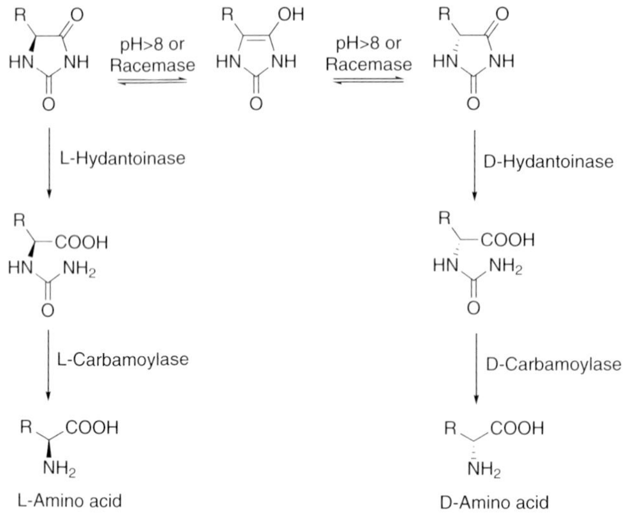

Figure 4.59 Synthesis of L- or D-amino acids using a combination of hydantoinase and carbamoylase. Complete conversion of racemic D,L-hydantoin can be achieved by racemization at alkaline pH with specific racemases.

Two different strategies have been described for the hydantoinase-based synthesis of D- or L-amino acids. In one approach, the carbamoyl formed is chemically hydrolyzed (with chemical racemization of the unwanted enantiomer). Alternatively, a combined one-pot process ("hydantoinase process") using a hydantoinase, a carbamoylase, and a racemase can be employed (Figure 4.59). This approach has the major advantage that an insufficient enantioselectivity of a hydantoinase can be overcome by the use of a more selective carbamoylase, racemization during the process enables directly up to 100% yield, and the use of chemicals and solvents can be avoided or considerably reduced.

The hydantoinase process for the production of L-methionine was improved by directed evolution (see Section 3.2.2) for the hydantoinase from *Arthrobacter* sp. DSM 9771 expressed in *E. coli*. A combination of error-prone PCR and saturation mutagenesis led to the identification of a variant with higher L-selectivity and fivefold increased specific activity. Recombinant technology was also used to clone and express together a D-hydantoinase from *Bacillus stearothermophilus* SD1 and a D-*N*-carbamoylase from *Agrobacterium tumefaciens* in *E. coli*. Both enzymes form approximately 20% of the total cell protein, and both proteins were expressed at a comparable level (ratio 1 : 1.2). Thus, D,L-*p*-hydroxyphenyl hydantoin ($30\,g\,l^{-1}$) was efficiently converted by this recombinant whole-cell catalyst to D-*p*-hydroxyphenylglycine – an important precursor in the synthesis of semisynthetic β-lactam antibiotics – within 15 h at 96% yield (Park, Kim, and Kim, 2000). A hydantoinase process using a genetically engineered *E. coli* expressing all three enzymes under the control of a rhamnose-inducible promoter was also described. With this system, L-tryptophan was produced with sixfold higher efficiency compared to the wild-type strain *A. aurescens* (Wiese *et al.*, 2001). In another example, D-hydantoinase, D-carbamoylase, and two different racemases all originating from *A. tumefaciens* strains were cloned and expressed in *E. coli* in different combinations. With these recombinant systems, complete conversion and excellent optical purity were reported for D-methionine, D-leucine, D-norleucine, D-norvaline, D-aminobutyric acid, and others starting from the corresponding racemic hydantoins (Martinez-Gomez *et al.*, 2007).

4.2.4 Lyases (EC 4)

4.2.4.1 Hydroxynitrile Lyases (EC 4.1.2.-)

Hydroxynitrile lyases (often also named oxynitrilases) catalyze the reversible stereoselective addition of HCN to aldehydes and ketones (Figure 4.60). Thus, HNLs allow the synthesis of optically pure compounds from prostereogenic substrates at (theoretically) quantitative yield. In addition, many HNLs show excellent stereoselectivity (for reviews, see Andexer *et al.*, 2009; Effenberger, Förster, and Wajant, 2000; Griengl, Schwab, and Fechter, 2000; Purkarthofer *et al.*, 2007). The resulting α-hydroxynitriles are versatile intermediates for a broad variety of chiral synthons, which can be obtained by subsequent chemical synthesis (Figure 4.61).

One of the earliest reports on biocatalysis involved the use of an HNL, when Rosenthaler used a hydroxynitrile lyase-containing extract (emulsin) for the preparation of (*R*)-mandelonitrile from benzaldehyde and HCN (Rosenthaler, 1908). However, little attention was paid to this discovery until the 1960s, when the corresponding enzyme was isolated, characterized, and used for the production of enantiomerically enriched (*R*)-cyanohydrins.

More than 3000 plant species are known to release HCN from their tissues, and for approximately 300 plants the HCN is released from a cyanogenic glycoside or lipid. These cyanide donors can be cleaved either spontaneously or by the action of an enzyme such as hydroxynitrile lyase.

Until the early 1990s, all HNLs were isolated from different plant sources, and about 11 enzymes with either (*S*)- or (*R*)-stereoselectivity were described. The oxynitrilase from *Prunus* sp. contains the cofactor FAD, but this is not involved in redox reactions. Instead, it seems to have a structure-stabilizing effect. Some of the HNLs are glycosylated and most of them are composed of subunits. The most thoroughly studied enzymes are listed in Table 4.6. As only small amounts of HNLs were available by extraction from plant sources, the enzymes from *Hevea brasiliensis* (rubber tree), *Manihot esculenta* (cassava), and *Linum usitatissimum* (flax) have been cloned and overexpressed in *E. coli* or *P. pastoris*. For instance, the gene encoding

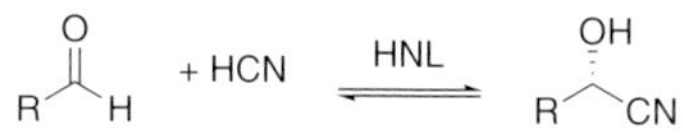

R–CHO + HCN ⇌(HNL) R–CH(OH)–CN

Figure 4.60 Example for the synthesis of optically active cyanohydrins from aldehydes and hydrogen cyanide catalyzed by a hydroxynitrile lyase (HNL).

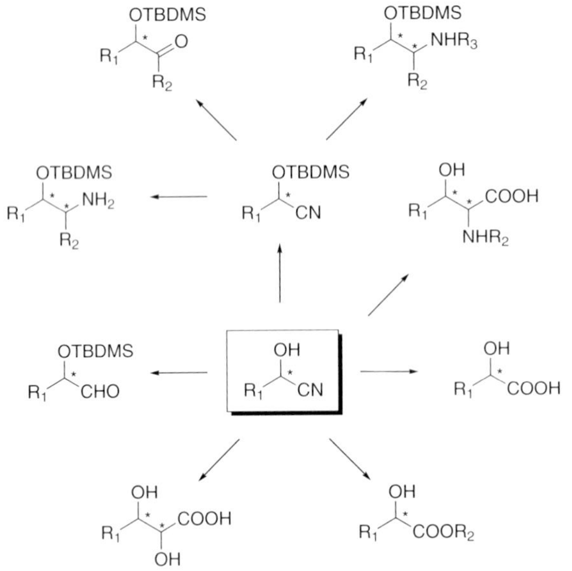

Figure 4.61 Examples of building blocks, which can be obtained by chemical synthesis from chiral cyanohydrins. Products with two stereocenters are accessible by these routes.

the enzyme from *H. brasiliensis* was expressed at high levels in the yeast *P. pastoris* under the control of the AOX1 promoter. On laboratory scale, a production of $23\,g\,l^{-1}$ of pure HNL was achieved in a high cell density fermentation, and the protein could be recovered by an one-step ion-exchange chromatography (Griengl, Schwab, and Fechter, 2000).

The crystal structures have been elucidated for enzymes from *H. brasiliensis*, *Sorghum bicolor*, *M. esculenta*, and *Prunus amygdalus* (almond). Interestingly, the enzymes share an α/β-hydrolase fold and also contain an active-site serine that is usually embedded in a GXSXG motif, similar to lipases and esterases (Jochens *et al.*, 2011). This might indicate an evolutionary relationship between these two

Table 4.6 Hydroxynitrile lyases for organic synthesis.

Enzyme	Specificity	Molecular weight (kDa)	pH optimum	Substrate spectrum
S. bicolor	*S*	105	n.d.	Aromatic aldehydes and ketones
M. esculenta	*S*	92–124	5.4	Broad
H. brasiliensis	*S*	58	5.5–6.0	Broad
L. usitatissimum	*R*	82	5.5	Aliphatic aldehydes and ketones
Prunus sp.	*R*	55–80	5.5	Broad

enzyme classes, hydrolases and lyases, although recent studies have suggested that HNLs obey a different mechanism. Indeed, it was possible to convert a plant esterase, SABP2, into a hydroxynitrile lyase using just two amino acid substitutions (G12T-M239K) even if the catalytic efficiency in the release of cyanide from mandelonitrile (20 mU mg^{-1}, $k_{cat}/K_m = 70\ M^{-1}\ min^{-1}$) was rather low (Padhi *et al.*, 2010).

The following mechanism was proposed for the HNL from *S. bicolor*. In contrast to a histidine residue serving as a general base in serine peptidases and lipases, the carboxylate group of a C-terminal tryptophan (Trp_{270} in *S. bicolor*) abstracts a proton from the cyanohydrin hydroxyl group. A water molecule bound to the active site appears to be involved in proton transfer. Thus, the entering cyanohydrin is hydrogen bonded to Ser_{158} and Trp_{270}, which abstracts a proton from the OH group of the substrate. Next, a proton is transferred from the tryptophan via the active-site water to the nitrile leaving group. Protonation of the cyanide ion results in the products 4-hydroxybenzaldehyde and HCN (Figure 4.62). This model can also explain the (*S*)-stereoselectivity observed for the reverse reaction (Lauble *et al.*, 2002).

Besides the availability of HNLs, two further problems had to be solved to allow an efficient application of HNLs in organic synthesis: (1) performing the reactions in water-immiscible organic solvents; and (2) reaction at low pH, typically in the range of pH 4–5. Both methods suppress the competing chemical reaction, which results in lower optical purities of the product. This is shown in Table 4.7 for the synthesis of several aliphatic α-hydroxynitriles using HNL from *P. amygdalus*.

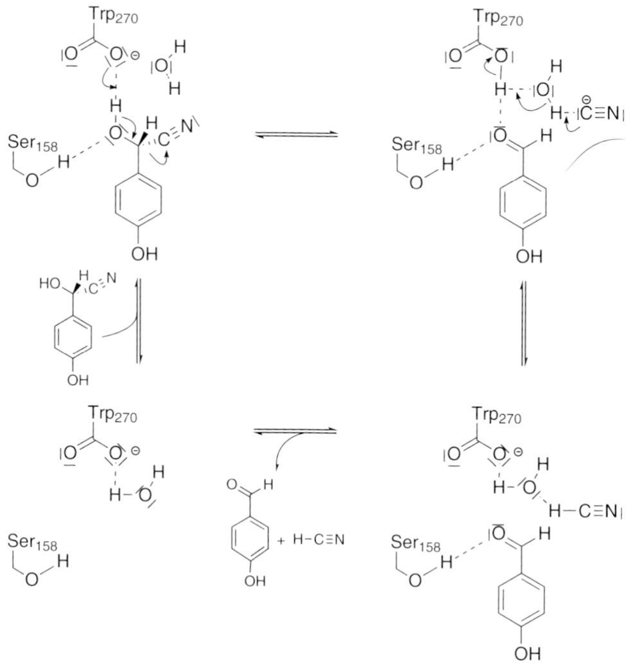

Figure 4.62 Suggested reaction mechanism for hydroxynitrile lyase from *S. bicolor*.

Table 4.7 Preparation of (*R*)-cyanohydrins from aldehydes using HNL from *P. amygdalus* (PaHNL) in organic solvent or buffer (modified from Fessner, 2000).

R–CHO + HCN → [(*R*)-PaHNL; *i*Pr_2O/Avicel or H_2O/EtOH] → R–C(OH)(H)–CN

R	H_2O/EtOH		$^iPr_2O$/Avicel	
	Yield (%)	ee (%)	Yield (%)	ee (%)
Ph	99	86	96	99
3-PhO-C_6H_4	99	11	99	98
C_3H_7	75	69	99	98
$(CH_3)_3C$	56	45	84	83

In contrast to the enzymes from *H. brasiliensis* and *M. esculenta*, where aliphatic and aromatic aldehydes function as substrates, the HNL from *S. bicolor* only catalyzes the formation and cleavage of aromatic (*S*)-cyanohydrins. The other enzymes listed in Table 4.6 show (*R*)-selectivity.

A broad range of aldehydes and ketones were converted into the corresponding α-hydroxynitriles using HNLs. Figure 4.63 shows some products obtained at high yields and with good to excellent optical purities using the enzymes listed in Table 4.6. As mentioned above, the use of water-immiscible solvents is

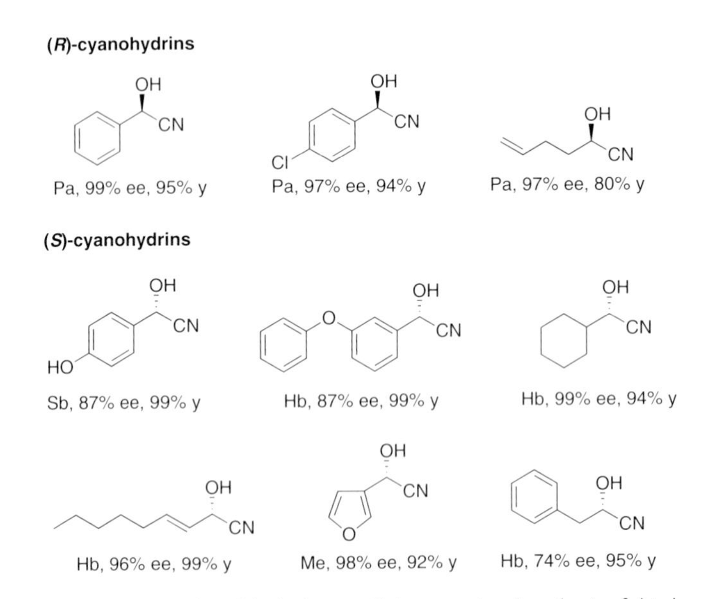

Figure 4.63 Selected examples of the hydroxynitrile lyase-catalyzed synthesis of chiral cyanohydrins. Pa, HNL from *P. amygdalus*; Sb, HNL from *S. bicolor*; Hb, HNL from *H. brasiliensis*; Me, HNL from *M. esculenta*.

Table 4.8 Preparation of (*R*)-cyanohydrins from ketones using HNL from *P. amygdalus* (PaHNL) in organic solvent or buffer (modified from Fessner, 2000).

R–CO–CH_3 + HCN —(*R*)-PaHNL, *i*Pr_2O or citrate buffer→ R–C(OH)(CH_3)CN

R	${}^{i}Pr_2O$/Avicel		Citrate buffer	
	Yield (%)	ee (%)	Yield (%)	ee (%)
C_3H_7	70	97	78	95
C_4H_9	90	98	94	98
C_5H_{11}	88	57	56	96
$(CH_3)_2CHCH_2$	57	98	40	98

recommended for organic synthesis with HNLs in order to avoid a nonstereoselective chemical reaction. As seen in Table 4.7, higher yields and excellent optical purities were only possible using PaHNL immobilized on Avicel (a cellulose membrane) in diisopropyl ether. In contrast, reactions in water/ethanol mixtures gave products with inferior optical purity, for example, only 11% ee for the *m*-substituted phenyl derivative. This effect, however, was less pronounced for reactions using ketones (Table 4.8).

An alternative to using the highly toxic HCN is to perform the reaction as a transcyanation (or transhydrocyanation), in which aromatic or aliphatic aldehydes are reacted with acetone cyanohydrin as shown for the (*R*)-HNL from almond (*P. amygdalus*) (Figure 4.64). However, the optical purity of the resulting products was slightly lower compared to reactions involving free HCN.

Very recently, processes for the production of (*S*)-cyanohydrins catalyzed by HNLs from *H. brasiliensis* and *M. esculenta* and the subsequent chemical hydrolysis to (*S*)-hydroxycarboxylic acids have been developed. The synthesis of (*S*)-*m*-phenoxybenzaldehyde cyanohydrin – an intermediate in the synthesis of pyrethroids – in a biphasic system has been commercialized (Figure 4.65) by DSM Chemie Linz and Nippon Shokubai.

4.2.4.2 Aldolases (EC 4.1.2.-, EC 4.1.3.-)

Aldolases catalyze the biological equivalent to the chemical aldol reaction, the formation of carbon–carbon bonds by (reversible) stereocontrolled addition of a nucleophilic ketone to an electrophilic aldehyde acceptor. Aldolases are usually classified

ketone or aldehyde (R_1–CO–R_2) + acetone cyanohydrin ⇌(HNL) acetone + chiral cyanohydrin (R_1R_2C(OH)CN)

Figure 4.64 The principle of transhydrocyanation.

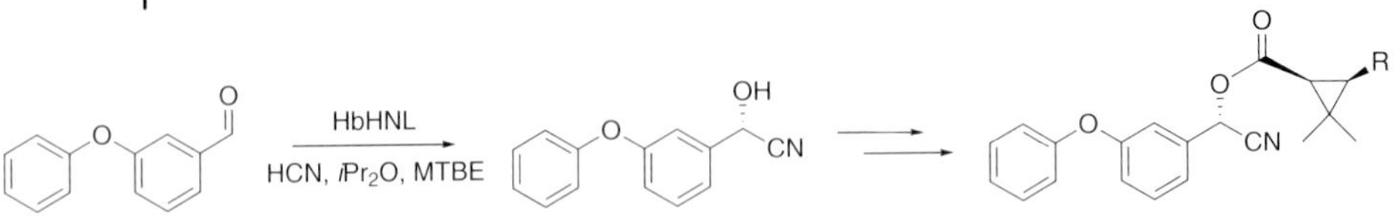

Figure 4.65 Synthesis of (S)-m-phenoxybenzaldehyde cyanohydrin as an intermediate of pyrethroids using a recombinant HNL from *H. brasiliensis*.

according to the nature of the nucleophilic component into (1) pyruvate (and phosphoenolpyruvate)-, (2) dihydroxyacetone phosphate-, (3) acetaldehyde-, and (4) glycine-dependent enzymes. The most important ones in organic synthesis use dihydroxyacetone phosphate (DHAP) as they allow the formation of two new stereocenters in a single reaction. A range of corresponding enzymes was also identified. The ability of aldolases to accept a variety of unnatural acceptor substrates and to generate new stereocenters of known absolute and relative stereochemistry makes them powerful tools for asymmetric synthesis.

Depending on their mechanism, aldolases are classified into type I and type II enzymes. Type I enzymes are predominantly found in higher plants and animals and are metal cofactor independent. The free amino group of a lysine residue in the active site reacts with DHAP under the formation of a Schiff base intermediate. An enamine is formed after deprotonation, which then attacks the aldehyde. Finally, the Schiff base intermediate product decomposes after reaction with water and the aldol and enzyme are released (Figure 4.66). Type II aldolases occur mostly in bacteria and fungi, and are Zn^{2+} dependent. The zinc ion acts as a Lewis acid, which polarizes the carbonyl group in DHAP (Figure 4.66).

As pointed out above, aldolases are highly specific and the stereochemical outcome of an aldol reaction can be usually predicted independently of the substrate

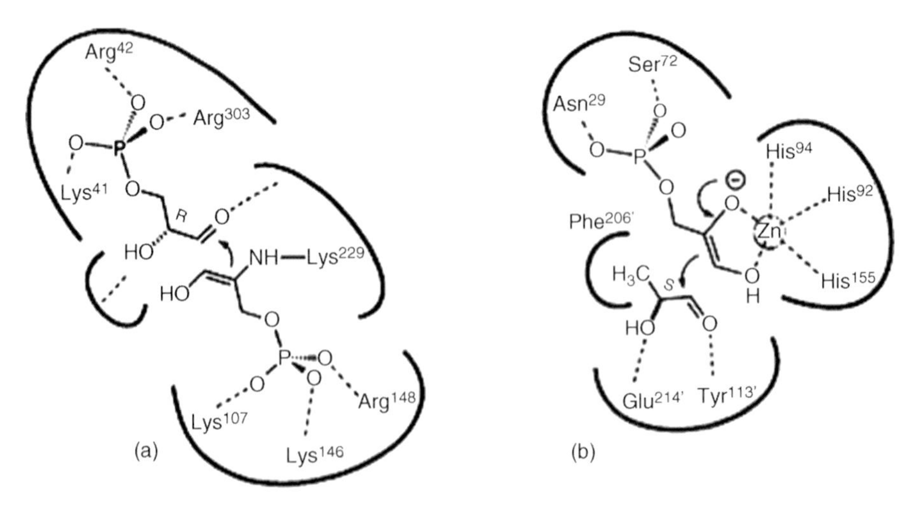

Figure 4.66 Mechanism of type I (a) and type II (b) aldolases. The example shows dihydroxyacetone phosphate (DHAP) as nucleophilic ketone.

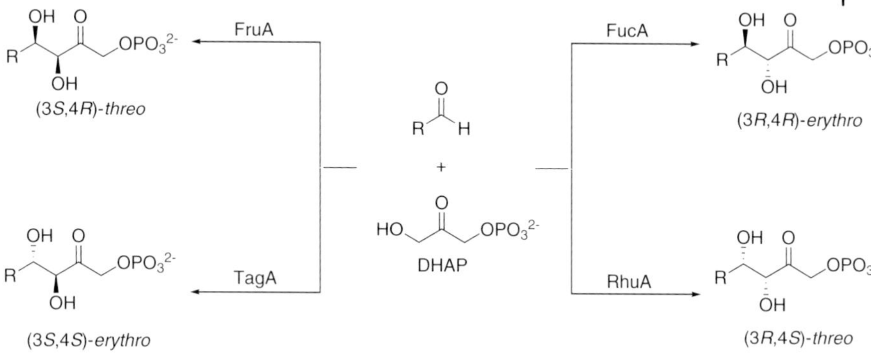

Figure 4.67 Aldolase reactions catalyzed by the four stereocomplementary aldolases FruA (fructose-1,6-diphosphate aldolase), FucA (fuculose-1-phosphate aldolase), TagA (tagatose-1,6-diphosphate aldolase), and RhuA (rhamnulose-1-phosphate aldolase).

structure. Consequently, the synthesis of all four diastereomers accessible from DHAP and an aldehyde is possible by using four different aldolases (Figure 4.67).

Numerous examples for the application of aldolases can be found in a number of excellent reviews (Clapes *et al.*, 2010; Mlynarski and Paradowska, 2008) and book chapters. Only two types of aldolases are described in the following paragraphs.

DHAP-Dependent Aldolases The most often used DHAP-dependent enzymes are Fru-aldolases (FruA, often also abbreviated to FDP aldolases; EC 4.1.2.13), which catalyze the reaction between DHAP and D-glyceraldehyde-3-phosphate to form D-fructose diphosphate (FDP). The equilibrium constant for this reaction is $\sim 10^4\,M^{-1}$ in favor of FDP formation. The enzyme has been isolated from various eukaryotic and prokaryotic sources, and both type I and type II biocatalysts have been described. Type I enzymes are usually tetramers of 160 kDa molecular weight, while type II FDP aldolases are dimers (~80 kDa). Sequence homologies are usually very low between type I and II aldolases, and especially for in-between type II enzymes. The crystal structures of the FDP aldolase from rabbit muscle (RAMA) and others have been determined. RAMA and several type II aldolases from microbial sources have been cloned and overexpressed, which substantially facilitates access to the enzymes. Aldolases have been widely employed in carbohydrate synthesis, and a few examples of products are shown in Figure 4.68.

The phosphorylated ketone DHAP must be available for these aldolase-catalyzed reactions. Although DHAP is, in principle, available via the reverse reaction – that is, from a retro-aldol reaction with FDP aldolase from FDP – this approach requires triosephosphate isomerase and can hardly be used, if a different aldolase is used in the forward aldol reaction, which then leads to a complex product mixture. The kinase-catalyzed phosphorylation of dihydroxyacetone is hampered by the requirement for ATP regeneration. A chemical alternative is a multistep synthesis, which usually provides a DHAP dimer that is stable and can easily be converted into DHAP by acid hydrolysis. Alternatively, arsenate derivatives of DHAP can be used, albeit at lowered aldolase activity.

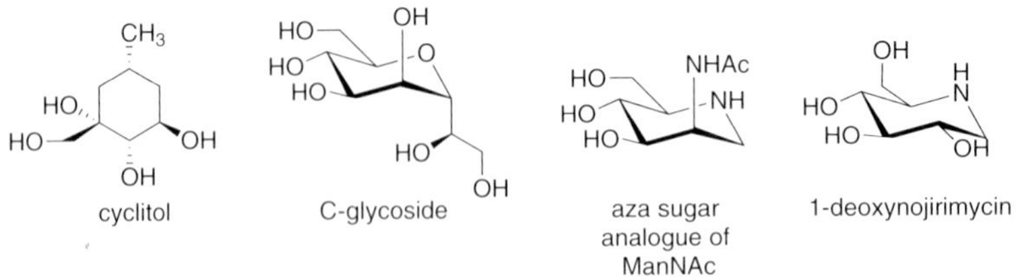

Figure 4.68 Selected examples of molecules synthesized with FDP aldolase (ManNAc, *N*-acetyl-D-mannosamine).

An alternative approach based on phosphorylation of glycerol using a phytase – a cheap and readily available enzyme – and inorganic pyrophosphate was also described. The feasibility of this reaction was demonstrated for the synthesis of 5-deoxy-5-ethyl-D-xylulose in a one-pot reaction combining four enzymatic steps with four different enzymes starting from glycerol (Figure 4.69). First, phosphorylation of glycerol by reaction with pyrophosphate in the presence of phytase at pH 4.0 in 95% glycerol afforded racemic glycerol-3-phosphate quantitatively. The L-enantiomer is then oxidized with glycerol phosphate oxidase (GPO) to DHAP under aerobic conditions at pH 7.5. Hydrogen peroxide is removed with the aid of catalase. *In situ*-generated DHAP reacts with butanal using FruA, followed by dephosphorylation of the aldol adduct, again using phytase at pH 4. Overall, 5-deoxy-5-ethyl-D-xylulose was obtained in 57% yield from L-glycerol-3-phosphate. The phytase "on/off switch," by changing the pH value, was the key to controlling phosphorylation and dephosphorylation (Schoevaart *et al.*, 2000).

Pyruvate/Phosphoenolpyruvate-Dependent Aldolases The best-studied enzyme utilizing pyruvate is *N*-acetylneuraminate (NeuAc) aldolase (EC 4.1.3.3). This enzyme catalyzes the reversible condensation of pyruvate with *N*-acetylmannosamine (ManNAc) to form sialic acid (α-NeuAc). The initial products of aldol cleavage are α-ManNAc and pyruvate. Although *in vivo* the equilibrium favors the

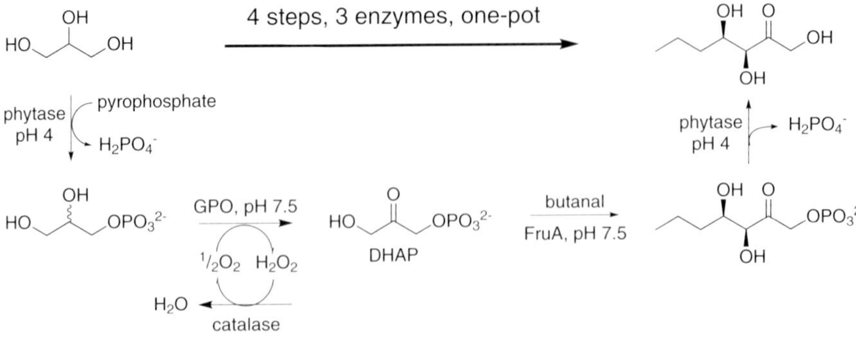

Figure 4.69 Enzymatic total synthesis of 5-deoxy-5-ethyl-xylulose with *in situ* formation of dihydroxyacetone phosphate. The key to success was the switch in pH from 4.0 (phytase reaction) to 7.5 (glycerol phosphate oxidase (GPO) and aldolase (FruA) reaction).

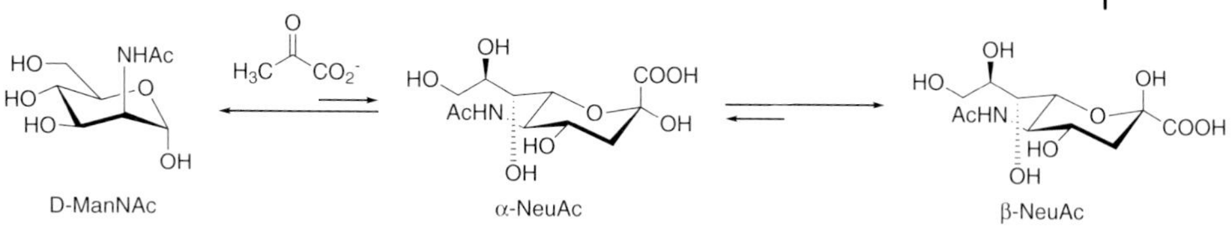

Figure 4.70 *In vivo* aldol reaction catalyzed by NeuAc aldolase.

retro-aldol reaction ($K_{eq} \sim 12.7\ M^{-1}$), the aldol reaction for organic synthesis can be achieved using excess pyruvate, which yields the β-anomer of NeuAc (Figure 4.70). NeuAc aldolase is a Schiff base type I aldolase, and has been isolated from both bacteria and animals. The enzymes from *Clostridium* and *E. coli* are commercially available, and the *E. coli* enzyme has been cloned and overexpressed. The optimum pH is 7.5, but activity is retained between pH 6 and 9. NeuAc aldolase can be used in solution, in immobilized form, or enclosed in a membrane system. With this aldolase, glycoconjugates – which play important roles in cell–cell interactions and cell adhesion – have been synthesized on the multigram scale.

4.2.5
Isomerases (EC 5)

The most important enzymes for organic synthesis in this class are racemases and epimerases (both EC 5.1.-.-), both of which catalyze the inversion of stereocenters. Racemases convert an enantiomer in a racemate, while epimerases convert one diastereomer selectively to another diastereomer. Well-studied enzymes in this class are glucose–fructose isomerase (EC 5.3), which is used in the synthesis of high-fructose corn syrup (see Section 12.1.2) and *cis–trans* isomerases.

Although racemases "destroy" the optical purity of a given chiral compound, they are very important for biocatalysis. If in a kinetic resolution one enantiomer is not needed, it can be racemized and again subjected to a resolution that eventually yields 100% product after several iterative cycles (recycling), or directly in a dynamic kinetic resolution. This has already been exemplified for the hydantoinase process in Section 4.2.3.7. In addition, racemases have also been used to increase access to the D-enantiomers of amino acids.

Most racemases described in the literature act on α-amino carboxylic acids, with the prominent exception of mandelate racemase (Kenyon *et al.*, 1995; St. Maurice and Bearne, 2004). The majority of racemases and epimerases act at a stereocenter adjacent to a carbonyl functionality, and reversibly cleave a C—H bond by lowering the pK_a of the hydrogen through stabilization of the resulting anion. As this anionic intermediate is planar and reprotonation can occur from both sides at identical probability, a racemic (or epimeric) product is formed.

Racemases are classified into two different types, depending on whether they require the cofactor pyridoxal-5′-phosphate, or not. The reaction mechanism for PLP-dependent enzymes is based on the formation of an imine linkage between the PLP cofactor and the substrate. This greatly acidifies the α-proton, resulting in

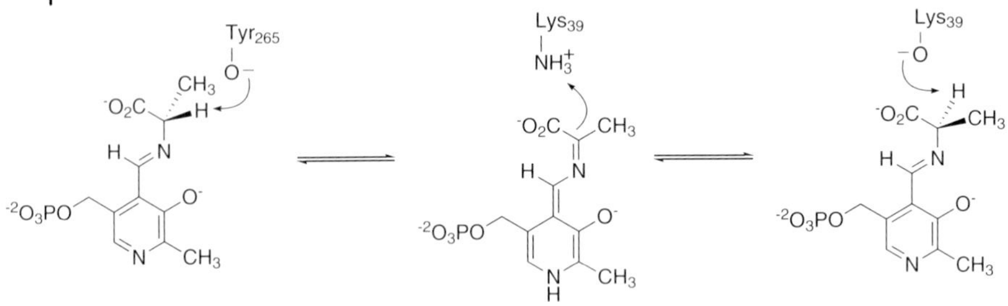

Figure 4.71 Mechanism of the PLP-dependent alanine racemase.

a lower pK_a and, consequently, an easier abstraction of the proton. This proton is then returned to the imine, the result being a 1 : 1 mixture of both enantiomers (Figure 4.71).

For PLP-independent racemases (e.g., glutamate racemase), a two-base mechanism (Figure 4.72) was suggested based on mechanistical studies with deuterated substrates. Mutational studies on glutamate racemase from *Lactobacillus fermenti* identified two cysteine residues (Cys_{73} and Cys_{184}) as catalytic acid/base residues.

Alanine racemase (EC 5.1.1.1) is a bacterial enzyme that catalyzes the racemization of alanine and requires PLP. It plays an important role in the bacterial growth by providing D-alanine used in peptidoglycan assembly. Several enzymes have been identified from different bacterial sources, and these have also been cloned. The enzyme has been used in biocatalysis for the production of various D-amino acids by combination of L-alanine dehydrogenase, D-amino acid aminotransferase, and formate dehydrogenase.

Glutamate racemases have been identified in various microorganisms, and the genes from *Lactobacillus* sp. and *Bacillus* sp. have been cloned and overexpressed. The structure of the enzyme from *Aquifex pyrophilus* has been determined using X-ray crystallography, and revealed that the enzyme is composed of two α/β-fold domains. The enzyme was described for the production of D-valine, D-alanine, D-aspartate, and other D-amino acids at high yields.

N-Acylamino Acid Racemases *N*-Acylamino acids racemize more easily than the corresponding amino acids. Therefore, the combination of chemical racemization and enantioselective hydrolysis of *N*-acylamino acids by aminoacylase (EC 3.5.1.14) has been used for the synthesis of L-tryptophan. Unfortunately, the reaction conditions for the chemical and the enzymatic step are very different, and therefore the

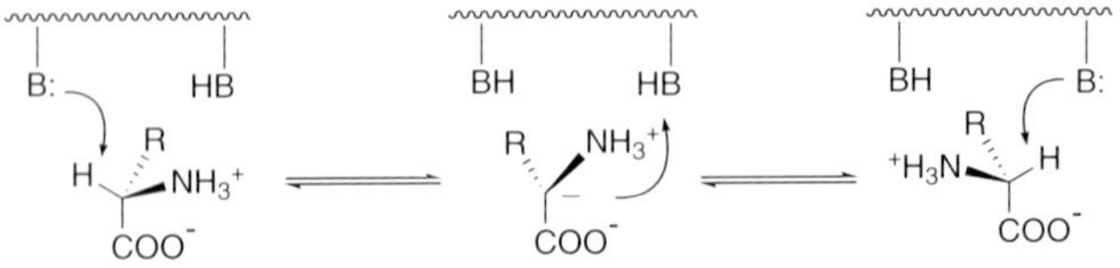

Figure 4.72 Mechanism of a PLP-independent racemase, that is, glutamate racemase.

resolution step and the racemization must be performed separately. Researchers at Takeda Chemical Industries (Japan) succeeded in identifying a racemase from *Streptomyces atratus* Y-53 that acts only on the *N*-acylamino acid but not on the free amino acid; this facilitates a dynamic kinetic resolution starting from D,L-*N*-acylamino acids. This racemase acts on *N*-acetyl D,L-methionine, *N*-acetyl L-valine, *N*-acetyl L-tyrosine, and *N*-chloroacetyl L-valine, but not on methyl or ethyl esters of *N*-acylamino acids.

Exercises

4.1 Explain the concept of a dynamic kinetic resolution.
4.2 Discuss the pros and cons of a process for the synthesis of an optically pure alcohol using either a dynamic kinetic resolution with a hydrolase as key enzyme or an asymmetric synthesis using the reduction of a ketone by an alcohol dehydrogenase.
4.3 Discuss the pros and cons of a P450-catalyzed hydroxylation in a whole-cell system or with isolated enzymes. What determines which process is the method of choice?
4.4 Which enzymes can be used for the synthesis of optically pure α-amino acids?
4.5 It is known that acetaldehyde released in the lipase-catalyzed resolution using vinyl esters as acyl donors can inactivate the biocatalyst. How can you circumvent this problem?
4.6 How would you screen a culture collection for the presence of racemase activity?
4.7 Describe possible enzymatic routes toward optically pure mandelic acid (α-hydroxyphenylacetic acid) and discuss the pros and cons of each route.

Literature

Textbooks

These provide general information relating to the use of enzymes in organic synthesis.

Bommarius, A.S. and Riebel, B.R. (2004) *Biocatalysis*, Wiley-VCH Verlag GmbH, Weinheim.

Bornscheuer, U.T. and Kazlauskas, R.J. (2006) *Hydrolases in Organic Synthesis – Regio- and Stereoselective Biotransformations*, 2nd edn, Wiley-VCH Verlag GmbH, Weinheim.

May, O., Gröger, H. and Drauz, K. (eds) (2012) *Enzyme Catalysis in Organic Synthesis*, 3rd edn, vols. 1–3, Wiley-VCH Verlag GmbH, Weinheim.

Faber, K. (2004) *Biotransformations in Organic Chemistry*, 5th edn, Springer, Berlin.

Patel, R.N. (ed.) (2000) *Stereoselective Biocatalysis*, Marcel Dekker, New York.

Patel, R.N. (ed.) (2006) *Biocatalysis in the Pharmaceutical and Biotechnology Industries*, CRC Press, Boca Raton, FL.

References

Alexeeva, M., Enright, A., Dawson, M.J., Mahmoudian, M., and Turner, N.J. (2002) Deracemization of alpha-methylbenzylamine using an enzyme obtained by *in vitro* evolution. *Angew. Chem., Int. Ed.*, **41**, 3177–3180.

Altenbuchner, J., Siemann-Herzberg, M., and Syldatk, C. (2001) Hydantoinases and related enzymes as biocatalysts for the synthesis of unnatural chiral amino acids. *Curr. Opin. Biotechnol.*, **12**, 559–563.

Andexer, J.N., Langermann, J.V., Kragl, U., and Pohl, M. (2009) How to overcome limitations in biotechnological processes – examples from hydroxynitrile lyase applications. *Trends Biotechnol.*, **27**, 599–607.

Archelas, A., Furstoss, R. (1998) Epoxide hydrolases: new tools for the synthesis of fine chemicals. *Trends Biotechnol.*, **16**, 108–116.

Azzolina, O., Vercesi, D., Collina, S., and Ghislandi, V. (1995) Chiral resolution of methyl 2-aryloxypropionates by biocatalytic stereospecific hydrolysis. *Il Farmaco*, **50**, 221–226.

Baeyer, A. and Villiger, V. (1899) Einwirkung des Caro'schen Reagens auf Ketone. *Chem. Ber.*, **32**, 3625–3633.

Bailey, K.R., Ellis, A.J., Reiss, R., Snape, T.J., and Turner, N.J. (2007) A template-based mnemonic for monoamine oxidase (MAO-N) catalyzed reactions and its application to the chemo-enzymatic deracemisation of the alkaloid (±)-crispine A. *Chem. Commun.*, **44**, 3640–3642.

Baillargeon, M.W. and McCarthy, S.G. (1991) *Geotrichum candidum* NRRL Y-553 lipase: purification, characterization and fatty acid specificity. *Lipids*, **26**, 831–836.

Balke, K., Kadow, M., Mallin, H., Saß, S., and Bornscheuer, U.T. (2012) Discovery, application and protein engineering of Baeyer-Villiger monooxygenases in organic synthesis. Org. Biomol. Chem., **10**, 6249–6265.

Balkenhohl, F., Ditrich, K., Hauer, B., and Ladner, W. (1997) Optically active amines via lipase-catalyzed methoxyacetylation. *J. Prakt. Chem.*, **339**, 381–384.

Bartsch, S., Kourist, R., and Bornscheuer, U.T. (2008) Complete inversion of enantioselectivity towards acetylated tertiary alcohols by a double mutant of a *Bacillus subtilis* esterase. *Angew. Chem., Int. Ed.*, **47**, 1508–1511.

Bdjìc, J.D., Kadnikova, E.N., and Kostic, N.M. (2001) Enantioselective aminolysis of an α-chloroester catalyzed by *Candida cylindracea* lipase encapsulated in sol-gel silica glass. *Org. Lett.*, **3**, 2025–2028.

Biermann, U., Bornscheuer, U., Meier, M.A., Metzger, J.O., and Schäfer, H.J. (2011) Oils and fats as renewable raw materials in chemistry. *Angew. Chem., Int. Ed.*, **50**, 3854–3871.

Bommarius, A.S. (2002) *Enzyme Catalysis in Organic Synthesis*, 2nd edn, vol. 3 (eds K. Drauz and H. Waldmann), Wiley-VCH Verlag GmbH, Weinheim, pp. 1047–1063.

Bommarius, A.S., Drauz, K., Groeger, U., and Wandrey, C. (1992) *Chirality in Industry* (eds A.N. Collins, G.N. Sheldrake, and J. Crosby), John Wiley & Sons, Ltd, London, pp. 371–397.

Bonsor, D., Butz, S.F., Solomons, J., Grant, S., Fairlamb, I.J.S., Fogg, M.J., and Grogan, G. (2006) Ligation independent cloning (LIC) as a rapid route to families of recombinant biocatalysts from sequenced prokaryotic genomes. *Org. Biomol. Chem.*, **4**, 1252–1260.

Bordusa, F. (2002) Proteases in organic synthesis. *Chem. Rev.*, **102**, 4817–4867.

Bornscheuer, U.T. (2002) Microbial carboxyl esterases: classification, properties and application in biocatalysis. *FEMS Microbiol. Rev.*, **26**, 73–81.

Bornscheuer, U.T. and Kazlauskas, R.J. (2005) *Hydrolases in Organic Synthesis – Regio- and Stereoselective Biotransformations*, Wiley-VCH Verlag GmbH, Weinheim.

Bougioukou, D.J., Kille, S., Taglieber, A., and Reetz, M.T. (2009) Directed evolution of an enantioselective enoate-reductase: testing the utility of iterative saturation mutagenesis. *Adv. Synth. Catal.*, **351**, 3287–3305.

Breuer, M., Ditrich, K., Habicher, T., Hauer, B., Keßeler, M., Stürmer, R., and Zelinski, T. (2004) Industrial methods for the production of optically active intermediates. *Angew. Chem., Int. Ed.*, **43**, 788–824.

Britton, L.N., Brand, J.M., and Markovetz, A.J. (1974) Source of oxygen in the conversion of 2-tridecanone to undecyl acetate by *Pseudomonas cepacia* and *Nocardia* sp. *Biochim. Biophys. Acta*, **369**, 45–49.

Brundiek, H.B., Evitt, A.S., Kourist, R., and Bornscheuer, U.T. (2012) Creation of a lipase highly selective for trans fatty acid by protein engineering, *Angew. Chem. Int. Ed.*, **51**, 412–414.

Carr, R., Alexeeva, M., Dawson, M.J., Gotor-Fernandez, V., Humphrey, C.E., and Turner, N.J. (2005) Directed evolution of an amine oxidase for the preparative deracemisation of cyclic secondary amines. *ChemBioChem*, **6**, 637–639.

Carr, R., Alexeeva, M., Enright, A., Eve, T.S.C., Dawson, M.J., and Turner, N.J. (2003) Directed evolution of an amine oxidase possessing both broad substrate specificity and high enantioselectivity. *Angew. Chem., Int. Ed.*, **42**, 4807–4810.

Chen, C.S., Fujimoto, Y., Girdaukas, G., and Sih, C.J. (1982) Quantitative analyses of biochemical kinetic resolutions of enantiomers. *J. Am. Chem. Soc.*, **104**, 7294–7299.

Chen, C.S., Wu, S.H., Girdaukas, G., and Sih, C.J. (1987) Quantitative analyses of biochemical kinetic resolution of enantiomers. 2. Enzyme-catalyzed esterifications in water–organic solvent biphasic systems. *J. Am. Chem. Soc.*, **109**, 2812–2817.

Cirino, P.C. and Arnold, F.H. (2002) Protein engineering of oxygenases for biocatalysis. *Curr. Opin. Chem. Biol.*, **6**, 130–135.

Clapes, P., Fessner, W.D., Sprenger, G.A., and Samland, A.K. (2010) Recent progress in stereoselective synthesis with aldolases. *Curr. Opin. Chem. Biol.*, **14**, 154–167.

Criegee, R. (1948) Die Umlagerung der Dekalin-peroxydester als Folge von kationischem Sauerstoff. *Justus Liebigs Ann. Chem.*, **560**, 127–141.

Cripps, R.E. (1975) The microbial metabolism of acetophenone. *Biochem. J.*, **152**, 233–241.

Cripps, R.E., Trudgill, P.W., and Whateley, J.G. (1978) The metabolism of 1-phenylethanol and acetophenone by *Nocardia* T5 and an *Arthrobacter* species. *Eur. J. Biochem*, **86**, 175–186.

Crosby, J.A., Parratt, J.S., and Turner, N.J. (1992) Enzymic hydrolysis of prochiral dinitriles. *Tetrahedron: Asymmetry*, **3**, 1547–1550.

de Gonzalo, G., Mihovilovic, M.D., and Fraaije, M.W. (2010) Recent developments in the application of Baeyer–Villiger monooxygenases as biocatalysts. *ChemBioChem*, **11**, 2208–2231.

Desai, A.A. (2011) Sitagliptin manufacture: a compelling tale of green chemistry, process intensification, and industrial asymmetric catalysis. *Angew. Chem., Int. Ed.*, **50**, 1974–1976.

DeSantis, G., Wong, K., Farwell, B., Chatman, K., Zhu, Z., Tomlinson, G., Huang, H., Tan, X., Bibbs, L., Chen, P., Kretz, K., and Burk, M.J. (2003) Creation of a productive, highly enantioselective nitrilase through gene site saturation mutagenesis (GSSM). *J. Am. Chem. Soc.*, **125**, 11476–11477.

DeSantis, G., Zhu, Z., Greenberg, W.A., Wong, K., Chaplin, J., Hanson, S.R., Farwell, B., Nicholson, L.W., Rand, C.L., Weiner, D. P., Robertson, D.E., and Burk, M.J. (2002) An enzyme library approach to biocatalysis: development of nitrilases for enantioselective production of carboxylic acid derivatives. *J. Am. Chem. Soc.*, **124**, 9024–9025.

de Vries, E.J. and Janssen, D.B. (2003) Biocatalytic conversion of epoxides. *Curr. Opin. Biotechnol.*, **14**, 414–420.

Donoghue, N.A., Norris, D.B., and Trudgill, P.W. (1976) The purification and properties of cyclohexanone oxygenase from *Nocardia globerula* CL1 and *Acinetobacter* NCIB 9871. *Eur. J. Biochem.*, **63**, 175–192.

Dunsmore, C.J., Carr, R., Fleming, T., and Turner, N.J. (2006) A chemo-enzymatic route to enantiomerically pure cyclic tertiary amines. *J. Am. Chem. Soc.*, **128**, 2224–2225.

Effenberger, F., Förster, S., and Wajant, H. (2000) Hydroxynitrile lyases in stereoselective catalysis. *Curr. Opin. Biotechnol.*, **11**, 532–539.

Faber, K., Mischitz, M., Kroutil, W. (1996) Microbial epoxide hydrolases. *Acta Chem. Scand.*, **50**, 249–258.

Fessner, W.-D., *Enzymatic synthesis using aldolases in* Stereoselective Biocatalysis (Ed.: R. N. Patel), Marcel Dekker, New York, 2000, pp. 239–265.

Forney, F.W., Markovetz, A.J., and Kallio, R.E. (1967) Bacterial oxidation of 2-tridecanone to 1-undecanol. *J. Bacteriol.*, **93**, 649–655.

Fraaije, M.W., Wu, J., Heuts, D.P.H.M., van Hellemond, E.W., Spelberg, J.H.L., and Janssen, D.B. (2005) Discovery of a thermostable Baeyer–Villiger monooxygenase by genome mining. *Appl. Microbiol. Biotechnol.*, **66**, 393–400.

Fried, J., Thoma, R.W., and Klingsberg, A. (1953) Oxidation of steroids by microorganisms. III. Side chain degradation, ring D-cleavage and dehydrogenation in ring A. *J. Am. Chem. Soc.*, **75**, 5764–5765.

Gadler, P. and Faber, K. (2007) New enzymes for biotransformations: microbial alkyl sulfatases displaying stereo- and enantioselectivity. *Trends Biotechnol.*, **25**, 83–88.

Gagnon, R., Grogan, G., Levitt, M.S., Robets, S.M., Wan, P.W.H., and Willetts, A.J. (1994) Biological Baeyer–Villiger oxidation of some monocyclic and bicyclic ketones using monooxygenases from *Acinetobacter calcoaceticus* NCIMB 9871 and *Pseudomonas putida* NCIMB 10007. *J. Chem. Soc., Perkin Trans. 1*, 2537–2543.

Glieder, A., Farinas, E.T., and Arnold, F.H. (2002) Laboratory evolution of a soluble, self-sufficient, highly active alkane hydroxylase. *Nat. Biotechnol.*, **20**, 1135–1139.

Goldstein, J.A. and Faletto, M.B. (1993) Advances in mechanisms of activation and deactivation of environmental chemicals. *Environ. Health Perspect.*, **100**, 169–176.

Griengl, H., Schwab, H., and Fechter, M. (2000) The synthesis of chiral cyanohydrins by oxynitrilases. *Trends Biotechnol.*, **18**, 252–256.

Grogan, G., Roberts, S., Wan, P., and Willetts, A.J. (1993) Camphor-grown *Pseudomonas putida*, a multifunctional biocatalyst for undertaking Baeyer–Villiger monooxygenase-dependent biotransformations. *Biotechnol. Lett.*, **15**, 913–918.

Hartmans, S. and deBont, J.A.M. (1986) Acetol monooxygenase from *Mycobacterium* Py1 cleaves acetol into acetate and formaldehyde. *FEMS Microbiol. Lett.*, **36**, 155–158.

Heinze, B., Kourist, R., Fransson, L., Hult, K., and Bornscheuer, U.T. (2007) Highly enantioselective kinetic resolution of two tertiary alcohols using mutants of an esterase from *Bacillus subtilis*. *Protein Eng. Des. Sel.*, **20**, 125–131.

Henke, E., Bornscheuer, U.T., Schmid, R.D., and Pleiss, J. (2003) A molecular mechanism of enantiorecognition of tertiary alcohols by carboxylesterases. *ChemBioChem*, **4**, 485–493.

Henke, E., Pleiss, J., and Bornscheuer, U.T. (2002) Activity of lipases and esterases towards tertiary alcohols: insights into structure–function relationships. *Angew. Chem., Int. Ed.*, **41**, 3211–3213.

Hermann, M., Kietzmann, M.U., Ivancic, M., Zenzmaier, C., Luiten, R.G., Skranc, W., Wubbolts, M., Winkler, M., Birner-Gruenberger, R., Pichler, H., and Schwab, H. (2008) Alternative pig liver esterase (APLE) – cloning, identification and functional expression in *Pichia pastoris* of a versatile new biocatalyst. *J. Biotechnol.*, **133**, 301–310.

Hosoya, N., Hatayama, A., Irie, R., Sasaki, H., Katsuki, T. (1994) Rational design of Mn-Salen epoxidation catalysts: Preliminary results. *Tetrahedron*, **50**, 4311–4322.

Höhne, M. and Bornscheuer, U.T. (2009) Biocatalytic routes to optically active amines. *ChemCatChem*, **1**, 42–51.

Höhne, M., Kühl, S., Robins, K., and Bornscheuer, U.T. (2008) Efficient asymmetric synthesis of chiral amines by combining transaminase and pyruvate decarboxylase. *ChemBioChem*, **9**, 363–365.

Höhne, M., Schätzle, S., Jochens, H., Robins, K., and Bornscheuer, U.T. (2010) Rational assignment of key motifs for function guides *in silico* enzyme identification. *Nat. Chem. Biol.*, **6**, 807–813.

Hulley, M.E., Toogood, H.S., Fryszkowska, A., Mansell, D., Stephens, G.M., Gardiner, J.M., and Scrutton, N.S. (2010) Focused directed evolution of pentaerythritol tetranitrate reductase by using automated anaerobic kinetic screening of site-saturated libraries. *ChemBioChem*, **11**, 2433–2447.

Hummel, A., Brüsehaber, E., Böttcher, D., Trauthwein, H., Doderer, K., and Bornscheuer, U.T. (2007) Isoenzymes of pig-liver esterase reveal striking differences in enantioselectivities. *Angew. Chem., Int. Ed.*, **46**, 8492–8494.

Hummel, W., Schütte, H., Schmidt, E., Wandrey, C., and Kula, M.-R. (1987) Isolation of L-phenylalanine dehydrogenase from *Rhodococcus* sp. M4 and its application for the production of L-phenylalanine. *Appl. Microbiol. Biotechnol.*, **26**, 409–416.

Janssen, D.B. (2004) Evolving haloalkane dehalogenases. *Curr. Opin. Chem. Biol.*, **8**, 150–159.

Jones, J.B., Hinks, R.S., Hultin, P.G. (1985) Enzymes in organic synthesis. 33. Stereoselective pig liver esterase-catalyzed hydrolyses of *meso* cyclopentyl-, tetrahydrofuranyl-, and tetrahydrothiophenyl-1,3-diesters. *Can. J. Chem.*, **63**, 452–456.

Jones, J.B. (1990) Esterases in organic synthesis: present and future. *Pure Appl. Chem.*, **62**, 1445–1448.

Jochens, H., Hesseler, M., Stiba, K., Padhi, S.K., Kazlauskas, R.J., and Bornscheuer, U.T. (2011) Protein engineering of alpha/beta-hydrolase fold enzymes. *ChemBioChem*, **12**, 1508–1517.

Joo, H., Lin, Z., and Arnold, F.H. (1999) Laboratory evolution of peroxide-mediated cytochrome P450 hydroxylation. *Nature*, **399**, 670–673.

Kadow, M., Loschinski, K., Saß, S., Schmidt, M., and Bornscheuer, U.T. (2012) Completing the series of BVMOs involved in camphor metabolism of Pseudomonas putida NCIMB 10007 by identification of the two missing genes, their functional expression in E. coli and biochemical characterization. *Appl. Microb. Biotechnol.*, doi: 10.1007/s00253-011-3859-1.

Kadow, M., Saß, S., Schmidt, M., and Bornscheuer, U.T. (2011) Recombinant expression and purification of the 2,5-diketocamphane 1,2-monooxygenase from the camphor metabolizing *Pseudomonas putida* strain NCIMB 10007. *AMB Express*. doi: 10.1186/2191-0855-1181-1113.

Kamerbeek, N.M., Mooen, M.J.H., van der Ven, J.G.M., van Berkel, W.J.H., Fraaije, M.W., and Janssen, D.B. (2001) 4-Hydroxyacetophenone monooxygenase from *Pseudomonas fluorescens* ACB. *Eur. J. Biochem.*, **268**, 2547–2557.

Kasche, V. (2001) *Proteolytic Enzymes. A Practical Approach* (eds M. Beynon and J. Bond), Oxford University Press, Oxford, pp. 265–292.

Katsuki, T., Martin, V.S. (1996) Asymmetric epoxidation of allylic alcohols: The Katsuki-Sharpless epoxidation reaction. *Org. React.*, **48**, 1–299.

Kayser, M., Chen, G., and Stewart, J. (1999) Designer yeast: an enantioselective oxidizing reagent for organic synthesis. *Synlett*, **1**, 153–158.

Kazlauskas, R.J., Weissfloch, A.N.E., Rappaport, A.T., and Cuccia, L.A. (1991) A rule to predict which enantiomer of a secondary alcohol reacts faster in reactions catalyzed by cholesterol esterase, lipase from *Pseudomonas cepacia*, and lipase from *Candida rugosa*. *J. Org. Chem.*, **56**, 2656–2665.

Kenyon, G.L., Gerlt, J.A., Petsko, G.A., and Kozarich, J.W. (1995) Mandelate racemase: structure–function studies of a pseudosymmetric enzyme. *Acc. Chem. Res.*, **28**, 178–186.

Kirschner, A. and Bornscheuer, U.T. (2006) Kinetic resolution of 4-hydroxy-2-ketones catalyzed by a Baeyer–Villiger monooxygenase. *Angew. Chem., Int. Ed.*, **45**, 7004–7006.

Kobayashi, M. and Shimizu, S. (1998) Metalloenzyme nitrile hydratase: structure, regulation, and application to biotechnology. *Nat. Biotechnol.*, **16**, 733–736.

Kobayashi, M., Nagasawa, T., Yamada, H. (1988) Regiospecific hydrolysis of dinitrile compounds by nitrilase from *Rhodococcus rhodochrous* J1. *Appl. Microbiol. Biotechnol.*, **29**, 231–233.

Koszelewski, D., Clay, D., Rozzell, D., and Kroutil, W. (2009) Deracemisation of α-chiral primary amines by a one-pot, two-step cascade reaction catalysed by ω-transaminases. *Eur. J. Org. Chem.*, 2289–2292.

Koszelewski, D., Lavandera, I., Clay, D., Guebitz, G.M., Rozzell, D., and Kroutil, W. (2008a) Formal asymmetric biocatalytic reductive amination. *Angew. Chem., Int. Ed.*, **47**, 9337–9340.

Koszelewski, D., Lavandera, I., Clay, D., Rozzell, D., and Kroutil, W. (2008b) Asymmetric synthesis of optically pure pharmacologically relevant amines employing ω-transaminases. *Adv. Synth. Catal.*, **350**, 2761–2766.

Kourist, R. and Bornscheuer, U.T. (2011) Biocatalytic synthesis of optically active tertiary alcohols. *Appl. Microbiol. Biotechnol.*, **91**, 505–517.

Kourist, R., Dominguez de Maria, P., and Bornscheuer, U.T. (2008) Enzymatic synthesis of optically active tertiary alcohols: expanding the biocatalysis toolbox. *ChemBioChem*, **9**, 491–498.

Kroutil, W., Mischitz, M., Plachota, P., Faber, K. (1996) Deracemization of (±)-*cis*-2,3-epoxyheptane *via* enantioconvergent biocatalytic hydrolysis using *Nocardia* EH1-epoxide hydrolase. *Tetrahedron Lett.*, **37**, 8379–8382.

Kroutil, W., Genzel, Y., Pietzsch, M., Syldatk, C., Faber, K. (1998) Purification and characterization of a highly selective epoxide hydrolase from *Nocardia* sp. EH1. *J. Biotechnol.*, **61**, 143–150.

Kroutil, W., Orru, R.V.A., Faber, K. (1998) Stabilization of *Nocardia* EH1 epoxide hydrolase by immobilization. *Biotechnol. Lett.*, **20**, 373–377.

Lam, L.K.P., Hui, R.A.H.F., Jones, J.B. (1986) Enzymes in organic synthesis. 35. Stereoselective pig liver esterase catalyzed hydrolyses of 3-substituted glutarate diesters. Optimization of enantiomeric excess via reaction conditions control. *J. Org. Chem.*, **51**, 2047–2050.

Lam, L.K.P., Brown, C.M., Jeso, B.d., Lym, L., Toone, E.J., and Jones, J.B. (1988) Enzymes in organic synthesis. 42. Investigation of the effects of the isozymal composition of pig liver esterase on its stereoselectivity in preparative-scale ester hydrolyses of asymmetric synthetic value. *J. Am. Chem. Soc.*, **110**, 4409–4411.

Lange, S., Musidlowska, A., Schmidt-Dannert, C., Schmitt, J., and Bornscheuer, U.T. (2001) Cloning, functional expression, and characterization of recombinant pig liver esterase. *ChemBioChem*, **2**, 576–582.

Lauble, H., Miehlich, B., Förster, S., Wajant, H., and Effenberger, F. (2002) Crystal structure of hydroxynitrile lyase from *Sorghum bicolor* in complex with the inhibitor benzoic acid: a novel cyanogenic enzyme. *Biochemistry*, **41**, 12043–12050.

Layh, N., Hirrlinger, B., Stolz, A., and Knackmuss, H.-J. (1997) Enrichment strategies for nitrile-hydrolysing bacteria. *Appl. Microbiol. Biotechnol.*, **47**, 668–674.

Leisch, H., Morley, K., and Lau, P.C. (2011) Baeyer–Villiger monooxygenases: more than just green chemistry. *Chem. Rev.*, **111**, 4165–4222.

Lewis, D.F.V. (1996) *Cytochromes P450: Structure, Function and Mechanism*, Taylor & Francis, London.

Li, Q.-S., Schwaneberg, U., Fischer, F., Schmitt, J., Pleiss, J., Lutz-Wahl, S., and Schmid, R.D. (2001) Rational evolution of a medium chain-specific cytochrome P-450 BM-3 variant. *Biochim. Biophys. Acta*, **1545**, 114–121.

Liese, A., Seelbach, K., and Wandrey, C. (eds) (2006) *Industrial Biotransformations*, Wiley-VCH Verlag GmbH, Weinheim.

Linker, T. (1997) The Jacobsen-Katsuki epoxidation and its controversial mechanism. *Angew. Chem. Int. Ed. Engl.*, **36**, 2060–2062.

Malito, E., Alfieri, A., Fraaije, M.W., and Mattevi, A. (2004) Crystal structure of a Baeyer–Villiger monooxygenase. *Proc. Natl. Acad. Sci. USA*, **101**, 13157–13162.

Martinez-Gomez, A.I., Martinez-Rodriguez, S., Clemente-Jimenez, J.M., Pozo-Dengra, J., Rodriguez-Vico, F., and Las Heras-Vazquez, F.J. (2007) Recombinant polycistronic structure of hydantoinase process genes in *Escherichia coli* for the production of optically pure D-amino acids. *Appl. Environ. Microbiol.*, **73**, 1525–1531.

Martin-Matute, B. and Bäckvall, J.E. (2007) Dynamic kinetic resolution catalyzed by enzymes and metals. *Curr. Opin. Chem. Biol.*, **11**, 226–232.

Matcham, G., Bhatia, M., Lang, W., Lewis, C., Nelson, R., Wang, A., and Wu, W. (1999) Enzyme and reaction engineering in biocatalysis. Synthesis of (*S*)-methoxyisopropylamine (= (*S*)-1-methoxypropan-2-amine). *Chimia*, **53**, 584–589.

Matcham, G.W. and Bowen, A.R.S. (1996) Biocatalysis for chiral intermediates: meeting commercial and technical challenges. *Chim. Oggi*, **14**, 20–24.

Mihovilovic, M.D., Müller, B., Kayser, M.M., Stewart, J.D., Fröhlich, J., Stanetty, P., and Spreitzer, H. (2001) Baeyer–Villiger oxidations of representative heterocyclic ketones by whole cells of engineered *Escherichia coli* expressing cyclohexanone monooxygenase. *J. Mol. Catal. B*, **11**, 349–353.

Mirza, I.A., Yachnin, B.J., Wang, S., Grosse, S., Bergeron, H., Imura, A., Iwaki, H., Hasegawa, Y., Lau, P.C.K., and Berghuis, A. M. (2009) Crystal structures of cyclohexanone monooxygenase reveal complex domain movements and a sliding cofactor. *J. Am. Chem. Soc.*, **131**, 8848–8854.

Mlynarski, J. and Paradowska, J. (2008) Catalytic asymmetric aldol reactions in aqueous media. *Chem. Soc. Rev.*, **37**, 1502–1511.

Moree, W.J., Sears, P., Kawashiro, K., Witte, K., and Wong, C.H. (1997) Exploitation of subtilisin BPN′ as catalyst for the synthesis of peptides containing noncoded amino acids, peptide mimetics and peptide conjugates. *J. Am. Chem. Soc.*, **119**, 3942–3947.

Musidlowska, A., Lange, S., and Bornscheuer, U.T. (2001) By overexpression in the yeast *Pichia pastoris* to enhanced enantioselectivity: new aspects in the application of pig liver esterase. *Angew. Chem., Int. Ed.*, **40**, 2851–2853.

Nagasawa, T., Takeuchi, K., and Yamada, H. (1991) Characterization of a new cobalt-containing nitrile hydratase purified from urea-induced cells of *Rhodococcus rhodochrous* J1. *Eur. J. Biochem.*, **196**, 581–589.

Nagasawa, T. and Yamada, H. (1995) Microbial production of commodity chemicals. *Pure Appl. Chem.*, **67**, 1241–1256.

Nardini, M., S., R.I., Rozeboom, H.J., Kalk, K. H., Rink, R., Janssen, D.B., Dijkstra, B.W. (1999) The X-ray structure of epoxide hydrolase from *Agrobacterium radiobacter* AD1. *J. Biol. Chem.*, **274**, 14579–14596.

Narhi, L.O. and Fulco, A.J. (1986) Characterization of a catalytically self-sufficient 119,000 dalton cytochrome P-450 monooxygenase induced by barbiturates in *Bacillus megaterium*. *J. Biol. Chem.*, **261**, 7160–7169.

Nelson, D.R. (1999) Cytochrome P450 and the individuality of species. *Arch. Biochem. Biophys.*, **369**, 1–10.

Nishizawa, M., Shimizu, M., Ohkawa, H., and Kanaoka, M. (1995) Stereoselective production of (+)-*trans*-chrysanthemic acid by a microbial esterase: cloning, nucleotide sequence, and overexpression of the esterase gene of *Arthrobacter globiformis* in *Escherichia coli*. *Appl. Environ. Microbiol.*, **61**, 3208–3215.

Öhrner, K., Mattson, A., Norin, T., and Hult, K. (1990) Enantiotopic selectivity of pig liver esterase isoenzymes. *Biocatalysis*, **4**, 81–88.

Ohshima, T. and Soda, K. (2000) *Stereoselective Biocatalysis* (ed. R.N. Patel), Marcel Dekker, New York, p. 877.

Oliver, C.F., Modi, S., Sutcliffe, M.J., Pimrose, W.U., Lian, L.Y., and Roberts, G.C. (1997) A single mutation in cytochrome P450 BM3 changes substrate orientation in a catalytic intermediate and the regiospecificity of hydroxylation. *Biochemistry*, **36**, 1567–1572.

Ollis, D.L., Cheah, E., Cygler, M., Dijkstra, B., Frolow, F., Franken, S., Harel, M., Remington, S.J., and Silman, I. (1992) The α/β hydrolase fold. *Protein Eng.*, **5**, 197–211.

Omura, T. and Sato, R.J. (1964) The carbon monoxide-binding pigment of liver microsomes. I. Evidence for its hemoprotein nature. *J. Biol. Chem.*, **239**, 2370–2378.

Orru, R.V.A., Kroutil, W., Faber, K. (1997) Deracemization of (±)-2,2-disubstituted epoxides *via* enantioconvergent chemoenzymic hydrolysis using *Nocardia* EH1 epoxide hydrolase and sulfuric acid. *Tetrahedron Lett.*, **38**, 1753–1754.

Orru, R.V.A., Osprian, I., Kroutil, W., Faber, K. (1998) An efficient large-scale synthesis of (*R*)-(-)-mevalonolactone using simple biological and chemical catalysts. *Synthesis*, 1259–1263.

Oyama, K. (1992) *Chirality in Industry* (eds A.N. Collins, G.N. Sheldrake, and J. Crosby), John Wiley & Sons, Ltd, Chichester, pp. 237–247.

Padhi, S.K., Bougioukou, D.J., and Stewart, J. D. (2009) Site-saturation mutagenesis of tryptophan 116 of *Saccharomyces pastorianus* old yellow enzyme uncovers stereocomplementary variants. *J. Am. Chem. Soc.*, **131**, 3271–3280.

Padhi, S.K., Fujii, R., Legatt, G.A., Fossum, S.L., Berchtold, R., and Kazlauskas, R.J. (2010) Switching from an esterase to a hydroxynitrile lyase mechanism requires only two amino acid substitutions. *Chem. Biol.*, **17**, 863–871.

Pàmies, O. and Bäckvall, J.-E. (2004) Chemoenzymatic dynamic kinetic resolution. *Trends Biotechnol.*, **22**, 130–135.

Park, J.H., Kim, G.J., and Kim, H.S. (2000) Production of D-amino acid using whole cells of recombinant *Escherichia coli* with separately and coexpressed D-hydantoinase and N-carbamoylase. *Biotechnol. Prog.*, **16**, 564–570.

Pedragosa-Moreau, S., Archelas, A., Furstoss, R. (1996) Microbiological transformations. 32. Use of epoxide hydrolase mediated biohydrolysis as a way to enantiopure epoxides and vicinal diols: Application to substituted styrene oxides. *Tetrahedron*, **52**, 4593–4606.

Peters, M.W., Meinhold, P., Glieder, A., and Arnold, F.H. (2003) Regio- and enantioselective alkane hydroxylation with engineered cytochromes P450 BM-3. *J. Am. Chem. Soc.*, **125**, 13442–13450.

Peters-Wendisch, P., Stolz, M., Etterich, H., Kennerknecht, N., Sahm, H., and Eggeling, L. (2005) Metabolic engineering of *Corynebacterium glutamicum* for L-serine production. *Appl. Environ. Microbiol.*, **71**, 7139–7144.

Pleiss, J., Fischer, M., Peiker, M., Thiele, C., and Schmid, R.D. (2000) Lipase engineering database understanding and exploiting sequence–structure–function relationships. *J. Mol. Catal. B*, **10**, 491–508.

Purkarthofer, T., Skranc, W., Schuster, C., and Griengl, H. (2007) Potential and capabilities of hydroxynitrile lyases as biocatalysts in the chemical industry. *Appl. Microbiol. Biotechnol.*, **76**, 309–320.

Quax, W.J. and Broekhuizen, C.P. (1994) Development of a new *Bacillus* carboxyl esterase for use in the resolution of chiral drugs. *Appl. Microbiol. Biotechnol.*, **41**, 425–431.

Reetz, M.T. and Wu, S. (2008) Greatly reduced amino acid alphabets in directed evolution: making the right choice for saturation mutagenesis at homologous enzyme positions. *Chem. Commun.*, 5499–5501.

Reetz, M.T. and Wu, S. (2009) Laboratory evolution of robust and enantioselective Baeyer–Villiger monooxygenases for asymmetric catalysis. *J. Am. Chem. Soc.*, **131**, 15424–15432.

Rehdorf, J. and Bornscheuer, U.T. (2010), Monooxygenases, Baeyer-Villiger oxidations in organic synthesis in: *Encyclopedia of Industrial Biotechnology, Bioprocess, Bioseparation and Cell Technology* (Flickinger, M.C. Ed.)., John Wiley & Sons, Hoboken.

Rehdorf, J., Mihovilovic, M.D., and Bornscheuer, U.T. (2010) Exploiting the regioselectivity of Baeyer–Villiger monooxygenases for the formation of beta-amino acids and beta-amino alcohols. *Angew. Chem., Int. Ed.*, **49**, 4506–4508.

Rink, R., Fennema, M., Smids, M., Dehmel, U., Janssen, D.B. (1997) Primary structure and catalytic mechanism of the epoxide hydrolase from *Agrobacterium radiobacter* AD1. *J. Biol. Chem.*, **272**, 14650–14657.

Rosenthaler, L. (1908) Durch Enzyme bewirkte asymmetrische Synthese. *Biochem. Z.*, **14**, 238–253.

Ruettinger, R.T., Wen, L.-P. and Fulco, A.J. (1989) Coding nucleotide, 5 regulatory, and deduced amino acid sequences of P-450 BM-3, a single peptide cytochrome P-450: NADPH-P-450 reductase from *Bacillus megaterium*. *J. Biol. Chem.*, **264**, 10987–10995.

Savile, C.K., Janey, J.M., Mundorff, E.C., Moore, J.C., Tam, S., Jarvis, W.R., Colbeck, J. C., Krebber, A., Fleitz, F.J., Brands, J., Devine, P.N., Huisman, G.W., and Hughes, G.J. (2010) Biocatalytic asymmetric synthesis of chiral amines from ketones applied to sitagliptin manufacture. *Science*, **329**, 305–309.

Savjalov, W.W. (1901) Zur Theorie der Eiweissverdauung. *Pflügers Arch. Ges. Physiol.*, **85**, 171.

Schätzle, S., Steffen-Munsberg, F., Höhne, M., Robins, K., and Bornscheuer, U.T. (2011) Enzymatic asymmetric synthesis of enantiomerically pure aliphatic, aromatic and arylaliphatic amines with (*R*)-selective amine transaminases. *Adv. Synth. Catal.*, **353**, 2439–2445.

Schellenberger, V. and Jakubke, H.D. (1991) Protease-catalyzed kinetically controlled peptide synthesis. *Angew. Chem., Int. Ed. Engl.*, **30**, 1437–1449.

Schmid, R.D. and Urlacher, V. (eds) (2007) *Modern Biooxidation*, Wiley-VCH Verlag GmbH, Weinheim.

Schmid, R.D., Verger, R. (1998) Lipases - interfacial enzymes with attractive applications. *Angew. Chem. Int. Ed.*, **37**, 1608–1633.

Schneider, S., Wubbolts, M.G., Sanglard, D., and Witholt, B. (1998) Biocatalyst engineering by assembly of fatty acid transport and oxidation activities for *in vivo* application of cytochrome P-450 BM-3 monooxygenase. *Appl. Environ. Microbiol.*, **64**, 3784–3790.

Schoevaart, R., Rantwijk, F.v., and Sheldon, R.A. (2000) A four-step cascade for the one-pot synthesis of non-natural carbohydrates from glycerol. *J. Org. Chem.*, **65**, 6940–6943.

Schwaneberg, U. and Bornscheuer, U.T. (2000) *Enzymes in Lipid Modification* (ed. U.T. Bornscheuer) Wiley-VCH Verlag GmbH, Weinheim.

Schwaneberg, U., Schmidt-Dannert, C., Schmitt, J., and Schmid, R.D. (1999a) A continuous spectrophotometric assay for P450 BM-3, a fatty acid hydroxylating enzyme, and its mutant F87A. *Anal. Biochem.*, **269**, 359–366.

Schwaneberg, U., Sprauer, A., Schmidt-Dannert, C., and Schmid, R.D. (1999b) P450 monooxygenase in biotechnology. I. Single-step, large-scale purification method for cytochrome P450 BM-3 by anion-exchange chromatography. *J. Chromatogr. A*, **848**, 149–159.

Secundo, F., Carrea, G., Riva, S., Battistel, E., and Bianchi, D. (1993) Cyclohexanone monooxygenase catalyzed oxidation of methyl phenyl sulfide and cyclohexanone with macromolecular NADP in a membrane reactor. *Biotechnol. Lett.*, **155**, 865–870.

Shin, J.-S. and Kim, B.-G. (1999) Asymmetric synthesis of chiral amines with ω-transaminase. *Biotechnol. Bioeng.*, **65**, 206–211.

Smit, M.S. (2004) Fungal epoxide hydrolases: new landmarks in sequence-activity space. *Trends Biotechnol.*, **22**, 123–129.

Steinreiber, A., Faber, K. (2001) Microbial epoxide hydrolases for preparative biotransformations. *Curr. Opin. Biotechnol.*, **12**, 552–558.

Spelberg, J.H.L., Rink, R., Kellogg, R.M., Janssen, D.B. (1998) Enantioselectivity of a recombinant epoxide hydrolase from *Agrobacterium radiobacter. Tetrahedron: Asymmetry*, **9**, 459–466.

Spelberg, J.H., Tang, L., van_Gelder, M., Kellogg, R.M., and Janssen, D.B. (2002) Exploration of the biocatalytic potential of a halohydrin dehalogenase using chromogenic substrates. *Tetrahedron: Asymmetry*, **13**, 1083–1089.

Spelberg, J.H., van Hylckama Vlieg, J.E., Tang, L., Janssen, D.B., and Kellogg, R.M. (2001) Highly enantioselective and regioselective biocatalytic azidolysis of aromatic epoxides. *Org. Lett.*, **3**, 41–43.

Stampfer, W., Kosjek, B., Moitzi, C., Kroutil, W., and Faber, K. (2002) Biocatalytic asymmetric hydrogen transfer. *Angew. Chem., Int. Ed.*, **41**, 1014–1017.

Stewart, J.D., Reed, K.W., and Kayser, M.M. (1996) Designer yeast: a new reagent for enantioselective Baeyer–Villiger oxidations. *J. Chem. Soc., Perkin Trans. 1*, 755–757.

St. Maurice, M. and Bearne, S.L. (2004) Hydrophobic nature of the active site of mandelate racemase. *Biochemistry*, **43**, 2524–2532.

Strohmeier, G.A., Pichler, H., May, O., and Gruber-Khadjawi, M. (2011) Application of designed enzymes in organic synthesis. *Chem. Rev.*, **111**, 4141–4164.

Stuermer, R., Hauer, B., Hall, M., and Faber, K. (2007) Asymmetric bioreduction of activated C=C bonds using enoate reductases from the old yellow enzyme family. *Curr. Opin. Chem. Biol.*, **11**, 203–213.

Sugiura, Y., Kuwahara, J., Nagasawa, T., and Yamada, H. (1987) Nitrile hydratase: the first non-heme iron enzyme with a typical low-spin Fe(III)-active center. *J. Am. Chem. Soc.*, **109**, 5848–5850.

Syldatk, C., May, O., Altenbuchner, J., Mattes, R., and Siemann, M. (1999) Microbial hydantoinases – industrial enzymes from the origin of life? *Appl. Microbiol. Biotechnol.*, **51**, 293–309.

Szolkowy, C., Eltis, L.D., Bruce, N.C., and Grogan, G. (2009) Insights into sequence–activity relationships amongst Baeyer–Villiger monooxygenases as revealed by the intragenomic complement of enzymes from *Rhodococcus jostii* RHA1. *ChemBioChem*, **10**, 1208–1217.

Tanner, A. and Hopper, D.J. (2000) Conversion of 4-hydroxyacetophenone into 4-phenyl acetate by a flavin adenine dinucleotide-containing Baeyer–Villiger-type monooxygenase. *J. Bacteriol.*, **182**, 6565–6569.

Tao, J., Lin, G.-Q., and Liese, A. (eds) (2009) *Biocatalysis for the Pharmaceutical Industry*, Wiley-VCH Verlag GmbH, Weinheim.

Taylor, D.G. and Trudgill, P.W. (1986) Camphor revisited: studies of 2,5-diketocamphane 1,2-monooxygenase from

Pseudomonas putida ATCC 17453. *J. Bacteriol.*, **165**, 489–497.

Toogood, H.S., Gardiner, J.M., and Scrutton, N.S. (2010) Biocatalytic reductions and chemical versatility of the old yellow enzyme family of flavoprotein oxidoreductases. *ChemCatChem*, **2**, 892–914.

Torres Pazmino, D.E., Snajdrova, R., Bass, B.-J., Ghobrial, M., Mihovilovic, M.D., and Fraaije, M.W. (2008) Self-sufficient Baeyer–Villiger monooxygenase: effective coenzyme regeneration for biooxygenation by fusion engineering. *Angew. Chem., Int. Ed.*, **47**, 2275–2278.

Turfitt, G.E. (1948) The microbiological degradation of steroids. 4. Fission of the steroid molecule. *Biochemistry*, **42**, 376–383.

Urlacher, V.B. and Eiben, S. (2006) Cytochrome P450 monooxygenases: perspectives for synthetic application. *Trends Biotechnol.*, **24**, 324–330.

Urlacher, V.P., Lutz-Wahl, S., and Schmid, R.D. (2004) Microbial P450 enzymes in biotechnology. *Appl. Microbiol. Biotechnol.*, **64**, 317–325.

Walsh, C.T. and Chen, Y.C.J. (1988) Enzymatic Baeyer–Villiger oxidations by flavin-dependent monooxygenases. *Angew. Chem., Int. Ed. Engl.*, **27**, 333–343.

Weissfloch, A.N.E. and Kazlauskas, R.J. (1995) Enantiopreference of lipase from *Pseudomonas cepacia* toward primary alcohols. *J. Org. Chem.*, **60**, 6959–6969.

Wendisch, V.F., Bott, M., and Eikmanns, B.J. (2006) Metabolic engineering of *Escherichia coli* and *Corynebacterium glutamicum* for biotechnological production of organic acids and amino acids. *Curr. Opin. Microbiol.*, **9**, 268–274.

Wiese, A., Wilms, B., Syldatk, C., Mattes, R., and Altenbuchner, J. (2001) Cloning, nucleotide sequence and expression of a hydantoinase and carbamoylase gene from *Arthrobacter aurescens* DSM 3745 in *Escherichia coli* and comparison with the corresponding genes from *Arthrobacter aurescens* DSM 3747. *Appl. Microbiol. Biotechnol.*, **55**, 750–757.

Wieser, M. and Nagasawa, T. (1999) *Stereoselective Biocatalysis* (ed. R. Patel), Marcel Dekker, New York, pp. 461–486.

Woo, J.C.G. and Silverman, R.B. (1995) Monoamine oxidase B catalysis in low aqueous medium. Direct evidence for an imine product. *J. Am. Chem. Soc.*, **117**, 1663–1664.

Yamamoto, K., Oishi, K., Fujimatsu, I., and Komatsu, K.-I. (1991) Production of *R*-(−)-mandelic acid from mandelonitrile by *Alcaligenes faecalis* ATCC 8750. *Appl. Environ. Microbiol.*, **57**, 3028–3032.

Zaks, A. and Dodds, D.R. (1995) Chloroperoxidase-catalysed asymmetric oxidations: substrate specificity and mechanistic study. *J. Am. Chem. Soc.*, **117**, 10419–10424.

Zhang, X.M., Archelas, A., Furstoss, R. (1991) Microbiological transformations. 19. Asymmetric dihydroxylation of the remote double bond of geraniol: a unique stereochemical control allowing easy access to both enantiomers of geraniol-6,7-diol. *J. Org. Chem.*, **56**, 3814–3817.

Zhang, J., Reddy, J., Roberge, C., Senanayake, C., Greasham, R., Chartrain, M. (1995) Chiral bio-resolution of racemic indene oxide by fungal epoxide hydrolase. *J. Ferment. Bioeng.*, **80**, 244–246.

5
Cells Designed by Metabolic Engineering as Biocatalysts for Multienzyme Biotransformations

Why designed cells for biotransformations?	• Complex multistep reactions easier to implement • No need for enzyme isolation/immobilization • No need for isolation of intermediate products • Use of abundant carbon sources possible • Cofactor recycling facilitated
Tools needed	• Suitable expression system/tools for genetic manipulation • Metagenomics/bioinformatics to access biodiversity and genes • Metabolic engineering to control and adapt pathways • "omics" technologies to understand complex pathways • Protein engineering to improve/tailor enzymes • Methods for product isolation/downstream processing

5.1
Introduction

The previous chapters dealt with the identification, characterization, and improvement of isolated enzymes to be used in biocatalysis in single-step reactions. The next chapters cover the importance of isolated or immobilized biocatalysts in the production of various products. However, if multistep reactions are required to synthesize the target compound, a decision must be made whether several single

Biocatalysts and Enzyme Technology, Second Edition. Klaus Buchholz, Volker Kasche, and Uwe T. Bornscheuer

enzymatic steps are combined in a process or whether the use of a whole-cell system is advantageous. One example in which a process initially developed with isolated enzymes was then changed to a whole-cell biotransformation is the synthesis of D- or L-amino acids starting from racemic hydantoins (Section 4.2.3.7). The biocatalytic route required production, isolation, and immobilization of the hydantoinase, the carbamoylase, and the racemase to run the reaction in consecutive fixed-bed reactors. The use of genetically engineered *Escherichia coli* cells expressing all three enzymes under the control of a rhamnose-inducible promoter allowed the transfer of the cascade reaction to a whole-cell system. This is a rather simple example for the use of designed cells as no undesired side reactions occurred and excellent yields and optical purities were possible. Furthermore, all enzymatic reactions do not require cofactors to be regenerated and there was no need to control and manipulate the metabolic flux within the *E. coli* cells.

This situation becomes much more complex if entire pathways need to be engineered to establish a novel biotransformation. Targets can be divided into two subgroups: (i) engineering of an existing pathway or (ii) introduction of new pathways (Figure 5.1).

In both cases, the host strains must be easy to manipulate on a genetic level (introduction and overexpression of novel genes, ability to delete genes, enhanced overexpression of existing genes). Furthermore, for all enzymes involved in a reaction cascade, the biochemical properties must be known in detail. This includes knowledge about their nucleotide and protein sequence, oligomeric structure, regulation of expression, details about kinetic constants, level of substrate or product inhibition, cofactor requirement, and so on (see Chapter 2). When necessary, their properties must be improved by protein engineering (see Chapter 3). To solve these problems, metabolic engineering, whose goal is defined as "targeted manipulation of enzymatic, transport, and regulatory functions of the cell with the use of recombinant DNA technology," has become an important tool in the past two decades (Bailey, 1991).

5.2
A Short Introduction to Metabolic Engineering

Around 1970, biochemists and cell and molecular biologists developed the metabolic control theory that described the regulation of the metabolism in cells to control the concentrations of metabolites (Rapoport, Heinrich, and Rapoport, 1976; Kacser and Burns, 1981).[1] Biochemical engineers found this theory useful to develop metabolic engineering as a tool to optimize the yield and productivity (or space–time yield; see Sections 2.8 and 10.2) of useful metabolic products as shown

1) Readers who are interested in the development of the metabolic control theory that also includes controversial discussions can learn more about this in the Discussion Forum of the January and June issues of *Trends in Biochemical Sciences* from 1987.

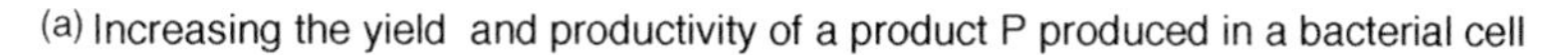

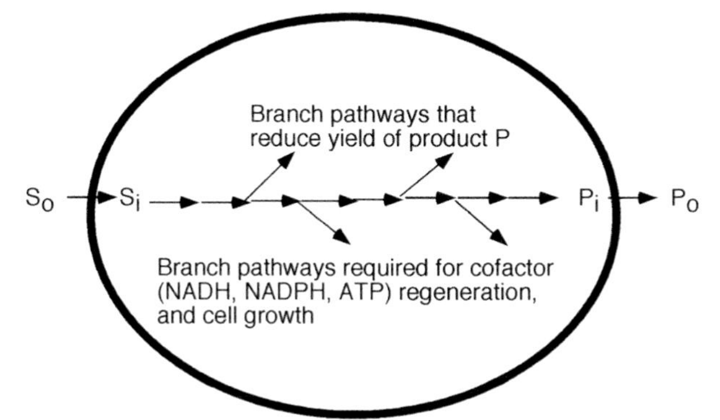

(b) Increasing the yield and productivity of a product P that is not produced in a wild-type bacterial cell by adding a pathway from a common metabolite from another cell (**red**)

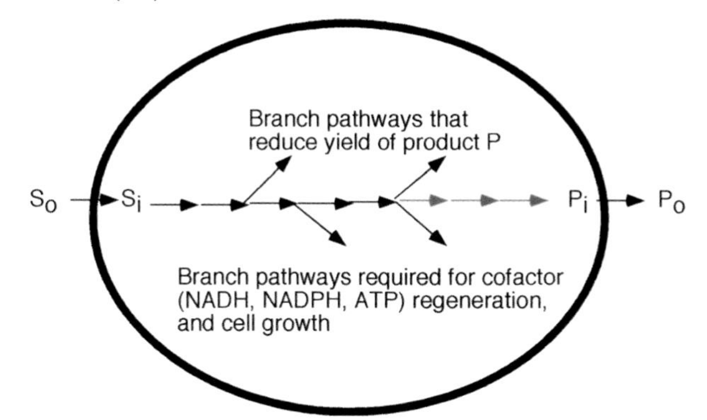

Figure 5.1 Problems solved by metabolic engineering. Improving the production of a product in a cell along an existing pathway (a) or by adding a new pathway (b) (given by red arrows). It may also include improvement of existing transport systems in and out of the cell membrane or the addition of new transport systems. Each arrow within the cell is a reaction catalyzed by an enzyme. The horizontal arrows in the cell show the shortest pathway from substrate S to product P. The arrows through the cell membrane represent transport systems for S and P; subscripts o and i denote their extra- and intracellular concentrations, respectively. The substrate S is generally a 6- or 5-C sugar derived from polysaccharides.

in Figure 5.1 (Bailey, 1991; Stephanopoulos and Vallino, 1991; Edwards *et al.*, 1999; Segre *et al.*, 2002).

How the yield and productivity can be influenced along the pathways in this figure can be analyzed for a part of the pathway with a branch out of the pathway to the desired product as shown in the following scheme for the first reactions of the pathway from the cytosolic substrate S_i:

$$S_i \xrightarrow[E_1]{J_1} M_1 \xrightarrow[E_2]{J_2} M_2 \xrightarrow[E_3]{J_3} M_3, \quad M_2 \xrightarrow[E_4]{J_4} M_4 \tag{5.1}$$

where M refers to a metabolite, J is the flux, and E is the enzyme that catalyzes the reaction to the metabolite with the same subscript. The dashed line indicates feedback inhibition.

The quantitative relations for the fluxes J_i in metabolic control theory and metabolic engineering are based on the assumption of a steady state in all metabolites between the substrate and the product (Stephanopoulos, Aristidou, and Nielsen, 1998). Then the following relation applies for the above equation (5.1):

$$J_1 = J_2 = J_3 + J_4. \tag{5.2}$$

This relation shows that both the yield and the productivity are reduced by branches out of the pathway. Assuming that the enzyme-catalyzed reactions in the pathway of Eq. (5.1) can be described by simple Michaelis–Menten kinetics (see Section 2.7), the maximum productivity and the product yield are given by the following relations:

$$\text{maximum productivity} = \frac{k_{cat,1}[E_1][S_i]}{K_{m,1} + [S_i]} = \frac{k_{cat,2}[E_2][M_1]}{K_{m,2} + [M_1]}, \tag{5.3}$$

$$\text{maximum product yield} = \frac{k_{cat,3}[E_3][M_2]/(K_{m,3} + [M_2])}{k_{cat,3}[E_3][M_2]/(K_{m,3} + [M_2]) + k_{cat,4}[E_4][M_2]/(K_{m,4} + [M_2])}. \tag{5.4}$$

The fluxes J_i and the maximal and realized productivity along the pathway in Eq. (5.1) can be measured using ^{13}C-labeled substrate, by determining the rate of labeling in the metabolites of the pathway (Marx *et al.*, 1996). In practice $[S_i] \gg K_{m,1}$; thus, the first enzyme of the pathway is saturated. From Eq. (5.3) follows that the productivity then can only be increased by higher concentrations of E_1. When it is inhibited by a metabolite M_i in the further pathway (end product inhibition), the productivity can be increased by protein engineering of E_1 to reduce this inhibition (see Chapter 3). Equation (5.4) shows that the product yield can be increased by reducing the concentration of E_4 or increasing the concentration of E_3. This can be achieved by deletion or downregulation of the gene for E_4, and increasing the expression of the gene for E_3. The same can be achieved when k_{cat}/K_m for E_3 is increased, and decreased for E_4. Thus, the yield and the productivity can be increased by changing the properties and/or concentrations of the enzymes in the pathway and at its branching points.

The successful application of metabolic engineering of cells requires sophisticated analytical methods. Not only the concentration of key metabolites must be known in a time-resolved manner (metabolomics), but also the protein level must be analyzed, for example, by transcriptomics or proteomics. Consequently, metabolic engineering is a highly multidisciplinary area, in which expertise from microbiology, molecular biology, biochemistry, metabolic flux analysis, fermentation, analytics, and bioinformatics must be combined to successfully establish a process.

The integration of all these aspects, and transport phenomena in bioreactors, is the topic of biosystems engineering, or systems biotechnology, in order to gain a holistic view of the complex phenomena (Deckwer *et al.*, 2006; Papini, Salazar, and Nielsen, 2010). The interplay of these is visualized in Figure 5.2.

Furthermore, protein engineering can be important if key enzymes need to be identified and adapted to overcome bottlenecks within the metabolic flux.

The above discussion can be summarized as follows: Once the pathway to be manipulated or established is known, biochemical engineers can direct the metabolic flux toward product formation by deletion of genes, reduction or increase of expression levels of certain genes, or introduction of new (to establish reactions not existing in the host strain) or alternative protein encoding genes (to replace enzymes existing in the host strains by other ones with, for example, higher activity, alternative selectivity, or less inhibition). For the overall process, it is also important that the uptake of carbon (e.g., glucose) and nitrogen sources is known and that preferentially a secretion mechanism for the product either exists naturally or can be established by the introduction of export systems. It is also required to ensure that the fluxes in branch pathways for cofactor regeneration and cell growth are at a sufficient level.

In the past 10 years, much more complex metabolic engineering examples have been established as given for those in the next sections. The processes are mainly carried out with suspended cells. These usually involve not only the establishment of new pathways in the host cell by cloning and expression of heterologous genes,

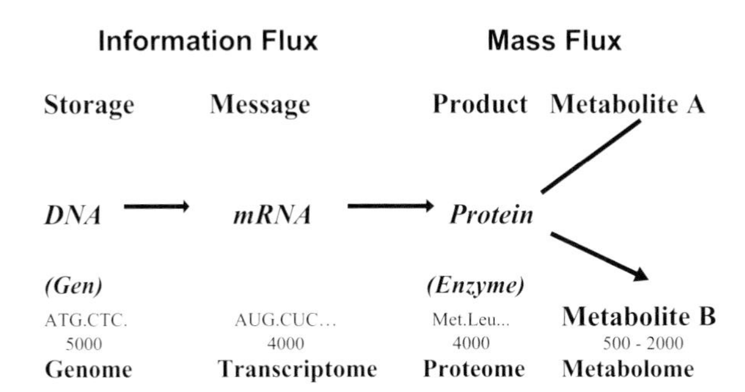

Figure 5.2 Basic molecular structure of biological processes. The information stored in DNA is transcribed into messenger molecules (mRNA), which in turn encode the synthesis of proteins on the ribosomes. Most of the proteins produced act as catalysts for the reactions of the metabolic network. The entirety of the DNA information is termed genome; for microorganisms it contains typically about 5000 genes. The proteome comprises about 4000 proteins, and roughly 2000 metabolites can be identified as comprising the metabolome (Deckwer *et al.*, 2006).

but also sophisticated analysis and redirection of the entire metabolic flux. This is important for two reasons: (i) to ensure that the key precursor substrate is quantitatively converted into the product, which means that cofactors are present or regenerated in stoichiometric amounts, and (ii) to ensure that cheap and renewable carbon sources can be used to establish a cost-efficient and environment-friendly process, especially if large-scale products ($>$10 000 t year^{-1}) are targeted.

For a further insight into the broad area of metabolic engineering, readers are referred to reviews (de Boer and Schmidt-Dannert, 2003; Lee *et al.*, 2008; Mijts and Schmidt-Dannert, 2003) and excellent books (Lee and Papoutsakis, 1999; Smolke, 2010; Stephanopoulos, Aristidou, and Nielsen, 1998; Klein-Marcuschamer *et al.*, 2010; Papini, Salazar, and Nielsen, 2010).

5.3
Examples

5.3.1
1,3-Propanediol

1,3-Propanediol (1,3-PD) is an industrial chemical and a key component in polyester/polymer production. In order to establish a production independent of petrol price and supply, an alternative microbiological process was developed. Initially, optimization of cultivation conditions for natural 1,3-PD-producing strains (i.e., *Klebsiella* sp., *Citrobacter* sp., *Clostridium* sp., all grown anaerobically) especially with glycerol as carbon source was performed (for a review, see Zeng and Biebl, 2002). These organisms are able to convert glycerol in a two-step reaction with glycerol dehydratase first to the aldehyde and then with an NADH-dependent oxidoreductase to 1,3-PD (Figure 5.3b). What appears to be a rather simple pathway proved to be more difficult as the glycerol dehydratase is a B_{12}-dependent coenzyme composed of three polypeptides and inactivation can occur, which can be circumvented by a reactivase. Due to these issues, alternative glycerol dehydratases were identified with higher resistance to inactivation by glycerol and less inhibition by 1,3-PD. The corresponding genes were cloned into *E. coli* to serve as production strain. As glucose is a much more abundant and cheaper carbon source, a pathway needed to be established to guide the metabolic flux from glucose to glycerol, the key substrate for the formation of 1,3-PD. The key metabolite derived from glucose is dihydroxyacetone phosphate (DHAP), which is then reduced to glycerol-3-phosphate using a DHAP dehydrogenase followed by hydrolysis of the phosphate with a glycerol-3-phosphatase to generate glycerol (Figure 5.3a). Consequently, the overall process requires two equivalents of NADH and its regeneration must be ensured. Furthermore, additional problems had to be taken into account to maintain the redox balance and the necessity of phosphorylation of the sugar by ATP or phosphoenolpyruvate. Furthermore, it needed to be avoided that glycerol enters the central carbon metabolism and that part of the carbon flux enters the TCA cycle.

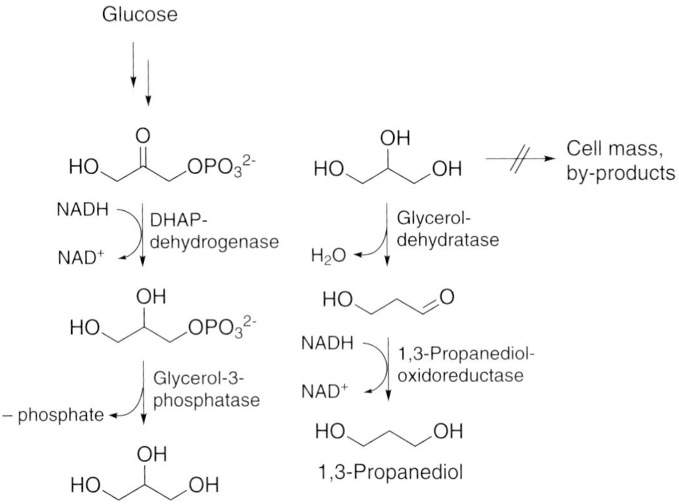

Figure 5.3 Enzymatic reactions that were implemented in *E. coli* to direct carbon flow from glucose to glycerol (a) and from glycerol to 1,3-propanediol (b).

The companies DuPont and Genencor International Inc. in a joint effort were able to address all these issues and constructed an engineered *E. coli* strain able to produce 1,3-PD in an aerobic fermentation with titers up to $130\,g\,l^{-1}$ of 1,3-PD and a yield of 0.63 mol 1,3-PD per mol glycerol. Overall, 36 enzymes needed to be examined and optimized (for an excellent review, see Nakamura and Whited, 2003). A more recent example for the production of a commodity chemical is the production of 1,4-butanediol (1,4-BD) using an engineered *E. coli*. Key to success was an algorithm to identify most suitable pathways for 1,4-BD biosynthesis from common metabolic intermediates such as succinate or α-ketoglutarate. The recombinant microorganism is capable of producing 1,4-BD from glucose, xylose, or sucrose with titers up to $18\,g\,l^{-1}$ (Yim *et al.*, 2011).

5.3.2 Synthesis of "Biodiesel" and Other Fatty Acid Derivatives

Biofuels such as ethanol and fatty acid methyl esters ("biodiesel") are still produced mostly from raw materials used in human and animal nutrition, for example, sugars and plant oils, which results in a competition between food/feed supply and fuel production. To avoid this conflict, more sustainable routes for biofuel production are investigated relying on alternative carbon sources such as lignocellulose (Lee *et al.*, 2008; Zhang, Rodriguez, and Keasling, 2011). Again, *E. coli* was engineered to produce fatty acid ethyl esters (FAEE), fatty alcohols, and wax esters from simple sugars including the utilization of plant biomass-derived hemicelluloses

(Steen *et al.*, 2010). Key to direct production of FAEE was cytosolic expression of a thioesterase, control of fatty acid chain length profile by introduction of plant-derived enzymes, elimination of several side reactions to deregulate fatty acid biosynthesis, coexpression of a wax ester synthase, and ethanol formation from pyruvate (Figure 5.4). Furthermore, fatty alcohol production was achieved by introduction of two reductases. The yields reported were, however, rather low (up to 674 mg l^{-1} FAEE, 9.4% of theoretical yield) and at least one order of magnitude too low for commercial production.

5.3.3
Conversion of Cellulosics to Ethanol

Future options for biofuel manufacture envisage fermentation of mixtures of glucose, xylose, and arabinose in order to obtain high ethanol yields from cellulosic substrates. Strain optimization has been successfully performed to utilize such substrates. A large number of rational metabolic engineering strategies have been directed toward sugar transport, initial pentose conversion, the pentose phosphate pathway, and cellular redox metabolism in *Saccharomyces cerevisiae*. The xylose pathway enzymes of, for example, *Pichia stipitis* have been introduced in *S. cerevisiae*, which, however, turned out to be not sufficient for efficient ethanol fermentation. Further metabolic engineering strategies to improve xylose

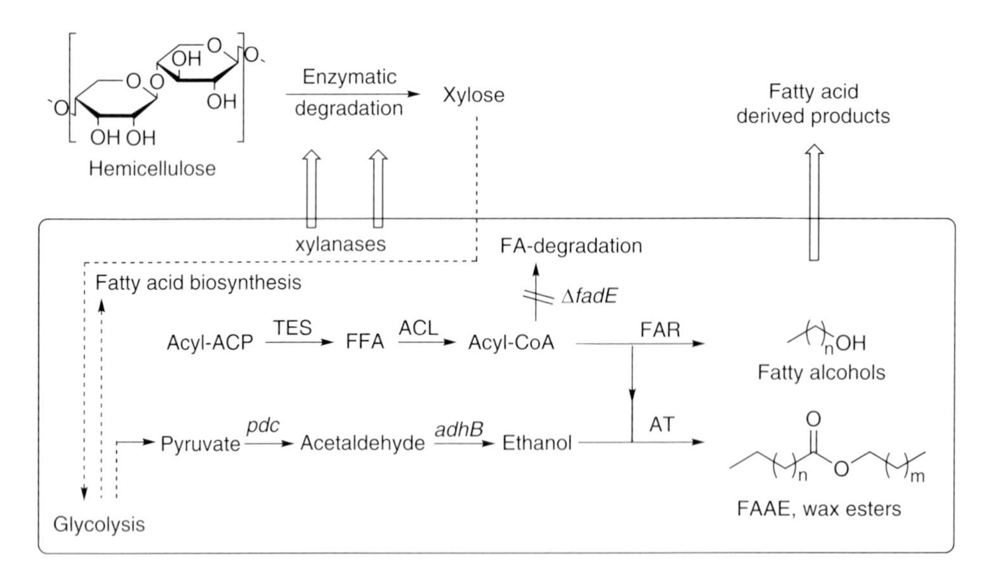

Figure 5.4 Engineered pathways for production of fatty acid alkyl esters (FAAE), fatty alcohols, and wax esters in *E. coli*. Overexpression of thioesterases (TES), acyl-CoA ligases (ACL), and deletion of β-oxidation (ΔfadE) enhanced free fatty acid (FFA) and fatty alcohol production. FAR: fatty acid reductase AT: acyl transferase; *pdc*: pyruvate decarboxylase; *adhB*: alcohol dehydrogenase.

fermentation comprised the introduction of transporter and xylulokinase genes, improving the conversion of xylose to xylulose, the pentose phosphate pathway, and engineering the redox metabolism. Chromosomal integration of the genes is necessary for industrial application of yeast strains, since plasmids are generally unstable (Hahn-Hägerdal *et al.*, 2007; Matsushika *et al.*, 2009; Karhumaa *et al.*, 2006).[2] In another example, an engineered *S. cerevisiae* strain has been developed that is capable of fermenting mixtures of glucose, xylose, and arabinose without formation of the side products xylitol and arabinitol. Batch cultivation with repeated cycles of consecutive growth in three media with different compositions (glucose, xylose, and arabinose; xylose and arabinose; and only arabinose) allowed rapid selection of an evolved strain exhibiting improved specific rates of xylose and arabinose consumption. This evolution strategy resulted in a 40% reduction in the time required to completely ferment a mixture containing $30\,g\,l^{-1}$ glucose, $15\,g\,l^{-1}$ xylose, and $15\,g\,l^{-1}$ arabinose, with an ethanol yield of $0.43\,g\,g^{-1}$ (total sugar) (Wisselink *et al.*, 2009).

Among other strains that have been engineered for ethanol fermentation are *Zymomonas mobilis* and *E. coli*; they ferment a variety of sugars (pentoses and hexoses), as well as tolerate stress conditions, and performed well in pilot studies (Zaldivar, Nielsen, and Olsson, 2001; Brethauer and Wyman, 2010; Rogers *et al.*, 2007; Jarboe *et al.*, 2007).

The potential of consolidated bioprocessing (CBP) has been outlined – featuring cellulase production, cellulose hydrolysis, and fermentation in one step for the conversion of lignocellulosic biomass into fuels, using both hexose and pentose sugars. Approaches comprise both engineering naturally occurring cellulolytic microorganisms to improve yield and titer and engineering noncellulolytic organisms that exhibit high product yields and titers to express a heterologous cellulase system enabling cellulose utilization, for example, by *S. cerevisiae* (Figure 5.5) (Lynd *et al.*, 2005; van Zyl *et al.*, 2007).

A whole-cell biocatalyst with the ability to induce synergistic and sequential cellulose degradation reaction was constructed through codisplay of three types of cellulolytic enzyme on the cell surface of the yeast *S. cerevisiae* (Fujita *et al.*, 2004). *Trichoderma reesei* endoglucanase II, cellobiohydrolase II, and *Aspergillus aculeatus* β-glucosidase 1 were simultaneously codisplayed as individual fusion proteins with the C-terminal half region of α-agglutinin. Codisplay of the three enzymes enabled the yeast strain to directly produce ethanol from amorphous cellulose. The yield (in grams of ethanol produced per gram of carbohydrate consumed) was $0.45\,g\,g^{-1}$, which corresponds to 88.5% of the theoretical yield. A critical aspect should be taken into account: the display of the cellulolytic enzymes on the cell surface makes attack of the macromolecular substrate difficult, or slow.

2) Reports on xylose fermentation by recombinant strains with industrial substrates are relatively few, and the rate of xylose fermentation, for example, for spruce hydrolysate, was lower by an order of magnitude than that in mineral medium. The major future challenge remains to translate the knowledge acquired from laboratory strains to industrial production strains (Hahn-Hägerdal *et al.*, 2007; Matsushika *et al.*, 2009).

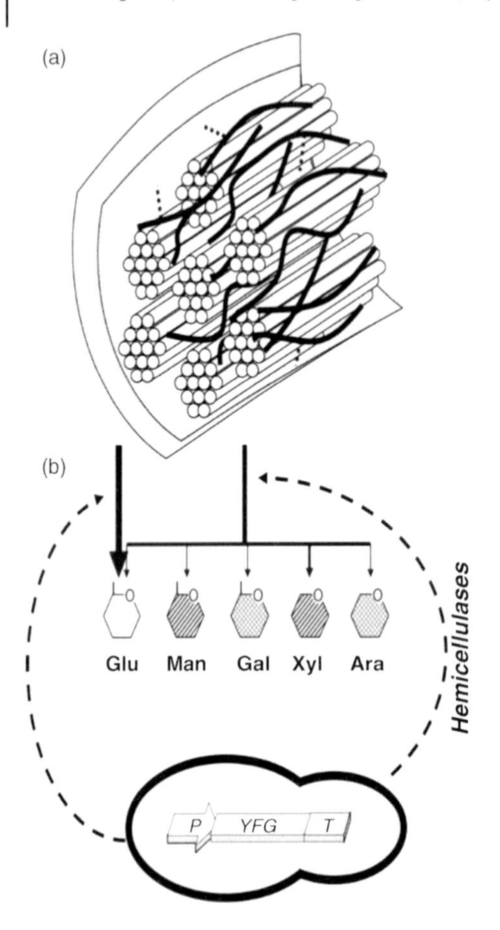

Figure 5.5 Graphic illustration of a CBP microorganism (van Zyl *et al.*, 2007, reproduced with permission from Springer Berlin/Heidelberg).

5.3.4
Conversion of D-Fructose to D-Mannitol

A whole-cell biotransformation system for the conversion of D-fructose to D-mannitol was developed in *E. coli* by constructing a recombinant oxidation/reduction cycle (Kaup *et al.*, 2003). First, the *mdh* gene, encoding mannitol dehydrogenase of *Leuconostoc pseudomesenteroides* ATCC 12291 (MDH), was expressed. To provide a source of reduction equivalents needed for D-fructose reduction, the *fdh* gene from *Mycobacterium vaccae* N10 (FDH), encoding formate dehydrogenase, was functionally coexpressed. FDH generates the NADH used for D-fructose reduction by dehydrogenation of formate to carbon dioxide. The introduction of a further gene, encoding the glucose facilitator protein of *Z. mobilis* (GLF), allowed the cells to efficiently take up D-fructose (Figure 5.6). Biotransformations conducted under pH control by formic acid addition yielded D-mannitol at a concentration of 362 mM

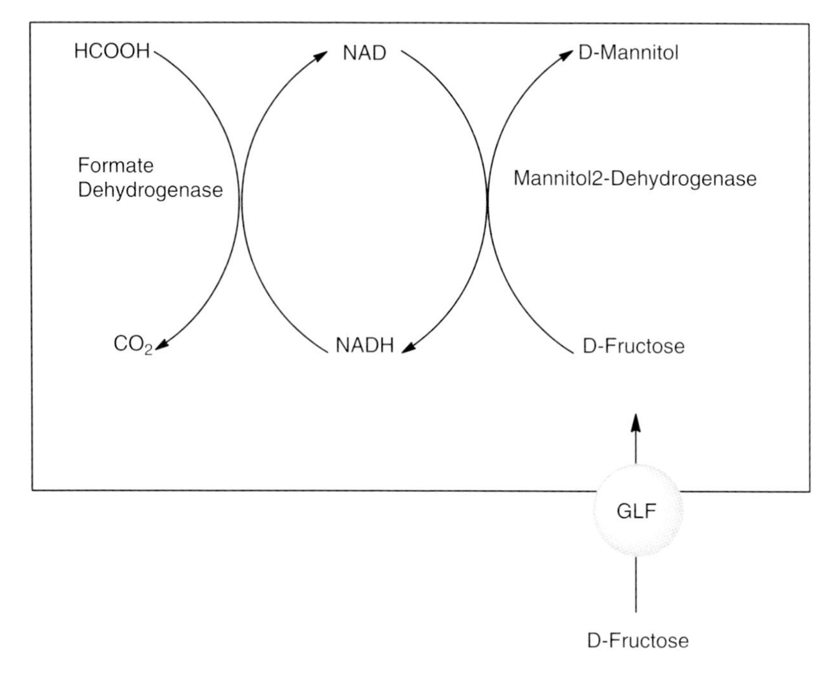

Figure 5.6 Biocatalyst for D-mannitol formation from fructose in a whole-cell biotransformation (GLF: glucose facilitator) (Kaup *et al.*, 2003).

within 8 h. The yield of mannitol (relative to fructose) was 84 mol%. An extension of this system may envisage the integration of glucose isomerase in order to use glucose as a substrate (Kaup *et al.*, 2003).

5.3.5
Synthesis of L-Ascorbic Acid

Much effort has been devoted to new routes for L-ascorbic acid (vitamin C) synthesis, and particularly by construction of designed recombinant microorganisms. Thus, an efficient way was developed with a recombinant strain of *Erwinia* sp., which oxidizes D-glucose to 2,5-diketo-D-gluconic acid. This strain furthermore encodes a recombinant reductase from *Corynebacterium* sp. producing 2-keto-L-gulonic acid, which easily can be rearranged by cyclization to vitamin C (Figure 5.7). When a culture of the *Erwinia* strain was fed with glucose to a total of 40 g l^{-1}, 49.4% of the glucose was converted to 2-keto-L-gulonate during a 72 h bioconversion (Grindley *et al.*, 1988). Anderson *et al.* (1985) succeeded in constructing a similar concept. The company BASF has realized a different concept on the industrial scale using a mixed culture of two microorganism. Sorbitol as the substrate is oxidized to L-sorbose followed by oxidation to 2-keto-L-gulonic acid, which then is rearranged to vitamin C; however, details have not been published.

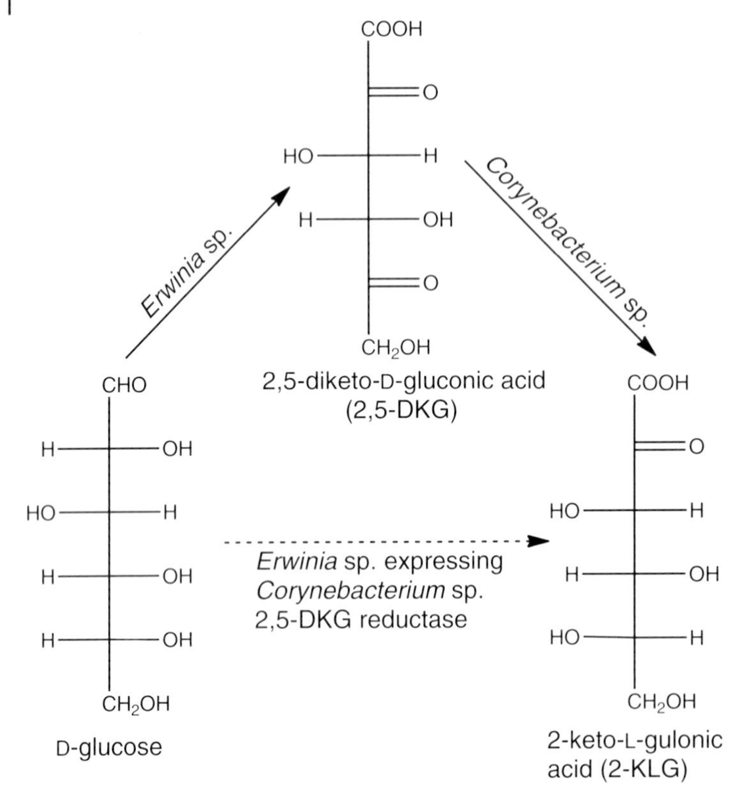

Figure 5.7 Whole-cell biotransformation of D-glucose to 2,5-diketo-D-gluconic acid and 2-keto-L-gulonic acid (serving as precursor for L-ascorbic acid, vitamin C) by a recombinant strain of *Erwinia* sp. (in solution, glucose and sugar acids are present in equilibrium as cyclic pyranose and lactones, respectively) (Grindley *et al.*, 1988).

5.3.6
Other Examples

Another product of commercial interest is indigo, one of the world's largest selling textile dyes. A biotransformation route has been designed by cloning genes for a multicomponent hydroxylase from *Acinetobacter* sp. in a solvent-resistant *E. coli* mutant encoding a NADH regeneration system. This strain was able to produce indigo from indole in an organic solvent–water two-phase culture system (Doukyu *et al.*, 2003).

Steroid pathway intermediates, formed during sterol catabolism, are widely recognized as pharmaceutically important precursors for drug synthesis. Different approaches have been developed to overcome difficulties due to the limited capabilities of microorganisms in such pathways. The group of Dijkhuizen focused on the bioengineering of molecularly defined mutant strains of *Rhodococcus* species blocked at the level of steroid Δ^1-dehydrogenation and steroid 9α-hydroxylation.

A molecular toolbox has been developed including characterization of the steroid catabolic pathway, cloning and expression vectors for *Rhodococcus* sp. for use in genomic library construction, functional complementation, and gene expression. The application of this toolbox has allowed rational construction of *Rhodococcus* strains with optimized properties for sterol/steroid bioconversions (van der Geize *et al.*, 2002).

Nucleotide-activated sugars (e.g., UDP-glucose, UDP-galactose, GlcNAc, GDP-mannose, GDP-fucose) are produced in Japan using the combination and reaction coupling between *Corynebacterium ammoniagenes* with a strong activity to synthesize nucleoside 5′-triphosphates and *E. coli* that is metabolically engineered and expresses genes of sugar nucleotide biosynthesis (www.kyowa.co.jp/bio/glyco_e/production/production.htm). Recombinant *E. coli* harboring five overexpressed enzymes were optimized and utilized for the production of UDP-galactose (UDP-Gal) from UMP, galactose, ATP (that was regenerated by acetate kinase), and acetyl phosphate. The yield of UDP-Gal reached 95% from 10 mM UMP (Lee *et al.*, 2010).

An engineering strategy for establishing new pathways with an improved NADH regeneration to utilize methanol as an unconventional cheap substrate, and to synthesize new high added-value (chiral) products, for example, drug intermediates, has been presented by Hartner (2007) (Figure 5.8). Studying the kinetic properties of alcohol oxidase (AOX), formaldehyde dehydrogenase (FLD), and formate dehydrogenase (FDH) and using the derived kinetic data for subsequent kinetic simulations of NADH formation rates led to the identification of FLD activity to

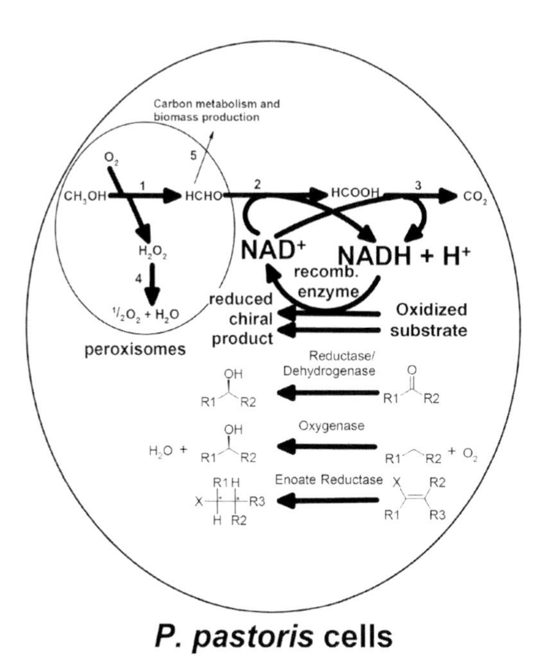

Figure 5.8 Engineered whole cells of *P. pastoris* can utilize methanol as substrate for carbon metabolism and for redox equivalents required for recombinant enantioselective redox enzymes from different sources to produce chiral intermediates (Hartner, 2007; Schroer *et al.*, 2010).

constitute the main bottleneck for efficient NADH recycling via the methanol dissimilation pathway. The simulation results were confirmed constructing a recombinant *Pichia pastoris* strain overexpressing *P. pastoris* FLD and the highly active NADH-dependent 1,4-butanediol dehydrogenase from *S. cerevisiae*. With this engineered strain, significantly improved 1,4-butanediol production rates were achieved in whole-cell biotransformations (Schroer *et al.*, 2010).

Exercises

5.1 When the flux J_1 in Eqs. (5.2) and (5.3) is increased, how can the fluxes J_2 and J_3 be increased without changing the concentrations of E_2, E_3, and E_4? (*Hint*: See Section 2.11.1 about the metabolite concentration and the K_m values.)

5.2 Are the processes to produce cephalosporin C or adipoyl-7-ADCA shown in Figure 12.12 examples of successful metabolic engineering?

Literature

Klein-Marcuschamer, D., Yadav, V., Ghaderi, A., and Stephanopoulos, G.N. (2010) *De novo* metabolic engineering & the promise of synthetic DNA. *Adv. Biochem. Eng. Biotechnol.*, **120**, 101–139.

Lee, S.Y. and Papoutsakis, E.T. (eds) (1999) *Metabolic Engineering*, Marcel Dekker, New York.

Papini, M., Salazar, M., and Nielsen, J. (2010) Systems biology of industrial microorganisms. *Adv. Biochem. Eng. Biotechnol.*, **120**, 51–99.

Stephanopoulos, G. (1999) Metabolic fluxes and metabolic engineering. *Metab. Eng.*, **1**, 1–11.

Stephanopoulos, G.N., Aristidou, A.A., and Nielsen, J. (1998) *Metabolic Engineering: Principles and Methodologies*, Academic Press, San Diego, CA.

References

Anderson, S., Berman Marks, C., and Lazarus, R., *et al.*, (1985) Production of 2-keto-L-gulonate, an intermediate in L-ascorbate synthesis, by a genetically modified Erwinia herbicola. *Science*, **230**, 144–149.

Bailey, J.E. (1991) Towards a science of metabolic engineering. *Science*, **252**, 1668–1675.

Brethauer, S. and Wyman, C.E. (2010) Continuous hydrolysis and fermentation for cellulosic ethanol production (Review). *Bioresour. Technol.*, **101**, 4862–4874.

de Boer, A.L. and Schmidt-Dannert, C. (2003) Recent efforts in engineering microbial cells to produce new chemical compounds. *Curr. Opin. Chem. Biol.*, **7**, 273–278.

Deckwer, W.D., Jahn, D., Hempel, D., and Zeng, A.D. (2006) Systems biology approaches to bioprocess development. *Eng. Life Sci.*, **6**, 455–469.

Doukyu, N., Toyoda, K., and Aono, R., (2003) Indigo production by Escherichia coli carrying the phenol hydroxylase gene from Acinetobacter sp. strain ST-550 in a water-organic solvent two-phase system. *Appl. Microbiol. Biotechnol.* **60**, 720–725.

Edwards, J.S., Ramakrishna R., Schilling C.H., Palsson B.Ø., (1999) Metabolic flux balance analysis. in Metabolic engineering, (eds Lee S.Y., Papoutsakis E.T.) (Marcel Dekker Inc. New York, NY), pp 13–57.

Fujita, Y., Ito, J., Ueda, M., Fukuda, H., and Kondo, A. (2004) Synergistic saccharification, and direct fermentation to ethanol, of amorphous cellulose by use of an engineered yeast strain codisplaying three types of cellulolytic enzyme. *Appl. Environ. Microbiol.*, **70**, 1207–1212.

Grindley, J.F., Payton, M.A., van den Pol, H., and Hardy, K.G., (1988) Conversion of glucose to 2-keto-L-gulonate, an intermediate in L-ascorbate synthesis, by a recombinant strain of Erwinia citreus, *Appl. Environ. Microbiol.*, **54**, 1770–1775.

Hahn-Hägerdal, B., Karhumaa, K., Jeppsson, M., and Gorwa-Grauslund, M. (2007) Metabolic engineering for pentose utilization in *Saccharomyces cerevisiae. Adv. Biochem. Eng. Biotechnol.*, **108**, 147–177.

Hartner, F.S. (2007) Engineering *Pichia pastoris* for whole cell biotransformations. Ph.D. thesis, Graz University of Technology.

Jarboe, L., Grabar, T., Yomano, L., Shanmugan, K., and Ingram, L. (2007) Development of ethanologenic bacteria. *Adv. Biochem. Eng. Biotechnol.*, **108**, 237–261.

Kacser, H. and Burns, J.A. (1981) The molecular basis of dominance. *Genetics*, **97**, 639–666.

Karhumaa, K., Wiedemann, B., Hahn-Hagerdal, B., Boles, E., and Gorwa-Grauslund, M.F. (2006) Co-utilization of L-arabinose and D-xylose by laboratory and industrial *Saccharomyces cerevisiae* strains. *Microb. Cell Fact.*, **5**, 18.

Kaup, B., Bringer-Meyer, S., and Sahm, H. (published online: 28 October, 2003) Metabolic engineering of Escherichia coli: construction of an efficient biocatalyst for D-mannitol formation in a whole-cell biotransformation, *Appl. Microbiol. Biotechnol.*, 1432–0614.

Lee, J.H., Chung, S.W., Lee, H.J., Jang, K.S., Lee, S.G., and Kim, B.G. (2010) Optimization of the enzymatic one pot reaction for the synthesis of uridine 5′-diphosphogalactose. *Bioprocess Biosyst. Eng.*, **33**: 71–78.

Lee, S.K., Chou, H., Ham, T.S., Lee, T.S., and Keasling, J.D. (2008) Metabolic engineering of microorganisms for biofuels production: from bugs to synthetic biology to fuels. *Curr. Opin. Biotechnol.*, **19**, 556–563.

Lynd, L., Zyl, W., McBride, J., and Laser, M. (2005) Consolidated bioprocessing of cellulosic biomass: an update. *Curr. Opin. Biotechnol.*, **16**, 577–583.

Marx, A., de Graaf, A.A., Wiechert, W., Eggeling, L., and Sahm, H. (1996) Determination of the fluxes in the central metabolism of *Corynebacterium glutamicum* by nuclear magnetic resonance spectroscopy combined with metabolite balancing. *Biotechnol. Bioeng.*, **49**, 111–129.

Matsushika, A., Inoue, H., Kodaki, T., and Sawayama, S. (2009) Ethanol production from xylose in engineered *Saccharomyces cerevisiae* strains: current state and perspectives. *Appl. Microb. Biotechnol.*, **84**, 37–53.

Mijts, B.N. and Schmidt-Dannert, C. (2003) Engineering of secondary metabolite pathways. *Curr. Opin. Biotechnol.*, **14**, 597–602.

Nakamura, C.E. and Whited, G.M. (2003) Metabolic engineering for the microbial production of 1,3-propanediol. *Curr. Opin. Biotechnol.*, **14**, 454–459.

Rapoport, T.A., Heinrich, R., and Rapoport, S.M. (1976) The regulatory principles of glycolysis in erythrocytes *in vivo* and *in vitro*. *Biochem. J.*, **154**, 449–469.

Rogers, P., Jeon, Y., Lee, K., and Lawford, H. (2007) *Zymomonas mobilis* for fuel ethanol and higher value products. *Adv. Biochem. Eng. Biotechnol.*, **108**, 263–288.

Schroer, K., Luef, K.P., Hartner, S.F., Glieder, A., and Pscheidt, B. (2010) Engineering the *Pichia pastoris* methanol oxidation pathway for improved NADH regeneration during whole-cell biotransformation. *Metab. Eng.*, **12**, 8–17.

Segre, D., Vitkup, D., and Church, G.M. (2002): Analysis of optimality in natural and perturbed metabolic networks, Proc. *Natl. Acad. Sci.* USA, **99**, 15112–15117.

Smolke, C.D. (ed.) (2010) *The Metabolic Pathway Engineering Handbook*, CRC Press, Boca Raton, FL.

Steen, E.J., Kang, Y., Bokinsky, G., Hu, Z., Schirmer, A., McClure, A., Del Cardayre, S. B., and Keasling, J.D. (2010) Microbial production of fatty-acid-derived fuels and chemicals from plant biomass. *Nature*, **463**, 559–562.

Stephanopoulos, G. and Vallino, J.J. (1991) Network rigidity and metabolic engineering in metabolite overproduction. *Science*, **252**, 1675–1681.

Van der Geize, R, Hessels, G I, and van Gerwen, R. *et al.*, (2002) Molecular and functional characterization of kshA and kshB, encoding two components of 3-ketosteroid 9-alpha-hydroxylase, a class IA monooxygenase, in Rhodococcus erythropolis strain SQ1. *Mol. Microbiol.* **45**, 1007–1018

van Zyl, W., Lynd, L., den Haan, R., and McBride, J. (2007) Consolidated bioprocessing for bioethanol production using *Saccharomyces cerevisiae*. *Adv. Biochem. Eng. Biotechnol.*, **108**, 205–235.

Wisselink, H.W., Toirkens, M.J., Wu, Q., Pronk, J.T., and van Maris, A.J.A. (2009) Novel evolutionary engineering approach for accelerated utilization of glucose, xylose, and arabinose mixtures by engineered *Saccharomyces cerevisiae* strains. *Appl. Environ. Microbiol.*, **75**, 907–914.

Yim, H., Haselbeck, R., Niu, W., Pujol-Baxley, C., Burgard, A., Boldt, J., Khandurina, J., Trawick, J.D., Osterhout, R.E., Stephen, R., Estadilla, J., Teisan, S., Schreyer, H.B., Andrae, S., Yang, T.H., Lee, S.Y., Burk, M.J., and Van Dien, S. (2011) Metabolic engineering of *Escherichia coli* for direct production of 1,4-butanediol. *Nat. Chem. Biol.*, **7**, 445–452.

Zaldivar, J., Nielsen, J., and Olsson, L. (2001) Fuel ethanol production from lignocellulose: a challenge for metabolic engineering and process integration. *Appl. Microb. Biotechnol.*, **56**, 17–34.

Zeng, A.P. and Biebl, H. (2002) Bulk chemicals from biotechnology: the case of 1,3-propanediol production and the new trends. *Adv. Biochem. Eng. Biotechnol.*, **74**, 239–259.

Zhang, F., Rodriguez, S., and Keasling, J.D. (2011) Metabolic engineering of microbial pathways for advanced biofuels production. *Curr. Opin. Biotechnol.*, **22**, 775–783.

6
Enzyme Production and Purification

For enzyme processes and for the use of enzymes in therapy and for analytical purposes, a steady and safe enzyme supply is required. In technical applications, the enzyme costs must be less than 1–10% of the product costs, with the low and high values applying to low- and high-value products, respectively. In order to achieve this, the production, purification, and formulation of an enzyme that is suitable as biocatalyst for a specific application can be improved as shown below.

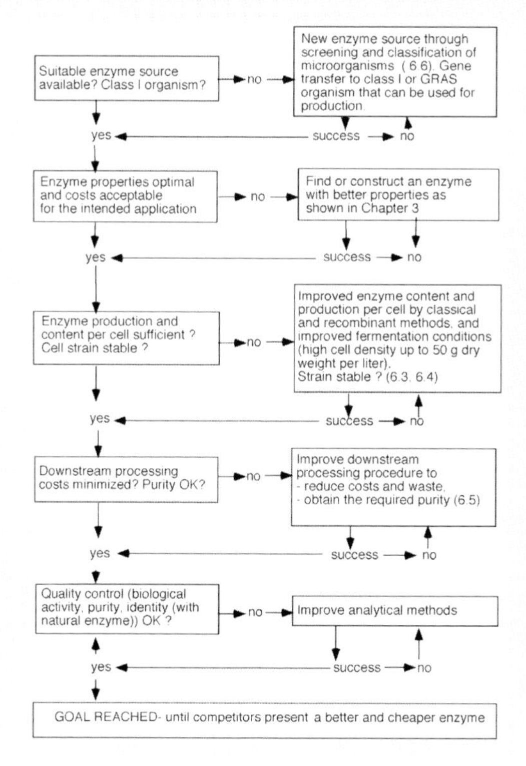

Biocatalysts and Enzyme Technology, Second Edition. Klaus Buchholz, Volker Kasche, and Uwe T. Bornscheuer

6.1 Introduction

The use of isolated enzymes as biocatalysts in enzyme processes is limited by their unavoidable inactivation. The enzyme cost in a process can be reduced by using more stable enzymes with the same or better activity or vice versa, provided that they can be produced with the same or higher yield as the original enzyme. From this it follows that enzyme processes must be coupled with the production and purification of enzymes. The latter must be designed to minimize the enzyme production costs, and to provide a stable supply for the enzyme with the required purity. The first step to reach this aim is to find a suitable enzyme source. Generally, the natural sources (animal and plant tissues, wild-type microorganisms) produce insufficient quantities of the desired enzyme, as natural systems have not been optimized for high enzyme contents during evolution. Therefore, the next step in the production of an enzyme in wild-type organisms that can be used for the safe production of enzyme (GRAS or class I organisms) is the optimization of the enzyme production.

When the enzyme-producing organism is not a GRAS or class I organism, it should not be used for the production of the enzyme, based on safety criteria. In this case, the gene for the desired enzyme can be transferred to and expressed in a GRAS or class I organism. The classification is based on the risks of the use of these organisms for human health and the environment. Their use is therefore governed by national and international laws and regulations. This legislation is currently in a process of revision. Therefore, an overview of the legislation that must be considered in bio- and enzyme technology is given in Section 6.6.

The optimization of enzyme production in wild-type or recombinant microorganisms involves manipulation of the transcriptional, translational, and post-translational processes that influence the yield of the active enzyme. This can be achieved with and without gene technological methods. Finally, the enzyme produced must be purified to the desired purity for its application as a biocatalyst. The purification is much easier for extracellular enzymes as they can be isolated directly from the medium without the disruption of cells, as is required for intracellular enzymes. The purification costs decrease with the enzyme content in the medium or homogenate, and increase with the purity of the enzyme that is required for its final use. Estimated costs of technical enzymes used as biocatalysts are between $\approx 10\,€\,kg^{-1}$ (amylases, pectinases) and $>1000\,€\,kg^{-1}$. These costs are lower than those for enzymes used for analytical and therapeutic purposes, where a much higher purity is required than for technical enzymes.

This chapter will provide an overview of how enzyme costs can be reduced by increasing the yield in the production and downstream processing to reach the desired purity. For more detailed information, the interested reader can consult the textbooks and reviews given in the literature list at the end of the chapter.

6.2
Enzyme Sources

6.2.1
Animal and Plant Tissues

Classical enzyme sources are animal and plant tissues from the processing of animals and plants, especially those that are residues not used for food. To these belong the pancreas (used earlier in the treatment of hides in the production of leather; see Section 1.3), calf stomach (used in cheese making), kidney, liver, parts of plants (papaya, pineapple) used for the tenderization of meat, or sodom apple leaves used for milk clotting in the production of cheese in Central Africa (Aworh, Kasche, and Apampa, 1994). These processes, except the use of pancreas, are up to 5000 years old. They were developed empirically, long before it was known that enzymes were essential as biocatalysts in these processes. The enzymes involved in such processes have been identified during the development of biochemistry and enzymology over the past 150 years. Some of the enzyme sources used here have large enzyme contents and can produce up to 1% enzyme of the wet tissue weight per day (Table 6.1). To these belong tissues producing digestive enzymes (pancreas, seeds) or expressing large quantities of enzyme for tissue-specific metabolism (liver, or muscle such as heart). Due to problems caused by difficulties in the controlled recovery of enzymes from these tissues or their alternative uses, and in order to avoid shortages in the enzyme supply, other enzyme sources had to be found. This is illustrated by the use of pancreas as an enzyme source. Until it was discovered (in 1921) that the pancreas also produces insulin, the tissue was mainly used as a source of technical enzymes (see Section 1.3) and digestive enzymes for substitution therapy. However, after this time the tissue became too expensive as a source of technical enzymes. The enzymes still produced from the pancreatic tissue (chymotrypsin and trypsin) were by-products in the production of insulin. Today, insulin is mainly produced in recombinant organisms (*E. coli* or yeast cells), and therefore the price of those enzymes that have established applications in therapy and animal cell culture technology has increased. For other technical enzymes – for which the pancreas was once a source – alternative supplies, mainly from microorganisms, have been found in the meantime. The traditional source for rennet (a mixture of the peptidases chymosin (EC 3.4.23.4) and pepsin (EC 3.4.23.1)), used to curdle milk as a first step in the production of cheese, is one of the stomachs of young calves. This source is limited, and could not meet the demand. Therefore, other chymosin/rennet sources had to be found. First wild-type microbial rennet was used as an additional source. Now also a recombinant chymosin produced in GRAS microorganisms is used in the production of cheese. It is still prepared with wild-type enzymes for consumers that only consume these cheeses (Soethaert and Vandamme, 2010).

Today, animal and plant tissues can be used for the production of heterologous proteins by transfer of genes (*pharming*) (Rudolph, 1999; Karg and Kallio, 2009). Until now, this approach has mainly been considered for therapeutic proteins, but

Table 6.1 Biologically active enzyme content and enzyme productivity in animal and plant tissues and microorganisms (Albee, Butler, and Wrigtht, 1990; Amneus, 1977; Deshpande *et al.*, 1994; Rúa *et al.*, 1998; Koller, Riess, and Aretz, 1998; Pilone and Pollegioni, 2002; Christiansen, Michaelsen, and Wümpelmann, 2003; Kasche *et al.*, 2005; Huang *et al.*, 2010).

Enzyme source	Enzyme/protein	Enzyme content (g g^{-1} cell dry weight) (% of total protein in tissue or fermenter)	Enzyme productivity (g l^{-1} tissue or fermenter per day)
Animal tissues			
Pancreas	Digestive enzymes such as chymotrypsin, lipase, nuclease, trypsin (e)[a]	The pro-enzymes of these enzymes are activated outside the pancreatic cells; pro-enzyme content ≈0.04	Up to 10
Liver	Aldolase (i)[a]	0.001	
Muscle	Aldolase (i)	0.03	
Porcine kidney	D-Amino acid oxidase (i)	0.001	
Plant tissues			
Papaya	Papain (e)	<0.01	
Eukaryotic cells			
S. cerevisiae	Aldolase (i)	0.016	
Recombinant CHO[b] cells	Monoclonal antibody (e)	0.4 (>50)	0.5
Prokaryotic cells			
E. coli			
Wild type	Penicillin amidase (p)[a]	0.004 (1)	
Recombinant	Penicillin amidase (p)	0.04–0.1 (8–20)	1–2
Recombinant	Glutaryl amidase (p)	0.05 (10)	0.5
Recombinant	Lipase (i + p)[c]	0.05 (10)	
Recombinant	D-Amino acid oxidase (i)	0.04 (8)	
B. clausii			
Recombinant	Savinase (a subtilisin)	0.016	0.4

a) (i) intracellular; (p) periplasmic (outside cell membrane inside cell wall); (e) extracellular (outside cell membrane and cell wall).
b) Chinese hamster ovary.
c) *B. thermocatenulatus* lipase is extracellular but was mainly produced as insoluble enzyme in *E. coli*.

the potential advantage lies in the reduced cost of protein production compared to that by microorganisms in fermenters. Pharming might be used to produce therapeutic proteins, though its use to supply technical enzymes that are nonnatural for these tissues remains questionable. In animals, this would lead to the production of antibodies, and in both animal and plant tissues these enzymes might catalyze unwanted reactions that can have negative influences on animal and plant growth. Another problem with enzymes derived from animal tissues is their possible contamination with prions (e.g., bovine spongiform encephalitis, BSE) or viruses that are harmful to humans. To avoid this, these enzymes can be produced recombinantly in microorganisms or prion- and virus-free mammalian cells.

6.2.2 Wild-Type Microorganisms

Microorganisms that have long been used safely in food production, mainly as preserving agents – that is, GRAS organisms and class I organisms – are ideal sources for the production of enzymes for enzyme processes. Enzymes from these sources are used mainly in food processing, where the demand to use nonrecombinant enzymes still is and is expected to be high in the future, and regulations are strict (see Section 6.6.6).

Wild-type microorganisms are also used to screen for "new" enzymes that would catalyze the desired conversion or degradation of a compound, A. When this conversion involves only one enzyme, it can also be used to catalyze the reverse reaction, that is, the synthesis of A from the products. Different methods exist to screen for such enzymes:

- from a sample with different microorganisms (Figure 6.1);
- from a sample with the same microorganisms but with a different enzyme content per cell (Figures 6.2 and 6.3).

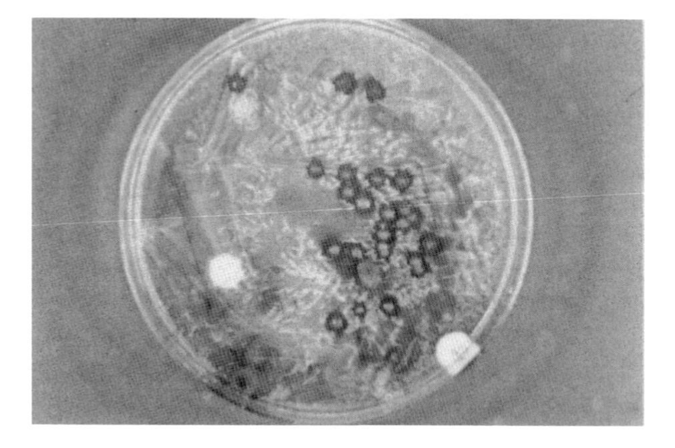

Figure 6.1 Screening for microorganisms from a soil sample that can degrade pyrene. The cells are grown isolated on the Petri dish with pyrene as the only C source. The clones that can degrade pyrene and divide are directly observed (Kästner, 1994, personal communication).

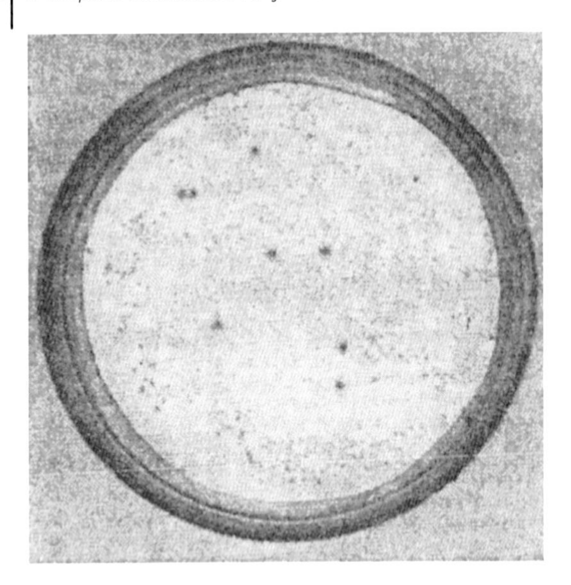

Figure 6.2 Screening for *E. coli* cells that produce penicillin amidase. A filter paper with a substrate selectively hydrolyzed by this enzyme that yields a yellow product is placed on the Petri dish with isolated cell clones. The clones producing penicillin amidase color the filter paper yellow (here seen as dark spots) above the clone. The cells that produce more enzyme develop the yellow color more rapidly than those producing less of the enzyme.

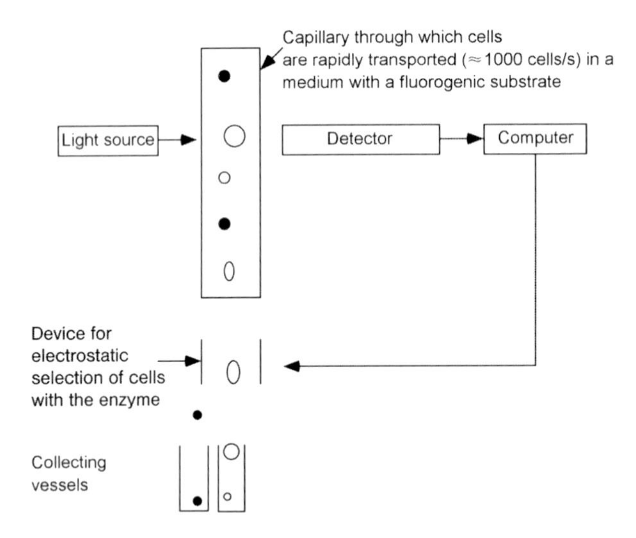

Figure 6.3 A fluorescence-activated cell sorter (FACS) for high-throughput screening (HTS) of cells with respect to different properties such as size, DNA content, or content of an intracellular or periplasmic enzyme, E. In the latter case, single cells are transported through a capillary in a solution with a fluorogenic substrate that can only be converted to a fluorescent product by the enzyme E. The cells with the enzyme (marked black) become fluorescent and are detected by the measuring device (detector and computer); this gives a signal to a device for electrostatic selection of these cells, and they are collected into a separate vessel.

These methods can also be used to screen for enzymes that are active under desired process conditions (pH and temperature). For processes to be carried out under alkaline or acidic pH conditions, the screening must be performed with organisms or metagenomes gathered from alkaline or acidic environments, respectively (see Section 2.11). "Metagenomes" are DNA from entire microbial consortia that cannot be cultivated directly (see Section 3.1). Genes from the metagenomes are transferred to microorganisms, which in turn express the enzymes encoded by the metagenome genes (see the next section). For processes that are optimal at higher temperatures (up to 100 °C), it is necessary to screen organisms from hot springs (extremophiles) (Niehaus *et al.*, 1999). The organisms producing the desired enzyme must then be classified.

6.2.3 Recombinant Microorganisms

When the enzyme yield is low in wild-type class I organisms, or the desired enzyme is not found in class I organisms, then alternative enzyme sources must be found. This can easily be achieved by transferring the enzyme gene to, and expressing it as a recombinant enzyme in, a suitable class I organism. For this purpose, frequently used microorganisms include bacteria and yeasts such as *Bacillus* sp., *Escherichia coli* (only class I strains), *Streptomyces* sp., *Aspergillus* sp., *Pichia pastoris*, and *Saccharomyces* sp. Nowadays, most enzymes used in enzyme processes are produced using recombinant technology. More details can be found in Chapter 3.

6.3 Improving Enzyme Yield

Enzymes are produced and purified up to a gram scale in laboratories where

- its 3D structure is determined or its function will be studied in detail for a longer time;
- its potential use as biocatalyst in an enzyme process is studied in detail.

When the enzyme source is not a class I organism or when such amounts cannot be easily produced in the wild-type class I organism enzyme source, such amounts can easily be obtained when the enzyme gene is transferred to and expressed in a suitable class I microorganism (host cell). Two possibilities exist here:

- **Homologous expression:** the host cell is a wild-type class I cell that produces the enzyme. The host cell has been transformed and contains extra enzyme genes in suitable plasmids.
- **Heterologous expression:** the host class I cell does not produce the enzyme, unless it contains plasmids with the enzyme gene from a different microorganism.

The class I microorganisms that can be used for homogeneous and heterogeneous expression of enzymes in laboratories are found in national lists for organism classification (see Section 6.6).

Most technical enzymes and pharmaceutical proteins (monoclonal antibodies, enzymes, insulin, etc.) are now produced in recombinant cells in quantities from less than 1 ton to more than 100 tons per year (≈100 kg penicillin amidase, ≈100 ton glucose isomerase). The number of class I cells that can be used as hosts for homologous or heterologous enzyme expression in production facilities is much lower than those that can be used for laboratory-scale production (see Section 6.6.). For enzymes that must be glycosylated, eukaryotes must be used as host cells. When this is not the case, bacteria are suitable hosts. Companies that produce enzymes on an industrial scale use a limited number of host cells, especially those with which they have obtained extensive proprietary know-how, concerning the optimal fermentation conditions. This knowledge is generally never published.

A high yield of active enzyme per cell, obtained as fast as possible, is desired when the enzyme is produced for laboratory or industrial uses. This is especially important in the latter case where the enzyme costs must be minimized. About 50% of the dry weight of a cell is protein. The cells require their enzymes and other proteins to produce the desired enzyme. Generally, the yield as active soluble enzyme is increased, compared to its nonrecombinant expression in its wild-type host, when it is expressed as a pre-(pro-)enzyme, that is, extracellular (or periplasmic), in suitable host cells. The maximal yields that have been observed are up to ≈20% of the total cell proteins, or ≈100 mg active enzyme per gram cell dry weight (Table 6.1; Schmidt, 2004). The content of the pre-pro-enzyme in the host cells is small compared to its content outside the cytoplasmic cell membrane. The cytoplasm is highly viscous and a crowded 20% protein solution (Luby-Phelps, 2000; McGuffee and Elcock, 2010). Producing a high content of foreign soluble and active protein in this environment disturbs the normal cell physiology, and results in the hydrolysis or precipitation of the foreign enzyme (protein) into insoluble inclusion bodies. Higher yields up to 40% have been obtained when the enzyme (protein) is produced as inactive inclusion bodies (Quyen, Schmidt-Dannert, and Schmid, 1999). Their renaturation and purification require additional processing steps and chemicals that increase the enzyme costs, compared to the production and purification of a soluble active extracellular or periplasmic enzyme (Figure 6.10 and Section 6.5). Therefore, technical enzymes are mainly produced by the latter procedure.

After the induction of the expression of an enzyme gene, a consecutive reaction scheme starts, where several intermediates are formed until the final active enzyme is formed. Processes that reduce the concentration of the intermediates, from which the active enzyme is formed, can influence the yield (Figure 6.4). The factors that influence the homologous and heterologous expression of enzymes in bacteria and yeast cells, mainly on the level of expression vectors (plasmid construction) and transcription, have been extensively reviewed (Jana and Deb, 2005; Burgess and Deutscher, 2009). Here, we will focus on the yield-reducing processes starting from the mRNA, that is, on the translational and the posttranslational level.

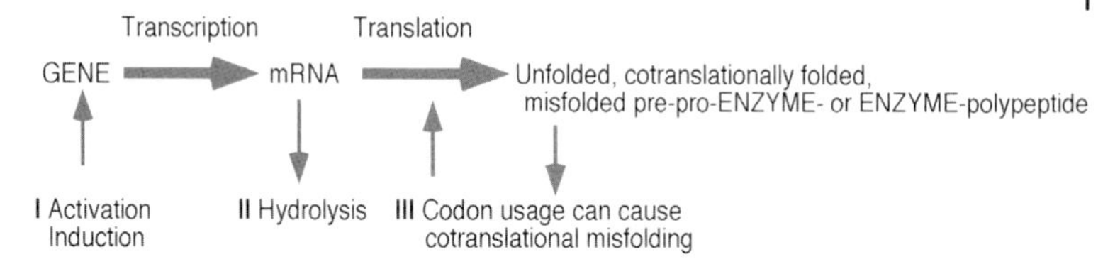

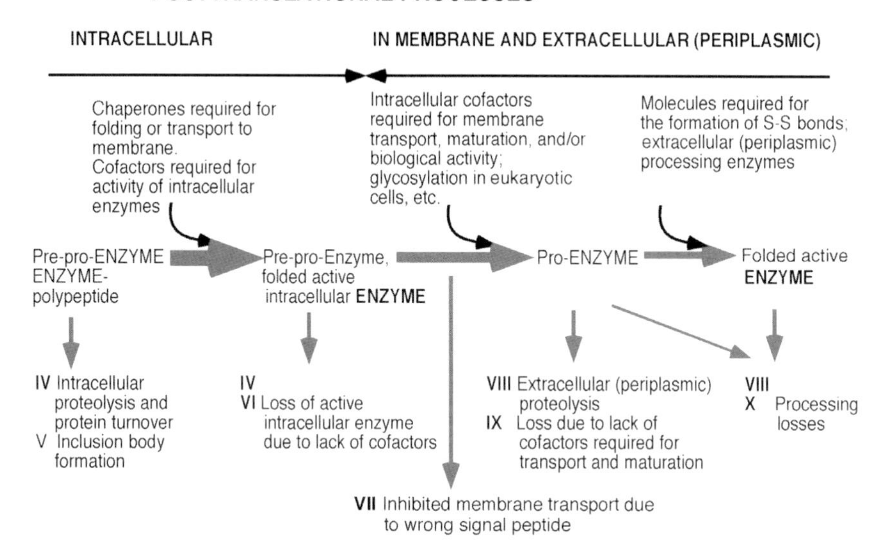

Figure 6.4 Transcriptional, translational, and posttranslational processes that influence the yield of intracellular, extracellular, and periplasmic enzymes. The latter two are synthesized as pre-pro-enzymes. A signal peptide (pre-) is required for the transport through the cell membrane, and is cleaved off by signal peptidases. The biologically active enzyme is written in boldface letters.

6.3.1
Processes that Influence the Enzyme Yield

Posttranscriptional, translational, and posttranslational processes that influence the yield of active enzyme are shown schematically in Figure 6.4. These processes include the following (where stages VII to X apply only to extracellular and periplasmic enzymes that are transported through the cell membrane):

I) For transcription to occur, the gene must first be activated. This requires an external inducer that in wild-type organisms is normally a molecule, the conversion of which is catalyzed by the enzyme encoded by the activated gene. The repressor protein that inhibits the binding of RNA polymerase to the gene in the double-stranded DNA, required for transcription, loses this property when it binds to an inducer molecule. These activation mechanisms are also used in the production of recombinant enzymes.

II) In bacteria, mRNA has a very short half-life time (range 2–10 min) and is rapidly degraded in the cells by different RNases. The half-life of mRNA can be increased, and the enzyme yield increased, by reducing the expression of specific RNases (Viegas *et al.*, 2005). In eukaryotes, mRNA undergoes additional processing steps, including RNA splicing, capping, and polyadenylation. In addition, the availability of mRNA for translation is controlled by microRNAs or RNA silencing (short interference RNA, siRNA) in eukaryotic hosts. In these, antisense nucleotides that can inhibit the translation or accelerate the degradation of the complementary mRNA are formed by transcription (Scherer and Rossi, 2003). Consequently, these processes reduce the yield of (pre-pro-)enzyme polypeptides by decreasing the concentration of the mRNA that can be translated.

III) In the heterologous expression of an enzyme that is cotranslationally folded in its wild-type host cell, the tRNA concentration of the new host modulates the rate of mRNA translation. Codon usage and tRNA concentrations determine the rate of translation of a single codon, which in turn strongly influences cotranslational folding (Komar, 2009; Zhang, Hubalewska, and Ignatova, 2009). Codon usage and tRNA concentrations have coevolved, and differ from organism to organism and their substrains. This may result in an incorrect folding of the enzyme in the host that reduces the active enzyme yield (Jana and Deb, 2005; Angov *et al.*, 2008; Zhang and Ignatova, 2009). Therefore, a host cell with a codon usage similar to the wild-type host cell must be selected for an optimal expression of the enzyme. Codon usage in different cell strains is found at http://www.kazusa.or.jp/codon/.

IV) Intracellular proteolysis is a selective highly regulated process that plays an important role in cellular physiology (Enfors, 1992; Wickner, Maurizi, and Gottesman, 1999; Henge and Bukau, 2003). Many intracellular enzymes are synthesized to catalyze specific processes during the cell cycle. When their activity is no more required, it can disturb the cell cycle, and must be inhibited. This is achieved by intracellular proteolysis by intracellular peptidases (protein turnover), a natural process used to recycle amino acids. Thus, many intracellular enzymes have biological half-life times less than the time required for a cell cycle (<100 min for prokaryotes) (Grünenfelder *et al.*, 2001). Other intracellular peptidases are involved in "quality control" of the synthesized (pre-pro-)enzymes. Some of these may be misfolded or form soluble aggregates, that is, be "aberrant" proteins that may disturb the cell cycle or form toxic aggregates. As they cannot be transformed to biologically active enzymes in the cell, they must be degraded by such intracellular peptidases. Heterologous enzymes may also be recognized as "aberrant" proteins in the host cell. Their yield can thus be reduced by intracellular proteolysis involved in protein turnover and quality control. This can be minimized using peptidase-deficient host cells (see Section 6.2.3).

V) When the concentration of (pre-pro-)enzyme polypeptides that are synthesized is high, neighboring polypeptides may interact with each other during folding. This can lead to the formation of soluble protein aggregates or insoluble protein aggregates (inclusion bodies). The latter must be solubilized, unfolded, and renatured to become biologically active (Mitraki *et al.*, 1991). This process is prevented by molecular chaperones that act as folding catalysts of intracellular enzymes or extracellular pre-pro-enzymes that must be folded before they are transported through the cell membrane. For extracellular pre-pro-enzymes that must be unfolded during the membrane transport, other proteins that assist their transport to the membrane prevent the folding in the cytoplasm, intracellular proteolysis, and inclusion body formation.

VI) Binding of cofactors to intracellular enzymes required for their biological activity. They are bound covalently (such as phosphorylation) or noncovalently, such as Zn^{2+} to hydantoinases (Werner *et al.*, 2004) or FAD to D-amino acid oxidase (Pilone and Pollegioni, 2002). An insufficient supply of these cofactors reduces the yield of active enzyme. The intracellular concentrations of ions such as Ca^{2+} and Zn^{2+} is $<1\ \mu M$; the binding constants for these ions to the enzymes that require them for their activity must therefore be much lower than $1\ \mu M$ in order to be saturated with the required metal ion (Jones, Holland, and Campbell, 2002).

VII) Extracellular (periplasmic) pre-pro-enzymes are transported through the cell membrane either folded or unfolded conformation. Folded proteins are transported through the Tat translocation system that is found only in archea, bacteria, and chloroplasts (Bagos *et al.*, 2009). It is used by enzymes with bound cofactors (redox or metal ions such as Ca^{2+}) in the cytoplasm (Bendtsen *et al.*, 2005; Kasche *et al.*, 2005). Unfolded proteins are translocated through the Sec translocation system present in archea, bacteria, and eukaryotic cells (Rapoport, 2007). A signal peptide (an N-terminal pre-peptide) guides the precursor protein to the corresponding translocation machinery; the signal peptide is removed by signal peptidases during the translocation process. Sec and Tat signal peptides differ from organism to organism, even for the same pre-pro-enzyme. In the heterologous expression of such enzymes, the yield of active enzyme can be improved by using a signal peptide of the host cell (Mergulhão, Summers, and Monteiro, 2005; Idiris *et al.*, 2010). Then, however, the membrane translocation system in the wild-type cell must be known. When this is not known, it can be predicted based on known signal peptides for these translocation systems from different organisms (Bendtsen *et al.*, 2004; Bagos *et al.*, 2009). Enzymes that are intracellular in the source cell can become extracellular in the host cell, when a suitable signal peptide has been attached to the N-terminal end of this enzyme (Kjærulff and Jensen, 2005). For the translocation through the Tat system for folded proteins, this requires that this end is located on the surface of the folded enzyme.

VIII) Extracellular (periplasmic) enzymes can be degraded by extracellular (periplasmic) peptidases. Here the use of peptidase (periplasmic and extracellular)-deficient strains can increase the yield of active enzymes.

IX) For pre-pro-enzymes that require cofactors (such as metal ions; see stage VI) for the transport through the membrane, maturation, and/or biological activity, a lack of these can lead to the accumulation of misfolded pro-enzyme or pro-enzyme in the cytoplasm (Kasche *et al.*, 2005).

X) Pro-enzymes that are proteolytically processed generally exist as several different processed forms that differ in activity (Figures 6.5 and 6.7).

How these processes can be influenced to increase the enzyme yield with and without gene technology is shown in Table 6.2. In addition, they can also be influenced by temperature, pH, and the medium composition. When the yield has been optimized in shake flask cultures, as indicated in Table 6.2, it can only be improved by increasing the cell density during the fermentation. When the cells grow exponentially, medium with necessary energy, C, N, S, and P sources, and different ions required as cofactors must be added exponentially in order to achieve this. In fermenters, the aggregation of cells to larger particles can influence the rate of enzyme production, as mass transfer limitations increase with aggregate size (see Chapter 10). This applies especially for filamentous fungi or bacteria, such as class I (or GRAS) *Aspergillus* sp. or *Streptomyces* sp. that are or can become industrially important enzyme producers, respectively. They can form pellets with diameters larger than 200 μm. In these, oxygen is depleted in the inner of the pellets, and the enzyme is produced only in the outer part of these (Nielsen, 1996; see also Figure 10.2). The size of the pellets can be reduced and the enzyme production per cell increased by selecting mutants that form smaller pellets, mechanical forces such as stirring, or adding inert microparticles (Driouch, Sommer, and Wittmann, 2009). In practice, cell densities up to 50–100 g cell dry weight per liter can be obtained in such high cell density fermentations. The upper limit for the concentration of an active enzyme synthesized under these conditions is about $10\,g\,l^{-1}$, or about 40% of the total cell protein. This is about the same order of magnitude as in the best enzyme-producing tissues (e.g., pancreas; see Table 6.1). How this has been achieved will be illustrated mainly for the posttranslational processes III–X, based on recent studies, for extracellular (lipases) and periplasmic (penicillin amidases) enzymes.

6.4
Increasing the Yield of Periplasmic and Extracellular Enzymes

Most technical enzymes are currently produced in unicellular organisms (bacteria, yeasts, fungi), and this applies to both nonrecombinant and recombinant enzymes. For the latter, *E. coli*, *Bacillus* sp., *Aspergillus* sp., and yeast cells (only class I strains) are frequently used hosts. The major hydrolases used in technical processes are either extracellular or periplasmic – that is, they have been transported through the

Table 6.2 How the yield of biologically active enzyme can be increased on the transcriptional, translational, and posttranslational level by influencing processes I–VIII in Figure 6.4.

Process influencing the yield per cell (Figure 6.4)		Yield can be increased without enzyme engineering by	Yield can be increased with enzyme engineering by[a)]
I	Gene activation by induction and rate of mRNA synthesis	Screening for improved enzyme-producing wild-type cells after mutagenesis (chemical, physical); selecting optimal inducer concentration	Increasing the number of genes per cell and the rate of mRNA synthesis
II	mRNA hydrolysis	Selecting cells with reduced rate of mRNA hydrolysis	Reducing the content of different RNases
III	Codon usage	Decreasing the temperature	Using host cells with suitable codon usage in heterologous enzyme expression
IV	Intracellular proteolysis	Selecting cells with lower intracellular peptidase content	Increasing chaperone content to reduce proteolysis losses; reducing intracellular enzyme content in host cells
V	Inclusion body formation	Reducing rate of polypeptide synthesis by reducing the temperature; using non-rich synthetic media to decrease the synthesis rate	Increasing chaperone content by coexpression of chaperones; reducing rate of polypeptide synthesis by reducing the temperature
VI	Cofactor binding to intracellular enzymes	Increasing extracellular cofactor concentration (metal ions)	Stimulating intracellular cofactor synthesis and increasing extracellular cofactor concentration (metal ions)
VII	Membrane transport	Selecting cells with higher membrane transport rate	Using host cell signal peptides in heterologous enzyme expression
VIII	Extracellular (periplasmic) proteolysis	Selecting cells with a low content of extracellular (periplasmic) peptidases	Reducing the content of extracellular and periplasmic peptidases
IX	Cofactor binding to extracellular and periplasmic enzymes and membrane transport	Increasing cofactor content; selecting cells with a high membrane transport rate	Increasing cofactor concentration and number of protein translocation systems in the membrane; using host cell signal peptides
X	Processing losses		Site-directed mutagenesis of peptide bond in proenzyme to prevent further processing

a) In all gene transformations with plasmids to microorganisms used for large-scale production of enzymes, the use of antibiotic resistance selection markers must be reduced, by developing alternative selections markers, to minimize the risk of selecting antibiotic-resistant microorganisms.

cell membrane. How their yields can be improved in wild-type and recombinant *E. coli* cells (as outlined in Table 6.2) will be illustrated for the hydrolases penicillin amidase and lipase, both of which are used as biocatalysts in large-scale enzyme processes (Makrides, 1996; Jana and Deb, 2005). This illustration is based on the yield of active enzyme, and not for inclusion bodies of the (pre-)pro-enzyme. Larger yields (in terms of protein content per cell) can be obtained for inclusion bodies, and these could be renatured to active enzyme after purification. However, this requires additional processing steps such as refolding to the native pro-enzyme and proteolytic processing *in vitro* that not only increase the processing costs but also reduce the yield of the active enzyme (Middelberg, 2002).

6.4.1 Penicillin Amidase

The penicillin amidases (EC 3.5.1.11) from *E. coli, A. faecalis, P. rettgeri, K. citrophila, B. megaterium,* and *A. viscosus* are used for the hydrolysis of about 30 000 tons of penicillin G (see Figure 1.7) each year. Recently, a new application of this enzyme as a biocatalyst for the kinetically controlled synthesis of semisynthetic penicillins and cephalosporins has been industrialized (see Sections 2.6 and 12.3; Bruggink, 2001). The penicillin amidases are produced as periplasmic enzymes in *E. coli* cells. The active enzyme contains a tightly bound Ca^{2+} ion as cofactor that is not directly involved in the biological function, but required for membrane translocation and maturation of the enzyme (McDonough, Klei, and Kelly, 1999; Kasche *et al.*, 2005). When produced in *E. coli*, the inactive pro-enzyme is first activated by intramolecular autoproteolysis, and then by intermolecular (auto)proteolysis, as shown in Figure 6.5 for the *E. coli* enzyme (PA_{EC}) (Kasche *et al.*, 1999). This activation occurs after or during translocation into the periplasm. Penicillin amidases produced from other microorganisms have similar processing schemes (Ignatova *et al.*, 1998).

The biological function of this enzyme in wild-type *E. coli* and other bacteria is still unknown. As its synthesis can be induced by phenylacetic acid, it has been postulated that it is involved in the metabolism of aromatic compounds (Prieto, Diaz, and Garcia, 1996).

Already in the 1960s it was observed that the synthesis of PA_{EC} was optimal at 28 °C, and decreased rapidly at higher temperatures. At the optimal temperature, its yield was reduced with increasing glucose content in the nutrient medium (Szentimai, 1964). At that time the explanation for this was considered to be on the transcriptional level, by repression of the transcription of the pre-pro-enzyme (ppPA) gene. Then the methods (SDS gel electrophoresis and isoelectric focusing with immunoblotting and activity stain techniques) now used to study posttranslational effects on enzyme production were not available. Using these methods it has been found that these early observed effects of temperature and C sources on the yield of active PA are mainly due to posttranslational processes, especially intracellular proteolysis, that reduce the yield

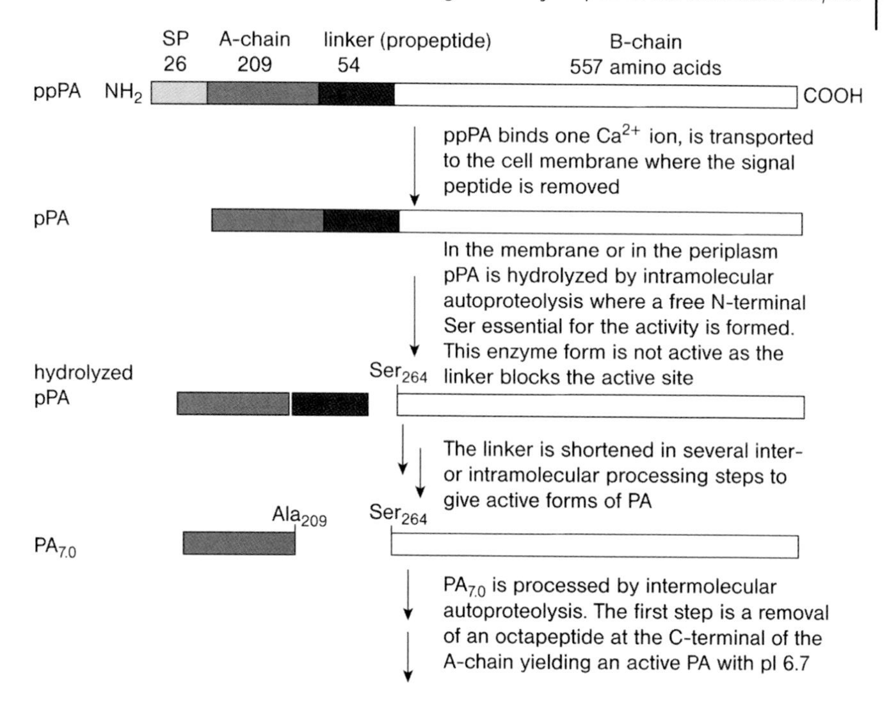

Figure 6.5 The proteolytic processing of penicillin amidase from *E. coli* (PA_{EC}, EC 3.5.1.11). The pre-pro-enzyme (ppPA) is one polypeptide chain that from the N-terminal consists of the signal peptide (SP) that directs the periplasmic enzyme to the membrane translocation system, the A-chain, the linker or pro-peptide that must be removed to obtain an active enzyme, and the B-chain. The number of amino acids in these peptides is given in the second line from above. The dominating active PA form in samples of technical enzyme ($PA_{7.0}$) from wild-type cells consists of the A- and the B-chain and has an isoelectric point (p*I*) of 7.0 (Kasche *et al.*, 1999, 2005).

of active enzyme (Figures 6.6 and 6.7 and Table 6.3). The yield of active PA_{EC} in wild-type *E. coli* could be increased by a factor of more than 20 by reducing the temperature from 37 to 28 °C, as shown in Table 6.3. However, in the cells cultivated at 28 °C, a large fraction of the pre-pro-enzyme is still lost by intracellular proteolysis.

The intracellular proteolysis is an energy-requiring process that can be influenced by the content of intracellular proteases and the medium. To study this, the influence of the medium on the yield of PA_{EC} produced in wild-type and recombinant *E. coli* cells (especially the peptidase-deficient BL21(DE3) strain) at the optimal temperature (28 °C) was studied (Table 6.3 and Figure 6.6). In all cases, the yield was lower with the energy-rich LB medium. A change to a minimal medium M9 increased the yield by a factor of 2, in the wild-type case, mainly due to reduced intracellular proteolysis (Figure 6.6). Using the peptidase-deficient strain, the yield could be increased additionally by a factor of ≈10. It should be emphasized that the above results were obtained in shake flask cultures where only a low cell density (up to 1–2 g dry cell weight) can be achieved.

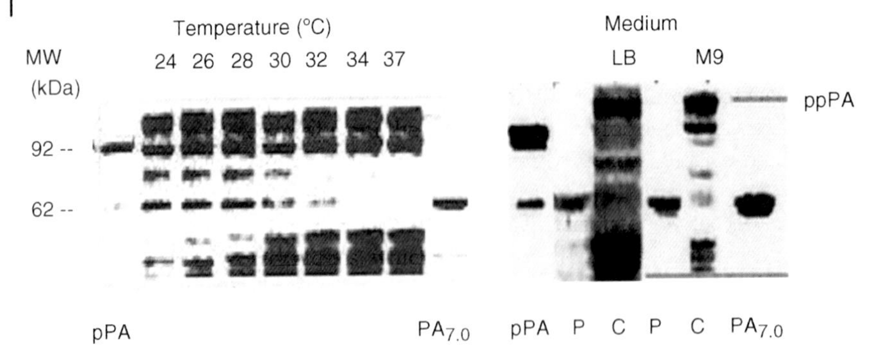

Figure 6.6 The influence of temperature and medium on the biosynthesis of PA_{EC} in the wild-type parental strain *E. coli* ATCC 11105 studied with SDS gel electrophoresis (12% gel) and immunoblotting using a monoclonal antibody against an epitope on the B-chain. In this denaturing electrophoresis, unfolded polypeptides are separated with respect to molecular weight. The immunoblotting visualizes $ppPA_{EC}$ and polypeptides derived from it with the epitope to which the antibody is bound. The temperature dependence was studied with LB medium; the medium influence (LB and a synthetic M9 medium) with cells grown at 28 °C. Cells were grown for 6 h after the induction with 6 mM phenylacetic acid, homogenized (temperature) or separated into cytoplasmic (C) and periplasmic (P) fractions. In each lane, the same number of cells was analyzed. Pure $PA_{7.0}$ and a slowly processing pro-PA (pPA) were used as standards. The intensity of the same polypeptide band increases with its content in the sample. The intensity of the B-chain (Figure 6.5) with MW 62 kDa is a measure for the synthesis of biologically active PA_{EC}, and shows that it has a temperature optimum at 26–28 °C and depends on the medium composition. The presence of the pPA and ppPA bands gives no information whether these represent correctly folded structures that can be processed to biologically active PA_{EC} that is localized in the periplasm. The other immunostained bands represent B-chain fragments with the epitope produced by intracellular proteolysis. The synthesis of the pre-pro-enzyme is hardly influenced by temperature in the range studied. With increasing temperature, the amount lost by misfolding, inclusion body formation, and intracellular processing approaches 100%.

In order to increase the volumetric PA_{EC} concentration further, the cells must be cultivated to a high cell density in a fermenter. For the recombinant BL21(DE3) cells, this can be achieved with an exponential feed starting from a medium that was optimal in the shake flask cultures. The feed contains a C source (glucose), N sources, and ions required for growth, especially Ca^{2+} that is required as cofactor. The glucose content in the medium must be kept low to avoid formation of acetic acid that reduces the yield of active PA_{EC}. The penicillin amidases from the above-mentioned organisms contain one tightly bound Ca^{2+} per molecule. The intracellular concentration of this ion is very low, and the question arises as to where this ion is added – before or after membrane translocation? In both wild-type and recombinant cells producing PA_{EC}, it has been found that only $ppPA_{EC}$ or pPA_{EC} having a bound Ca^{2+} ion can be translocated through the membrane (Figure 6.7) (Kasche *et al.*, 2005). Thus, the intracellular concentration of the cofactor markedly influences the yield of PA_{EC}. Without Ca^{2+}, the pro-enzymes, probably misfolded or aggregated, accumulated in the cytoplasm (Figure 6.7).

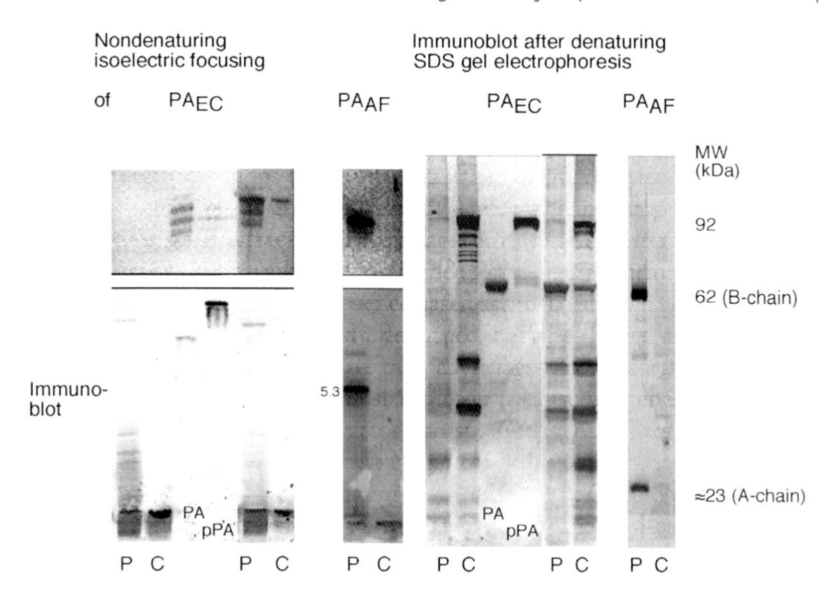

Figure 6.7 The influence of the Ca^{2+} content in the medium on the yield of biologically active homogeneously expressed PA_{EC} (*E. coli* penicillin amidase) and heterogeneously expressed PA_{AF} (*A. faecalis* penicillin amidase) in *E coli* (BL21(DE3)) cells exponentially grown at 28 °C. Aliquots were removed 9 h (PA_{EC}) and 12 h (PA_{AF}) after induction with ITPG, fractionated into cytoplasmic (C) and periplasmic (P) fraction, and separated as native folded proteins with respect to isoelectric point (p*I* = pH of the focused band) by nondenaturing isoelectric focusing or molecular weight by denaturing SDS gel electrophoresis (12.5% gel). The same standards for PA_{EC} and pPA_{EC} were used as in Figure 6.6. After IEF, a chromogenic substrate (NIPAB) was layered on the gel. The colored bands indicate the existence of more than one active enzyme forms, as is expected from the processing scheme in Figure 6.5. In this case, the intensity of the band increases with the amount of the enzyme form, as long as not all substrate near the band has been consumed. The dominating form of PA_{EC} produced in these cells has a p*I* of 7.3 and a higher specific activity than the dominating form produced by wild-type cells in shake flasks that has a p*I* of 7.0. After IEF and gel electrophoresis, the gels were immunostained with a monoclonal antibody against an epitope on the B-chain of PA_{EC} or a polyclonal antibody against PA_{AF}. This visualizes correctly folded enzymes in IEF, and in SDS gel electrophoresis unfolded polypeptides with epitopes recognized by the antibodies. Without Ca^{2+}, no correctly folded pre-pro- or pro-enzyme is found and the yield of active enzyme was very low (Table 6.3). The cytoplasmic band with MW ≈ 92 kDa was identified as pPA_{EC} by N-terminal sequencing. At optimal Ca^{2+} content, much less intracellular proteolysis was observed in the cells producing PA_{AF}. Unlike the pre-pro-PA_{EC}, pre-pro-PA_{AF} is transported in an unfolded conformation through the Sec translocation system. Its content per cell is higher than that for the Tat translocation system, through which folded $ppPA_{EF}$ is transported. The residence time in the cytoplasm for $ppPA_{EC}$ is probably longer than that for $ppPA_{AF}$. Thus, the yield of PA_{EC} is lower than that for PA_{AF}, due to loss of precursor by intracellular proteolysis.

The results of such fermentations with and without Ca^{2+} in the feed are shown in Figures 6.7 and 6.8. At optimal Ca^{2+} concentrations, up to ≈10% of the protein in the fermenter was active PA_{EC}, while for the PA from *A. faecalis* the corresponding yield was 20%. These results demonstrate the necessity of supplying sufficient

Table 6.3 Effect of various factors that influence the transcriptional, translational, and posttranslational processes in Table 6.2 and Figure 6.4 on the active PA_{EC} content in wild-type and recombinant *E. coli* cells cultivated in shake flasks and a fermenter (Ignatova, Taruttis, and Kasche, 2000; Ignatova *et al.*, 2003; Kasche *et al.*, 2005).

Factor	Process in Figure 6.4 mainly influencing yield of active PA_{EC}	*E. coli* cell strain/cultivation conditions/medium	Active PA_{EC} content (mg g^{-1} CDW)
Temperature (°C)	IV, V: inclusion body formation, intracellular proteolysis	Wild-type/shake flask LB medium[a)]	
24			1.7
28			1.9
32			0.5
36			<0.1
***E. coli* cell strains**: wild-type without plasmid PA_{EC}, others with plasmid PA_{EC}	All processes	Shake flask/28 °C, M9 minimal medium or LB medium[a,b)]	M9 minimal medium (LB medium)
Wild-type ATCC 11105			3.5 (2.0)
DH5			3.1 (1.2)
K5			9.0 (3.2)
JM109			11.3 (4.8)
BL21(DE3)			30 (2.7)
High cell density: fermentation with exponential glucose and Ca^{2+}, Mg^{2+}, and Fe^{2+} in feed with different final Ca^{2+} **content (mM)**	IV, IX: cofactor addition in cytoplasm required for membrane translocation and maturation of PA	BL21(DE3) with plasmid PA_{EC}/2 l fermenter; 28 °C/M9 minimal medium[b)]	
0			1
0.48			16
2.4			36
4.8			32

a) The Ca^{2+}content in the medium was 100 μM that was found to be optimal for the production of PA_{EC} and PA_{AF} in shake flasks.
b) The glucose content in the medium at and after the induction was <100 μM, found to be optimal for the production of PA_{EC} and PA_{AF}. It minimizes intracellular proteolysis and acetate formation, but hardly influences the synthesis of the pre-pro-enzymes.

amounts of Ca^{2+} to the pre-pro-enzymes or pro-enzymes that require this ion for membrane transport and the proteolytic processing to yield the active enzyme.[1)] The different yields of active enzyme are probably due to the different translocation systems used by $ppPA_{EC}$ (Tat) and $ppPA_{AF}$ (Sec) (Ignatova *et al.*, 2002). In the latter

1) Other periplasmic or extracellular enzymes important in enzyme technology that have tightly bound Ca^{2+} that is not directly involved in enzyme function include α-amylases (Nielsen and Borchert, 2000), subtilisins (Bryan, 2000), and some lipases (see Section 6.4.2).

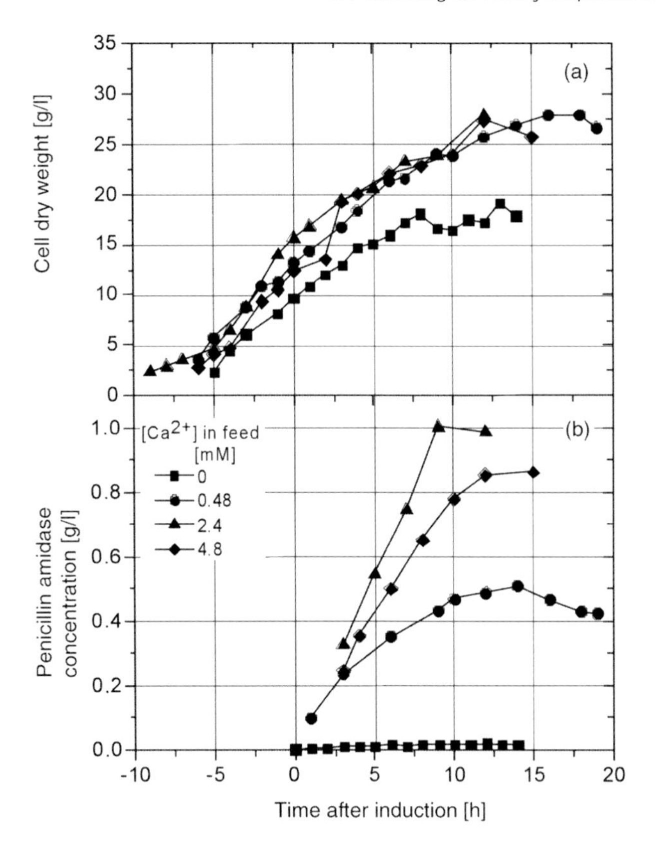

Figure 6.8 Production of penicillin amidase from *E. coli* in recombinant *E. coli* BL21(DE) cells grown to high cell density (CDW, g l^{-1}). The cells were fed exponentially with glucose and different concentrations of Ca^{2+} at 10 h before induction with ITPG. The cells were induced to produce the pre-pro-enzyme at time 0 (Kasche *et al.*, 2005).

case, no pro-enzyme was found in the cytoplasm indicating that almost 100% of pre-pro-enzyme has been processed to active PA_{AF}. The corresponding figure for PA_{EC} is estimated to be about 50% (Figure 6.7). Fermentations with a higher glucose content resulted in higher acetate (117 mM) concentrations than those observed in the fermentations shown in Figure 6.8 (≈1 mM). This did not influence the yield of the pre-pro-enzyme, but decreased the yield of active PA (Kasche *et al.*, 2005; Eiteman and Altman, 2006).

6.4.2
Lipase

Lipases (EC 3.1.1.3) from different microorganisms are used as biocatalysts for a large number of different enzyme processes (see Section 4.2.4). Lipases are all

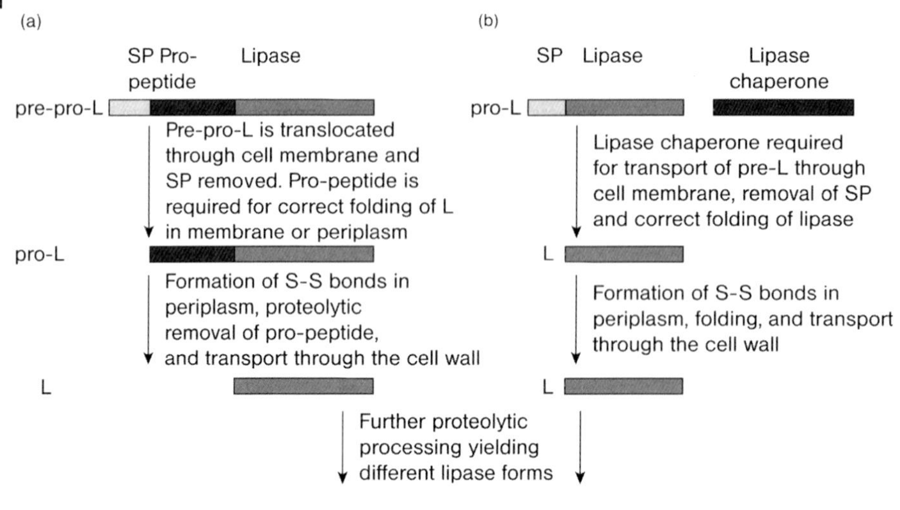

Figure 6.9 Mechanism for the posttranslational processing of extracellular lipases (L, EC 3.1.1.3) from different microorganisms. SP = signal peptide. The schemes apply for (a) eukaryotic lipases (such as *Rh. miehei*, *Rh. oryzae*) and (b) prokaryotic lipases (such as *Pseudomonas* sp., *Burkholderia* sp., *C. viscosum*) (Jaeger, Dijkstra, and Reetz, 1999; Ueda *et al.*, 2002).

expressed homogeneously (i.e., produced in their natural host), are extracellular, and are, in contrast to penicillin amidase, transported through the outer cell wall of their wild-type hosts into the medium. Different mechanisms for the post-translational processing and maturation of lipases are shown in Figure 6.9. Some prokaryotic lipases (from *Pseudomonas aeruginosa*, *B. glumae*, and *C. viscosum*) that are produced as pre-enzymes require simultaneously synthesized proteins such as lipase-specific chaperones (lipase-specific foldases; Rosenaua *et al.*, 2004). These protect the enzymes against intracellular proteolysis and the formation of inclusion bodies, and are also required for membrane transport and folding in the periplasm (Figure 6.9b). The regulation of their gene expression, folding, and secretion has been reviewed (Rosenaua *et al.*, 2004). Some of these lipases also contain a tightly bound Ca^{2+} ion (Nardini *et al.*, 2000), though at what stage this ion is added and whether it is required for the activity has not yet been elucidated.

Eukaryotic lipases from filamentous fungi (*Rh. miehei*, *Rh. oryzae*) are, as in the case of penicillin amidase, produced as pre-pro-enzymes (Figure 6.9a). The pro-enzyme of these lipases has been shown to be more active than the mature enzyme (Beer *et al.*, 1998). For these enzymes, the pro-peptide acts as an internal folding catalyst.

The influence of certain posttranslational factors that influence enzyme yield (see Table 6.2) has been studied for some lipases. The prokaryotic lipase from *P. cepacia* expressed in *E. coli* was mainly produced as inclusion bodies. The expression of the active enzyme could be increased when the wild-type signal peptide was exchanged to an *E. coli* signal peptide, and the pro-lipase inclusion bodies could be renatured in the presence of the lipase-specific foldase that had been separately expressed as inclusion bodies in *E. coli*. The final yield of active lipase was $\approx 0.2\,g\,g^{-1}$ cell dry weight – that is, about 40% of the cellular protein (Quyen, Schmidt-Dannert, and Schmid, 1999). This was somewhat larger than the yields for recombinant active enzymes given in Table 6.1, although the production required more processing steps than did the direct production of active enzymes.

The eukaryotic lipase from *Rh. oryzae* has been expressed in *E. coli* and *S. cerevisiae* (Beer *et al.*, 1998; Ueda *et al.*, 2002). At 42 °C, mainly inclusion bodies were produced in *E. coli*, but by reducing the temperature to 39.4 °C some active enzyme was formed. The amount formed was larger when the wild-type signal peptide was exchanged to an *E. coli* signal peptide. Later, the yield of active soluble extracellular pro-lipase could be improved using *E. coli* Origami (DE3) with the plasmid pET-11d, grown at 20 °C (Di Lorenzo *et al.*, 2005).

6.5 Downstream Processing of Enzymes

The costs for the downstream processing of enzymes – that is, purification to a desired degree of purity and formulation for final use – generally account for more than 50% of the total enzyme production costs. The downstream costs are much higher for enzymes used for therapy and diagnostics than for technical enzymes; hence, in order to reduce these costs, it is important to improve the downstream processing.

Following their production, the enzymes are located either inside the cell membrane (intracellular), outside the cell membrane and inside the cell wall (periplasmic), or outside the cell wall (extracellular). The downstream processing steps for the recovery and purification of enzymes are shown in Figure 6.10. The goal is to optimize the yield of active enzyme as fraction of the enzyme produced in the fermentation, with the required purity, with minimal number of processing steps. To achieve this, it may be necessary to add buffer (adjusting the pH), and add or remove salt (change the ionic strength *I*). These additions must be selected so that they do not interfere in the further processing steps, interfere with the final use of the enzyme, or increase the wastewater treatment costs. Ideally, minimal amounts must be used to achieve this. The clarification step in Figure 6.10 can be omitted by using expanded bed adsorption and desorption (EBA) that is mainly used for large-scale production of intracellular and periplasmic enzymes. The homogenate, that is, solution after the disruption step, is introduced from below into an expanded bed with porous adsorbent particles. They have the same properties as those used for column chromatography (Table 6.5), except that they have a higher density to allow the formation of a stable bed. In these, the target enzyme is adsorbed. Depending on the properties of the adsorbent, other proteins, DNA/RNA, and low molecular weight compounds can also be adsorbed. The cells or disrupted cell particles flow between the particles out of expanded bed. One downstream processing step is saved with the EBA step, as the target enzyme already is partly purified, compared to the medium or solutions after the clarification step (Figure 6.10). Following the clarification or EBA step, the solutions with intracellular, periplasmic, or

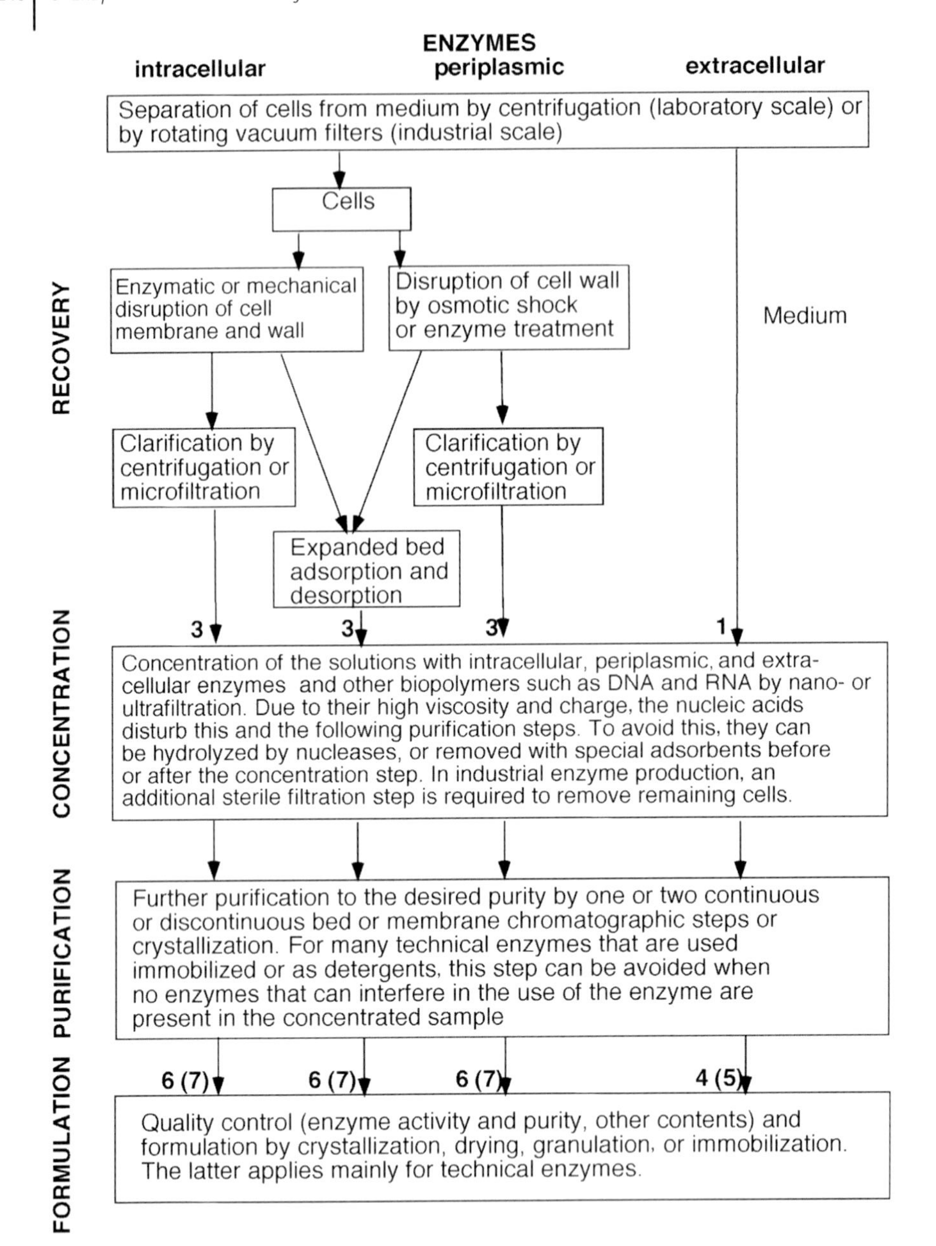

Figure 6.10 Downstream processing steps for the recovery, concentration, purification, formulation, and quality control of soluble biologically active intracellular, extracellular, and periplasmic enzymes. The number of processing steps before the concentration step and the formulation and quality control is given without and (with) removal of nucleic acids.

extracellular enzymes should be concentrated using ultra- (or nano-)filtration. For this step, it is important to know the molecular weight of the desired enzyme. The filters used for this step should be selected to give a maximal activity yield for this enzyme. Therefore, the enzyme activity in the original solution and the concentrate must be measured. The cutoff of the filters should also be selected so that >90% of the subject enzyme is in the concentrate. The filter producers give data, from which the cutoff can be selected. Then also the concentration of contaminating proteins with lower molecular weight than the enzyme is reduced in the concentrate. The previously frequently used concentration step by precipitation of the proteins, including the desired enzyme, with $(NH_4)_2SO_4$, should be avoided. Especially in large-scale enzyme production this implies a high salt content and N load in the wastewater that can be avoided by using the ultrafiltration step.

In the concentration and sterile filtration steps, the cells are practically completely removed from the concentrate, and the amount of low molecular weight impurities and ions that can interfere in the further purification steps reduced. In addition to the target enzyme, the concentrate contains other biopolymers (enzymes, proteins, nucleic acids, soluble polysaccharides, and endotoxins) (Table 6.4). The biopolymer composition in the concentrates after whole cell disruption (intra- or periplasmic

Table 6.4 Weight fraction (in % of total biopolymer dry weight) in concentrated and sterile filtered concentrates after recovering the subject enzyme or mAb, after whole cell disruption, osmotic shock, or separation of the medium from the cells after high cell density fermentations to produce periplasmic penicillin amidase in *E. coli* (Kasche *et al.*, 2005), using published data for the whole-cell biopolymer content (Carta and Jungbauer, 2010), extracellular phytase in *Hansenula polymorpha* (Mayer *et al.*, 1999), or mAb in CHO cells (Huang *et al.*, 2010).

	Biopolymer (% of total biopolymer dry weight)			
	Prokaryotic cell, *E. coli*, penicillin amidase after disruption of whole cells[a]	**Prokaryotic cell, *E. coli*, periplasmic penicillin amidase after isolation of the periplasmic cell fraction by osmotic shock**[a]	**Eukaryotic yeast cell, *H. polymorpha*, extracellular phytase**[a]	**Eukaryotic animal cell, CHO, extracellular monoclonal antibody**[a]
Subject enzyme/protein	10	≤ 65	≤ 95	≤ 80
Other proteins	50	≥ 35	-5	≥ 10[b]
DNA/RNA	<30	2	ND	≤ 10[b]
Polysaccharides	<10	ND	ND	ND
Endotoxin[c]	<1	<0.1		

a) Cell, strain, and produced enzyme or mAb.
b) Estimated from the number of dead cells.
c) Endotoxins, also known as pyrogens (cause fever), are lipopolysaccharides bound in the outer cell wall of Gram-negative bacteria (Carta and Jungbauer, 2010). They are toxic for animals and humans. Their contents are influenced by the fermentation conditions and vary from strain to strain (Svensson *et al.*, 2005).

enzyme), osmotic shock to recover a periplasmic enzyme, or recovering an extracellular enzyme or monoclonal antibody (mAb) from the medium is given in Table 6.4. This table also illustrates that the purity of the subject enzyme after the concentration and sterile filtration steps is much higher for extracellular and periplasmic enzymes than for intracellular enzymes. Therefore, recombinant or wild-type enzymes should preferably be expressed as extracellular or periplasmic enzymes. Some of the impurities are due to cells that have been disrupted during the fermentation or the osmotic shock step. They are then present in the medium. Their concentration is higher in the concentrates with intracellular and extracellular enzymes or after the EBA desorption step compared to periplasmic enzymes. In the latter case, part of the medium is separated from the cells before the osmotic shock step. After this, the periplasmic cell fraction is separated from the cells with the cell membrane by centrifugation or microfiltration. Of these impurities, especially the nucleic acids DNA and RNA can interfere in the further processing steps – mainly chromatographic – required to obtain the desired purity of the target enzyme. They increase the viscosity of the concentrates, and are adsorbed to anion exchangers. To avoid this, the nucleic acids can be adsorbed to specific nucleic acid adsorbents (that also can adsorb proteins), or hydrolyzed by nucleases (such as Benzonase™, Merck, Germany) before or after the concentration step. Unfortunately, the nucleic acid content in the homogenates or concentrates is rarely determined. They are not negligible. DNA concentrations of up to 100 mg l^{-1} have been observed in the medium after high cell density fermentation of *E. coli* (Castan and Enfors, 2000). This is also illustrated by the fact that concentrated solutions of technical extracellular enzymes or the medium of animal cells producing monoclonal antibodies exhibit an ultraviolet absorption spectrum that more closely resembles that of DNA/RNA than that of protein (Scheidat, 1999).

Whether the impurities in Table 6.4 must be separated from the target enzyme depends on the final use and the required specific activity of the enzyme on a weight basis, and whether the impurities disturb the process in which the enzyme is to be used as a catalyst. The table also gives information for the rational design of the further purification steps. The further purification steps are mainly chromatographic using suitable adsorbents. For these and the previous filtration steps, it is essential to know the following properties of the target enzyme:

1) The isoelectric point (pI) – that is, the pH at which the net charge of the protein is 0. This can be determined by isoelectric focusing (IEF), and is essential for selecting the separation conditions for ion-exchange or mixed-mode chromatography. This separation is, in contrast to SDS gel electrophoresis, carried out under nondenaturing conditions. The pI can be determined as shown in Figure 6.7, or by measuring the pH on the surface of the gel with a planar pH electrode. In IEF, the proteins are focused there where their pI equals the pH in the gel, that is, where their net charge is zero. The pH scale is determined by standards with different pI. After separation, the pI of the enzyme can be determined either with an enzyme-specific chromogenic substrate that is converted to a colored product on the gel or immunologically with an enzyme-specific

antibody, as in the Western blots or activity stains in Figure 6.7. It is not recommended to use pI calculated by software from the protein sequence as the true pI also depends on the folded protein structure.

2) Molecular weight (MW): it can be determined by SDS gel electrophoresis under denaturing conditions (see Figure 6.6). The best sample for this is the band with the active enzyme from the isoelectric focusing. Before separation, the enzymes are unfolded in a SDS solution and the S–S bonds reduced. It is then separated with respect to molecular weight in the gel and the MW scale is obtained with MW standards. For oligomeric enzymes, the MW of the monomer is determined. Many technical enzymes are active as oligomers (alcohol dehydrogenase, arginase, catalase, β-D-galactosidase (lactase), glucose isomerase, etc.). For enzymes consisting of several polypeptide chains formed by proteolytic processing (e.g., penicillin amidase), the MWs of the different peptides are determined. After separation, the proteins are visualized either by protein staining or with immunological methods (Western blot) that visualizes only the target enzyme and its processing products that interact with the antibody against the protein (see Figure 6.6).
3) Cofactors, especially tightly bound FAD or PLP (pyridoxal-5′-phosphate) or metal ions required for optimal enzyme activity and stability, such as Ca^{2+} for penicillin amidase and lipase discussed in Section 6.4 that may be lost during the purification procedures. This must be avoided.
4) The pH range where the enzyme is stable, within which the purification must occur.

Besides these the following static and dynamic properties of the chromatographic adsorbents must be considered, for a rational design of the chromatographic purification steps.

6.5.1 Static and Dynamic Properties of Chromatographic Adsorbents that Must Be Known for a Rational Design of Chromatographic Protein Purification

6.5.1.1 Static Properties

Type of Interaction of the Enzyme with the Adsorbent in the Adsorption and Desorption Step The enzyme properties used for their separation from impurities with different chromatographic adsorbents are listed in Table 6.5. The chemical and physical properties of the porous adsorbent particles are similar to those used to immobilize enzymes (see Section 8.2.3). For membrane adsorbers, usually a cellulose-based matrix is used (Fraud, 2008). These macroporous particles and monoliths (macroporous cylindrical structures) have large perfusable pores, with diameters $>1\ \mu m$, through which the mobile phase flows. This results in shorter diffusion distance for the molecules that are adsorbed, and a more rapid adsorption than in the porous particles (Carta and Jungbauer, 2010). To the inner surfaces of these adsorbents, functional groups, enzyme inhibitors, or antibodies are covalently bound to

Table 6.5 Adsorbents that can be used for the chromatographic separation based on different properties of enzymes, and the conditions for adsorption and desorption to/from the adsorbents.

Chromatographic adsorbents for	Enzyme properties	Adsorption conditions	Desorption conditions
Size-exclusion chromatography	Molecular size and weight	Not applicable	Not applicable
Anion-exchange chromatography	Surface charge	pH > pI and low ionic strength	Increasing ionic strength
Cation-exchange chromatography	Surface charge	pH < pI[a] and low ionic strength	Increasing ionic strength
Hydrophobic chromatography	Hydrophobicity of surface	High ionic strength	Reduced ionic strength
Metal affinity chromatography	Histidines at the surface; or His tags[b]	At the pH optimum for the interaction between metal ion on adsorbent and His tag	By free metal ions that bind to the His tag as Ni^{2+}
Bifunctional or mixed-mode chromatography as in Figure 6.11	Surface charge and hydrophobicity of the surface	Hydrophobic adsorption at pH > pI, where support is practically uncharged	Desorption by electrostatic repulsion at pH < pI
Biospecific chromatography (enzyme–inhibitor; enzyme–antibody; biotin–streptavidin; peptide–peptide antibody)	Specific hydrophobic and electrostatic binding interactions with inhibitor or antibody or biospecific tags (streptavidin or peptide)[b]	At the pH optimum for the interaction	Change in pH provided that this is within the pH stability of the enzyme

a) In cation- *but not in anion*-exchange chromatography, metal ions bound to the enzyme that are essential for the activity may be removed by the adsorbent.

b) A His tag with up to eight His is added to the N- or C-terminal of the enzyme. This tag binds to the metal ions (Co^{2+}, Ni^{2+}) complexed to chelates covalently immobilized on the surface of the adsorbent. It is only successful when the C- or N-terminal is located on the enzyme surface, and when the tag allows a correct folding of the enzyme. This and other tags fused to the enzyme allow a one-step isolation of enzymes from homogenates of enzymes that can be "tagged" without loss of activity. The metal ions used for the desorption and the tags are to be removed, when the enzyme must be identical with the wild-type enzyme in the final use (Terpe, 2003). This requires at least two additional purification steps, compared to the purification with the other adsorbents. One- or two-step chromatographic isolations of pure enzymes can be obtained without tags using biospecific, ion-exchange, or bifunctional adsorbents (Figure 6.15).

give them the different properties of the adsorbents in Table 6.5. Of these, the mixed-mode adsorbents were introduced for large-scale applications around 2000 (Kasche *et al.*, 1990; Burton and Harding, 1998; Zhao, Dong, and Sun, 2009; Voitl, Müller-Späth, and Morbidelli, 2010). Suppliers of such adsorbents are given in the

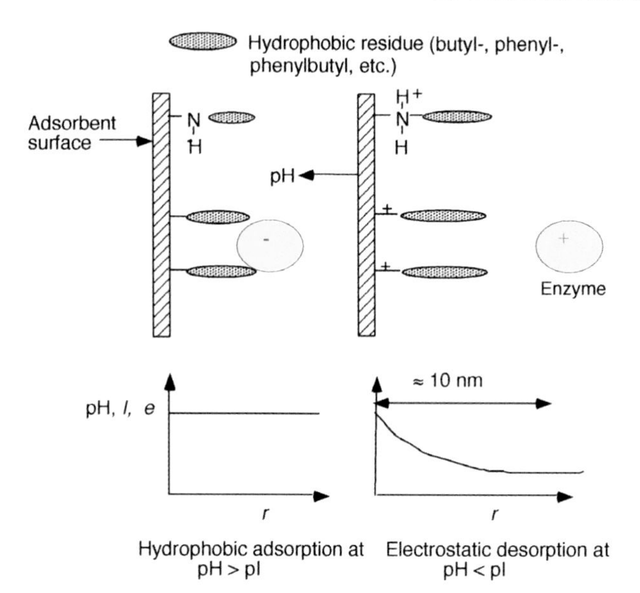

Figure 6.11 Mixed-mode adsorbents use different interactions for the adsorption and desorption. Enzymes are bound by hydrophobic interactions to an uncharged adsorbent at a pH > pI. Then the pH, ionic strength *I*, and the electric field strength *e* are constant in the electric double layer on the adsorbent surface. When the bulk pH is lowered, the enzyme and the surface charge on the adsorbent become positive, and the physical properties depend on the distance from the surface. With increasing charge on the enzyme and the surface, the repulsion force becomes larger than the hydrophobic interaction, and the enzyme is desorbed. The desorption pH depends on the p*I* of the enzyme.

review by Zhao, Dong, and Sun (2009). These and hydrophobic adsorbents can also be used for a rapid immobilization of technical enzymes after the filtration steps in Figure 6.10 (see Chapter 8). Their function is illustrated in Figure 6.11.

Mechanical Stability Adsorbents are available with a mechanical stability that allows their use, without being compressed from atmospheric to high-pressure liquid chromatography conditions (see also Section 8.2.3).

Maximal Adsorption Capacity (*n*) and the Dissociation Constant (*K*) for the Interaction of the Enzyme with the Functional Groups of the Adsorbent They are determined from the adsorption isotherm. In its simplest form as a Langmuir adsorption isotherm, it is given by

$$x = \frac{nc}{K + c}, \tag{6.1}$$

where n is the static capacity of the adsorbent, c is the concentration of free enzyme in solution, x is the concentration of adsorbed enzyme in the adsorbent volume,

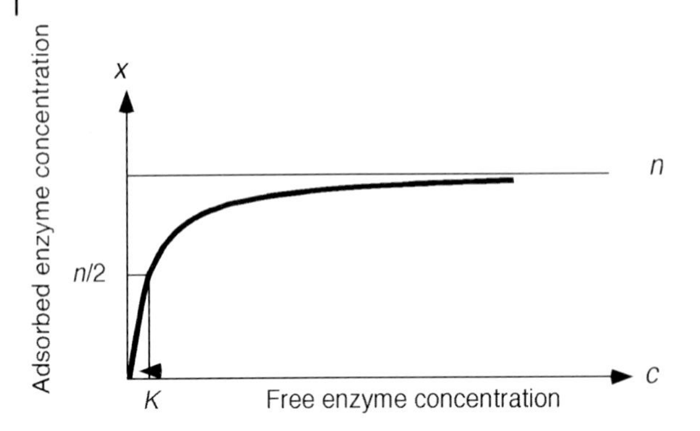

Figure 6.12 The Langmuir adsorption isotherm for the binding of an enzyme to an adsorption site on the surface of a porous particle or a porous membrane. The static capacity n and the apparent dissociation constant K are determined as shown.

and K is the apparent dissociation constant for the binding of the enzyme to the adsorbent. The graphical representation of Eq. (6.1) is given in Figure 6.12.

The manufacturers of the adsorbents generally give values for the static capacity, but not always data for the dissociation constants. Table 6.6 gives ranges for these properties based on these and literature data for different adsorbents.

6.5.1.2 Dynamic Properties

Pore Diffusion Coefficient The effective diffusion coefficient of a molecule in porous adsorbents is (Conder and Hayek, 2000; Kasche, Galunsky, and Ignatova, 2003; Kasche *et al.*, 2003)

$$D_{\mathrm{p,eff}} = D_{\mathrm{p}}/[\varepsilon_{\mathrm{p}}(1 + nK/[K + c(r)]^2)], \tag{6.2}$$

where D_{p} is the pore diffusion coefficient without adsorption, ε_{p} is the particle porosity, and $c(r)$ is the free molecule (unadsorbed) concentration at distance r

Table 6.6 Range of observed static capacities and order of magnitude for binding (dissociation) constants for different adsorbents (porous particles and membranes).

Adsorbent	Static capacity for an enzyme with MW 100 kDa, n (mM or g l^{-1})	Binding constant, K (mM)
Ion exchanger	0.1–1 (10–100)	0.01
Hydrophobic	0.1–1 (10–100)	0.01
Mixed-mode	0.1–1 (10–100)	0.01
Biospecific	0.01–0.1 (1–10)	0.001

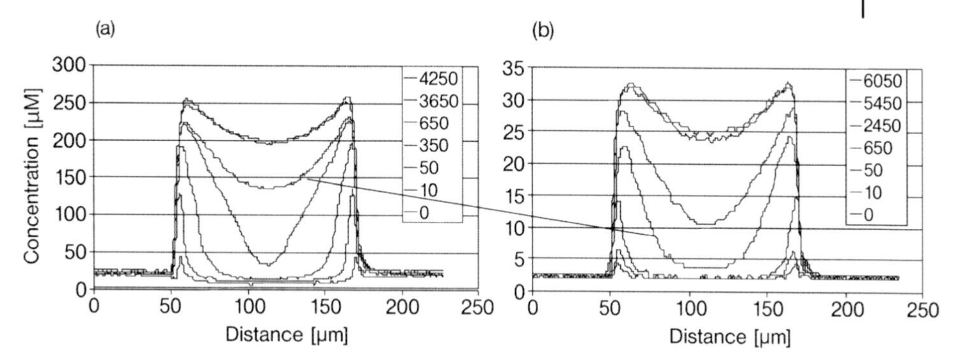

Figure 6.13 The time dependence of the adsorption of fluorescein-labeled α-chymotrypsin to soybean trypsin inhibitor covalently immobilized in porous Sepharose particles ($n = 700\ \mu M$; $K = 12\ \mu M$) as a function of the enzyme concentration outside the particle. The concentration profiles within the particle were determined at different times increasing from the bottom at times in seconds given to the right, with CLSM (confocal laser scanning microscopy) (Kasche, Galunsky, and Ignatova, 2003; Kasche *et al.*, 2003): (a) outside fluorescein-labeled α-chymotrypsin concentration $22\ \mu M > K$; (b) outside fluorescein-labeled α-chymotrypsin concentration $2.2\ \mu M < K$. As expected from Eqs. (6.1) and (6.2), the adsorption is much faster, and the adsorbed protein amount per unit adsorbent volume is higher at the higher concentration, as follows from the line connecting the profiles after 650 s.

from the particle center. From this equation follows that $D_{p,eff}$ increases with $c(r)$ when this concentration is larger than K, that is, in the nonlinear part of the adsorption isotherm. Analytical chromatography is carried out in the linear part of the adsorption isotherm where $c(r) \ll K$. $D_{p,eff}$ is then independent of concentration. The goal of laboratory- and preparative-scale enzyme purification is to purify as much as possible in the shortest time, that is, high throughput. To achieve this, it should be carried out in the nonlinear part of the adsorption isotherm. This requires that the concentration of the target protein is larger than K for the used adsorbent found in Table 6.6 (Figure 6.13). This is one reason for the nano(ultra)filtration of the homogenate, periplasmic fraction, or homogenate before the chromatographic step in Figure 6.10. Another reason is that the concentration of low molecular weight compounds that can interfere in the adsorption to the adsorbents is reduced in this concentration step. It is also favorable for any further chromatographic purification, as less adsorbent is required to adsorb the enzymes in the sample when it is concentrated. This also follows from Figure 6.13.

In the desorption step, the mobile phase composition is changed so that K is increased and becomes larger than the free enzyme concentration c outside and inside the porous particles. Then the effective pore diffusion constant (Eq. (6.2)) increases. The shorter diffusion distances in perfusable membranes, monoliths, and macroporous particles result in a more rapid adsorption than in the porous particles.

Mobility in a Packed Bed The linear velocity or mobility of a molecule A, u_A, in a packed bed, where only the molecules in the mobile phase (volume V_M) are moving, is

$$u_A = \frac{u}{1 + [1 + n/(K + c_A)](V_S/V_m)}, \tag{6.3}$$

where u is the linear flow rate of the solvent in the column and V_S is the volume of the stationary phase of the column. When $K \gg c_A$, that is, in the linear part of the adsorption isotherm, u_A is concentration independent. This is used in analytical chromatography to measure the concentration of a compound, as its retention time in a column is kept constant. In preparative chromatography, where c_A should be $>K$, that is, in the nonlinear part of the adsorption isotherm, u_A increases with c_A.

Dynamic Capacity and Column Efficiency From Figure 6.13 follows that the dynamic capacity decreases with the linear flow rate of the mobile phase u in a packed bed with porous adsorbent particles or a porous membrane. The contact (residence) time between the adsorbent particles and the protein sample decreases with increasing u.

The linear flow rate also influences the column efficiency that is the peak width of an eluted target enzyme (protein). This determines its dilution in the column and resolution from contaminating proteins. A measure for this is the peak width that should be as small as possible. It has been shown that the peak width as a function of u has a minimum (Carta and Jungbauer, 2010). Then the column efficiency is optimal, and has a maximal resolution. The latter is expressed by the plate number given by the relation

$$N = \frac{L}{H} = c\frac{LD_0}{ud_p^2}, \tag{6.4}$$

where L is the column length, H is the height equivalent to a theoretical plate (HETP),[2)] d_p is the diameter of the porous adsorbent particles, D_0 is the diffusion coefficient of the enzyme (protein) in free solution, and c is constant for an isocratic elution that depends on the interaction of the protein with the adsorbent. The last expression for N in Eq. (6.4) applies for proteins in porous adsorbent particles, where intraparticle mass transfer determines the column efficiency (Carta and Jungbauer, 2010). Chromatographic enzyme (protein) purification is carried out under nonisocratic conditions, using a pH or ionic strength gradient. Then c in Eq. (6.4) is not constant. The equation can, however, be used for the scale-up of such purifications, where the column efficiency must be kept constant. For the purification of a specific enzyme, this implies that L/ud_p^2 must be kept constant (Rathore and Velayudhan, 2003; Carta and Jungbauer, 2010). With given column and adsorbent particles, the optimal linear flow rate must be found or chosen for the purification of a target enzyme. Hints for this can be found in the data for protein separations given by the producers of adsorbent particles, or from published data for optimal ratios of L/u.

2) It can be calculated from experimental data using the relation $H = (W/t_{peak})^2 L/16$, where W is the baseline peak width and t_{peak} is the time for the elution of the peak maximum.

6.5.2 Chromatographic Purification of Enzymes: Problems and Procedures

6.5.2.1 Problems

From Table 6.4 follows that the purification of an extracellular or periplasmic enzyme after the filtration step in Figure 6.10 is a separation of three (or four when nucleic acids are present) components: the target enzyme from proteins with higher or lower p*I*, and the nucleic acids. From Table 6.5 follows that this requires a different mobile phase composition during the adsorption and desorption step (nonisocratic chromatography). This does not apply for SEC that is not an adsorption chromatographic procedure but a filtration procedure. Smaller proteins have larger accessible pore volume in the porous particles. Therefore, they have larger residence times in these than larger proteins. SEC is suitable to desalt or change buffers of protein solutions. This can, however, also be partly achieved by ultrafiltration.

It is favorable to avoid the presence of nucleic acids in chromatographic purification of enzymes. They increase the viscosity that results in increased pressures in packed beds (Section 11.3.5). They can also influence the adsorption of the target enzyme to the adsorbent, especially to positively charged adsorbents (anion exchanger, mixed-mode adsorbents that are positively charged during the desorption step (Figures 6.11 and 6.15), and positively charged biospecific adsorbents). For the mixed-mode adsorbent, this is minimized as the adsorbent is practically uncharged during the adsorption step, and when this is carried out using a high salt content in the sample and adsorption buffer. The best way to remove the nucleic acids is to hydrolyze them with protease-free nucleases as Benzonase^{TM} (Merck, Germany), before the concentration steps in Figure 6.10. The concentration of the resulting low molecular weight hydrolysis products is then reduced in the ultrafiltration step.

For the large number of enzymes that require tightly bound metal ions, anion exchanger or mixed-mode adsorbents should preferably be used for the chromatographic purification. In these, the bound metal ions cannot be adsorbed that can reduce the activity yield of the enzyme in the purification procedure.

6.5.2.2 Procedures

Previously, the purification of an enzyme from a concentrated clarified homogenate involved up to more than five separation steps involving different types of chromatography. In each step, more than 10% of the enzyme was lost, and this led to a low recovery of the enzyme. Hence, to reduce the enzyme costs it was necessary to lower the number of processing steps in order to increase the recovery (Figure 6.10). The design of process engineering for discontinuous column and membrane chromatography of enzymes has been improved considerably during the past 10–20 years (Rathore and Velayudhan, 2003; Carta and Jungbauer, 2010). Such processes are suitable for the laboratory- and large-scale separation of three- or four-component systems discussed in Section 6.5.2.1. For the separation of two-component systems, continuous chromatographic processes such as simulated moving bed (SMB), continuous separation (CSEP), and continuous annular

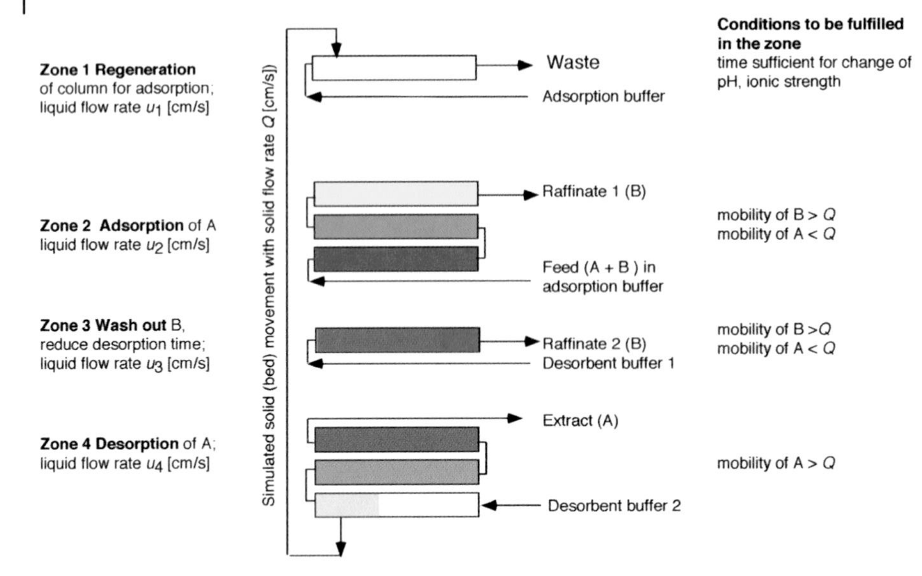

Figure 6.14 Continuous chromatography with a simulated moving bed for the nonisocratic separation of a two-component protein system (blue = A, the target protein that is adsorbed in the adsorbent particles; yellow = B, other proteins that are not adsorbed in the adsorbent). The SMB consists of eight identical columns filled with adsorbent particles whose inlets can be connected to outlets of adjacent columns in the zone or the feed and adsorbent and desorbent buffers. The outlets can be connected to inlets of adjacent columns in the zone or the containers for pure A, pure B, and the waste. The figure shows the distribution of A and B in the zones 1–4 between two consecutive switches of the last column in each zone one step downward. The time between the switches is constant and gives Q, the mobility of the solid phase that moves opposite to the liquid mobile phase. The feed to the first column in zone 2 continues until it is filled with adsorbed A. As the mobility of A in this section must be $< Q$, the adsorption isotherms (Eqs. (6.1) and (6.3)) of A (and B) must be known to design the process. The column with adsorbed A then moves downward one step and is connected to desorbent 1, and replaces the column from zone 3 without B that is moved to the desorption zone and connected to desorbent 2. At the same time, the regenerated column from zone 1 is connected to the feed, and the first column in the desorption zone, from which A has been desorbed, is moved to zone 1 and connected to the adsorbent buffer. The inlet and outlet ports of the columns that are moved to the next zone must be changed as shown in the figure. This is achieved using multiport valves that are controlled by a computer (Gottschlich and Kasche, 1997).

chromatography have been introduced into the downstream processing of proteins (Gottschlich and Kasche, 1997; Imamoglu, 2002; Uretschlager and Jungbauer, 2002; Carta and Jungbauer, 2010). They are mainly used for large-scale purification of low molecular weight two-component systems (see Section 12.3) and proteins. The design criteria for a SMB process are shown in Figure 6.14. Continuous chromatography for the separation of three-component systems for protein purification has recently been developed using multicolumn countercurrent solvent gradient purification (MCSGP) (Müller-Späth *et al.*, 2008). For the separation of the same amount of two or three components, the purified components are less diluted and less solvent and adsorbent are required for the continuous chromatographic process, compared with the discontinuous chromatographic process. The rational design of these continuous and discontinuous processes for the purification of enzymes using different adsorbents requires the knowledge of the p*I*, MW, and biospecific interactions of the enzyme, and the dynamic and static properties of the adsorbents discussed in Section 6.5.1.

6.5.3
Chromatographic Purification and Conditioning of Technical and Therapeutic Enzymes

6.5.3.1 Technical Enzymes

When the concentrate of the clarified homogenate (Figure 6.10) containing a technical enzyme does not contain contaminants that disturb the enzyme's use as a biocatalyst, it must not be purified further. This applies for most of the technical enzymes, especially those that are immobilized (see Chapter 8). However, when the final process requires a DNA- and RNA-free enzyme, or an enzyme preparation with a higher specific activity (in U g^{-1} protein), the clarified homogenate must undergo further purification. This applies especially for enzymes, whose enzyme properties for its possible application in an enzyme process, primary and 3D structure, are to be determined. Here very pure enzyme preparations are required. The enzyme amounts required for this are generally <1 g. DNA and RNA can be hydrolyzed in the clarified homogenate by adding nucleases, after which the hydrolysis products and contaminating proteins with molecular weights lower than that of the enzyme can be removed by ultrafiltration using a membrane with a cutoff below the MW of the target enzyme.

For laboratory- and large-scale enzyme purifications, ion-exchange and mixed-mode adsorbents are the first choice. They can be used for all enzymes, are very stable, and are available for low- and high-pressure column and membrane chromatography. The adsorbent producers also present data for different protein separations that give hints for the design of a purification of an enzyme from a concentrated homogenate. As stated before, this requires that the p*I* of the target enzyme is known. Then the laboratory-scale enzyme purification is optimized as follows. Using a small column (volume <10 ml) and small amounts of the concentrate with the target enzyme, the optimal gradients (ionic strength for ion-exchange adsorbents or pH for mixed-mode adsorbents) that give a pure enzyme fraction must be found. Then the sample size can be increased as long as a pure enzyme fraction still can be obtained with the column.

Figure 6.15 shows such a laboratory ion-exchange purification for the wild-type penicillin amidase from *A. faecalis* produced in recombinant *E. coli* cells fermented to a high cell density (Figures 6.7 and 6.8). Samples were purified with and without hydrolysis of nucleic acids in the concentrate (activity yield 90%) before the anion-exchange chromatography. The hydrolysis products were removed from the nuclease-treated concentrate by dialysis and buffer exchange with SEC chromatography. With nucleic acids in the sample the enzyme cannot be purified in one step, as the nucleic acids are adsorbed,

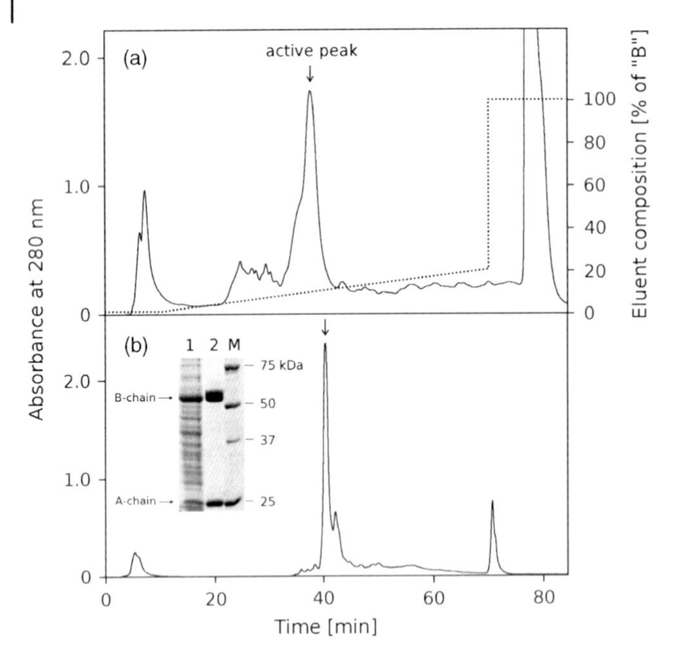

Figure 6.15 Chromatographic purification of wild-type *A. faecalis* penicillin amidase (PA) produced in recombinant *E. coli* at optimal Ca^{2+} content as shown in Figure 6.7, with (b) and without (a) hydrolysis of the nucleic acids in the concentrated clarified homogenate. An anion-exchange Mono Q 10/10 column (GE Healthcare) was used. After equilibration with 30 mM Tris–HCl, pH 7.5 at 2 ml min^{-1}, and application of the sample with ≈3 mg active enzyme in the same buffer, the column was eluted as follows: 0–70 min with a linear salt gradient 0–20% 1 M NaCl in 30 mM Tris–HCl; 60–90 min with 30 mM Tris–HCl containing 1 M NaCl. The inset in (b) shows a Coomassie-stained SDS electrophoretic gel (lane 1: the homogenate; lane 2: the enzyme peak shown by the arrow; lane 3: molecular weight markers). The last peak in (a) contains mainly nucleic acids (Yuryev, 2010, and unpublished results).

reduce the column capacity for the enzyme, and interfere with the resolution of the column. Without nucleic acids, a pure enzyme could be obtained in one chromatographic step, with an activity yield of >90%. The overall activity yield from the homogenate, its concentrate, treatment with nuclease, buffer exchange, and chromatography was >70%. This demonstrates that pure extracellular and periplasmic enzymes in homogenates as shown in Table 6.4 can be obtained in one ion-exchange chromatographic step with a high activity yield, after hydrolysis of the nucleic acids in the homogenate. After sample scale-up, up to 30 mg pure penicillin amidases from *A. faecalis* and *E. coli* could be isolated per separation. By repeating this, 100–1000 mg pure penicillin amidase could be prepared to study its application for the hydrolysis and synthesis of β-lactam antibiotics, kinetic resolution of racemates, and the proteolytic processing of the enzymes, their enzyme kinetic properties, stability, and 3D structure (Kasche *et al.*, 1999; Ignatova *et al.*, 1998; Yuriev *et al.*, 2010).

To isolate a larger enzyme amount in one step, a larger column with the same resolution (or plate height) as in the smaller column is required. Then the following quantity

$$\frac{L}{ud_{\mathrm{p}}^{2}}, \tag{6.5}$$

where L is the column length, d_{p} is the adsorbent particle diameter, and u is the linear flow rate, must be kept constant. At constant column length, linear flow rate, and the same adsorbent particles, the column diameter D for the large (subscript l) scale separation can be calculated from the ratio of the sample volumes V for the large (subscript l) and small (subscript s) samples using the relation (Rathore and Velayudhan, 2003; Carta and Jungbauer, 2010)

$$\frac{V_{\mathrm{sample,\,l}}}{V_{\mathrm{sample,\,s}}} = \frac{D_{\mathrm{l}}^{2}}{D_{\mathrm{s}}^{2}}. \tag{6.6}$$

Other scale-up relations for change of particle size or column length can be found in the cited literature. They are important for the scale-up of the purification of technical enzymes to a desired purity on an industrial scale.

Technical enzymes are distributed to the end users as either solutions or powders, or immobilized to different porous supports (see Chapter 8). The solutions must have a defined active enzyme content that can be obtained by concentration (ultrafiltration) or dilution. To avoid bacterial growth and to maintain a desired activity over a given time, preservation and stabilization agents may be added to these solutions. The powder can be either freeze-dried (producing small particles that can be inhaled) or dried in a spray dryer and covered with a protecting layer with particle sizes that cannot be inhaled. All of these conditioning steps must be carried out in hoods or closed production units in order to avoid exposure of employees or humans in the environment to small enzyme particles that might cause an allergic response when inhaled.

6.5.3.2 Enzymes for Therapy and Diagnostics

These enzymes must be very pure. In particular, therapeutic enzymes must be free from pyrogens and have a very low DNA content to guarantee that they are free from viruses and unwanted genes. Many of these enzymes are glycosylated, as lactase (EC 3.2.1.108) or β-galactosidase (EC 3.2.1.23) used to hydrolyze lactose in the food for humans that suffer from lactose intolerance, and tissue plasmin activator (tPA, EC 3.4.21.68) used to dissolve blood clots in the treatment of stroke. To achieve this, they should be produced in virus-free eukaryotic cells that in contrast to prokaryotes can glycosylate enzymes, and more purification steps are required than for technical enzymes. Here the production and purification methods of monoclonal antibodies may be applied (Shukla and Thömmes, 2010). Their production yields in animal cells (CHO, Chinese hamster ovary, the main animal cell used for the production of therapeutic proteins) have been increased from 0.05 to $5\,\mathrm{g\,l^{-1}}$ in the past 20 years (Huang *et al.*, 2010; Walsh, 2010). This should also be possible for glycosylated therapeutic enzymes. For these, however, the increase in the production yields has not been published.

Table 6.7 Methods to determine the purity, biological activity, and identity of therapeutic enzymes (Burgess and Deutscher, 2009).

Quality criteria	Method
Purity (this includes the determination of impurities such as DNA and other compounds)	PCR (DNA) Electrophoresis (IEF and SDS) Chromatography Mass spectrometry
Biological activity	Determination of enzyme activity Immunological methods
Identity (with the human enzyme)	Chromatography N-terminal sequencing Mass spectrometry Peptide mapping after hydrolysis with a peptidase

As for mAbs the first purification step after the concentration step in Figure 6.10 is often a biospecific chromatography, using immobilized protein A for mAb or antibodies for therapeutic enzymes. Then additional mixed-mode or ion-exchange chromatographic steps follow to remove impurities. Here also continuous chromatography can be applied (Müller-Späth *et al.*, 2008). The goal here is to reduce the number of purification steps in order to obtain the required purity, to reduce processing costs.

Finally, the purified enzyme solutions are sterilized by filtration that also removes viruses, and generally distributed to the final users as freeze-dried powders that are either ingested (digestive enzymes for substitution therapy as lactase or pancreatic enzymes) or dissolved in sterile buffers before use.

The quality of these enzymes must be checked with respect to three criteria: purity, biological activity, and identity (with the human enzyme). The methods available for performing this validation are detailed in Table 6.7.

6.6 Regulations Based on Risk Assessments/Safety Criteria that Influence the Production of Enzymes and Their Use for Analytical, Pharmaceutical, Scientific, and Technical Purposes

6.6.1 Regulations Governing the Use of Genetically Modified Microorganisms for the Production of Enzymes in Laboratories and Production Facilities

As shown in Section 1.3, enzymes have been used for pharmaceutical and technical purposes, mainly in food processing, for more than 150 years. The enzyme sources were edible animal and plant tissues or microorganisms that had been used safely in food processing for more than 1000 years. With the increasing demand of enzymes, new enzyme sources – mainly from wild-type microorganisms – were found. Whether the enzymes from these new sources are safe for humans when they are used for food processing or for therapeutic purposes had to be regulated. For this aim, the US Food and Drug Administration (FDA) introduced the GRAS (Generally Recognized as Safe) concept that applies for food additives. *Under Sections 201(s) and 409 of the US Federal Food, Drug, and Cosmetic Act (from 1938), any substance that is intentionally added to food is a food additive, that is subject to premarket review and approval by FDA, unless the substance is generally recognized, among qualified experts, as having been adequately shown to be safe under the conditions of its intended use, or unless the use of the substance is otherwise excluded from the definition of a food additive* (FDA, 2010b). This involves the safety evaluation of both the microorganism and the produced enzyme used for food processing. Enzymes are essential for all living species, but can in the wrong place be harmful. They can act as allergens when inhaled or ingested (Section 1.5). The producers of enzymes added to washing powder neglected this when these products were introduced around 1970. This caused an increase in enzyme allergies in the employees of the enzyme producers and among the end users of the enzyme. However, the enzyme producers solved these problems rapidly, by a contained enzyme production in closed units, and by covering the enzymes with a wax layer to increase the particle size so that they could not be inhaled.

With the development of genetic engineering techniques in the beginning of the 1970s, new recombinant enzyme sources became possible. This necessitated the development of new legislation to minimize the risks for humans and the environment due to the application of genetic engineering in laboratories and production facilities. For this aim, the microorganisms that can be used for the production of recombinant enzymes (or other bioproducts) were classified by their risks for human health and the environment as shown in Table 6.8.

Based on this risk classification, the work with biological agents using gene technological methods in laboratories and production facilities had to be regulated in order to protect human health and the environment. This is also required by an international agreement: the *Cartagena Protocol on Biosafety to the Convention on Biological Diversity* is an international treaty governing the movements of living modified organisms (LMOs) resulting from modern biotechnology from one country to another. It was adopted on January 29, 2000 as a supplementary agreement to the Convention on Biological Diversity and entered into force on September 11, 2003. It has now been ratified by 159 states and the European Union (not yet by the United States and Russia) (http://bch.cbd.int/protocol/text/), and requires that these activities must be contained. WHO has developed safety rules to assist member states to formulate the necessary regulation (WHO, 2004). The contained use is achieved by the following:

- **Physical containment:** The requirements how the use of LMOs should be separated from other activities in the same building and the environment by physicals means (laboratory equipment, requirements on buildings and plants,

Table 6.8 Classification of wild-type microorganisms by risk group used by WHO and US and EU authorities.

Risk group and definition	WHO Laboratory Biosafety Manual, Chapter 1, Table 3 (WHO, 2004)	NIH (US National Institutes of Health) Guidelines for Research Involving Recombinant DNA Molecules, Section II-A-1 (NIH, 2009)	European Union (EU, 2000)*
1. No or low individual and community risk	A microorganism that is unlikely to cause human or animal disease	Agents not associated with disease in healthy adult humans	A biological agent** that is unlikely to cause human disease
2. Moderate individual risk; low community risk	A pathogen that can cause human or animal disease but is unlikely to be a serious hazard to laboratory workers, the community, livestock, or the environment. Laboratory infections may cause serious infection, but effective treatment and preventive measures are available. The risk of spread of infection is limited.	Agents associated with human disease that is rarely serious and for which preventive or therapeutic interventions are often available	A biological agent that can cause human disease and might be a hazard to workers; it is unlikely to spread to the community; there is usually effective prophylaxis or treatment available
3. High individual risk; low community risk	A pathogen that usually causes serious human or animal disease but does not ordinarily spread from one individual to another. Effective treatment and preventive measures are available	Agents associated with serious or lethal human disease that is rarely serious for which preventive or therapeutic interventions may be available	A biological agent means one that can cause severe human disease and present a serious hazard to workers; it may present a risk of spreading to the community, but there is usually effective prophylaxis or treatment available

4. High individual and community risk	A pathogen that usually causes serious human or animal disease and that can be readily transmitted from one individual to another, directly or indirectly. Effective treatment and preventive measures are not usually available	Agents likely to cause serious or lethal human disease that is rarely serious and for which preventive or therapeutic interventions are not usually available	A biological agent means one that causes severe human disease and is a serious hazard to workers; it may present a high risk of spreading to the community; there is usually no effective prophylaxis or treatment available
Comments	This classification is a recommendation to WHO's member states	Valid guidelines for the United States; no effects on animals and environment considered	*Detailed procedures for the classification are given in this directive that must be followed in the legislation in all member states of the EU **"Biological agents" shall mean microorganisms (including viruses and viroids), including those that have been genetically modified, cell cultures (of animal and plant tissues), and human endoparasites, which may be able to provoke any infection, allergy, or toxicity. The attached list of biological agents does only contain those in risk classes 2–4. The corresponding list from German authorities contains biological agents in risk classes 1–4. It specifically lists the biological agents of risk class 1 that can be used for production of bioproducts. It is only a fraction of those in risk class 1 that can be used in laboratories (ZKBS, 2010)

Table 6.9 Risk classes for the contained use of the biological agents classified in Table 6.8.

Risk class (the same risk definitions as in Table 6.1)	Risk class definition given in Article 4 of EU Directive 2009/41 (EU, 2009)*,**
1. No or low individual and community risk	Activities of no or negligible risk, that is to say activities for which level 1 containment is appropriate to protect human health and the environment
2. Moderate individual risk; low community risk	Activities of low risk, that is to say activities for which level 2 containment is appropriate to protect human health and the environment
3. High individual risk; low community risk	Activities of moderate risk, that is to say activities for which level 3 containment is appropriate to protect human health and the environment
4. High individual and community risk	Activities of high risk, that is to say activities for which level 4 containment is appropriate to protect human health and the environment

*This directive gives detailed information on the minimal requirements for the different physical containment levels in laboratories and production facilities, and other requirements that must be fulfilled before the start of the activities. They must be adhered to when the corresponding regulations are formulated and decided by the national authorities in the member states in the European Union. The cost for the containment increases with the risk class

**Note the following exceptions: Techniques or methods of genetic modification yielding microorganisms to be excluded from this directive on condition that they do not involve the use of recombinant nucleic acid molecules or GMMs other than those produced by one or more of the techniques/methods listed below:

1. Mutagenesis
2. Cell fusion (including protoplast fusion) of prokaryotic species that exchange genetic material by known physiological processes
3. Cell fusion (including protoplast fusion) of cells of any eukaryotic species, including production of hybridomas and plant cell fusions
4. Self-cloning consisting in the removal of nucleic acid sequences from a cell of an organism that may or may not be followed by reinsertion of all or part of that nucleic acid (or a synthetic equivalent), with or without prior enzymatic or mechanical steps, into cells of the same species or into cells of phylogenetically closely related species that can exchange genetic material by natural physiological processes where the resulting microorganism is unlikely to cause disease to humans, animals, or plants. Self-cloning may include the use of recombinant vectors with an extended history of safe use in the particular microorganisms.

treatment of exhaust air and waste, etc.). The requirements on physical containment increase with the risk class of the activity, and differ for laboratory and production facilities (EU, 2009; WHO, 2004).

- **Biological containment:** The requirements for this are concisely given in the NIH Guidelines Appendix I-I (NIH, 2009) as "In consideration of biological containment, the vector (plasmid, organelle, or virus) for the recombinant DNA and the host (bacterial, plant, or animal cell) in which the vector is propagated in the laboratory will be considered together. Any combination of vector and host which is to provide biological containment shall be chosen or constructed so that the following types of 'escape' are minimized: (i) survival of the vector in its host outside the laboratory, and (ii) transmission of the vector from the propagation host to other non-laboratory hosts. The following levels of biological containment (host–vector systems) for prokaryotes are established. Appendices I-I-A through I-II-B describe levels of biological containment (host–vector systems) for prokaryotes. Specific criteria will depend on the organisms to be used." This also requires applying Good Microbiological Techniques (GMT). The NIH Guidelines (NIH, 2009) and the WHO Laboratory Biosafety Manual (WHO, 2004, Chapter 16) apply mainly for laboratory use.

The classification of the risk of the contained uses involves four classes, one for each risk class given in Table 6.8. The formulations applying for the 27 member states of the European Union are given in Table 6.9. Information on the actual state of the regulations required in the Cartagena Protocol on Biosafety in the states that have ratified or signed the protocol is provided by the Biosafety Clearing-House, an information exchange institution for the protocol (http://bch.cbd.int/about/).

6.6.2 Regulations Governing the Use of Enzymes Produced in Wild-Type or Recombinant Organisms

Pharmaceutical and technical enzymes should be produced under conditions that minimize the risks for human health, the environment, and the costs of the containment in Table 6.9. Therefore, only risk class I or GRAS microorganisms (Table 6.8) that can be used for production of bioproducts should be used for this aim. The above regulations also restrict the use of vectors used for the transfer of enzyme genes to these organisms. This does not yet apply for the selection markers used to select the microorganism to which the gene has been successfully transferred. This applies especially to antibiotic resistance (see footnote to Table 6.3). It is to be expected that the above regulations will be amended frequently, as has been done since the first edition of this book, also in the future. The reader should therefore consult the Internet pages of the national authorities to follow this.

Besides these regulations, the use of enzymes is also regulated depending on their final use. Such regulations recommended by international organizations (WHO or FAO (Food and Agricultural Organization)) or decided by US and EU authorities are given in Table 6.10. From this table follows that the regulatory and legal system for chemicals in general and specifically enzymes is, nationally and on an international level, in a transition situation. This has to be considered when developing new enzyme processes or changing older processes to comply with the new regulations. GLP (Good Laboratory Practice) and GMP (Good Manufacturing Practice) apply for all of the above activities.

Table 6.10 International and national authorities that regulate the final use of enzymes.

Final enzyme use	WHO/FAO	United States	EU
For food, food additive, and feed processing (enzyme is or can be a food or feed additive)	JECFA (Joint FAO/WHO Expert Committee on Food Additives) plans to update its current guidance document on enzymes to include enzymes produced in recombinant microorganisms (JECFA, 2009). JECFA has a database with its evaluations (up to 2009) of enzymes and other compounds used as food additives and for other purposes: http://www.fao.org/ag/agn/jecfa-additives/index.html	FDA regulates based on the GRAS concept. It has published a list of GRAS microorganisms (also some recombinant) and microbial-derived ingredients (also enzymes) that are used in foods (FDA 2010): http://www.fda.gov/Food/FoodIngredientsPackaging/ucm078956.htm. For new microorganisms or enzymes, the GRAS Notification Program applies (FDA, 2010a)	Regulated in EC Regulation 1332/2008 on food enzymes (EU, 2008). Based on this, the European Food Safety Authority (ESFA) will evaluate enzymes. This will result in a positive list of approved food enzymes that can be used for food processing and food additives or ingredients in the EU. The enzyme-producing companies are integrated in this process (AMFEP, 2010).
As a drug	—	FDA regulations used for drugs	EU regulations used for drugs
As a biocatalyst in industrial processes and for processing of waste	The International Programme on Chemical Safety (http://www.who.int/ipcs/en/) gives recommendations on the assessment and use of chemicals (also enzymes)	The TSCA (Toxic Substances Control Act of 1976) provides EPA (US Environmental Protection Agency) with authority to require reporting, record keeping, and testing requirements, and restrictions relating to chemical substances and/or mixtures. Certain substances are generally excluded from TSCA, including, among others, food, drugs, cosmetics, and pesticides. EPA currently plans a revision of the TSCA similar to REACH in the EU (see the next column)	The REACH (Registration, Evaluation, Authorization, and Restriction of Chemicals) of the EU (EU, 2006) requires that all existing and new chemicals (also enzymes) in use must be registered and assessed (based on different risks) until about 2020. This is done by the European Chemical Agency (http://echa.europa.eu/home_en.asp), and can result in the restriction of the use of existing or new chemicals. See this page for REACH legislation amendments
In washing powder			This also applies for enzymes and household cleaning products used in detergents. Here the involved industry performs an interesting and informative risk assessment project (http://www.heraproject.com)

Exercises

6.1 Is *E. coli* generally a class I organism?

6.2 *S. cerevisiae* is a class I and GRAS organism that is used by some companies as a host for technical enzymes such as chymosin (used in the production of cheese) and savinase (used in washing powder; see Table 6.1). In order to select the cells that have been transformed, the enzyme gene is transferred with a selection marker. Previously (and for laboratory use), the main selection marker was and is antibiotic resistance (in bacteria a gene for β-lactamase, an enzyme that hydrolyzes β-lactam antibiotics). Why should this be avoided in the recombinant production of technical enzymes and enzymes for therapy? What policy – and alternatives – do the enzyme-producing companies have here?

6.3 *E. coli* cells contain ≈70% H_2O and 15% protein. The cells can be considered to be cylindrical. What is the maximum protein amount that can be produced per liter in fermentation? Is this smaller or larger than that for the real cells that are spherical and not planar at the end of the cylinder?

6.4 How much penicillin amidase could be produced when the cells are grown to the maximum density in Exercise 6.3, based on the data given in Figure 6.8? Discuss whether this is possible to achieve.

6.5 What processes that influence the enzyme yield shown in Figure 6.4 and Table 6.2 have been studied for penicillin amidases and lipases in Section 6.4?

6.6 Why must enzymes that bind cofactors such as Ca^{2+} or Zn^{2+} in the cell have very low equilibrium constants (dissociation constants) for the binding of these ions?

6.7 Is it sustainable to produce inclusion bodies?

6.8 Technical enzymes from extreme thermophilic organisms can be produced in *E. coli*. Discuss how they can be easily separated from a homogenate, based on information provided in Table 2.9. Heat-stable proteins tend to be more slowly hydrolyzed by peptidases than less stable proteins. This frequently correlates with allergenic potential, caused by the uptake of the whole unhydrolyzed protein in the intestine. What must therefore be studied for heat-stable enzymes before they are used in the food industry?

6.9 Why does the medium with extracellular enzymes contain DNA? How can it be separated from the enzyme? When the medium is separated from the cells by filtration, must the cells be washed to increase the enzyme yield? Why should the solution with the enzyme be concentrated by ultrafiltration before chromatographic purification steps?

6.10 In Figure 6.5, the processing scheme is given for penicillin amidase (PA) from *E. coli*. A similar scheme applies for the processing of wild-type PA from *A. faecalis* expressed in *E. coli*. In both cases, the processing involves successive hydrolysis of oligopeptides from the C-terminal of the A-chain. It has been found that the shorter the length of the A-chain, the lower the specific activity of the enzyme. The hydrolysis, after the first intramolecular autoproteolysis, has been shown to be intermolecular autoproteolysis or catalyzed by periplasmic peptidases. In the high cell density fermentations using the peptidase-deficient *E. coli* BL21(DE3) cells, the A-chain of the enzymes was longer than that found for the PA produced in wild-type *E. coli*, or the PA from *A. faecalis* produced *E. coli* cells that are not peptidase deficient (Kasche *et al.*, 2005).

a. How would you explain these observations?
b. What consequences does this have for the immobilization of these PAs with a high activity yield?
c. When you filter the purified PA (Figure 6.15) with the longer A-chain, how would you expect that the activity yield will change with incubation time of the enzyme before the filtration?

6.11 You have optimized a chromatographic purification of an enzyme in a column (diameter 1 cm, length 10 cm). Now you want to scale-up to purify 10 times of the enzyme in one step, by changing either the diameter or length of the column.

a. What would the dimensions of the columns be, keeping the resolution of the column constant?
b. Discuss the pros and cons of changing the diameter or the length!

6.12 Explain how the concentration profiles given by the intensity of the colors and the separation of A and B, given by different colors, are developed in the SMB chromatography shown in Figure 6.14.
6.13 The yield of PA_{AF} was about twice the yield of PA_{EC} in high cell density fermentations (Figures 6.7 and 6.8). Try to explain this based on the results in these figures. How would you try to increase the yield of PA_{EC}?
6.14 What laws and regulations exist in your country that influence the production and use of enzymes?

Literature

Textbooks and Internet Resources

The following are general sources on enzyme biosynthesis, production, and purification, classification of microorganisms, and on safety aspects of enzyme production and use.

AMFEP (Association of Manufacturers and Formulators of Enzyme Products) home page (http://www.amfep.org) contains important information on safety aspects of enzyme production and use, and has links to the main producers of enzymes and to international and national organizations that formulate the regulations on the use of enzymes, especially in the food sector.

Burgess, R.R. and Deutscher, M.P. (eds) (2009) *Methods in Enzymology*, vol. **463**, Academic Press, San Diego, CA; this volume covers all topics of this chapter – except large-scale production and purification of enzyme in detail.

Carta, G. and Jungbauer, A. (2010) *Protein Chromatography*, Wiley-VCH Verlag GmbH, Weinheim.

Cartagena Protocol on Biosafety. An information exchange institution, the Biosafety Clearing-House (http://bch.cbd.int/about/), gives information of laws and regulation on biosafety in the countries that have signed this protocol.

Rathore, A.S. and Velayudhan, A. (2003) *Scale-up and Optimization in Preparative*

Chromatography, Marcel Dekker, New York.

Subramanian, G. (ed.) (1998) *Bioseparation and Bioprocessing*, vols. I and II, Wiley-VCH Verlag GmbH, Weinheim,.

References

Albee, K.R., Butler, M.H., and Wrigtht, B.E. (1990) Cellular concentration of enzymes and their substrates. *J. Theor. Biol.*, **143**, 163–195.

AMFEP (Association of Manufacturers and Formulators of Enzyme Products) (2010) *Establishment of the First EU Positive List of Food Enzymes*, AMFEP, www.Amfep.org.

Amneus, H. (1977) On the use of pancreatic proteases for indication of genetic mutation. Dissertation, Uppsala University, Sweden.

Angov, E., Hillier, C.J., Kincaid, R.L., and Lyon, J.A. (2008) Heterologous protein expression is enhanced by harmonizing the codon usage frequencies of the target gene with those of the expression host. *PLoS ONE*, **3**, e2189.

Aworh, O.C., Kasche, V., and Apampa, O.O. (1994) Purification and some properties of sodom-apple latex proteinases. *Food Chem.*, 359–362.

Bagos, P.G., Tsirigos, K.D., Plessas, S.K., Liakopoulos, T.D., and Hamodrakas, S.J. (2009) Prediction of signal peptides in archea. *Protein Eng. Des. Sel.*, **22**, 27–35.

Beer, H.D., McCarthy, J.E.G., Bornscheuer, U.T., and Schmid, R.D. (1998) Cloning, expression, characterization and role of the leader sequence of a lipase from *Rhizopus oryzae*. *Biochim. Biophys. Acta*, **1399**, 173–180.

Bendtsen, J.D., Nielsen, H., von Heijne, G., and Brunak, S. (2004) Improved prediction of signal peptides. *J. Mol. Biol.*, **340**, 783–795.

Bendtsen, J.D., Nielsen, H., Widdick, D., Palmer, T., and Brunak, S. (2005) Prediction of twin-arginine signal peptides. *BMC Bioinform.*, **6**, 167.

Bruggink, A. (ed.) (2001) *Synthesis of β-Lactam Antibiotics*, Kluwer Academic Publishers, Dordrecht.

Bryan, P.N. (2000) Protein engineering of subtilisin. *Biochim. Biophys. Acta*, **1543**, 203–222.

Burton, S.C. and Harding, D.R.K. (1998) Hydrophobic charge induction chromatography: salt independent protein adsorption and facile elution with aqueous buffers. *J. Chromatogr. A*, **814**, 71–81.

Castan, A. and Enfors, S.O. (2000) Characterization of a DO-controlled fed-batch culture of *Escherichia coli*. *Bioprocess Eng.*, **22**, 509–515.

Christiansen, T., Michaelsen, S., and Wümpelmann, M. (2003) Production of savinase and population viability of *Bacillus clausii* during high cell density fermentations. *Biotechnol. Bioeng.*, **83**, 344–352.

Conder, J.R. and Hayek, B.O. (2000) Adsorption kinetics and equilibria of bovine serum albumin on rigid ion-exchange and hydrophobic interaction chromatography matrices in a stirred cell. *Biochem. Eng. J.*, **6**, 215–223.

Deshpande, S.S., Ambedkar, V.K., Sudhakaran, V.K., and Shewale, J.G. (1994) Molecular biology of β-lactam acylase. *World J. Microbiol. Biotechnol.*, **10**, 129–138.

Di Lorenzo, M., Hidalgo, A., Haas, M., and Bornscheuer, U.T. (2005) Heterologous production of functional forms of *Rhizopus oryzae* lipase in *Escherichia coli*. *Appl. Environ. Microbiol.*, **71**, 8974–8977.

Driouch, H., Sommer, B., and Wittmann, C. (2009) Morphology engineering of *Aspergillus niger* for improved enzyme production. *Biotechnol. Bioeng.*, **105**, 1058–1068.

Eiteman, M.A. and Altman, E. (2006) Overcoming acetate in *Escherichia coli* recombinant protein fermentations. *Trends Biotechnol.*, **24**, 530–536.

Enfors, S.O. (1992) Control of proteolysis in fermentation of recombinant proteins. *Trends Biotechnol.*, **10**, 310–315.

EU (2000) Directive 2000/54/EC of the European Parliament and of the Council of 18 September 2000 on the protection of workers from risks related to exposure to biological agents at work. *Off. J. Eur. Communities*, **L262**, 22.

EU (2006) Regulation (EC) No. 1907/2006 of the European Parliament and of the Council of 18 December 2006 concerning the Registration, Evaluation, Authorisation and Restriction of Chemicals (REACH), establishing a European Chemicals Agency. *Off. J. Eur. Union*, **L396**, 1–849.

EU (2008) Regulation (EC) No. 1332/2008 of the European Parliament and of the Council of 16 December 2008 on food enzymes. *Off. J. Eur. Union*, **L354**, 7–15.

EU (2009) Directive 2009/41/EC of the European Parliament and of the Council of 6 May 2009 on the contained use of genetically modified micro-organisms. *Off. J. Eur. Union*, **L125**, 75–97.

FDA (2010) Partial List of Microorganisms and Microbial-Derived Ingredients that Are Used in Foods, http://www.fda.gov/Food/FoodIngredientsPackaging/ucm078956.htm (last updated in 2009).

FDA (2010a) Generally Recognized as Safe, http://www.fda.gov/Food/FoodIngredientsPackaging/GenerallyRecognizedasSafeGRAS/default.htm.

FDA (2010b) Revised Recommendations to Industry (Not Binding!) on Enzyme Preparations Used as Food Additives, http://www.fda.gov/downloads/Food/GuidanceComplianceRegulatoryInformation/GuidanceDocuments/FoodIngredientsandPackaging/UCM217735.pdf.

Fraud, N. (2008) Membrane chromatography: an alternative to polishing column chromatography. *Bioprocess. J.*, **7**, 34–37.

Gottschlich, N. and Kasche, V. (1997) Purification of monoclonal antibodies by simulated moving-bed chromatography. *J. Chromatogr. A*, **765**, 201–206.

Grünenfelder, B., Rummel, G., Vohradsky, J., Roder, D., Langen, H., and Jenal, U. (2001) Proteomic analysis of the cell cycle. *Proc. Natl. Acad. Sci. USA*, **98**, 4681–4686.

Henge, R. and Bukau, B. (2003) Proteolysis in prokaryotes: protein quality control and regulatory principles. *Mol. Microbiol.*, **49**, 1451–1462.

Huang, Y.M., Hu, W., Rustandi, E., Chang, K., Yusuf-Makagiansar, H., and Ryll, T. (2010) Maximizing productivity of CHO cell based fed-batch culture using chemically defined media conditions and typical manufacturing equipment. *Biotechnol. Prog.*, **26**, 1400–1410.

Idiris, A., Tohda, H., Kumagai, H., and Takegawa, K. (2010) Engineering of protein secretion in yeast: strategies and impact on protein production. *Appl. Microbiol. Biotechnol.*, **86**, 403–417.

Ignatova, Z., Hörnle, C., Nurk, A., and Kasche, V. (2002) Unusual signal peptide directs penicillin amidase from *Escherichia coli* to the Tat translocation machinery. *Biochem. Biophys. Res. Commun.*, **291**, 146–149.

Ignatova, Z., Mahsunah, A., Geogieva, M., and Kasche, V. (2003) Improvement of posttranslational bottlenecks in the production of penicillin amidase in recombinant *Escherichia coli* strains. *Appl. Environ. Microbiol.*, **69**, 1237–1245.

Ignatova, Z., Stoeva, S., Galunsky, B., Hörnle, C., Nurk, A., Piotraschke, E., Voelter, W., and Kasche, V. (1998) Proteolytic processing of penicillin amidase from *A. faecalis* in *E. coli* yields several active forms. *Biotechnol. Lett.*, **20**, 977–982.

Ignatova, Z., Taruttis, S., and Kasche, V. (2000) Role of the intracellular proteolysis in the production of the periplasmatic penicillin amidase in *Escherichia coli*. *Biotechnol. Lett.*, **22**, 1727–1732.

Imamoglu, S. (2002) Simulated moving bed chromatography (SMB) for application in bioseparation. *Adv. Biochem. Eng. Biotechnol.*, **76**, 211–231.

Jaeger, K.-E., Dijkstra, B.W., and Reetz, M.T. (1999) Bacterial biocatalysts: molecular biology, three-dimensional structures, and biotechnological applications of lipases. *Annu. Rev. Microbiol.*, **53**, 315–351.

Jana, S. and Deb, J.K. (2005) Strategies for efficient production of heterologous proteins in *Escherichia coli*. *Appl. Microbiol. Biotechnol.*, **67**, 289–298.

JECFA (Joint FAO/WHO Expert Committee on Food Additives) (2009) Evaluation of Certain Food Additives, http://whqlibdoc.who.int/trs/WHO_TRS_952_eng.pdf; their latest report, where also recombinant enzymes used in food production or as food additives are evaluated. References are given in previous reports on the same topic.

Jones, H.E., Holland, I.B., and Campbell, A. K. (2002) Direct measurement of free Ca^{2+} shows different regulation of Ca^{2+} between the periplasm and the cytosol of *Escherichia coli*. *Cell Calcium*, **32**, 183–192.

Karg, S.R. and Kallio, P.T. (2009) The production of biopharmaceuticals in plant systems. *Biotechnol. Adv.*, **27**, 879–894.

Kasche, V., de Boer, M., Lazo, C., and Gad, M. (2003) Direct observation of intraparticle equilibration and the rate-limiting step in adsorption of proteins in chromatographic adsorbents with confocal laser scanning microscopy. *J. Chromatogr. B*, **790**, 115–129.

Kasche, V., Galunsky, B., and Ignatova, Z. (2003) Fragments of pro-peptide activate mature penicillin amidase of *Alcaligenes faecalis*. *Eur. J. Biochem.*, **270**, 4721–4728.

Kasche, V., Ignatova, Z., Märkl, H., Plate, W., Punckt, N., Schmidt, D., Wiegandt, K., and Ernst, B. (2005) Ca^{2+} is a cofactor required for membrane transport and maturation and is a yield-determining factor in high cell density penicillin amidase production. *Biotechnol. Prog.*, **21**, 432–438.

Kasche, V., Lummer, K., Nurk, A., Piotraschke, E., Riecks, A., Stoeva, S., and Voelter, W. (1999) Intramolecular autoproteolysis initiates the maturation of penicillin amidase from *E. coli*. *Biochim. Biophys. Acta*, **1433**, 76–86.

Kasche, V., Scholzen, T., Boller, Th., Krämer, D.M., and Löffler, F. (1990) Rapid protein purification using PBA-Eupergit™: a novel method for large-scale procedures. *J. Chromatogr.*, **510**, 149–154.

Kjærulff, S. and Jensen, M.R. (2005) Comparison of different signal peptides for secretion of heterologous proteins in fission yeast. *Biochem. Biophys. Res. Commun.*, **336**, 974–982.

Koller, K.-P., Riess, G.J., and Aretz, W. (1998) Process for the preparation of glutarylacylase in large quantities. US Patent 5,830,743.

Komar, A.A. (2009) A pause for thought along the co-translational folding pathway. *Trends Biochem. Sci.*, **34**, 16–24.

Luby-Phelps, K. (2000) Cytoarchitecture and physical properties of cytoplasm: volume, viscosity, diffusion, intracellular surface area. *Int. Rev. Cytol.*, **192**, 189–209.

Makrides, S.C. (1996) Strategy for achieving high-level expression of genes in *Escherichia coli*. *Microbiol. Rev.*, **60**, 512–538.

Mayer, A.F., Hellmuth, K., Schlieker, H., Lopez-Ulibarri, R., Oertel, S., Dahlems, U., Strasser, A.W.M., and van Loon, A.P. G.M. (1999) An expression system matures: a highly efficient and cost-effective process for phytase production by recombinant strains of *Hansenula polymorpha*. *Biotechnol. Bioeng.*, **63**, 373–381.

McDonough, M.A., Klei, H.E., and Kelly, J.A. (1999) Crystal structure of penicillin G acylase from the Bro1 mutant strain of *Providencia rettgeri*. *Protoc. Sci.*, **8**, 1971–1981.

McGuffee, S.R. and Elcock, A.H. (2010) Diffusion, crowding & protein stability in a dynamic molecular model of the bacterial cytoplasm. *PLOS Comput. Biol.*, **6**, e1000694.

Mergulhão, F.J.M., Summers, D.K., and Monteiro, G.A. (2005) Recombinant protein secretion in *Escherichia coli*. *Biotechnol. Adv.*, **23**, 177–202.

Middelberg, A.P.J. (2002) Preparative protein refolding. *Trends Biotechnol.*, **20**, 437–443.

Mitraki, A., Fane, B., Haase-Pettigell, C., Sturtevant, J., and King, J. (1991) Global suppression of protein folding defects and inclusion body formation. *Science*, **290**, 54–58.

Müller-Späth, T., Aumann, L., Melter, L., Ströhlein, G., and Morbidelli, M. (2008) Chromatographic separation of three monoclonal antibody variants using multicolumn countercurrent solvent gradient purification (MCSGP). *Biotechnol. Bioeng.*, **100**, 1166–1177.

Nardini, M., Lang, D.A., Liebeton, K., Jaeger, K.-E., and Dijkstra, B.W. (2000) Crystal structure of *Pseudomonas aeruginosa* lipase in the open conformation. *J. Biol. Chem.*, **275**, 31219–31225.

Niehaus, F., Bertoldo, C., Kähler, M., and Antranikian, G. (1999) Extremophiles as a source of novel enzymes for industrial application. *Appl. Microbiol. Biotechnol.*, **51**, 711–729.

Nielsen, J. (1996) Modelling the morphology of filamentous microorganisms. *Trends Biotechnol.*, **14**, 438–443.

Nielsen, J.E. and Borchert, T.V. (2000) Protein engineering of bacterial α-amylases. *Biochim. Biophys. Acta*, **1543**, 253–274.

NIH (2009) Guidelines for Research Involving Recombinant DNA Molecules, http://oba.od.nih.gov/rdna/nih_guidelines_oba.html.

Pilone, M.S. and Pollegioni, L. (2002) D-Amino acid oxidase as an industrial biocatalyst. *Biocatal. Biotransform.*, **20**, 145–159.

Prieto, I., Diaz, E., and Garcia, J.L. (1996) Molecular characterization of the 4-hydroxyphenylactate catabolic pathway in *Escherichia coli*. *J. Bacteriol.*, **178**, 111–120.

Quyen, D.T., Schmidt-Dannert, C., and Schmid, R.D. (1999) High-level formation of active *Pseudomonas cepacia* lipase after heterologous expression of the encoding gene and its modified chaperone in *Escherichia coli* and rapid *in vitro* refolding. *Appl. Environ. Microbiol.*, **65**, 787–794.

Rapoport, T.A. (2007) Protein translocation across the eukaryotic endoplasmic reticulum and bacterial plasma membranes. *Nature*, **450**, 663–669.

Rosenaua, F., Tommasen, J., and Jaeger, K.-E. (2004) Lipase-specific foldases. *ChemBioChem*, **5**, 152–161.

Rúa, M.L., Atomi, H., Schmidt-Dannert, C., and Schmid, R.D. (1998) High-level expression of the thermoalkalophilic lipase from *Bacillus thermocatenulatus* in *Escherichia coli*. *Appl. Microbiol. Biotechnol.*, **49**, 405–410.

Rudolph, N.S. (1999) Biopharmaceutical production in transgenic livestock. *Trends Biotechnol.*, **17**, 367–374.

Scheidat, B. (1999) Bioverfahrenstechnische Aspekte zum Einsatz von technischen Enzymen am Beispiel der kommunalen Abwasserreinigung. Thesis, TU Hamburg-Harburg.

Scherer, L.J. and Rossi, J.J. (2003) Approaches for the sequence-specific knockdown of mRNA. *Nat. Biotechnol.*, **21**, 1457–1465.

Schmidt, F.R. (2004) Recombinant expression systems in the pharmaceutical industry. *Appl. Microbiol. Biotechnol.*, **65**, 363–372.

Shukla, A.A. and Thömmes, J. (2010) Recent advances in large-scale production of monoclonal antibodies and related proteins. *Trends Biotechnol.*, **28**, 253–261.

Soethaert, W. and Vandamme, E.J. (2010) *Industrial Biotechnology*, Wiley-VCH Verlag GmbH, Weinheim, p. 354.

Svensson, M., Han, L., Silfversparre, G., Häggström, L., and Enfors, S.O. (2005) Control of endotoxin release in *Escherichia coli* fed-batch cultures. *Bioprocess Biosyst. Eng.*, **27**, 91–97.

Szentimai, A. (1964) Production of penicillin acylase. *Appl. Microbiol.*, **12**, 185–187.

Terpe, K. (2003) Overview of tag protein fusions: from molecular and biochemical fundamentals to commercial systems. *Appl. Microbiol. Biotechnol.*, **60**, 523–533.

Ueda, M., Takahashi, S., Washida, M., Shiraga, S., and Tanaka, A. (2002) Expression of *Rhizopus oryzae* lipase gene in *Saccharomyces cerevisiae*. *J. Mol. Catal. B*, **17**, 113–124.

Uretschlager, A. and Jungbauer, A. (2002) Preparative continuous annular chromatography (P-CAC), a review. *Bioprocess Biosyst. Eng.*, **25**, 120–140.

Viegas, S.C., Schmidt, D., Kasche, V., Arraiano, C.M., and Ignatova, Z. (2005) Effect of increased stability of the penicillin amidase mRNA on the protein expression levels. *FEBS Lett.*, **579**, 5069–5073.

Voitl, A., Müller-Späth, T., and Morbidelli, M. (2010) Application of mixed mode resins for the purification of antibodies. *J. Chromatogr. A*, **1217**, 5753–5760.

Walsh, G. (2010) Biopharmaceutical benchmarks 2010. *Nat. Biotechnol.*, **28**, 917–924.

Werner, M., Las Heras Vasques, F.J., Fritz, C., Vielhauer, O., Siemann-Herzberg, M., Altenbuchner, J., and Syldatk, C. (2004) Cloning of D-specific hydantoinase utilization genes from *Arthrobacter crystallopoietes*. *Eng. Life Sci.*, **4**, 563–572.

WHO (World Health Organization) (2004) *Biosafety Manual*, 3rd edn, WHO, Geneva.

Wickner, S., Maurizi, M.R., and Gottesman, S. (1999) Posttranslational quality control: folding, refolding, and degrading proteins. *Science*, **286**, 1888–1893.

Yuryev, R., Kasche, V., Ignatova, Z. Galunsky, B. (2010) Improved A. faecalis Penicillin Amidase Mutant Retains the Thermodynamic and pH Stability of the Wild Type Enzyme, *Protein J.*, **29**, 181–187.

Zhang, G., Hubalewska, M., and Ignatova, Z. (2009) Transient ribosomal attenuation coordinates protein synthesis and co-translational folding. *Nat. Struct. Mol. Biol.*, **16**, 274–280.

Zhang, G. and Ignatova, Z. (2009) Generic algorithm to predict the speed of translational elongation: implications for protein biogenesis. *PLoS ONE*, **4**, e5036.

Zhao, G., Dong, X.Y., and Sun, Y. (2009) Ligands for mixed-mode protein chromatography: principles, characteristics, and design. *J. Biotechnol.*, **144**, 3–11.

ZKBS (Zentrale Kommission für die biologische Sicherheit) (2010) Bekanntmachung der Liste risikobewerteter Spender- und Empfängerorganismen für gentechnische Arbeiten (The German Classification of Microorganisms), http://www.bvl.bund.de/SharedDocs/Downloads/06_Gentechnik/register_datenbanken/organismenliste.pdf?__blob=publicationFile&v=4.

7
Application of Enzymes in Solution: Soluble Enzymes and Enzyme Systems

Topics/purposes/problems	Solutions to problems
Reaction/transformation • Technical application • Preparative purpose	Enzymes in solution, or immobilized biocatalyst (see below for criteria)
Insoluble high molecular substrates	Enzymes in solution
Enzyme amount required at • Given substrate concentration • Time • Temperature	Calculations based on kinetics, productivity, inactivation kinetics
Cost of enzymes relating to cost of product	Calculation based on productivity • Low cost: – Application of enzymes in solution • High cost: – Optimization of enzyme production (Chapter 6) – Membrane reactors (Section 7.4) – Immobilization (Chapters 8 and 9)
Further criteria • Residual enzyme in product	
• Reaction dependent on cosubstrates	Membrane reactors (Section 7.4)
Multienzyme reaction sequences	Engineered whole-cell biocatalysts (Chapter 5)
• Substrate insoluble in water	Reaction engineering, application of solid/solution equilibria

Biocatalysts and Enzyme Technology, Second Edition. Klaus Buchholz, Volker Kasche, and Uwe T. Bornscheuer

7.1
Introduction and Areas of Application

Soluble enzymes continue to dominate technical applications, most notably in the areas of

- detergent formulations with proteases, lipases, and cellulases;
- food manufacturing, with starch hydrolysis and further processing, bread, cheese, and fruit juice manufacture;
- biofuel production from starch (for production from lignocellulose, see Section 12.2.2); and
- synthesis of chiral compounds such as Sitagliptin (Section 4.2.3).

These applications represent the most important ones with respect to volume and turnover, and a general survey of the situation is presented in Table 7.1. (For an overview of products and processes in biotechnology, see Buchholz and Collins, 2010). Applications have extended into the areas of paper and pulp processing, textile manufacture, and the degumming of oil, with one-step reactions and hydrolytic enzymes continuing to dominate the scene.

However, in recent years newer areas of application and perspectives of synthesis in organic and pharmaceutical chemistry have gained much interest, and these are described in more detail in Chapters 4 and 12.

The impact of genetic engineering, notably with regard to recombinant techniques, has played a major role with regard to both significantly increased enzyme yields with consequent reductions in cost and improved performance in terms of stability and productivity. In addition, the range of applications for enzymes has also been significantly extended, and their (regio-, chemo-, and stereo-) selectivity greatly modified. In this respect, one important factor is that of safety aspects, notably in food manufacture, where the GRAS (generally recognized as safe) status plays a key role (see Sections 6.6 and 7.1.3).

Several conditions govern the potential for the application of soluble enzymes, and their advantages and/or shortcomings as compared to chemical reactions (cf. in starch hydrolysis and in washing) or to the application of immobilized enzymes (Table 7.2).

Soluble enzymes in production processes are limited to singular application, with the condition of low price (5–20 € kg^{-1} or € L^{-1} of concentrate, respectively), or their application in small amounts, as in analytical testing or medical/pharmaceutical use.

The hydrolysis of high molecular weight, or insoluble, or adsorbed substrates proceeds efficiently only with the use of soluble enzymes. With immobilized biocatalysts, slow diffusion and poor accessibility of macromolecules or adsorbed substances would imply an extremely low efficiency. Thus, the application areas of proteases, cellulases, and lipases for detergents in laundry and textile manufacture, as well as glycosidases in starch hydrolysis, for biofuel production, and in baking, brewing, and fruit juice and vegetable processing, remain the major areas for the utilization of soluble enzymes (see Section 7.3) (Aehle, 2004; Schäfer *et al.*, 2007; Soetaert and Vandamme, 2010; Wenda *et al.*, 2011).

Table 7.1 Industrial applications of enzymes: major selected areas (Aehle, 2004; Schäfer *et al.*, 2007; Soetaert and Vandamme, 2010).

Market share[a]	Enzyme	Purpose, application in solution	Membrane systems	Immobilized systems
	Hydrolases			
Detergents: about 34%	Proteases	Detergents		
	Cellulases	Detergents		
	Lipases	Textile, oil		
Food and feed: 31%	Rennin, chymosin	Cheese manufacture		
	Glycosidases			
(Starch: 11–15%)[b]	Amylases, amyloglucosidases	Starch hydrolysis, biofuel production, detergents, baking, brewing		(Dextrin hydrolysis)
	Cellulases	Pulp and paper manufacture, textiles, biofuels		
	Xylanases, β-glucanases, pectinases[c]	Juice manufacture, feed processing		
	Lactase	Milk products		Lactose hydrolysis
	Esterases			
	Acylases		Amino acid synthesis	Amino acid synthesis
	Penicillin amidase			Penicillin hydrolysis/synthesis
	Oxidoreductases			
	Glucose Oxidase	Drinks manufacture, Analytics		Glucose analyzer
	Lipoxygenase	Baking		
	Transferases			
	Aminotransferase			Amino acid manufacture
	Cyclodextrin transferase			Cyclodextrin manufacture
	Isomerases			
(12%)[b]	Glucose isomerase			Manufacture of high-fructose corn syrup
2300 million €[d]	*Total turnover* Worldwide			

a) Technical enzymes, worldwide, estimates.
b) Included in food.
c) Including other activities.
d) 2009 (1 € ≈ 1.25 US $).

Table 7.2 Advantages and disadvantages of the application of enzymes in solution and immobilized enzymes, respectively.

Advantage for soluble enzymes	Disadvantage for soluble enzymes	Advantage for immobilized enzymes	Disadvantage for immobilized enzymes
Low enzyme price	High enzyme price; allergies due to enzymes in product	High enzyme price requires immobilization	High cost of immobilization
Singular application (analytical, therapeutic)	Limited productivity	Continuous processes, high productivity	
Low amounts (analytical, therapeutic application)			
Efficiency with insoluble, or adsorbed, or high molecular weight substrates[a]		Fewer by-products[b]	More by-products[c]
Complex systems with coupled reaction path			Low efficiency with complex systems
Coenzyme-dependent reactions easy to implement			No coenzyme retention
Application of membrane reactor systems (retention or recycling of enzymes)		Easy reuse	

a) Enzyme remains in the product.
b) Example: less trisaccharides due to decreasing lactose gradient in the carrier in lactose hydrolysis.
c) Example: more reversion products due to increasing product gradient in the carrier with amyloglucosidase (dextrin hydrolysis).

Membrane processes allow for the continuous use of soluble enzymes in the conversion of low molecular weight substrates, with these being notably favorable for systems requiring coenzymes and/or multienzyme systems (see Section 7.4).

The total turnover in industrial enzyme sales has increased from about 1 billion € in 1995 (Godfrey and West, 1996) to 2.5 billion € in 2010 (www.novozymes.com). The most important areas of this market were the food and feed sector (market share 31%) including 11–15% for starch processing and 12% for glucose isomerization, and technical enzymes (market share 35%), while enzymes in detergents had a market share of 34% (Table 7.1; see also Section 1.3.3 and Figure 1.5). Further, major amounts of cellulases and other glycan hydrolases are required for the hydrolysis of lignocellulosic biomass for biofuel production (see Section 12.2.2). The total value of enzymes produced for enzyme processes is much larger, as many companies produce the technical enzymes that they use.

7.1.1
The Impact of Genetic Engineering

The impact of genetic engineering, with recombinant techniques, has played a major role in extension of the fields of enzyme applications, the majority of enzymes being modified and/or produced using recombinant DNA methods. Yields for enzymes in fermentation processes have increased tremendously, by factors up to 100, and this in turn has led to much reduced prices, typically lower by a factor of 10 as compared to enzymes from conventional sources (see Chapter 6).

Thermal stability could be improved by site-directed and random mutagenesis, as well as by directed evolution. Although these are now established techniques, they must be worked out for each enzyme individually. Site-directed mutagenesis may also be used to exchange amino acids that are unstable at elevated temperature. Examples of this include asparagine and glutamine that undergo deamidation, and lysine in amylases and glucose isomerase that tends to form Schiff bases with reducing sugars (Sicard, Leleu, and Tiraby, 1990; Quax *et al.*, 1991). Recent approaches of protein engineering, including rational, combinatorial, and data-driven designs, have been covered in Chapter 3.

The modification of regions involved in unfolding has also been a target. The exchange of between 4 and 12 amino acids may be required for technically relevant effects, such as shifting the application range by >5 °C toward a higher operation temperature. Thermostable amylases obtained from hyperthermophilic microorganisms are active up to 130 °C, and are used industrially in the range of 105–110 °C for about 5 min during starch processing (Section 12.1.1). Other important improvements concern ion requirements (e.g., amylases that no longer need Ca^{2+} ions to be added to substrate solutions) or shifts in the pH application range (e.g., lowering the pH optimum of amylases and approaching the operational range of amyloglucosidases) so that ion-exchange operations to adjust the pH of substrate solutions may be omitted. A range of successful modifications has been summarized in Chapter 3. Despite these successes, however, native enzymes are generally required for food processing in several European countries, mainly due to customer preferences.

7.1.2 Medium Design

Medium design is essential in order to improve enzyme stability. The protective effect of salts and polyols is well known, and applied to improve the storage stability of commercial enzyme preparations. Under operating conditions, the stabilizing effects may also be effective at both elevated temperatures and pressures. A stabilizing effect was seen to be optimal for salts, including ammonium or potassium and sulfuric acid ions, and is outlined in the following lyotropic series (Foster, Frackman, and Jolly, 1995; Curtis *et al.*, 2002; Morgan and Clark, 2004):

- **Cations:** $NH_4^+ > K^+ > Na^+ > Mg^{2+} > Ca^{2+}$ (in some cases, Ca^{2+} is essential for the enzyme function (Section 6.4)).
- **Anions:** $SO_4^{2-} > Cl^- > NO_3^-$.

However, when they are added they may increase wastewater treatment costs. The protective effect of polyols has also been investigated by several authors (Monsan and Combes, 1984; Ye, Combes, and Monsan, 1988). Prominent among these is the example of sorbitol stabilizing β-galactosidase by more than two orders of magnitude at elevated concentrations (Athes and Combes, 1998; see also data below). These effects are most significant in industrial processes for hydrolyzing starch, and for the isomerization, synthesis, and transformation of sugars, where high concentrations of substrates and products are favorable for product recovery (Cheetham, 1987). Polymers have also been used as enzyme stabilizers; thus, polysaccharides such as dextran and synthetic polymers such as polyvinyl alcohol, polyvinylpyrrolidone, and polyvinylsaccharides were investigated with respect to their protecting and stabilizing effects, by the addition of up to 1% to the solution (Gibson, Hulbert, and Woodward, 1993; Bryjak and Noworyta, 1994).

Enzyme modification – for example, at the external protein surface – has been applied to both stabilization and specific applications. Certain reagents – for example, acyl derivatives or epoxides – may react with amino acid functional groups, thus modifying the surface charge and/or the hydrophobic character of the protein.

The formulation of detergent enzymes aims at improved stability (notably against high pH) and safety in handling. Liquid detergent enzyme products often have an active substance content similar to the powder products (1–6%). They are generally stabilized by "preferential exclusion" of water from the protein surface by the addition of water-associated compounds such as polyols (propylene glycol), sugar, and lyotropic salts. Thus, the unfolding of the protein molecule is minimized. For powder products, there are little to none stability issues. At present, most of the enzyme granulates are produced by high shear mixing, spraying on a core particle in fluidized bed, or by extrusion (Maurer, 2010). In many applications, granulation by agglomeration of the enzyme concentrate in the presence of different salts and polysaccharides is a well-established method for making particles of appropriate size in the range of 0.5–1 mm, mainly in order to avoid allergic reactions (Eriksen, 1996).

7.1.3
Safety Aspects

Safety aspects are very important, not only for food manufacture but also for products where contact with the protein is possible (e.g., detergents). They have been and are considered by a joint FAO/WHO Expert Committee on Food Additives (JECFA; see Section 6.6). Enzymes obtained from the edible tissues of animals or plants commonly used as foods, and from microorganisms traditionally accepted as constituents of, or for the preparation of foods, were all regarded as foods. For enzymes produced by nonpathogenic microorganisms as well as by genetically modified organisms, safety regulations are compiled in Chapter 6 (Section 6.6, Table 6.10). For the production of recombinant enzymes, microorganisms with GRAS status (see also Section 6.1) should be used preferentially as host, including, for example, *Bacillus* sp., *A. niger*, or yeast used in food manufacture. For detergent application, guidelines for safe handling and use of enzymes have been developed in order to avoid allergies caused by inhaled enzymes and skin contact irritation (coated or granulated formulations) (Maurer, 2010).

7.2
Space–Time Yield and Productivity

The kinetics of enzyme processing were outlined in Chapter 2, and form the basis for calculating or estimating the amount of enzyme and/or the reaction time (and thus the reactor volume) required for a given transformation on the technical scale. These aspects are summarized by the space–time yield (STY), giving the substrate amount reacted or product produced per unit of time and volume (see Sections 2.7 and 10.6).

Catalyst productivity is the second economic key parameter, and indicates the amount of product that can be obtained per unit amount of biocatalyst (this will be dealt with in more detail at the end of this paragraph). It is important to bear in mind that transformations on the technical scale normally must be performed at nonoptimal conditions with respect to the reaction rate and stability, such as high conversion, elevated temperature, and nonideal mixing in technical reactors (thus far from natural conditions; see Section 11.3). The important process conditions are summarized in Table 7.3.

The selection of reaction conditions aims at maximizing the yield and minimizing the overall cost of the process: the cost for raw material, investment, energy, enzyme, disposal of residual material, and environmental protection (wastewater, exhaust gas, etc.). Enzyme costs form a minor part of the total in general (1–10%), whereas raw material expenditure dominates in processes at the large scale with bulk products (e.g., 30–50% for products prepared from starch). In these cases, as well as in detergent applications, the enzyme costs typically comprise about 1%. *Expense and investment costs* for upstream and downstream operations – for example, for product isolation and purification – are in general significantly higher than those for the transformation step; thus, the latter must be optimized with respect to these steps (high product yield and concentration, minimal side products).

The basic kinetic equations that allow for the calculation of process data and parameters discussed previously were indicated in Chapter 2, and are summarized in Box 7.1. It must be remembered that a transformation can proceed towards an equilibrium (e.g., in glucose isomerization) or under kinetic control (e.g., synthesis of antibiotics, peptides, and other condensation products). For consistent and

Table 7.3 Selection of reaction conditions in technical processes.

Reaction conditions	Reasons	Consequences
High concentrations	Simplified, efficient product recovery; avoiding contaminations; high STY	Substrate inhibition
High conversion	As before; minimizing substrate cost	Product inhibition
Temperature	Optimization of reaction rate and yield; avoiding contaminations	Enzyme inactivation
pH	Shift of equilibrium towards optimal yield; adaptation to side and subsequent reactions; corrosion	Nonoptimal conditions for reaction rate and stability

precise calculations, the corresponding equations must be applied (see Section 2.7). The kinetics of synthetic reactions with more than one substrate and different reaction pathways – as well as the hydrolysis of high molecular weight substrates – are much more complex. An example of this is the oligosaccharide synthesis with glucansucrases from sucrose and another sugar as acceptor (Demuth, Jördening, and Buchholz, 1999; Seibel *et al.*, 2010).

For the application of enzymes in solution, batch (discontinuous) processes are standard, and the concentration of educts and products is then a function of time (Figure 7.1). Hence, for the calculation of conversion rates, the integrated equations given in Box 7.1 must be applied.

Box 7.1:

Equations for the calculation or estimation, respectively, of the biocatalyst amount expressed as an activity V_{max} per unit reactor volume required for a defined substrate conversion in a given time.[1] From this, the biocatalyst productivity (amount of substrate converted to product)/V_{max} can be calculated (for derivations, see Chapter 2).

Without inhibition and inactivation of the biocatalyst. The rate equation is (Eq. (2.8))

$$v = -\frac{d[S]}{dt} = \frac{k_{cat}[E][S]}{K_m + [S]} = \frac{V_{max}[E][S]}{K_m + [S]}. \tag{7.1}$$

The dependence of V_{max} and K_m on pH, T, and inhibitors I is shown in Section 2.7.2. Integration of Eq. (7.1) gives

$$V_{max} = \frac{([S]_0 - [S]) + K_m \ln([S]_0/[S])}{t}, \tag{7.2}$$

where the subscript 0 denotes the initial substrate concentration or biocatalyst activity.

The biocatalyst productivity is then

$$([S]_0 - [S])/V_{max}. \tag{7.3}$$

Inhibition but no inactivation of the biocatalyst. For *noncompetitive substrate inhibition* (Eq. (2.25)), Eq. (7.1) is modified to

$$v = \frac{V_{max}[S]}{K_m + [S](1 + K_m/K_i) + ([S]^2/K_i)}. \tag{7.4}$$

Integration gives

$$V_{max} = \frac{([S]_0 - [S])(1 + K_m/K_i) + K_m \ln([S]_0/[S]) + ([S]_0^2 - [S]^2)/2K_i}{t}. \tag{7.5}$$

For *competitive* product inhibition (Eq. (2.36)), Eq. (7.1) is modified to

$$v = \frac{V_{max}[S]}{K_m + (K_m/K_i)[S]_0 + [S](1 - K_m/K_i)}. \qquad (7.6)$$

Integration gives

$$V_{max} = \frac{([S]_0 - [S])(1 - K_m/K_i) + (K_m + (K_m/K_i)[S]_0)\ln([S]_0/[S])}{t}. \qquad (7.7)$$

These equations apply, for instance, to penicillin hydrolysis by penicillin amidase (see Section 2.7.2.3).

Inhibition and inactivation of the biocatalyst. The inactivation of the biocatalyst can be either a first-order process

$$V_{max} = V_{max,0}\, e^{-k_i t} \qquad (7.8)$$

or a second-order process. The latter occurs, for instance, in the autoproteolysis of peptidases, or when peptidases are present in the biocatalyst used. With first-order biocatalyst inactivation, the denominator t in Eqs. (7.2), (7.5), and (7.7) is replaced by

$$\frac{1 - e^{-k_i t}}{k_i} \geq t, \qquad (7.9)$$

that is, the biocatalyst productivity decreases when the biocatalyst is inactivated.

1) The biocatalyst amount for free enzymes is usually expressed in U (see Section 2.7.1), that is, the biocatalyst amount that converts 1 μmol substrate per minute at defined conditions. The calculated V_{max} values can be easily converted to U.

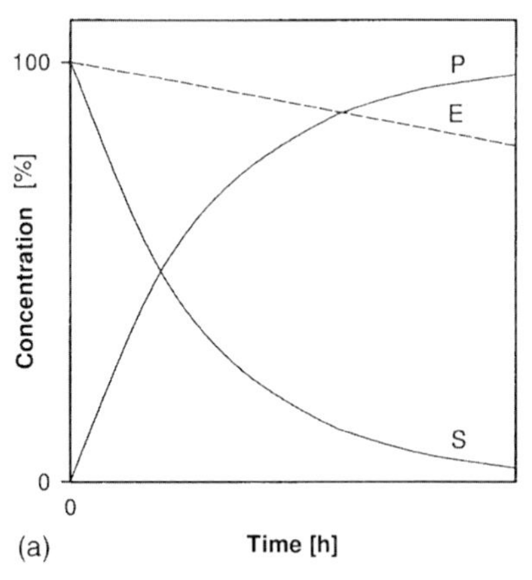

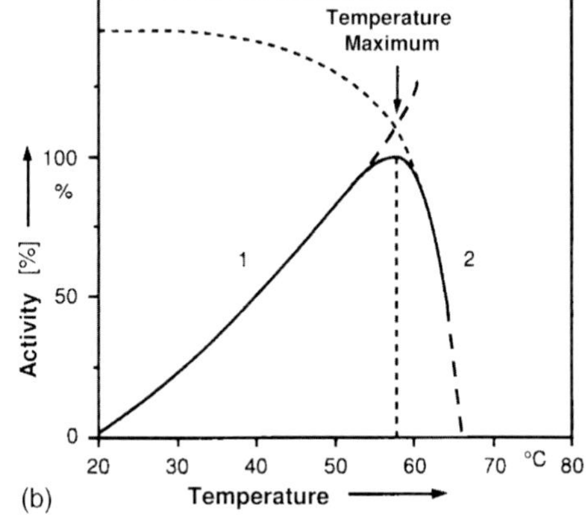

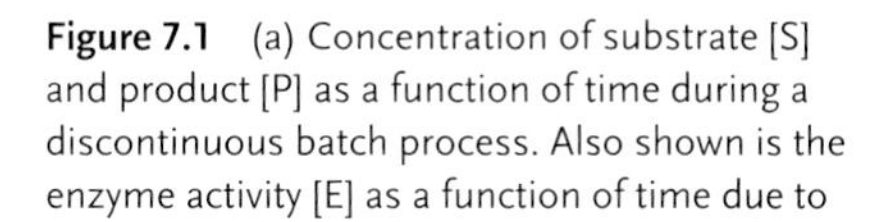

Figure 7.1 (a) Concentration of substrate [S] and product [P] as a function of time during a discontinuous batch process. Also shown is the enzyme activity [E] as a function of time due to inactivation. (b) Enzyme activity as a function of temperature with the different effects of increasing reaction rate (curve 1) and increasing inactivation rate (curve 2).

Enzyme inactivation is of major importance. The enzyme is lost after the transformation (this is different for membrane systems; see Section 7.4). Thus, the optimal temperature must be selected for a rapid transformation, but taking into account that enzyme inactivation should not lead to low reaction rates prior to achievement of the desired conversion. In general, activation energies for enzyme catalysis are in the range of 40–80 kJ mol^{-1}, while those for enzyme inactivation are distinctly higher, in the range of 60–400 kJ mol^{-1} (see also Section 2.10 and Exercise 2.10). Rather simple estimations with simplified kinetics allow for the identification of an appropriate compromise (see Exercise 2).

In order to calculate an enzymatic conversion, the procedure is as follows: the kinetic constants are taken from the literature (see Table 2.4 for important technical data; BRENDA; Internet resources in Appendix A; take into account thermodynamic data for the catalyzed reaction, exo- or endothermic reaction). If no data are available, they must be determined experimentally. The conversion required is fixed, for example, 98%, which corresponds to $([S_0] - [S]/[S_0]) = 0.98$, then the reaction time (and thus the STY) is also fixed (e.g., 24 h). These figures are then introduced into the appropriate integrated equation and the required amount of enzyme $V_{max,0}$ is calculated. *Note*: It is important to take both substrate and/or product inhibition into account, or the results may be incorrect.

Finally, the loss of enzyme activity by inactivation must be calculated. Few data are available in the literature, so that this parameter – inactivation kinetics, a key for economic application – often must be determined experimentally (see Prenosil *et al.*, 1987). It must be taken into account that enzyme stability under process conditions often differs from that in model solutions. In many cases, stabilizing conditions can be identified, notably in food manufacture, where carbohydrates are processed in concentrated solutions.

Thus, at high sucrose concentration (>50%) a considerable stabilization was observed for (immobilized) invertase in the temperature range of 65–70 °C with a productivity of 10 tons (product) per kg (biocatalyst) (Monsan and Combes, 1984). An analogous effect was found for a protease developed for detergents (Adler-Nissen, 1986). For the application at elevated temperatures, enzymes from thermophiles exhibiting high thermostability are available, and these can be cloned into appropriate enzyme-producing organisms. Experimental results for a protease are shown in Figure 7.2a and b. The influence of pH on stability is obvious. For a thermostable α-amylase from *Bacillus licheniformis*, the following kinetic data for inactivation rates were obtained (de Cordt *et al.*, 1992) (all subsequent data for k_i reported refer to a reaction following first-order with the dimension of min^{-1}): $k_i = 0.02–0.06$ (when the commercial enzyme preparation was diluted by factors of 1: 100 to 1: 500, 95 °C, pH 8,5, Ca^{2+}: 70 ppm); activation energy EA = 426 kJ mol^{-1} (for further data on kinetics of inactivation, including temperature and pressure as parameters, see Weemaes *et al.*, 1996). Further data published by Feng *et al.* (2002) for inactivation kinetics of two thermostable amyloglucosidases, including stabilization by sorbitol, are provided in Exercise 2.

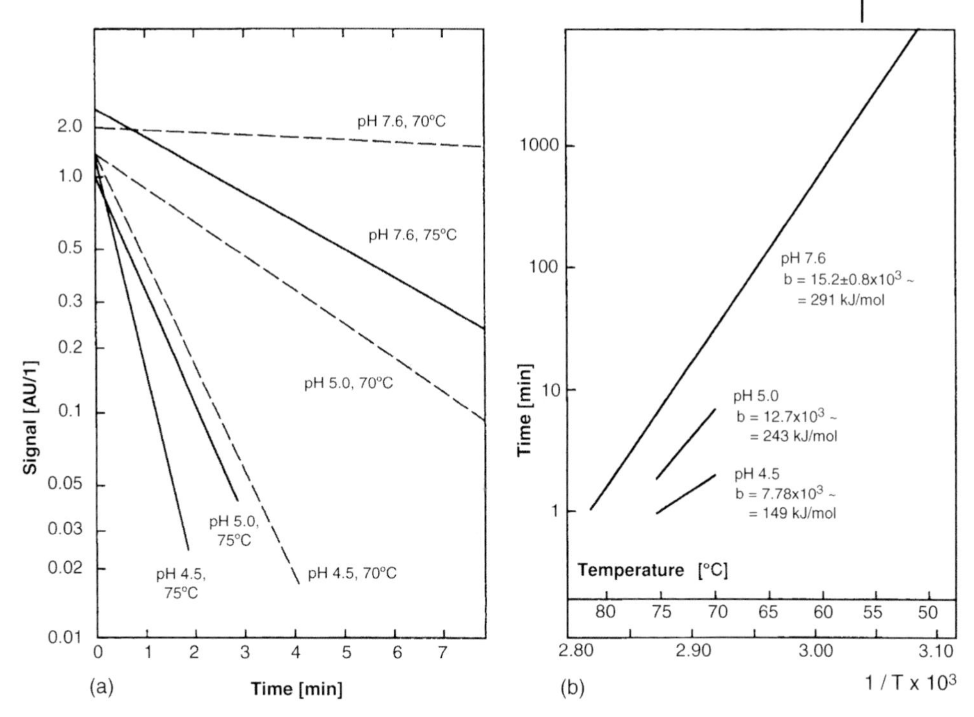

Figure 7.2 (a) Inactivation of a protease (Alcalase) that is stable under alkaline conditions at different pH and temperatures (activity given in logarithmic scale as a function of time in min) (Adler-Nissen, 1986). (b) Correlation according to Arrhenius for t_D (t_D in min corresponds to the time during which the activity decreases by a factor of 10, thus the *practical maximal time of operation*) (Adler-Nissen, 1986).

Additional data on the kinetics of inactivation were summarized by Fullbrock (1996), and referred to proteases, β-galactosidase, α-amylases, and glucose isomerase. Among the highest stabilities reported so far was that of an esterase from *Pyrococcus furiosus* that was cloned in *E. coli* and exhibited a half-life of 50 min at 126 °C (Ikeda and Clark, 1998). A two-stage series-type mechanism of inactivation was observed for immobilized β-galactosidase (Illanes, Altamirano, and Aillapán, 1998). Recent approaches for stabilization comprise glycosidation with functionalized β-cyclodextrin derivatives, and the addition of glycol chitosan and fumed silica. However, data on inactivation kinetics remain scarce (Polizzi *et al.*, 2007). The biocatalyst (or enzyme) productivity can be calculated using Eq. (7.3). Based on this calculation, the enzyme costs as part of the product costs can be deduced. Examples of the enzyme amount required, the productivity, and half-life (together with simple calculations) are provided in the exercises Section.

7.3 Examples for the Application of Enzymes in Solution

7.3.1 Survey

Since the first technical enzyme preparations were introduced during the nineteenth century, a broad range of diverse industrial applications has been developed. With respect to economic relevance, however, two areas have dominated: (1) enzymes in detergents, where proteases (and more recently cellulases), lipases, and amylases are important activities; and (2) amylases and amyloglucosidases in starch processing (see Table 7.1). Some major companies that supply enzymes are listed in Table 7.4.

More extensive details of technical enzymes and their commercial applications are listed in Table 7.5. One major field is that of detergents, and notably of washing powders and *liquid preparations*, where proteases, cellulases, lipases, and amylases are all applied (see Section 7.3.3.1). The success of lipase application evolved with genetically modified preparations that were active in aqueous solution (native lipases are primarily active at interfaces, such as oil/water). Most applications have been – and continue to be – within the food sector, mainly in areas of food and feed processing, modification, and upgrading (see Table 7.5).

Improved economics of processing, with enhanced yield, fewer by-products, and improved environmental protection (e.g., less wastewater) are also of

Table 7.4 Selected major enzyme supplying/selling companies.

Company	Country
AB Enzymes	Germany
Amano Pharmaceutical Co.	Japan
BBI Enzymes Ltd	Great Britain
Biocatalysts Ltd	Great Britain
Christian Hansen AS	Denmark
Danisco (including Genencor Enzymes)[a]	Denmark
DSM	The Netherlands
Enzyme Development Corporation (with Röhm Enzymes and others)	United States
Fluka-Sigma-Aldrich	Switzerland
Iogen Co.	Canada
Meiji Seika Kaisha Ltd	Japan
National Enzyme Company Inc.	United States
Novozymes	Denmark
Roche Penzberg	Germany

a) Acquired by DuPont (USA) in 2011 (CHEManager, 1–2/2011, p. 3; CHEManager 21–22/2011, p. 5).

Table 7.5 Applications of technical enzymes (selected according to economic relevance) (Aehle, 2004; Schäfer *et al.*, 2007).

Enzyme	Source	Reaction type	Application
Proteases			
Alkaline protease	*Bacillus, Aspergillus*	Protein degradation	Detergents, leather manufacturing
Rennin	Calf	Specific casein hydrolysis	Cheese manufacturing
Pepsin and others	*Mucor*; pancreas of pig, cattle	Protein degradation	Meat production
Papain	*Carica papaya*	Protein hydrolysis	Bread manufacture; brewing; meat production
Glycosidases[a]			
α-Amylases	*Bacillus, Aspergillus*	Endohydrolysis of starch to dextrins	Starch processing, dextrin production; biofuel production; brewing; bread manufacture; detergents
Amyloglucosidases	*Aspergillus*	Exohydrolysis of dextrins to glucose	Glucose syrup production
Pullulanase, isoamylase		Hydrolysis of α-1,6-bonds in amylopectin	Starch processing, brewing
Cellulases	*Aspergillus, Trichoderma*	Hydrolysis of cellulose	Detergents; pulp and paper, textile manufacture; biofuel production
Pectinases,[b] including galactanases, xylanases, arabanases	*Aspergillus*	Hydrolysis of corresponding polysaccharides in plant cell walls	Processing of fruits, vegetables, production of wine, juices, pureés, and so on; brewing; baking; feed processing
Pentosanases (similar as before, except pectinases)	*Humicola, Trichoderma*	Hydrolysis of pentosan fractions	Bread manufacture; bakery products
Lactase	*Aspergillus*	Hydrolysis of lactose	Milk products; whey processing
Invertase	*Saccharomyces*	Hydrolysis of sucrose	Production of invert sugar syrup; bakery, confectionery
Esterases, amidases			
Lipases	*Pseudomonas, Aspergillus, Candida*		Detergents; fat processing, modification, fatty acid production; milk products; baking

(*continued*)

Table 7.5 (Continued)

Enzyme	Source	Reaction type	Application
		Fat, triglyceride hydrolysis, transesterification, organic synthesis	
Acetylesterases			Amino acid, speciality production
Hydantoinases/carbamoylases		Hydrolysis of hydantions	Amino acid production
Transferases			
Cyclodextrin transferases	*E. coli*	Glycosyl transfer/cyclization	Cyclodextrin production
Transglutaminase			Meat processing
Oxidoreductases			
Glucose oxidase	*Aspergillus, Penicillium*	Oxidation of glucose	Glucose analysis; elimination of oxygen in juices, beer, wine; baking
Catalase	*Aspergillus, Micrococcus*	Degradation of hydrogen peroxide	(As before, together with glucose oxidase)
Lipoxygenase	Soybeans		Bread manufacturing

a) In many cases, mixtures with different enzymes or activities.
b) Mixtures, in general, offered as pectinases.

major importance, and in this respect one or more of the following aspects may play a role:

- modified or new products (e.g., sweeteners, bakery ingredients, confectionery);
- nutritive value, functional food;
- taste, texture;
- physical properties (water activity, viscosity); and
- economics.

Starch hydrolysis and processing is one of the most important processes and incorporates a wide range of different products (see Section 7.3.2). Furthermore, microbial amylases and other glucanases are used in brewing and alcohol production, in addition to malt enzymes.

7.3.1.1 Food Applications

Bread manufacture has become an important field for enzymes such as α-amylases, xylanases as dough conditioners, glucose oxidase, lipases, and lipoxygenase. The different activities act synergistically, as is the case with most complex substrates (e.g., cellulolytic enzymes, pectinases). They affect dough quality, bread texture, volume and crumb structure, antistaling, and so on (Aehle, 2004). The motive for this is the improved processing of doughs, a better or more pleasant structure and volume of the bread, and also improved storage quality (antistaling effect, maltogenic α-amylase). The amounts applied are low, in the range of 0.1–10 g enzyme per 100 kg of flour. Furthermore, proteases are applied when there is a need to partially hydrolyze the gluten present – this makes the dough less viscous, and its processing more easy. Lipases, acting as dough conditioners, can replace emulsifiers by promoting emulsion formation. Glucose oxidase (including catalase) is used for strengthening doughs by oxidation of the free sulfhydryl units and the formation of disulfide linkages in gluten protein. Granulation of the enzyme preparations prevents dust formation during application. All of these enzymes are inactivated at elevated temperatures during the baking process. Similar enzyme mixtures, including amylases, glucoamylases, cellulases, xylanases, and proteases, are applied in brewing processes. In addition, a rapidly growing market with a considerable volume (>100 million €) has recently developed for enzymes applied to feed production, especially β-glucanases, endoxylanases, and phytases. In cheese manufacturing, rennet or chymosin (or related proteases) is used to coagulate the milk protein, while further proteases and lipases may be added in order to improve the taste. Lactase (EC 3.2.1.108) provides the hydrolysis of lactose in milk and soft ice products for lactose-intolerant consumers (O'Rourke, 1996; Aehle, 2004; Schäfer *et al.*, 2007; Soetaert and Vandamme, 2010; Monsan and O'Donohue, 2010). Further to be mentioned are enzymes for flavor production (Cheetham, 2010).

For fruit and vegetable processing, a broad range of pectinases including hemicellulases (glycanases) are used in order to improve yields and/or to produce pureés by maceration and accelerating the filtration and pressing stages (cf. Grassin and Fauquembergue, 1996; Benen and Visser, 2003; Benen, Voragen, and Visser, 2003; Benen *et al.*, 2003). Pectic enzymes in fruit processing provide higher juice yields, increased filtration rates, and improved color (e.g., in red wine). Furthermore, they assist in the clarification of juices or the cloud stabilization of pulpy drinks. Other applications are in processing of vegetables for maceration products, pastes or pureés of tomatoes, carrots, or potatoes, and the preparation of pectins for specific applications and further fields of the food industries (Benen and Visser, 2003; Benen, Voragen, and Visser, 2003; Benen *et al.*, 2003).

For application, it is advisable to have an insight into the substrate's composition and structure. The structure of pectic material is highly complex. The fibrils of the plant cell wall are linked to other polysaccharides, such as xylan, xyloglucan, araban, arabinogalactan, and galactan (often termed hemicelluloses). These in turn are cross-linked to pectic substances, highly heterogeneous polymers (Schols and Voragen, 2003). As is typical for many plant substrates processed in the food industries, a range of different enzyme activities is required to act synergistically – that is, to function much more efficiently in complex than in singular activity. The most important activities of pectinase preparations comprise pectin methyl esterases, endo-pectinases, endo-pectin lyases, endo-polygalacturonases, endo-pectate lyases, and rhamnogalacturonan hydrolases (process details have been provided by Uhlig (1998)).

Saccharide and oligosaccharide synthesis by aldolases is dealt with in Section 4.2.4.2. For Leloir glycosyltransferases, see Wong (2005). The modification and synthesis of sugars, oligosaccharides, and derivatives relevant in food and nutrition, and in pharmaceutical industries (synthesis of vaccines, cancer proliferation) will be discussed in Section 8.4.1; industrially established processes with immobilized enzymes such as glucose isomerization and oligosaccharide and cyclodextrin production with glucosyltransferases will also be discussed in Section 8.4.1.

Further important enzymatic routes to different products are dealt with in other sections of this book, including amino acid synthesis (Sections 4.2.1.1, 4.2.3.7, 4.2.5, and 8.4.3.1), semisynthetic antibiotics (Sections 2.7 and 11.2), and a wide range of synthetic routes (Chapter 4).

7.3.1.2 **Other Industrial Applications**

A remarkable amount of enzymes – some 10% of the market share – is purchased by the *textile industries*. For example, adhesive sizes are used in order to protect fabrics during weaving, and these are starch-based in most applications. Desizing after machining requires that the starch is solubilized, and this is done using amylases of different origin. Surfactants must also be added during desizing, and these must be largely nonionic with small amounts of cationic types in order not to damage the enzymes; $CaCl_2$ must also be added to provide an appropriate enzyme performance. Generally, both conventional and thermostable bacterial amylases are applied in amounts of 0.025 up to 0.25% (w/v), during 2–4 h, in the temperature range of 75–80 °C (Godfrey, 1996; Schäfer *et al.*, 2007).

Another major application of enzymes is in *paper and pulp processing*. The motives here are to produce a cost-effective and environmentally benign technology, one precondition being an improved understanding of the interactions between enzymes, materials (substrate), and the process. Mixtures of cellulases and hemicellulases are used to increase the rate of pulp dewatering by the partial solubilization of fines and the cleaning of small fibrils on the cellulose fiber surface, thus improving the process economics. Furthermore, xylanases are used to aid bleaching of kraft pulp, where lignin is removed from the cellulose fiber, and to reduce the use of chlorine-based bleaching compounds. Lipases are used for pitch removal, and levan hydrolases for slime removal in paper making (Tolan, 1996, 2010; Bajpai, 1999).

A much larger number of enzymes are applied in analytical procedures and diagnostics, but these are not included in the statistics of technical enzymes. Although the total amount produced is perhaps rather low, their purity and high added value means that their economic relevance is rather high. The leading company for diagnostics, Roche AG, Division Diagnostics (Basel, Switzerland), had sales of some $9.7 billion in 2008 (Roche, 2008; Kresse, 1995).

Following this survey, a few selected examples will be treated in more detail. Aspects concerning reaction mechanisms are referred to in order to highlight their role with respect to selectivity and by-product formation. The reaction conditions (concentrations, pH, temperature, reaction or residence time, and effectors such as metal ions) are described for some examples as these play a key role in process and product stability, and also in the process economics. These parameters reflect the complexity of the process, while their treatment and optimal design require insight into basic scientific aspects as well as a broad empirical knowledge and experimental experience.

7.3.2 Starch Processing

Starch is the most important carbon and energy source among plant carbohydrates, and it is second following cellulose, at 10^{10} tons per annum, in total biosynthesis. Its industrial production is about 48×10^6 t a^{-1} (EU, about 10×10^6 t a^{-1}). The most important starch sources are corn (maize, 3.6×10^6 t a^{-1} starch in the EU), wheat, and potato (Vorwerg, Radosta, and Dijksterhuis, 2005). Starch is composed of two polysaccharides: 15–30% is amylose, a linear polyglucan with α-D-(1→4)-linked glucopyranosyl units (up to 6000) and a molar mass ranging from 3×10^5 up to 10^6 Da (Figure 7.3a); special corn types ("high-amylose") have been developed that contain up to 80% amylose. Conventional starch is composed of 70–85% amylopectin, a polyglucan with main chains as in amylose, with α-D-(1→4)-linked glucopyranosyl units, but with side chains linked by α-D-(1→6)-glycosidic bonds to the main chain, which has about 5% branch points (Figure 7.3b). The molar mass of amylopectin ranges from 16 to 160×10^6 Da, corresponding to about 10^6 glucose units. Seeds or tubers contain starch as compact granules of several μm diameter, with partially crystalline amylose and amylopectin, which are essentially inaccessible to enzymes (Robyt, 1998). However, when heated in water to about 60 °C, these granules take up large amounts of water and begin to swell, being transformed into a gel, and strongly increasing the viscosity of the suspension.

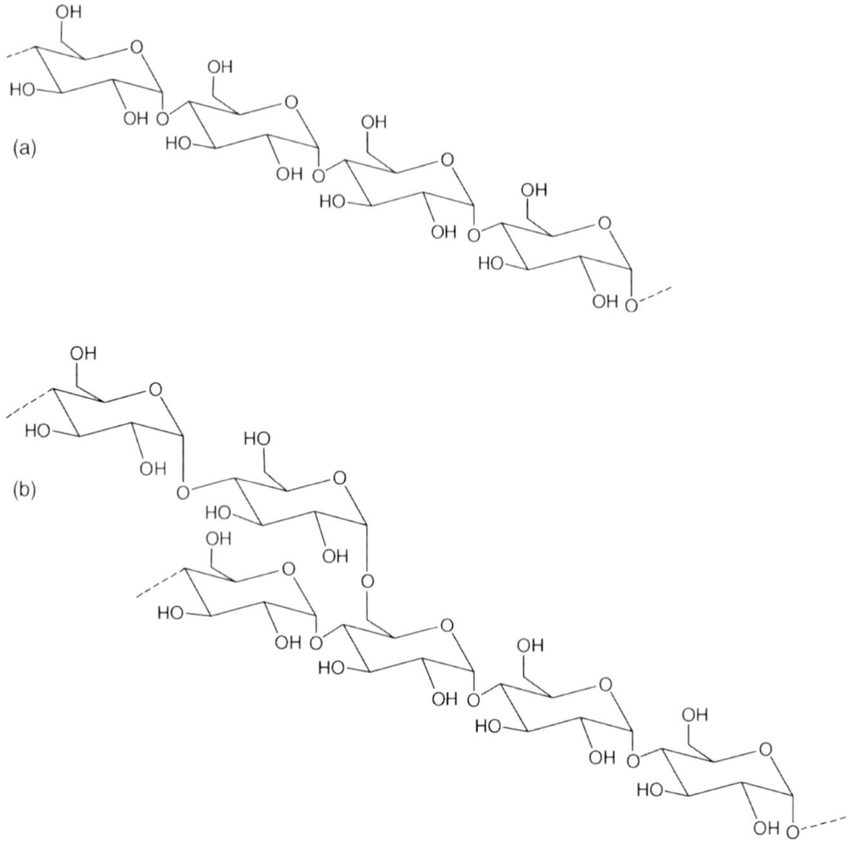

Figure 7.3 Structure of starch molecules: (a) amylose; (b) amylopectin.

Eventually, under shear forces the granules disintegrate, and it is only after this transformation that starch can be efficiently hydrolyzed by enzymes. The degree of hydrolysis is expressed as the DE (dextrose equivalent) value, which is determined by the reducing groups (equivalent to the amount of hydrolyzed bonds) taken as glucose equivalents per dry mass. Hence, DE 100 is pure glucose, and DE 0 is unhydrolyzed long-chain starch.

A low degree of hydrolysis (DE 10–15) results in so-called "soluble starch," with a considerably lower viscosity than that of the original solution. Oligosaccharide fractions with 4–20 glucosyl units are called dextrins. Such products are obtained by acid or α-amylase hydrolysis in one step from starch. Further hydrolysis requires two enzymatic steps involving α-amylase and glucoamylase. Syrups with higher degree of DE up to those with essentially pure glucose, which can be crystallized to yield crystalline glucose, can thus be produced. However, these require additional downstream processing, since besides low amounts of oligosaccharides from starch (i.e., maltose), other oligosaccharides are formed by condensation. The formation of such "reversion" (condensation) products is favored at high sugar concentrations, with preferential formation of an α-(1→6)-glycosidic bond, leading to isomaltose and higher isomaltooligosaccharides. This is due to the high concentration of the reaction solutions (>30%) required for economic reasons.

Thus, a wide range of different products such as starch hydrolysates, maltose (with β-amylase), glucose, high-fructose corn syrups (HFCS), and cyclodextrins are available commercially (Tramper and Poulson, 2000; Schäfer *et al.*, 2007; Bentley, 2010). Enzymes for starch processing, encompassing α-amylase, glucoamylase, β-amylase, pullulanase, isoamylase, and glucose isomerase, comprise about 30% of the world's industrial enzyme production. Isoamylases and pullulanases exclusively hydrolyze α-(1→6) bonds (van der Maarel *et al.*, 2002). Cyclomaltodextrin glucanotransferases are remarkable enzymes (as other glucosyltransferases) insofar as they exhibit only very minor hydrolase activity, allowing only sugars as nucleophiles for attack of the covalently bound glucosyl group to synthesize oligosaccharides (cyclodextrins, etc.; see Section 8.4.1). Glucoamylase is produced by filamentous fungi, mainly *Aspergillus* sp., in higher tonnage than almost any other industrial enzyme, and thus it is available at low price. Data on enzyme properties, structure, mechanism, and kinetics are available (Sauer *et al.*, 2000; Reilly, 2003; Kelly *et al.*, 2009).

In starch processing, the first step, starch liquefaction, combines thermal treatment at temperatures beyond 105 °C (gelatinization) and hydrolysis by thermostable α-amylases (liquefaction) active in the range of 85–120 °C (for details see Section 12.1). This is followed by liquefaction by α-amylases at pH 6–7, 85–95 °C, giving maltodextrins of DE 10–15 as trade products. Further hydrolysis to produce a glucose syrup (saccharification) is performed with glucoamylase and additional pullulanase or isoamylase at pH 4.3–4.5 and 55–60 °C.

Major efforts have been devoted to research on thermostable α-amylases with much success, revealing structural determinants responsible for the high thermostability of *Bacillus* enzymes (Declerck *et al.*, 2000). The results of stability studies on barley α-amylase isoenzymes suggested that stabilization by electrostatic interactions might play a role (Svensson, Bak-Jensen, and Mori, 1999). Two major problems were the motives for extensive research to be conducted into the genetic modification of glucoamylase. First, the formation of reversion side products (see Section 2.4) at high dissolved solid concentrations typically applied in technical starch processing, leading to the formation of di-, tri-, and tetrasaccharides, the most important being isomaltose. Second, glucoamylase is not as stable as α-amylase and glucose isomerase, being used at 60 °C in starch processing. Combining favorable mutations was successful in decreasing the kinetics of the enzyme toward the formation of isomaltose, increasing its glucose yield from 96 to 97.5% (equivalent to 200 000 extra tons at actual production scale) (Reilly, 1999). One condition for the successful design of mutations of single amino acids in glucoamylase is advanced knowledge of its structure and function. Thus, stiffening α-helices by Gly→Ala and Ser→Pro mutations, as well as creating disulfide bonds across two loops, contributed to a fourfold increase in the thermostability of *A. niger* glucoamylase (Reilly, 1999; Suvd *et al.*, 2001).

Genetic engineering and recombinant technologies thus had significant impact on enzyme technology, favored by new insights on structures and mechanisms (Kelly *et al.*, 2009) (see Section 2.4 and Chapter 3):

- creating a tremendous increase in the productivity of enzyme fermentation with recombinant organisms;
- creating a significant improvement in thermostability (α-amylases, glucoamylases), by screening of thermophilic organisms, directed evolution, and rational design via site-directed mutagenesis; and
- providing process optimization by improved selectivity (fewer side products) and reduced or no Ca^{2+} dependence (α-amylase).

7.3.3
Detergents

The economically most important application of technical enzymes – besides starch hydrolysis – is that in washing powder (or liquid) by the detergent industries, worth about 850 million € (see Figure 1.5 and Table 7.1). Eight enzymes are typically included in one detergent formulation, essential for removal of complex stains: two proteases, two amylases, two cellulases, one lipase, and one mannanase. Also the formula of automatic dishwasher detergents was adjusted to allow the use of enzymes, mostly proteases and amylases, that lead to improved performance (Maurer, 2010; Novozymes, 2011, www.novozymes.com/, www.biotimes.com).

The enzymes are applied, with optimal activity under alkaline conditions (up to pH 11) at moderate or elevated temperature (30–90 °C), corresponding to the requirements of washing procedures (Figure 7.4). They hydrolyze proteins from

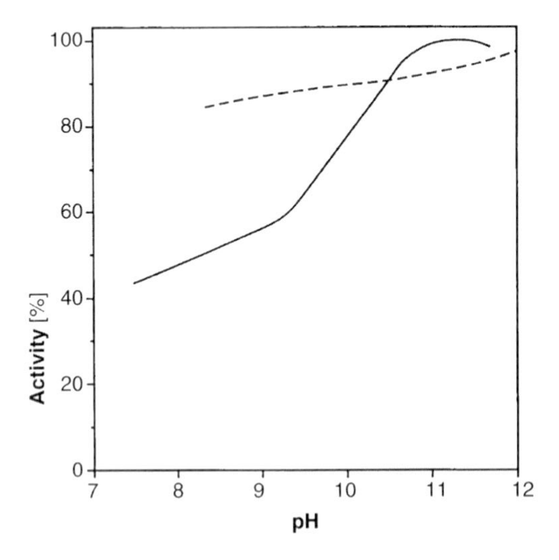

Figure 7.4 Activity (%) of an alkaline protease as a function of pH (substrate: casein, reaction at 40 °C, solid line; hemoglobin (denatured), reaction at 25 °C, broken line). (after International Bio-Synthetics, 1988).

milk, starch, fat from butter, and other components of food, thereby exhibiting stain removing effects. Typical protein-associated stains are, for example, blood and serum stains, egg stains, and milk-containing stains, including chocolate and coffee/milk mixtures; starch acts as an adhesive for colored stains such as sauce, pudding, or mousse au chocolat.[2)]

Practically, all proteases actually used until today belong to the group of subtilisins (E.C. 3.4.21.62), originating from *Bacillus* species. These extracellular serine peptidases are characterized by broad pH profiles from neutral to highly alkaline conditions, broad substrate specificity, and sufficient process and storage stability in the presence of surfactants. Cellulases exhibit fiber effects that can be described as polishing effects on the cotton fibrils resulting in color freshening, and removal of pilling (hydrolyzing the ends of broken fibers that scatter light) resulting in anti-graying effects.[3,4)] Lipases could be introduced only after recombinant production systems were available; they have been used with success in laundry detergents, where they contribute significantly to the removal of stains containing natural fats and oils, for example, triglyceride lipstick stains and serum. Mannanases are used to remove mannan in colored food components such as sauces, ice cream, and chocolate (Maurer, 2010).

All relevant technologies for enzyme engineering have been applied for development, and recombinant systems are used for production, including site-directed mutagenesis, error-prone PCR, gene shuffling, and directed evolution, synthesis of genes with randomization at certain positions, and metagenomic screening, with dramatic improvements in the bacterial or fungal production systems[5)] (Shaw and Bott, 1996; Maurer, 2010) (see also Chapter 3). The principal requirement is a low specificity of the enzyme for acting on the broad diversity of the substrates, and at the same time a high specific performance over the full range of temperature applied from cold to high-temperature washing conditions. Efforts to modify the substrate specificity or the flexibility of proteases have shown good success (Chaparro Riggers *et al.*, 2006). Destabilizing effects by anionic surfactants must be reduced by addition of nonionic surfactants, which exhibit a stabilizing effect.

A major success relates to lipases that hydrolyze fat (triglycerides) into a mixture of free fatty acids, di- and monoglycerides, and glycerol – all of which can be removed much more easily than fat. Lipases normally are not active during washing in the aqueous phase, due to a "lid" that must be opened to give access to a

2) In addition to stain removal, proteases can also show lytic activity on bacteria, viruses, or prions. These activities are considered as side effects with increasing hygienic relevance (Maurer, 2010).

3) Amylases and subtilisin proteases are coexpressed in *Bacillus* strains, and exhibit synergies in acting in laundry detergents. Most cellulases used in detergents are from fungal origin and are produced in fungal production strains (Maurer, 2010).

4) Novozymes claimed also the color refreshing effect of its cellulase product. Some cellulases are also being used in textile technology, like enzymatic stone washing (Novozymes, 2009, 2010; www.novozymes.com/, www.biotimes.com).

5) The success of the shuffling technology was not as overwhelming as expected, one reason being that the resulting enzyme molecules have to be free of third party's intellectual property – in a patent landscape as dense as in the field of detergent enzymes (Maurer, 2010).

substrate; opening occurs at nonpolar interfaces, and not in the aqueous phase (Eriksen, 1996). Removal of this "lid" by genetic engineering of lipases significantly improved their efficiency during the washing procedure (Svendsen, 2000). Further improvements were introduced by modified electrostatic surface charge, including a N-terminal extension with positive charges, in order to have activity in the presence of anionic detergents. The problem was solved by "classical" protein engineering. For improved thermostability, a rather broad spectrum of methods was applied, comprising computational chemistry with substrate docking, molecular dynamics, and localized random mutagenesis (error-prone PCR of relevant gene fragments, combined with doped DNA oligonucleotides) and family shuffling (see Chapter 3). The approach was successful and led to the production of a variant that was active up to and over 80 °C (Vind, 2002).

Detergent enzymes are produced in a small number of microorganisms. These are *Bacillus* species for bacterial amylases and proteases, and *Aspergillus oryzae* and *Trichoderma* species for fungal cellulases and lipases. Production is normally based on genetically modified production strains.[6] The enzymes are typically fermented, filtered (usually rotating vacuum filters), ultrafiltered for concentration of the enzyme, filtered again (sterile filtration to remove remaining cells from the production), and finally formulated (liquid enzymes typically in a salt/sugar formulation, dry preparations are granulated) (Figure 6.10). Detergent enzymes are sold as stabilized liquid concentrates or dust-free, coated (e.g., by polyethylene glycol) granulates.[7] Concentrations in granular preparations are as follows (%, w/w): proteases 0.04–0.06, amylases 0.013–0.04, cellulases 0.004–0.007, and lipases 0.002 (upper limits) (Maurer, 2010). Granulates are dust-free preparations, with an inclusion formulation of the protein inside a water-soluble inert coating. This technique is one of the most important unit operations of enzyme confectioning, where the formation of protein dust and contact by inhalation or with the skin, and thus allergic reactions, are avoided.

A recent trend is to create benefits in sustainability that can be seen in low-temperature washing performance, which allows to save heating energy, and reduction of chemicals, as is obvious from life cycle assessments. Reduced detergent consumption and washing at lower temperature have a significant effect on the carbon footprint and the energy consumption of washing.[8] Expectations are even targeting at a 50% replacement of surfactants (Maurer, 2010).

6) The use of recombinant high expression systems is essential for the economic production of enzymes with higher specificity and purity. By achieving high titers, the sustainability of enzyme production is increased in the terms of economy, consumption of energy and renewable resources, reduction of waste, and easier control of occupational health hazards (Maurer, 2010).

7) Several companies, including Novozymes (Denmark), BRAIN AG (Germany), and Henkel (Germany) found and applied enzymes with sufficient activity at lower temperature to allow washing at 40 and 30 °C, even at 15 °C, instead of 60 °C, thus with considerably lower energy requirement (Schäfer *et al.*, 2007; Maurer, 2010) (Novozymes, 2010; www.novozymes.com/, www.biotimes.com; information on production by genetically modified microorganisms).

8) Recent results suggest the potential of reducing the washing at lower temperature to even 15 °C with improved performance and significantly reduced environmental impact (CO_2 emissions) (Novozymes, 2010; www.novozymes.com/, www.biotimes.com).

7.4
Membrane Systems and Processes

Enzymes can be retained in, or recycled to, a reactor by means of semipermeable membranes, corresponding to immobilization (Figures 7.5 and 7.6). Ultrafiltration membranes are commercially available for such purposes with 5 or 10 kDa cutoffs, and these are impermeable to proteins of higher molar masses.

Membrane systems offer advantages for cofactor-dependent reactions, when the coenzyme must be regenerated. In general, two enzymes (E_1 and E_2) are required, which catalyze a coupled redox reaction, such as

$$AH_2 + NAD^+ \xrightarrow{E_1} A + NADH + H^+$$
$$B + NADH + H^+ \xrightarrow{E_2} BH_2 + NAD^+$$

Both enzymes and the cofactor must be retained in, or recycled to, the reactor for cost reasons, with a cutoff for enzymes >>99%. The efficiency of cofactor regeneration and recycling should be, for NADH (with a price in the range of 1 € g^{-1} and a product of moderate price, <25 € kg^{-1}), better than 99.9%. The total turnover number of such systems must be >1000 for high-price products, such as L-*tert*-leucine, and >10 000 for low-price products, such as L-leucine, in order to provide for economic conditions (Kula and Wandrey, 1987; Kragl *et al.*, 1996). The covalent

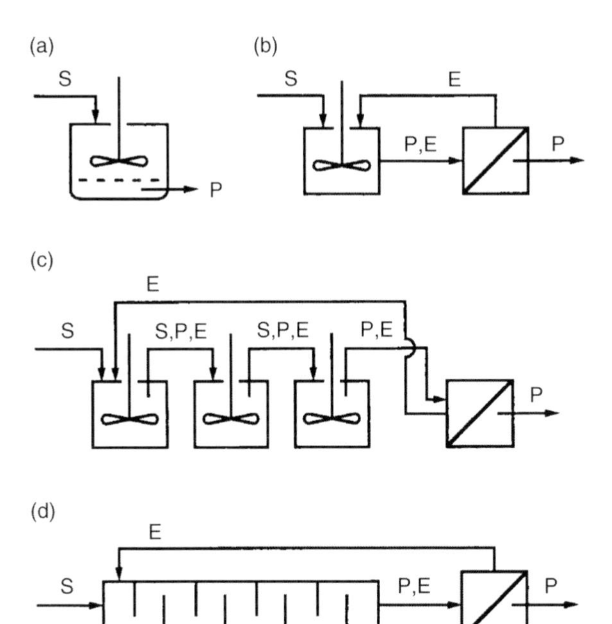

Figure 7.5 Principles of membrane systems. (a) Membrane reactor with integrated semipermeable membrane; (b) continuous stirred tank reactor; (c) cascade; (d) tubular reactor, each system with membrane module/unit and recycling of enzyme.

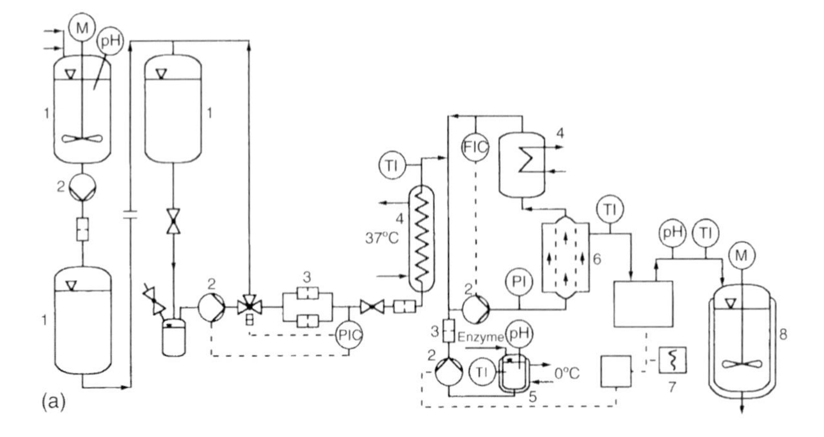

Figure 7.6 (a) Scheme of production system with membrane unit, with mixing and storage tanks (1), pumps (2), filters (3), heat exchangers (4), enzyme tank (5), ultrafiltration module (6), analytical monitors (7), product vessel (8), and control units for pH, temperature (TI), pressure (PIC), and volumetric flow (FIC) (regulated control: dashed and solid lines). (b) Unit for amino acid production. (reproduced with kind permission of Degussa, Hanau, Germany).

immobilization of such systems with two enzymes and a cofactor, which must access both active centers, is difficult and tedious, and thus economically not feasible.

The method of choice for coenzyme-dependent, two- or multienzyme systems is the application of membrane reactors, which retain or recycle all catalytically active components. The low molecular weight cofactors must be coupled to polymers of sufficiently high molecular weight. Polyethylene glycol (PEG) with a mean molar mass of 20 kDa serves as the soluble macromolecular carrier (Riva *et al.*, 1986). All types of reactor systems with membrane module may be applied (see Figure 7.5). In order to provide for maximum conversion rates and favorable enzyme efficiency, batch operations are favored; as alternatives, a cascade of continuous stirred tanks or a tubular reactor with a membrane module, which recycles the enzymes and the cofactor, may be used (Figure 7.5c and d). Combinations of stirred and tubular reactors with a membrane module should also be considered. A scheme of a production unit for continuous operation with peripheral vessels, valves, pumps, and control units is shown in Figure 7.6. Further, membrane systems, mostly membrane filtration units, are used with whole-cell biocatalysts, which perform two- or multistep reactions including cofactor regeneration (Table 7.6b).

Several membrane systems are available, with different membrane structures and module configurations (Figure 7.7). Details of these were discussed by Kula and Wandrey (1987), Tutunjian (1985), and Giorno and Drioli (2010). Usually composite membranes are used, with an extremely thin layer of homogeneous polymer

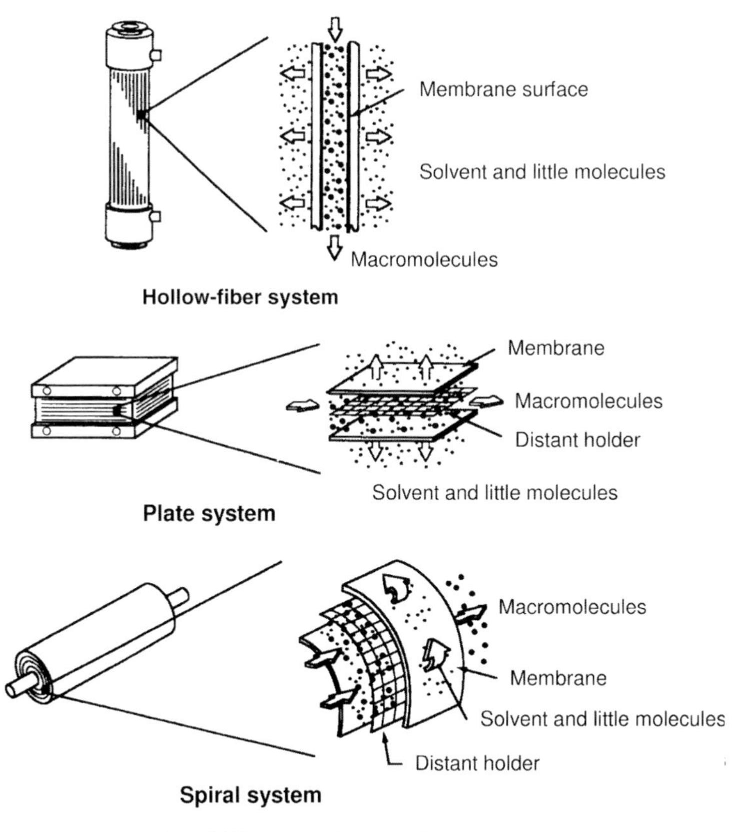

Figure 7.7 Examples of different membrane configurations (Tutunjian, 1985).

Table 7.6 Examples for industrial processes with membrane systems, from Liese, Seelbach, and Wandrey (2000, 2006).

(a) Processes with soluble enzymes and ultrafiltration modules

Enzyme system	Substrate	Product	Application	Company
Alcohol dehydrogenase/formate dehydrogenase[a)]	1-Phenyl-2-propanone	*S*-1-Phenyl-2-propanol	Intermediate for synthesis of amphetamines[b)]	
Lactate dehydrogenase/formate dehydrogenase[a)]	2-Oxo-4-phenylbutyric acid	2-Hydroxy-4-phenylbutyric acid	Precursor for different ACE inhibitors[c)]	Ciba-Geigy
Aminoacylase	D,L-*N*-Acetylmethionine[d)]	L-Methionine	Feed and food additive[b)]	Evonik (former Degussa)
Leucine dehydrogenase/formate dehydrogenase[a)]	Trimethylpyruvic acid	L-*tert*-Leucine	Building block for drug synthesis[b)]	Evonik (former Degussa)
Sialic aldolase[e)]	*N*-Acetyl-D-glucosamine + pyruvic acid	*N*-Acetyl-D-neuraminic acid	Building block for synthesis of oligosaccharides[f)]	Marukin Shoyu

(b) Processes with suspended cells and membrane filtration systems (see also Chapter 5)

Catalyst: whole cells	Substrate	Product	Application	Company
E coli (engineered)/styrene monooxygenase	Styrene (and derivatives)	(*S*)-Styrene oxide (and derivatives)	Building blocks for a range of products	DSM
Brevibacterium flavum/L-aspartate ammonia lyase	Fumaric acid	L-Aspartic acid	Synthesis of aspartame	Mitsubishi Chem. Corp.
Achromobacter xylosoxidans/nicotinic acid hydrolase	Nicotinic acid	6-Hydroxynicotinate	Building block for insecticides	Lonza AG

Arthrobacter sp./catechol dioxygenase[g]	Benzoic acid	*cis,cis*-Muconic acid[g]	Raw material for pharmaceuticals, agrochemicals	Mitsubishi Chem. Corp.
Nocardia sp./monooxygenase	Simvastatin	6-β-Hydroxymethyl simvastatin	Cholesterol-lowering drug	Merck & Co.
E coli/carnitine:NAD^+ 3-oxidoreductase[h]	4-Butyrobetaine	(*R*)-L-Carnitine (3-hydroxy-4-(trimethylamino)butanoate[h]	Infant, health, sport, geriatric nutrition	Lonza AG

a) Second enzyme for cofactor regeneration.
b) Also other products are manufactured via this route (Liese, Seelbach, and Wandrey, 2000, 2006).
c) ACE: angiotensin converting enzyme.
d) *N*-Acetyl-D,L-amino acid precursors are conveniently accessible through acetylation of chemically synthesized D,L-amino acids. After enzymatic deacylation, L-methionine is separated by ion-exchange chromatography and purified by crystallization; the D-component is recycled after racemization (Wöltinger *et al.*, 2005).
e) *N*-Acetyl-D-neuraminic acid aldolase.
f) Also for glycoprotein synthesis, important for studies on biological recognition processes.
g) Three-step reaction by catechol-oxygen 1,2-oxidoreductase with NADH and NAD^+ regeneration.
h) Multistep reaction.

supported on a thick supporting substructure. Reaction systems with more hydrophobic enzymes normally exhibit better filtration characteristics with hydrophilic membrane materials, such as regenerated cellulose or hydrophilized polyethersulfones; in contrast, more hydrophobic membrane materials such as polysulfone are preferred with hydrophilic enzymes (they do not easily form protein layers on the membrane since they are repelled). With respect to low investment cost and simple scale-up, hollow-fiber modules are often preferred for homogeneous reaction systems; for reaction mixtures containing high solid concentrations, for example, whole cells, tubular membranes or plate and frame modules are used. Scale-up may be performed by numbering up the filtration modules (Wöltinger *et al.*, 2005).

A considerable number of developments, notably with oxidoreductases or dehydrogenases (DH), have been published (Table 7.6a) (see also Section 4.2). Cofactor regeneration is essential. The cosubstrate used for regeneration should be soluble, nontoxic, noninhibitory, and readily available at low price. The product should be easy to separate and/or represent a valuable by-product. For the regeneration of NADH, formic acid or its salts are suited as it is cheap, the enzyme required (formate DH, FDH from *Candida boidinii*) is readily available, and the product (CO_2) can be easily separated (see Section 4.2.1.1, Figure 4.6, and Table 4.2).

Examples from a range of processes that are currently used both with soluble enzymes and ultrafiltration modules and with suspended cells and membrane (micro-)filtration systems, which provide for recycle of the catalyst, are given in Table 7.6a and b (Liese, Seelbach, and Wandrey, 2000, 2006). Most are two- or multistep reactions including cofactor regeneration.

As an example, L-*tert*-leucine is produced with LeuDH and FDH using a fed batch mode (compare below, Figure 7.8). Enzymes have been optimized by evolutionary or rational protein design, for example, site-directed mutagenesis of FDH, replacing oxidation-sensitive cysteines by serine and alanine (Wöltinger *et al.*, 2005). Considerable data have been published relating to the production of L-leucine from a keto acid (α-ketoisocaproate) with LeuDH in a membrane reactor (Kula and Wandrey, 1987; Kragl *et al.*, 1996) (Figures 7.8–7.10 and Table 7.7): the derivative of the coenzyme (PEG-NADH) remains in the reactor with high degree of retention ($R >$ 0.999%), where R is defined as

$$R = (C_r - C_f)/C_r,$$

where C_r and C_f are the concentrations in the retentate and filtrate, respectively (index 0 for $t = 0$); the loss by washout follows from

$$C_r/C_{r,0} = \exp - [(1 - R)/\tau),$$

where τ is the mean residence time.

The amounts of enzyme consumed and product obtained as a function of process time are shown in Figure 7.10. Results including concentrations, conversion, space–time yield, turnover number, and consumption of enzymes per product unit are listed in Table 7.7.

For further examples, see Section 4.2.1.1. Hydroxy acids can be produced following the same principle. (R)-2-Hydroxy-4-phenylbutyric acid is produced industrially

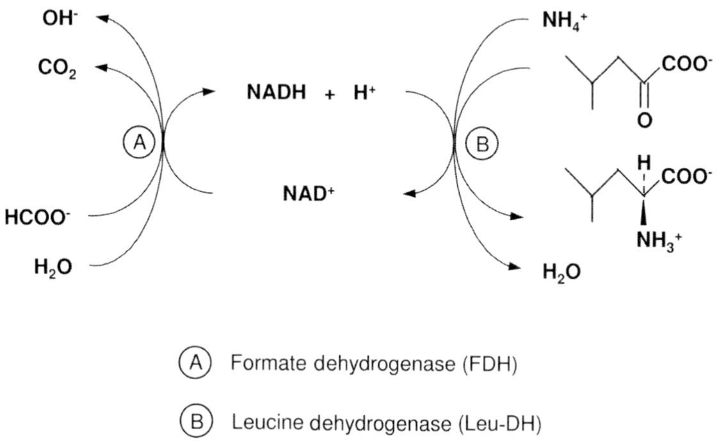

OH^-
CO_2
NADH + H^+
NH_4^+
COO^-
O
(A)
(B)
H
COO^-
NAD^+
$HCOO^-$
NH_3^+
H_2O
H_2O

(A) Formate dehydrogenase (FDH)

(B) Leucine dehydrogenase (Leu-DH)

Figure 7.8 Reaction scheme for continuous production of L-leucine from a keto acid (α-ketoisocaproate) with leucine DH (B), formate DH (A), and polymer-bound cofactor.

from 2-oxo-4-phenylbutyric acid using (*R*)-lactate-NAD oxidoreductase. Formate DH and formic acid are used for cofactor regeneration. The reaction is carried out in a continuous stirred tank reactor with an ultrafiltration membrane (cutoff 10 kDa; cf. Figure 7.5b) with 4.6 h residence time. At a conversion of 91% with 99.9% ee, the enzyme consumption is 150 U kg^{-1} lactate DH and 150 U kg^{-1} FDH

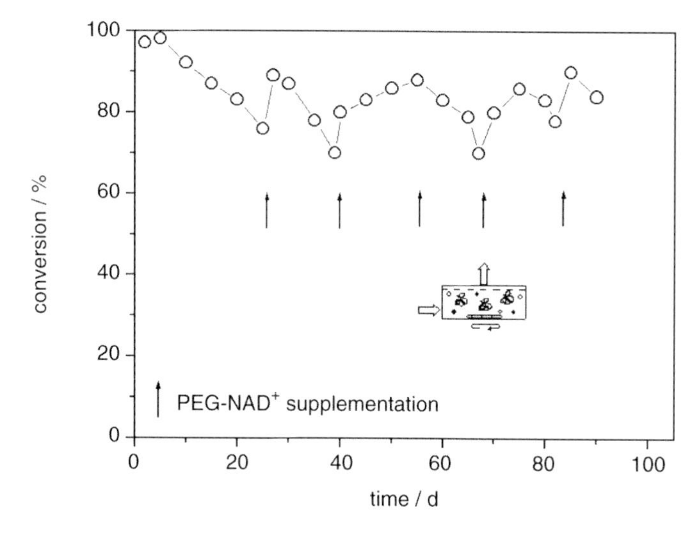

Figure 7.9 Continuous production of L-leucine in a single enzyme membrane reactor; conditions (feed concentrations): 100 mM α-ketoisocaproate, 400 mM ammonia, and 400 mM formic acid; LeuDH 2.5 U ml^{-1}, FDH 2.5 U ml^{-1}; 0.2 mM PEG-NAD$^+$; residence time 1.25 h. The arrows indicate supplementation of PEG-NAD$^+$ when the cofactor concentration decreased to 0.1 mM (Kragl *et al.*, 1996).

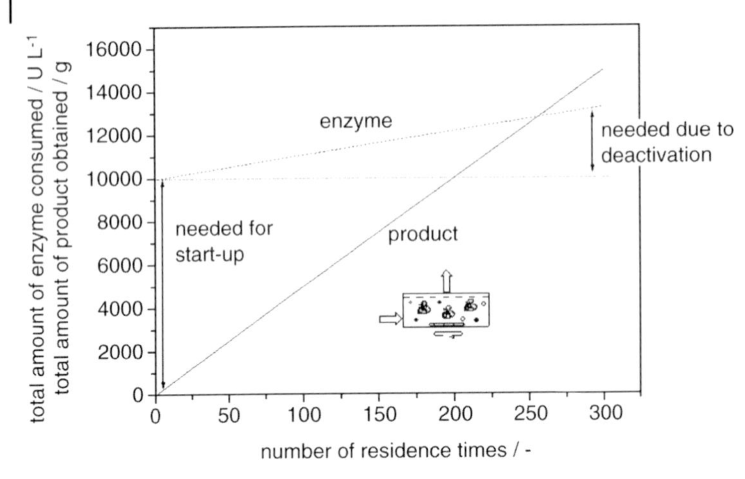

Figure 7.10 Total amounts of enzyme consumed and product obtained as a function of process time; example used for calculation: substrate 500 mM, enzyme deactivation 3% per day, molecular weight of product 100 g mol^{-1}. Residence time 1 h (Kragl *et al.*, 1996).

(Liese, Seelbach, and Wandrey, 2000, 2006). The stereochemistry, empirical rules, and further examples have been discussed earlier in Section 4.2. The advantages and disadvantages of membrane processes are summarized in Table 7.8.

However, for cases of high-cost enzymes, limited availability, and systems of high complexity (more than two enzymes with cofactors), the use of recombinant microorganisms or cells, optimized by metabolic engineering, may be more convenient as compared to membrane systems with isolated enzymes (Chapter 5).

Table 7.7 Conditions and results for the production of L-leucine (Kragl *et al.*, 1996).

Feed concentrations	
• Ketoisocaproate	100 mM
• Ammonia	400 mM
• Formic acid	400 mM
Reactor concentrations	
• PEG-NAD^+	0.2 mM
• LeuDH	2.5 U ml^{-1}
• FDH	2.5 U ml^{-1}
Reactor volume	10 ml
pH	8.0
Temperature	25 °C
Residence time	1.25 h
Mean conversion	85%
Space–time yield	214 g l^{-1} day^{-1}
Total turnover number	80 000
Consumption LeuDH	261 U kg^{-1} product
Consumption FDH	298 U kg^{-1} product

Table 7.8 Advantages and disadvantages of membrane processes.

Advantages	Disadvantages
No loss and cost for immobilization	Higher investment and more complex periphery
Production with constant conversion due to continuous enzyme dosage	Continuous pumping required
Easy exchange of substrate/enzyme systems	Tubular reactor characteristics require cascade reactor or column with internals
Homogeneous catalysis	Soluble enzymes are in general less stable as compared to immobilized ones
No mass transfer limitation	
Efficient use of multienzyme systems	

Exercises

Exercise 1: Production of Invert Sugar by Invertase

Invert sugar is to be produced in a technical process from sucrose by hydrolysis using invertase (E):

$$\text{Sucrose} + H_2O \xrightarrow{E} \text{Glucose} + \text{Fructose}$$

This is an exothermal reaction where the equilibrium is shifted to the right with increasing temperature (Table 2.7). No equilibrium constant for this reaction has been published. It can be assumed that the equilibrium is near 100% conversion for both temperatures.

A) Calculate the amount of enzyme required in terms of V_{max} for 95% conversion. Reaction temperature: 20 °C (no consideration of enzyme inactivation).

Conversion of 600 kg sucrose per m^3 ($1.75 \times 10^3\,\text{mol m}^{-3}$) by 95% to give a final concentration of $30\,\text{kg m}^{-3}$ ($0.0877 \times 10^3\,\text{mol m}^{-3}$) in 10 h ($3.6 \times 10^4$ s). Calculation of the conversion by means of Eq. (7.2):

$$[S]_0 - [S] + K_m \ln([S]_0/[S]) = V_{max}t,$$

$$K_m = 0.01\,\text{mol l}^{-1} \cong 10\,\text{mol m}^{-3}.$$

B) Compare the amount of enzyme required in terms of V_{max} at 20 and 50 °C. Reaction temperature: 50 °C (consideration of inactivation). Reaction conditions (Westphal, Vogel, and Pusch, 1988):

- Activation energy of the reaction: $E_A = 39.1\,\text{kJ mol}^{-1}$.
- Activation energy of the inactivation reaction: $E_I = 404.4\,\text{kJ mol}^{-1}$.
- Rate constant of inactivation: $\ln k_i = -48\,610(1/T) + 143.33$ (temperature, K; T (K) $= \vartheta$ (°C) $+ 273.15$).

Calculation of the mean enzyme activity considering inactivation (integral over reaction time t, divided by t):

$$V_{max} = V_{max,0}\frac{1 - e^{-k_i t}}{k_i t}.$$

From this follows the activity required at the start of the reaction:

$$V_{max,0} = V_{max} \frac{k_i t}{1 - e^{-k_i t}}.$$

The rate constant k_2 at 50 °C is higher as compared to k_1 at 20 °C, corresponding to

$$k_2/k_1 = (\exp - E_A/RT_2)/(\exp - E_A/RT_1).$$

C) Calculate the productivity of the enzyme catalyst in terms of kg (products) per kg (enzyme) (enzyme activity assumed: 300 U mg^{-1}).

Additional Questions

1. Would a further increase of temperature to 55 °C be advantageous? An increased temperature would be favorable in order to reduce the risk of contamination by infections.
2. Calculate the amount of enzyme required (kat m^{-3}) in order to convert sucrose (60 kg m^{-3}) to dextran by dextransucrase with 90% yield in 7 days; must substrate inhibition (K_i = 1.5 mol l^{-1}) be taken into account (cf. Demuth, Jördening, and Buchholz, 1999)?

$$K_m = 2.4 \times 10^{-2}\ \text{mol l}^{-1};$$
$$k_i = 7.1 \times 10^{-4}\ \text{h}^{-1}\quad \text{(inactivation, 20 °C)}.$$

Activation energies: reaction: 40 kJ mol^{-1}; inactivation: 200 kJ mol^{-1} (values assumed). Reaction temperatures: (a) 20 °C; (b) 40 °C.

Exercise 2: Calculation of Enzyme Inactivation under Different Conditions

Data for inactivation kinetics were obtained for two thermostable amyloglucosidases (GA), including stabilization by sorbitol as a model polyol (Feng *et al.*, 2002): *Thermoanaerobacterium thermosaccharolyticum* GA was stable up to 65 °C, k_i (min^{-1}) being 0.0043 (7.17×10^{-5} s^{-1}) at 70 °C. For *A. niger*, GA k_i values at 60, 65, and 70 °C were 0.018, 0.062, and 0.088 (3.0×10^{-4}, 1.03×10^{-3}, and 1.47×10^{-3} s^{-1}), respectively.

Questions

1. Compare the stability of the two enzymes at 70 °C, in terms of inactivation time for a 20% residual activity; give results in minutes, or hours, as appropriate.

   Sorbitol stabilized *A. niger* GA considerably at 65 °C, with k_i values of 0.020, and 0.0080 (3.33×10^{-4} and 1.33×10^{-4} s^{-1}) at 1 and 2 M sorbitol, respectively, while 4 M sorbitol gave complete stability over 40 min of incubation.
2. Calculate the half-life times and the factor of stabilization of *A. niger* GA by 2 M sorbitol at 65 °C.
3. For a peptidase ("glutenase"), inactivation data were given with the product description, where times for 90% inactivation in the presence of wheat

gluten were ~220 min at 60 °C and 15 min at 70 °C, whereas inactivation was rapid at 80 °C, as required in the baking process (Novo Nordisk, 1995).

What does "inactivation was rapid at 80 °C" mean in quantitative terms (minutes)?

Calculate k_i for 60 and 70 °C, the respective activation energy, k_i at 80 °C, and the time for 90% inactivation at that temperature.

Exercise 3: Hydrolysis of Penicillin with Penicillin Amidase

Penicillin G (300 mmol) is hydrolyzed by penicillin amidase (PA) at 37 °C and pH 7.8 ($K_m = 10$ μM) in 120 min to yield 95% of 6-APA and phenylacetic acid. Enzyme inactivation may be neglected if the pH is kept constant. Phenylacetic acid is a competitive inhibitor ($K_i = 20$ μM). It is an endothermal reaction; thus, the product yield increases with temperature (see Chapter 2.8). The optimal reaction temperature is limited by the temperature stabilities of the enzyme, substrate, and product.

Questions

1. How many units (U) of PA are required per kg of 6-APA? (prices: 6-APA ≈ 20 € kg^{-1}; ≈ 1 € per 100 U PA).
2. Is the process meaningful for biotechnical purposes? How many times must the enzyme be reused in order to bring the enzyme cost below 5% of the production cost?

The amount of enzyme required for a given conversion of substrate in a given time can be calculated from Eqs. (2.34) and (2.36) for different cases (integrated Michaelis–Menten equation for competitive product inhibition to be used).

Literature

General Information: Books, Reviews, and Encyclopedias

Aehle, W. (ed.) (2004) *Enzymes in Industry*, Wiley-VCH Verlag GmbH, Weinheim.

Atkinson, B. and Mavituna, F. (1991) *Biochemical Engineering and Biotechnology Handbook*, 2nd edn, Macmillan Publishers Ltd, New York, pp. 529–546.

Ballesteros, A., Plou, F.J., Iborra, J.L., and Halling, P.J. (eds) (1998) *Stability and Stabilization of Biocatalysts*, Elsevier, Amsterdam.

BRENDA, http://www.brenda-enzymes.org, www.empproject.com/links/.

Encyclopedia of Industrial Biotechnology, John Wiley & Sons, Inc., New York, http://tinyurl.com/wileyEIB.

Godfrey, T. and West, S.I. (eds) (1996) *Industrial Enzymology*, Macmillan Press, London, pp. 1–8.

Laane, C., Tramper, J., and Lilly, M.D. (eds) (1987) *Biocatalysis in Organic Media: Studies in Organic Chemistry*, vol. **29**, Elsevier, Amsterdam.

Liese, A., Seelbach, K., and Wandrey, C. (2000) *Industrial Biotransformations*, Wiley-VCH Verlag GmbH, Weinheim.

Liese, A., Seelbach, K., and Wandrey, C. (2006) *Industrial Biotransformations*, 2nd edn, Wiley-VCH Verlag GmbH, Weinheim.

Schäfer, T., Borchert, T.W., Nielsen, V.S., Skagerlind, P. *et al.* (2007) Industrial enzymes. *Adv. Biochem. Eng. Biotechnol.*, **105**, 59–131.

Soetaert, W. and Vandamme, E. (eds) (2010) *Industrial Biotechnology*, Wiley-VCH Verlag GmbH, Weinheim.

Straathof, A.J.J. and Adlercreutz, P. (2000) *Applied Biocatalysis*, Harwood Academic Publishers, Amsterdam.

Uhlig, H. (1998) *Industrial Enzymes and Their Applications*, John Wiley & Sons, Inc., New York.

Ullmann's Encyclopedia of Industrial Chemistry, Biocatalysts, Enzymes, Wiley-VCH Verlag GmbH, Weinheim, www.mrw.interscience.wiley.com/emrw/.

Whitaker, J.R., Voragen, A.G.J., and Wong, D.W.S. (eds) (2003) *Handbook of Food Enzymology*, Marcel Dekker, New York.

References

Adler-Nissen, J. (1986) *Enzymic Hydrolysis of Food Proteins*, Elsevier Publishers, Barking, UK, Appendix, p. 355.

Aehle, W. (2004) *Enzymes in Industry: Production and Applications*, Wiley-VCH Verlag GmbH, Weinheim.

Athes, V. and Combes, D. (1998) Effect of high hydrostatic pressure on enzyme stability, in *Stability and Stabilization of Biocatalysts* (eds A. Ballesteros, F.J. Plou, J.L. Iborra, and P.J. Halling), Elsevier, Amsterdam, pp. 205–210.

Bajpai, P. (1999) Application of enzymes in the pulp and paper industry. *Biotechnol. Prog.*, **15**, 147–157.

Benen, J.A.E., Alebeek, G.W.M., Voragen, A.G.J., and Visser, J. (2003) Pectic esterases, in *Handbook of Food Enzymology* (eds J.R. Whitaker, A.G.J. Voragen, and D.W.S. Wong), Marcel Dekker, New York, pp. 849–856.

Benen, J.A.E., and Visser, J. (2003) Polygalacturonases, in *Handbook of Food Enzymology* (eds J.R. Whitaker, A.G.J. Voragen, and D.W.S. Wong), Marcel Dekker, New York, pp. 857–866.

Benen, J.A.E., Voragen, A.G.J., and Visser, J. (2003) Pectic enzymes, in *Handbook of Food Enzymology* (eds J.R. Whitaker, A.G.J. Voragen, and D.W.S. Wong), Marcel Dekker, New York, pp. 845–848.

Bentley, I.S. (2010) Enzymes, starch conversion, in *Encyclopedia of Industrial Biotechnology*, John Wiley & Sons, Inc., New York, http://tinyurl.com/wileyEIB.

Bryjak, J. and Noworyta, A. (1994) Storage stabilization and purification of enzyme by water-soluble synthetic polymers. *Enzyme Microb. Technol.*, **16**, 616–621.

Buchholz, K. and Collins, J. (2010) Chapter 16, in *Concepts in Biotechnology – History, Science and Business*, Wiley-VCH Verlag GmbH, Weinheim.

Chaparro Riggers, J.F., Breves, R. *et al.* (2006) Modulation of infectivity in phage display as a tool to determine the substrate specificity of proteases. *ChemBioChem*, **7** (6), 965–970.

Cheetham, P.S.J. (1987) Production of isomaltulose using immobilized microbial cells, in *Methods in Enzymology*, vol. **136** (ed. K. Mosbach), Academic Press, New York, pp. S432–S454.

Cheetham, P.S.J. (2010) Enzymes for flavor production, in *Encyclopedia of Industrial Biotechnology*, John Wiley & Sons, Inc., New York.

Curtis, R.A., Ulrich, J., Montaser, A., Prusnitz, J.M., and Blanch, H.W. (2002) Protein–protein interactions in concentrated electrolyte solutions. *Biotechnol. Bioeng.*, **79**, 367–380.

Declerck, N. *et al.* (2000) Probing structural determinants specifying high thermostability in *Bacillus licheniformis* alpha-amylase. *J. Mol. Biol.*, **301**, 1041–1057.

de Cordt, S., Vanhoof, K., Hu, J., Maesmans, G., Hendrickx, M., and Tobback, P. (1992) Thermostability of soluble and immobilized α-amylase from *Bacillus licheniformis*. *Biotechnol. Bioeng.*, **40**, 396–402.

Demuth, B., Jördening, H.-J., and Buchholz, K. (1999) Modelling of oligosaccharide synthesis by dextransucrase. *Biotechnol. Bioeng.*, **62**, 583–592.

Eriksen, N. (1996) Detergents, in *Industrial Enzymology* (eds T. Godfrey and S.I. West), Macmillan Press, London, pp. 187–207.

Feng, P-.H., Berensmeier, S., Buchholz, K., and Reilly, P.J. (2002) Production, purification, and characterization of *Thermoanaerobacterium thermosaccharolyticum* glucoamylase. *Starch/Stärke*, **54**, 328–337.

Foster, K.A., Frackman, S., and Jolly, J.F. (1995) Production of enzymes as fine chemicals, in *Biotechnology*, vol. **9** (eds G.

Reed and T.W. Nagodawithana), Wiley-VCH Verlag GmbH, Weinheim, pp. 73–120.

Fullbrock, P.D. (1996) Practical limits and prospects (kinetics), in *Industrial Enzymology* (eds T. Godfrey and S.I. West), Macmillan Press, London, pp. 503–540.

Gibson, T.D., Hulbert, J.N., and Woodward, J.R. (1993) Preservation of shelf life of enzyme based analytical systems using a combination of sugars, sugar alcohols and cationic polymers or zinc ions. *Anal. Chim. Acta*, **279**, 185–192.

Giorno, L. and Drioli, E. (2010) Biocatalytic membrane reactors, in *Encyclopedia of Industrial Biotechnology*, John Wiley & Sons, Inc., New York, http://tinyurl.com/wileyEIB.

Godfrey, T. (1996) Textiles, in *Industrial Enzymology* (eds T. Godfrey and S.I. West), Macmillan Press, London, pp. 359–371.

Grassin, C. and Fauquembergue, P. (1996) Fruit juices, in *Industrial Enzymology* (eds T. Godfrey and S.I. West), Macmillan Press, London, pp. 225–264.

Ikeda, M. and Clark, D.S. (1998) Molecular cloning of extremely thermostable esterase gene from hyperthermophilic archeon *Pyrococcus furiosus* in *E. coli*. *Biotechnol. Bioeng.*, **57**, 624–629.

Illanes, A., Altamirano, C., and Aillapán, A. (1998) Packed bed reactor performance with immobilized lactase under thermal inactivation. *Enzyme Microb. Technol.*, **23**, 3–9.

Kelly, R.M., Dijkhuizen, L. *et al.* (2009) Starch and α-glucan acting enzymes, modulating their properties by directed evolution. *J. Biotechnol.*, **140** (3–4), 184–193.

Kragl, U., Kruse, W., Hummel, W., and Wandrey, C. (1996) Enzyme engineering aspects of biocatalysis: cofactor regeneration as examples. *Biotechnol. Bioeng.*, **52**, 309–319.

Kresse, G.B. (1995) Analytical uses of enzymes, in *Biotechnology* (ed. G.R.H.-J. Rehm), VCH Verlagsgesellshaft GmbH, Weinheim.

Kula, M.-R. and Wandrey, C. (1987) *Methods in Enzymology* (ed. K. Mosbach), vol. **136**, Academic Press, New York, pp. S9–S21.

Liese, A., Seelbach, K., and Wandrey, C. (2000) *Industrial Biotransformations*, Wiley-VCH Verlag GmbH, Weinheim.

Maurer, K.H. (2010) Enzymes, detergent, in *Encyclopedia of Industrial Biotechnology*, John Wiley & Sons, Inc., New York, http://tinyurl.com/wileyEIB.

Monsan, P. and Combes, D. (1984) Effect of water activity on enzyme action and stability. *Ann. N. Y. Acad. Sci.*, **434**, 48–60.

Monsan, P. and O'Donohue, M.J. (2010) *Industrial Biotechnology in the Food and Feed Sector*, Wiley-VCH Verlag GmbH, Weinheim.

Morgan, J.A. and Clark, D.S. (2004) Salt-activation of nonhydrolase enzymes for use in organic solvents. *Biotechnol. Bioeng.*, **85**, 456–459.

Novo Nordisk (1995) *Glutenase*, B 836a-D 2000.

O'Rourke, T. (1996) Brewing, in *Industrial Enzymology* (eds T. Godfrey and S.I. West), Macmillan Press, London, pp. 103–131.

Polizzi, K.M., Bommarius, A.S. *et al.* (2007) Stability of biocatalysts. *Curr. Opin. Chem. Biol.*, **11** (2), 220–225.

Prenosil, J.E., Dunn, I.J., and Heinzle, E. (1987) *Biotechnology*, vol. **7a** (eds H.-J. Rehm and G. Reed), Verlag Chemie, Weinheim, pp. 489–545.

Quax, W.J., Mrabet, N.T., Luiten, R.G.M., Schuuhuizen, P.W., Staussens, P., and Lasters, I. (1991) Enhancing the thermostability of glucose isomerase by protein engineering. *Biotechnology*, **9**, 723–742.

Reilly, P.J. (1999) Protein engineering of glucoamylase to improve industrial performance – a review. *Starch/Stärke*, **51**, 269–274.

Reilly, P. (2003) Glucoamylase, in *Handbook of Food Enzymology* (eds J.R. Whitaker, A.G.J. Voragen, and D.W.S. Wong), Marcel Dekker, New York, pp. 727–738.

Riva, S., Carrea, C., Veronese, F.M., and Bückmann, A.F. (1986) Effect of coupling site and nature of the polymer on the coenzymatic properties of water-soluble macromolecular NAD derivatives with selected dehydrogenase enzymes. *Enzyme Microb. Technol.*, **8**, 556–567.

Robyt, J.F. (1998) *Essentials of Carbohydrate Chemistry*, Springer, New York.

Roche (2008) Geschäftsbericht Business Report. F. Hoffmann La-Roche AG, Basel.

Sauer, J. *et al.* (2000) Glucoamylase: structure/function relationships, and protein engineering. *Biochim. Biophys. Acta*, **1543**, 275–293.

Schäfer, T., Borchert, T.W., Nielsen, V.S., Skagerlind, P. *et al.* (2007) Industrial enzymes. *Adv. Biochem. Eng. Biotechnol.*, **105**, 59–131.

Schols, H.A. and Voragen, A.G.J. (2003) Pectic polysaccharides, in *Handbook of Food Enzymology* (eds J.R. Whitaker, A.G.J. Voragen, and D.W.S. Wong), Marcel Dekker, New York, pp. 829–843.

Seibel, J., Jördening, H.J. *et al.* (2010) Extending synthetic routes for oligosaccharides by enzyme, substrate and reaction engineering. *Adv. Biochem. Eng. Biotechnol.*, **120**, 1–31.

Shaw, A. and Bott, R. (1996) Engineering enzymes for stability. *Curr. Opin. Struct. Biol.*, **6** (4), 546–550.

Sicard, P.J., Leleu, J.-B., and Tiraby, G. (1990) Toward a new generation of glucose isomerase through genetic engineering. *Starch*, **42**, 23–27.

Suvd, D., Fujimoto, Z., Takase, K., Matsumura, M., and Mizuno, H. (2001) Crystal structure of *Bacillus stearothermophilus* alpha-amylase: possible factors of the thermostability. *J. Biochem. (Tokyo)*, **129**, 461–468.

Svendsen, A. (2000) Lipase protein engineering. *Biochim. Biophys. Acta*, **1543**, 223–238.

Svensson, B., Bak-Jensen, K.S., and Mori, H. (1999) The engineering of specificity and stability in selected starch-degrading enzymes, in *Recent Advances in Carbohydrate Bioengineering* (ed. H.J. Gilbert), The Royal Society of Chemistry, Cambridge, pp. 272–281.

Tolan, J.S. (1996) Pulp and paper, in *Industrial Enzymology* (eds T. Godfrey and S.I. West), Macmillan Press, London, pp. 327–338.

Tolan, J.S. (2010) Enzymes, pulp and paper processing, in *Encyclopedia of Industrial Biotechnology*, John Wiley & Sons, Inc., New York, http://tinyurl.com/wileyEIB.

Tramper, J. and Poulson, P.B. (2000) Enzymes as processing aids and final products, in *Applied Biocatalysis* (eds A.J. J. Straathof and P. Adlercreutz), Harwood Academic Publishers, Amsterdam, pp. 55–91.

Tutunjian, R.S. (1985) *Comprehensive Biotechnology*, vol. **2** (ed. M. Moo-Young), Pergamon Press, Oxford, pp. S411–S437.

Uhlig, H. (1998) *Industrial Enzymes and Their Applications*, John Wiley & Sons, Inc., New York.

van der Maarel, M.J., van der Veen, B., Uitdehaag, J.C., Leemhuis, H., and Dijkhuizen, L. (2002) Properties and applications of starch-converting enzymes of the α-amylase family. *J. Biotechnol.*, **94**, 137–155.

Vind, J. (2002) Optimization of fungal lipase for diverse applications, *Engineering Enzymes*, Preprints, Paris, p. 31.

Vorwerg, W., Radosta, S., and Dijksterhuis, J. (2005) Technologie der Kohlenhydrate – 2: Stärke, in *Chemische Technik: Prozesse und Produkte*, vol. **8** (eds R. Dittmeyer, W. Keim, G. Kreysa, and A. Oberholz), Wiley-VCH Verlag GmbH, Weinheim, pp. 355–406.

Weemaes, C. *et al.* (1996) High pressure, thermal, and combined pressure–temperature stabilities of α-amylases from *Bacillus* species. *Biotechnol. Bioeng.*, **50**, 49–56.

Wenda, S. Illner, S. *et al.* (2011) Industrial biotechnology – the future of green chemistry? *Green Chem.*, **13** (11), 3007–3047.

Westphal, G., Vogel, J., and Pusch, D. (1988) Prozeßberechnung der enzymatischen Saccharosehydrolyse in Abhängigkeit von der Temperatur. *Acta Biotechnol.*, **8**, 357–365.

Wöltinger, J. Karau, A. *et al.* (2005) Membrane reactors at Degussa. *Adv. Biochem. Eng. Biotechnol.*, **92**, 289–316.

Wong, C.H. (2005) Protein glycosylation: new challenges and opportunities. *J. Org. Chem.*, **70** (11), 4219–4225.

Ye, W.N., Combes, D., and Monsan, P. (1988) Influence of additives on the thermostability of glucose oxidase. *Enzyme Microb. Technol.*, **10**, 498–502.

8
Immobilization of Enzymes (Including Applications)

Why immobilize biocatalysts?	Method for reuse and stabilization of biocatalysts
Immobilization of enzymes: in cells or as isolated enzymes?	For high enzyme density in a cell or when cofactor regeneration is required: immobilization in cells (see Chapter 9); otherwise immobilization of isolated enzymes
Important enzyme properties for immobilization	Protein surface properties (functional groups, ionic charge, and hydrophobic groups; see Sections 2.3 and 8.1.1)
Carriers for immobilization, properties	Porous systems, high internal surface area for adsorption or covalent binding; functionalization of carrier surface
Methods for immobilization on different carriers	• Adsorption: equilibrium (T, pH, I), prevention of desorption by cross-linking • Covalent binding: activation of the carrier, reaction with functional enzyme groups
Relevant aspects	Yield of active immobilized enzyme, kinetics and efficiency, stability
Examples	Selected technical applications

8.1
Principles

Insoluble enzymes offer all the advantages of classical heterogeneous catalysis: convenient separation for reuse after the reaction by filtration, centrifugation, and so on; application in continuous processes, in fixed-bed, fluidized bed, and stirred tank reactors provided with a filter system for retention. Continuous processes combine the generally more simple technical equipment with the potential for

Biocatalysts and Enzyme Technology, Second Edition. Klaus Buchholz, Volker Kasche, and Uwe T. Bornscheuer

easier process control, automation, and convenient coordination with up- and downstream processing, including product recovery and purification.

Today, more than 15 processes of major importance are in operation, in addition to over a hundred of further, specialized applications. Selected examples of industrial application are listed in Table 8.1, and these range from a production scale of several million down to a few hundred tons per year, albeit with high value-added, notably chiral, products (further examples are provided in Section 1.3.3, Table 1.2). The most important of these processes are to be found in the food and pharmaceutical industries, although the production of basic chemicals such as acrylamide and biodiesel is also included. For details of sales and costs, the reader is advised to consult Godfrey and West (1996), Cheetham (2000), Liese, Seelbach, and Wandrey (2006), Ghisalba *et al.* (2010), and Breuer *et al.* (2004).

The most obvious reason for immobilization is the need to reuse enzymes, if they are expensive, in order to make their use in industrial processes economic. The benefits and shortcomings of continuous processing are summarized in Table 8.2. Limitation by mass transfer in heterogeneous biocatalysts, and thus reduced efficiency of the enzyme, is another important topic that requires consideration and optimization (see Section 10.4).

The contact of an enzyme with the carrier surface may induce changes due to forces that cause interaction of surface-active groups, both of the matrix and of the protein. Conformational changes and inactivation have been observed in many cases. Detailed analyses including adsorption isotherms have been published (Norde and Zoungrana, 1998; Hartmann and Jung, 2010).

Numerous investigations have been reported on the stabilization of enzymes by their immobilization on insoluble carriers. Not all effects are understood on a molecular basis, but two are clear: (1) reduction of inactivation by protease hydrolysis, as immobilization decreases the access of peptidases to the targets of proteolysis; and (2) provision of stabilization of the enzyme's tertiary structure by a rigid matrix, which restricts unfolding to a certain degree. Otherwise, this subject is governed by empirical analysis and procedures. In recent years, continued efforts have resulted in commercial biocatalysts of high operational stability, such as glucose isomerases with productivities in the range of 12–15 tons (product) kg^{-1} (biocatalyst), and penicillin acylases with productivities of 200–2000 kg (product) kg^{-1} (biocatalyst) (corresponding to 400 000 kg product per kg enzyme immobilized), and operating lifetimes in excess of 3–6 months.

Enzymes may be immobilized by transformation into an insoluble form and retained – as a heterogeneous biocatalyst – inside a reactor by mechanical means, or by inclusion in a definite space (Figure 8.1). The second principle – inclusion by a semipermeable membrane – has been discussed in Section 7.4. The principles applied most frequently – binding to carriers – will be dealt with here in more detail. One-step reactions are the domain of immobilized enzymes, if not other factors, such as high molecular weight substrates, represent serious problems. The classical principles of binding enzymes onto carriers include physical adsorption, ionic binding to ion exchangers, and eventually cross-linking, as well as covalent coupling to an insoluble matrix (Figure 8.1). The last two procedures dominate in technical applications. Important examples applied on the large scale are glucose and sucrose

Table 8.1 Examples of immobilized enzymes used in major commercial processes (estimates) (Poulson, 1984; Cheetham, 2000; Liese, Seelbach, and Wandrey, 2006; Schmid *et al.*, 2002; Ghisalba *et al.*, 2010; Biermann *et al.*, 2011).

Enzyme (amount of immobilized biocatalyst used)	Product (amount)	Companies (enzyme suppliers; applicants)
Glucose isomerase (xylose isomerase) (about 1000 $t\,a^{-1}$)	Glucose–fructose syrup, about 12 million $t\,a^{-1}$	Novozymes, DSM, DuPont/Danisco/Genencor, Nagase; Archer Daniels Midland, A.E. Staley, Cargill, CPC International, and others
Sucrose mutase	Isomaltulose, about 100 000 $t\,a^{-1}$	Cerestar, Mitsui Seito Co., Südzucker
β-Galactosidase (lactase)	Glucose–galactose syrup, >6000 $t\,a^{-1}$	Central del Latte, Snow Brand Milk Prod., Sumimoto Chemical Industries, Valeo/Alko
Penicillin acylase (or amidase)[b)] (several $t\,a^{-1}$)	6-APA, >10 000 $t\,a^{-1}$	Asahi Chemical Ind. Co., DSM, Recordati, Sandoz, Toyo Jozo
(1) D-Amino acid oxidase; (2) glutarylamidase[b,c)]	7-ACA, 4000 $t\,a^{-1}$	Asahi Chemical Ind. Co., Sandoz, DSM, Sanofi Aventis, Toyo Jozo
Thermolysin	Aspartame,[d)] >10 000 $t\,a^{-1}$	Holland Sweetener Company (Tosoh and DSM)
Nitrilase	Acrylamide, 400 000 $t\,a^{-1}$; nicotinamide, 15 000 $t\,a^{-1}$[e)]	Nitto Chemicals, Lonza
Aminoacylase	L-Amino acids, for example, alanine, leucine, phenylalanine, L-DOPA, *tert*-leucine; several thousands of tons annually	Amino, Evonik, DSM, Fluka, Tanabe Seiyaku, and others
Hydantoinases	D-Amino acids, for example, *p*-hydroxy phenylglycine,	Ajinomoto, Bayer, Evonik, DSM, Kanegafuchi

(*continued*)

Table 8.1 (*Continued*)

Enzyme (amount of immobilized biocatalyst used)	Product (amount)	Companies (enzyme suppliers; applicants)
	phenylalanine, tryptophan, valine, about 1000 t a^{-1}	
Lipases	Modified oils and fats	Novozymes and others
	Diglycerides, >30 000 t a^{-1}	Kao Corp./ADM
	Structured triglycerides (cocoa butter equivalent)	Unilever, Fuji Oil, Novozymes
	Biodiesel, 20 000 t a^{-1}	(China)
	Pharmaceutical, cosmetic and agrochemical intermediates, for example, (*S*)-alcohols, (*R*)-1-phenylethylamine, multi-kg; >1000 t a^{-1} [g]	Novozymes, DSM, Genzyme Co., BASF, Bristol Myers Squibb, Fluka, Sumitomo Chemical Co., Tanabe Seiyaku Co., and others

a) Sucrose isomer: α-D-glucosyl-1,6-D-fructoside.
b) Mainly for semisynthetic penicillin and cephalosporin antibiotics (ampicillin, amoxicillin) with sales $>3 \times 10^9$ € a^{-1} (Cheetham, 2000, pp. 120–122); see Section 12.3.
c) For 7-ACA, a traditional two-step process is still in operation; a new one-step process using a modified glutarylamidase has been introduced, see Section 12.3.
d) Sales about 850×10^6 € a^{-1} (Cheetham, 2000, pp. 113–117).
e) See Section 4.2.3.6.
f) Most amino acids are produced now using whole-cell processes.
g) See Section 4.2.3.1.

Table 8.2 Reasons and limitations for enzyme immobilization.

Reasons	Limitations
Reuse of enzyme, reducing cost	Cost of carriers and immobilization
Continuous processing	Mass transfer limitations
• Facilitated process control	Problems with cofactors and regeneration
• Low residence time (high volumetric activity)	Problems with multienzyme systems
• Optimization of product yield	
Easy product separation and recovery	Changes in properties (selectivity)
Stabilization by immobilization	Activity loss during immobilization

isomerization, the manufacture of modified oils and fats, acrylamide production, penicillin and cephalosporin hydrolysis, synthesis of semisynthetic β-lactam antibiotics, and the production of some enantiomerically pure amino acids (see Table 8.1). Enzymes are either soluble or membrane bound. Immobilization of the cells clearly includes the enzymes listed (see Chapter 9), and this provides advantages over the immobilization of isolated enzymes for complex transformations that require more than one active enzyme and/or cofactor regeneration.

8.1.1 Parameters of Immobilization

As for other physical and chemical processes, the rate and yield of immobilization depend on the parameters involved, notably the type of carrier, method of immobilization, concentration, pH, temperature, and reaction time (Monsan, 1978; Buchholz, Duggal, and Borchert, 1979; Mosbach, 1987; Hartmann and Jung, 2010; Guisan, Betancor, and Fernandez-Lorente, 2010). Binding to insoluble porous carriers is the standard method for both laboratory and industrial applications. The

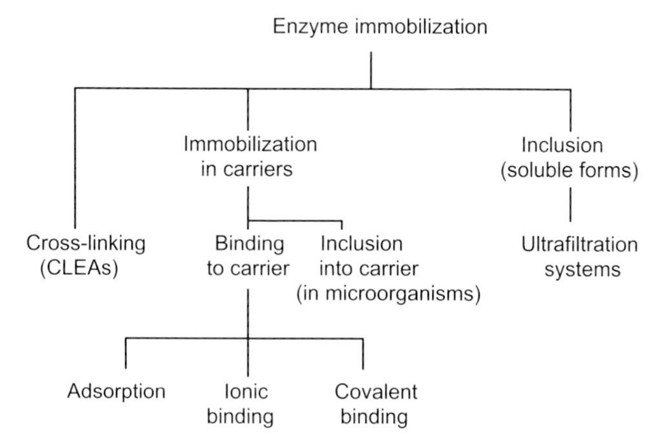

Figure 8.1 Principles of enzyme immobilization.

Table 8.3 Functional groups in proteins relevant for immobilization.

Functional groups	pK_a[a)]	Mean content in proteins (%)	Trypsin, tot. (access.)[b)] number	Chymotrypsin, tot. (access.)[b)]	Glucoamylase
Amino-					
Guanidino-					
Lysine-NH_2	10.5		13 (>11)	13 (>11)	8
Arginine-NH_2		7	3.8	2	3
Carboxyl-					
Aspartate-COO^-	3.9	4.8	24 (7)	21 (7–8)	38
Glutamate-COO^-	4.1	4.8			
Other					
Tyrosine-OH		3.4	10 (3)	4	12
Tryptophan-NH		1.2	4	7–8	12
Histidine-	About 7	3.4/2.2			4
Mercapto-					
Cystine-SH					
Oligosaccharides[c)]					

a) Values depend to some degree on the microenvironment.
b) tot: total content; (access.): accessible at the protein surface.
c) For enzymes from eukaryotes only.

properties of the external protein surface and the functional groups accessible play a major role in binding to carriers. Adsorption is dependent upon the hydrophilic and hydrophobic characteristics of the surface regions. The dominating ionic groups and their interaction depend on the amino acids, with pH-dependent charge and their density, which in turn determines the overall surface charge, which is influenced by the pH of the solution (Table 8.3). Most (up to 95%) of the titratable residues are located on the surface of soluble proteins, and about 40% are hydrophobic residues (Petersen *et al.*, 1998; Creighton, 1993). For covalent binding, several functional groups may be utilized, if they are accessible at the protein surface (Table 8.3). However, only a few are used in practice, notably the amino groups of lysine and arginine, and the carboxyl groups of aspartic and glutamic acid.

Stabilization of enzymes by immobilization is an important aspect, and common half-lives for industrial biocatalysts are in the range of 3–6 months, or more. However, kinetic data are rare.[1)] However, it is important to consider the strong interactions between the enzyme and the carrier surface with regard to ionic, hydrophilic and/or hydrophobic, and hydrogen bonding, and also with respect to enzyme

1) Inactivation rate constants for cephalosporin amidase gave roughly twofold reduced rate constants for the immobilized enzyme (from Hartmann and Jung (2010)). Stabilizing effects of immobilization, using multipoint covalent attachment, were determined fort D-amino acid oxidase by Dib and Nidetzky (2008). Dramatic effects for stabilization of pyruvate decarboxylase by adding glycerol and trehalose and by immobilization onto an ion-exchange polymer have been claimed by Matsuda *et al.* (2010).

stability. Several strong interactions (and their cooperative effects) may result in unwanted irreversible adsorption at the carrier surface and loss of enzyme activity. This may also be due in part to induced conformational changes of the tertiary structure of the protein. These effects have been observed notably for multiple interactions at rigid carrier surfaces.

The introduction of surface groups – known as "spacers" – in an appropriate density is common practice (see below, Figure 8.4), as these protect the enzyme against an aggressive surface that is rendered inaccessible and thus avoids inactivation. Similarly, the adsorption of a certain amount of inactive, cheap protein (e.g., bovine serum albumin) prior to enzyme immobilization has been found to be useful as a protective step. The specific functionality of the amino and carboxyl groups, and notably binding due to their reactivity and density at many protein surfaces (provided that they are not involved in the catalytic step), is apparent from the data listed in Table 8.3.

8.2 Carriers

Here, only a few examples taken from the overwhelming body of literature will be presented, according to their practical relevance. Methods have been extensively reported and summarized earlier (see General Literature, notably Mosbach, 1987; Bickerstaff, 1997; Buchholz, Duggal, and Borchert, 1979; see also Sheldon, 2007; Guisan, Betancor, and Fernandez-Lorente, 2010). Very few methods have found broad application both in the laboratory and on a technical scale.

The first option for the immobilization of enzymes may be realized by cross-linking, and this includes the techniques of cross-linking enzyme aggregates (CLEAs) (Tischer and Kasche, 1999; Sheldon, 2007; Sheldon, 2011). In addition, cross-linking of the microorganism that might provide the carrier, thereby leaving the singular enzyme required in an active form, has been successfully applied to glucose isomerase. Immobilization by matrix inclusion into polymeric networks is effective only with whole microorganisms, and is one of the most convenient methods available (see Chapter 9). (A single enzyme has a too small diameter and generally leaks out of the particles.)

The standard method for enzyme immobilization is binding to insoluble porous carriers, and this will form the focus of this section. Adsorption on the (internal) carrier surface, binding to ion exchangers, and eventually cross-linking to prevent desorption during operation, and covalent binding are all options used in the laboratory as well on the industrial scale. Some examples will be discussed in more detail in order to illustrate the potential of these methods.

The immobilization of an enzyme requires certain specific functions, technical applications, and on the other hand further qualities. A summary of the main properties of matrices that must be considered and serve specific purposes is listed in Table 8.4. A range of commonly used carriers was characterized with respect to particle size distribution, porosity, surface area, and skeletal density (Barros *et al.*, 2000; www.resindion.com).

Table 8.4 Properties of matrices relevant for enzyme immobilization.

Properties	Desired	
Chemical		
• Hydrophilicity/hydrophobicity		
• Swelling properties	Low	
• Chemical stability	High	
• Microbial stability	Good	
Morphological		
• Particle diameter, particle distribution	0.2–1 mm/narrow distribution	
• Pore size[a)]	30–60 nm	
• (Inner) surface for adsorption/binding	Large	
Mechanical		
• Resistance to pressure/compressibility	Good/low	
• Elasticity	Sufficient (no abrasion by stirrer)	
General	Food grade (for application in food manufacture)	
	Low cost	
	Lower limit	**Upper limit**
Pore diameter	20–40 nm	80–160 nm
Inner surface (corresponding)	250–125 $m^2 g^{-1}$	45–25 $m^2 g^{-1}$

a) Pore diameter and corresponding inner surface (example: porous glass): Trypsin with a molar mass of 23 800 Da requires a surface of about 15 nm^2.

The accessible surface inside the pores must be sufficiently large in order to accommodate the amount of enzyme required in a monomolecular layer providing for good activity of the biocatalyst. On the other hand, the pore diameter must be sufficiently large in order to allow the protein to diffuse into the internal volume. As the diameter of technical enzymes is in the range of 4–8 nm, pore sizes >20 nm are required for unrestricted diffusion. Since increasing pore diameter correlates with decreasing internal surface (Table 8.4), a compromise is required with respect to the capacity and transport by diffusion (see Section 10.4). The capacity of a carrier for the immobilization of an enzyme by adsorption can be determined by measuring the adsorption isotherm under optimal conditions (pH, T, buffer, ionic strength) (Chapter 6). In many cases, a Langmuir-type isotherm, suggesting a monomolecular protein layer on the internal matrix surface, has been observed.

Furthermore, the particles should preferentially exhibit a regular shape (in general spherical, Figure 8.3) and a narrow particle size distribution. This provides for optimal flow in a fixed-bed reactor or suspension in a fluidized bed reactor. Fluid flow and pressure drop in fixed-bed reactors depend strongly on these properties, as the pressure drop clearly increases as the void volume decreases, this being due either to the presence of small particles and/or to a broad particle size distribution (see Section 11.3.5).

An until now neglected property of all particles, except ion-exchange particles, is the residual stationary charges (mainly negative) on the pore surfaces that have been determined to be in the range from ≈1 mM for polysaccharides to ≈100 mM for synthetic methacrylate carriers (Kasche, 1973). They influence the rate of immobilization of negatively charged enzymes (Section 8.3), and can influence their selectivities in reactions with charged substrates (Section 10.6.2.3).

The different types of carriers used may be classified according to either the basic material, origin or source, or their structure. Furthermore, the functional groups accessible on the inner (pore) surface also play a major role. In this respect, frequently used carriers are (Table 8.5)

- inorganic;
- organic from natural sources; and
- organic synthetic materials.

8.2.1 Inorganic Carriers

Inorganic carriers exhibit high pressure stability, but may undergo abrasion in stirred vessels. An example of highly efficient adsorption of a lipase onto a silica carrier and its application was provided by Pedersen and Christensen (2000) (see Section 8.3.1). SiO_2-based carriers can be simply functionalized by introducing amino groups bound to spacers, for example, by treating with aminopropyltriethoxysilane (APTS) (Weetall, 1976) (see Section 8.3.2). A complete functionalization leading to a high density of functional groups is essential to produce a stable carrier in aqueous solutions. The covalent binding of enzymes is carried out using glutaraldehyde activation of the carrier. These materials, such as porous glass and silica, are commercially available and have been used mainly at the laboratory scale.

Celite is a convenient carrier used for the adsorption and stabilization of enzymes in organic media (Balco *et al.*, 1996; Andersson *et al.*, 1999). Bentonite also has excellent adsorption capacity (in the range of 0.3–1.5 g protein g^{-1} bentonite) for enzymes such as penicillin amidase, without substantial loss of activity. It is used for enzyme isolation by adsorption and desorption on a large scale. Cross-linking with glutaraldehyde prevents desorption, while the carrier may be entrapped in alginate to provide biocatalyst particles of an appropriate size (Dauner-Schütze *et al.*, 1988; Ralla *et al.*, 2010).

8.2.2 Polysaccharides

Organic materials from natural sources in most cases offer favorable compatibility with proteins. A range of polysaccharides and derivatives have been used for enzyme immobilization (Figure 8.2) (Sheldon, 2007). Polysaccharides exhibit a typical wide network structure, but, due to their hydrophilic properties, they exert weak interactions with proteins with rather no inactivation.

Cellulose derivatives have been used on both laboratory and technical scales. Typical functional groups in commercial carriers are DEAE (diethylaminoethyl)

Table 8.5 Common carriers for enzyme immobilization.

		Application[a]	Functional groups	Manufacturer[b]
Inorganic carriers				
Porous glass	SiO_2	I	–OH	Grace (USA)
Porous silica (Aerolyst®)[c]	SiO_2	I	–OH	Evonic (former Degussa) (Germany)
Porous silica	SiO_2	L, I	–OH	Align Chemical (UK)
Porous titan dioxide (Aerolyst®)	TiO_2	I		Evonic (former Degussa) (Germany)
Organic carriers, natural origin				
Polysaccharides: Sepharose™ (agarose derivatives)		L	(Polyol functions) –OH	GE Healthcare (USA) (former Amersham-Pharmacia (UK))
Cross-linked dextrans (Sephadex®)		L	–OH	GE Healthcare (USA)
Porous organic synthetic carriers[d]				
Methacrylate derivatives: Sepabeads®		L, I	–CH(–O–)CH2 (epoxide); others: diol, amino, alkyl	Resindion (Japan)
Polyacrylic acid derivatives		L, I	Carboxylic acid, epoxide	Align Chemical (UK)
Polystyrene derivatives		L, I	Aromatic residues	Align Chemical (UK)
Polypropylene		L, I	Quaternary ammonia: $-NR_3^+$	Align Chemical (UK)
Polypropylene (Accurel®)		L, I	Alkyl	Membrana GmbH (Germany)

a) Application: L: laboratory mainly; I: industrial scale.
b) Selected commercial suppliers.
c) Based on Degussa's Aerosil®.
d) See also Table 8.7.

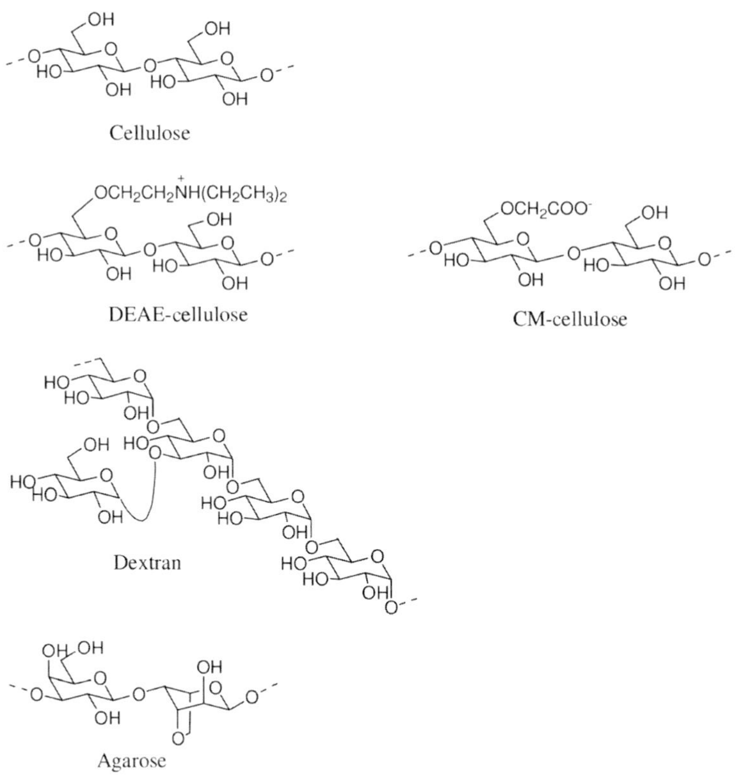

Cellulose

DEAE-cellulose

CM-cellulose

Dextran

Agarose

Figure 8.2 Structures of polysaccharides used as carriers for enzymes.

and CM (carboxymethyl) groups (Figure 8.2). Thus, a composite material of polystyrene, titanium dioxide, and DEAE-cellulose has been applied on an industrial scale for the immobilization of glucose isomerase (see below, Figure 8.8) (Antrim and Auterinen, 1986). The enzyme is bound by ion exchange with high stability. After inactivation of the biocatalyst (having produced up to 15 t of product syrup per kg biocatalyst), the carrier can be regenerated and will bind new enzyme molecules.

Dextran and agarose (Figure 8.2) have been used widely for enzyme immobilization, and in gel chromatography of proteins. Dextran is water-soluble and is cross-linked for application as an insoluble carrier material with either a narrow or a wide network structure, depending on the range of application. A wide network structure exhibits a rather low mechanical and pressure stability. A wide range of commercial preparations of dextran and agarose (e.g., Sephadex®; Sepharose™; Amersham-Pharmacia (UK), now part of GE Healthcare (USA)) are available. Activation for protein binding is conventionally performed using bromocyanide (cyanogen bromide), but the limited stability and the toxicity of reagents used for carrier activation have restricted the use of dextran and agarose mainly to the laboratory (see Table 8.6) (Scouten, 1987; Kohn and Wilchek, 1982; Guisan, Betancor, and Fernandez-Lorente, 2010).

Table 8.6 Carriers, functional groups, protein functional groups, and reactions.

Original functional group of matrix	Activation reagent	Active intermediate	Reactive group of matrix	Reactive group of enzyme	Coupling reaction, binding
SiO_2 |/\/\/NH_2 Aminoalkyl	$OHC(CH_2)_3CHO$	|–$N{=}CH(CH_2)_3CHO$[a]	–CHO Aldehyde	$—NH_2$	Schiff base[a]
|–$CONH_2$ Acrylamide	$OHC(CH_2)_3CHO$	|–$CON{=}CH(CH_2)_3CHO$[a]	–CHO Aldehyde	$—NH_2$	Schiff base[a]
|–CH_2–CH–CH_2 (epoxide, O) Glycidyl-methacrylate			–O–CH–CH_2 (epoxide, O)	$—NH_2$, —OH, —SH	Alkylation
|–COOH Acrylic acid Methacrylic acid	R—N=C=N—R′, H^+		–COO–C(–NH–R)(=$^+$NH–R′)	$—NH_2$	Peptide bond
SiO_2 |/\/\/COOH Acid	R—N=C=N—R′, H^+		O-Acyliso-urea	$—NH_2$	Peptide bond
|–CO–O–CO–| (cyclic) Methacrylic acid anhydride			–C(=O)–O–C(=O)– Acid anhydride	$—NH_2$	Peptide bond

⊢OH, ⊢OH Polysaccharide	CNBr	$-O$, $-O$ $>C{=}NH$[a]	$-O$, $-O$ $>C{=}NH$ Imidocarbonate		Isourea, imidocarbonate, carbamate
Polystyrene[b]			$\vert-C_6H_4-N_2^+Cl^-$ Diazonium salt	$-NH_2$, $-SH$, $-C_6H_4-OH$	Diazo bond
⊢$CONH_2$ Acrylamide	H_2NNH_2; HNO_2		$-CH_2CON_3$ Acyl azide	$-NH_2$, $-SH$, $\vert-C_6H_4-NH_2$	Peptide bond
$\vert-C_6H_4-NH_2$	Cl_2CO		$-C_6H_4-NCO$ Isocyanate	$-NH_2$	Urea derivative
⊢$R-NH_2$	Cl_2CS		$-R-NCS$	$-NH_2$	Thiourea derivative

a) Simplified.
b) Copolymer with amino groups.

8.2.3 Synthetic Polymers

Organic synthetic carriers in general exhibit high chemical stability and can be adapted to requirements of proteins to be immobilized. Ion-exchange materials have proven to be an economic and technically appropriate solution, as a wide range of carriers is currently available offering good capacity for enzyme immobilization, as well as properties that are relevant to industrial-scale processing. Most enzymes (exhibiting a low isoelectric point) may be adsorbed on anion exchangers, including products commercially available (Guisan, Betancor, and Fernandez-Lorente, 2010). These polymers have also found widespread application in commercial processes, mainly due to the cost effectiveness of the carriers and simplicity of the method (see Table 8.7). They have also been shown to be very

Table 8.7 Commercial ion-exchange resins (www.rohmhass.com/; www.ionexchange.com/ion/de/; www.diaion.com).

Chemical basis[a]	Functional groups
Polystyrene derivatives, cross-linked	Weak anion exchangers: $—NR_2H^+$
Polyacrylic ester derivatives, cross-linked	Weak cation exchangers: $—COO^-$
Polysaccharide based (cellulose, dextran, agarose)	All weak anion (DEAE, TEAE) or cation (CM, SP) exchangers[b]
Commercial suppliers of ion-exchange resins[a]	
Amberlite (Rohm & Haas)	Types[c]
Methacrylic acid polymers	Weak cation exchangers: $—COO^-$
Styrene–divinylbenzene polymer	Weak anion exchangers: $—NR_3^+$
Duolite (polystyrene and polyacrylic ester derivatives, Rohm and Haas Company)	
Lewatit (Lanxess)	VP OC 1600[d]
Dowex (styrenic polymers, Dow Chemical)	
Diaion (styrenic, acrylic, methacrylic polymers, Mitsubishi)	Weak anion exchangers: $—NR_3^+$; weak cation exchangers: $—COO^-$
GE Healthcare (USA) (former Amersham-Pharmacia (UK))	Sephacel (cellulose-based), Sephadex (dextran-based), agarose-based (functional groups see above)
Bio-Rad	Cellulose-based, agarose-based (functional groups see above)
Recommended parameters for immobilization	
Carrier particle size	0.1–0.8 mm
Pore size	10–200 nm
Internal surface	$10–100\ m^2\ g^{-1}$
Enzyme load	0.1–10 g (protein) l^{-1} wet resin
Adsorption time; temperature	0.5–20 h; *T*: 10–60 °C

a) Dardel and Arden (2008) and Shaeiwitz and Henry (2000); see also Guisan, Betancor, and Fernandez-Lorente (2010).
b) DEAE: diethylaminoethyl; TEAE: triethylaminoethyl; CM: carboxymethyl; SP: sulfopropyl.
c) Price range: 5–25 € l^{-1}.
d) For enzyme immobilization

useful in the immobilization of lipases for reactions in nonaqueous systems (Balcao, Paiva, and Malcata, 1996).

Polystyrene, cross-linked with divinylbenzene, and acrylic-type polymers are the most common chemical basis for ion-exchange and adsorbent materials. Derivatives with functional groups are available within a wide range of commercial products, as well as particles exhibiting both gel or macroporous structures (Table 8.7). Carriers with a macroporous structure are particularly suited as they offer pores that are sufficiently wide for protein diffusion inside the particle, and the high mechanical strength that is required in technical reactors. Functional groups may be introduced either with the monomers (acrylic type) or by derivatization of polystyrene via nitration and reduction to yield amino groups, or by sulfonation. A survey on ion exchange and basic structural and biophysical aspects (equilibria, kinetics, etc.) has been published (Dardel and Arden, 2008). It must be considered that weak ion exchangers will be best suited and have been used for enzyme immobilization in industrial processes. Strong ion exchangers bear the risk of enzyme inactivation due to strong electrostatic interaction. Polysaccharide-based ion exchangers used in chromatography of enzymes may prove appropriate for immobilization, since they have good adsorption capacities and exhibit mild interactions (see Table 8.7) (Shaeiwitz and Henry, 2000).

Special carriers for enzyme immobilization have also been developed, mainly based on materials such as polymethacrylate copolymers (developed by the Röhm GmbH, Germany) and copolymers of vinylacetate and divinyl-ethylene-urea (developed by HOECHST AG, Germany), both exhibiting macroporous structure and excellent binding capacity (> 60 mg of protein per g of carrier) (Figure 8.3) (Burg *et al.*, 1988). A broad range of copolymers with different characteristics, including hydrophilicity and hydrophobicity balance, functional groups with carboxyl-, amino-, epoxy- (oxiran-), and alkyl-functions, also including spacers, have been designed, and are currently available commercially (e.g., Sepabeads®, Figure 8.4); see www.resindion.com; Table 8.5). Immobilization to carriers with an epoxide function has been investigated in detail, thus, different modifications and particle sizes of Sepabeads® for the immobilization of cephalosporin C (CPC) amidase (Katchalski-Katzir and Krämer, 2000; Chikere *et al.*, 2001; Boniello, Mayr *et al.*, 2010; Guisan 2010). An example for continuing optimization is the process for 7-ACA production developed at the (former) Hoechst company, where different carriers have been applied (see Section 12.3). Since an amidase for a one step process could not be found in the early eightees, a two-step process had been developed first using a D-amino acid oxidase and a glutarylamidase (see section 12.3, Figure 12.14). Eupergit® (former Röhm GmbH, Germany; no longer available) was used until 2004, for D-amino acid oxidase, an open pore variant that was stable over >150 cycles; for the glutarylamidase standard Eupergit® could be applied, which was stable over >400 cycles in the scale of >20 m^3. Later, the cheaper Sepabeads® epoxy-activated carrier (Resindion/Mitsubishi, Japan) was used for immobilization of glutarylamidase. Diffusion limitation, due to proton diffusion, was observed in CPC hydrolysis using CPC amidase immobilized on Sepabeads® with standard pore diameter (30–40 nm), highlighting the importance of appropriate carrier choice (Boniello, Mayr *et al.*, 2010; Boniello *et al.*, 2012). In parallel

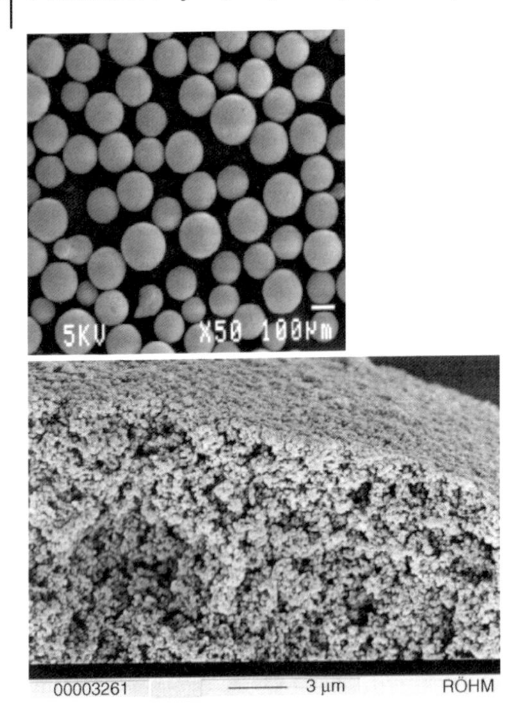

Figure 8.3 Synthetic polymer carrier; macroporous particle structure. Scanning electron microscopy (SEM) pictures of Eupergit® C 250L. The spherical shape and porous structure is easily recognizable. (Reproduced courtesy of S. Menzler, Degussa/Röhm GmbH.)

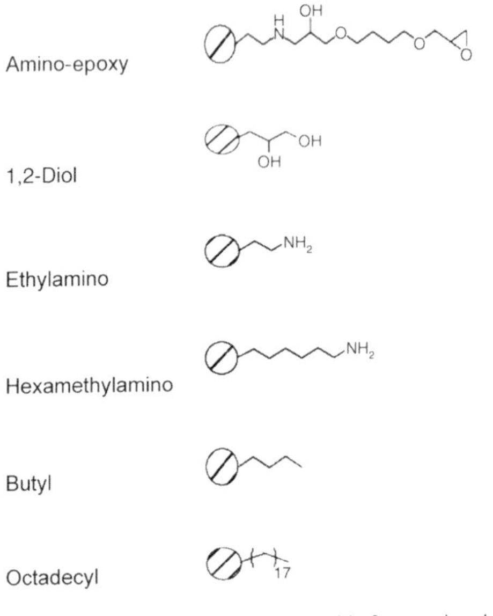

Figure 8.4 Functional groups available for Sepabeads®.

to the existing process, the one-step process research was resumed by Aventis (Fritz Wolf *et al.*, 2002) and Sandoz (Austria) and finally an amidase mutant from Pseudomonas sp. SE 83 (WO 2005) was implemented. For immobilization of the amidase, an anion exchange carrier, with subsequent cross-linking, proved superior, cheaper, and considerably more stable. For this remarkable success, processing of CPC to 7-ACA in one step, two major achievements by genetically engineering glutarylamidase were essential. First, the enzyme activity toward the substrate CPC had to be improved by several orders of magnitude. Second, the stability of the biocatalyst had as well to be improved dramatically with respect to process conditions (K. Sauber, 2012, personal communication).

Synthetic resins, including the XAD adsorbents (Amberlite™ types, for example, XAD-7) and Sepabeads® with octadecyl chains as functional groups, exhibiting hydrophobic characteristics, are suitable for the immobilization of enzymes with high yields of activity, since their surface is hydrophobic to ≈40–50% (Section 2.3). A wide range of immobilization systems and their application in synthetic reactions has been reported. Results including carriers and binding methods used and reactions with optimal conditions have been summarized (Balcao, Paiva, and Malcata, 1996; Sheldon, 2007; Guisan, Betancor, and Fernandez-Lorente, 2010). Porous polypropylene, for example, Accurel® (MP1001, Membrana GmbH, Wuppertal, Germany), is used industrially for lipase immobilization (Liese, Seelbach, and Wandrey, 2000, 2006). It was shown that immobilization of lipases or esterases can be significantly improved by "hydrophilization" of the highly hydrophobic surface using plasma functionalization with an oxygen plasma (creating, for example, hydroxyl groups), resulting in much enhanced immobilization yield, activity, and high stability (Vorhaben *et al.*, 2010). Optimal conditions for enzyme adsorption and, potentially, cross-linking must be established empirically by trial and error, screening of different carriers, and parameters for immobilization (for examples of appropriate immobilization conditions, see Exercises).

Microchannel reactors with immobilized enzymes may be emerging tools for biocatalytic processes and their development. They offer potential in process intensification for the production of fine chemicals exhibiting highly efficient heat exchange and mass transfer as a consequence of the much larger surface area to volume ratios compared to conventional reactors[2] (Schwarz *et al.*, 2009).

2) In an example, the enzyme is covalently immobilized in multiple linear flow channels of the reaction plate supported by, for example, a macroporous layer of γ-aluminum oxide. Thus, the thermostable β-glycosidase CelB from *Pyrococcus furiosus* is used for the production of β-glucosylglycerol (βGG), in a kinetically controlled enzymatic transglucosylation from a suitable donor substrate to glycerol as acceptor that has potential applications as a moisturizing agent in cosmetic products. As donors 2-nitrophenyl-β-D-glucoside or cellobiose have been used to give high or moderate yields, respectively (Schwarz *et al.*, 2009). In another concept, the enzyme is immobilized on a honeycomb, ceramic monolith, analogous to those used in catalytic exhaust systems in automobiles. Cordierite monoliths were functionalized, by coating with polyethylene imine or different types of carbon, in order to create adsorption sites for the enzyme (CALB) (De Lathouder *et al.*, 2005).

Enzymes immobilized on *nanoparticles* dispersed in solution are believed to offer specific advantages with respect to activity, stability, and mass transfer (Fei *et al.*, 2010). Various nanostructures have been examined as hosts for enzyme immobilization via approaches including enzyme adsorption, covalent attachment, enzyme encapsulation, and sophisticated combinations of methods. Materials include nanoparticles, nanofibers, mesoporous materials, and single enzyme nanoparticles[3] (Kim *et al.*, 2006). Further, *nanostructured magnetic materials* may find applications as support matrices for enzyme immobilization (as readily recyclable biocatalysts), immunoassays, drug delivery, and biosensors (to be magnetically separated from reaction mixtures). Thus, the bioconjugation of histidine-tagged enzymes and other proteins to the surface of composite "magnetomicelles" consisting of encapsulated magnetic γ-Fe_2O_3 nanoparticles has been described (Herdt *et al.*, 2007; Lee *et al.*, 2005). However, the advantages of such systems for catalysis still have to be demonstrated, and nanodispersed biocatalysts may remain dwarfed in practical applications due to their tiny size and difficulties in recycling.[4] *Magnetic silica beads* with different surface functional groups (including alkyl, carboxyl, and amino groups) are available from MoBiTec (Germany) (www.mobitec.com).

8.3 Binding Methods

8.3.1 Adsorption

The first step in immobilization, the adsorption of proteins (e.g., chymotrypsin and penicillin amidase) on adsorbents (Sepharose™ and Eupergit® types), has been investigated in detail by Kasche *et al.* (2003), including profiles of protein density as a function of carrier radius and time. The rate of adsorption increases with the enzyme concentration. To obtain high immobilized enzyme concentrations, the adsorption must be carried out in the nonlinear part of the isotherm (Section 6.5.1). It has also been shown that the profiles of enzyme density within carriers can be fixed by quick immobilization in order to improve enzyme efficiency (by reducing the maximal enzyme loading and shortened diffusion depth inside the carrier) (Borchert and Buchholz, 1984; Scharer *et al.*, 1992). A sophisticated technique to analyze the enzyme distribution inside a carrier has been developed using

3) Bruns and Tiller (2005) have reported a novel method for the immobilization of enzymes in a prefabricated, nanophase separated amphiphilic network, whereby the enzyme is situated in its hydrophilic domains that consist of poly(2-hydroxyethyl acrylate) (PHEA). The substrate that diffuses into the hydrophobic polydimethylsiloxane (PDMS) phase can access the enzyme via the large interface, owing to its nanophase separation and peculiar swelling properties.

4) Therefore, nanoparticle-based biocatalysts were first constructed and then assembled as cell-like microreactors, using porous microcapsules as containers, creating a rigid/porous capsule structure by suspension polymerization. Such systems are claimed to provide also effective protection for delicate enzymes. The biocatalyst was recycled using a cellulose acetate membrane (Gao *et al.*, 2009; Fei *et al.*, 2010).

confocal laser microscopy (Spiess and Kasche, 2001). Recently, new methods for the analysis of molecular details of surface derivatization and immobilization steps have been used including AFM (atomic force microscopy), transmission FTIR, and XPS (X-ray photoelectron spectroscopy). Thus, enzyme loading can be observed depending on conditions (e.g., functionalization and density of functional groups); also multilayer formation has been observed at high density of functional groups (Libertino *et al.*, 2008; Miletic *et al.*, 2010).

The adsorption of enzymes onto silica or clay particulate materials is both simple and cost-effective. Pedersen and Christensen (2000) developed a straightforward process for lipase adsorption, agglomeration, drying, and sieving to produce an active and stable biocatalyst of defined mechanical strength and particle diameter (Pedersen and Christensen, 2000).

Adsorption onto ion-exchange materials has proven to be an efficient, technically appropriate, and economic method since a broad range of carriers is currently available at a reasonable price (5–25 € l^{-1}). A selection of commercial ion exchangers is listed in Table 8.7. The selection of an appropriate carrier for an enzyme must proceed experimentally, as well as the identification of optimal conditions for immobilization, such as the ratio of enzyme and carrier amounts (or concentration, respectively), pH, buffer, and temperature. In order to stabilize an adsorbed enzyme, cross-linking with glutaraldehyde is a favored option (adsorption is reversible in general) (see below and Exercises).

The adsorption of lipases onto synthetic resins has been successfully used (see Section 8.2.3). Adsorption was reported to be tight, with desorption not occurring to any significant extent when using established procedures. The effect of cross-linking on adsorption has also been investigated. Mechanisms of partial unfolding in the context of multiple binding at the carrier surface have been discussed, and a wide range of investigations including binding yield, operational conditions, stability, and water activity have been reported and summarized (Balco *et al.*, 1996; Bornscheuer, 2003).

8.3.2 Covalent Binding

An analysis of 125 protein families revealed that most (up to 95%) of polar, titratable residues are located at the surface of soluble proteins (Table 8.3) (Petersen *et al.*, 2000). The carrier functional groups, as well as their density, play a major role in binding the overall protein, and in the activity yield. When the immobilization on a carrier with negative stationary charges is carried out at a pH > p*I* of the enzyme, the rate of immobilization is increased when it is carried out at a high ionic strength. This applies also for the immobilization by adsorption to such carriers. Unreacted functional groups should be inactivated after immobilization.

An example of the functionalization of a silica carrier and enzyme binding is shown in Figure 8.5. Porous silica or glass is treated with aminopropyltriethoxysilane in order to introduce amino groups. Subsequently, the carrier is activated with glutaraldehyde (GA), followed by enzyme binding. A detailed study of the

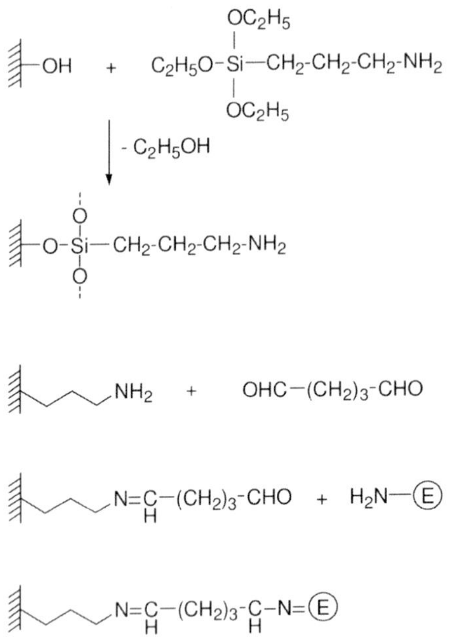

$$\text{carrier–OH} + C_2H_5O\text{–}Si(OC_2H_5)_2\text{–}CH_2\text{-}CH_2\text{-}CH_2\text{-}NH_2$$

$$\downarrow -C_2H_5OH$$

$$\text{carrier–O–}Si(O\text{–})_2\text{–}CH_2\text{-}CH_2\text{-}CH_2\text{-}NH_2$$

$$\text{carrier}\sim NH_2 + OHC\text{–}(CH_2)_3\text{-}CHO$$

$$\text{carrier}\sim N{=}\underset{H}{C}\text{–}(CH_2)_3\text{-}CHO + H_2N\text{–}(E)$$

$$\text{carrier}\sim N{=}\underset{H}{C}\text{–}(CH_2)_3\text{-}\underset{H}{C}\text{–}N{=}(E)$$

Figure 8.5 Derivatization of a silica carrier; activation with glutaraldehyde, and binding of an enzyme by a free amino group (e.g., lysine).

parameters involved in and optimal conditions for coupling to an amino group activated carrier via GA was published by Monsan (1978) (Table 8.8). Until now, this method has been the most popular due to its simplicity, low cost, and effectiveness. It has been applied manifold both at laboratory and technical scales. However, it must be kept in mind that GA in solution consists of a mixture of oligomers that may also react.[5] The reaction conditions essentially allow for good yields (>90% of

Table 8.8 Immobilization of trypsin via glutaraldehyde, range of conditions (from Monsan, 1978).

Carrier: silica (SiO_2) amino-functionalized (Sherosil)	
Amount	100 g
Internal surface range	6–180 $m^2\,g^{-1}$
Pore size	75–2800 nm
Enzyme: trypsin	Dissolved under slightly acidic conditions in order to prevent autolysis
Conditions:	
Activation reaction	pH 8.6 (phosphate buffer, 50 mM), 15 h, 25 °C
Concentration range of glutaraldehyde	0.5–5%
Coupling reaction	2 $g\,l^{-1}$ protein in phosphate buffer pH 8.6, 50 mM, 20 ml, 2 h, 4 °C

5) Different reaction mechanisms may occur to an extent that is not quite clear. The condensation reaction yields double bonds that can be recognized by a light color; hydrogenation is advisable providing for bonds that are chemically more stable.

active immobilized enzyme can be achieved with trypsin, for example), at optimal pH, *T*, *t*, amount of carrier, and concentrations of GA and enzyme. A two-step procedure with activation of the carrier in the first step and subsequent enzyme immobilization is recommended. An example is given with Exercises. The reaction is rather quick, with coupling proceeding in the range of minutes (at 25 °C) or a few hours (at 4 °C) (Buchholz, Duggal, and Borchert, 1979, pp. 169–181). GA activated supports are commercially available (see also Sheldon, 2007). Binding of an enzyme (L-aminoacylase) to different carriers has been described by Toogood *et al.* (2002). The data include derivatization of XAD resins with amino groups and activation by GA, as well as optimal conditions for high activity yield after immobilization. Spacers for application in organic solvents should provide hydrophobic characteristics. As an example, a polytryptophan tether ($n \approx 78$) has been developed for covalent immobilization of *Candida antarctica* lipase B (CALB) on nonporous, functionalized 1 μm silica microspheres. The biocatalyst exhibited much enhanced esterification activity (Schilke and Kelly, 2008).

A reaction that has been used successfully at both laboratory and technical levels is protein binding by epoxide groups. These react with nucleophilic functions such as amino, thiol, and hydroxyl groups (which are less reactive) under formation of stable C—N—, C—S—, or C—O— bonds, respectively (Figure 8.6). Yields of actively bound enzymes are high when the procedure is optimized. The reaction rates are low, such that effective coupling may require many hours or even days (Krämer, 1979). A carrier of this type is shown in Figure 8.3. Various enzymes, including hydrolases, penicillin amidase, cephalosporin C amidase, hydantoinase, β-galactosidase, lipoxygenases, and lipases, have successfully been immobilized and, in part, applied

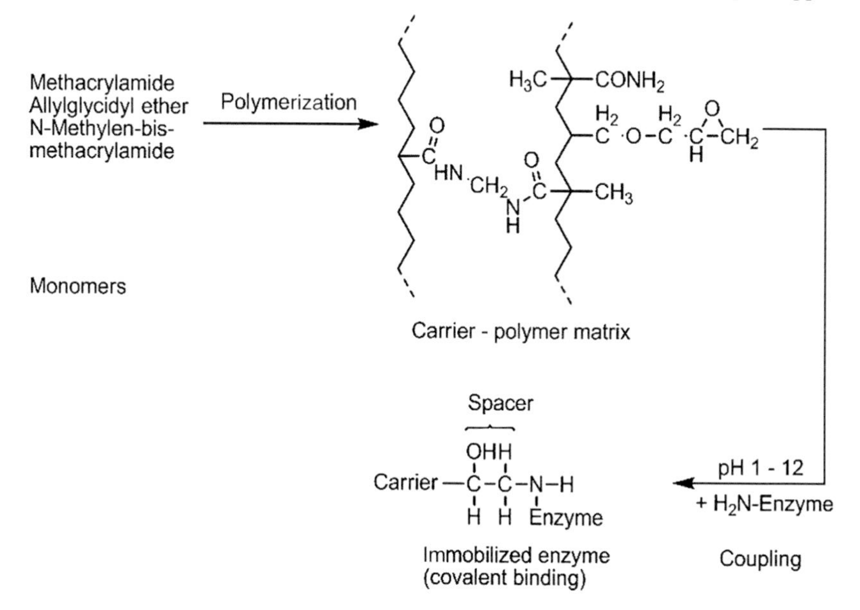

Figure 8.6 Covalent binding of penicillin amidase to a carrier matrix by epoxide groups (Katchalski-Katzir and Krämer, 2000).

at the technical scale (for carriers see Table 8.5) (Chikere *et al.*, 2001; Sheldon, 2007; Boniello *et al.*, 2010; Guisan, Betancor, and Fernandez-Lorente, 2010).

Multipoint covalent immobilization has been designed to significantly stabilize enzymes; it has been claimed that it promotes rigidification and stabilization of enzymes against distorting agents (heat, organic cosolvents, pH, etc.)[6] (Fernandez-Lafuente *et al.*, 1995; Allard, Cheynet, and Oriol, 2002; Guisan, Betancor, and Fernandez-Lorente, 2010).

A laboratory method has been the activation of polysaccharide carriers with cyanogen bromide and subsequent coupling to yield different types of rather instable bonds, resulting in substantial leakage of the enzyme from the matrix. Also the activation of carboxyl groups by carbodiimide derivatives and binding to amino groups to yield amide bonds has been used in many laboratories. The reagents used for these methods are expensive and highly toxic, and scale-up is difficult. Consequently, these approaches were not applied industrially to any major degree. More extensive overviews of the many now traditional methods have been published by Scouten (1987) and Buchholz and Kasche (1997, pp. 141–166).[7]

Cross-linked enzyme aggregates (CLEAs) have been developed by Sheldon (2007, 2011). They are prepared by precipitation as physical aggregates of protein molecules from an aqueous medium, by addition of ammonium sulfate or polyethylene glycol and subsequent cross-linking, while retaining most (up to 100%) activity. It essentially combines purification and immobilization into a single unit operation that does not require a highly pure enzyme. Thus, penicillin amidase could be immobilized with about 80% hydrolytic activity. In addition, the preparation maintained its synthesis activity in a wide range of organic solvents, whereas the native enzyme exhibited a very low stability (Cao, van Rantwijk, and Sheldon, 2000). CLEAs were prepared from seven commercially available lipases and the effects of various parameters on their activities were investigated. It has been claimed that lipase CLEAs exhibit activities up to 12 times those of free enzymes. Cascade processes can be achieved by immobilizing two or more enzymes in "combi-CLEAs" by coupling steps together, to drive unfavorable equilibria towards the product[8] (Sheldon, 2011).

Cross-linking of the whole microorganism containing an enzyme in sufficiently high concentrations, where the (dead) cells provide the matrix, or carrier, respectively, has been applied to glucose isomerase immobilization (see Section 8.4).

6) Thus, the adsorption of enzymes by anionic exchange on a heterofunctional amino-epoxy support (e.g., Sepabeads®) may be achieved first, followed by alkaline incubation that may promote an intense multipoint covalent attachment. Stabilization of multimeric enzymes has been described by additional cross-linking of all subunits by using polyfunctional flexible polymers (Guisan, Betancor, and Fernandez-Lorente, 2010).

7) Several different methods have been developed for coupling in enzyme sensors, including affinity interaction, an example being biotinylated protein binding to adsorbed avidin (Polzius *et al.*, 1996).

8) Thus, a triple combi-CLEA has been developed, comprising a hydroxynitrile lyase, a nitrilase, and an amidase in the one-pot conversion of benzaldehyde to *S*-mandelic acid in >99% ee at 96% conversion, as well as a combi-CLEA with a hydroxynitrile lyase and an alkaliphilic nitrile hydratase, to catalyze the bienzymatic cascade process for the conversion of aldehydes to *S*-*R*-hydroxycarboxylic acid amides (Sheldon, 2011).

For the immobilization procedure, it is essential to have a balance of enzyme activity introduced and recovered in the immobilized biocatalyst. The role of a spacer at the matrix surface may be important in order to provide flexibility for the enzyme (Manecke, Ehrenthal, and Schlüsen, 1979; see also Buchholz and Kasche, 1997, pp. 141–166). Furthermore, the efficiency of immobilized enzymes plays a crucial role, as discussed in Section 10.4.2. Special procedures that provide improved efficiency by coupling enzymes in the shell of a particle near the external surface have been developed and applied; an example is penicillin hydrolysis (Carleysmith, Dunnil, and Lilly, 1980; Borchert and Buchholz, 1984).

8.4 Examples: Application of Immobilized Enzymes

The first successful applications of immobilized enzymes were glucose isomerization for application as a sweetener in food and the hydrolysis of penicillin and cephalosporin in the pharmaceutical sector. Both of these processes will be treated in more detail in Sections 12.1 and 12.3; hence, only some general information on them will be presented in this section. For application of enzymes in biosensors and binding of biomolecules to transducer surfaces, see Nakamura and Karube (2003), and Schmidt *et al.* (2008).

8.4.1 Hydrolysis and Biotransformation of Carbohydrates

High-fructose corn syrup is currently produced at a scale of about 12 million tons per year, using about 1000 tons of immobilized enzyme worth 50 million €. The product is a syrup containing about 53% glucose, 42% fructose, and 5% other products, or a syrup containing about 40% glucose and 55% fructose (it is upgraded in sweetness to that of sucrose by addition of fructose from chromatographic separation). This has the advantage that at an equal sweetener level it is some 10–20% cheaper than sucrose and, as a technical advantage, exhibits a lesser tendency to crystallize in a wide range of food products. It is largely used in drinks such as cola, lemonades, nectars, fruit juices, confectionary, and sauces, mainly in the United States. Sweetness and taste are almost equal to those of sucrose.

The isomerization of glucose to fructose is an equilibrium-controlled reaction (Figure 8.7b), with fructose equilibrium concentrations of 48% at 45 °C and 55% at 85 °C. Processing at 85 °C, however, is not possible due to the limited enzyme operational stability and the instability of fructose at temperatures beyond 65 °C due to the Maillard reaction. The mechanism of isomerization by glucose isomerase (GI) consists of ring opening, isomerization through a hydride shift from C2 of the open-form glucose to C1 of the product fructose mediated by two Mg^{2+} ions, followed by ring closure to give the α-ketol (Figure 8.7a) (Lavie *et al.*, 1994; Reilly and Antrim, 2003).

Many GIs (which basically is a D-xylose ketol-isomerase, EC 5.3.1.5) have been characterized and more than 70 tertiary structures are now known. The structure

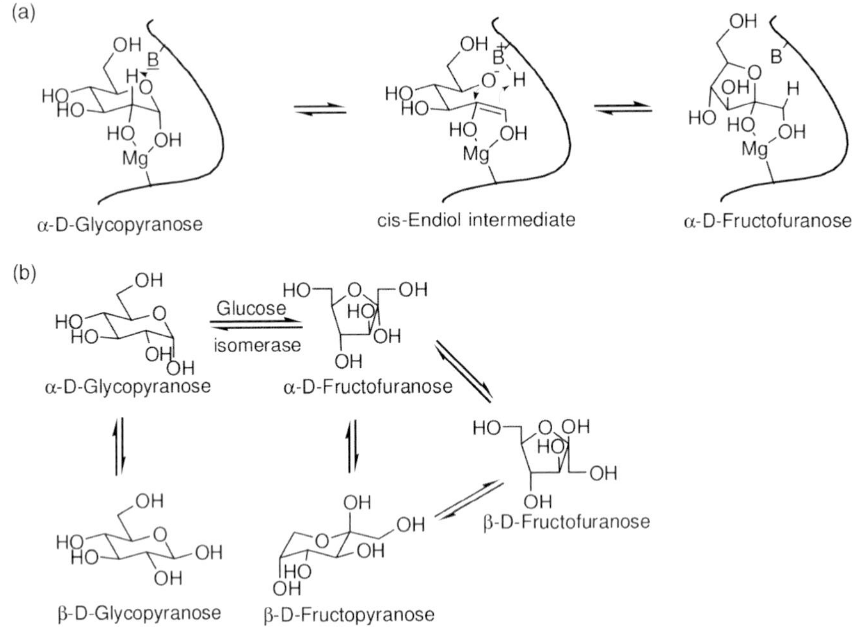

Figure 8.7 (a) Mechanism of glucose isomerization (Makkee, Kieboom, and van Bekkum, 1984). (b) Chemical equilibria in solution.

of the GI from *Streptomyces rubiginosus* can be found under the pdb entry 1OAD (http://www.rcsb.org/pdb/). The amino acid sequence alignment has been presented by Misset (2003). Most GIs are homotetramers composed of two tightly bound dimers. Monomers have $(\beta, \alpha)_8$ barrels, with eight β-strands being surrounded by eight helices (Lavie *et al.*, 1994; see also overviews: Misset, 2003; Reilly and Antrim, 2003). GI has been successfully genetically engineered with respect to thermal stability (see Chapter 3) and tight binding of Mg^{2+}. This provides a significant advantage insofar as two steps up- and downscale – to provide Mg^{2+} ions to and remove them from – the product syrup by ion exchange are no longer necessary. Furthermore, GI has been modified with reference to substrate binding and catalytic activity on glucose and xylose, exhibiting a higher V_{max} and a lower K_m on glucose, which provides advantages for glucose isomerization. Rational protein design has succeeded in increased flexibility of the active site to accommodate the larger substrate (Genencor International, 1990).

The productivity of GI is in the range of 12–15 t (dry substance, d.s.) kg^{-1} biocatalyst, with a half-life of 80–150 days (Misset, 2003; Swaisgood, 2003). Reactor dimensions typically are 1.5 m diameter and 5 m height for fixed-bed bioreactors, and the operation is carried out at 58–60 °C. Several different procedures have been developed, and different commercial biocatalysts are utilized at the technical scale. One established method of immobilization is cross-linking of the cells of *Streptomyces murinus* with GA while maintaining almost all of the GI activity (Figure 8.8a). In another procedure, the enzyme is isolated and purified by chromatography and

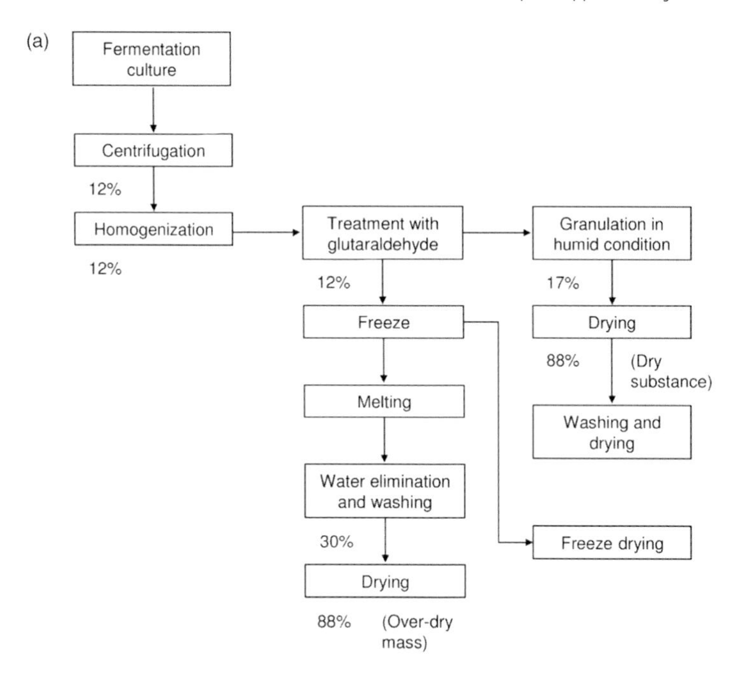

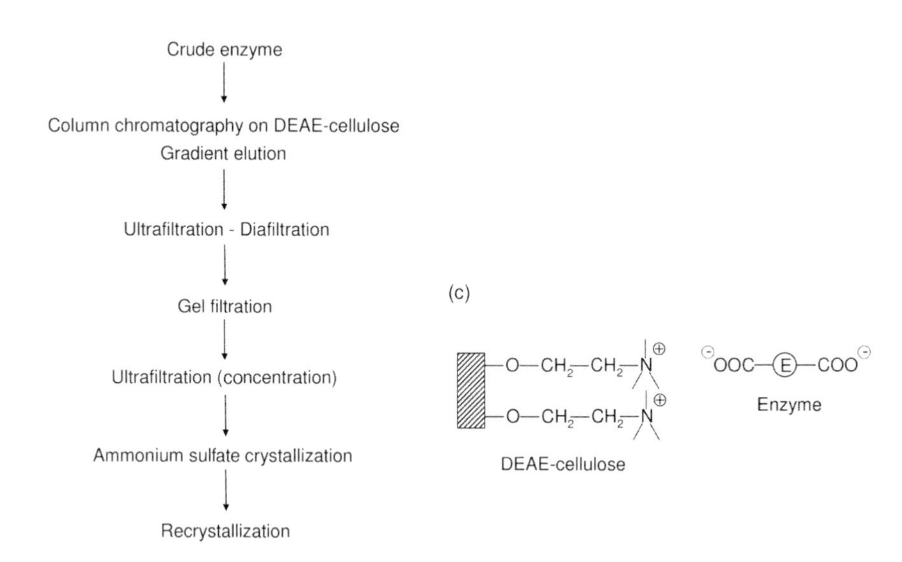

Figure 8.8 Scheme of procedures for immobilization of glucose isomerase. (a) Novo: cross-linking of cells (Pedersen and Christensen, 2000). (b) Genencor: principle of isolation of crystalline enzyme. (c) Adsorption onto a composite ion-exchange matrix formed by extrusion with 50% polystyrene, 20% TiO_2, 30% DEAE-cellulose; d_p 400–800 μm (Antrim and Auterinen, 1986).

Table 8.9 Examples of commercial immobilized glucose isomerases (Pedersen and Christensen, 2000).

Manufacturer[a)]	Trade name	Enzyme source	Immobilization method	Typical values of initial space velocities at 60 °C (bed vol. h^{-1})
Corn Products International	G-zyme G 994	*S. olivochromogenes*	Purified GI adsorbed on an anion-exchange resin	6
Genencor International (Danisco)	Spezyme (600 IGIU g^{-1})	*S. rubiginosus*	Purified GI adsorbed on an anion-exchange resin consisting of DEAE-cellulose agglomerated with polystyrene and TiO_2	3.9
Novozymes A/S	Sweetzyme T (300 IGIU g^{-1})	*S. murinus*	Cross-linking of cell material with glutaraldehyde, extruded	2.1

a) Enzyme systems have been developed earlier, not all are currently available.

crystallization, and finally adsorbed to an ion-exchange matrix (Figure 8.8b) (Pedersen and Christensen, 2000; Antrim and Auterinen, 1986). Commercial biocatalysts are listed in Table 8.9. Process data are provided in Section 12.1.

A considerable number of further processes using immobilized biocatalysts to convert sugars and polysaccharides into new products, and to synthesize oligosaccharides (OS) and derivatives, mostly as sweeteners and functional foods, have been applied industrially, and several more are currently under development. These comprise hydrolytic as well as synthetic reactions by hydrolases and glycosyltransferases (Buchholz and Seibel, 2008). Enzymatic processes fully exhibit their potential in the carbohydrate field in terms of regio- and stereospecificity, and this makes them superior to chemical and classical catalytic reactions. The potential to build up different linkages via α- and β-bonds and to different regio positions (e.g., 1 to 6 in hexose sugars) makes the synthesis of OS dramatically complex. In general, enzymes easily form a single bond, or one main product in high yield (mostly >80%) with low by-product formation of isomers. (For regulatory issues in the EU and the United States, see Praaning (2003) and Whitaker *et al.* (2003, pp. 67–76), respectively.)

Figure 8.9 provides some selected examples of established processes, and further examples and data for glycosyltransferases are listed in Table 8.10. The hydrolysis of sucrose is accomplished by invertase (immobilized, for example, on an epoxy-activated carrier) (Uhlig, 1991, p. 204). This process competes with others based on

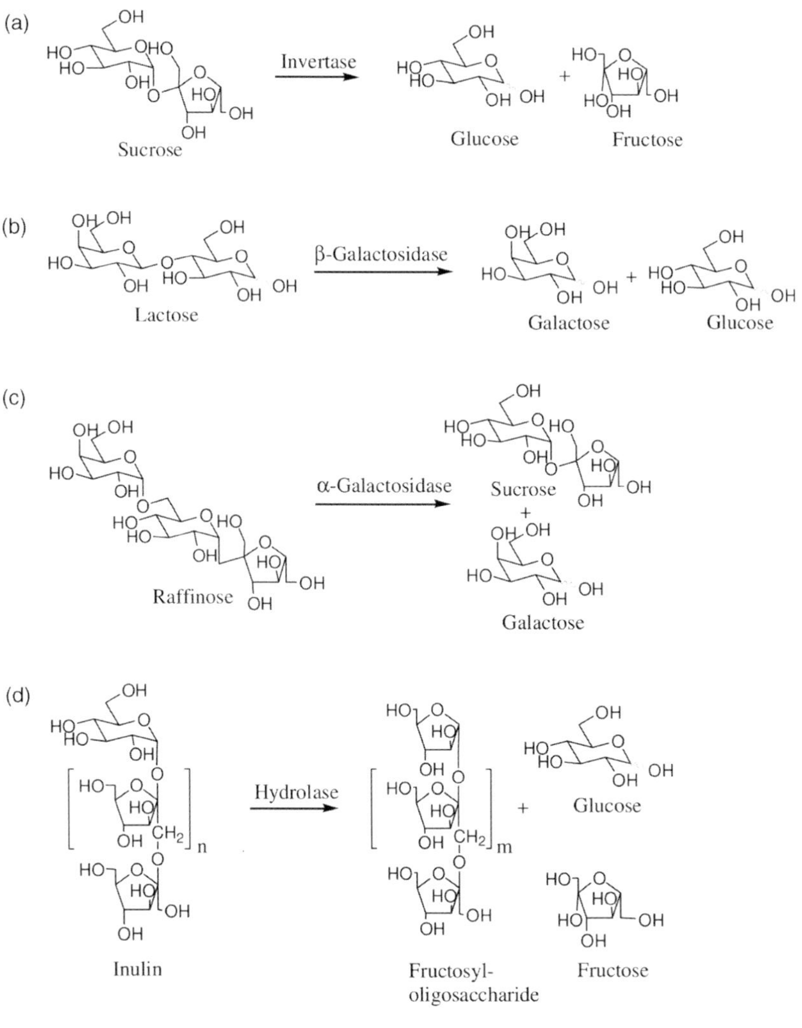

Figure 8.9 Hydrolytic carbohydrate reactions. (a) Sucrose with invertase; (b) lactose with β-galactosidase; (c) raffinose with α-galactosidase; (d) inulin with inulinase.

hydrolysis by ion-exchange resins, which are well established. The enzymatic process offers the advantage of less by-product (e.g., hydroxymethylfurfural) formation. Lactose in milk and whey is hydrolyzed by β-galactosidase (EC 3.2.1.23) or lactase (EC 3.2.1.62) into glucose and galactose, exhibiting a greater sweetness and improved fermentation behavior when used as a substrate. The principal reason for hydrolyzing lactose in milk and whey products, however, is to overcome the problem of lactose intolerance, which is widespread among certain populations. Whey syrup with 60% solids is treated at a scale of several thousands of tons annually. A typical biocatalyst has been based on Eupergit®, with a half-life of 20 months, and a productivity of 2000 kg per kg biocatalyst. Other companies purchasing

Table 8.10 Applications of immobilized glycosyltransferases (Nakakuki, 2005; Buchholz and Seibel, 2008).

Enzyme (reference)	Product	Market (estimated, $t\,a^{-1}$)	Company
Sucrose mutase (1)	Isomaltulose	100 000	Südzucker (Germany)
Cyclodextrin transferase	Malto-oligosylsaccharides	5000	Hayashibara (Japan)
α-Glucosidase and glucosyltransferase (4)	Isomalto-oligosaccharides	4800	Hayashibara (Japan) and others
Dextransucrase	Isomalto-oligosaccharides		BioEurope, Solabio (France)
Dextransucrase[a)]	Dextran	200	Amersham Biosciences (UK)
β-Galactosidase	Galacto-oligosaccharides	7000	Yacult (Japan)
Fructosyltransferase (2)	Neosugar	4000	Beghin-Meji (France)
Cyclodextrin transferase (3)	β-Cyclodextrin and other cyclodextrins	3000	Wacker-Chemie (Germany)

Reference sources: (1) Schiweck *et al.* (1991); (2) Ouarne and Guibert (1995); (3) Wimmer, 1999, personal communication, Wacker-Chemie, München; (4) Buchholz and Monsan (2003).
a) Single example where the product is not made by an immobilized enzyme.

biocatalysts, or applying this process, include Valio and the former Snamprogetti (Uhlig, 1991, 1998; Swaisgood, 2003). Galacto-oligosaccharides (OS) are applied as prebiotic ingredients for infant formulas with 90% short-chain galacto-OS and 10% long-chain fructo-OS. They are produced by transgalactosylation using β-galactosidase (lactase) and lactose as the substrate (Boehm *et al.*, 2005; Nakayama and Amachi, 2010). Industrial applications of α-D-galactosidases, mainly in the sugar industry and medicine, have been reviewed, including genetic engineering that has enabled the creation of glycosynthases from the hydrolytic glycosidases (Weignerova *et al.*, 2009). α-Galactosidase has been used in Japan on a large scale for the hydrolysis of raffinose into sucrose and galactose in concentrated sugar syrups and molasses from sugar beet processing (Figure 8.9) (Linden, 1982).

A range of oligosaccharides is produced by glycosyltransferases, including both glucosyl- and fructosyltransferases, and using sucrose or starch, or dextrins, respectively, as convenient, high-quality, and low-price commercial substrates. The synthesis of gluco- and fructo-OS based on sucrose by glycosyltransferases furthermore utilizes the advantage of the high glycosidic bond energy that is almost equal to that of nucleotide-activated sugars ($-23\,kJ\,mol^{-1}$). This drives the synthesis to high yields when water is used as the solvent, which is a precondition for food-grade and economic processing (Seibel *et al.*, 2006b; Desmet and Soetaert, 2011). Sucrose phosphorylase has been used for glucoside synthesis making as well use of sucrose as a high-energy glucosyl donor. Its glucosyl acceptor specificity, focusing on applications, has been recently reviewed (Goedl *et al.*, 2010). The

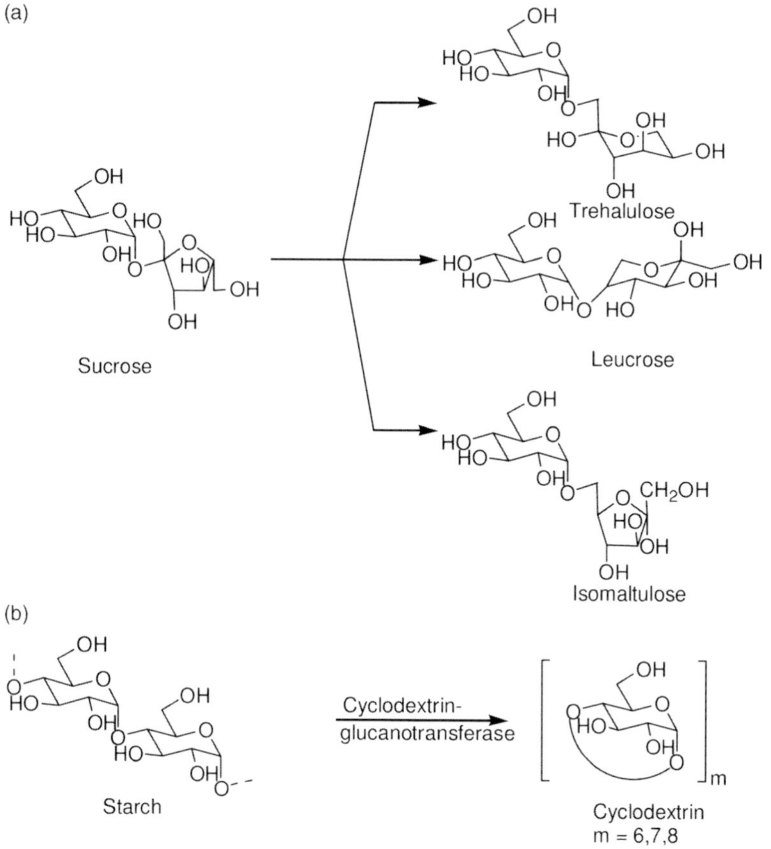

Figure 8.10 Transglycosidations by glycosyltransferases. (a) Sucrose isomerization; (b) cyclodextrin synthesis.

alternative of using glycosyl hydrolases at high concentration in order to favor the condensation reaction over hydrolysis has not succeeded in application (except for galactosyl-OS), mainly due to limited yields and regioselectivity.

The production of isomaltulose (α-D-glucosyl-1,6-D-fructose) by isomerization of sucrose (α-D-glucosyl-1,2-β-D fructose) (Figure 8.10a) is established on the large scale, with about 100 000 tons produced annually (Südzucker AG, Cerestar, Germany; Mitsui Seito Co., Japan). The enzyme sucrose mutase is a glycosyltransferase that effects isomerization by an intramolecular rearrangement. For enzyme immobilization, the cells (*Protaminobacter rubrum*) with enzymatic activity are cross-linked and entrapped in alginate beads. Most remarkable is the much increased stability of the biocatalyst at high sugar concentrations, this being further improved under operational conditions in sucrose solution (1.6 M; half-life of over 8000 h) (Cheetham, 1987). The yield of isomaltulose critically depends on the kinetic control of the reaction, with increasing formation of trehalulose as a

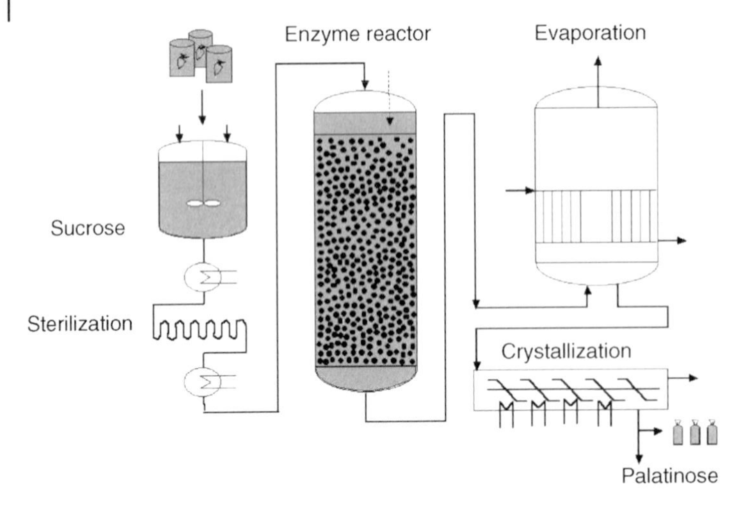

Figure 8.11 Scheme of industrial isomaltulose production (Rose and Kunz, 2002).

by-product at extended reaction (residence) time to be controlled in the fixed-bed reactor. A scheme of the process is shown in Figure 8.11. Most of the product is further processed by catalytic hydrogenation to a mixture of α-D-glucosylsorbitol and -mannitol (Isomalt®) that is used as an alternative sweetener in many food products (Schiweck *et al.*, 1991; Rose and Kunz, 2002).

Isomalto-oligosaccharides are produced by companies in Japan and Europe for use in food, feed, and dermocosmetics (Buchholz and Monsan, 2003; Monsan *et al.*, 2010). Raw materials are starch or sucrose with an acceptor, such as glucose or maltose, to which a glucosyl unit from sucrose is transferred.

Fructo-oligosaccharides (FOS) have gained major attention during recent years as functional food ingredients and medical foods (Rastall and Hotchkiss, 2003; Tungland, 2003). They are manufactured either by partial hydrolysis of inulin or by glycosyltransferases using sucrose as the substrate (Ouarne and Guibert, 1995; Yun and Kim, 1999; Yun and Song, 1999; Tungland, 2003). An alternative method to produce FOS is by fructosyltransferases acting on sucrose, with the transfer of a fructosyl unit from sucrose to another sucrose linking the fructosyl units by β-(2→1) bonds to the fructosyl moiety of sucrose. Several fructosyltransferases – mainly from lactic acid bacteria – have recently been cloned, expressed, and characterized (Van Hijum *et al.*, 2006). These may be both of the inulosucrase type forming β-(2→1) bonds and of the levansucrase type, forming β-(2→6) bonds. The product mixture obtained with inulosucrases is composed mainly of glucose, sucrose, and tri- to pentasaccharides, with one glucose and two to four fructosyl units linked by β-(2→1) bonds (kestose, etc.) exhibiting prebiotic properties. Physiological functionalities attributed to FOS are prevention of dental caries, dietary fiber, stimulation of bifidobacteria, prevention of diarrhea and constipation, reduction of serum cholesterol, further provision of texture, mouthfeel, taste improvement, and fat replacement (Yun and Song, 1999; Tungland, 2003; Rastall and

Hotchkiss, 2003). The production of total oligosaccharides (except isomaltulose) is estimated to be in the range of several tens of thousands of tons annually (Nakakuki, 2005; Buchholz and Seibel, 2008). A selection of commercial products and their producing companies is listed in Table 8.10. The price of oligosaccharides in the food sector ranges from 2 to 5 € kg^{-1}; thus, these products (except for cyclodextrins) represent bulk low-price commodities requiring cheap raw materials (sucrose, starch, inulin), together with economic processing with low cost for the biocatalyst, which must be immobilized and exhibit high productivity.

Cyclodextrins (Figure 8.10b) are products of major importance in the fields of food, pharmaceuticals (drug protection, slow release), and commodities; for example, they are used in textile drying, for odor removal, and as a perfume carrier. In food, their function may be aroma complexation and slow release, stabilization of flavors, and the reduction of bitterness. They are also used to remove cholesterol from milk and egg products. Cyclodextrins are torus-shaped molecules of cyclic α-(1→4)-linked gluco-oligosaccharides. They form inclusion complexes with a wide variety of hydrophobic guest molecules, leading to the applications mentioned. The enzymes used for their manufacture from starch or dextrins are cyclodextrin glycosyltransferases; several have been crystallized, and their structures elucidated and characterized in detail, including the reaction mechanism; they have also been engineered to improve selectivity (Dijkhuizen and van der Veen, 2003; Kelly *et al.*, 2009).

Recent developments made further OS and polysaccharides, as well as derivatives, available industrially. These include malto-, galacto-, xylo-OS, for example, trehalose, gentio- and nigero-OS, lactosucrose, branched cyclodextrins, and pullulan. They exhibit, in addition to sweetness, physiological properties such as gastrointestinal functions, promoting bifidobacteria, increasing absorption and bioavailability of minerals, and cariogenic prophylaxis. Ascorbic acid 2-glucoside is produced by binding glucose to conventional vitamin C; it is blended into many whitening products in Japan (Whitaker *et al.*, 2003; Eggleston and Côté, 2003; Nakakuki, 2005; Buchholz and Seibel, 2008) (www.Hayashibara.co.jp/english/main.html/).

The modification and synthesis of sugars, oligosaccharides, and derivatives has gained much interest since the recognition of their role in biological communication, recognition, and cell–cell interactions, relevant notably in infection and cancer proliferation. New and diverse applications have been derived in the food and pharmaceutical industries and markets, as food additives, prebiotics, and so-called "neutraceuticals" with protective and immune-stimulating functions (Rastall and Hotchkiss, 2003). Chemical synthesis is extremely laborious due to the need for regio- and stereoselectivity in products with multiple stereocenters, whereas enzymes in general exhibit high selectivity in synthesis. Glycosyltransferases represent a most useful group of synthetic enzymes that exhibit only very minor hydrolase activity – that is, they exclude water from the active site, allowing only sugars as nucleophiles for attack of a covalently bound glycosyl group, and thus synthesis of oligosaccharides. Using sucrose or starch as substrates, they transfer glucosyl or fructosyl groups on acceptors to efficiently synthesize OS, for example, trisaccharides such as kestoses, or cyclodextrins, or polysaccharides, such as dextran and fructans. Several overviews of current research and application are available (Buchholz and Seibel, 2008; Homann and Seibel, 2009; Seibel and Buchholz, 2010; Monsan *et al.*, 2010).

New approaches, aiming at the design of tailor-made biocatalysts by enzyme engineering and substrate engineering, have generated new routes to efficient OS synthesis. Among the goals are the chemoenzymatic synthesis of vaccines or vaccine building blocks, such as antigenic surface OS, for example, against *Plasmodium falciparum*, causing malaria, or bacterial pathogens, or against ovarian cancer (Kamena *et al.*, 2008; Champion *et al.*, 2009; Zhu *et al.*, 2009). In order to achieve OS synthesis, avoiding hydrolysis, *glycosynthases* have been conceived via mutagenesis, leading to the development of endo- and exoglycosynthases, which have been summarized in various reviews. Glycosyl fluorides were chosen as activated glycosyl donors, as these mimic the enzyme–substrate intermediate, and aryl glycosides as acceptors, leading to transglycosylation with inverted configuration (Hancock *et al.*, 2006; Jahn and Withers, 2010). Another new approach for the synthesis of oligo- and polysaccharides uses sucrose analogues that are nonnatural activated substrates, derived from sucrose and monosaccharides. They can function as activated donor substrates, providing a convenient and rather low-cost access to synthesis and glycosidation (Kralj *et al.*, 2008; Seibel and Buchholz, 2010). An exofructosyltransferase from *B. subtilis* transfers the fructosyl residue of the substrate sucrose to a series of monosaccharide acceptors (D-mannose, etc.) to yield the β-D-fructofuranosyl α-D-glycopyranosides (D-mannosyl-, D-galactosyl-, D-fucosyl-, D-xylosylfructoside) (Seibel *et al.*, 2006c). They are mostly formed in good or high yields. Sucrose analogues thus open a gate, for example, for the synthesis of new glycopyranosyl oligofructosides using fructosyltransferases (FTFs, fructansucrases) (Seibel *et al.*, 2006a). Thus, modified FTFs from *A. niger* and *B. subtilis* transformed sucrose analogues efficiently and with high yield into the 1-kestose and 1-nystose, or the 6-kestose analogues functionalized with different monosaccharides of potential interest (Figure 8.12) (Zuccaro *et al.*, 2008; Beine *et al.*, 2008).[9] Aiming at improved oligosaccharide versus polysaccharide synthesis, Hellmuth *et al.* (2008) performed a mutagenesis screen yielding a glucosyltransferase variant with favored oligosaccharide formation while exhibiting negligible polymer formation.

The modification of mono- and oligosaccharides to produce derivatives via enzymatic routes has also gained much interest since they operate in aqueous solution, with good or high regio- and stereoselectivity and good yields in general. The modification of sucrose is of major interest as it is available in large quantities, at high purity, and at a low price. Sucrose can be oxidized by *Agrobacterium tumefaciens* with high regioselectivity to 3-ketosucrose (α-D-ribo-hexopyranosyl-3-ulose-β-D-fructofuranoside) with up to 75% yield (Stoppok and Buchholz, 1999). Likewise, other disaccharides such as maltose, lactose, and glucosylsorbitol can be oxidized via this

9) Using L-glycopyranosides as acceptors led to the formation of β-D-fructofuranosyl-β-L-glycopyranosides (L-Glc-Fru, L-Gal-Fru, L-Fuc-Fru, L-Xyl-Fru, Rha-Fru) (Seibel *et al.*, 2006a). The solution of structures of several fructosyltransferases and, recently, of a glucosyltransferase has significantly facilitated the construction of modified and optimized enzymes (Meng and Fütterer, 2008; Vujicic-Zagara *et al.*, 2010).

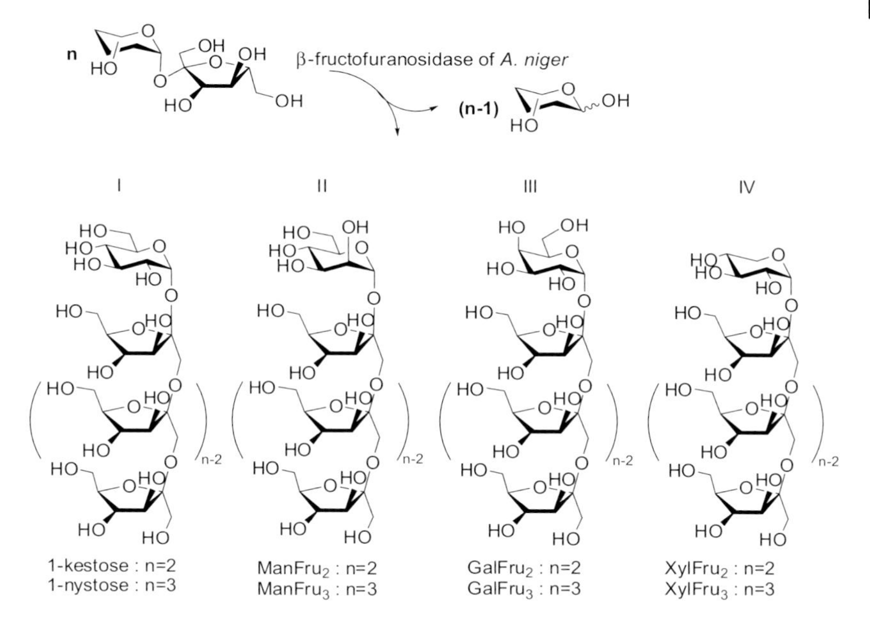

Figure 8.12 Different products with sucrose analogues as substrates. Enzymatic synthesis of 1-kestose, 1-nystose, and their analogues by *β*-fructofuranosidase of *A. niger*. Structures of fructo-oligosaccharides: (I) commercial products, (II) mannose-, (III) galactose-, and (IV) xylose-substituted analogues (Zuccaro *et al.*, 2008; Seibel and Buchholz, 2010).

route, and can be used as building blocks for a range of subsequent chemical steps, including catalytic reduction and coupling with amino acids (Timme *et al.*, 1998; Pietsch *et al.*, 1994). Pyranose oxidases provide access to various keto sugars and keto sugar acids that may serve as chiral intermediates and synthons in chemoenzymatic synthesis. Enzymes acting on 2-, 3-, 4-, or 6-position of different sugars such as D-glucose, D-galactose, D-xylose, or D-ribose have been identified (Röper, 1990, Giffhorn, 2000).

For the synthesis with nucleotide-activated sugars, see Sauerzapfe and Elling (2008). For high-throughput screening of new and engineered carbohydrate-active enzymes, a NMR-based structure determination method has been developed recently (Irague *et al.*, 2011). *Glycochips*, carbohydrate-array types that impact the understanding of glycan-mediated biological processes and give rise to new medical applications, as well as immobilization and readout techniques, have been reviewed by Hillringhaus and Seibel (2010).

8.4.2
Hydrolysis and Synthesis of Penicillins and Cephalosporins

The second example of high economic importance is the hydrolysis and synthesis of penicillin and cephalosporin antibiotics and derivatives. The enzyme mechanism is discussed in Section 2.4 (Figure 2.5); process details are provided in Section 12.3.

The first large-scale enzyme processes developed using immobilized biocatalysts were the hydrolysis of penicillin G and V to yield 6-aminopenicillanic acid (6-APA), during the 1970s. Most of the immobilized biocatalysts used in these processes are produced by the companies that use them for hydrolysis, and few data have been published on these proprietary supports (Liese, Seelbach, and Wandrey, 2006, pp. 386–402). Some companies that have produced (or still produce) immobilized enzymes for commercial purpose did, however, publish more extensive data on their immobilized biocatalysts (e.g., Katchalski-Katzir and Krämer, 2000).

The penicillin amidases used are from *Alcaligenes faecalis*, *Bacillus megaterium*, and *Escherichia coli*. The enzyme has been covalently immobilized to different supports (for carriers used recently see Section 8.2.3). The data published for the process (pH and T) are within the process window derived in Section 12.3 and Chapter 2 (Exercise 2.13). pH control is essential since at pH <7.5 the equilibrium conversion is too low, and at pH >9 both the substrate and product degrade too rapidly. Efficient mixing in technical reactors is important in order to prevent enzyme inactivation by the acids produced in the hydrolytic reaction that are titrated with base (ammonia, or carbonate) to keep the pH constant. The conversion is 98–99%, and the yield is in the range of 86–93% (enzyme productivity decreases with increasing yield). The enzyme productivity is in the range of 1000–2000 kg product kg^{-1} biocatalyst (or in the range of 5 g product U^{-1}). The enzyme cost per product is about 2.5 € kg^{-1} (Poulson, 1984; Cheetham, 2000; Liese, Seelbach, and Wandrey, 2006, pp. 386–402). The hydrolysis of cephalosporin C to 7-aminocephalosporanic acid (7-ACA), which was developed and introduced on an industrial scale during the 1990s, is detailed in Section 12.3. Semisynthetic cephalosporins and penicillins are produced from the hydrolysis products 6-APA, 7-ACA, and 7-ADCA by the condensation of other side chains to the amino group mainly by enzymatic routes (see Section 12.3) (Bruggink, 2001).

8.4.3 Further Processes

8.4.3.1 Amino Acid, Peptide, and Amide Synthesis

Amino acid synthesis is a major sector to which continued research and development efforts have been devoted, both to enzymatic and to fermentation routes (for these see Leuchtenberger *et al.*, 2005; Drauz *et al.*, 2007; Becker *et al.*, 2011). The main reason is that enantiopure products are required, as is generally the case in the food and pharmaceutical sectors (for more details, see Section 4.2.1.1). Chemical synthesis is straightforward, flexible, and economic to provide a range of precursors of different amino acids, such as hydantoins and caprolactam derivatives, including nonnatural varieties (Syldatk *et al.*, 1999; Wenda *et al.*, 2011) (see Section 4.2.3.7). For the industrial production of enantiopure amino acids from these precursors, and also from mixtures of L- and D-isomers produced by chemical synthesis, immobilized enzyme systems are applied in order to keep enzyme costs low. These processes were first introduced by Chibata and coworkers in Japan in 1969 (Chibata *et al.*, 1987). The classical route is acetylation of the D,L-mixture followed

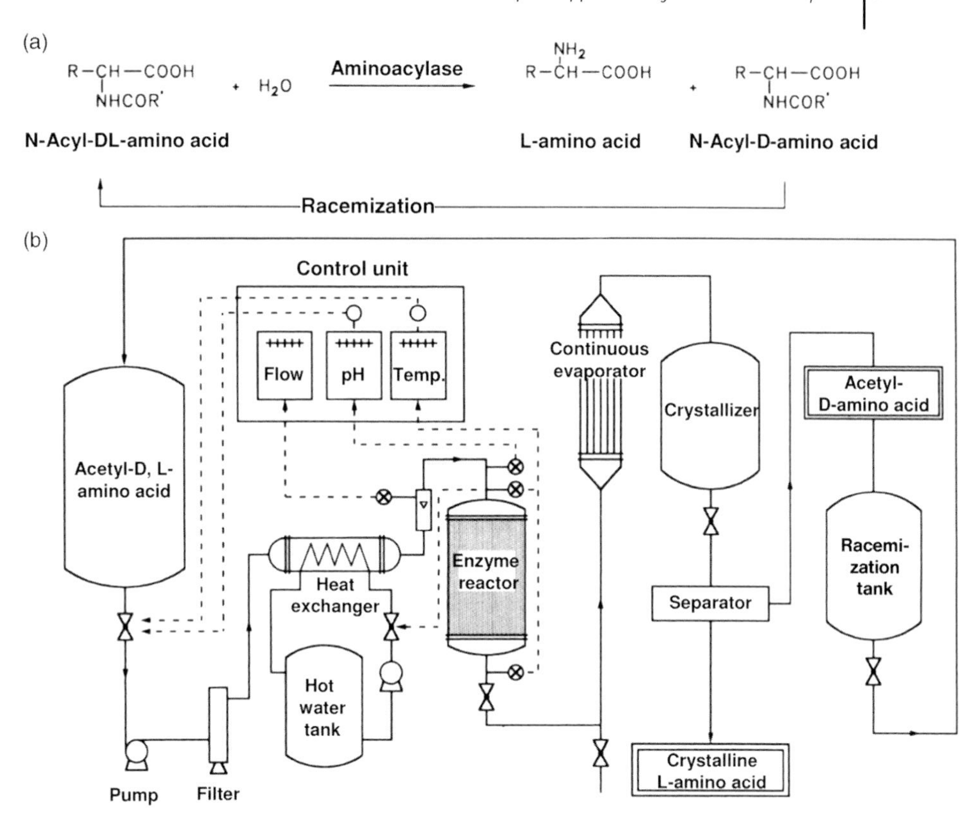

Figure 8.13 Scheme of the production of L-amino acids by amino acylases (Chibata *et al.*, 1987). (a) Reaction; (b) reactor system and peripheral instrumentation.

by stereospecific hydrolysis by an amino acylase and subsequent separation by ion exchange and crystallization. The remaining acetylated isomer is recycled after racemization (Figure 8.13). The amino acylase has been immobilized by adsorption onto an ion-exchange resin, though other methods have also been applied.

A success story has been the production of acrylamide using nitrile hydratase on a scale of 400 000 t a^{-1} (Wenda *et al.*, 2011). After successive optimization rounds, Lonza developed a process with continuous feed of 3-cyanopyridine of 10–20 wt% concentration to an immobilized *Rhodococcus rhodochrous* strain with active nitrile hydratase. Another economically highly relevant peptide synthesis is that of aspartame (Figure 8.14), a high-intensity sweetener that is used to replace sugar and reduce the caloric content in foods, for example, in soft drinks. The synthesis uses protected (*N*-benzylcarboxy-)aspartic acid (*Z*-Asp) and cheap racemic phenylalanine methyl ester (PheMe) (see also Section 4.2.3.3). The residual D-PheMe is racemized and recycled. Thermolysin was found after a broad screening effort to catalyze the synthesis both regio- and enantioselectively, without hydrolyzing the methyl ester. It is applied after immobilization in stirred tank reactors using ethyl acetate:water

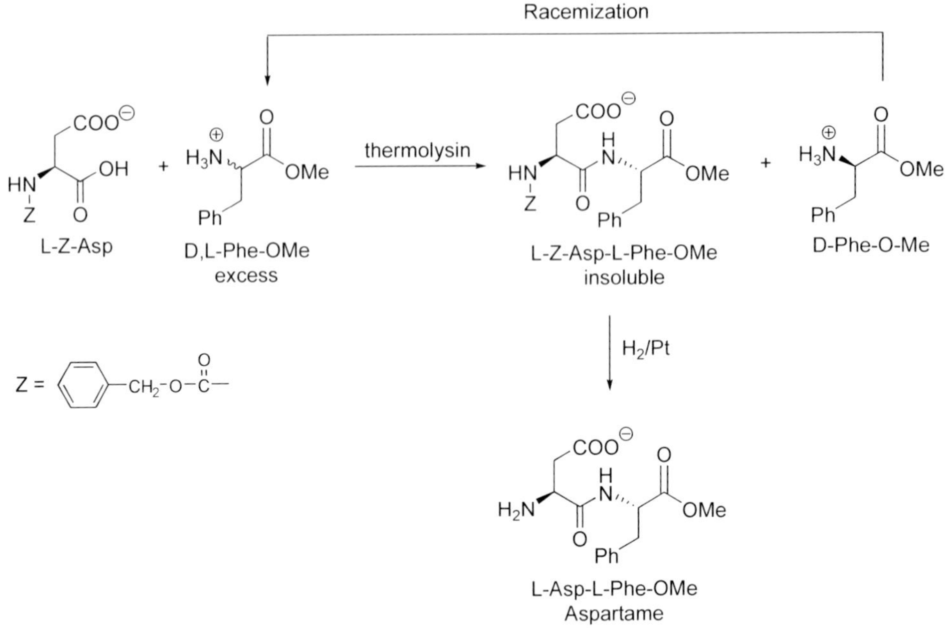

Racemization

thermolysin

L-Z-Asp

D,L-Phe-OMe
excess

L-Z-Asp-L-Phe-OMe
insoluble

D-Phe-O-Me

H_2/Pt

Z = C_6H_5–CH_2–O–C(=O)–

L-Asp-L-Phe-OMe
Aspartame

Figure 8.14 Synthesis of aspartame using thermolysin.

as the solvent. The yield is >95% at 55 °C, and annual sales are in the range of 850×10^6 € (Cheetham, 2000; Hanzawa, 2010).

Reactions using dehydrogenases are discussed in Section 4.2.1.1. Yet another route to the production of chiral amino acids is via transamination with α-keto or α-hydroxy acids as readily available substrates (Höhne and Bornscheuer, 2010). An example has been discussed in Section 7.4.

8.4.3.2 **Application of Lipases**

Immobilized lipases are used in a range of industrial processes including the manufacture of *structured triglycerides* (TAGs), specific products with enriched polyunsaturated fatty acids, and biofuels (biodiesel). Further phospholipases are used on a large scale for degumming – the removal of phospholipids – of natural fats and oils. TAGs with a defined distribution of fatty acids are important compounds for applications in human nutrition, for example, to treat patients with pancreatic insufficiency, in infant nutrition (products with selected types of fatty acids), and for rapid energy supply, for example, in sports. They are manufactured using, for example, lipases from *Rhizomucor miehei* (Novozyme RMIM) or the lipase from *C. antarctica* (CAL-B) that allow for the synthesis of TAGs due to their fatty acid type specificity (in case of CAL-B) and *sn*1.3-regioselectivity (in case of RMIM) (Biermann *et al.*, 2011). As an example for a TAG to be applied in infant nutrition, 1,3-oleoyl-2-palmitoylglycerol was synthesized by a two-step process in high yields and purity using *sn*1,3-regiospecific lipases. The best results were achieved with immobilized lipases from *R. miehei* and *Rhizopus delemar.* Emphasis was given to develop the process in solvents allowed to be used in foodstuffs and to perform the second step in a solvent-free system (Schmid *et al.*, 1999). Further examples of industrial processes are the production of hardstocks for margarines (ADM/Novozymes) without *trans*-fatty acids, the selective synthesis of 1,3-diglycerides with properties as low-calorie fat replacements, of diglyceride-based cooking and frying oils (Kao Corp./ADM) on a $>30\,000\,\mathrm{t\,a^{-1}}$ production scale, and further the enrichment by selective transesterification to make products with enhanced polyunsaturated fatty acids (PUFAs) that exhibit positive effects on human health. Unilever and Fuji Oil manufacture cocoa butter equivalent from palm oil using 1,3-selective lipases for transesterification or acidolysis using tristearin or stearic acid as acyl donors. Lipases have also been used on the industrial scale to produce esters for cosmetic applications, for example, myristyl myristate and cetyl ricinoleate, and specialty oils such as oleochemicals from specialty esters and amides. Enzymatic interesterification is a more cost-efficient process since it allows for higher yields and gives purer products and savings in energy (ambient temperature instead of 160–180 °C), as compared to chemical synthesis (see also Chapter 4.2.3.1) (Schäfer *et al.*, 2007; Biermann *et al.*, 2011) (www.novozymes.com/; www.amano-enzyme.co.jp/).

Enzymatic *biodiesel* transesterification has found much attention. Currently, lipases are the enzymes of choice for the synthesis of fatty acid esters (FAEs) from fats and oils, yielding biodiesel with the methyl esters (FAME) as the most important product. Whole-cell *Rhizopus oryzae* immobilized within biomass support particles (BSPs) as catalyst for biodiesel production in *tert*-butanol exhibited enhanced stability and were found to give good yields (>80%)[10] (Li *et al.*, 2007). More recently, the direct production of FAE using engineered whole-cell microorganism has also been described (Du *et al.*, 2008; Adamczak *et al.*, 2009). Still, however, the cost of enzymes remains a barrier for its industrial implementation (Nielsen *et al.*, 2008). Nevertheless, it has been reported recently that the first large-scale biodiesel plant has started operation in China, to produce fatty ester alkyl esters using a lipase and *tert*-butanol as cosolvent (capacity $20\,000\,\mathrm{t\,a^{-1}}$) (Fjerbaek *et al.*, 2009; Biermann *et al.*, 2011).

Exercises

Immobilization of Enzymes on Ion Exchangers

- Lipase (Novo, European Patent 0382 767, 1988)

  Adsorption on ion exchanger: Lewatit E2001/85, or equivalent (e.g., LEWATIT® VP OC 1600) (Lanxess, D) (see Table 8.7).
  Enzyme: *Mucor miehei* lipase 2.6 g (112 lipase-U $\mathrm{mg^{-1}}$) in 25 ml solution, addition to carrier: 8.5 g, pH 6.1, 21 h.

10) Critical aspects were feedstocks (e.g., rapeseed oil), the contents of free fatty acids, phospholipids, and water, which preferentially should be removed by adsorbents (Li *et al.*, 2007).

Immobilized activity: 171 BIU g^{-1} (batch interesterification unit, 60 °C). Application in molten substrates without water or solvents.

- Transglycosidase and pullulanase (Cerestar, European Patent 875 585, 1997; further examples: European Patent 336 376, Mitsubishi Kasei Corp., 1988)

  Coimmobilization of two enzymes on Duolite A 568, or quivalent (e.g., Amberlite™ styrene-DVB, MR, weak base, FPA 51, or acrylic-DVB, weak base, FPA 53, see also Table 8.7); adsorption and cross-linking.
  Enzymes:

1) Transglycosidase (Amano Pharmaceutical Co.) (www.amano-enzyme.co.jp/eng/productuse/starch.html) 11 mg ml^{-1}, 103 TGU ml^{-1} (solution).
   Addition of 1 ml solution to 10 ml Duolite, or Amberlite™, 4 h, 20 °C.
2) Pullulanase (400 PU ml^{-1} solution), addition of 15 ml solution; 4 h.
   Cross-linking: addition of 5 ml glutaraldehyde solution (1%; 0.2% final concentration), 12 h, 20 °C, washing.

Immobilization of Enzymes on Porous Glass or Silica

- Amyloglucosidase, porous silica as carrier

Reagents and equipment

- Amyloglucosidase (AG) (may be kept for some time as dry preparation in the refrigerator; test for activity before use).
- Porous silica, or porous glass, derivatized with amino groups (see Table 8.5 and Section 8.3.2 for different properties).
- Solution of glutaraldehyde (GA), 5%.
- Sörensen phosphate buffer, pH 7, 0.01 M (add a droplet of toluene as antimicrobial agent).
- Sörensen phosphate buffer, pH 5, 0.01 M (add a droplet of toluene as antimicrobial agent).
- Thermostated glass vessel, about 50 ml, with frit, or sieve, and stirrer.
- Thermostat, 50 °C.
- UV photometer, cuvette (10 mm).
- Peristaltic pump.
- Glucose and/or ethanol analysis by enzymatic test, or GC, HPLC, or glucose analyzer.

Immobilization
Suspend 3 g of carrier in 20 ml buffer (pH 7), degas; add 5 ml of GA solution (5%), if needed add another 10 ml of buffer; the reaction should proceed for 30 min while stirring and pumping the solution through a bypass (from the lower outlet of the glass vessel and reintroducing it from the top); take off the solution, wash two times with water.

Add solution of AG (corresponding to 0.2 g, or 1000 U) in 5 ml of buffer (pH 7); measurement of extinction at 280 nm every 15 min (dilution by 1: 20), let reaction proceed for 45 min; take off the solution (test for residual enzyme activity), wash three times with water, suspend in 50 ml of buffer (pH 5).

Measurement of activity in suspended solution of immobilized enzyme (with a sample of suspended carrier) and with a solution of dextrin as substrate (low molecular weight; high molecular weight substrate hydrolysis is limited by diffusion); glucose analysis by enzymatic test, or HPLC, or glucose analyzer. The concentrations of glucose and/or ethanol are graphically recorded as a function of time; the catalytic activity (g (glucose/ethanol) per g (yeast) per h) determined from the slope; compare the reaction with the same amount (activity) of free enzyme; calculate balance of active immobilized enzyme.

Immobilization of Enzymes on a Synthetic Carrier

- Amyloglucosidase, Sepabeads® amino-epoxy as carrier (see Figure 8.4)

Reagents and equipment

- Amyloglucosidase (AG).
- Sepabeads® amino-epoxy, from Resindion (Japan).
- Potassium phosphate buffer, pH 5, 0.01 M.
- Citrate buffer 0.1 M (pH 4.5).
- Glass vessels, 10 or 20 ml volume.

Immobilization

Prepare a solution of 50 mg (or 300 U) of AG in 50 ml of buffer (pH 6) in four glass vessels; add different amounts of carrier (see the following table) and let react while shaking the closed vessels for 24 h at 25 °C. After sedimentation of the carrier, the supernatant is removed with a pipette. Residual epoxy groups may be inactivated using ethanol or glycine (Guisan, Betancor, and Fernandez-Lorente, 2010). The carrier with immobilized enzyme is washed three to four times with citrate buffer 0.1 M (pH 4.5).

The supernatants of the different preparations are tested for residual enzyme activity. The immobilized enzyme preparations are suspended in 25 ml of citrate buffer (pH 4,5); samples are tested for activity in shaked glass vessels (see above).

Amounts for immobilization

Number	Amyloglucosidase (g)	Sepabeads® (g)	Ratio of enzyme to carrier
1	0.05	0.25	0.20
2	0.05	0.50	0.10
3	0.05	0.75	0.067
4	0.05	1.00	0.05

As an alternative, the procedure published and well documented (Experimental, Table 2) by Hilterhaus *et al.* (2008) may be applied using Sepabeads® EC-EP with epoxide functional groups.

Literature

General Literature

Bickerstaff, G.F. (ed.) (1997) Methods in biotechnology, in *Immobilization of Enzymes and Cells*, vol. **1**, Humana Press Inc., Totowa, NJ.

Buchholz, K. (ed.) (1979) *Characterization of Immobilized Biocatalysts, DECHEMA Monograph*, vol. **84**, Wiley-VCH Verlag GmbH, Weinheim.

Godfrey, T. (ed.) (1996) *Industrial Enzymology*, Macmillan Press, London.

Liese, A., Seelbach, K., and Wandrey, C. (2006) *Industrial Biotransformations*, Wiley-VCH Verlag GmbH, Weinheim.

Mosbach, K. (ed.) (1987) *Methods in Enzymology*, vol. **135**, Academic Press, Orlando, FL.

Pedersen, S. and Christensen, M.W. (2000) Immobilized biocatalysts, in *Applied Biocatalysis* (eds A.J.J. Straathof and P. Adlercreutz), Harwood Academic Publishers, Amsterdam, pp. 213–228.

Prenosil, J.E. *et al.* (2009) Biocatalysis. 2. Immobilized biocatalysts, in *Ullmann's Encyclopedia of Industrial Chemistry*, Wiley-VCH Verlag GmbH, Weinheim, www.mrw.interscience.wiley.com/emrw/.

Swiss-Prot, http://www.kr.expasy.org/enzyme/.

Tischer, W. and Kasche, V. (1999) Immobilization enzyme crystals. *Trends Biotechnol.*, **17**, 326–335.

References

Adamczak, M., Bornscheuer, U.T. *et al.* (2009) The application of biotechnological methods for the synthesis of biodiesel. *Eur. J. Lipid Sci. Technol.*, **111** (8), 800–813.

Allard, L., Cheynet, V., Oriol, G. *et al.* (2002) Versatile method for production and controlled polymer-immobilization of biologically active recombinant proteins. *Biotechnol. Bioeng.*, **80**, 341–348.

Andersson, M., Samra, B., Holmberg, H., and Adlercreutz, P. (1999) Use of Celite-immobilised chloroperoxidase in predominantly organic media. *Biocatal. Biotransform.*, **17**, 293–303.

Antrim, R.L. and Auterinen, A.-L. (1986) A new regenerable immobilized glucose isomerase. *Stärke*, **38**, 132–137.

Balcao, V.M., Paiva, A.L., and Malcata, F.X. (1996) Bioreactors with immobilized lipases: state of the art. *Enzyme Microb. Technol.*, **18**, 392–416.

Barros, R.J., Wehtje, E., Garcia, F., and Adlercreutz, P. (2000) Physical characterization of porous materials. *Biocatal. Biotransform.*, **16**, 67–85.

Becker, J., Zelder, O. *et al.* (2011) From zero to hero – design-based systems metabolic engineering of *Corynebacterium glutamicum* for L-lysine production. *Metab. Eng.*, **13**, 159–168.

Beine, R., Moraru, R. *et al.* (2008) Synthesis of novel fructooligosaccharides by substrate and enzyme engineering. *J. Biotechnol.*, **138** (1–2), 33–41.

Biermann, U., Bornscheuer, U. *et al.* (2011) Oils and fats as renewable raw materials in chemistry. *Angew. Chem., Int. Ed.*, **50**, 3854–3871.

Boehm, G., Stahl, B. *et al.* (2005) Prebiotic carbohydrates in human milk and formulas. *Acta Paediatr.*, **94**, 18–21.

Boniello, C., Mayr, T. *et al.* (2010) Intraparticle concentration gradients for substrate and acidic product in immobilized cephalosporin C amidase and their dependencies on carrier characteristics and reaction parameters. *Biotechnol. Bioeng.*, **106** (4), 528–540.

Boniello, C., Mayr, T., Juan M. Bolivar and Nidetzky, B. (2012) Dual-lifetime referencing (DLR): a powerful method for on-line measurement of internal pH in carrier-bound immobilized biocatalysts. *BMC Biotechnology*, **12**:11.

Borchert, A. and Buchholz, K. (1984) Improved biocatalyst effectiveness by controlled immobilization of enzymes. *Biotechnol. Bioeng.*, **26**, 727–736.

Bornscheuer, U.T. (2003) Immobilizing enzymes: how to create more suitable biocatalysts. *Angew. Chem., Int. Ed.*, **42**, 3336–3337.

Breuer, M., Ditrich, K. *et al.* (2004) Industrial methods for the production of optically active intermediates. *Angew. Chem., Int. Ed.*, **43** (7), 788–824.

Bruggink, A. (ed.) (2001) *Synthesis of β-Lactam Antibiotics*, Kluwer Academic Publishers, Dordrecht.

Bruns, N. and Tiller, J.C. (2005) Amphiphilic network as nanoreactor for enzymes in organic solvents. *Nano Lett.*, **5** (1), 45–48.

Buchholz, K., Duggal, S.K., and Borchert, A. (1979) *Characterization of Immobilized Biocatalysts, DECHEMA Monograph*, vol. **84**, Wiley-VCH Verlag GmbH, Weinheim, pp. 1–48, 169–181.

Buchholz, K. and Kasche, V. (1997) *Biokatalysatoren und Enzymtechnologie*, Wiley-VCH Verlag GmbH, Weinheim.

Buchholz, K. and Monsan, P. (2003) Dextransucrase, in *Handbook of Food Enzymology* (eds J.R. Whitaker, A.G.J. Voragen, and D.W.S. Wong), Marcel Dekker, New York, pp. 589–603.

Buchholz, K. and Seibel, J. (2008) Industrial carbohydrate biotransformations. *Carbohydr. Res.*, **343** (12), 1966–1979.

Burg, K., Mauz, O., Noetzel, S., and Sauber, K. (1988) Neue synthetische Träger zur Fixierung von Enzymen. *Angew. Makrom. Chem.*, **157**, 105–121.

Cao, L., van Rantwijk, F., and Sheldon, R.A. (2000) Cross-linked enzyme aggregates: a simple and effective method for the immobilization of penicillin acylase. *Org. Lett.*, **2**, 1361–1364.

Carleysmith, S.W., Dunnil, P., and Lilly, M.D. (1980) Kinetic behaviour of immobilized penicillin acylase. *Biotechnol. Bioeng.*, **22**, 735–756.

Champion, E., Andre, I. *et al.* (2009) Design of α-transglucosidases of controlled specificity for programmed chemoenzymatic synthesis of antigenic oligosaccharides. *J. Am. Chem. Soc.*, **131** (21), 7379–7389.

Cheetham, P.S.J. (ed.) (1987) Production of isomaltose using immobilized microbial cells, in *Methods in Enzymology*, vol. **136** (ed. K. Mosbach), Academic Press, Orlando, FL, pp. 432–454.

Cheetham, P. (2000) Case studies in the application of biocatalysts, in *Applied Biocatalysis* (eds A.J.J. Straathof and P. Adlercreutz), Harwood Academic Publishers, Amsterdam, pp. 93–152.

Chibata, I., Tosa, T., and Sato, T. (1987) Application of immobilized biocatalysts in pharmaceutical and chemical industries, in *Biotechnology*, vol. **7a** (eds H.J. Rehm and G. Reed), Verlag Chemie, Weinheim, pp. 653–684.

Chikere, A.C., Galunsky, G., Schünemann, V., and Kasche, V. (2001) Stability of immobilised soybean lipoxygenases: influence of coupling conditions of the ionization state of the active site Fe. *Enzyme Microb. Technol.*, **28**, 168–175.

Creighton, T.E. (1993) *Proteins: Structures and Molecular Properties*, W.H. Freeman, New York.

Dardel, F. and Arden, T.V. (2008) Ion exchangers, in *Ullmann's Encyclopedia of Industrial Chemistry*, Wiley-VCH Verlag GmbH, Weinheim, www.mrw.interscience.wiley.com/emrw/.

Dauner-Schütze, C., Brauer, E., Borchert, A., and Buchholz, K. (1988) Development of a high capacity adsorbent for enzyme isolation and immobilization, in *Bioreactor Immobilized Enzymes and Cells* (ed. M. Moo-Young), Elsevier Applied Science, London, pp. 63–70.

De Lathouder, K.M., Marques Fló, T. *et al.* (2005) A novel structured bioreactor: development of a monolithic stirrer reactor with immobilized lipase. *Catal. Today*, **105** (3–4), 443–447.

Desmet, T. and Soetaert, W. (2011) Enzymatic glycosyl transfer: mechanisms and applications. *Biocatal. Biotransform.*, **29**, 1–18.

Dib, I. and Nidetzky, B. (2008) The stabilizing effects of immobilization in D-amino acid oxidase from *Trigonopsis variabilis*. *BMC Biotechnol.*, **8** (1), 72.

Dijkhuizen, L. and van der Veen, B. (2003) Cyclodextrin glycosyltransferase, in *Handbook of Food Enzymology* (eds J.R. Whitaker, A.G.J. Voragen, and D.W.S. Wong), Marcel Dekker, New York, pp. 615–627.

Drauz, K., Grayson, I., Kleemann, A. *et al.* (2007) Amino acids, in *Ullmann's Encyclopedia of Industrial Chemistry*, Wiley-VCH Verlag GmbH, Weinheim.

Du, W., Li, W. *et al.* (2008) Perspectives for biotechnological production of biodiesel and impacts. *Appl. Microbiol. Biotechnol.*, **79** (3), 331–337.

Eggleston, G. and Côté, G.L. (2003) Oligosaccharides in food and agriculture. ACS Symposium Series 849, American Chemical Society, Washington, DC.

Fei, G., Ma, G.H. *et al.* (2010) Enzyme immobilization, biocatalyst featured with nanoscale structure, in *Encyclopedia of Industrial Biotechnology*, John Wiley & Sons, Inc., New York, http://tinyurl.com/wileyEIB.

Fernandez-Lafuente, R., Rosell, C.M., Rodriguez, V., and Guisan, J.M. (1995) Strategies for enzyme stabilisation by intramolecular crosslinking with bifunctional reagents. *Enzyme Microb. Technol.*, **17**, 517–523.

Fjerbaek, L., Christensen, K.N., and Norddahl, B. (2009) A review of the current state of biodiesel production using enzymatic transesterification. *Biotechnol. Bioeng.*, **102**, 1298–1315.

Fritz Wolf, K., Koller, K.P. *et al.* (2002) Structure based prediction of modifications in glutarylamidase to allow single step enzymatic production of 7-aminocephalosporanic acid from cephalosporin C. *Protein Sci.*, **11** (1), 92–103.

Gao, F., Su, Z.G. *et al.* (2009) Double emulsion templated microcapsules with single hollow cavities and thickness-controllable shells. *Langmuir*, **25** (6), 3832–3838.

Genencor International (1990) Glucose isomerases having altered substrate specificity. European Patent 1 264 883.

Ghisalba, O., Meyer, H.P. *et al.* (2010) Industrial biotransformation, in *Encyclopedia of Industrial Biotechnology*, John Wiley & Sons, Inc., New York, http://tinyurl.com/wileyEIB.

Giffhorn, F. (2000) Fungal pyranose oxidases: occurrence, properties and biotechnical applications in carbohydrate chemistry. *Appl. Microbiol. Biotechnol.*, **54**, 727–740.

Godfrey, T. and West, S. (1996) Introduction to industrial enzymology, in *Industrial Enzymology* (eds T. Godfrey and S. West), Macmillan Press, London, pp. 1–8.

Goedl, C., Sawangwan, T. *et al.* (2010) Sucrose phosphorylase: a powerful transglucosylation catalyst for synthesis of α-D-glucosides as industrial fine chemicals. *Biocatal. Biotransform.*, **28**, 10–21.

Guisan, J.M., Betancor, L., and Fernandez-Lorente, G. (2010) Immobilized enzymes, in *Encyclopedia of Industrial Biotechnology*, John Wiley & Sons, Inc., New York, http://tinyurl.com/wileyEIB.

Hancock, S.M., Vaughan, M.D. *et al.* (2006) Engineering of glycosidases and glycosyltransferases. *Curr. Opin. Chem. Biol.*, **10** (5), 509–519.

Hanzawa, S. (2010) Aspartame, in *Encyclopedia of Industrial Biotechnology*, John Wiley & Sons, Inc., New York, http://tinyurl.com/wileyEIB.

Hartmann, M. and Jung, D. (2010) Immobilization of proteins and enzymes: mesoporous supports, in *Encyclopedia of Industrial Biotechnology*, John Wiley & Sons, Inc., New York, http://tinyurl.com/wileyEIB.

Hellmuth, H., Wittrock, S. *et al.* (2008) Engineering the glucansucrase GTFR enzyme reaction and glycosidic bond specificity: toward tailor-made polymer and oligosaccharide products. *Biochemistry*, **47** (25), 6678–6684.

Herdt, A.R., Kim, B.S. *et al.* (2007) Encapsulated magnetic nanoparticles as supports for proteins and recyclable biocatalysts. *Bioconjug. Chem.*, **18** (1), 183–189.

Hillringhaus, L. and Seibel, J. (2010) Glycochips, in *Encyclopedia of Industrial Biotechnology*, John Wiley & Sons, Inc., New York, http://tinyurl.com/wileyEIB.

Hilterhaus, L., Minow, B. *et al.* (2008) Practical application of different enzymes immobilized on Sepabeads. *Bioprocess Biosyst. Eng.*, **31** (3), 163–171.

Homann, A. and Seibel, J. (2009) Towards tailor-made oligosaccharides – chemo-enzymatic approaches by enzyme and substrate engineering. *Appl. Microbiol. Biotechnol.*, **83** (2), 209–216.

Höhne, M., Bornscheuer, U.T. (2009), Biocatalytic routes to optically active amines, *ChemCatChem*, **1**, 42–51.

Irague, R., Massou, S. *et al.* (2011) NMR-based structural glycomics for high-throughput screening of carbohydrate-active enzyme specificity. *Anal. Chem.* **83**, 1202–1206.

Jahn, M. and Withers, S.G. (2010) New approaches to enzymatic oligosaccharide synthesis: glycosynthases and thioglycoligases. *Biocatal. Biotransform.*, **21** (4–5), 159–166.

Kasche, V. (1973) Effects of the microenviroment on the specific interaction between alpha-chymotrypsin and immobilized soybean trypsin inhibitor. *Studia Biophysica*, **35**, 45–56.

Kamena, F., Tamborrini, M. *et al.* (2008) Synthetic GPI array to study antitoxic malaria response. *Nat. Chem. Biol.*, **4** (4), 238–240.

Kasche, V., de Boer, M. *et al.* (2003) Direct observation of intraparticle equilibration and the rate-limiting step in adsorption of proteins in chromatographic adsorbents with confocal laser scanning microscopy. *J. Chromatogr. B*, **790** (1–2), 115–129.

Katchalski-Katzir, E. and Krämer, D.M. (2000) Eupergit[R] C, a carrier for immobilization of enzymes. *J. Mol. Catal. B*, **10**, 157–176.

Kelly, R.M., Dijkhuizen, L. *et al.* (2009) The evolution of cyclodextrin glucanotransferase product specificity. *Appl. Microbiol. Biotechnol.*, **84** (1), 119–133.

Kim, J., Grate, J.W. *et al.* (2006) Nanostructures for enzyme stabilization. *Chem. Eng. Sci.*, **61** (3), 1017–1026.

Kohn, J. and Wilchek, M. (1982) A new approach (cyano-transfer) for cyanogen bromide activation of Sepharose at neutral pH, which yields activated resins free of interfering nitrogen derivatives. *Biochem. Biophys. Res. Commun.*, **107**, 878–884.

Kralj, S., Buchholz, K. *et al.* (2008) Fructansucrase enzymes and sucrose analogues: a new approach for the synthesis of unique fructo-oligosaccharides. *Biocatal. Biotransform.*, **26** (1–2), 32–41.

Krämer, D.M. (1979) *Characterization of Immobilized Biocatalysts, DECHEMA Monograph*, vol. **84** (ed. K. Buchholz), Wiley-VCH Verlag GmbH, Weinheim, p. 168.

Lavie, A., Allen, K.N., Petsko, G.A., and Ringe, D. (1994) X-ray crystallographic structures of D-xylose isomerase–substrate complexes position the substrate and provide evidence for metal movement during catalysis. *Biochemistry*, **33**, 5469–5480.

Lee, J., Lee, D. *et al.* (2005) Preparation of a magnetically switchable bio-electrocatalytic system employing cross-linked enzyme aggregates in magnetic mesocellular carbon foam. *Angew. Chem., Int. Ed.*, **44** (45), 7427–7431.

Leuchtenberger, W., Huthmacher, K. *et al.* (2005) Biotechnological production of amino acids and derivatives: current status and prospects. *Appl. Microbiol. Biotechnol.*, **69** (1), 1–8.

Li, W., Du, W. *et al.* (2007) Optimization of whole cell-catalyzed methanolysis of soybean oil for biodiesel production using response surface methodology. *J. Mol. Catal. B*, **45** (3–4), 122–127.

Libertino, S., Giannazzo, F. *et al.* (2008) XPS and AFM characterization of the enzyme glucose oxidase immobilized on SiO_2 surfaces. *Langmuir*, **24** (5), 1965–1972.

Liese, A., Seelbach, K., and Wandrey, C. (2000) *Industrial Biotransformations*, Wiley-VCH Verlag GmbH, Weinheim.

Liese, A., Seelbach, K., and Wandrey, C. (2006) *Industrial Biotransformations*, 2nd edn, Wiley-VCH Verlag GmbH, Weinheim.

Linden, J.C. (1982) Immobilized α-D-galactosidase in the sugar beet industry. *Enzyme Microb. Technol.*, **4**, 130–136.

Makkee, M., Kieboom, A.P.G., and van Bekkum, H. (1984) Glucose isomerase catalyzed D-glucose–D-fructose interconversion. *Recl. Trav. Chim. Pays-Bas*, **103**, 361–364.

Manecke, G., Ehrenthal, E., and Schlüsen, J. (1979) *Characterization of Immobilized Biocatalysts*, in *DECHEMA Monograph*, vol. **84** (ed. K. Buchholz), Wiley-VCH Verlag GmbH, Weinheim, pp. S49–S52.

Matsuda, T., Nakayama, K. *et al.* (2010) Stabilization of pyruvate decarboxylase under a pressurized carbon dioxide/water biphasic system. *Biocatal. Biotransform.*, **28** (3), 167–171.

Meng, G. and Fütterer, K. (2008) Donor substrate recognition in the raffinose-

bound E 342 A mutant of fructosyltransferase *Bacillus subtilis* levansucrase. *BMC Struct. Biol.*, **8** (1), 16.
Miletic, N., Nguyen, L.T.T. *et al.* (2010) Formation, topography and reactivity of *Candida antarctica* lipase B immobilized on silicon surface. *Biocatal. Biotransform.*, **28**, 357–369.
Misset, O. (2003) Xylose (glucose) isomerase, in *Handbook of Food Enzymology* (eds J.R. Whitaker, A.G.J. Voragen, and D.W.S. Wong), Marcel Dekker, New York, pp. 1057–1077.
Mitsubishi Kasei Corp. (1988) European Patent 336 376.
Monsan, P.J. (1978) Optimization of glutaraldehyde activation of a support for enzyme immobilization. *J. Mol. Catal.*, **3**, 371–384.
Monsan, P., Remaud-Siméon, M. *et al.* (2010) Transglucosidases as efficient tools for oligosaccharide and glucoconjugate synthesis. *Curr. Opin. Microbiol.*, **13** (3), 293–300.
Nakakuki, T. (2005) Present status and future prospects of functional oligosaccharide development in Japan. *J. Appl. Glycosci.*, **52** (3), 267–271.
Nakamura, H. and Karube, I. (2003) Current research activity in biosensors. *Anal. Bioanal. Chem.*, **377** (3), 446–468.
Nakayama, T. and Amachi, T. (2010) *β*-Galactosidase, in *Encyclopedia of Industrial Biotechnology*, John Wiley & Sons, Inc., New York, http://tinyurl.com/wileyEIB.
Nielsen, P.M., Brask, J. *et al.* (2008) Enzymatic biodiesel production: technical and economical considerations. *Eur. J. Lipid Sci. Technol.*, **110** (8), 692–700.
Norde, W. and Zoungrana, T. (1998) Activity and structural stability of adsorbed enzymes, in *Stability and Stabilization of Biocatalysts* (eds A. Ballesteros, F.J. Plou, J.L. Iborra, and P.J. Halling), Elsevier, Amsterdam, pp. 495–504.
Ouarne, F. and Guibert, A. (1995) Fructo-oligosaccharides: enzymic synthesis from sucrose. *Zuckerindustrie*, **120**, 793–798.
Pedersen, S. and Christensen, M.W. (2000) Immobilized biocatalysts, in *Applied Biocatalysis* (eds A.J.J. Straathof and P. Adlercreutz), Harwood Academic Publishers, Amsterdam, pp. 213–228.
Petersen, S.B., Jonson, P.H., and Fojan, P. (1998) Protein engineering the surface of enzymes. *J. Biotechnol.*, **66**, 11–26.
Pietsch, M., Walter, M., and Buchholz, K. (1994) Regioselective synthesis of new sucrose derivatives via 3-ketosucrose, *Carb. Res.* **254** 83–194.
Polzius, R., Schneider, T., Bier, F.F., Bilitewsky, U., and Koschinski, W. (1996) Optimization of biosensing using grating couplers: immobilization on tantalum oxide waveguides. *Biosens. Bioelectron.*, **11**, 503–514.
Poulson, P.B. (1984) Current applications of immobilized enzymes for manufacturing purposes. *Biotechnol. Genet. Eng. Rev.*, **1**, 121.
Praaning, D.P. (2003) Regulatory issues of enzymes used in foods from the perspective of the E.U. market, in *Handbook of Food Enzymology* (eds J.R. Whitaker, A.G.J. Voragen, and D.W.S. Wong), Marcel Dekker, New York, pp. 59–65.
Ralla, K., Sohling, U. *et al.* (2010) Adsorption and separation of proteins by a smectitic clay mineral. *Bioprocess Biosyst. Eng.*, **33**, 847–861.
Rastall, R.A. and Hotchkiss, A.T., Jr., (2003) Potential for the development of prebiotic oligosaccharides form biomass. *ACS Symp. Ser.*, **849**, 44–53.
Reilly, P. and Antrim, R.L. (2003) Enzymes in grain wet milling, in *Ullmann's Encyclopedia of Industrial Chemistry*, Wiley-VCH Verlag GmbH, Weinheim.
Röper, H. (1990) Selective Oxidation of D-Glucose: Chiral Intermediates for Industrial Utilization. starch/stärke, **42**, 342–349.
Rose, T. and Kunz, M. (2002) Production of isomalt. *Landbauforschung*, **241**, 75–80.
Sauerzapfe, B. and Elling, L. (2008) Multienzyme systems for the synthesis of glycoconjugates, in *Multi-Step Enzyme Catalysis: Biotransformations and Chemoenzymatic Synthesis* (ed. E. Garcia-Junceda), Wiley-VCH Verlag GmbH, Weinheim.
Schäfer, T., Borchert, T.W. *et al.* (2007) Industrial enzymes. *Adv. Biochem. Eng. Biotechnol.*, **105**, 59–131.
Scharer, R., Hossain, M.M. *et al.* (1992) Determination of total and active

immobilized enzyme distribution in porous solid supports. *Biotechnol. Bioeng.*, **39** (6), 679–687.

Schilke, K.F. and Kelly, C. (2008) Activation of immobilized lipase in non aqueous systems by hydrophobic poly-DL-tryptophan tethers. *Biotechnol. Bioeng.*, **101** (1), 9–18.

Schiweck, H., Munir, M., Rapp, K.M., Schneider, B., and Vogel, M. (1991) New developments in the use of sucrose as an industrial bulk chemical, in *Carbohydrates as Organic Raw Materials* (ed. F.W. Lichtenthalter), Wiley-VCH Verlag GmbH, Weinheim, pp. 57–94.

Schmid, U., Bornscheuer, U.T. *et al.* (1999) Highly selective synthesis of 1,3-oleoyl-2-palmitoylglycerol by lipase catalysis. *Biotechnol. Bioeng.*, **64** (6), 678–684.

Schmid, A., Hollmann, F., Park, J.B., and Bühler, B. (2002) The use of enzymes in the chemical industry in Europe. *Curr. Opin. Biotechnol.*, **13**, 359–366.

Schmidt, H.L., Schuhmann, W. *et al.* (2008) Specific features of biosensors, in *Sensors: Chemical and Biochemical Sensors*, John Wiley & Sons, Inc., New York.

Schwarz, A., Thomsen, M.S. *et al.* (2009) Enzymatic synthesis of glucosylglycerol using a continuous flow microreactor containing thermostable glycoside hydrolase CelB immobilized on coated microchannel walls. *Biotechnol. Bioeng.*, **103** (5), 865–872.

Scouten, W.H. (1987) A survey of enzyme coupling techniques, in *Methods in Enzymology*, vol. **135** (ed. K. Mosbach), Academic Press, Orlando, FL, pp. 30–65.

Seibel, J. and Buchholz, K. (2010) Tools in oligosaccharide synthesis: current research and application. *Adv. Carbohydr. Chem. Biochem.*, **63**, 101–138.

Seibel, J., Beine, R. *et al.* (2006a) A new pathway for the synthesis of oligosaccharides by the use of non-Leloir glycosyltransferases. *Biocatal. Biotransform.*, **24** (1–2), 157–165.

Seibel, J., Jördening, H.J. *et al.* (2006b) Glycosylation with activated sugars using glycosyltransferases and transglycosidases. *Biocatal. Biotransform.*, **24** (5), 311–342.

Seibel, J., Moraru, R. *et al.* (2006c) Synthesis of sucrose analogues and the mechanism of action of *Bacillus subtilis* fructosyltransferase (levansucrase). *Carbohydr. Res.*, **341** (14), 2335–2349.

Shaeiwitz, J.A. and Henry, J.D., Jr., (2000) Biochemical separations, in *Ullmann's Encyclopedia of Industrial Chemistry*, Wiley-VCH Verlag GmbH, Weinheim, www.mrw.interscience.wiley.com/emrw/.

Sheldon, R.A. (2007) Enzyme immobilization: the quest for optimum performance. *Adv. Synth. Catal.*, **349** (8–9), 1289–1307.

Sheldon, R.A. (2011) Cross-linked enzyme aggregates as industrial biocatalysts. *Org. Process Res. Dev.*, **15**, 213–223.

Spiess, A. and Kasche, V. (2001) Direct measurement of pH profiles in immobilised enzymes during kinetically controlled synthesis using CLSM. *Biotechnol. Prog.*, **17**, 294–303.

Stoppok, E., Buchholz, K. (1999) The production of 3-keto derivatives of disaccharides. In: C. Bucke ed, Methods in Biotechnology - Carbohydrate Biotechnology Protocols, 277–289, Humana Press, Totowa, New Jersey, USA, 1999.

Swaisgood, H.E. (2003) Use of immobilized enzymes in the food industry, in *Handbook of Food Enzymology* (eds J.R. Whitaker *et al.*), Marcel Dekker, New York, pp. 359–366.

Syldatk, C., Altenbuchner, J., Mattes, R., and Siemann-Herzberg, M. (1999) Microbial hydantoinases. *Appl. Microbiol. Biotechnol.*, **51**, 293–309.

Timme, V., Buczys, R., Buchholz, Kinetic, K., (1998) investigations on the hydrogenation of 3-ketosucrose, Starch/Stärke, **50**, 29–32.

Tischer, W. and Kasche, V. (1999) Immobilization enzyme crystals. *Trends Biotechnol.*, **17**, 326–335.

Toogood, H.S., Taylor, I.N., Brown, R.C., Taylor, S.J.C., McCague, R., and Littlechild, J.A. (2002) Immobilisation of the thermostable L-aminoacylase from *Thermococcus litoralis* to generate a reusable industrial biocatalyst. *Biocatal. Biotransform.*, **20**, 241–249.

Tungland, B.C. (2003) Fructooligosaccharides and other fructans, in *Oligosaccharides in Food and Agriculture, ACS Symposium Series 849* (eds G. Eggleston and G.L. Côté), American Chemical Society, Washington, DC.

Uhlig, H. (1991) *Enzyme arbeiten für uns*, Hanser, München.

Uhlig, H. (1998) *Industrial Enzymes and Their Applications*, John Wiley & Sons, Inc., New York.

Van Hijum, S.A.F.T., Kralj, S. *et al.* (2006) Structure–function relationships of glucansucrase and fructansucrase enzymes from lactic acid bacteria. *Microbiol. Mol. Biol. Rev.*, **70** (1), 157.

Vorhaben, T., Böttcher, D. *et al.* (2010) Plasma modified polypropylene as carrier for the immobilization of *Candida antarctica* lipase B and *Pyrobaculum calidifontis* esterase. *ChemCatChem*, **2** (8), 992–996.

Vujicic-Zagara, A., Pijninga, T. *et al.* (2010) Crystal structure of a 117 kDa glucansucrase fragment provides insight into evolution and product specificity of GH70 enzymes. *Proc. Natl. Acad. Sci. USA*, **107**, 21406–21411.

Weetall, H.H. (1976) Covalent coupling methods for inorganic support materials, in *Methods in Enzymology*, vol. **44** (ed. K. Mosbach), Academic Press, Orlando, FL, pp. 134–148.

Weignerova, L., Simerska, P. *et al.* (2009) α-Galactosidases and their applications in biotransformations. *Biocatal. Biotransform.*, **27** (2), 79–89.

Wenda, S., Illner, S. *et al.* (2011) Industrial biotechnology – the future of green chemistry? *Green Chem.*, **13** (11), 3007–3047.

Whitaker, J.R., Voragen, A.G.J., and Wong, D.W.S. (eds) (2003) *Handbook of Food Enzymology*, Marcel Dekker, New York.

WO 2005/ 014821, Cephalosporin C acylase mutant and methods for preparing 7-ACA using same.

Yun, J.W. and Kim, D.H. (1999) Enzymatic production of inulooligosaccharides from inulin, in *Methods in Biotechnology – Carbohydrate Biotechnology Protocols* (ed. C. Bucke), Humana Press Inc., Totowa, NJ, pp. 153–163.

Yun, J.W. and Song, S.K. (1999) Enzymatic production of fructooligosaccharides from sucrose, in *Methods in Biotechnology – Carbohydrate Biotechnology Protocols* (ed. C. Bucke), Humana Press Inc., Totowa, NJ, pp. 141–151.

Zhu, J., Wan, Q. *et al.* (2009) Biologics through chemistry: total synthesis of a proposed dual-acting vaccine targeting ovarian cancer by orchestration of oligosaccharide and polypeptide domains. *J. Am. Chem. Soc.*, **131** (11), 4151–4158.

Zuccaro, A., Götze, S. *et al.* (2008) Tailor made fructooligosaccharides by a combination of substrate and genetic engineering. *ChemBioChem*, **9** (1), 143–149.

9
Immobilization of Microorganisms and Cells

Why immobilize microorganisms?	• No enzyme isolation required • Application for complex reactions • Application of multienzyme systems • Intracellular cofactor regeneration • Continuous processing
Methods of immobilization	• Organism as carrier only (see Chapter 8) • Inclusion in matrices • Adhesion onto surfaces
Technical applications	• Acetic acid (vinegar) production • Biopharmaceuticals production • Chemicals production • Anaerobic wastewater treatment • Exhaust gas purification
Examples and exercises	

9.1
Introduction

The focus of this chapter is on the immobilization of living or viable microorganisms and cells. A definition is as follows: "The physical confinement or localization of cells to a certain defined region of space with preservation of activity." The background of the application of such systems is the observation that the enzyme activities of immobilized cells remain active for longer periods compared to that in isolated cells, this being due to the (re)synthesis of enzymes and cofactors as well as their regeneration, and due to protection against toxic or noxious effects. This is also essentially true for resting cells. Remarkably, as long ago as 1823, the cells of *Acetobacter*, when adsorbed onto wood chips, were used in the technical production of acetic acid from ethanol (Knapp, 1847, pp. 480–486). The basic principles of this process were later laid down by Hattori and Fusaka in 1959, and by K. Mosbach and R. Mosbach in 1966. Living bacteria in mixed cultures have been applied

Biocatalysts and Enzyme Technology, Second Edition. Klaus Buchholz, Volker Kasche, and Uwe T. Bornscheuer

successfully to environmental technologies since the 1970s and 1980s (Jördening and Buchholz, 1999; Lettinga *et al.*, 1999; van Lier, 2008). A range of important biopharmaceuticals is produced using mammalian cell culture, mostly in suspension culture; some processes are in operation using immobilized adherent cell cultures (Wurm, 2005; Walsh, 2010). Currently, cells with multienzyme systems and/or intracellular cofactor regeneration – notably recombinant cells with designed reaction pathways (Chapter 5 and Section 9.7.1) – have gained much interest for the production of intermediates and high added-value compounds that otherwise might be difficult to produce. Recently, microbial fuel cells for the generation of electricity have attracted considerable attention (Section 9.7.2). When compared to classical fermentation processes, immobilized microorganisms offer the advantage of short residence time, and high volumetric productivity in continuous processes in different types of reactors (e.g., tubular and fluidized bed reactors) can be achieved. (For manufacture of chemicals and pharmaceuticals by fermentation, see Buchholz and Collins, 2010, Chapters 16 and 17.) The immobilized systems may function with resting cells that do not grow (e.g., due to growth limitation by controlling C, N, or P sources), but nevertheless can regenerate cofactors; on the other hand, the growth of microorganisms may be possible under appropriate conditions. The immobilization of growing cells is applicable when the formation or regeneration of enzyme activities is dependent upon growth, and if new biomass is formed and/or delivered to the medium and appropriately removed.

In addition to microorganisms, plant or animal cells can also be immobilized. All of these biocatalysts offer the advantages of reuse and continuous application. Further advantages of immobilized microorganisms and cells include

- no requirements for enzyme isolation and purification;
- large amounts of catalyst easily available;
- high cell density and productivity in some cases;
- reaction sequences or transformations requiring multienzyme systems;
- cofactor regeneration in the native or engineered system;
- formation/production of secondary metabolites (eventually coupled to growth);
- application of anchorage-dependent cells that grow only when attached to surfaces (e.g., mammalian cells);
- application of immobilized mixed cultures with syntrophic growth or metabolism;
- reuse of biocatalysts and operation in continuous mode; and
- protection against shear forces, and in part also against toxic substances (e.g., oxygen, metal ions, aromatics).

Thus, complex reaction sequences can only be performed by viable microorganisms (bacteria, yeasts, or mammalian cells) that are capable of synthesizing enzymes and regenerating cofactors. The immobilization of microorganisms is also applicable if the enzyme(s) are difficult to isolate or show low stability and activity outside the cell, as is the case for nitrile hydratase in acrylamide production. (The topic of enzyme immobilization is discussed in Chapter 8.)

Disadvantages and problems relating to the application of viable, immobilized microorganisms and cells result from a variety of different conditions, including

- limited volumetric activity in many cases;
- cost of immobilization;
- insufficient stability;
- mass transfer limitation due to growing cells in the catalyst matrix;
- side reactions, degradation of products; and
- by-products from the lysis of cells or traces of toxic metabolites (notably in pharmaceutical and food applications).

Indeed, the above-mentioned problems have considerably limited the use of immobilized microorganisms and cells. Thus, the immobilization of *Gluconobacter oxydans* for oxidative biotransformations was hindered by the high oxygen demand, which led to mass transfer limitations and a loss in activity after only one biotransformation cycle (Schedel, 2000).

Therefore, the main areas for the application of immobilized microorganisms and cells include transformations that require complex reaction sequences, the regeneration of enzymes and cofactors, and/or cell growth. Wastewater treatment, notably with anaerobic cultures, and exhaust gas purification have been established on a very large scale (see Sections 9.6.2.2 and 9.6.2.4).

A review on the research and development of immobilized microorganisms has been written by Brodelius and Vandamme (1987), while basic research on their use has been compiled by Wijffels *et al.* (1996), Wijffels (2001), Verica *et al.* (2010), and Willaert (2009) (the last one focusing on engineering aspects). This has incorporated topics such as adhesion, details of immobilization procedures, viable cells, physiology, influence of the microenvironment, mass transfer and dynamic modeling, oxygen concentration profiles, and mechanical stability. Junter *et al.* (2002) later presented data on the physiology and protein expression that support the existence of a specific metabolic behavior in the immobilized state. These authors focused on the proteomic approach as a complementary tool to gene level investigations. Perspectives include new routes generated by applying metabolic engineering and combinatorial strategies that modify central metabolism, aromatic biosynthetic pathways, transport, and regulatory functions, complemented with heterologous gene expression and protein engineering. Microorganisms used include engineered *Escherichia coli* and *Pseudomonas putida* strains, enabling the development of sustainable processes for the manufacture of 2-phenylethanol, *p*-hydroxycinnamic acid, *p*-hydroxystyrene, and many others (Gosset, 2009). A new concept developed makes use of biofilms as robust, self-immobilized, and self-regenerating catalysts in productive catalysis, particularly in cases for which substrates and/or products affect cell viability. This biofilm approach includes synthetic chemistry, ranging from specialty to bulk chemicals, the food industry, and bioenergy using microbial fuel cells (see Sections 9.7.1 and 9.7.2) (Rosche *et al.*, 2009; Schröder, 2007). For details of immobilized mammalian cells, the reader is referred to reviews by Tokayashi and Yokoyama (1997), Doyle and Griffiths (1998), and Waugh (1999), for biopharmaceuticals production by adherent mammalian cells to Wurm

(2005), and for immobilized plant cells to reviews by Dörnenburg and Knorr (1995) and Ishihara *et al.* (2003).

The most important principles for the immobilization of microorganisms and cells are their inclusion in polymer matrices, adsorption or adhesion onto carriers, and flocculation and/or aggregation. The most relevant method for inclusion into ionotropic gels is discussed in detail in Section 9.4.2, while adhesion to carriers – based on the interaction of surfaces of microorganisms and solid surfaces – is detailed in Section 9.6.[1)]

In industrial applications, three types of immobilized viable cell systems are used currently: (1) nicotinamide production using immobilized resting cells of *Rhodococcus rhodochrous* (see Ghisalba *et al.*, 2010), and beer maturation with immobilized yeast cells (however, with problems of unbalanced flavor) (Verbelen *et al.*, 2006; Brányik *et al.*, 2005; for trends in food processing, see Willaert and Nedovic, 2006; Nedovic *et al.*, 2011); (2) anchorage-dependent mammalian cells immobilized on microcarriers (for the production of biopharmaceuticals; see also Section 9.6.2.1); and (3) environmental technology using mixed cultures (see Sections 9.6.2.2–9.6.2.4). Despite much effort being made into research investigations (up to pilot plant level for the continuous production of alcohol), few other relevant industrial applications have yet been reported that utilize immobilized viable cells. The reasons for this may be, first, that the immobilized systems seem to be unstable under conditions of continuous processing, as experienced in beer and alcohol fermentation, and, second, that fine chemicals are produced at rather low (but high price) scale, which make efforts for immobilization less attractive. The alternative of using membrane systems for cell recycle has been applied in a range of processes (see Section 7.4 and Table 7.6).

9.2 Fundamental Aspects

The immobilization of microorganisms leads to modifications in their microenvironment, with consequential concerns relating to mass transfer phenomena that affect substrate and product gradients and metabolism. At present, limited details are available of these phenomena, mainly due to their complexity and analytical problems. For structural analysis, transmission electron microscopy and visualization of messenger RNA with respect to position and dynamics in both fixed and living yeast cells have been used recently (Lee *et al.*, 2011; Gallardo and Chartrand, 2011). *Mass transfer phenomena* are described more fully in Section 10.4. The performance of immobilized cells is influenced by mass transport where both external

1) Microencapsulation technologies, including cell immobilization by spray coating, have been a useful tool to improve the delivery of bioactive compounds into foods, particularly probiotics, minerals, vitamins, phytosterols, and antioxidants. Moreover, these technologies could promote the successful delivery of bioactive ingredients to the gastrointestinal tract (Champagne and Fustier, 2007). Further, encapsulation is being applied in bioindustry and biomedicine. It is clinically applied for the treatment of a wide variety of endocrine diseases (de Vos *et al.*, 2009).

and internal mass transfer limitations have to be considered. Details were treated by Willaert for gel systems, microcapsules, and dense cell masses, that is, biofilms, bioflocs, and mammalian cell aggregates (Willaert, 2009). Systematic studies with varying particle diameters and densities of cells within the matrix have been reported by Klein and Vorlop (1985).

As a general rule, and if no cell growth occurs, when microorganisms are immobilized in ionotropic gels, the diffusion of low molecular weight substrates is not significantly slower than diffusion in water (Table 9.1). In bioflocs, mixed microbial aggregates, effective diffusion is reduced significantly. Relative effective diffusion coefficients D_e/D_a in bioflocs typically are in the range of 53–58% (for oxygen in *Aspergillus niger* pellets), 11% (acetate in methanogenic flocs), and 31–36% (oxygen in mixed microbial flocs), and relative permeabilities (also D_e/D_a, %) in the range of 46–68 (for nitrate) in denitrifying cultures and 27–52 (for glucose) in mixed microbial culture (data given as %, with effective diffusion coefficient, D_e, related to the diffusion coefficient in water, D_a). Further data concerning diffusion in mammalian cell aggregates are also given (Willaert, 2009). Similar conditions may be valid for biofilms growing on surfaces. However, under conditions of significant growth and increasing biofilm thickness (>100 μm), mass transfer can be reduced severely, the consequence being a much lower effective diffusion of substrates into the biofilm. Omar (1993) has reported on oxygen diffusion into gels in the presence of microorganisms, while mass transfer and coupled intracellular phenomena have been analyzed in pellets of *A. niger* and biofilms (Bössmann *et al.*, 2003).

Table 9.1 Comparison of diffusion coefficients (D) of substrates in water and in particles of Ca^{2+} alginate at 30 °C (Tanaka *et al.*, 1984; Berensmeier *et al.*, 2004), and in different matrices at 37 °C (ag: agarose; low a: low-density alginate; high a: high-density alginate) (Lundberg and Kuchel, 1997).

Substrate	**MW ($g\ mol^{-1}$)**	**D ($10^{-6}\ cm^2\ s^{-1}$ or $10^{-10}\ m^2\ s^{-1}$) in water**	**D ($10^{-6}\ cm^2\ s^{-1}$ or $10^{-10}\ m^2\ s^{-1}$) in Ca^{2+} alginate (2%)**
Tanaka *et al.* (1984)			
L-Tryptophan	204	6.7	6.7
α-Lactalbumin	15 600	1.0	1.0
Albumin	69 000	0.70	No diffusion
Berensmeier *et al.* (2004)			
Glucose	180	7.0	5.6 ± 0.2
Sucrose	342	5.23	4.7 ± 0.2
Lundberg and Kuchel (1997)			
Glucose	180	9.6 ± 0.2	
ag			6.8 ± 0.1
low a			7.6 ± 0.1
high a			6.7 ± 0.3
Glycine	75	13.8 ± 0.4	
ag			10.4 ± 0.3
low a			12.5 ± 0.3
high a			9.7 ± 0.1

Diffusion–reaction–growth correlations were investigated both experimentally and by modeling using *E. coli* as a model organism (Lefebvre and Vincent, 1995). Heterogeneity in the biomass distribution inside gel-immobilized cell systems was examined in detail, and the phenomena responsible included substrate diffusion, relevant notably at low concentration, and accumulation of inhibiting products (Lefebvre and Vincent, 1997).

Immobilized cell physiology, which relates to the fundamental aspects of correlation of immobilization and the response of cell growth and productivity, was reviewed by Junter *et al.* (2002) and Verica *et al.* (2010). Information has been collected on a variety of organisms such as *Pseudomonas* sp., *Bacillus* sp., *Saccharomyces cerevisiae*, and *Candida* sp., together with details of the matrices used for immobilization (e.g., alginate, agarose, carrageenan beads). In most cases, growth was limited due to nutrient and/or oxygen mass transfer, with the immobilized biomass concentrating in the peripheral areas of the particles. Growth-promoting effects were attributed to protection against inhibiting or toxic substances or metabolites (e.g., ethanol, see below). Many examples have shown unchanged or lower specific productivities compared to those of suspended cultures. More detailed analysis revealed a high resistance of several bacteria (e.g., *Pseudomonas aeruginosa*) against antibiotics, and parameters other than diffusion limitation were clearly involved in these phenomena (Junter *et al.*, 2002).[2)]

9.3 Immobilization by Aggregation/Flocculation

Several methods for the immobilization of microorganisms and cells are presented in Figure 9.1. For most applications, aggregation or flocculation (Figure 9.2), adsorption, and/or adhesion as well as entrapment in polymeric networks have gained broad acceptance, and these methods will be treated in more detail in the following sections. Those methods in which the whole microorganism is immobilized, but only a singular enzyme activity is used and the organism itself serves as the matrix (in general after cross-linking with glutaraldehyde), are described in Chapter 8.

The aggregation or flocculation of microorganisms and cells – that is, the formation of agglomerates – can be regarded as a natural and simple method of immobilization. The basic phenomena and physiology of these processes have

2) In contrast, an enhanced rate of ethanol formation – by 40–50% – has been found for immobilized yeast as compared to suspended cells, and a growth rate reduction of 45% and lower intracellular pH values were also observed (Galazzo and Bailey, 1990). It could also be shown for *E. coli* that cell permeability was changed due to an alteration in the porins as channels for nonspecific diffusion. Subsequent two-dimensional electrophoresis (2DE) of the total cellular proteins from suspended and immobilized *E. coli* cells revealed notable qualitative and quantitative differences. Those proteins for which amounts varied according to the growth mode represented about 20% of the total cellular proteins detected. In addition, indications were identified that related some of these differences to stress response (Junter *et al.*, 2002).

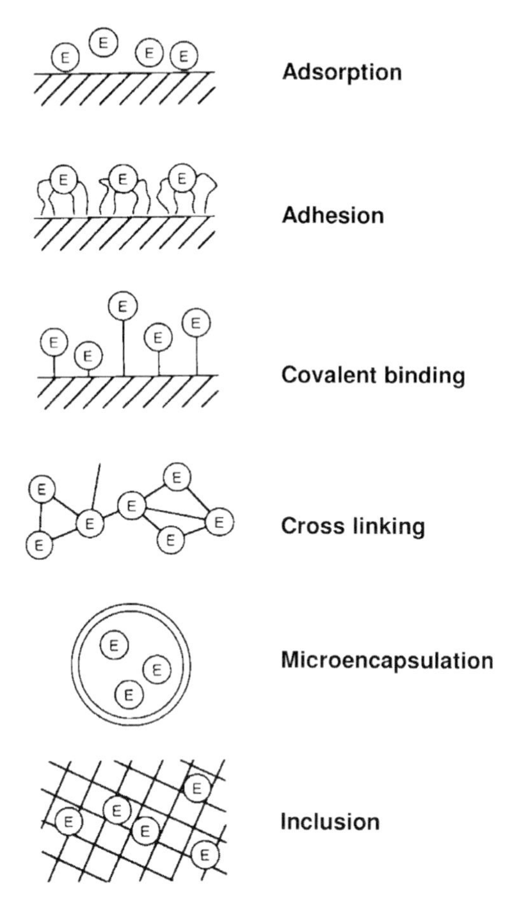

Figure 9.1 Methods of immobilization of microorganisms (E signifies the enzyme system inside the immobilized cells).

been discussed in detail elsewhere. Bioflocs and biofilms usually contain mixed populations of bacteria, where filamentous morphologies may play a major role, but they also represent a habitat for eukaryotic microorganisms. Maximal concentrations of bacteria range from 10^{10} to 10^{12} cells ml^{-1}, and this results from both direct and indirect interactions of the organisms. Cell surface polymers and extracellular polymeric substances are of major importance for the development and structural integrity of flocs and biofilms (Marshall, 1984; Wingender and Flemming, 1999).

Flocculating or aggregating anaerobic bacteria (including other material such as $CaCO_3$) have been applied successfully in biological wastewater treatment (Sam-Soon *et al.*, 1987; Lettinga *et al.*, 1999; van Lier, 2008). Under appropriate conditions, pellets with favorable settling (sedimentation) properties are formed, which contain mixed bacterial populations of high density, but which may also include inorganic material. These systems are well known as "upflow anaerobic sludge

Table 9.2 Characteristics, parameters, and criteria for flocculation or granulation of anaerobic mixed cultures (Hulshoff *et al.*, 1983; Sam-Soon *et al.*, 1987; van Lier, 2008).

Parameters	Criteria
Type and availability of substrates	Beneficial substrates: • Low fatty acids (besides acetic acid) • N source (not limited) • High partial pressure of hydrogen Unfavorable substrates: • Lipids
Temperature	30–37 °C (practical application range)
pH	6.5–7.8
Ionic strength, ratio of mono- and divalent ions	Ratio of Ca^{2+} to K^{+} ions: >5: 1
Control of start-up	Selection pressure for aggregating species, short residence time

bed" (UASB) and "expanded granular sludge bed" (EGSB). The aggregates or pellets usually have a diameter of 0.15–4 mm and, under favorable conditions (e.g., inclusion of $CaCO_3$), exhibit high sedimentation velocities (typically 15–50 m h^{-1}). In mixed cultures, filamentous (e.g., *Methanothrix soehngenii*) and rod-type morphological forms have been observed; basically, the aggregate should contain all species required for degradation of the substrate, including hydrolytic, acetogenic, and methanogenic bacteria.

At the laboratory scale, reaction rates of chemical oxygen demand (COD) or degradation, respectively, of up to 50 kg m^{-3} day^{-1} could be obtained, where the biomass exhibited a specific activity of about 2 kg COD kg^{-1}organic dry matter (odm, essentially biomass) day^{-1} (Hulshoff *et al.*, 1983). In industrial-scale reactors with volumes of 2000–5000 m^3, loading rates of 20–30 kg m^{-3} day^{-1} are being applied, and criteria for flocculation or granulation have been established for such systems (Table 9.2) (van Lier, 2008). The low cost, enhanced efficiency, and operational stability for certain types of wastewater make this method of immobilization particularly attractive (see Section 9.6.2.2).

Flocculating yeast species have been investigated for the continuous production of ethanol, with improved productivity in tubular reactors and stirred tank reactors fitted with a subsequent settler system (Netto, Destruhaut, and Goma, 1985). However, the use of this method on a technical scale exhibited problems of operational stability of the biocatalyst.

The conditions for *stable* flocculation have been investigated for several examples (Table 9.2), and they are known in detail for anaerobic mixed cultures. Multivalent ions (e.g., Ca^{2+}) and polyelectrolytes (e.g., chitosan) may favor the aggregation of microorganisms. The substrate, pH, ionic strength, and temperature each play a major role; relevant characteristics are listed in Table 9.3.

Aggregation is also an important phenomenon in fungi, an example being *A. niger*. The morphological development is thought to start with the aggregation of

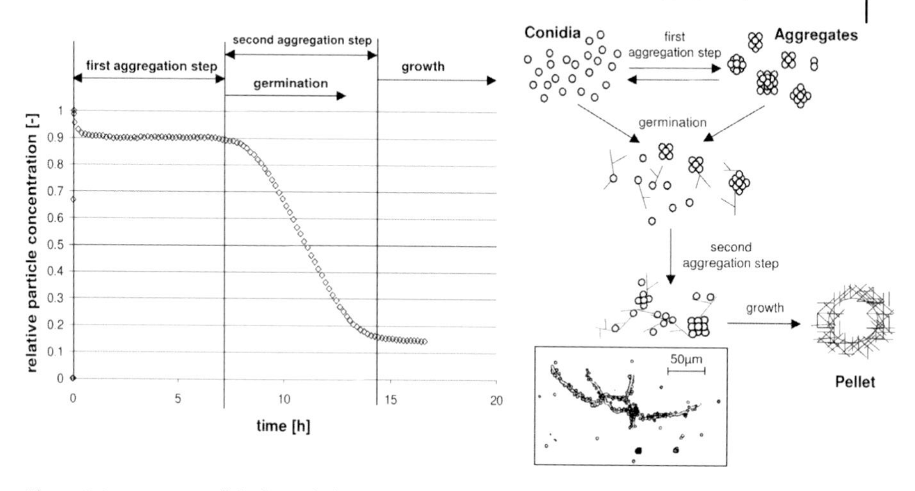

Figure 9.2 Kinetic model of conidial aggregation of *A. niger* AB 1.13 based on experimental results (Grimm *et al.*, 2004).

conidia, immediately after inoculation. A model for conidial aggregation has been proposed, which consists of two separate aggregation steps. The first takes place immediately after inoculation (timescale 5–8 h), and the second is triggered by germination and hyphal growth (timescale about 10 h). A model of these steps is shown in Figure 9.2 [3)] (Grimm *et al.*, 2004).

Table 9.3 Characteristics of life in microbial aggregates (Wingender and Flemming, 1999).

Characteristics	Associated phenomena
Formation of stable microconsortia	High cell density
Cell-to-cell communication	Facilitated gene transfer
Restriction of mass transfer	Formation of gradients in pH, oxygen, nutrients
	Aerobic, microaerophilic, and anaerobic habitats in close proximity
	Improved tolerance to toxic substances

3) Product formation of mycelial organisms is intimately linked with their morphology. Pellet morphology and glucoamylase formation were investigated under different agitation intensities of *A. niger* AB 1.13. For pellet formation, it is necessary to apply low energy dissipation rates, which also favor product formation (Kelly *et al.*, 2004). In further studies, *A. niger* has been used as an expression system for glucose oxidase (GOD), and a pellet model has been developed, assuming both suspended hyphae and a fixed number of pellets. Although pellet density was identified as a key parameter, total pellet number was also seen to be important as it influenced the pellet surface and hence the turnover of substrate and product formation (Rinas *et al.*, 2005).

9.4
Immobilization by Entrapment

9.4.1
Entrapment in Polymeric Networks

This method of immobilization had dominated research and development studies, while adhesion has been applied mainly to pharmaceuticals production by mammalian cells and in environmental technologies. The many and diverse methods for producing polymeric porous networks allow the design of different solutions with complete retention of the cells to be immobilized, yet with almost unlimited transport of substrates and products (except for polymers) (Verica *et al.*, 2010). It is important to select nontoxic components under appropriate conditions (pH, temperature, solvents) in order to maintain the catalytic potential of the cells. Thus, methods using organic solvents have not been widely used.

Ionotropic gel formation with polymeric polyelectrolytes is most important when compared to other inclusion methods, as it combines the use of low-cost materials (most are based on natural polysaccharides) and gentle conditions, together with a broad range of matrices and designs (see Section 9.4.2).

A simple and gentle method is that of *gelation*, which involves a phase transition in a polymer–solvent mixture due to a temperature shift. For this purpose, the cells are suspended at elevated temperature in the polymer (e.g., agar or agarose) solution; subsequent cooling causes the formation of a three-dimensional network – the gel – which includes the cells. The nature of the polysaccharides used allows for a smooth immobilization, and this is particularly suited for sensitive plant cells (Hulst and Tramper, 1989). Polymers based on polyvinyl alcohol (PVA) have been used successfully as matrices for the immobilization of different organisms (Jahnz *et al.*, 2001; Sheldon, 2007). Gel formation can be performed at room temperature under gentle conditions, with high rates of survival of active organisms. The polymer is heated to produce a solution, which is then cooled to room temperature and mixed carefully with the suspension of microorganisms. Droplets are deposited onto a surface, followed by partial drying (15–20 min); a stabilizer solution is then overlaid to complete the hardening. The resultant particles have a flat, lens-like shape, with a diameter of 3–4 mm and a thickness of 200–400 μm – and therefore short diffusion path for substrates. The process has been automated to yield reproducible catalysts (LentiKats®). Typical microorganisms immobilized in this way include *S. cerevisiae* and *Leuconostoc oenos* for alcohol production and *E. coli* for the production of tryptophan. The system can also be operated under nonsterile conditions, as the matrix is not degraded by fungi or other organisms (Jahnz *et al.*, 2001).[4] Sol–gel hybrid materials, using polyol-modified silanes and organic

4) Chemically resistant and mechanically stable networks can be formed via polycondensation and polyaddition reactions. Thus, bifunctional oligomeric prepolymers were reacted with appropriate multifunctional components. Polyurethanes based on prepolymers with two terminal isocyanate groups have proven to be flexible systems, notably for the conversion of hydrophobic substrates (Fukui and Tanaka, 1982; Kawamoto and Tanaka, 2003).

polymers, have been developed for the immobilization of living cells in biocompatible matrices (Desimone *et al.*, 2009).

Basic considerations of processing, including the effect of processing parameters, relevant for high productivity and operational stability as well as examples of biotransformations have been summarized by Freeman and Lilly (1998). They include the immobilization method, the mode of operation and bioreactor configuration, aeration and mixing, medium composition, temperature, pH, and *in situ* product and/or excess biomass removal. The authors stress the relevance of understanding cell physiology and correlations with productivity and operational stability, which has limited application for viable immobilized microorganisms in synthesis (Freeman and Lilly, 1998).

9.4.2 Entrapment in Ionotropic Gels

9.4.2.1 **Principle**

Water-soluble polyelectrolytes can form solid polymeric networks (gels) by cross-linking with either polycations or polyanions. Polysaccharides with carboxyl or sulfonyl groups (alginate, pectin, carrageenan) or amino groups (chitosan obtained from chitin by deacetylation) are specifically suited for this purpose. These materials are available at low or reasonable cost, and they are nontoxic. Furthermore, this method offers diverse optimization strategies with different procedures and conditions; several counterions can be applied for cross-linking to form a variety of secondary structures. The most frequently used ions are Ca^{2+} and K^+ as cations, and polyphosphates as anions (Table 9.4). In principle, Al^{3+} and Fe^{3+} can also be used, though restrictions with regard to their use in the food and pharmaceutical industries limit their application.

The structural elements of the most popular polysaccharides are shown in Figure 9.3. The secondary and tertiary structures that lead to the formation of networks (gels) are shown schematically in Figure 9.4. Inside these networks, the microorganisms, cells, and parts of cells (e.g., microsomes) can be entrapped, both gently and efficiently. Enzymes usually cannot be immobilized using this method as their dimensions are smaller by several orders of magnitude, and consequently they easily diffuse out of the gel (Reischwitz *et al.*, 1995). The structural elements

Table 9.4 Ionic polysaccharides and counterions for cross-linking.

Fixed matrix ion	Carrier matrix	Counterion
$R{-}COO^-$	Alginate	Ca^{2+} (Al^{3+}, Zn^{2+}, Fe^{2+}, Fe^{3+})
	Pectin	Ca^{2+} (Al^{3+}, Zn^{2+}, Mg^{2+})
	Carboxymethyl cellulose	Ca^{2+} (Al^{3+})
$R{-}SO_3^{2-}$	Carrageenan	K^+, Ca^{2+}
$R{-}NH_3^+$	Chitosan	Polyphosphates

(a) Alginate

(b)

(c)

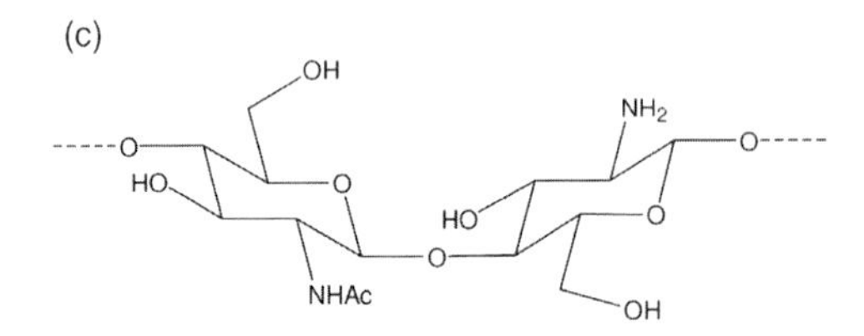

Figure 9.3 Structures of ionic polysaccharides. (a) Alginates with D-mannuronic acid, L-guluronic acid, and mixed structural elements; (b) κ-carrageenan (R = OH) and ι-carrageenan (R = OSO_3^-); (c) chitosan.

are of major importance with regard to the mechanical properties of the gel and particles formed.

Alginic acid is a polyuronic acid extracted from seaweeds, and is composed of varying proportions of (1→4)-linked β-D-mannuronic (M) and α-L-guluronic acids (G) (see Figure 9.3). The residues occur in varying proportions depending on the source, and are arranged in block patterns comprised of homopolymeric regions

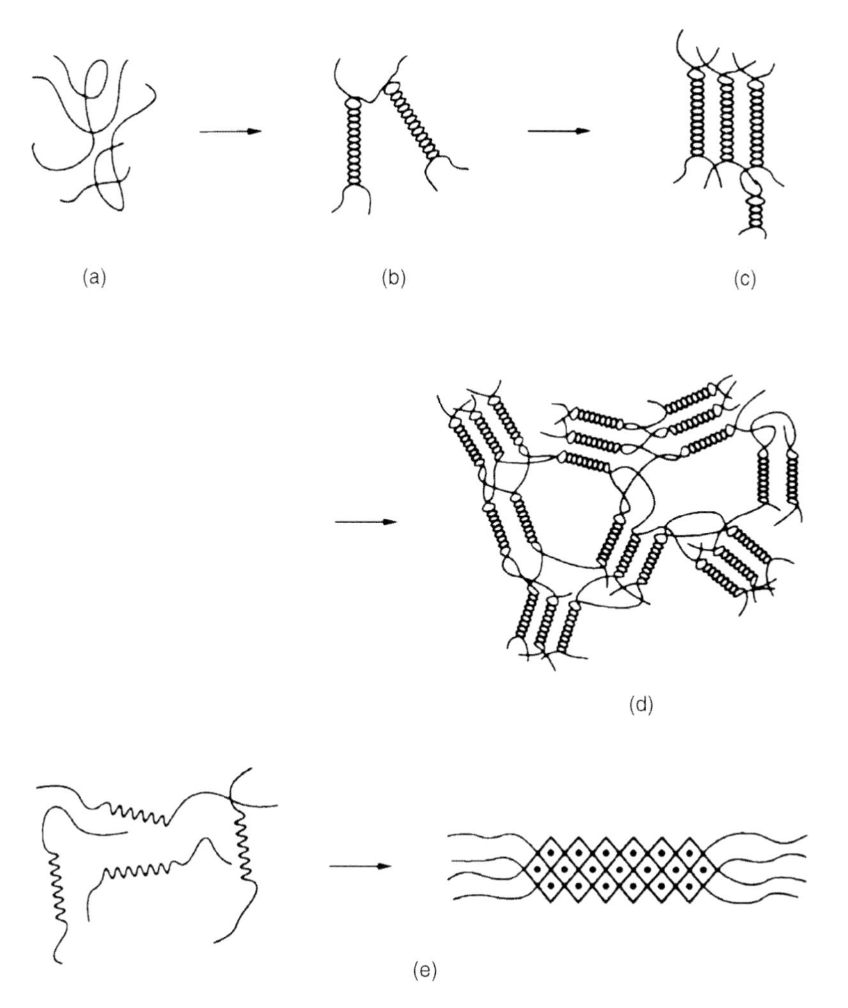

Figure 9.4 Mechanisms of gel formation. (a) Statistical polymer network in solution; (b) double/twin helices; (c) bundle/packet of double helices and (d) their tertiary/supramolecular structure; (e) network formation in alginates due to cross-linking by multivalent cations (filled circles).

(MM blocks and GG blocks) interspersed with alternating regions of heteropolymeric regions (MG blocks). The gel strength is related to the G content, which varies from 20 to 75% depending on the seaweed source. The GG blocks have preferential binding sites for divalent cations, such as Ca^{2+}. The resultant gel is biochemically inert and mechanically stable, with interstitial spaces that are suitable for cell immobilization (Fraser and Bickerstaff, 1997).

Carrageenans are able to build different networks with other types of binding, which are formed also with K^+ ions. These polyanion systems are unstable at elevated pH, whereas the polycation systems tend to dissolve at low pH (Brodelius and Vandamme, 1987).

A considerable number of publications have demonstrated the merits and shortcomings as well as the flexibility of gel entrapment methods, notably with alginate and carrageenan (see surveys in Mosbach, 1987; Nagashima, Azuma, and Nogushi, 1987; Freeman and Lilly, 1998; Dörnenburg and Knorr, 1995; Bickerstaff, 1997; Junter *et al.*, 2002).

9.4.2.2 Examples

The entrapment of cells in alginate is one of the simplest methods of immobilization. Alginates are commercially available as water-soluble sodium alginates, and have been used for more than 65 years in the food and pharmaceutical industries as thickening, emulsifying, film-forming, and gelling agents. Entrapment in insoluble calcium alginate gel is recognized as a rapid, nontoxic, inexpensive, and versatile method for the immobilization of cells (Fraser and Bickerstaff, 1997). This has also been the most popular method, with numerous applications for immobilizing both microorganisms and plant cells. Major developments of this procedure were reported by Klein and Vorlop (1985), Kierstan and Bucke (1977), Jahnz *et al.* (2001), and Verica *et al.* (2010).

To prepare immobilized cells, a solution of sodium alginate, typically 2–4% (w/v), and an appropriate buffer solution are mixed with the cells to be immobilized and passed dropwise through a nozzle (syringe) into a vessel containing a solution of calcium chloride (200 mM) under gentle stirring (Figure 9.5). In this way, particles of about 2 mm diameter (range: 1–4 mm) are obtained. The particle size can be reduced in part by drying. The density

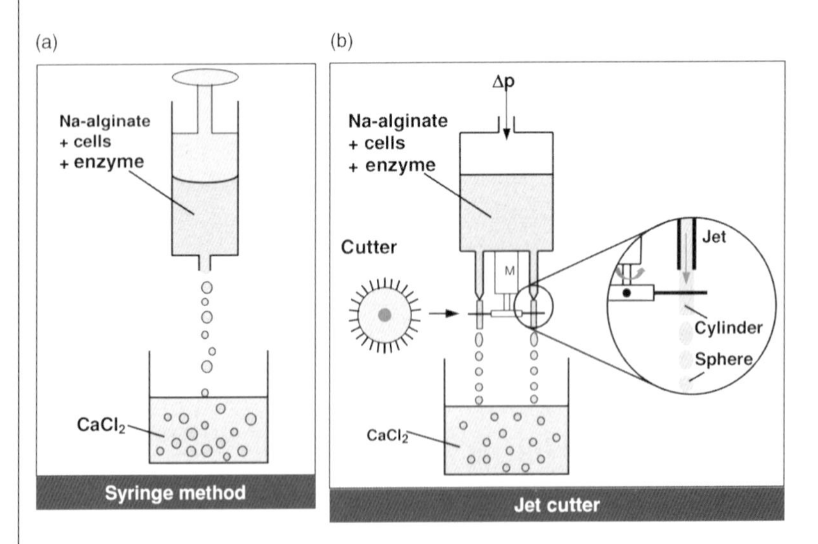

Figure 9.5 Immobilization in alginate. (a) Syringe method; (b) jet cutter method with alginate–cell suspension, pressure application Δp, rotating cutter, jet nozzle, cylindrical solution leaving the nozzle, cutting to yield spherical droplets falling down into the $CaCl_2$ solution for cross-linking (Prüße *et al.*, 2000).

of the biomass can thus be increased significantly by up to 30% (relative to wet weight) (see also Exercises).

Specific methods have been developed in order to control the particle diameter, and notably to prepare small particles with optimal mass transfer characteristics. In a system developed by Prüße *et al.* (1998), the fluid is pressed through a nozzle as a liquid jet. Beneath the nozzle, the jet is cut into cylindrical segments by a rotating cutting tool made of small wires fixed in a holder. The segments form spherical beads while falling into a 200 mM $CaCl_2$ solution (Figure 9.5), the particle diameter being dependent on the speed of rotation. In this way, particles of 0.5 mm and narrow size distribution can be obtained (Figure 9.6). Scale-up to produce biocatalysts in the kilogram scale has been achieved (Prüße *et al.*, 1998). Furthermore, particles could be obtained that are appropriate for use in fluidized bed reactors by

(a)

(b)

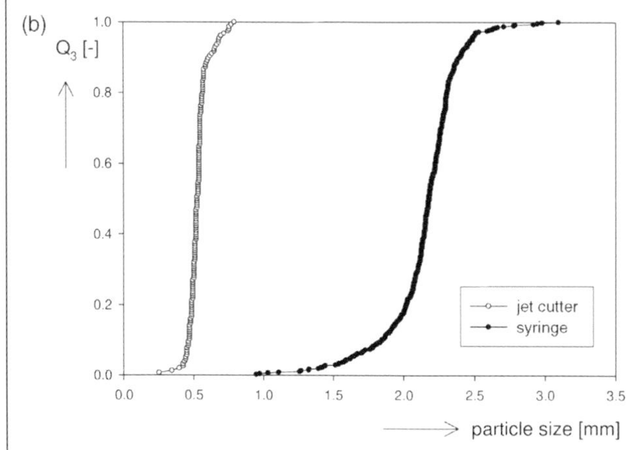

Figure 9.6 Particles (a) and particle size distribution (b) of alginate beads (with 30% (w/w) sand for improved settling properties) prepared by syringe or jet cutter (Reproduced with permission of Elsevier © 2004, from Berensmeier *et al.*, 2004).

the additional entrapment of titanium dioxide or sand particles, thus providing improved mechanical stability and high sedimentation rates due to an enhanced density (Berensmeier *et al.*, 2004).

The destabilization of calcium alginate is promoted by chelating agents (phosphate, citrate, EDTA) that remove calcium ions and by other mono- (K^+, Na^+) or divalent cations that can exchange with Ca^{2+}. Operational stability during the conversion of substrate solutions requires a ratio of monovalent and Ca^{2+} ions of <5 or 20 (maximal ratio) in general. This is due to the exchange and stationary equilibrium of all ions present in solution. Leakage can occur from the gel beads, and this is influenced by initial alginate concentration, mechanical treatment of the beads, and cell productivity, if dividing cells have been immobilized. Electron microscopy of a 2% calcium alginate bead showed pores that ranged from 5 to 200 nm in diameter. After gelation with Ca^{2+} ions, alginate particles may be further cross-linked with chitosan, polyacrylic acid, polyvinyl alcohol, or polyethylenimine to produce more stable, lower porosity complexes with improved leakage characteristics (Fraser and Bickerstaff, 1997).[5]

Owing to the importance of this system, several studies have focused on the metabolic responses (Galazzo *et al.*, 1990; Junter *et al.*, 2002). A stimulation of active metabolism on immobilization, with increased specific rates of substrate (essentially glucose) uptake and product (essentially ethanol) secretion, was observed for different types of immobilized systems. Enhanced production efficiencies as compared to suspended cells were frequently reported, but careful interpretation should be made of published data on such effects.

A survey summarizing the developments of gel-immobilized microorganisms is provided in Table 9.5. For specific production rates and/or yields with immobilized as compared to suspended cells, see Junter *et al.* (2002). The first industrial utilization of immobilized microorganisms was based on the method with entrapment in κ-carrageenan developed by Chibata *et al.* (1987) for the production of amino acids (single enzyme step, nonviable cells).[6] Redox reactions that require multienzyme systems with cofactors and their regeneration are carried out mainly with suspended whole-cell

5) Many investigations and major developmental efforts have been devoted to ethanol production by immobilized yeast or bacteria (e.g., *Zymomonas*) up to the pilot scale with several m^3 reactor volumes (Nagashima *et al.*, 1987; Junter *et al.*, 2002). However, despite numerous efforts, no industrial process for ethanol production with immobilized yeast or bacteria has yet been established.

6) In principle, gelation occurs by dissolving the potassium salt of κ-carrageenan (2–5% solution) at 70–80 °C, suspending the cells at 40 °C in the polysaccharide solution, and subsequently cooling. For particle stability, however, it proved advantageous to drop the suspension into a solution of potassium chloride (e.g., 0.1 M). Particles with carrageenan can be further stabilized and hardened by several methods, such as cross-linking by hexamethylenediamine, polyethyleneimine, or glutaraldehyde.

Table 9.5 Examples of immobilized microorganisms in different supports (Chibata *et al.*, 1987; Freeman and Lilly, 1998; Junter *et al.*, 2002).

Product	Organism	Carrier	Reactor[a]
Ethanol	*S. cerevisiae*	Alginate	
Ethanol	*Zymomonas mobilis*	Alginate	
2,3-Butanediol	*Enterobacter aerogenes*	Alginate	
L-Isoleucine	*Serratia marcescens*	Alginate	
11β-Hydroxyprogesterone	*Acetobacter phoenicis*	Alginate	
Prednisolone	*Arthrobacter globiformis*	Polyacrylamide	RB
L-DOPA	*Erwinia herbicola*	Carrageenan	PB
Thienamycin	*Streptomyces cattleya*	Celite	BC
Glucoamylase	*A. niger*	Alginate	SF/RB
(Conversion of toluene)	*Pseudomonas fluorescens*	Anion-exchange resin	
	Bacillus subtilis	Carrageenan	

a) RB: repeated batch; PB: packed bed; SF: shake flask.

systems and batch-type reactions (see Liese *et al.*, 2000, 2006; Cheetham, 2000; Kirst, Yeh, and Zmijewski, 2001).

Plant cells can also be immobilized in alginate or carrageenan. However, *few applications have been reported* for the formation of products of high added value, such as pharmaceutical specialties, components of aromas and fragrances, and secondary metabolites (Rosevear, 1988; Hulst and Tramper, 1989; Roisin *et al.*, 1996; Tang and Mavituna, 2001; Ishihara *et al.*, 2003).

9.5 Adsorption

Acetic acid (vinegar) has been produced by adsorbing *Acetobacter* sp. onto wood chips since 1823, and even today 60 m^3 reactors filled with curled beechwood shavings are used as so-called "trickling generators" with ethanol and oxygen as substrates. A pump circulates an ethanol/water/acetic acid mixture from a storage reservoir to a distributor at the top of the tank. A blower is used to force air through the packing. A good conversion efficiency is 88–90% of the theoretical value, and a correctly operated generator has a packing life of 20 years (Atkinson and Mavituna, 1991, pp. 1171–1173).

Although the use of immobilized yeast cells has been widely studied for beer production, the only industrial production process known to date with viable cells has been the final stage in beer manufacture, where aroma formation occurs during maturation. Immobilized yeast cells (*S. cerevisiae*) were prepared by adsorption onto porous glass or silica particles with a high internal surface. During maturation and aroma formation, the essential step of α-acetolactate transformation into acetone and diacetyl has been optimized with respect to a much reduced residence time (2 h compared to 10–20 days in the conventional process) and continuous processing (Breitenbücher, 1996; Iersel *et al.*, 1999; Cashin, 1996).[7]

General requirements for cell immobilization carriers have been summarized as follows:

- Material where cells can easily attach to proliferate on the surface, innoxious to the cells.
- Density and size adequate for fluidization by mild mixing.
- Large surface area per unit volume.
- Mechanical strength and durability.
- In porous support: appropriate dimensions for cell accommodation and nutrient and oxygen diffusion, the latter being one of the critical parameters.
- Scale-up potential and commercially available at a reasonable price (Tokayashi and Yokoyama, 1997).[8]

9.6
Adhesion

The adsorption of microorganisms at, and their growth on, surfaces is a well-known and common phenomenon in nature. Typical matrices include stones, sand, clay minerals, plant materials, teeth, and metal and polymer/plastic surfaces. Adsorption is based on a physical – and, in principle, reversible – interaction. Often, however, a subsequent step follows when microorganisms produce and secrete secondary metabolites that cause a strong, partially irreversible adhesion to the primary matrix, in many cases by formation of a secondary matrix of polymers formed as metabolites. In the context of this chapter, this process is termed "adhesion" and it is most often encountered with biofilm-producing bacteria.

For technical applications, immobilization based on adhesion is important because it forms a very stable and biocompatible attachment of the microorganisms to the carrier. It represents an elegant and powerful method of immobilization, as simple, convenient, and cheap carriers can be used in general (e.g., silicone, other materials with appropriate coatings, sand, pumice, charcoal, lava), without any special additives. As reactors, tubular configurations, packed beds, as well as fluidized beds with carriers exhibiting good settling properties, are convenient. For biopharmaceuticals production, mostly suspended cell culture technology is being applied, but also adherent cell culture systems using microcarriers, for example, based on cross-linked dextran, are used, suspended in stirred bioreactors (Wurm, 2005; Buchholz and Collins, 2010, pp. 383–385).

7) Originally "Siran" (Schott, Germany) with an internal surface (up to $0.4\,m^2\,g^{-1}$, d_p 2–3 mm) was used. Two reactors, each of $2.5\,m^3$, were installed for a production volume of 400 000 hl (1 hl = 100 l) of beer. Actually, a similar carrier (Immopore) is offered by EasyProof (www.easyproof.de).

8) Scale-up has been performed using reactors of 1–5 m^3 size, albeit with low cell densities, usually about 2×10^6 cells ml^{-1} (Tokayashi and Yokoyama, 1997). Stirred, fixed-bed, and fluidized bed reactors have also been investigated (Lüllau *et al.*, 1994; Pörtner *et al.*, 2007; Waugh, 1999).

A range of processes has gained importance in the field of environmental technology; indeed, this sector is considered to be most relevant inasmuch as it has considerable economic importance for the application of living immobilized microorganisms. For example, the annual overall cost for biological wastewater treatment in Germany alone is estimated to be over 10 billion €. Diverse mixed microorganism cultures are used, after being adapted to the conditions of the different wastewaters (which obviously cannot be closed systems operating under sterile conditions). Both aerobic cultures for wastewater and exhaust air purification and anaerobic cultures for the treatment of high organic load wastewater of the food and chemical industries are used, and in this respect several hundred functional plants are now on stream worldwide.

9.6.1 Basic Considerations

The interaction of microorganisms or cells and a carrier depends essentially on the surface properties of each. This is notably true for the first step, the primary adsorption – that is, before extracellular metabolites, which may modify the surface, are formed and secreted.

Whereas many matrices (e.g., silicone, glass, silica, clay minerals, cellulose) have been well characterized, less is known about the surface properties of microorganisms, which in turn depend on the growth conditions and/or the surrounding medium. X-ray photoelectron spectroscopy has proven to be the most promising method to provide information on the elements and types of bonds in the external molecular layers down to a 10 nm depth from the surface (Brodelius and Vandamme, 1987; Amory and Rouxhet, 1988; Mozes, Léonhard, and Rouxhet, 1988). This method allowed for an analysis of the elemental composition of the cell surfaces of different bacteria and yeasts. Furthermore, the electrophoretic mobility provides information on the surface charge, and hydrophobic chromatography on the hydrophobicity of the surface (van Loosdrecht *et al.*, 1987; Amory and Rouxhet, 1988).[9]

9) Correlation of these physicochemical data led to the following conclusions (van Loosdrecht *et al.*, 1987; Amory and Rouxhet, 1988):

- Phosphate groups play a dominant role with respect to the surface charge; the N/P concentration of the surface correlates with the surface charge.
- The bacteria investigated exhibited higher N and P concentrations as compared to yeasts. It must be considered in this respect that the elemental composition for a certain strain depends on the fermentation conditions (e.g., the substrates) and the fermentation time (cell age).
- Proteins of the cell surface correlate with the hydrophobicity. For yeasts, the hydrophobicity is proportional to the N/P ratio at the cell surface. In contrast, the hydrophobicity of bacteria exhibits an inverse correlation with the N/P ratio and the oxygen concentration; here, hydrophobicity is proportional to the hydrocarbon percentage of the cell surface.
- The main interactions of the surfaces of microorganisms and carriers are van der Waals forces, dipole–dipole, and electrostatic interactions. In many cases, repulsive forces must also be taken into account as both microorganisms (e.g., *Saccharomyces, Acetobacter, Bacillus* sp.) and carrier materials (such as glass or silica surfaces) carry negative surface charges. In such cases, the surface charge of the carrier can be modified or changed (reversed) by appropriate ionic components (Fe^{3+}, chitosan, etc.) in order to facilitate the primary adsorption.

Flocculation or granulation, where microorganisms aggregate to form larger particles, may be considered as a special form of immobilization by adsorption and/or adhesion. In such cases, bacteria and yeasts form granules that may include other materials such as calcium carbonate, as is the case in anaerobic wastewater treatment. These granular particles can be readily retained, for example, in fluidized bed reactors. For flocculating yeasts, further specific interactions than those mentioned before could be identified (Kihn, Masy, and Mestdagh, 1988).

Adhesion, flocculation, and granulation in general require selection processes and also conditions for operational stability that are not fully known. Granulation occurs preferentially when proper growth nuclei, that is, inert organic and inorganic bacterial carrier materials as well as bacterial aggregates, are already present in the seed sludge. Aggregate formation automatically occurs when applying dilution rates higher than the bacterial growth rates under the prevailing environmental conditions (van Lier, 2008). Processes occurring during the adsorption of microorganisms may be divided phenomenologically into the following steps (Rouxhet, 1990; Wanda *et al.*, 1990):

1) (Primary) adsorption.
2) Adhesion (secondary steps, building of bridges, for example, by cell organelles, pili, or secondary metabolites such as extracellular polymeric substances (EPS), including polysaccharides and/or proteins).
3) Growth of the adhering cells (formation of a biofilm and/or a secondary matrix of microorganisms, secondary metabolites, and further components such as calcium carbonate) up to a limiting condition where growth and detachment by shearing reach a stationary equilibrium.

Figure 9.7 illustrates the single steps of adsorption, adhesion, and biofilm growth cycle taken from an example of growth of *Pseudomonas* sp. (Rosche *et al.*, 2009). Another example shows the growth of an anaerobic mixed culture on glass surfaces. It is clear that the primary adsorption proceeds quickly (within several hours), whereas the formation of a biofilm takes a much longer time (several days or weeks) (Figure 9.8a and b). Formation of the biofilm depends critically on the available carbon source; this provides the stimulus for the formation of extracellular products (EPS, such as polysaccharides; Figure 9.8c) that form the secondary matrix on the carrier surface. For anaerobic organisms, complex media with different carbon sources – notably medium-chain fatty acids – have proven to be superior with respect to the formation of a polysaccharide matrix and thus a biofilm with high stability and activity, as compared to media with a single carbon source (Wanda *et al.*, 1990).

The (secondary) adhesion can provide for mechanically stable biofilms that, in fluidized bed reactors, resist abrasion. It can be initiated by selection pressure on the bacterial mixed culture. For this, the carrier is kept in the reactor (e.g., with a fixed or fluidized bed) that is provided with wastewater carrying a mixed culture of broad species distribution. The medium residence time should be below the reciprocal growth rate of the microorganisms, so that all organisms that are unable to adhere to the surface are eluted. Both aerobic (with a hydraulic residence time τ of about 45 min) and anaerobic ($\tau = 1$–4 h) biomass can be immobilized by adhesion

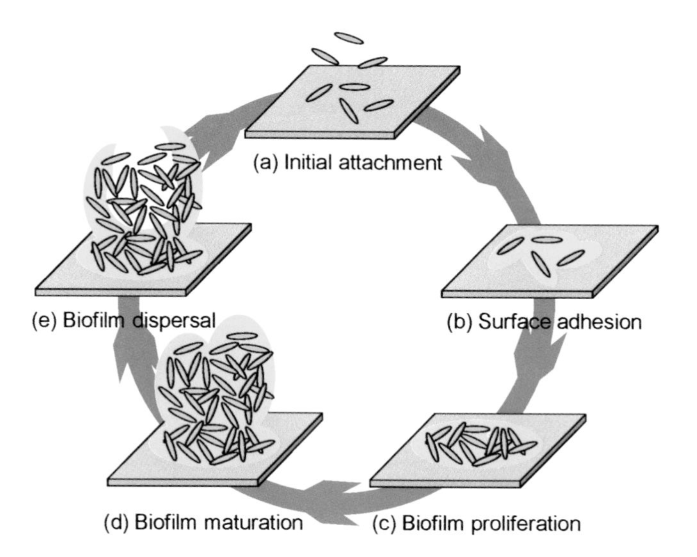

Figure 9.7 Biofilm growth cycle. Initially, free-swimming cells settle on a surface via reversible attachment (a), which triggers the production of extracellular polymeric substances for surface adhesion (b). Biofilm proliferates within the EPS matrix (c) and exhibits typical biofilm morphology on maturation (d) in a microorganism such as *Pseudomonas*. During biofilm dispersal (e), individual free-swimming cells are released from mature biofilms to colonize new surfaces, which completes the biofilm growth cycle (Reproduced with permission of Elsevier © 2009, from Rosche *et al.*, 2009).

via this method, thereby implementing a selection pressure (Heijnen *et al.*, 1991). The growth to high density of active biomass on the carrier may require between a few days and several weeks, but it can be accelerated by the application of an inoculum with carrier-fixed (preselected) microorganisms (which are transferred by a low rate of detachment of the inoculum and subsequent attachment to the new carrier) (Jördening *et al.*, 1991). The analysis of profiles in bioaggregates and biofilms has been developed considerably in recent years (Bössmann *et al.*, 2003).

Several conditions (which are known only partially for different applications) must be maintained in order to obtain an immobilized bacterial culture with high activity and stability, notably with respect to anaerobic wastewater and exhaust gas treatment:

- **Stable pH.** in large reactors this requires sufficient mixing and buffer capacity.
- **Control of temperature.** a convenient and cheap source of energy ($<50\,^{\circ}C$) and heat exchangers are required in order to maintain the optimal temperature.
- **Minimal substrate requirements.** carbon source, minerals (N, P, and sources of other growth requirements); constant concentrations are favorable, large variations may cause problems; starvation and sudden changes of substrates can lead to a breakdown of cultures that are not adapted; notably, anaerobic cultures may require long time (weeks) in order to form new active biomass.
- No substances that are toxic to the microorganisms.

The most important criteria for carrier materials are their mechanical, chemical, and biological stability, the specific weight (or density), the surface per reactor

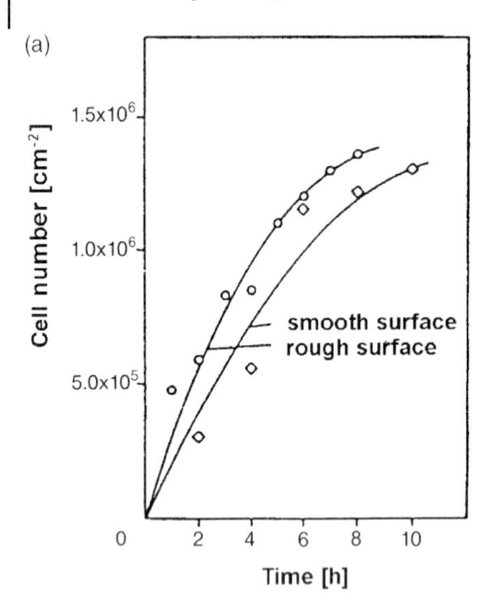

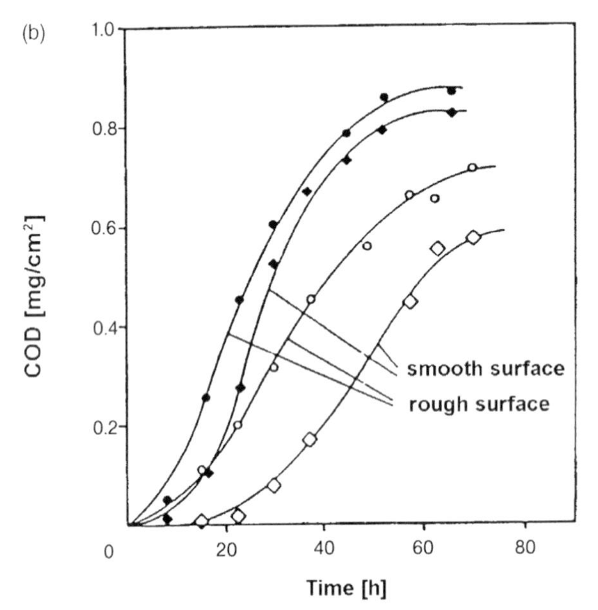

Figure 9.8 Processes during growth of an anaerobic mixed culture on either smooth and even or, in contrast, raw glass surfaces (Wanda *et al.*, 1990). (a) Rate of primary adsorption (measurement of cell number per cm^2) with acetic acid as carbon source. (b) Formation of a biofilm on a glass surface (measurement via COD per cm^2) with different carbon sources (open circles, open diamonds, reactor A: acetic acid; closed circles, closed diamonds, reactor B: acetic and butyric acid). (c) Polysaccharide content of the system shown in (b).

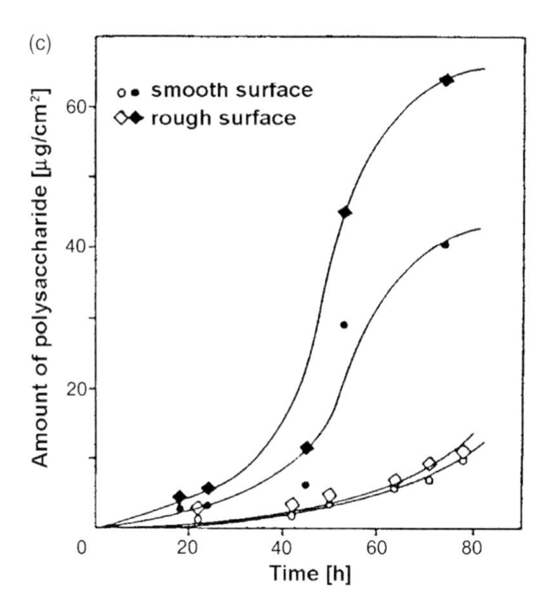

Figure 9.8 (*Continued*)

volume (m^2 per m^3) available for attachment of microorganisms, and the void volume available for fluid flow. In addition, it should be noted that several additives can favor immobilization, notably calcium carbonate precipitates or ions, which either ease or enhance adsorption. Many different materials have been applied as carriers for anaerobic mixed cultures, for example, particles (granules, spherical, porous, or compact carriers) can be used in fluidized beds preferentially. When compared to nonporous carriers (e.g., sand), porous materials (e.g., porous silica, pumice) allow for a much higher density of active biomass per volume, which in turn provides for a much higher degradation potential, and they require, due to lower density, lower energy input for maintaining fluidization (Keim *et al.*, 1989, 1990; Jördening *et al.*, 1991; Aivasidis and Wandrey, 1992).[10)]

The formation of a biofilm in the initial phase depends on the inoculum and favorable conditions (substrate, pH, temperature). The growth rates for anaerobic mixed cultures, including methane bacteria, are in the range of 0.04–0.15 day^{-1}. In pilot reactors, an increase in performance of 10–20% per day during the exponential growth phase could be obtained (Jördening *et al.*, 1991), while the thickness of the biofilm on a smooth surface may be in the range of 0.02–0.2 mm. Extended investigations on the dynamics in the biofilm have been undertaken by Trulear and Characklis (1982) and van Lier (2008). Data for characteristics of granular microbial aggregates and carriers for fluidized systems with anaerobic mixed cultures and their properties are presented in Table 9.6a and b (fixed-bed reactors have not been used recently).

10) The fluidization of pumice requires only about 5 $W\ m^{-3}$, which is one-tenth of that required for sand, (50 $W\ m^{-3}$, both at 50% bed expansion and similar particle size; see Table 9.6) (Jördening *et al.*, 1991).

Table 9.6 (a) Characteristics of granular microbial aggregates in wastewater treatment systems (van Lier, 2008) and (b) support material for anaerobic fluidized bed systems (Jördening and Buchholz, 1999).

(a)

Diameter range (mm)	Density (kg m^{-3})	Shape	Settling velocities (m h^{-1})
0.1–8 (typically 0.15–4)	1.0–1.05	Spherically formed and well-defined surface	2–100 (typically 15–50)

(b)

Support	Diameter (mm)	Density (kg m^{-3})	Surface (m^2 m^{-3})	Porosity	Upflow velocity (m h^{-1})	Biomass (kg m^{-3})	References
Sand	0.5	2540	7100[a)]	0.41	30	4–20	Anderson *et al.* (1990)
Sepiolite	0.5	1980	20 300[b)]			32	Balaguer *et al.* (1992)
GAC	0.6					34	Chen *et al.* (1995)
Biomass granules					2–6.5		Franklin *et al.* (1992)
Sand	0.1–0.3	2600			16	40	Heijnen (1985)
Biolite	0.3–0.5	2000			5–10	30–90	Ehlinger (1994), Holst *et al.* (1997)
Pumice	0.25–0.5	1950	2.2×10^6	0.85	10		Jördening (1996), Jördening and Küster (1997)

a) Calculated from the given data with the assumption of total sphericity.
b) Calculated with data given in Sanchez *et al.* (1994).

9.6.2 Applications

Microorganisms immobilized by adhesion are applied mainly in areas of biopharmaceuticals production and environmental technology. (For perspectives in chemicals production and microbial fuel cells, see Sections 9.7.1 and 9.7.2.) The advantages of systems with immobilized microorganisms are a high density of organisms and thus a high biocatalyst activity, when compared to free suspended cells or microorganisms. Hence, high-capacity reactors may have a much smaller volume and lower costs. Difficult or cost-intensive recycling of slow-growing *cells* or microorganisms (e.g., anaerobic) is not necessary due to immobilization. The dominating technologies for anaerobic wastewater treatment use sludge granulation (UASB and EGSB systems). Several fluidized bed reactors with viable anaerobic mixed cultures either with carrier or as granular bacterial aggregates are utilized in the food industry for the purification of high-load wastewater. Likewise, trickle-bed systems to immobilize bacteria performing nitrification for the elimination of nitrogen-containing substances, as well as biofilters, are used for exhaust gas purification, mainly for the elimination of foul-smelling organic pollutants, in a large number of different production facilities.

9.6.2.1 Adherent Mammalian Cells for Biopharmaceuticals Production

Most of the biologics, which are on the market, are produced with suspension culture and stirred tank technology (Walsh, 2010). A limited number is produced using carrier systems for adherent (anchorage-dependent) cell lines, notably for vaccines manufacture, for example, with Madin-Durby canine kidney (MDCK) and Vero cell lines (GE Healthcare, 2011, www.gelifesciences.com/, Application Note 28-9576-34 AB; Behrend, 2011, personal communication). The immobilization of animal and human cells with high cell densities and productivity has been realized. Microcarrier-based systems are frequently used, that is, the growth of adherent cells on small particles (usually spheres with a diameter range of 100–300 μm) suspended in stirred culture medium. They are well established at both laboratory and production scales in stirred tank bioreactors, and they can be operated in batch or perfusion modes (providing continuously perfused medium). Widely used microcarriers are cross-linked dextran, for example, Cytodex™ microcarrier (cross-linked dextran with triethylamine groups, particle size 190 μm, surface area 0.44 $m^2\,g^{-1}$ dry weight) (GE Healthcare, www.gelifesciences.com/). Others are based on gelatin, collagen, modified cellulose, or silica. The key criteria are to provide an extended surface and a stable environment, both chemically and electrostatically appropriate for cell attachment, spreading, and growth. The continuous perfused production process aims at the highest cell concentration possible. Perfused cultures can be maintained for many weeks or months, whereby product harvesting is carried out repeatedly throughout that period (Pörtner *et al.*, 2007; Tokayashi and Yokoyama, 1997; Doyle and Griffiths, 1998; Wurm, 2005). The medium composition has a significant effect on cell attachment. An appropriate microcarrier concentration is 3 $g\,l^{-1}$ (Cytodex) at an inoculum concentration of MDCK cells of

3×10^5 cells ml^{-1}; cells grew to a final concentration of 3×10^6 cells ml^{-1} (GE Healthcare, 2011, www.gelifesciences.com/, Application Note 28-9576-34 AB). Several processes have been developed in the human and animal vaccine industry, as well as for tissue engineering, using the microcarrier concept. In vaccine production, the cells serve as substrates for the multiplication of viruses such as measles, polio, or mumps. CHO (Chinese hamster ovary) cells are used for the production of several human recombinant proteins on microcarriers in stirred bioreactors (most notably at Merck-Serono). The antihemophilic protein factor VIII has reliably been manufactured using perfusion technology with BHK (baby hamster kidney) cells (Becker *et al.*, 2007, Chapter 9.1; Wurm, 2005).

9.6.2.2 **Anaerobic Wastewater Treatment**

There are considerable advantages of anaerobic wastewater treatment as compared to aerobic treatment. Worldwide over 2000 industrial plants are in operation, about 36% in the food, and some 30% in the brewing and beverage industries (van Lier, 2008; Austermann-Haun, 2008). Typically, the specific productivity of an anaerobic system is much higher, and the engineering is simple. Dilution rates, and thus the residence time in reactors, are independent of bacterial growth rates due to high settling rates when either granular bacterial aggregates or fluidized immobilized cultures are utilized, with high applicable COD loading rates of some 20–35 kg COD per m^3 of reactor per day, and highly efficient transformation of COD into biogas (>90%). Whereas much energy is needed for aeration in aerobic wastewater treatment plants, anaerobic treatment produces energy in the form of usable biogas with an energy output in the range of 55–390 MJ m^{-3} day^{-1}, or about 13.5 MJ energy (CH_4 kg^{-1} COD removed), and electric power output of 0.25–1.7 kW m^{-3} (40% electric conversion efficiency assumed) (van Lier, 2008; van Lier *et al.*, 2008). On the other hand, this also means that the growth of anaerobic bacteria is slow, and for this reason immobilization is the preferential technology. In particular, the start-up of an anaerobic plant needs special attention. An overview of the most important reaction pathways of anaerobic degradation of biomass (polysaccharides, proteins, lipids) is presented in Figure 9.9. The so-called "acidifying bacteria" convert carbohydrates, for example, to fatty acids, alcohols, carbon dioxide, and hydrogen, and this in general is a rather rapid reaction. These primary fermentation products are further converted by acetogenic bacteria to yield acetate, formate, CO_2, and H_2 that ultimately are used by methanogenic bacteria to form biogas, mainly CH_4 and CO_2, which are the slowest reaction steps (McInerny, 1999).

Both thermodynamics and kinetics of anaerobic systems are discussed in detail by van Lier, and stoichiometric values of COD and TOC (total organic carbon) per unit mass for different pure organic compounds are given (van Lier, 2008). The kinetics of this reaction network are complex. For the growth of new biomass, the essential step during the slow start-up of an anaerobic reactor, the concentration of biomass x can accumulate with a rate μ corresponding to exponential growth:

$$\mu = \frac{1}{x}\frac{dx}{dt}. \tag{9.1}$$

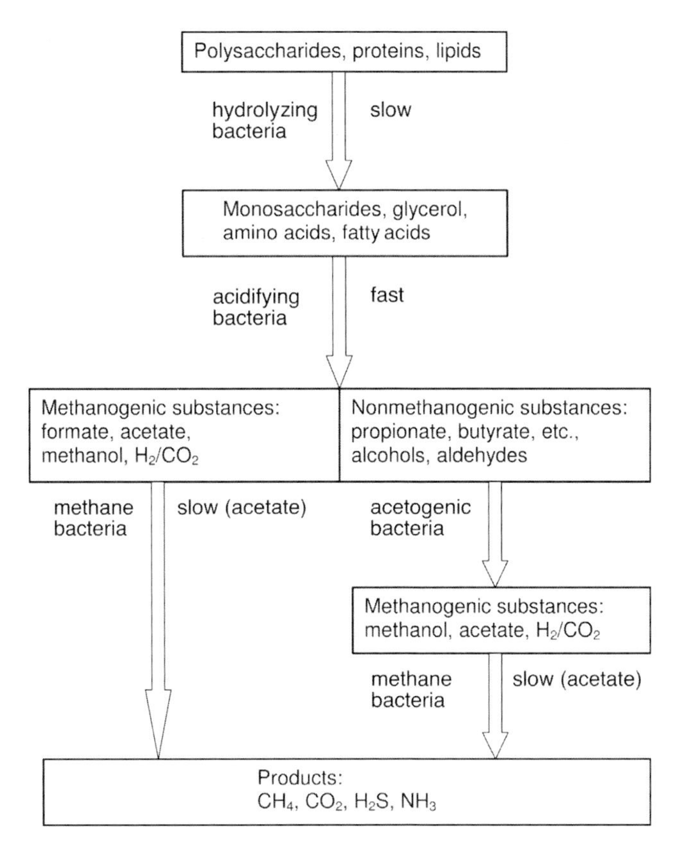

Figure 9.9 Scheme of anaerobic degradation reactions.

Integration gives the concentration of biomass x at time t:

$$x = x_0\, e^{\mu \cdot t}. \tag{9.2}$$

The growth rate μ depends on the substrate concentration [S] in a saturation type curve – that is, μ depends on [S] as follows:

$$\mu = \mu_{\max}\left(\frac{[S]}{K_m + [S]}\right), \tag{9.3}$$

where K_m corresponds to the substrate concentration where the growth rate reaches half of the maximal value.

From Eqs. (9.1) and (9.3), one obtains for the growth of microorganisms (with $\mu_{\max}$ typically around 0.12 day^{-1}) (for start-up with high substrate concentration, that is, $[S] \gg K_m$, as is used conventionally):

$$\frac{dx}{dt} = \mu_{\max}\left(\frac{x\,[S]}{K_m + [S]}\right) \sim \mu_{\max} x, \quad \text{if} \quad [S] \gg K_m. \tag{9.4}$$

For substrate conversion, the simple equation of Monod, or Michaelis and Menten, can often be used, corresponding to the assumption of one rate-limiting step (in most cases methane formation):

$$-\frac{d[S]}{dt} = \frac{V_{max}[S]}{K_m + [S]} \quad \text{(approximation)}. \tag{9.5}$$

The stationary effectiveness factor η for a system with microorganisms immobilized by adhesion is

$$\eta = \frac{V_{Biofilm}}{V_{Suspension}} \tag{9.6}$$

and can be calculated or estimated according to Section 10.4.2. In most cases, this value is near 1, but at a high film thickness with an active biomass η may be significantly lower. In such cases, inactivation and detachment of the microorganisms from the carrier surface may occur due to insufficient nutrient supply. Kinetics and mass transfer in anaerobic granular aggregates of microorganisms (sludge) have been investigated by Gonzales-Gil *et al.* (2001). It was found that external mass transfer can be neglected, whereas an increase in apparent K_m was observed due to intraparticle mass transfer.

Integration of Eq. (9.5) gives

$$[S]_o - [S] + K_m \ln\frac{[S]_o}{[S]} = V_{max}t. \tag{9.7}$$

With this equation an estimation of the conversion ($[S]_o - [S]$), or the exit concentration [S], can be obtained that depends on the concentration at the reactor inlet $[S]_o$, the residence time $\tau = t$, and the values of V_{max} and K_m determined empirically. For this purpose, it is essential to control the pH (≥ 7) and the temperature at constant values. It must furthermore be considered that the (apparent) values of V_{max} and K_m may change with stationary conditions (e.g., the concentration or the substrate composition).

As analytical data for the wastewater charge COD values are conveniently taken. If one acid (mostly acetic acid) is dominating, the COD may serve as an approximation for the substrate concentration [S] (see Eq. (9.5)) (1 g acetic acid corresponds to 1.07 g COD). This simple approximation, however, can only be used if the stationary conditions are well defined and constant.

In cases where this approach cannot be used, the coupled differential equations corresponding to the most important substrates and intermediary products can be established based on simple Michaelis–Menten kinetics (Eq. (9.5)) and solved by numerical methods. As an example, the differential equations for the degradation of butyric acid (But) (a rather common case) have been given by Mösche and Jördening (1998, 1999).

The kinetic model does not describe external deviations such as the instationary occurrence or accumulation of lactic and/or propionic acid – a common phenomenon in wastewater from the food industries. The conversion of these may become a serious bottleneck (if the mixed culture is not adapted), with a drop in pH and inactivation of the methanogens. Typical data for industrial substrates and kinetics

Table 9.7 Data for common industrial substrates (wastewater from the sugar industry) and (a) kinetics of immobilized anaerobic systems (laboratory data) (Jördening *et al.*, 1991; Buchholz *et al.*, 1992) and (b) industrial granular sludge (van Lier, 2008).

(a) Laboratory data

Wastewater composition	(kg m^{-3})	Kinetic data	
Organic substances		Active biomass	
COD (total)	5–15	(Determined as odm)[a]	15–80 g l^{-1}
Acetate	1–2	Max. growth rate, μ_{max}	0.08–0.12 day^{-1}
Propionate	0.5–1.5	Biofilm thickness	0.02–0.2 mm
Butyrate	1–2	V_{max} for acidification	Up to 100 g (COD) l^{-1} day^{-1}
(D,L)-Lactate	0.2–2.5	V_{max} for methanogenesis (porous glass as carrier)	30–150 g (COD) l^{-1} day^{-1}
		Total degradation rate	40–80 g (COD) l^{-1} day^{-1}

(b) Typical values for industrial plants

		Reaction rates	0.5–1.0 kg (COD) kg^{-1} (odm) day^{-1}

a) odm = organic dry matter.

are listed in Table 9.7. Data for industrial plants have been summarized by van Lier (van Lier, 2008; van Lier *et al.*, 2008).

Growth kinetics for anaerobic bacteria – including slow-growing methanogenic species – are in the range of 0.1–0.12 day^{-1}, corresponding to doubling rates (where the active biomass grows to twice the original mass) of 5–6 days. For the start-up of advanced systems, from one to a few weeks are required when sufficient inoculum is available (about 5% of the final active biomass is required) and optimal strategies are applied. For campaign-dependent systems (operating during certain seasons), which are common in many food industries, it means that a restart at the beginning of a campaign requires about 1 week to attain full capacity (assuming a common value of 50% activity loss during 6–8 months of intercampaign). Further data for UASB and EGSB systems have been published by Gonzales-Gil *et al.* (2001) and van Lier (van Lier, 2008; van Lier *et al.*, 2008).[11]

11) The anaerobic treatment of wastewaters includes acid formation and methane formation, which differ significantly in terms of nutritional needs, growth kinetics, and sensitivity with regard to the environmental conditions, notably pH (Figure 9.9) Hence, these two steps are preferentially physically separated, and two-stage systems are considered to be superior, as their performance in terms of stability and space–time yield will be superior to that of one-stage systems. The first step does not require immobilized bacteria, as those involved exhibit rapid growth, whereas the second step requires immobilized systems for high performance. The load of reactors for the second step with volatile fatty acids (in a system with a separate acidification reactor) may be higher by a factor of 4–5, compared to a one-step system and feeding with complex substrates (not acidified completely) (Henze and Harremoes, 1983).

The biocatalyst comprises granular aggregates of microorganisms, or the carrier with adhering microorganisms, extracellular secondary products (such as polysaccharides), and eventually precipitated inorganic compounds (such as $CaCO_3$). Data for microbial aggregates and carriers for fluidized bed reactors have been mentioned previously (see Table 9.6). Carriers (e.g., porous silica, pumice) should exhibit a large, rough, and – preferentially – a macroporous surface in order to accommodate a large number of bacteria and provide protection against shear forces. The microorganisms of the second, methanogenic step comprise complex mixed cultures (including propionate-, butyrate-, lactate-transforming, and methanogenic bacteria) that may secrete extracellular secondary metabolites such as polysaccharides and proteins to form a secondary matrix that is favorable for adhesion (see Figure 9.10). The complex reaction sequences (Figure 9.9) can proceed in a single particle as methanogenic and the other species mentioned that form acetate, formate, and hydrogen grow symbiotically, and utilize the products formed immediately, including interspecies hydrogen transfer (McInerny, 1999).

(a)

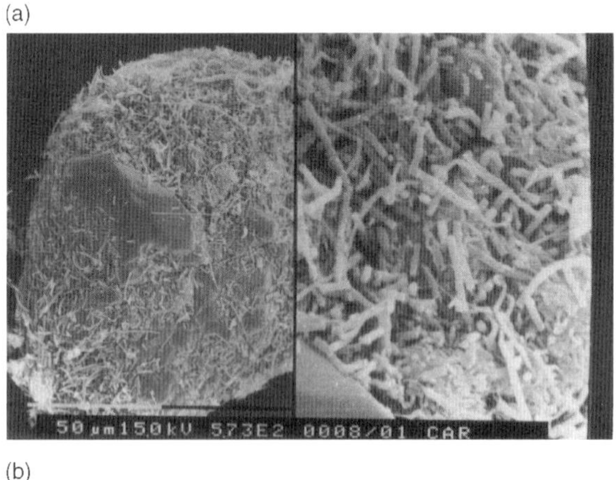

(b)

Figure 9.10 Microorganisms of an anaerobic mixed culture adhering to (a) sand and (b) pumice particles at two different magnifications (scale bar = 50 μm).

The most successful high-performance anaerobic treatment systems are UASB and EGSB reactors, relating to wastewater from industries based on agricultural and forestry products, with typically high concentrations of organic substrates that are readily degraded by anaerobic bacteria. These offer the advantage of high-load systems, require much less volume and space, and hence less investment as compared to conventional systems. Furthermore, these systems tend to operate more stable under transient conditions, such as fluctuations of substrates and pH. In fact, a large number of different types of plants with immobilized anaerobic bacteria are currently on stream, including fixed-bed and fluidized bed reactors with carriers.

Substrates typically result from raw material washing procedures, blanching, extraction, fermentation, or enzymatic processing. Original substrates are in most cases carbohydrates such as sugar, starch, cellulose, and hemicellulose, proteins, and fats, all of which readily undergo bacterial degradation to fatty acids, mainly acetic, propionic, and butyric and lactic acids. The majority of installations are in the potato, starch, and sugar industries, in fruit, vegetable, and meat processing, in beer and fruit juice, cheese, yeast, alcohol, and citric acid manufacturing, and in the paper and pulp industries. The concentrations of substrates are typically in the range of 5–50 kg COD m^{-3}, and these are diluted by recirculation (loop reactor for control of concentration, pH, and to provide suspension of particles; see Section 11.1) down to less than 2 kg COD m^{-3}.

The most widely used system worldwide has been the UASB system (about 50% out of a total of over 2000 plants); however, in recent years new installations have been preferentially the EGSB system (about 20%), a similar principle with a significantly higher loading potential. Both systems operate without a carrier, but rather utilize granular bacteria aggregates (Lettinga *et al.*, 1999; McCarty, 2001, 2001; Franklin, 2001; van Lier, 2008; Austermann-Haun, 2008). In the UASB system, the incoming wastewater is distributed equally over the cross section by a system of tubes. At low superficial upflow velocities (1–2 m h^{-1}), the wastewater flows vertically through a bed of granular particles of aggregated bacteria (sludge bed). In the upper part of the reactor, a three-phase separation unit is integrated, which avoids disintegration of the sludge pellets. The high liquid upflow velocities require specific separators (screens or modified internal lamella separators).

Typical loading rates are in the range of 50–90 kg COD m^{-3} day^{-1} for common wastewater from agricultural and food processes with a high conversion (over 80% of degradable COD) at laboratory and pilot scales (Jördening and Buchholz, 1999). The load rates reported for the performance on a technical scale are in a significantly lower range of 20–30 kg m^{-3} day^{-1}, as shown in Table 9.8.

Anaerobically treated wastewater usually cannot be released directly into the environment. The COD concentration in the effluent is generally not as low as can be achieved with aerobic treatment, and this often leads to an odorous product. For this reason, high-loaded wastewaters are often treated in a combination of anaerobic and aerobic processes. The second (aerobic) step is then, if necessary, constructed also for nitrogen elimination.

Table 9.8 Data for technical-scale anaerobic fluidized bed reactors, UASB and EGSB systems.[a]

Waste[b]	Company	Support	Reactor volume (m^3)	Ratio[c]	Load (kg m^{-3} day^{-1})	Concentration (kg COD m^{-3})	HRT[d] (h)	Removal (%)	Reference
Fluidized bed					(mean[e]: about 30)				
Soybean	Dorr–Oliver	Sand	304	2.0	14–21	0.8–10		75–80	Sutton *et al.* (1982)
Several	Degremont	Biolite	210–480		16–21	3.8–5	0.25	50–60	Ehlinger (1994)
Yeast and pharmaceuticals	Gist-Brocades[f]	Sand	400	4.4	8–30	1.9–4	0.14–0.41	95–98	Franklin *et al.* (1992)
Sugar beet	BMA	Pumice	500	5.0	20–30	1–5	1.5	90–95	Jördening (1996)
UASB					(mean[e]: 9–11)				
Aspartame	Nutrasweet Co.		2 × 600		7.8	22			Macarie (2001)
PET[g]	Eastman Chemical		144		12	12		90–95	
Nylon	Rhone Poulenc[h]		990		8	16		80	
EGSB					(mean[e]: 20 up to 30)				
Formaldehyde	Caldic Europort		275		17	40		98	Macarie (2001)
DMT[g]	Kosa		550		13	34			
DMT and PET[g]	Sasa		2 × 1000		13	6.5			

a) Companies offering plants include Paques, Biothane, Biotim, Enviroasia, Degremont, Envirochemie, and others (van Lier, 2008).
b) Source of wastewater.
c) Height/diameter.
d) HRT: hydraulic retention time.
e) Austermann-Haun (2008) and van Lier (2008).
f) Currently, DSM (NL).
g) PET: polyethylene terephthalate; DMT: dimethyl terephthalate.
h) Currently, Aventis.

9.6.2.3 **Nitrogen Elimination (Nitrification and Denitrification)**

Nitrogen elimination from wastewater is a standard requirement, as limiting values by authorities for total nitrogen-containing compounds (notably ammonia) require this step to be taken. It is usually a process subsequent to anaerobic degradation of organic compounds. During the anaerobic degradation of nitrogen-containing compounds, nitrogen is converted to ammonia. Subsequent nitrification and denitrification converts it to nitrogen (N_2) and, to a lesser degree, to N_2O, which should be avoided as far as possible. Two groups of bacteria catalyze the oxidation of ammonia to nitrate; the ammonium-oxidizing group (e.g., *Nitrosomonas* sp.) forms nitrite, which subsequently is oxidized to nitrate by nitrite-oxidizing bacteria (e.g., *Nitrobacter* sp.). Nitrate is finally reduced under anoxic conditions by denitrifying bacteria with acetic acid as a cosubstrate to yield CO_2, H_2O, and N_2 (for details, see Dorias *et al.*, 1999; Kayser, 2004).

Technical systems with immobilized bacteria in mixed cultures are well established, and two main systems are used: (1) trickling filters and (2) rotating biological contactors. Conditions for high process efficiency and general guidelines can be found in Dorias *et al.* (1999) and EPA (1993). In trickling filters, the main portion is the inserted packing material as carrier, which may be either rocks, lava, or plastic material. Rotating biological contactors with submerged drums, specially equipped with rotating disks made of plastic material placed halfway into the water, are used mainly in smaller wastewater treatment plants.

9.6.2.4 **Exhaust Gas Purification**

Immobilized mixed cultures are frequently used in exhaust gas treatments, as a wide variety of organic substances (Table 9.9) and even some inorganic odor compounds can be eliminated from the waste gas (VDI, 1991, 1996, 2000; Lu *et al.*, 2002; for recent reviews, see Iranpour *et al.*, 2005; Kennes *et al.*, 2009). Microbiological and engineering fundamentals have been summarized in Klein and Winter (2000). The principle has found successful application in the food industries, including animal, meat, fish, yeast, and aroma production, as well as in the chemical industries, with diverse examples such as the degradation of hydrocarbons (aliphatic and aromatic), phenols, sulfides, amines, amides, acrylonitrile, styrene, and even halogenated compounds, in foundries, and for spray cabins in automobile production (Table 9.9).

Table 9.9 Examples for groups of readily degradable substances in exhaust gas (VDI, 1991, 1996).

Group of substances	Examples
Aliphatic hydrocarbons	Hexane, ethylene
Aromatic hydrocarbons	Toluene, xylene
Alcohols	Methanol, butanol
Ethers	Tetrahydrofuran
Aldehydes	Formaldehyde, acetic aldehyde
Carboxylic acids	Butyric acid
Amines	Trimethylamine

Table 9.10 Ranges of pollutant concentrations and operating parameters for biofilters and trickling bed reactors (VDI, 1996; Waweru *et al.*, 2000).

	Biofilter	**Trickling bed reactor**
Pollutant concentration ($g\,m^{-3}$)	<1	<0.5
Henry coefficient	<10	<1
Surface loading rate ($m^3\,m^{-2}\,h^{-1}$)	50–200	100–1000
Mass loading rate ($g\,m^{-3}\,h^{-1}$)	10–160	<500
Empty bed contact time (s)	15–60	1–10[a]; 30–60[b]
Volumetric loading rate ($m^3\,m^{-3}\,h^{-1}$)	100–200	
Elimination capacity ($g\,m^{-3}\,h^{-1}$)	10–160	
Removal efficiency (%)	95–99	

a) For easy degradable compounds.
b) For slowly degradable compounds.

Reactor types with immobilized mixed cultures of microorganisms include the bioscrubber or biofilter, and trickling filter. Examples of their application have been summarized by Fischer (2000) and Plaggemeier and Lämmerzahl (2000), and in VDI (2002). The advantages of biological exhaust gas treatment are, first, operation at low temperature and, second, the efficiency of the elimination of toxic and odorous substances at low concentrations in the waste gas stream, at low cost.

The conditions for efficient treatment are good mass transfer – that is, the sufficiently rapid transfer of the components from the gas into the fluid phase – and their biological degradability. If these conditions are satisfied, then mixed cultures can adapt to the substrate if an inoculum with a sufficiently broad bacterial consortium is present. The characteristics and essential parameters of the waste gas stream are relative humidity (>95%), temperature (<60 °C), waste gas flow rate, chemical pollutant identity and concentration, odor concentration, surface loading rate ($m^3\,m^{-2}\,h^{-1}$), mass loading rate ($g\,m^{-3}\,h^{-1}$), elimination capacity ($g\,m^{-3}\,h^{-1}$), and removal efficiency (%) (Table 9.10). Further relevant parameters are pH and buffer capacity of the fluid phase, available surface, and the residence time distribution of the phases. Substances with toxic properties toward microorganisms can inhibit the system. For odor elimination, the olfactrometric measurement is essential (VDI, 1991, 2002; Waweru *et al.*, 2000).

The degradation routes are more or less complex, depending on the compounds and the microorganisms, which are mostly bacterial and fungal species. Part of the substrate (up to 50%) is utilized for the growth of biomass, and this must be removed from the system. The oxidative degradation in general yields CO_2 and water, for example, in the case of propionic aldehyde:

$$C_2H_5CHO + 4O_2 \rightarrow 3CO_2 + 3H_2O$$

Inorganic odorous substances can also be transformed in part, and thus be removed from the waste gas. Ammonia is oxidized after absorption to nitrite and nitrate, while hydrogen sulfide can be oxidized by some bacterial species to sulfuric acid.

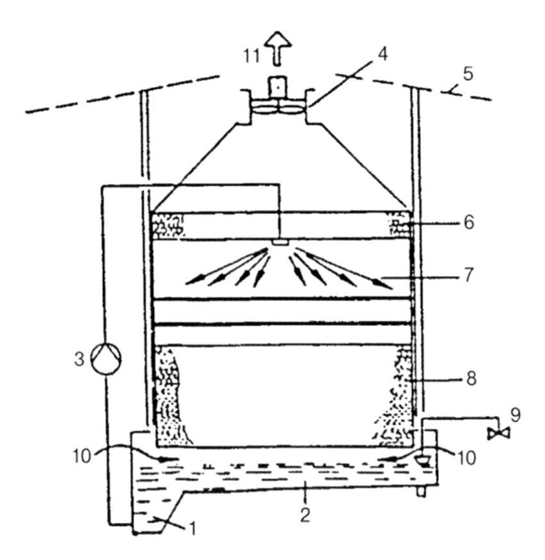

Figure 9.11 Scheme of a biological trickling filter, with water reservoir (1), collecting trough (2), pump (3), ventilator (4), root (5), packings for droplet removal (6), water distribution system (7), bed with packings (8), outflow control (9), waste gas inflow (10), and purified gas (11).

Reactors used frequently are the biofilter (this is the most common, with many hundreds of installations) and the trickling filter (trickle-bed air biofilter), which achieves higher specific rates of treatment. Biofilters are fixed-bed reactors where the waste gas flows through a bed of carrier material (mostly natural fiber material) with bacterial mixed cultures adhering to the surface.

Trickling filters are another standard method for the purification of waste gas, and are capable of degrading more contaminants per unit volume, as compared to biofilters, and treating up to 100 000 $m^3 h^{-1}$. In a trickle-bed reactor, an aqueous phase is continuously circulated through a bed of inert material. The packing can consist of inert bulk material (Raschig rings, saddles, etc.) or structured packings. The material can be plastics, such as polyethylene, polypropylene, or polyurethane foam. The surface should be appropriate for adhesion of the biofilm. The waste air flows either concurrently or countercurrently with a recirculated liquid through the packing (Figure 9.11). Ranges of pollutant concentrations and operating parameters are summarized in Table 9.10 (VDI, 1996, 2002; Waweru *et al.*, 2000).

9.7 Perspectives

9.7.1 Biofilm Catalysis

Biofilms are considered to exhibit great potential as biocatalysts for the production of chemicals because of their inherent characteristics of self-immobilization, high

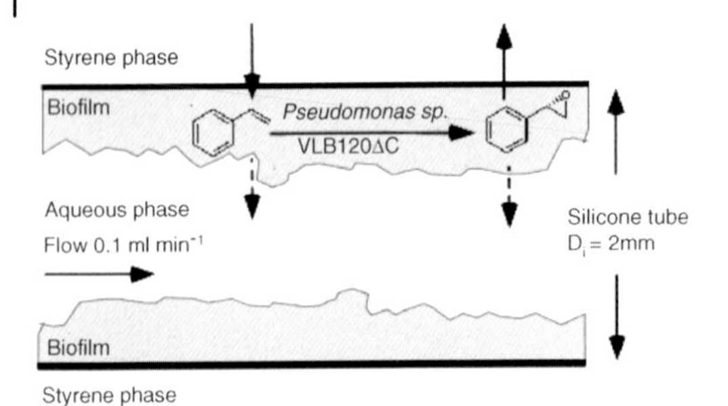

Figure 9.12 Principle of substrate supply and product extraction in the biofilm tube reactor (Gross *et al.*, 2007).

resistance to reactants and to cope with physiological stress, and long-term activity, which all facilitate continuous processing. Especially organisms used for bioremediation and for recombinant enzyme expression are interesting from the biotechnological point of view, since they tend to grow on surfaces, and as many key reactions in the degradation pathways yield important industrial products or derivatives thereof. Topics addressed include synthetic chemistry (ranging from specialty to bulk chemicals), bioenergy, and biologics.[12] In a major project, a large number of species of biosafety level 1 bacteria and yeasts from over 40 different genera and five phyla were screened by growing them in microtiter plates. Over 30 of these strains were classified as strong biofilm formers, potentially suitable as catalysts in biofilm applications, including strategies in whole-cell catalysis. A variety of biofilm reactor configurations have been explored for productive catalysis and some reactors have been operated continuously for months. The most critical issue identified was the volumetric oxygen mass transfer capability (k_{La}), responsible for a consistent productivity; this parameter was shown to directly influence biofilm growth and biotransformation performance. Further bottlenecks of biofilm-based processes include excessive biomass growth, which is difficult to control and often results in clogging of the reactor system (Alcalde *et al.*, 2006; Rosche *et al.*, 2009; Li *et al.*, 2007).

In an example, an engineered *Pseudomonas* sp. strain was chosen as a model organism due to its excellent biofilm forming capacity and its well-studied capability of catalyzing asymmetric epoxidations. A tubular reactor was used for the biotransformation of styrene to (*S*)-styrene oxide as a model reaction (Figure 9.12). Silicone tubing was chosen because the organism grew very well on this material. *In situ* product extraction prevented product inhibition of the catalyst. Biofilm physiology and dynamics have been characterized, as well as the influence of mass

12) Traditionally, only bulk chemicals such as acetic acid, ethanol, butanol, butane-2,3-diol, lactic acid, fumaric acid, and succinic acid have been produced in biofilm systems. Lactic acid fermentation by natural cell immobilization or biofilms has developed considerable interest over the past few decades. In a search for natural immobilization of lactic acid-producing microorganisms, optimum parameters and how they are affected by the physical, chemical, and biological variables of the process have been investigated (Dagher *et al.*, 2010).

transfer, specific membrane area, and tube wall thickness. The process was stable for at least 50 days; an average volumetric productivity of 24 g $l_{aq}{}^{-1}$ day^{-1} with a maximum of 70 g $l_{aq}{}^{-1}$ day^{-1} and biomass concentrations of 45 g CDW $l_{aq}{}^{-1}$ were achieved (CDW: cell dry weight) (Gross *et al.*, 2007, 2010). A further development suggested a microporous ceramic unit as an ideal microenvironment for biofilm growth of *Pseudomonas* sp. and for efficient oxygen transfer. A uniform and dense biofilm developed on this matrix. In this unit, an average productivity of 24 g $l_{aq}{}^{-1}$ day^{-1} by a dry biomass of 3.7 g CDW in a 10 ml aqueous phase was achieved by integrating an *in situ* substrate feed and an *in situ* product recovery technique, where styrene was both the substrate and product solvent phase, and (*S*)-styrene oxide enriched to 70% by vacuum distillation. The system was stable for more than 30 days (Halan *et al.*, 2010).

A special, sophisticated approach has been developed by Tsoligkas *et al.* (2011) using spin coating to control biofilm formation. To test this approach, *E. coli* PHL644, which overproduces curli and readily forms biofilms, was transformed with a high copy number plasmid expressing tryptophan synthase from *Salmonella* sp. After growth, the cells were spin coated onto a poly-L-lysine-coated flat glass substrate by using centrifugation. The topology and adhesive forces within the engineered and naturally deposited biofilms were explored by using environmental scanning electron microscopy (ESEM) and atomic force microscopy (AFM), which enabled high-resolution and global analysis of the structure and strength of the biofilm. The most dramatic change in biofilm development was observed between days 5 and 6, with a sharp increase in the adhesive force, as measured by AFM. This increase in strength corresponded to the observed production of an adhesive extracellular polymeric substance, visible by ESEM as a white fibrous material on the surface of the cells following 6 days maturation. At days 6 and 7, deep pores and channels were observed by ESEM. These confer a large catalytic surface area to the biofilm. Efficient conversion of haloindoles and serine to L-halotryptophans was achieved by the transformed *E. coli* spin-coated biofilm (e.g., 93 and 78% conversion to 5-F and 5-Cl tryptophan, respectively), as compared to a cell-free lysate containing tryptophan synthase under free and immobilized conditions or planktonic cells, respectively.

In another example, biofilm reactors of the rotating disk type have been used as niche-mimic bioreactors, which is essentially the cultivation of the producing microbe in reactor conditions that mimic its ecological niche. In this system, a marine *Streptomyces* sp. producing an antimicrobial compound has been investigated (Sarkar *et al.*, 2010). Biofilm cultures of *A. niger* grown on perlite as inert support grew less than other cultures (solid-state and submerged fermentation), but produced significantly higher cellulase yields and volumetric productivities, suggesting that fungal attached growth on perlite may favor better enzyme production (Gamarra *et al.*, 2010). In still another approach, poly(3-hydroxybutyrate) (PHB) biosynthesis in the biofilm of *Alcaligenes eutrophus* was achieved, using glucose enzymatically released from pulp fiber sludge. Almost 78% of the cell mass was converted to pure PHB. The anion exchanger DEAE-Sephadex A-25 was chosen as a microcarrier for packed bed biofilm cultures. Conditions for attachment, growth, and detachment for isolation of PHB were established by varying the ionic strength of the attachment medium (Zhang *et al.*, 2004).

Different analytical techniques have been applied and developed for control of biofilm formation and performance. Calorimetry represents an efficient tool, enabling real-time monitoring and data acquisition (Buchholz *et al.*, 2010). An advanced analytical approach was presented by Halan *et al.* (2011), enabling a non-invasive, quantitative approach to study biofilm development and its response to the toxic solvent styrene for *Pseudomonas*. Biofilm grown cells displayed stable catalytic activity, producing (*S*)-styrene oxide continuously. However, the cells experience severe membrane damage during styrene treatment, although they obviously are able to adapt to the solvent. Concomitantly, the fraction of concanavalin A (ConA)-stainable extrapolysaccharides increased, substantiating the assumption that those polysaccharides play a major role in structural integrity and enhanced biofilm tolerance toward toxic environments.[13)]

In order to gain new insight into molecular phenomena involved in control factors, biofilm dispersal and formation and EPS production were studied. 3,5-Cyclic diguanylic acid (c-di-GMP) is an intracellular messenger that controls cell motility by flagellar rotation and biofilm formation through synthesis of curli and cellulose. A novel c-di-GMP binding protein, BdcA, has been discovered and engineered and its influence on biofilm dispersal investigated (Ma *et al.*, 2011). Further, evidence has been presented that toxin/antitoxin (TA) systems are involved in biofilm and persister cell formation and that these systems may be important regulators of the switch from the planktonic to the biofilm lifestyle as a stress response by their control of secondary messenger c-di-GMP (Wang and Wood, 2011).

9.7.2
Microbial Fuel Cells

Microbial fuel cells (MFCs) represent a most promising concept where the microbial catabolic activity in the degradation of organic matter is converted directly into cheap and environment-friendly energy. An MFC is an electrochemical device in which microbially produced reduction equivalents are utilized to deliver electrons to a fuel cell anode. Substrates may be various types of organic materials in solution, such as methanol, ethanol, acetate (e.g., by-products from fermentation processes), glucose as well as other sugars (e.g., hydrolysates of hemicellulosics), and even wastewater constituents including organic acids. The energy conversion can be achieved with MFCs, in which anaerobic microorganisms serve as biocatalysts and from which electrons are diverted and transferred to an electrode to generate electricity. Finding an efficient way to "wire" the microbial activity is the key to success of the concept (Schröder *et al.*, 2003; Schröder, 2007). The current state of the

13) Compared to control experiments with planktonic (suspended) grown cells, the *Pseudomonas* biofilm adapted much better to toxic concentrations of styrene, as nearly 65% of biofilm cells were not permeabilized (viable), compared to only 7% in analogous planktonic cultures (Halan *et al.*, 2011).

art is presented in several reviews (Chae *et al.*, 2009; Rabaey *et al.*, 2009a, 2009b; Pandey *et al.*, 2010).[14)]

One approach uses electrocatalytic anode catalysts based on platinum–polyaniline and platinum–poly(tetrafluoroaniline) sandwich electrodes or on tungsten carbide that allow the efficient *in situ* oxidation of microbial hydrogen and a number of organic acids. Further three-dimensional carbon fiber nonwovens have been applied to serve as electrodes. Current densities in the range of 15–30 A m^{-2} have been achieved using a great versatility of fermentative, photofermentative, and even photosynthetic microbial activities (Chen *et al.*, 2011; Rosenbaum *et al.*, 2006). Connecting several microbial fuel cell units in series or parallel can increase voltage and current. Six units in a stacked configuration produced a power output of 258 W m^{-3} h^{-1} (Aelterman *et al.*, 2006). The effects of fluidic connections between stacks of MFC were investigated by Winfield *et al.* (2011). More recently, a power output of 1 kW m^{-3} h^{-1} has been achieved (Rabaey *et al.*, 2009b; Schröder, 2009, personal communication). Microbial fuel cells may be operated discontinuously or continuously – the continuous mode being the most appealing mode.

Most promising are systems of mixed bacterial biofilms that grow on and adhere to anode surfaces, not requiring sterile conditions, which is essential for the utilization of waste materials. Thus, the integration of MFCs into existing water or process lines has been discussed and a pilot test system has been run at a brewery in Brisbane, Australia (Rabaey *et al.*, 2009b). An important fact is that the microorganisms assemble spontaneously under appropriate conditions, and that inocula from wastewater systems and even soil may be used, which are rich in all types of microorganisms. An electrochemically driven selection in natural community-derived microbial biofilms using flow cytometry has been revealed (Harnisch *et al.*, 2011).[15)] Fermentative bacteria convert different types of substrates (carbohydrates, organic acids, etc.) into hydrogen, formate, or acetate; these are further oxidized by electrochemically active bacteria to yield H_2O and CO_2 and electrons, collected by the anode (Figure 9.13d). Procedures for performance evaluation of biofuel cells, including microbial fuel cells and enzyme fuel cells, with possible procedures and parameters, have been given by Watanabe (2010).

14) Biofuel cells may also be based on immobilized enzymes in nanostructured materials, such as mesoporous media, nanoparticles, nanofibers, and nanotubes, which have been demonstrated as efficient hosts of enzyme immobilization. It is evident that, when nanostructures of conductive materials are used, the large surface area of these nanomaterials can increase the enzyme loading and facilitate reaction kinetics, and thus improve the power density of biofuel cells (Kim *et al.*, 2006). However, enzyme fuel cells are less developed currently as are MFCs. The construction and characterization of a noncompartmentalized, mediator- and cofactor-free glucose–oxygen biofuel cell based on adsorbed enzymes exhibiting direct bioelectrocatalysis has been reported. Cellobiose dehydrogenase from *Dichomera saubinetii* and laccase from *Trametes hirsuta* as the anodic and cathodic bioelements, respectively, were used (Coman *et al.*, 2008).

15) It has been demonstrated that electrochemically active microbial biofilms and their enrichment at the anode of a microbial bioelectrochemical system (BES) can be quantitatively and easily characterized by flow cytometry. This analysis revealed that the anodic biofilm of a BES, formed from a highly diverse microbial community and fed with single substrate artificial wastewater, was dominated by only one phylotype (Harnisch *et al.*, 2011).

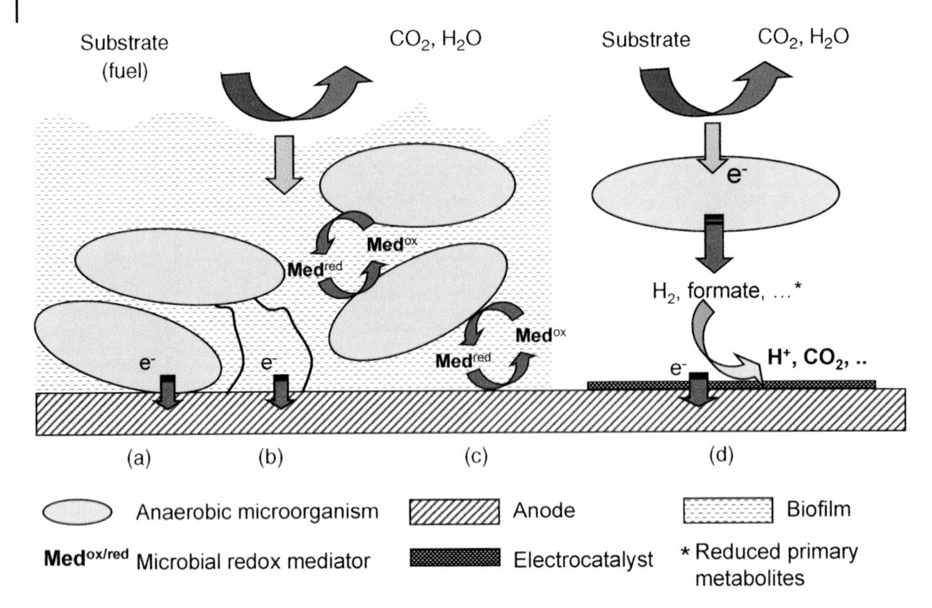

Figure 9.13 Identified electron transfer mechanisms in MFCs. Electron transfer via (a) cell membrane-bound cytochromes, (b) electrically conductive pili (nanowires), (c) microbial redox mediators, and (d) oxidation of reduced secondary metabolites (Rosenbaum *et al.*, 2006).

Different concepts and mechanisms for electron transfer from the biocatalyst to the anode of the fuel cell have been proposed. Electrontransfer in the biofilm is assumed to proceed by cell–cell transfer. Thus, conductive bacterial pili ("nanowires") have been observed. Further membrane-bound or excreted bacterial redox mediators (e.g., outer membrane redox proteins) are assumed to enhance the electron transport to the anode and between multiple cell layers (Figure 9.13) (Rabaey *et al.*, 2004, 2005; Rosenbaum *et al.*, 2006). Thus, *Geobacter* sp., known to evolve electronically conducting pili, has been identified. Remarkably, such systems have been found to operate even under conditions of stress, for example, with toxic substances introduced or at nonoptimal temperature, the heterogeneous biofilm obviously stabilizing the whole system.

Recent studies provided insight into molecular phenomena. A mixed culture-derived microbial biofilm, grown on silver electrodes, was studied *in vivo* with surface-enhanced resonance Raman scattering in combination with cyclic voltammetry. This experimental approach provides unprecedented structural information about the outer membrane cytochromes involved in the direct electron transfer across the bacterium–electrode interface (Millo *et al.*, 2011). In *Shewanella oneidensis*, electron transfer involves decaheme cytochromes that are located on the bacterial cell surface at the termini of *trans*-outer membrane electron transfer conduits. The first X-ray crystallography structure of a protein in the complex that is responsible for shuttling electrons from the microbe's innards to solution has been presented, and the spatial organization of 10 hemes visualized. The knowledge of this configuration should enable the more efficient use of the bacterium in MFC (Clarke *et al.*, 2011).[16] It has also been demonstrated that a genetic cassette can create a conduit for electronic communication from living cells to inorganic materials. Thus, a portion of the extracellular electron transfer chain of *S. oneidensis* MR-1 has been reconstituted into the model microbe *E. coli* (Jensen *et al.*, 2010).

Exercises

Entrapment of Yeast in Calcium Alginate for Ethanol Production

Equipment

- Syringe (20 or 100 ml); sieve or filter funnel; thermostated shaker; thermostated column bioreactor with mesh (about 10–100 ml).
- Glucose and/or ethanol analysis by enzymatic test, or GC, HPLC, or glucose analyzer.
- Sodium alginate solution: sodium alginate solutions can be sterilized by autoclaving, or by membrane filtration. Prepare well before required. A solution of alginate, typically 2–4% (w/v), is prepared by dissolving solid sodium alginate in distilled water or a buffer solution appropriate to the enzyme or cells to be immobilized. The dissolving process is slow, and normally requires stirring for up to 5 h. The solution may be stored at 4 °C, but should be prepared freshly each week.
- Calcium chloride ($CaCl_2 \cdot 6H_2O$) solution: prepare liter volumes in concentrations of 0.15 M. The solution may be stored at room temperature.
- Dried baker's yeast *S. cerevisiae*: suspend 1.0 g in 5 ml distilled water and mix in a small pestle and mortar until a smooth paste is formed.
- Yeast cell–alginate mixture: mix 1 ml of the yeast paste with 50 ml of alginate solution and stir the mixture thoroughly to ensure complete mixing. The maximum workable concentration of sodium alginate solution is about 4–5% (w/v), depending on the quality of sodium alginate (this is due to high viscosity).
- Buffer: 0.1 M sodium phosphate buffer, pH 7.0.

Immobilization

The setup for the formation of beads is shown in Figure 9.5. All the immobilization steps by ionotropic gel formation with sodium alginate may be carried out at 4 °C. Transfer the alginate mixture with yeast cells to a dropping device, such as a syringe (100 ml; a 0.55 mm internal diameter (i.d.) needle is appropriate; the number and

16) The cell surface cytochromes can potentially play multiple roles in mediating electron transfer directly to insoluble electron sinks, catalyzing electron exchange with flavin electron shuttles or participating in extracellular intercytochrome electron exchange along "nanowire" appendages. The hemes involved are organized across four domains in a unique crossed conformation, in which a staggered 65 Å octaheme chain transects the length of the protein and is bisected by a planar 45 Å tetraheme chain that connects two extended Greek key split β-barrel domains (Clarke *et al.*, 2011).

size of the alginate beads are influenced by the size of the opening through which the alginate is allowed to pass). Place a beaker containing 1 l of $CaCl_2$ solution below the dropping device and magnetically stir the solution to produce a light vortex. Adjust the flow control to allow dropwise flow, from a height of about 10 cm, into the $CaCl_2$ solution. Allow the flow to continue until the desired number of beads have been formed.

Allow a further 20–30 min stirring before collecting the beads using either a Büchner funnel or a large filter funnel. Care must be taken when using suction with Büchner or other funnels to ensure that the beads are maintained in liquid at all times. Wash the beads on the funnel with 1 l of buffer appropriate to the cells (for yeast cells, use distilled water). Store the beads in a minimal amount of distilled water until required (Fraser and Bickerstaff, 1997).

Supplements (titan dioxide or fine sand particles) can be added to the alginate solution prior to the gelation process in order to adjust the bead density, sedimentation rate, and mechanical strength (Berensmeier *et al.*, 2004).

Dissolution of the beads can be achieved by incubating them at room temperature with 0.1 M sodium phosphate buffer, pH 7.0, for a few hours, depending on the original alginate concentration, until complete.

Test: Conversion of Glucose and Formation of Ethanol

Discontinuous Experiment

Fractions of beads (5–10 g) with different diameters are placed into Erlenmeyer flasks; 50 ml of glucose solution (5 g l^{-1}) is added, and the flasks put into a shaker at 37 °C; concentrations of glucose and/or ethanol are determined at 10 and 20 min intervals during 2 h (conversion expected, about 80%). Concentrations of glucose and/or ethanol are recorded graphically as a function of time, the catalytic activity (g (glucose/ethanol) per g (yeast) per h) is determined from the slope; the activity is compared for biocatalyst beads of different diameter; from these data the efficiency may be estimated (see Sections 10.3 and 10.4).

The theoretical yield is 2 mol ethanol from 1 mol glucose (or 0.51 g ethanol from 1 g glucose). About 10% of the glucose is consumed for maintenance metabolism.

Continuous Experiment

Place 100 g biocatalyst (with 10–30% immobilized yeast) into a column with a sieve to obtain a packed bed. A glucose solution (5 g l^{-1}) is pumped through the bed of immobilized yeast at rates of 0.4 up to 4 l h^{-1} (residence time in the range of 30–3 min, calculated for 200 ml bed volume). Between three and five residence times are required until stationary conditions are approached. Take samples for the analysis of glucose and/or ethanol every 10 or 20 min. Calculate the conversion, space–time yield, and catalyst productivity and compare for different flow rate and conversion.

Note: A technical process should operate at near-maximal concentrations, which is about 10–12% for ethanol, due to toxic effects limiting the productivity of yeast. The glucose concentration therefore would be approximately 200 g l^{-1}. At high conversion – as would be required for economic reasons – product inhibition and inactivation of the biocatalyst must be taken into consideration.

Characterization of an Anaerobic Fluidized Bed Reactor
The technical properties of a fluidized bed reactor with two different carriers are to be investigated; the performance of an anaerobic fluidized bed reactor shall be characterized.

Equipment
Two glass reactors (tubes with a sieve at the bottom, volume 7 l) with recirculation and two pumps (for introducing the substrate and for recirculation), flow meter, conductometer, pH measurement and regulation, equipment for COD or TOC measurement, gas chromatograph (GC) or HPLC, or ion chromatograph.

Reagents
Sand (d_p 0.1–0.2), porous glass, or pumice (d_p 0.2–0.4), carrier with adhering anaerobic mixed culture, and substrate (acetate, propionate, butyrate, or a mixture of acids).

Pressure Drop in the Fluidized Bed
For a reactor without a carrier, with sand, and with porous glass or pumice, 20 different superficial upflow velocities shall be established; the pressure drop (Δp) and the bed expansion should be measured as a function of the flow rate (Q). The point of beginning expansion is determined from the correlation of Q and Δp.

Performance of an Anaerobic Fluidized Bed Reactor
The kinetic parameters V_{max} and $K_{m(apparent)}$ should be determined in a fluidized bed reactor with adhering anaerobic mixed culture. The carrier material is expanded by about 50%, with an appropriate flow rate. A substrate solution with a concentration of 1 g COD is introduced into the reactor. Samples are taken every 30 min, and COD or TOC are determined. The experiment is finished when no more gas is formed.

Anaerobic COD Degradation
The maximal reaction rate and the apparent Michaelis–Menten constant $K_{m(apparent)}$ as well as the transformation of different acids and intermediates shall be investigated by means of the degradation and intermediate formation of organic acids. Measurements are performed by taking samples from a continuous fluidized bed reactor (carrier: porous glass or pumice with adhering anaerobic mixed culture including acetogenic and methanogenic bacteria; T: 37 °C; pH 6.8–7.3; substrate: butyric acid at the start of the experiment; samples every 15 min). Concentrations of butyric and acetic acids and, if possible, intermediates such as isobutyric acid are determined by GC or HPLC. The kinetic parameters are determined according to Michaelis–Menten kinetics, where the reversible isomerization is considered (Eqs. (9.5) and (9.7)).

Microbial Fuel Cell Mechanisms – Elements
How do microorganisms organize electron transfer to media outside the cell? What are basic elements of electron transfer? Does a spatial organization play a role?

Update the recent literature relating to mechanisms of electron transfer.

References

Aelterman, P., Rabaey, K., Phan, H.T., Boon, N., and Verstraete, W. (2006) Continuous electricity generation at high voltages and currents using stacked microbial fuel cells. *Environ. Sci. Technol.*, **40**, 3388–3394.

Aivasidis, A. and Wandrey, C. (1992) Reaktionstechnische Optimierung und Maßstabsvergrö-ßerung von anaeroben Festbett-Umlauf- und Wirbelschichtreaktoren. *Chem.-Ing.-Tech.*, **64**, 374–375.

Alcalde, M., Ferrer, M. *et al.* (2006) Environmental biocatalysis: from remediation with enzymes to novel green processes. *Trends Biotechnol.*, **24** (6), 281–287.

Amory, D.E. and Rouxhet, P.G. (1988) Surface properties of *Saccharomyces cerevisiae* and *Saccharomyces carlsbergensis*: chemical composition, electrostatic charge and hydrophobicity. *Biochim. Biophys. Acta*, **938**, 61–70.

Anderson, G.K., Ozturk, I., Saw, C. B. (1990) Pilot-scale experiences on anaerobic fluidized bed treatment of brewery wastes, *Water Sci. Technol.* **22**, 157–166.

Anderson, S., Berman Marks, C., Lazarus, R. *et al.* (1985) Production of 2-keto-L-gulonate, an intermediate in L-ascorbate synthesis, by a genetically modified Erwinia herbicola. *Science*, **230**, 144–149.

Atkinson, B. and Mavituna, F. (1991) *Biochemical Engineering and Biotechnology Handbook*, 2nd edn, MacMillan Publisher Ltd, New York, pp. 529–546, 1171–1173.

Austermann-Haun, U. (2008) Anaerobverfahren – Uebersicht. *GWF Wasser Abwasser*, **149** (14), S6–S11.

Balaguer, M.D., Vicent, M.T., Paris, J.M. (1992) Anaerobic fluidized bed with sepiolite as support for anaerobic treatment of vinasse, *Biotechnol. Lett.* **5**, 333–338.

Becker, T., Breithaupt, D., Doelle, H.W. *et al.* (2007) Biotechnology, in *Ullmann's Encyclopedia of Industrial Chemistry*, Wiley-VCH Verlag GmbH, Weinheim.

Berensmeier, S., Ergezinger, M., Bohnet, M., and Buchholz, K. (2004) Design of immobilised dextransucrase for fluidised bed application. *J. Biotechnol.*, **114**, 255–267.

Bickerstaff, G.F. (ed.) (1997) *Immobilization of Enzymes and Cells*, Methods in Biotechnology, vol. **1**, Humana Press Inc., Totowa, NJ.

Bössmann, M., Staudt, C., Neu, T.R., Horn, H., and Hempel, D.C. (2003) Investigation and modeling of growth, structure and oxygen penetration in particle supported biofilms. *Chem. Eng. Technol.*, **26**, 219–222.

Brányik, T., Vicente, A.A. *et al.* (2005) Continuous beer fermentation using immobilized yeast cell bioreactor systems. *Biotechnol. Prog.*, **21**, 653–663.

Breitenbücher, K. (1996) Schott information 79.

Brodelius, P. and Vandamme, E.J. (1987) Immobilized cell systems, in *Biotechnology*, vol. 7a (eds H.J. Rehm, G. Reed, and J.F. Kennedy), VCH Verlagsgesellschaft, Weinheim, pp. 405–464.

Buchholz, K. and Collins, J. (2010) *Concepts in Biotechnology – History, Science and Business*, Wiley-VCH Verlag GmbH, Weinheim.

Buchholz, K., Diekmann, H., Jördening, H.-J., Pellegrini, A., and Zellner, G. (1992) Anaerobe Reinigung von Abwässern in Fließbettreaktoren. *Chem.-Ing.-Tech.*, **64**, 556–558.

Buchholz, F., Harms, H. *et al.* (2010) Biofilm research using calorimetry – a marriage made in heaven? *Biotechnol. J.*, **5** (12), 1339–1350.

Cashin, M.M. (1996) Comparative studies of five porous supports for yeast immobilisation by adsorption/attachment. *J. Inst. Brew.*, **102** (1), 5–10.

Chae, K.J., Choi, M.J. *et al.* (2009) Effect of different substrates on the performance, bacterial diversity, and bacterial viability in microbial fuel cells. *Bioresour. Technol.*, **100** (14), 3518–3525.

Champagne, C.P. and Fustier, P. (2007) Microencapsulation for the improved delivery of bioactive compounds into foods. *Curr. Opin. Biotechnol.*, **18** (2), 184–190.

Cheetham, S.J., Li, C.T., Shieh, W.K. (1995) Performance evaluation of the anaerobic fluidized bed system:I. substrate utilization and gas production, *J. Chem. Technol. Biotechnol.* **35B**, 101–109.

Cheetham, P.S.J. (2000) Case studies in the application of biocatalysts for the production of (bio)chemicals, in *Applied Biocatalysis* (eds A.J.J. Straathof and P. Adlercreutz), Harwood Academic Publishers, Amsterdam, pp. 93–152.

Chen, S., Hou, H., Harnisch, F., Patil, S.A., Carmona-Martinez, A.A., Agarwal, S., Zhang, Y., Sinha-Ray, S., Yarin, A.L., Greiner, A., and Schröder, U. (2011) Electrospun and solution blown three-dimensional carbon fiber nonwovens for application as electrodes in microbial fuel cells. *Energy Environ. Sci.*, **4**, 1417–1421.

Chibata, I., Tosa, T., Sato, T., and Takata, I. (1987) Immobilization of cells in carrageenan, in *Methods in Enzymology*, vol. 135 (ed. K. Mosbach), Academic Press, Orlando, FL, pp. 189–198.

Clarke, T.A., Edwards, M.J., Gates, A.J., Hall, A., White, G.F. *et al.* (2011) Structure of a bacterial cell surface decaheme electron conduit. *Proc. Natl. Acad. Sci. USA*, **108**, 9384–9389.

Coman, V., Vaz-Domínguez, C. *et al.* (2008) A membrane-, mediator-, cofactor-less glucose/oxygen biofuel cell. *Phys. Chem. Chem. Phys.*, **10** (40), 6093–6096.

Dagher, S.F., Ragout, A.L. *et al.* (2010) Cell immobilization for production of lactic acid: biofilms do it naturally. *Adv. Appl. Microbiol.*, **71**, 113–148.

Desimone, M.F., Alvarez, G.S. *et al.* (2009) Development of sol–gel hybrid materials for whole cell immobilization. *Recent Patents Biotechnol.*, **3** (1), 55–60.

de Vos, P., Bucko, M. *et al.* (2009) Multiscale requirements for bioencapsulation in medicine and biotechnology. *Biomaterials*, **30** (13), 2559–2570.

Dorias, B., Hauber, G., and Baumann, P. (1999) Design of nitrification/denitrification in fixed growth reactors, in *Biotechnology*, vol. 11a (ed. J. Winter), Wiley-VCH Verlag GmbH, Weinheim, pp. 335–348.

Dörnenburg, H. and Knorr, D. (1995) Strategies for the improvement of secondary metabolite production in plant cell cultures. *Enzyme Microb. Technol.*, **17**, 674–684.

Doyle, A. and Griffiths, J.B. (eds) (1998) Chapter 5.8: Microcarriers – basic techniques, in *Cell and Tissue Culture: Laboratory Procedures in Biotechnology*, John Wiley & Sons, Inc., New York, pp. 262–267.

Ehlinger, F. (1994) Anaerobic biological fluidized beds: operating experiences in France. Proceedings of the 7th International Symposium on Anaerobic Digestion, Cape Town, South Africa, pp. 315–323.

EPA (1993) Manual Nitrogen Control, EPA/625/R-93/010.

Fischer, K. (2000) Biofilters, in *Environmental Processes III. Biotechnology*, vol. 11c (eds J. Klein and J. Winter), Wiley-VCH Verlag GmbH, Weinheim, pp. 321–332.

Franklin, R.J. (2001) Full-scale experiences with anaerobic treatment of industrial wastewater. *Water Sci. Technol.*, **44**, 1–6.

Franklin, R.J., Koevoets, W.A.A., van Gus, W.M.A., and van der Pas, A. (1992) Application of the biobed upflow fluidized bed process for anaerobic waste water treatment. *Water Sci. Technol.*, **25**, 373–382.

Fraser, J.E. and Bickerstaff, G.F. (1997) Entrapment in calcium alginate, in *Immobilization of Enzymes and Cells, Methods in Biotechnology*, vol. 1 (ed. G.F. Bickerstaff), Humana Press Inc., Totowa, NJ.

Freeman, A. and Lilly, M.D. (1998) Effect of processing parameters on the feasibility and operational stability of immobilized viable microbial cells. *Enzyme Microb. Technol.*, **23**, 335–345.

Fukui, S. and Tanaka, A. (1982) Immobilized microbial cells. *Annu. Rev. Microbiol.*, **36**, 145–172.

Galazzo, J.L. and Bailey, J.E. (1990) Growing *Saccharomyces cerevisiae* in calcium alginate beads induces cell alterations which accelerate glucose conversion to ethanol. *Biotechnol. Bioeng.*, **36** (4), 417–426.

Gallardo, F. and Chartrand, P. (2011) Visualizing mRNAs in fixed and living yeast cells. *Methods Mol. Biol.*, **714**, 203.

Gamarra, N.N., Villena, G.K. *et al.* (2010) Cellulase production by *Aspergillus niger* in biofilm, solid-state, and submerged fermentations. *Appl. Microbiol. Biotechnol.*, **87** (2), 545–551.

Ghisalba, O., Meyer, H.P. *et al.* (2010) Industrial biotransformation, in *Encyclopedia of Industrial Biotechnology*, John Wiley & Sons, Inc., New York.

Gonzales-Gil, G., Seghezzo, L., Lettinga, R., and Kleerbezem, R. (2001) Kinetics and mass transfer in anaerobic sludge. *Biotechnol. Bioeng.*, **73**, 125–134.

Gosset, G. (2009) Production of aromatic compounds in bacteria. *Curr. Opin. Biotechnol.*, **20** (6), 651–658.

Grimm, L.H., Kelly, S., Hengstler, J., Göbel, A., Krull, R., and Hempel, D.C. (2004) Kinetic studies on the aggregation of *Aspergillus niger* conidia. *Biotechnol. Bioeng.*, **87**, 213–218.

Gross, R., Hauer, B. *et al.* (2007) Microbial biofilms: new catalysts for maximizing productivity of long term biotransformations. *Biotechnol. Bioeng.*, **98** (6), 1123–1134.

Gross, R., Lang, K. *et al.* (2010) Characterization of a biofilm membrane reactor and its prospects for fine chemical synthesis. *Biotechnol. Bioeng*, **105** (4), 705–717.

Halan, B., Schmid, A. *et al.* (2010) Maximizing the productivity of catalytic biofilms on solid supports in membrane aerated reactors. *Biotechnol. Bioeng.*, **106** (4), 516–527.

Halan, B., Schmid, A. *et al.* (2011) Real-time solvent tolerance analysis of *Pseudomonas* sp. strain VLB120ΔC catalytic biofilms. *Appl. Environ. Microbiol.*, **77** (5), 1563.

Harnisch, F., Koch, C., Patil, S.A., Hübschmann, T., Müller, S., and Schröder, U. (2011) Revealing the electrochemically driven selection in natural community derived microbial biofilms using flow-cytometry. *Energy Environ. Sci.*, **4**, 1265–1267.

Heijnen, J.J. (1985) *US Patent* 4560479.

Heijnen, J.J., Mulder, A., Weltevrede, R., Hols, J., and van Leeuwen, H.L.J.M. (1991) Large-scale anaerobic–aerobic treatment of complex industrial waste water using biofilm reactors. *Water Sci. Technol.*, **23**, 1427–1436.

Henze, M. and Harremoes, P. (1983) Anaerobic treatment of wastewater in fixed film reactors – a literature review. *Water Sci. Technol.*, **15**, 1–101.

Holst, T.C., Truc, A., and Pujol, R. (1997) Anaerobic fluidized beds: ten years of industrial experience, *Water Sci. Technol.* **36**, 415–422.

Hulshoff, L.W., de Zeeuw., W.J., Velzeboer, C. T.M., and Lettinga, G. (1983) Granulation in UASB reactors. *Water Sci. Technol.*, **15**, 291–304.

Hulst, A.C. and Tramper, J. (1989) Immobilized plant cells: a literature survey. *Enzyme Microb. Technol.*, **11**, 546–558.

Iersel, van M.F.M., van Dieren, B., Rombouts, F.M., and Abee, T. (1999) Flavor formation and cell physiology during the production of alcoholfree beer with immobilized *Saccharomyces cerevisiae*, *Enzyme Microb. Technol.*, **24**, 407–411.

Iranpour, R., Cox, H.H.J. *et al.* (2005) Literature review of air pollution control biofilters and biotrickling filters for odor and volatile organic compound removal. *Environ. Prog.*, **24** (3), 254–267.

Ishihara, K., Hamada, H., Hirata, T., and Nakajima, N. (2003) Biotransformation using plant cultured cells. *J. Mol. Catal. B*, **23**, 145–170.

Jahnz, U., Wittlich, P., Pruesse, U., and Vorlop, K.D. (2001) New matrices and bioencapsulation processes, in *Focus on Biotechnology*, vol. 4 (eds M. Hofmann and J. Anne), Kluwer Academic Publishers, Dordrecht, pp. 293–307.

Jensen, H.M., Albers, A.E., Malley, K.R., Londer, Y.Y. *et al.* (2010) Engineering of a synthetic electron conduit in living cells. *Proc. Natl. Acad. Sci. USA*, **107** (45), 19213–19218.

Jördening, H-.J. (1996) Scaling-up and operation of anaerobic fluidized bed reactors. *Zuckerindustrie*, **121**, 847–854.

Jördening, H.-J. and Buchholz, K. (1999) Fixed film stationary-bed and fluidized-bed reactors, in *Biotechnology*, vol. 11a (ed. J. Winter), Wiley-VCH Verlag GmbH, Weinheim, pp. 493–515.

Jördening, H.-J., Jansen, W., Brey, S., and Pellegrini, A. (1991) Optimierung des Fließbettsystems zur anaeroben Abwasserreinigung. *Zuckerindustrie*, **116**, 1047–1052.

Jördening, H.-J. and Küster, W. (1997) Betriebserfahrungen mit einem anaeroben Fließbettreaktor zur Behandlung von Zuckerfabriksabwasser. *Zuckerindustrie*, **122**, 934–936.

Junter, G-.A., Coquet, L., Vilain, S., and Jouenne, T. (2002) Immobilized-cell physiology: current data and the potentialities of proteomics. *Enzyme Microb. Technol.*, **31**, 201–212.

Kawamoto, T. and Tanaka, A. (2003) Entrapment of biocatalysts by prepolymer methods, in *Handbook of Food Enzymology* (eds J.R. Whitaker, A.G.J. Voragen, and D. W.S. Wong), Marcel Dekker, New York, pp. 331–341.

Kayser, R. (2004) Activated sludge processes, in *Environmental Biotechnology* (eds H.-J. Jördening and J. Winter), Wiley-VCH Verlag GmbH, Weinheim, pp. 79–120.

Keim, P., Luerweg, M., Aivasidis, A., and Wandrey, C. (1989) Entwicklung der Wirbelschichttechnik mit dreidimensional kolonisierbaren Trägermaterialien aus makroporösem Glas am Beispiel der anaeroben Abwasserreinigung. *Korresp. Abwasser*, **36**, 675–687.

Keim, P., Luerweg, M., Striegel, B., Aivasidis, A., and Wandrey, C. (1990) Einsatzbeispiele und Scale-up von Wirbelschichtreaktoren mit Kugeln aus porösem Sinterglas in der anaeroben Abwasserreinigung. *Chem.-Ing.-Tech.*, **62**, 336–337.

Kelly, S., Grimm, L.H., Hengstler, J., Schultheis, E., Krull, R., and Hempel, D. C. (2004) Agitation effects on submerged growth and product formation of *Aspergillus niger*. *Bioprocess Biosyst. Eng.*, **26**, 315–323.

Kennes, C., Rene, E.R. *et al.* (2009) Bioprocesses for air pollution control. *J. Chem. Technol. Biotechnol.*, **84** (10), 1419–1436.

Kierstan, M. and Bucke, C. (1977) The immobilization of microbial cells, subcellular organelles and enzymes in calcium alginate gels. *Biotechnol. Bioeng.*, **19**, 387–397.

Kihn, J.C., Masy, C.L., and Mestdagh, M.M. (1988) Yeast flocculation: competition between nonspecific repulsion and specific bonding in cell adhesion. *Can. J. Microbiol.*, **34**, 773–778.

Kim, J., Jia, H. *et al.* (2006) Challenges in biocatalysis for enzyme-based biofuel cells. *Biotechnol. Adv.*, **24** (3), 296–308.

Kirst, H.A., Yeh, W.-K., and Zmijewski, M.J., Jr. (eds) (2001) *Enzyme Technologies for Pharmaceutical and Biotechnological Applications*, Marcel Dekker, New York.

Klein, J. and Vorlop, K.-D. (1985) Immobilization techniques – cells, in *Comprehensive Biotechnology*, vol. 2 (ed. M. Moo-Young), Pergamon Press, Oxford, pp. 203–224.

Klein, J. and Winter, J. (eds) (2000) *Environmental Processes III. Biotechnology*, vol. **11c**, Wiley-VCH Verlag GmbH, Weinheim.

Knapp, F. (1847) *Lehrbuch der Chemischen Technologie*, vol. 2, F. Vieweg und Sohn, Braunschweig.

Lee, K.H., Choi, I.S. *et al.* (2011) Enhanced production of bioethanol and ultrastructural characteristics of reused *Saccharomyces cerevisiae* immobilized calcium alginate beads. *Bioresour. Technol.*, **102**, 8191–8198.

Lefebvre, J. and Vincent, J.-C. (1995) Diffusion-reaction-growth coupling in gel-immobilized cell systems: model and experiment. *Enzyme Microb. Technol.*, **17**, 276–284.

Lefebvre, J. and Vincent, J.-C. (1997) Control of the biomass heterogeneity in immobilized cell systems. Influence of initial cell and substrate concentrations, structure thickness, and type of bioreactors. *Enzyme Microb. Technol.*, **20**, 536–543.

Lettinga, G., Hulshof Pol, L.W., van Lier, J.B., and Zeemann, G. (1999) Possibilities and potential of anaerobic waste water treatment using anaerobic sludge bed (ASB) reactors, in *Biotechnology*, vol. 11a (ed. J. Winter), Wiley-VCH Verlag GmbH, Weinheim, pp. 517–526.

Li, X.Z., Hauer, B. *et al.* (2007) Single-species microbial biofilm screening for industrial applications. *Appl. Microbiol. Biotechnol.*, **76** (6), 1255–1262.

Liese, A., Seelbach, K., and Wandrey, C. (2000) *Industrial Biotransformations*, Wiley-VCH Verlag GmbH, Weinheim.

Liese, A., Seelbach, K., and Wandrey, C. (2006) *Industrial Biotransformations*, 2nd edn, Wiley-VCH Verlag GmbH, Weinheim.

Lu, C., Lin, M.-R., and Wey, I. (2002) Removal of acrylonitrile and styrene mixtures from waste gases by a trickle bed air biofilter. *Bioprocess Biosyst. Eng.*, **25**, 61–67.

Lüllau, E., Biselli, M., and Wandrey, C. (1994) Growth and metabolism of CHO-cells in porous glass carriers, in *Animal Cell Technology: Products of Today, Prospects of Tomorrow* (eds R.E. Spier, J.B. Griffiths, and W. Berthold), Butterworth-Heinemann, pp. 152–255.

Lundberg, P. and Kuchel, P.W. (1997) Diffusion of solutes in agarose and alginate gels: ^{1}H and ^{23}Na PFGSE and ^{23}Na TQF NMR studies. *Magn. Reson. Med.*, **37**, 44–52.

Ma, Q., Yang, Z., Pu, M. *et al.* (2011) Engineering a novel c-di-GMP-binding protein for biofilm dispersal. *Environ. Microbiol.*, **13**, 631–642.

Macarie, H. (2001) Overview of the application of anaerobic treatment to chemical and petrochemical wastewaters. *Water Sci. Technol.*, **44**, 201–214.

Marshall, K.C. (ed.) (1984) *Microbial Adhesion and Aggregation*, Springer, Berlin.

McCarty, P.L. (2001) The development of anaerobic treatment and its future. *Water Sci. Technol.*, **44**, 149–156.

McInerny, M.J. (1999) Anaerobic metabolism and its regulation, in *Biotechnology*, vol. 11a (ed. J. Winter), Wiley-VCH Verlag GmbH, Weinheim, pp. 455–478.

Millo, D., Harnisch, F., Patil, S.A., Ly, H.K., Schröder, U., and Hildebrandt, P. (2011) *In situ* spectroelectrochemical investigation of electrocatalytic microbial biofilms by surface-enhanced resonance Raman spectroscopy. *Angew. Chem., Int. Ed.*, **50**, 2625–2627.

Mosbach, K. (ed.) (1987) *Methods in Enzymology*, vol. **135**, Academic Press, Orlando, FL.

Mösche, M. and Jördening, H.J. (1998) Detection of very low saturation constants in anaerobic digestion: influences of calcium carbonate precipitation and pH. *Appl. Microbiol. Biotechnol.*, **49** (6), 793–799.

Mösche, M. and Jördening, H.J. (1999) Comparison of different models of substrate and product inhibition in anaerobic digestion. *Water Res.*, **33** (11), 2545–2554.

Mozes, N., Léonhard, A.J., and Rouxhet, P.G. (1988) On the relation between the elemental surface composition of yeast and bacteria and their charge and hydrophobicity. *Biochim. Biophys. Acta*, **945**, 324–334.

Nagashima, M., Azuma, M., and Nogushi, S. (1987) Large scale preparation of alginate immobilized yeast cells and its application to industrial ethanol production, in *Methods in Enzymology*, vol. **136** (ed. K. Mosbach), Academic Press, Orlando, FL, pp. 394–405.

Nedovic, V.A., Manojlovi, V. *et al.* (2011) State of the art in immobilized/encapsulated cell technology in fermentation processes. *Food Eng. Interfaces*, (Part 1), 119–146.

Netto, C.B., Destruhaut, A., and Goma, G. (1985) Ethanol production by flocculating yeast: performance and stability dependence on a critical fermentation rate. *Biotechnol. Lett.*, **7**, 359–360.

Omar, S.H. (1993) Oxygen diffusion through gels employed for immobilization. Part 2. In the presence of microorganisms. *Appl. Microbiol. Biotechnol.*, **40**, 173–181.

Pandey, A., Lee, D.J., and Logan, B.E. (2010) Algal biofuels and microbial fuel cells. *Bioresour. Technol.*, **102**, 1–426.

Plaggemeier, T. and Lämmerzahl, O. (2000) Treatment of waste gas pollutants in trickling filters, in *Environmental Processes III. Biotechnology*, vol. 11c (eds J. Klein and J. Winter), Wiley-VCH Verlag GmbH, Weinheim, pp. 333–344.

Pörtner, R., Platas, O. *et al.* (2007) Fixed bed reactors for the cultivation of mammalian cells: design, performance and scale-up. *Open Biotechnol. J.*, **1**, 41–46.

Prüße, U., Dallun, J., Breford, J., and Vorlop, K.-D. (2000) Production of spherical beads by jet cutting. *Chem. Eng. Technol.*, **23**, 1105–1110.

Prüße, U., Fox, B., Kirchhoff, M., Bruske, F., Breford, J., and Vorlop, K.-D. (1998) The jet cutting method as a new immobilization technique. *Biotechnol. Technol.*, **12**, 105–108.

Rabaey, K., Angenent, L., Schroeder, U., and Keller, J. (eds) (2009a) *Bioelectrochemical Systems*, IWA Publishing, London.

Rabaey, K., Boon, N., Siciliano, D., Verhaege, M., and Verstraete, W. (2004) Biofuel cells select for microbial consortia that self-mediate electron transfer. *Appl. Environ. Microbiol.*, **70**, 5373–5382.

Rabaey, K., Boon, N., Verstraete, W., and Höfte, M. (2005) Microbial phenazine production enhances electron transfer in biofuel cells. *Environ. Sci. Technol.*, **39**, 3401–3408.

Rabaey, K., Keller, J., Angenent, L., Lens, P., and Schroeder, U. (2009b) *Introduction – MFCs and BESs in the Context of Environmental and Industrial Biotechnology*, IWA Publishing, London.

Reischwitz, A., Reh, K.D., and Buchholz, K. (1995) Unconventional immobilization of dextransucrase with alginate. *Enzyme Microb. Technol.*, **17**, 457–461.

Rinas, U., El-Enshasy, H., Emmler, M., Hille, A., Hempel, D.C., and Horn, H. (2005) Model-based prediction of substrate conversion and protein synthesis and excretion in recombinant *Aspergillus niger* biopellets. *Chem. Eng. Sci.*, **60**, 2729–2739.

Roisin, C., Bienaimé, C., Nava Saucedo, J.E., and Barbotin, J.-N. (1996) Influence of the microenvironment on immobilised *Gibberella fujikuroi*, in *Immobilized Cells: Basics and Applications* (eds R.H. Wijffels, R.M. Buitelaar, C. Bucke, and J. Tramper), Elsevier, Amsterdam, pp. 189–195.

Rosche, B., Li, X.Z. *et al.* (2009) Microbial biofilms: a concept for industrial catalysis? *Trends Biotechnol.*, **27** (11), 636–643.

Rosenbaum, M., Zhao, F. *et al.* (2006) Interfacing electrocatalysis and biocatalysis with tungsten carbide: a high performance, noble metal free microbial fuel cell. *Angew. Chem., Int. Ed.*, **45** (40), 6658–6661; *Angew. Chem.*, **118**, 6810–6813.

Rosevear, A. (1988) Immobilized plant cells. *Food Biotechnol.*, **2**, 109–114.

Rouxhet, P.G. (1990) Biocatalysts/interfacial chemistry. *Ann. N. Y. Acad. Sci.*, **613**, 265–278.

Sam-Soon, P., Loewenthal, R.E., Dold, P.L., and Marais, G.R. (1987) Hypothesis for pelletisation in the upflow anaerobic sludge bed reactor. *Water SA*, **13**, 69–80.

Sanchez, J.M., Arijo, S., Munoz, M.A., Morinigo, M. A., Borrego, J. J. (1994) Microbial colonization of different support materials used to enhance the methanogenic process, *Appl. Microbiol. Biotechnol.* **41**, 480–486.

Sarkar, S., Roy, D. *et al.* (2010) Production of a potentially novel antimicrobial compound by a biofilm-forming marine *Streptomyces* sp. in a niche-mimic rotating disk bioreactor. *Bioprocess Biosyst. Eng.*, **33** (2), 207–217.

Schedel, M. (2000) Regioselective oxidation of aminosorbitol with *Gluconobacter oxydans*: key reaction in the industrial 1-deoxynojirimycin synthesis. *Biotechnology*, **8b**, 295–311.

Schmid, A., Hollmann, F., Park, J.B., and Bühler, B. (2002) The use of enzymes in the chemical industry in Europe. *Curr. Opin. Biotechnol.*, **13**, 359–366.

Schröder, U. (2007) Anodic electron transfer mechanisms in microbial fuel cells and their energy efficiency. *Phys. Chem. Chem. Phys.*, **9** (21), 2619–2629.

Schröder, U., Nießen, J., and Scholz, F. (2003) A generation of microbial fuel cells with current outputs boosted by more than one order of magnitude. *Angew. Chem., Int. Ed.*, **42** (25), 2880–2883; *Angew. Chem.*, **115**, 2986–2989.

Sheldon, R.A. (2007) Enzyme immobilization: the quest for optimum performance. *Adv. Synth. Catal.*, **349** (8–9), 1289–1307.

Sutton, P.M., Li, A., Evans, R.R., and Korchin, S. (1982) Dorr–Oliver's fixed film and suspended growth anaerobic systems for industrial wastewater treatment and energy recovery. 37th Industrial Waste Conference, Purdue University, West Lafayette, IN.

Tanaka, H., Matsumura, M., and Veliky, I.A. (1984) Diffusion characteristics of substrates in Ca-alginate gel beads. *Biotechnol. Bioeng.*, **26**, 53–58.

Tang, C.W. and Mavituna, F. (2001) Cell immobilisation of Taxus media, in *Novel Frontiers in the Production of Compounds for Biomedical Use* (eds A. Van Broekhovenet *al.*), Kluwer Academic Publishers, The Netherlands, pp. 401–407.

Tokayashi, M. and Yokoyama, S. (1997) Cell cultivation technology, in *Mammalian Cell Biotechnology in Protein Production* (eds H. Hauser and R. Wagner), Walter de Gruyter, Berlin.

Trulear, M.G. and Characklis, W.G. (1982) Dynamics of biofilm processes. *J. WPCF*, **54**, 1288–1301.

Tsoligkas, A.N., Winn, M. *et al.* (2011) Engineering biofilms for biocatalysis. *ChemBioChem*, **12** (9), 1391–1395.

van der Geize, R., Hessels, G.I., van Gerwen, R. *et al.* (2002) Molecular and functional characterization of *kshA* and *kshB*, encoding two components of 3-ketosteroid 9-alpha-hydroxylase, a class IA monooxygenase, in *Rhodococcus erythropolis* strain SQ1. *Mol. Microbiol.*, **45**, 1007–1018.

van Iersel, M.F.M., van Dieren, B., Rombouts, F.M., and Abee, T. (1999) Flavor formation and cell physiology during the production of alcohol-free beer with immobilized *Saccharomyces cerevisiae*. *Enzyme Microb. Technol.*, **24**, 407–411.

van Lier, J.B. (2008) High-rate anaerobic wastewater treatment: diversifying from end-of-the-pipe treatment to resource-oriented conversion techniques. *Water Sci. Technol.*, **57** (8), 1137–1148.

van Lier, J.B., Mahmoud, N., and Zeeman, G. (2008) Anaerobic wastewater treatment, in *Biological Wastewater Management: Principles, Modelling and Design* (eds M. Henze, M.C.M. van Loosdrecht, and G.A. Ekama), International Water Association.

van Loosdrecht, M.C.M., Lyklema, J., Norde, W., Schraa, G., and Zehnder, A.J.B. (1987) Electrophoretic mobility and hydrophobicity as a measure to predict the initial steps of bacterial adhesion. *Appl. Environ. Microbiol.*, **53**, 1898–1901.

VDI (1991) VDI-Richtlinien 3477: Biological Waste Gas/Waste Air Purification. Biofilters (German, English).

VDI (1996) VDI-Richtlinien 3478: Biological Waste Gas Purification. Bioscrubbers and Trickle Bed Reactors (German, English).

VDI (2002) VDI-Richtlinien 3477: Biologische Abgasreinigung. Biofilter (German).

Verbelen, P.J., De Schutter, D.P. *et al.* (2006) Immobilized yeast cell systems for continuous fermentation applications. *Biotechnol. Lett.*, **28** (19), 1515–1525.

Verica, M., Branko, B. *et al.* (2010) Immobilized cells, in *Encyclopedia of Industrial Biotechnology: Bioprocess, Bioseparation, and Cell Technology*, John Wiley & Sons, Inc., New York.

Walsh, G. (2010) Biopharmaceutical benchmarks 2010. *Nat. Biotechnol.*, **28** (9), 917–924.

Wanda, U., Wollersheim, R., Diekmann, H., and Buchholz, K. (1990) Adhesion of anaerobic bacteria on solid surfaces, in *Physiology of Immobilized Cells* (eds J.A.M. de Bont, J. Visser, B. Mattiassen, and J. Tramper), Elsevier, Amsterdam, pp. 109–114.

Wang, X. and Wood, T.K. (2011) Toxin–antitoxin systems influence biofilm and persister cell formation and the general stress response. *Appl. Environ. Microbiol.*, **77** (16), 5577–5583.

Watanabe, K. (2010) Biofuel cells, performance characterization, in *Encyclopedia of Industrial Biotechnology: Bioprocess, Bioseparation, and Cell Technology*, John Wiley & Sons, Inc., New York.

Waugh, A. (1999) Culturing animal cells in fluidized bed reactors, in *Animal Cell Biotechnology*, Methods in Biotechnology, vol. 8 (ed. N. Jenkins), Humana Press Inc., Totowa, NJ, pp. 179–185.

Waweru, M., Herrygers, V., Van Langenhove, H., and Verstraete, W. (2000) Process engineering of biological waste gas purification, in *Environmental Processes III. Biotechnology*, vol 11c (eds J. Klein and J. Winter), Wiley-VCH Verlag GmbH, Weinheim, pp. 259–273.

Weiland, P., Thomsen, H., and Wulfert, K. (1988) Entwicklung eines Verfahrens zur anaeroben Vorreinigung von Brennereischlempen unter Einsatz eines Festbettreaktors, in *Verfahrenstechnik Der Mechanischen, Thermischen, Chemischen und Biologischen Abwasserreinigung, Part 2: Biologische Verfahren*, VDI, Düsseldorf, pp. 169–186.

Wijffels, R.H. (2001) *Immobilized Cells*, Springer, Berlin.

Wijffels, R.H., Buitelaar, R.M., Bucke, C., and Tramper, J. (1996) *Immobilized Cells: Basics and Applications*, Elsevier, Amsterdam.

Willaert, R. (2009) Cell immobilization: engineering aspects, in *Encyclopedia of Industrial Biotechnology: Bioprocess, Bioseparation, and Cell Technology*, John Wiley & Sons, Inc., New York.

Willaert, R. and Nedovic, V.A. (2006) Primary beer fermentation by immobilised yeast – a review on flavour formation and control strategies. *J. Chem. Technol. Biotechnol.*, **81** (8), 1353–1367.

Winfield, J., Ieropoulos, I. *et al.* (2011) Investigating the effects of fluidic connections between stacks of microbial fuel cells. *Bioprocess Biosyst. Eng.*, **34**, 477–484.

Wingender, J. and Flemming, H.-C. (1999) Autoaggregation of microorganisms: flocs and biofilms, in *Biotechnology*, vol. **11a** (ed. J. Winter), Wiley-VCH Verlag GmbH, Weinheim, pp. 65–83.

Wurm, F.M. (ed.) (2005) Manufacture of recombinant biopharmaceutical proteins by cultivated mammalian cells in bioreactors, in *Modern Biopharmaceuticals: Design, Development and Optimization*, Wiley-VCH Verlag GmbH, Weinheim, Germany.

Zhang, S., Norrlow, O. *et al.* (2004) Poly(3-hydroxybutyrate) biosynthesis in the biofilm of *Alcaligenes eutrophus*, using glucose enzymatically released from pulp fiber sludge. *Appl. Environ. Microbiol.*, **70** (11), 6776.

10
Characterization of Immobilized Biocatalysts

The characterization is required to answer the following questions:	This is required to answer these questions: phenomena involved and required data
How much immobilized biocatalyst is required to obtain a desired space–time yield (*STY*)?	The required *STY* and the effectiveness factors (Sections 10.2, 10.3 and 10.7)
What other properties influence – in comparison with free biocatalysts – *STY* and productivity with immobilized biocatalysts?	The coupling of reaction and diffusion, particle size, and effectiveness factors (Sections 10.3 and 10.4)
How can effectiveness factors be calculated? How much can they differ from experimental data?	From relations that describe the coupling of reaction and diffusion (Sections 10.4 and 10.7)
Comparison of different continuous reactors (stirred tank or packed bed reactor) – which is better? Which effectiveness factor must be used?	Calculation of *STY* for these reactors (Section 10.5)
What properties of immobilized biocatalysts must be known to design an enzyme process? What properties of immobilized biocatalysts are important to improve them for a given enzyme process?	Determination of important mechanical, physical, chemical, and catalytic properties of immobilized biocatalysts for an enzyme process (Section 10.6)
Can immobilized biocatalysts be used to carry out enzyme processes with slightly soluble substrates or products in suspensions (emulsions)?	Section 10.10

Biocatalysts and Enzyme Technology, Second Edition. Klaus Buchholz, Volker Kasche, and Uwe T. Bornscheuer

How can negative effects on enzyme processes due to microenvironmental (concentration or pH gradients) and nanoenvironmental (pore surface charges) effects in immobilized biocatalysts be minimized?	Use suitable buffers; select suitable support with small surface charge and diffusion distance for substrates (Section 10.9)

10.1 Introduction

Immobilized biocatalysts are used in enzyme technology for analytical, preparative, and industrial purposes. They allow reuse of the enzyme and reduce the biocatalyst cost per unit of produced product (see Chapters 8 and 9). The first studies on the properties of immobilized biocatalysts appeared around 1960. Before this, the importance of immobilized enzymes in natural systems such as cells and tissues or in soil had been recognized and studied (McLaren and Packer, 1970; Weisz, 1972). Compared with systems using free enzymes, the reaction rate is determined not only by the catalytic properties but also by the mass transfer to, from, and inside the immobilized biocatalysts, and their micro- and nanoenvironment (pH, ionic strength on micrometer and nanometer scales). The first experimental and theoretical studies on such heterogeneous systems were performed on physiological systems during the 1920s (Warburg, 1923; Rashevsky, 1940). Phenomenologically similar systems – heterogeneous chemical catalysts – have been used in chemical engineering since the 1930s, and were initially extensively analyzed quantitatively at that time (Damköhler, 1937; Thiele, 1939; Zeldovich, 1939). The analytical description of both systems has – once the similarity was recognized – contributed much to the analytical and experimental characterization of immobilized biocatalysts.

These systems provide the basic knowledge for the rational design of enzyme processes with immobilized biocatalysts. For this aim, the quantitative relations for the time t required for a given substrate conversion with a given biocatalyst activity (or vice versa) as a function of different system properties must be derived (see Sections 10.2–10.5). From this, the space–time yield (*STY*), the biocatalyst productivity (the amount of substrate converted per unit biocatalyst amount, expressed as activity or weight), and the biocatalyst cost can be obtained. In order to improve the immobilized biocatalysts, properties that are important for their application must be determined, and how they can be influenced by the nano- and microenvironment around the immobilized enzymes must be studied in detail (see Sections 10.6–10.9).

10.2
Factors Influencing the Space–Time Yield of Immobilized Biocatalysts

The space–time yield of an enzyme process is (see Section 2.8)

$$STY = \frac{[\mathrm{S}]_0 - [\mathrm{S}]_t}{t}, \tag{10.1}$$

where t is the time required for the desired change in substrate concentration from $[\mathrm{S}]_0$ to $[\mathrm{S}]_t$. For free enzymes, the biocatalyst activity $V_{\max,0}$ required to obtain this STY can be calculated from Eq. (2.33):

$$V_{\max,0} = \frac{\int_{[\mathrm{S}]_t}^{[\mathrm{S}]_0} f([\mathrm{S}],[\mathrm{P}])\mathrm{d}[\mathrm{S}]}{(1 - \mathrm{e}^{k_i t})/k_i}, \tag{10.2}$$

where k_i is the rate constant for the first-order inactivation of the biocatalyst. In reactors with immobilized biocatalysts with the same catalyst activity, it cannot be assumed that all catalysts are used simultaneously. Some localized in the inner part of the particle are not accessible for substrate, when it has been converted in the outer part of the particle. To consider this, an effectiveness factor η (degree of catalyst utilization ≤ 1) must be introduced in Eq. (10.2). This gives the following relationship:

$$V_{\max,0} = \frac{\int_{[\mathrm{S}]_t}^{[\mathrm{S}]_0} f([\mathrm{S}],[\mathrm{P}])\mathrm{d}[\mathrm{S}]}{\eta(1 - \mathrm{e}^{-k_i t})/k_i} \tag{10.3}$$

from which $V_{\max,0}$ can be calculated, once η is known. The definition and calculation of effectiveness factors for different reactors is outlined in Sections 10.3–10.5.

The STY as a function of $V_{\max,0}$ is limited by the following system properties (Figure 10.1):

- The maximum amount of substrate that can be transported to the particle with immobilized biocatalysts per unit time.
- The maximal biocatalyst density that can be obtained in the particles with immobilized enzymes or cells.

They can be influenced by engineering (particle size, mass transfer rate) or biochemical/biological (catalyst content) means. Besides this, the STY can be influenced by

- Properties of the particle in which the biocatalyst is immobilized:
  - chemical, mechanical, and thermal stability;
  - concentration and type of functional groups (charges, hydrophobic residues) on the inner (in pores) and the outer surface per unit particle volume;
  - porosity, pore size, and structure;
  - particle density.

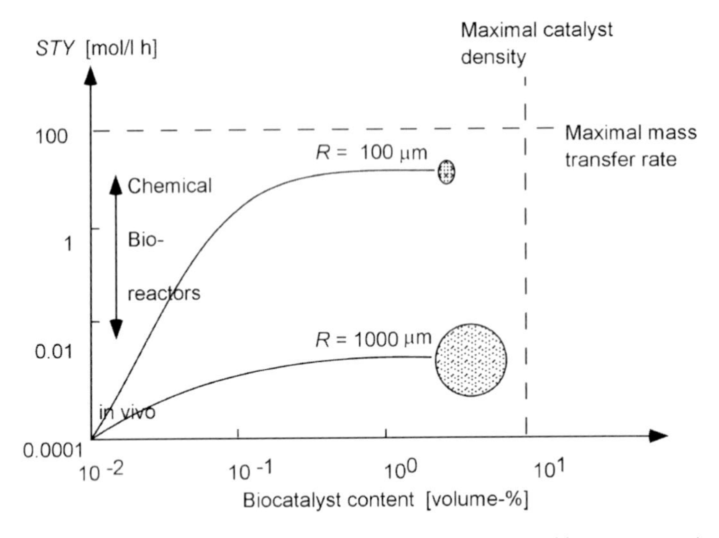

Figure 10.1 Limiting system properties for the maximal space–time yield in reactors with immobilized biocatalysts as a function of the biocatalyst content (as % of reactor volume) and particle radius *R*.

- Properties of the immobilized biocatalyst (enzyme or cell):
  - intrinsic properties of the immobilized biocatalyst (see Sections 2.6 and 2.7.1); maximum activity V'_{max}[1] or turnover number k'_{cat}, Michaelis–Menten and inhibitor binding constants K'_m and K'_i, stereoselectivity E', or synthesis/hydrolysis selectivity in kinetically controlled reactions $(k'_T/k'_H)_{app}$;
  - biocatalyst distribution within the particle;
  - conformation change of biocatalyst due to the immobilization;
  - stability.
- Concentration gradients of substrates and products inside the particles (especially pH gradients; Tischer and Kasche, 1999; Spieß and Kasche, 2001).

10.3
Effectiveness Factors for Immobilized Biocatalysts

The substrate concentration outside particles with immobilized enzyme is shown in Figure 10.2. At steady state, the substrate gradient is time independent, and the same amount of substrate that is transported to the particle per unit time is converted to product inside the particle.

Outside the particle (cell), a concentration gradient is formed, as substrate must be transported to the particle. This diffusion can only be driven by a concentration gradient that is characterized by the thickness δ of the unstirred diffusion layer (film).

1) In this chapter, all intrinsic properties of the immobilized enzyme will be primed as they generally differ from the properties of the free enzyme. The same applies for the diffusion coefficient of substrates and products in the particles with immobilized enzyme.

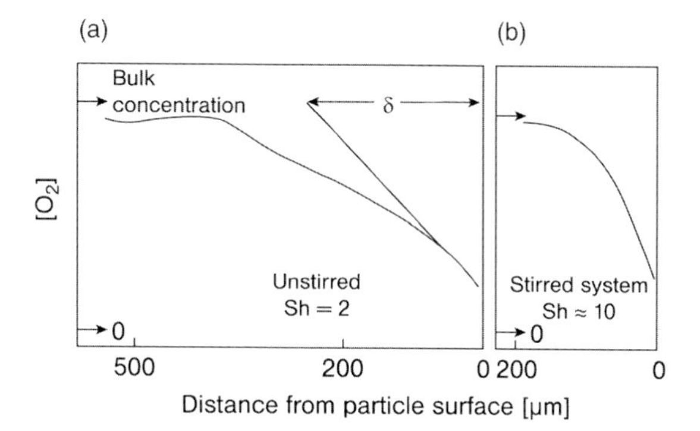

Figure 10.2 Oxygen concentration gradient outside a spherical particle (radius $R \approx 300\,\mu m$) with immobilized glucose oxidase that catalyzes the oxidation of glucose with oxygen: (a) without stirring ($Sh = 2$); (b) with stirring ($Sh \approx 10$). The oxygen concentration was measured using oxygen microelectrodes with a tip diameter of several micrometers. Note that the thickness of the diffusion layer given by the intersection of the gradient (tangent) at the particle surface and the bulk oxygen content approximately equals the particle radius in the unstirred system. This is in agreement with the theoretical analysis (Kasche and Kuhlmann, 1980).

In this layer, the mass transfer occurs only by diffusion. The thickness of the film can be reduced by stirring, which increases the velocity of the particles relative to the solvent (Figure 10.2). The stirring causes shear forces that may cause abrasion of the particles, and thus the stirring speed has an upper limit in order to avoid this abrasion.

From Figure 10.2 follows that the immobilized biocatalysts are surrounded by a substrate concentration that is lower than the bulk concentration. Two different effectiveness factors – the stationary and the operational – that account for this have been defined (Figures 10.3 and 10.4).

The *stationary effectiveness factor* η is the ratio of the initial rates (Figure 10.3), determined under equal conditions and with the same biocatalyst content (V_{max}) and bulk substrate concentration:

$$\eta = v_{imm}/v_f, \tag{10.4}$$

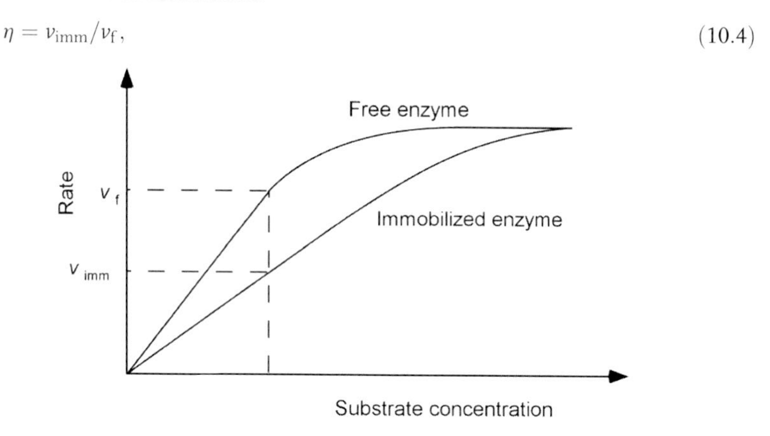

Figure 10.3 Determination of the stationary effectiveness factor η.

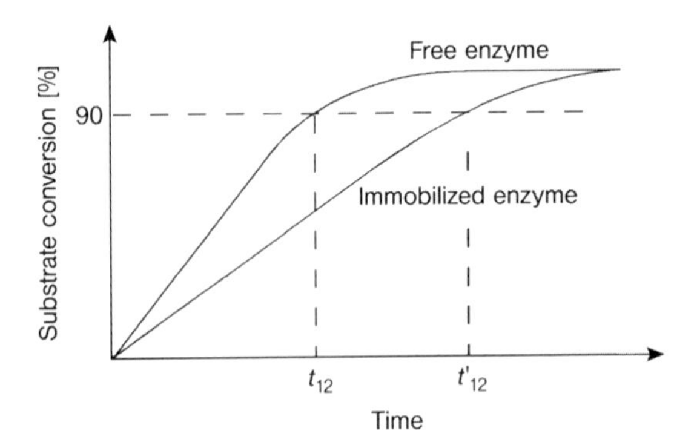

Figure 10.4 Determination of the operational effectiveness factor $\eta_o = t_{12}/t'_{12}$ from progress curves.

which is easy to determine experimentally. The stationary effectiveness factor can be subdivided into an external η_e (for the diffusion layer) and an internal η_i (inside the particle). These are, however, contrary to η, difficult to determine experimentally, and will not be used here.

The stationary effectiveness factor does not provide any information on how rapidly the desired end point of an enzyme process (see Figures 1.9 and 2.9) can be reached. During this process, the substrate concentration outside the particles and the stationary effectiveness factor are changed. Therefore, the *operational effectiveness factor* η_o was introduced (Kasche, 1983), and is defined as the ratio of the time required to reach the end point of a process with the same initial substrate concentration, and same amount of free and immobilized biocatalyst per reactor volume unit, under otherwise equal conditions (Figure 10.4):

$$\eta_o = t_{12}/t'_{12}. \tag{10.5}$$

It can be determined easily from progress curves (substrate concentration as a function of time) in Figure 10.4.

10.4 Mass Transfer and Reaction

10.4.1 Maximal Reaction Rate of Immobilized Biocatalysts as a Function of Particle Radius

The *STY* or observed rate of the catalyzed reaction v_{obs} as a function of the biocatalyst concentration in spherical particles is shown in Figure 10.1. It is limited by the maximum rate of substrate transfer to the particles and the biocatalyst content in the support. When the latter is rate determining, the rate can be increased by higher biocatalyst concentrations until v_{obs} becomes mass transfer limited.

The optimal biocatalyst concentration, expressed as the activity V'_{max}, can be estimated as follows. At the intersection of the linear part of the curve at low biocatalyst content and the maximal mass transfer rate, given by the horizontal part of the curves in Figure 10.1, the following ratio equals 1:

$$\frac{\text{maximal reaction rate in the particle}}{\text{maximal mass transfer rate to the particle}} = \frac{V_p \eta (V'_{max}/f([S],[P]))}{A_p k_L \alpha c_S(\infty)} \qquad (10.6)$$

and can be used to estimate V'_{max}, where V_p and A_p are the particle volume and outer surface area of the immobilized enzyme particle, respectively, η is the stationary effectiveness factor, primed quantities are properties of the immobilized biocatalyst, α is the concentration change in the diffusion layer as a fraction of the bulk substrate concentration $c_S(\infty)$, and k_L is the mass transfer coefficient. The latter, with the dimension velocity, is a function of the stirring speed in a batch reactor, or of the linear flow rate in a packed bed reactor. It can be expressed as

$$k_L = D_S\, Sh/2R = D_S/\delta, \qquad (10.7)$$

where D_S is the diffusion coefficient of the substrate and Sh is the dimensionless Sherwood number (the ratio of particle diameter to the thickness of the unstirred diffusion layer or film). This value is 2 for a spherical particle in an unstirred system (Kasche and Kuhlmann, 1980; for packed bed and stirred tank reactors, see Section 11.3.4). V'_{max} can be estimated from Eqs. (10.6) and (10.7) with the following assumptions: $c_S(\infty) \gg K'_m$, then $f([S],[P])$ and η in Eq. (10.6) are ≈ 1 (see Section 2.7). A high initial substrate concentration up to the molar range is desirable in enzyme processes, and this assumption is then justified (see Table 2.4). The expression for V'_{max} derived from Eqs. (10.6) and (10.7) is

$$V'_{max} = (1.5/R^2) D_S\, Sh\, \alpha c_S(\infty). \qquad (10.8)$$

For these, V'_{max} has been calculated as a function of the particle radius and different bulk substrate concentrations in Figure 10.5, for substrates with MW $\approx$ 500,

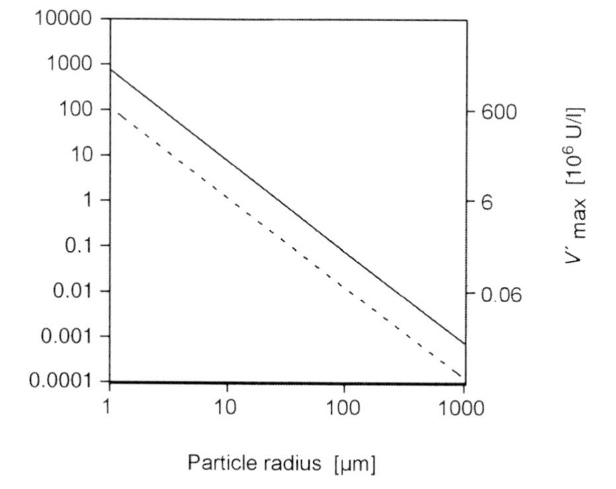

Figure 10.5 V'_{max} as a function of particle radius calculated from Eq. (10.10) for the initial bulk substrate concentrations of 1 M (solid line) and 0.1 M (dashed line).

$D_S \approx 6 \times 10^{-6}$ cm^2 s^{-1}, $Sh = 10$, and $\alpha = 0.1$. Immobilized biocatalysts that are used in packed bed reactors should have a radius of more than 100–200 μm to avoid too large backpressures in these reactors. For stirred batch reactors, the radius can be lower. When the radius is $\leq 10\,\mu$m, the particles are difficult to remove from the reaction mixture by filtration or sedimentation. The radius should not be much more than ~100–200 μm, to allow for high rates and *STY* (Figure 10.5). The question remains as to whether the V'_{max} calculated in this figure can be realized in available supports for biocatalyst immobilization (see Chapter 8). In these particles, up to ~10 g of enzyme can be immobilized per liter of wet support, which provides an enzyme concentration of 100–400 μM for enzymes in the MW range of 25–100 kDa. Many enzymes have a turnover number of 100 s^{-1} (see Tables 2.4 and 2.5), and this gives V'_{max} in the range of 0.01–0.04 M s^{-1} (or 6–24 × 10^5 U l^{-1}) – that is, almost of the same order of magnitude as calculated in Figure 10.5. The value may be smaller due to the assumptions $\alpha = 0.1$ (too large) and $\eta \approx 1$ (too large). In order to analyze this situation, the effectiveness factors and substrate concentration profiles both inside and outside the particles must be calculated (see Section 10.4.2).

10.4.2
Calculation of Effectiveness Factors and Concentration Profiles Inside and Outside the Particles

The concentration profile inside and outside a spherical particle with immobilized biocatalyst can be determined at steady state from the mass conservation relationship in the volume element between $r + dr$ and r (Figure 10.6). There, the net diffusion equals the conversion of the substrate (simple Michaelis–Menten kinetics

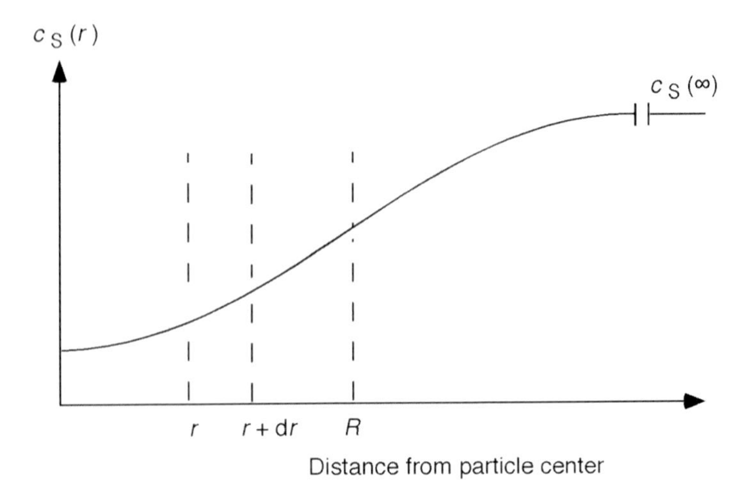

Figure 10.6 The substrate concentration (c_S) inside and outside a particle (radius R) with immobilized biocatalyst (enzyme, cell) at steady state.

without inhibition, properties of the immobilized biocatalyst are primed) or

$$D'_S\, A(r+\mathrm{d}r)\frac{\mathrm{d}c_S(r+\mathrm{d}r)}{\mathrm{d}r} - D'_S\, A(r)\frac{\mathrm{d}c_S(r)}{\mathrm{d}r} = \frac{V'_{\max}\, c_S(r)}{K'_m + c_S(r)} A(r)\,\mathrm{d}r, \qquad (10.9)$$

where $A(r + \mathrm{d}r)$ and $A(r)$ are areas perpendicular to the diffusional flow. From the Taylor series (see mathematical textbooks), it follows that

$$\frac{\mathrm{d}c_S(r+\mathrm{d}r)}{\mathrm{d}r} = \frac{\mathrm{d}c_S(r)}{\mathrm{d}r} + \frac{\mathrm{d}^2 c_S(r)}{\mathrm{d}r^2}\,\mathrm{d}r. \qquad (10.10)$$

With this, Eq. (10.9) is transformed to

$$D'_S\left(\frac{\mathrm{d}^2 c_S(r)}{\mathrm{d}r^2} + \frac{n}{r}\frac{\mathrm{d}c_S(r)}{\mathrm{d}r}\right) = \frac{V'_{\max}\, c_S(r)}{K'_m + c_S(r)}, \qquad (10.11)$$

where $n = 0$, 1, or 2 for planar, cylindrical, or spherical geometry, respectively. This equation can be solved with the following boundary conditions:

$$\frac{\mathrm{d}c_S(r)}{\mathrm{d}r} = 0 \quad \text{at} \quad r = 0,$$

$$c_S(R) = c_S(\infty) - \frac{D_S}{k_L}\frac{\mathrm{d}c_S(R)}{\mathrm{d}r} \quad \text{for} \quad r = R \quad \text{(particle surface)}$$

from Eq. (10.7) and Figure 10.6, after which $\mathrm{d}c_S(R)/\mathrm{d}r$ can then be determined. This provides the substrate transport rate into the particle that, at steady state, equals the rate of substrate conversion in the particle.

It is suitable to transform Eq. (10.11) into a dimensionless form using the variable transformations

$$\gamma = \frac{c_S(\infty)}{K'_m}, \qquad z = \frac{r}{R}, \qquad c_S(z) = \frac{c_S(r)}{c_S(\infty)}.$$

This results in

$$\frac{\mathrm{d}^2 c_S(z)}{\mathrm{d}z^2} + \frac{n}{z}\frac{\mathrm{d}c_S(z)}{\mathrm{d}z} = \frac{R^2 V'_{\max}}{D'_S K'_m}\frac{c_S(z)}{1 + c_S(z)\gamma}. \qquad (10.12)$$

The dimensionless quantity

$$\varphi^2 = \frac{R^2 V'_{\max}}{D'_S K'_m} \qquad (10.13)$$

in Eq. (10.12) is the square of the *Thiele modulus*, a dimensionless number. It is the ratio of the maximum rate of substrate conversion in the particle and the maximum rate of substrate transport to the particle. Such a ratio characterizing heterogeneous catalysts was first introduced by Damköhler (1937). It contains properties that influence the activity of immobilized biocatalysts in the support. Of these values, D'_S and R have no influence on the activity of the free biocatalyst. As φ^2 is introduced here, it applies for all substrate concentrations – that is, also for first-order ($c_S < K'_m$) and zero-order ($c_S \gg K'n_m$) reactions. (In some textbooks on heterogeneous catalysis, the dimensionless numbers are derived so that they only apply for first- or zero-order reactions.)

The differential equation (Eq. (10.12)) can only be solved analytically for first- and zero-order reactions, and must therefore generally be solved by numerical methods. A suitable method here is the collocation method (Villadsen and Michelsen, 1978; Kasche, 1983), with which the steady-state concentration profiles $c_S(z)$ (and $c_S(r)$) can be calculated. Then the effectiveness factors can be obtained as follows using the definitions in Section 10.3. The stationary effectiveness factor (Eq. (10.3)) for particles with radius R is

$$\eta = \frac{\text{rate of substrate diffusing into the particles}}{\text{rate of substrate conversion in the particles without mass transfer limitation } c_S = c_S(\infty)}$$

$$= \frac{4\pi R^2 D'_S (dc_S(R)/dr)}{(4/3)\pi R^3\, V(c_S(\infty))} = \frac{\text{rate of substrate conversion in the particles with mass transfer limitation}}{\text{rate of substrate conversion in the particles without mass transfer limitation } c_S = c_S(\infty)} \tag{10.14}$$

$$= \frac{\int_0^R 4\pi r^2 [V'_{max} c(r)/(K'_m + c(r))]\, dr}{(4/3)\pi R^3\, V(c_S(\infty))},$$

where $c_S(\infty)$ is the bulk substrate concentration (Figure 10.6) and $V(c_S(\infty)) = V'_{max}\, c_S(\infty)/(K'_m + c_S(\infty))$.

Then t'_{12} in Eq. (10.5) can be calculated from the integral derived from the expression

$$-d(c_S(\infty))/dt = \eta(c_S(\infty))V(c_S(\infty))n(4/3)\pi R^3,$$

that is, per unit volume the rate of change in the substrate concentration outside the particles equals the rate of the enzyme-catalyzed substrate conversion inside the particles; n is the particle number with immobilized enzyme per unit volume. By variable separation, this relation is transformed to

$$\int_{c_{S1}(\infty)}^{c_{S2}(\infty)} \frac{-dc_S(\infty)}{\eta c_S(\infty) V(c_S(\infty)) n(4/3)\pi R^3} = \int_0^{t'_{12}} dt = t'_{12}. \tag{10.15}$$

The integral can be integrated numerically using stationary effectiveness factors for the different substrate concentrations calculated with Eq. (10.14). This gives t'_{12}, and t_{12} in Eq. (10.5) is obtained when Eq. (10.15) is integrated with $\eta = 1$. Then η_o is calculated from Eq. (10.5).

The values of the effectiveness factors calculated for different parameters are given in Figures 10.7 and 10.8. As parameter for the external mass transfer, the Sherwood number (Eq. (10.7)) is used. In an unstirred system with spherical particles, the Sherwood number is 2, whereas in stirred tank or packed bed reactors it can be increased to 20–30. At higher values, where more energy is required for stirring and pumping at higher speeds, the effectiveness factors are marginally changed. The values calculated in these figures can be used to estimate the effectiveness factors for immobilized biocatalysts. The comparison with experimental data (see Figure 10.8) shows a good agreement within experimental error estimated to be in the range of ±20–30%. This large error results from the many quantities in Eqs. (10.7) and (10.13) that must be determined to estimate the square of the

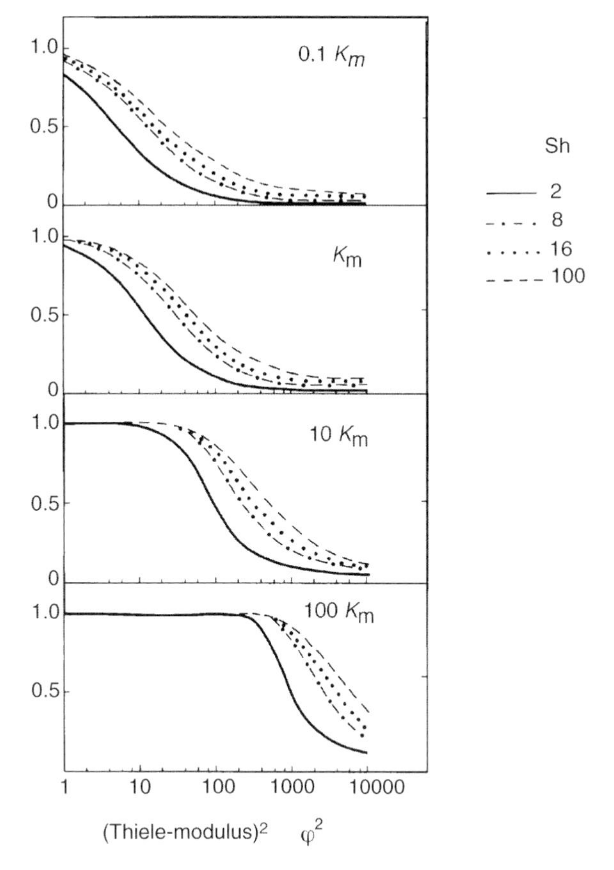

Figure 10.7 The stationary effectiveness factor η for simple Michaelis–Menten kinetics (one substrate, no inhibition) as a function of the square of the Thiele modulus, different Sherwood numbers (*Sh*), and for different substrate concentrations, given in units of K'_m (Kasche, 1983).

Thiele modulus and the Sherwood number. The error in determination of the single quantities is at best ±5%. For these estimations, the K_m value for the free enzyme is used. From Figure 10.7 follows that the intrinsic molecular property K'_m of the immobilized enzyme can only be determined in systems with $\varphi^2 < 1$, as the rates must also be determined for $[S] \leq K'_m$ (see Section 2.7.1).

The values in Figures 10.7 and 10.8 have been calculated for simple Michaelis–Menten kinetics (one substrate and no inhibition). This does generally not apply for enzyme processes. The effectiveness factors for other cases (one substrate with inhibition; two substrates without and with inhibition) can be calculated as shown here, when the rate equations covering these cases are used. Many enzyme processes, such as hydrolysis reactions, can be considered as reactions with one substrate. Here, however, competitive product inhibition cannot be neglected. For these, K'_m can be estimated to be

$$K'_m = K_m(1 + [P]/K_i) \qquad (10.16)$$

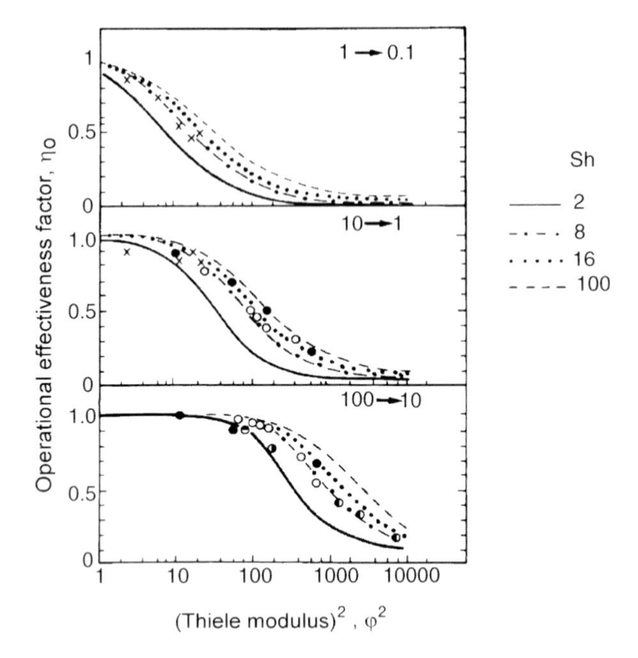

Figure 10.8 The operational effectiveness factor η_o for simple Michaelis–Menten kinetics (one substrate, no inhibition) as a function of the square of the Thiele modulus, different Sherwood numbers (*Sh*), and for different initial substrate concentrations, given as γ/K_m, for spherical particles and 90% substrate conversion. Calculated (curves) and experimental (circles) data for such processes are compared (Kasche, 1983).

where [P] increases with time (see Table 2.6). For such processes, the effectiveness factors can be estimated from Figures 10.7 and 10.8 using the average value $[P] = 0.5[S]_0$ – that is, half the initial substrate content.

Thus, these figures can be used to

- estimate the effectiveness factors for a given immobilized biocatalyst, and
- design immobilized catalysts where the effectiveness factors are ≈ 1.

10.5
Space–Time Yields and Effectiveness Factors for Different Reactors

In order to design enzyme processes with immobilized biocatalysts, and to evaluate the different reactors used in these, the amount of biocatalyst with a known activity (V_{max}) required to obtain a given *STY* must be calculated (Eq. (10.3)). The relationship between the *STY* and V_{max} will be derived for the most important continuous reactors – the continuous stirred tank (CST) reactor and the packed bed (PB) reactor – and this will be performed with the following (idealized) assumptions:

- for the CST reactor: complete mixing in the reactor;
- for the PB reactor: plug flow; no radial dispersion.

Both reactors are kept at the same temperature and contain the same immobilized biocatalyst, the inactivation of which is neglected. The catalyzed reaction is described by simple Michaelis–Menten kinetics (one substrate, no inhibition). This should also answer the question of which effectiveness factor applies to the different reactors.

10.5.1
Continuous Stirred Tank Reactor

The CST reactor (Figure 10.9) is characterized as follows: V_{CST} = reactor volume; α_{CST} = immobilized biocatalyst volume as a fraction of the reactor volume; and Q = flow rate. At steady state in the reactor ([S] in the reactor is time independent), the rate of net substrate flow into the reactor equals the rate of substrate conversion in the reactor. Then

$$[\mathrm{S}]_0 Q - [\mathrm{S}]Q = \frac{\eta([\mathrm{S}])\alpha_{\mathrm{CST}} V_{\mathrm{CST}} V'_{\max}[\mathrm{S}]}{K'_{\mathrm{m}} + [\mathrm{S}]}. \quad (10.17)$$

Here, the stationary effectiveness factor (Eq. (10.4)) must be used as the substrate concentration in the reactor is constant. With the relative substrate concentrations,

$$\gamma_2 = \frac{[\mathrm{S}]}{K'_{\mathrm{m}}}, \qquad \gamma_1 = \frac{[\mathrm{S}]_0}{K'_{\mathrm{m}}}. \quad (10.18)$$

Equation (10.17) can be rewritten as

$$(\gamma_1 - \gamma_2)Q = \frac{\eta(\gamma_2)\alpha_{\mathrm{CST}} V_{\mathrm{CST}} V'_{\max}\gamma_2}{(1+\gamma_2)K'_{\mathrm{m}}}. \quad (10.19)$$

The residence time in this reactor $t_{12,\mathrm{CST}}$ required for the substrate conversion from γ_1 to γ_2 is

$$t_{12,\mathrm{CST}} = \frac{V_{\mathrm{CST}}}{Q} = \frac{\gamma_1 - \gamma_2}{\gamma_2}\frac{(1+\gamma_2)K'_{\mathrm{m}}}{\alpha V'_{\max}\eta(\gamma_2)}. \quad (10.20)$$

The *STY* is

$$\begin{aligned} STY &= \frac{\text{amount of converted substrate(in moles)}}{\text{unit time} \times \text{unit reactor volume}} \\ &= \frac{(\gamma_1 - \gamma_2)K'_{\mathrm{m}}}{t_{12,\mathrm{CST}}} = \eta(\gamma_2)V'_{\max}\frac{\gamma_1\alpha_{\mathrm{CST}}}{1+\gamma_2} \end{aligned} \quad (10.21)$$

from which the amount of biocatalyst ($\alpha_{\mathrm{CST}} V_{\max}$) per unit reactor volume to obtain this *STY* can be calculated.

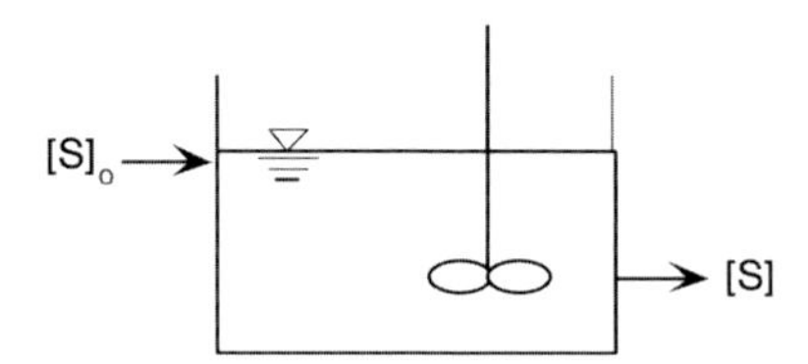

Figure 10.9 Continuous stirred tank reactor; operational details.

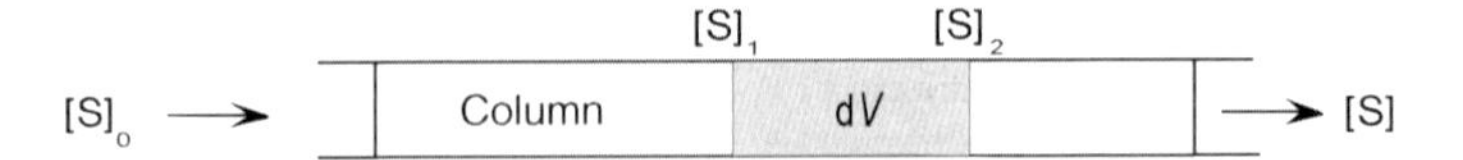

Figure 10.10 Packed bed reactor; operational details.

10.5.2 Packed Bed Reactor or Stirred Batch Reactor

The reactor (Figure 10.10) is characterized as follows. With the change in the subscripts from CST to PB, the same characteristics apply for this reactor. In the volume element dV, the following relationship applies for the steady state, where the rate of net flow of substrate into the volume element equals the rate of substrate conversion in this element. Then

$$([\mathrm{S}]_1 - [\mathrm{S}]_2)\frac{\mathrm{d}V}{\mathrm{d}t} = \frac{\eta([\mathrm{S}])\alpha_{\mathrm{PB}} V'_{\mathrm{max}}[\mathrm{S}]\mathrm{d}V}{K'_{\mathrm{m}} + [\mathrm{S}]}. \tag{10.22}$$

In the PB reactor, the stationary effectiveness factor is not constant, but varies with [S]. With the relative substrate concentrations in Eq. (10.18),(10.22) is changed to

$$\frac{\mathrm{d}\gamma}{\mathrm{d}t} = \frac{\eta([\mathrm{S}])\alpha_{\mathrm{PB}} V'_{\mathrm{max}}\gamma}{K'_{\mathrm{m}}(1+\gamma)} \quad \text{or} \quad \mathrm{d}t = \frac{K'_{\mathrm{m}}(1+\gamma)\mathrm{d}\gamma}{\eta([\mathrm{S}])\alpha_{\mathrm{PB}} V'_{\mathrm{max}}\gamma}.$$

Integration over the packed bed gives the residence time required for the substrate conversion from γ_1 to γ_2 in the packed bed reactor:

$$t'_{12,\mathrm{PB}} = \int_0^{t_{12,\mathrm{PB}}} \mathrm{d}t = \frac{K'_{\mathrm{m}}}{\alpha_{\mathrm{PB}} V'_{\mathrm{max}}} \int_{\gamma_2}^{\gamma_1} \frac{(1+\gamma)\mathrm{d}\gamma}{\eta(\gamma)\gamma}. \tag{10.23}$$

It can be determined either by numerical integration of the above integral, using the stationary effectiveness factors as shown in Figure 10.7, or from the definition of the operational effectiveness factor η_{o} (Eq. (10.5)). Then $t_{12,\mathrm{PB}}$ is calculated by integration of Eq. (10.23) with $\eta(\gamma) = 1$ (for the same amount of free enzyme):

$$t_{12,\mathrm{PB}} = \int_0^{t_{12,\mathrm{PB}}} \mathrm{d}t = \frac{K'_{\mathrm{m}}}{\alpha_{\mathrm{PB}} V'_{\mathrm{max}}} \{\ln(\gamma_1/\gamma_2) + (\gamma_1 - \gamma_2)\}. \tag{10.24}$$

Then $t'_{12,\mathrm{PB}}$ is given by the relation

$$t'_{12,\mathrm{PB}} = t_{12,\mathrm{PB}}/\eta_{\mathrm{o},12}(\gamma_1) \tag{10.25}$$

using operational effectiveness factors calculated as shown in Figure 10.8.

These results for the packed bed reactor also apply for the discontinuous stirred batch reactors, for which the operational effectiveness factor must be used. The *STY* for these reactors is analogous to the derivation of Eq. (10.21):

$$STY_{\mathrm{PB}} = \eta_{\mathrm{o},12}(\gamma_1) V'_{\mathrm{max}} \frac{\alpha_{\mathrm{PB}}(\gamma_1 - \gamma_2)}{\ln(\gamma_1/\gamma_2) + \gamma_1 - \gamma_2} \tag{10.26}$$

from which the required amount of immobilized biocatalyst ($\alpha_{PB}V'_{max}$) per unit reactor volume can be calculated.

10.5.3
Comparison of CST and PB Reactors

Equations (10.21) and (10.26) show that the *STY* values differ in the effectiveness factors used and the terms expressing the degree of substrate conversion. For the same degree of substrate conversion (≥90%), the ratio of Eq. (10.26) to Eq. (10.21) is

$$\frac{STY_{PB}}{STY_{CST}} = \frac{\eta_{o,12}(\gamma_1)\alpha_{PB}(\gamma_1 - \gamma_2)(1 + \gamma_2)}{\eta(\gamma_2)\alpha_{CST}\gamma_2(\ln(\gamma_1/\gamma_2) + \gamma_1 - \gamma_2)}, \quad (10.27)$$

a relationship that can be used to compare both reactors. Generally, $\gamma_1 \gg 1$, $\alpha_{PB} > \alpha_{CST}$, and $\eta_{o,12}(\gamma_1) \geq \eta(\gamma_2)$ (Figures 10.7 and 10.8). Then the ratio in Eq. (10.27) is >1 – that is, for this case the *STY* is larger for the PB than the CST reactor. This also implies that, for the same *STY*, less immobilized biocatalyst and a smaller reactor volume are required in the PB than in the CST reactor. In the latter reactor, α_{CST} must be less than 0.05 in order to avoid abrasion due to shear forces and particle–particle collisions. This limits the *STY* that can be obtained in this reactor. In the PB reactor, where α_{PB} is ≥ 0.5, a much higher *STY* can be obtained at equal reactor volume.

For processes with several substrates, inhibition, and enzyme inactivation, Eqs. (10.21) and (10.26) will be more complex. With product inhibition, the PB reactor is better than the CST reactor, as the latter works at a higher product concentration than the former. In some cases, the CST reactor is better than the PB reactor, especially for products that are gases or that must be neutralized, such as H^+. The disadvantageous pH gradient over the PB reactor can be reduced by dividing it into several consecutive reactors, where the pH is adjusted to the desired value between the reactors. CST and PB reactors with volumes in the range of 0.01–10 m^3 are used in enzyme processes.

10.6
Determination of Essential Properties of Immobilized Biocatalysts

Immobilized biocatalysts have been developed for a variety of applications in the laboratory, for analytical purposes, and in enzyme technology. In order to allow for their reproducible use, and to select the optimal biocatalyst for a specific application, they must be well characterized – that is, their essential physicochemical (including mechanical) properties, kinetic properties, and their stability under process conditions must be determined in standardized procedures. These have been proposed in the scientific literature (Buchholz *et al.*, 1979; Anonymous, 1983; Gardossi *et al.*, 2010). Properties that are important for large-scale enzyme processes are listed in Table 10.1, but not all of these are important for biocatalyst use on a laboratory or an analytical scale. This may explain why immobilized biocatalysts are often insufficiently characterized in the scientific literature, or inadequate information is provided by the immobilized biocatalyst manufacturer.

Table 10.1 Basic information or properties that should be given or determined for immobilized biocatalysts, and factors influencing these in order to allow for their reproducible production and to evaluate the support properties that are essential for an enzyme process (Buchholz *et al.*, 1979; Anonymous, 1983; Gardossi *et al.*, 2010).

		Information/property	Factors influencing the property
1.	**General information**		
	1.1	Reaction that is catalyzed	
	1.2	Enzyme (EC number) and enzyme source (microorganism or tissue); purity, p*I* (isoelectric point)	
	1.3	Support used for the immobilization Composition; functional groups	
2.	**Immobilization**		
	2.1	Immobilization method; conditions during immobilization (concentration, pH, I, T); how are unreacted reactive groups on the support inactivated?	pH, I, T, t
	2.2	Immobilization yield (with respect to enzyme amount and enzyme activity) and how it is determined	pH, I, T, t, [E] at the start of the immobilization, adsorption isotherm for immobilization by adsorption
	2.3	For immobilized biocatalysts used in dry solvents: pH and buffer used before they were transferred to the water-free solvent	
3.	**Physicochemical characterization of the immobilized biocatalyst**		
	3.1	Shape, average wet particle radius (R) and its distribution; surface area g^{-1} wet weight; average pore radius (R_p) and its distribution	pH, I
	3.2	Swelling (wet volume/dry volume)	pH, I
	3.3	*Compressibility/pressure drop in fixed beds*	Bed height H, R, pressure drop Δp; linear flow rate u
	3.4	*Abrasion in stirred tank or fluidized bed reactors*	R, particle density, stirrer diameter, and speed
	3.5	Biocatalyst distribution and conformation by fluorescence measurements	pH, R, I, immobilization time, stationary charge
	3.6	Stationary charges on support (positive or negative, even in non-ion-exchange supports!), n_c; pore surface: hydrophilic or hydrophobic	pH, I

		Property	Parameters
4.	**Kinetic characterization**		
	4.1	Intrinsic properties of the immobilized enzyme (V'_{max}; k'_{cat}) and active enzyme content n_E; K'_m, K'_i; selectivities (stereo- and $(k'_T/k_H)_{app}$)	pH, I, T, [S], n_E
	4.2	Stationary and operational effectiveness factors at process conditions	pH, I, R, T, V'_{max}, K'_m, K'_i, D'_S, [S], [P], u for PB, and n, d_i for CST and batch reactors
	4.3	Space–time yield at process conditions	pH, I, R, T, V'_{max}, K'_m, K'_i, D'_S, [S], [P]
	4.4	Substrate/product concentration and pH gradients	pH, I, R, T, V'_{max}, K'_m, K'_i, D'_S, [S], [P]
5.	**Stability and productivity**		
	5.1	Storage stability	pH, I, T, t
	5.2	*Stability under process conditions (operational stability)*	pH, I, T, k_i, t
	5.3	*Productivity* (space–time yield integrated over the time the biocatalyst is used) (amount of product/unit amount of immobilized enzyme)	pH, I, T, R, V'_{max}, K'_m, K'_i, D'_S, [S], k_i

Properties in italics must only be determined for immobilized biocatalysts used in industry.

The characterization of immobilized biocatalysts, as outlined above – and especially comparative studies with the same enzyme immobilized in different supports used for the same enzyme process – can provide important information on the properties of the supports that are important to improve the process. The recent reinvention of nanobiotechnology – that is, molecular interactions on a nanometer scale, as on the pore surface in immobilized biocatalysts or on natural or artificial membrane surfaces – has revived the importance of such studies on immobilized biocatalysts.

When immobilized biocatalysts are used in the food or pharmaceutical industries, health and environmental regulations must be considered. This information is given in Section 6.6.

10.6.1
Physicochemical Properties

The physicochemical properties of immobilized biocatalysts are listed in Table 10.1. The average particle radius of a swollen particle can be determined using microscopy, or with specialized instruments that determine particle size distribution. A narrow distribution is important for PB reactors as it determines the bed porosity and thus the pressure drop. The measurement of particle size distribution can also be used to measure the abrasion in stirred tank or fluidized bed reactors. To be of technical relevance, the abrasion as a function of particle concentration must be performed in reactors where high sheer forces can be realized. This requires baffled vessels and stirrers with a diameter $\geq$50 mm (see Section 11.3). As expected, abrasion increases with the concentration of immobilized particles in the reactor, and based on this the upper particle content in these reactors should be less than 5% of the reactor volume. This limits the space–time yield in these reactors. In enzyme technology, particles with radius R in the range of 50–500 μm are used, and these have an internal surface area of 10–100 $m^2\,g^{-1}$ wet particles. The average pore diameter is generally >20 nm. The pore size distribution can be measured only for dry particles using porosimeters, or from scanning electron micrographs of shock-frozen particles (see Chapter 8).

The particle radius influences the effectiveness factor (see Section 10.4) in all reactors. In PB reactors, the square of the radius determines the pressure drop over the reactor, and this is also influenced by the interstitial volume fraction, ε (or bed porosity). This is about 35–40% of the bed volume for optimally packed beds of spherical particles. In fluidized bed reactors, the particle radius and density determine the bed expansion as a function of the linear liquid flow rate. In general, the particle radius distribution is heterogeneous, and thus average values must be used to estimate the effectiveness factors. They are given as a function of the Thiele modulus (Eq. (10.13)) in Figures 10.7 and 10.8. When the intrinsic enzyme properties of the immobilized enzyme and the substrate diffusion coefficient in the particles do not depend on the particle size, the average Thiele modulus is

$$\overline{\varphi^2} = \frac{\sum_{i=1}^{n} \varphi_i^2}{n} = \frac{\sum_{i=1}^{n} R_i^2}{n} \cdot \text{constant}, \qquad (10.28)$$

where n is the number of particles. Thus, the average radius required for the estimation of the Thiele modulus for spherical particles with different sizes is the following average:

$$\bar{R} = \sqrt{\frac{\sum_{i=1}^{n} R_i^2}{n}}. \qquad (10.29)$$

The same average must be used for the estimation of the pressure drop over a packed bed reactor (see Section 11.3.5).

Some properties must be determined in the reactor used for the enzyme process. In order to minimize the immobilized biocatalyst amount required for this, suitable scaled-down reactors should be used. For packed and fluidized bed reactors, the same bed height but a much smaller cross section of the reactor can be used for this purpose. The compressibility is essential in PB reactors, when the bed height falls by more than 10% at higher flow rates; this indicates that the particles are deformed at higher pressures, and this causes an increase in the pressure drop. The pressure drop can be determined in a simple column with 1 m bed height, and at least 20 g of particles at different flow rates; a minimum pressure drop of 1 bar must be maintained for 1 h. Compressible particles are not suitable for use in PB reactors. For fluidization, see Chapter 11.

10.6.1.1 **Immobilized Biocatalyst Distribution and Conformation**

The fluorescence spectra of immobilized enzymes can be used to determine the enzyme distribution within the support, with conformational changes due to immobilization being monitored either by spectrofluorimetry or by confocal microscopy (Carleysmith, Dunnill, and Lilly, 1980; Kasche *et al.*, 1994; Heinemann *et al.*, 2002).

10.6.1.2 **Stationary Charge Density in the Support**

One important property of immobilized biocatalysts that frequently is not measured, *especially in non-ion-exchange supports*, is the stationary charge density, which can be determined by titration. When this has been carried out, net charge densities of more than 10 mM have been observed. This is much larger than net charge densities due to immobilized enzymes, where n_E is <0.1 mM. These charges either are formed in the deactivation of functional groups not used for the immobilization of the enzyme or are due to charged compounds used in the production of the supports (Bozhinova *et al.*, 2004; Kasche *et al.*, unpublished data). As these charges are localized on the surface of the pores in the particles, they are distributed in a much smaller volume than the total particle volume. The local charge density at the pore surface can then become more than 1 M. This gives rise to an electric double layer with dimensions in the nm range with pH and I-values that differ from the bulk values (Figure 10.11). This influences the intrinsic properties of the immobilized enzyme, especially when charged substrate or product molecules are involved in the enzyme process. It can also reduce the immobilization yield when the enzyme has the same charge as the surface. This requires that the isoelectric point (p I) of the enzyme is known, and this can be determined using isoelectric focusing (see Section 6.5). In this case, immobilization should be carried out at a high ionic

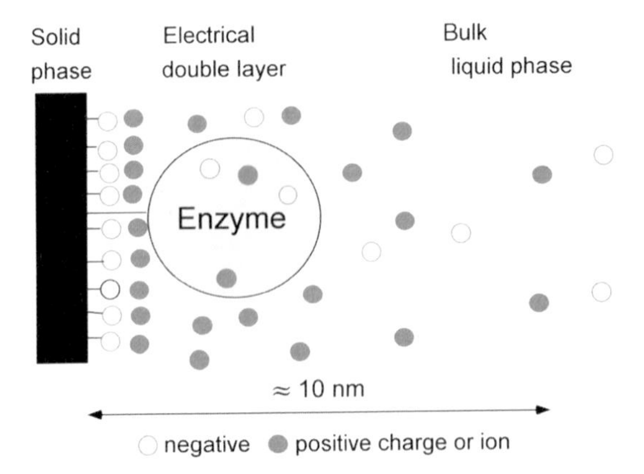

Figure 10.11 The electrical double layer formed by stationary charges on the surface of pores in particles with immobilized enzymes. In the double layer, the pH and ionic strength differ from the corresponding values in the bulk phase. This can change the intrinsic properties of the enzyme that is located in the double layer (Hunter, 1993; González-Caballero and Shilov, 2002).

strength, in order to reduce repulsion between the support surface and the enzyme, in order to increase the yield (Chikere *et al.*, 2001).

10.6.2
Kinetic Characterization of Immobilized Biocatalysts: Influence of Support Properties on the Nano- and Micrometer Level in Aqueous and Other Systems

The intrinsic properties listed in Table 10.1 are required to estimate the amount of biocatalyst needed to obtain a given space–time yield – that is, for the design of an enzyme process. Some of the immobilized enzyme molecules may be inactivated during the immobilization procedure. This can be determined from the mass balance for both concentration and activity (enzyme amount/activity before immobilization – enzyme amount/activity in wash solutions after immobilization) that for pure enzymes can be obtained using spectroscopic measurements. Whether all of the immobilized enzymes also are active can only be determined by measuring the immobilized active-site concentration, n_E. As the kinetic properties generally are changed when enzymes are immobilized, this is only possible for enzymes for which the active sites can be titrated. For this aim, substrates that react specifically with the active site irreversibly, and have a colored or fluorescent compound as releasing group, are used. When the substrate is in excess, the concentration of the leaving group equals the number of active sites in the sample of the immobilized enzyme. Alternatively, an inhibitor that irreversibly blocks the active site can be used. By increasing the concentration of this inhibitor until the enzyme becomes inactive, the number of active sites can be titrated. These active-site titrations are mainly available for hydrolases such as α-chymotrypsin, trypsin, penicillin amidase, and lipase (Gabel, 1974; Svedas *et al.*, 1977; Fujii *et al.*, 2003).

For covalent immobilization, almost quantitative binding yields can be reached. In this case, the concentration of immobilized enzyme is limited by the content of reactive groups on the support surface and the total surface area (see Section 8.2). For noncovalent adsorption, the amount of immobilized enzyme is determined by the adsorption isotherm and concentration of the free enzyme in solution. When peptidases are immobilized, a loss of active enzyme by autoproteolysis must be avoided; this is achieved by dissolving the enzyme in a buffer of a pH at which the enzyme activity is low, and by performing the immobilization at a low enzyme concentration (Kasche, 1983).

The maximal immobilized active enzyme concentration n_E is limited by the cross-sectional area of the enzyme ($\approx$1–5 $\times$ 10^{-17} m^2 for molecular weight in the range of 10–100 kDa) and the total surface area of the particles (10–100 $m^2\,ml^{-1}$ wet support). The concentration is obtained when the pore surface is covered by immobilized enzymes; the maximal values for n_E are then in the range of 1.6–16 mM (16–160 mg ml^{-1}) for a 10 kDa enzyme and 0.3–3 mM (30–300 mg ml^{-1}) for a 100 kDa enzyme. However, only 10% of these values are reached in practice (Katchalski-Katzir and Krämer, 2000; Janssen *et al.*, 2002).

10.6.2.1 Determination of V'_{max}, k'_{cat}, Substrate/Product Concentration, and pH Gradients

The intrinsic properties of immobilized enzymes can only be determined in systems that are not mass transfer limited – that is, where the stationary effectiveness factor η equals 1. This applies always when [S] $\gg$ K'_m – that is, for the determination of V'_{max} (see Figure 10.7). The turnover number k'_{cat} can only be determined for enzymes of which the active sites can be titrated.

A simple method is available to determine η. The particles can be disintegrated by sonication to reduce R and the Thiele modulus to a value where $\eta \approx 1$. By comparing the initial rate of the enzyme-catalyzed reaction with the same initial substrate content for whole and disintegrated particles, η for the whole particles can be determined (Figures 10.3 and 10.12). Even when $\eta \approx 1$, the turnover number determined may not be an intrinsic property (Figure 10.12b). This applies for reactions where acids or bases are either produced or consumed. A pH gradient may then be formed in the particles, which perturbs the pH-dependent intrinsic properties (see Section 2.7). To avoid this, a buffer with high buffering capacity at the pH where measurements are performed should be used. This implies that the pK value of the buffer should be near this pH value (Tischer and Kasche, 1999). For such measurements, a buffer with an ionic strength of 0.05 M is sufficient (Figure 10.12). The buffer must be inert; therefore, Tris should not be used as it can act as a nucleophile and react with acyl- or glycosyl-enzyme intermediates (see Section 2.6, Table 2.3).

It is well known that the intrinsic kinetic properties of an enzyme are changed when it has bound a ligand, or is chemically modified in solution. The same is expected for an enzyme that is bound (adsorbed) or chemically modified by covalent bonds to the surface of a support. This is shown in Figure 10.13, where the turnover number of penicillin amidase covalently (except for PBA-Eupergit where it is adsorbed and covalently cross-linked) immobilized in different supports is given for

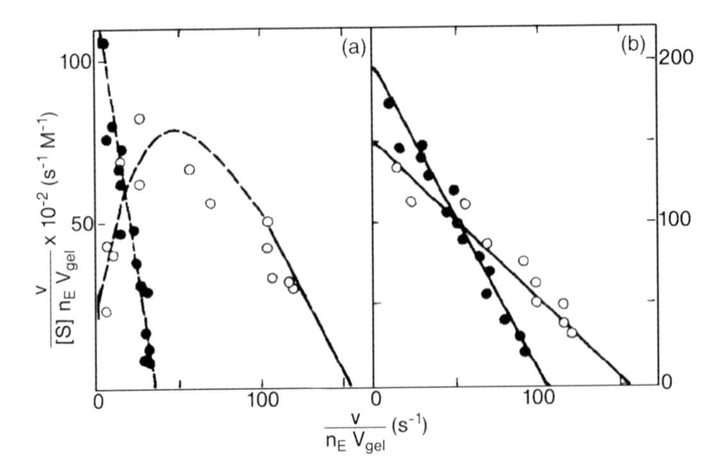

Figure 10.12 Effects of pH gradients in stirred batch reactors for enzyme processes where H^+ is produced. Influence of buffer capacity and particle radius on the determination of k'_{cat} and K'_m for α-chymotrypsin immobilized in Sepharose 4B particles ($n_E = 90\,\mu M$), from rate measurements using Eq. (2.13). (a) $R = 60\,\mu m$; (b) $R = 3\,\mu m$ (homogenized particles from (a)) at 25 °C and pH 10.0. Total ionic strength 0.25 M of which for buffer 0.05 M (open circles) or 0.001 M (closed circles). Substrate: *N*-acetyl-(*S*)-tyrosine ethyl ester (Kasche and Bergwall, 1973).

two substrates. For one substrate the immobilization led to an increase, and for the other a decrease in k_{cat}, compared with the value for the free enzyme. The observed changes were up to a factor of 2. Similar results have been observed by Kallenberg, van Rantwijk, and Sheldon (2005). The causes for this are still unclear, but these results show that – as expected – the intrinsic properties of enzymes are changed

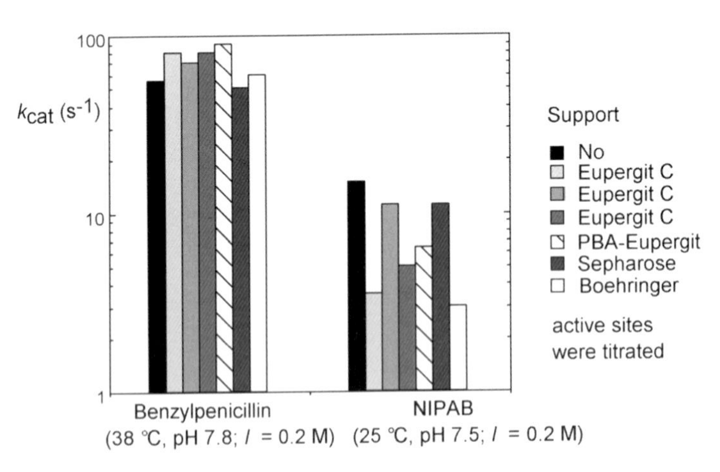

Figure 10.13 Turnover number for the hydrolysis of penicillin G and 6-nitro-3-(phenylacetamido)benzoic acid (NIPAB) by penicillin amidase from *E. coli* immobilized in different supports, determined in stirred batch reactors. The ionic strength is only due to the phosphate buffer used (B. Galunsky and V. Kasche, unpublished data).

when they are immobilized. This is also observed for enzymes adsorbed to hydrophobic supports (Kasche, Galunsky, and Michaelis, 1991). In some cases, the changes are within experimental error; for example, the turnover number determined for α-chymotrypsin in whole and homogenized particles at high buffer capacity in Figure 10.12 equals k_{cat} for the free enzyme (see Table 2.4).

In processes where H^+ is either formed or consumed, perturbation of the intrinsic constants due to pH gradients in the particles cannot be neglected. They can be determined when fluorochromes with a pH-dependent fluorescence intensity are coimmobilized with the enzymes (Spieß and Kasche, 2001; Bonielli *et al.*, 2010). This is shown for the equilibrium-controlled hydrolysis of penicillin and glutaryl-7-aminocephalosporanic acid in Figure 10.14. Without buffer, the formed acid

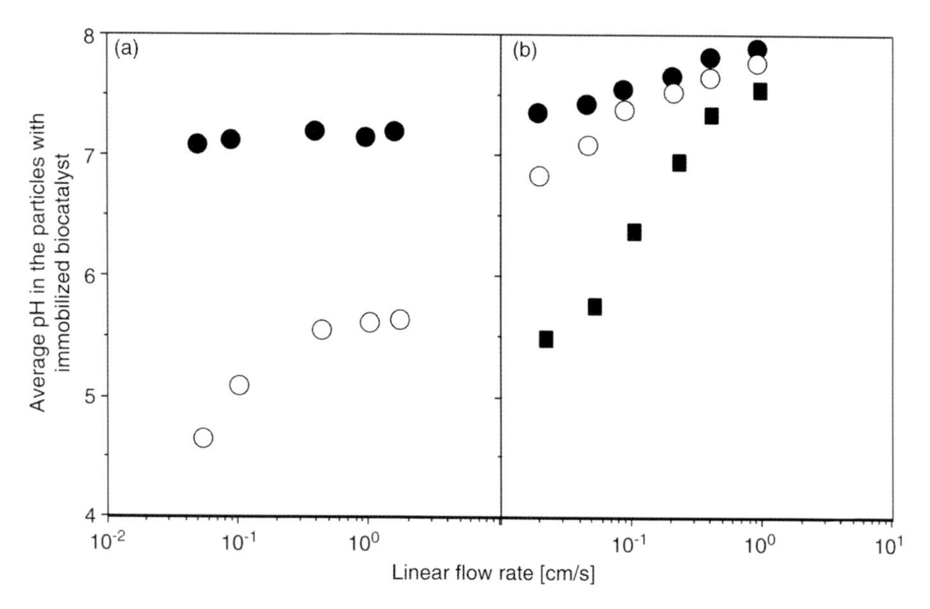

Figure 10.14 Effects of pH gradients in immobilized enzyme particles in packed bed reactors for an enzyme process where H^+ is produced. Influence of buffer capacity and linear flow rate on average pH in particles with immobilized biocatalysts. Filled symbols: with buffer; open symbols: without buffer. The average pH in the particles was determined from the fluorescence intensity of coimmobilized fluorescein, the fluorescence intensity of which is pH dependent below pH 9. A small packed bed reactor (length 3–5 mm) was used, where the pH decrease over the reactor was less than one pH unit. The stationary effectiveness factor at the initial substrate content was ≈ 1; that is, the systems used were not mass transfer controlled. (a) Hydrolysis of 300 mM penicillin by penicillin amidase from *E. coli* immobilized in Eupergit C ($n_E = 20\,\mu M$; average particle radius 100 μm) with and without phosphate buffer ($I = 0.05$) at 310 °C. The pH of the substrate solution at the reactor inlet was 7.9. (b) Hydrolysis of 40 mM (circles) glutaryl-7-aminocephalosporanic acid (Glu-7-ACA) with and without phosphate buffer ($I = 0.2$ M), and 135 mM Glu-7-ACA (squares) with buffering components ammonia, carbonate, and acetate ($I \approx 0.6$ M) at 25 °C, with glutaryl amidase immobilized in particles similar to Eupergit C produced by former Hoechst AG ($n_E = 100\,\mu M$; average particle radius 100 μm). The pH of the substrate solution at the reactor inlet was 10.0 (Spieß *et al.*, 1999).

(p$K \approx$ 4–5) and base (p$K \approx$ 4.5) cause a significant decrease in pH, to about 5 in the particles, but by increasing the linear flow rate to up to 1 cm s^{-1} this can be reduced. A further reduction is achieved by adding a suitable buffer with a pK near the optimum pH for the reaction.

Confocal microscopy can also be used to determine substrate concentration gradients in the particles (Bonielli *et al.*, 2010; Zavrel *et al.*, 2010).

10.6.2.2 K'_m and K'_i

In order to determine these properties, n_E must not itself be determined, but the effectiveness factor must be equal to 1. As for the turnover number, these intrinsic properties are expected to be changed when an enzyme is immobilized. For the equilibrium constants, a decrease is expected due to steric hindrance caused by the immobilization that reduces the rates of association reactions. Changes in rate constants for the dissociation reaction may counteract this. The K_m value for the enzyme in the homogenized particles in Figure 10.12 is fivefold larger than the value for the free enzyme (Kasche and Bergwall, 1973). Similar or opposite changes have been observed for other enzymes (Table 10.2).

These results show that the turnover number and the Michaelis–Menten constant can change considerably when an enzyme is immobilized. In most cases – except for the case of adsorption to ion exchangers shown in Table 10.2 – the causes for this remain unclear. It is probably partly due to conformational changes caused by the covalent binding or adsorption to the surface that, for free enzymes in solution, also leads to changes in k_{cat} and K_m. The nanoenvironment around the immobilized enzyme also causes changes in the apparent constants (Table 10.2). At low ionic strength, the positively charged substrate benzoyl-Arg ethyl ester accumulates at the negatively charged surface of the support, but this does not occur at a higher ionic strength.

Table 10.2 Intrinsic k_{cat} and K_m values for free and immobilized penicillin amidase (PA) from *E. coli* and bovine trypsin (TRY) immobilized in different supports for the hydrolysis of different substrates determined in phosphate buffer of different ionic strength (pH 7.5) at 25 °C.

Substrate/enzyme	Support	I (M)	k_{cat} (s^{-1})	K_m (mM)	E_{eq}
(*R*)-Phenylglycine amide/PA	None	0.20	35	19	
	Eupergit C	0.20	30	110	
	PBA-Eupergit C	0.20	16	7	
	Sepharose	0.20	22	7	
(*S*)-Phenylglycine amide/PA	None	0.20	6	11	0.30
	Eupergit C	0.20	11	26	0.25
	PBA-Eupergit C	0.20	17	34	0.20
	Sepharose	0.20	14	26	0.15
Benzoyl-Arg ethyl ester/TRY	None	0.04	n.d.	6.9	
	None	0.50		6.9	
	Ethylene malate	0.04		0.2	
	Copolymer (negative)	0.50		5.2	

The enantioselectivity (Eq. (2.15)) for PA is also given (Goldstein, Levin, and Katchalski, 1964; Wiesemann, 1991).

10.6.2.3 Selectivities

The stereoselectivity E_{kin} and selectivity in the kinetically controlled synthesis of free and immobilized penicillin amidase (see Section 2.7.1.2) using different supports are illustrated in Figure 10.15. It has been observed that most non-ion-exchange supports for immobilized biocatalysts contain stationary charges (mainly carboxyl or amino groups) whose concentrations are orders of magnitude larger than those due to the charges on the immobilized enzymes. With charged substrates, this can (as shown in Figure 10.15) cause large changes in the selectivities. The measurements in Figure 10.15 were carried out under conditions where $\eta \approx 1$. This is also verified by comparison of the whole and homogenized supports in Figure 10.15b. The total rates of synthesis and hydrolysis were not changed when the particles were homogenized. The hydrolysis of (*R*)-phenylglycine amide by penicillin amidase has been found to be independent of the ionic strength. The reduction in selectivity observed in this figure is due to the ionic strength in

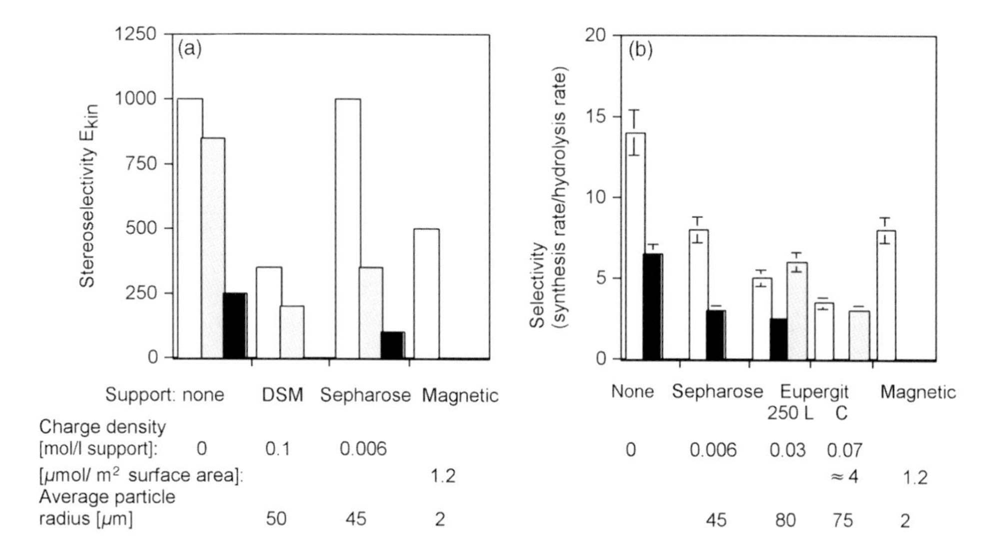

Figure 10.15 The influence of the support charge on (a) the stereoselectivity E_{kin} (Eq. (2.16)) and (b) selectivity in kinetically controlled synthesis (Eq. (2.1)) for reactions with charged substrates catalyzed by free and immobilized penicillin amidase (PA). The DSM support is based on gelatin (Spieß and Kasche, 2001) and the magnetic support on polyvinyl alcohol (Bozhinova *et al.*, 2004). (a) S_1' stereoselectivity in the kinetically controlled synthesis of (*R*)-phenylglycyl-(*R*,*S*)-Phe from 20 mM (*R*)-phenylglycine amide and 100 mM-(*R*,*S*)-Phe at pH 9.0 and 25 °C, catalyzed by PA from different sources (open bars: from *E. coli*; gray bars: from *A. faecalis*; black bars: from *B. megaterium*). *E*-values above 1000 cannot be accurately determined; the experimental error in the values below 1000 is ±10%. (b) Selectivity in the kinetically controlled synthesis of cephalexin from 200 mM (*R*)-phenylglycine amide and 50 mM 7-aminodesacetoxycephalosporanic acid in phosphate buffer ($I = 0.05$) at pH 7.5 and 5 °C, catalyzed by free and immobilized PA from *E. coli*. The measurements were performed without (open bars) and with (black bars) addition of 1 M NaCl. For the Eupergit support measurements were also carried without NaCl in homogenized supports (gray bars). After homogenization, the average particle size was 2.3 and 1.1 μm for Eupergit 250L and Eupergit C, respectively.

the electric double layer on the support surface where the enzyme is located. This reduces binding of the negatively charged nucleophile (7-ADCA) to its positively charged binding site on the enzyme (Kasche, 1986). The selectivity decreases with the concentration of stationary charges on the support for the immobilized biocatalyst, or the ionic strength for the free enzyme. These results show that the charge density of supports must be reduced in order to improve the selectivity of immobilized biocatalysts for kinetically controlled synthesis of charged substrates.

The changes in stereoselectivity caused by stationary charges on the support in Figure 10.15a are more complex to analyze. The binding of the negatively charged nucleophile (*R*,*S*)-Phe to the positively charged binding site is reduced by the ionic strength in the double layer. However, this alone cannot cause the reduction in E_{kin}, as this binding should not be stereoselective. The p*K* of the N-terminal Ser that is acylated in the enzyme-catalyzed reaction (see Figure 2.5) has been shown to be increased by the negatively charged carboxyl group of (*R*)-Phe (Lummer *et al.*, 1999). This effect can be reduced by the ionic strength in the double layer, and contributes to the decrease in stereoselectivity of the immobilized biocatalysts.

For uncharged substrates, the stereoselectivity has been found to both increase and decrease when the enzyme is immobilized in charged (non-ionic exchange) or hydrophobic supports where $\eta = 1$ (Kasche, Galunsky, and Michaelis, 1991; Palomo *et al.*, 2003; Mateo *et al.*, 2007). Further studies are required to explain the causes for these changes, but the above discussion indicates that nanoenvironmental effects can cause large changes in selectivities when the enzyme is immobilized.

10.6.2.4 Determinations of Effectiveness Factors

The *stationary effectiveness factor* η can be determined as shown in Figure 10.3, and can also be estimated from the Thiele modulus calculated using the values for the free enzyme, and the initial substrate content using Figure 10.7.

The *operational effectiveness factor* η_o can be determined as shown in Figure 10.4, or estimated from the Thiele modulus calculated as above, using Figure 10.8. It can also be calculated from experimental data for the times required for the same substrate conversion t_{12} and t'_{12} for free (amount per unit reactor volume V_{max}) and immobilized (amount per unit reactor volume V'_{max}) enzyme, respectively, using the relationship

$$\eta_{o.12} = (t_{12} V_{max})/(t'_{12} V'_{max}) \qquad (10.30)$$

assuming that $K_m = K'_m$.

10.6.3 Productivity and Stability under Process Conditions

The biocatalyst concentration V'_{max} and the space–time yield *STY* of immobilized biocatalysts cannot be kept constant under process conditions, due to inactivation of the biocatalyst. Besides the inactivation processes listed in Table 2.9, the following processes contribute to the inactivation of immobilized biocatalysts:

- Abrasion due to shear forces, stirrer, and particle–particle collisions in stirred reactors.

- Chemical instability of the support or the covalent bond between the support and the enzyme and desorption of adsorbed enzyme (Ulbrich, Golbik, and Schellenberger, 1991).
- Bimolecular reactions with substrates or products at the high concentrations used in enzyme processes, such as inactivation due to reaction between glucose and amino groups of Lys in glucose isomerase-catalyzed isomerization of glucose (see Section 12.1).
- Mass transfer limitations due to adsorption of molecules or microorganisms in or on the support.
- Desorption of metal ions required for activity and/or stability, such as Zn^{2+} for carboxypeptidase A (Figure 2.4) or hydantoinase or Ca^{2+} for α-amylase, some lipases, penicillin amidase, and subtilisin (see Section 6.4).

The storage, pH, and temperature stability of immobilized biocatalysts is generally higher than that for the free enzymes (Wiesemann, 1991; Aguade *et al.*, 1995; Monti *et al.*, 2000; Park *et al.*, 2002; Ferreira *et al.*, 2003; Mateo *et al.*, 2007). The inactivation processes can – as for free enzymes – be characterized by an apparent first-order rate constant, k_i, or a half-life time $t_{1/2}$ – that is, the process time when the biocatalyst activity is 50% of its initial value. This value is generally determined by the producers or users of immobilized biocatalysts, by determination of the *STY* as a function of time for the specific enzyme process. A high value of $t_{1/2}$ ($\geq$50–100 days) is essential to reduce the biocatalyst cost and to increase biocatalyst productivity. These values have now been reached for important enzyme processes, such as the hydrolysis of penicillin and the isomerization of glucose. The integral of the *STY* over time under process conditions until the immobilized biocatalyst has 20–50% of the initial *STY*, and is replaced by new biocatalyst, indicates the productivity of the immobilized biocatalyst. This is generally expressed as kg substrate converted per kg immobilized biocatalyst. For different enzyme processes, this value varies in the range of 100–15 000 $kg\,kg^{-1}$. Both the producers and users of immobilized biocatalysts make continuous attempts to increase productivity in order to reduce biocatalyst costs.

10.7 Comparison of Calculated and Experimental Data for Immobilized Biocatalysts

When designing an enzyme process, the immobilized biocatalyst activity V'_{max} required to obtain a desired *STY*, or the time needed for a given substrate conversion with a given V'_{max}, must be either calculated or estimated. This was discussed in Section 2.8 for free enzymes, but for immobilized enzymes the relationships derived in Section 10.5 can be used. The effectiveness factors calculated in Section 10.4 must also be considered.

Such calculations have been presented for some enzyme processes, such as the hydrolysis of penicillin G, penicillin V, and glutaryl-7-aminocephalosporanic acid (Carleysmith, Dunnill, and Lilly, 1980; Haagensen *et al.*, 1983; Tischer *et al.*, 1992; Spieß *et al.*, 1999). The influence of pH gradients (Figure 10.14) and effectiveness factors was only considered in the last of the above references, and the accuracy of these estimations will be illustrated for the hydrolysis of penicillin G (the process window is given in Chapter 2, Exercise 2.14), under conditions used in large-scale processes. This enzyme process, with competitive product inhibition, was previously mainly carried out in batch reactors with some added buffer at a pH ~10.0, and temperatures between 30 and 310 °C. Under these conditions, the end point of the reaction can be selected to be ≥90% substrate conversion. For an accurate calculation, all enzyme kinetic constants (V'_{max}, K'_m, and inhibition constants) for the hydrolysis and the back-reaction and their pH and temperature dependence of this equilibrium-controlled reaction must be known. For most enzyme processes, these constants (except for V'_{max}) are difficult and time-consuming to determine, and hence these data are rarely published. Such calculations must therefore be performed with simplifying assumptions. For this case, they are constant pH, biocatalyst inactivation, and the back-reaction neglected. The time t required for a substrate conversion from $[S]_0$ to $[S]_t$ for this reaction with competitive product inhibition in a stirred batch reactor is (Eqs. (10.3), (2.36,) and (10.24))

$$t = \frac{K'_m((1+[S]_0/K_i)\ln([S]_0/[S]_t) + (1-K'_m/K_i)([S]_0-[S]_t)}{\eta_o V'_{max,0}}. \quad (10.31)$$

With $V_{max,0}$ from the data in Figure 10.16, K'_m the value for the free enzyme from Table 2.4, K_i the value for the free enzyme = 20 μM for the competitive product

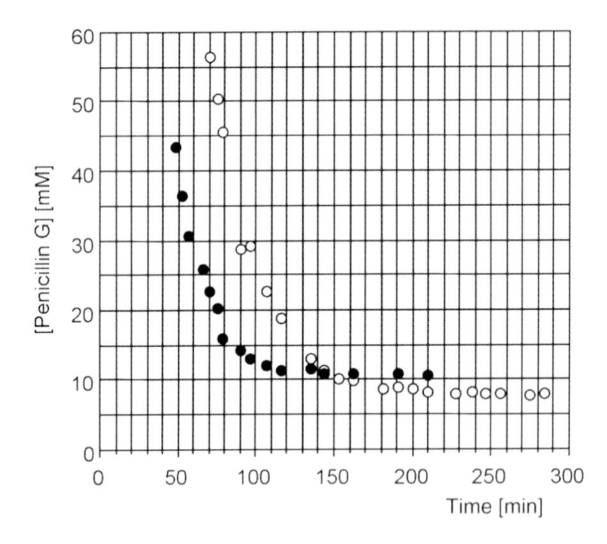

Figure 10.16 Hydrolysis of 300 mM penicillin G at 310 °C and pH 7.10 (phosphate buffer I = 0.05 M) with free (open circles, 9000 U l^{-1} reactor) and immobilized (closed circles, 13 000 U l^{-1} reactor) penicillin amidase (PA) from *E. coli* in a stirred batch reactor. The pH in the reaction solution was kept constant by titration with 3 M NH_3. The PA was adsorbed to the bifunctional support (see Table 6.4) phenylbutylamine-Eupergit (PBA-Eupergit, average particle radius 100 μm) and cross-linked with glutaraldehyde. The concentration of active immobilized enzyme, determined by active-site determination, was 3.4 mg ml^{-1} wet support (or 130 000 U l^{-1} wet support). The penicillin G concentration was determined by HPLC.

inhibition by phenylacetic acid, and $D'_S \approx 5 \times 10^{-6}$ cm^2 s^{-1}, the square of the Thiele modulus becomes <1, as for this process the apparent K_m value $K_m(1 + [S]_0/2K_i)$ can be used (see Section 10.4.2). From Figure 10.8 it follows that the operational effectiveness factor for 90% substrate conversion is ≈1. The times calculated with these values are

- for the free enzyme, 41 min (experimental value ≈90 min);
- for the immobilized biocatalyst, 29 min (experimental value ≈55 min).

In both cases, the calculated value is about a factor of 2 smaller than the experimental value. For the immobilized biocatalyst, this is partly due to the pH gradient inside the particles (Figure 10.14a) – that is, the V_{max} value used in the calculation is too high. Local pH gradients may also occur with the free enzyme. This can be avoided by using more buffer, but it is not recommended for enzyme processes as the buffer used must ultimately be removed from the product and the wastewater. Other causes for an underestimation of the time required for desired substrate conversion are

- experimental error in the kinetic data used (≥20% in K_m and K_i (Deranleau, 1969) and ≥10% for V_{max});
- the assumptions made for the calculation.

With fewer assumptions and more data, the discrepancy between calculated and experimental values can be reduced (Spieß *et al.*, 1999). However, in the practical design of enzyme processes, such estimations must be performed (as illustrated above) by using simplifying assumptions, together with personal and published data with considerable experimental errors. Thus, a difference of a factor of 2 in the calculated and experimental data is not unexpected. The additional studies required to determine all necessary data for reducing this difference may take longer than the adjustment of the design based on observed differences between calculated and experimental data.

Today, the hydrolysis of penicillin G is increasingly carried out in several sequential or recirculation PB reactors where the pH is adjusted between the beds or the vessel used in the recirculation reactor (see Sections 11.1.2 and 12.3). In this way, shorter hydrolysis times, in order to reduce loss of the unstable product 6-aminopenicillanic acid, can be obtained (see case study in Section 12.3). A higher conversion can be obtained when the hydrolysis is carried out at pH 9 with the more stable penicillin amidase from *A. faecalis* (Figure 2.28).

10.8 Application of Immobilized Biocatalysts for Enzyme Processes in Aqueous Suspensions

Many substrates and products used or produced in enzyme processes are slightly soluble in water. In Section 2.9.1, it was shown that processes with such compounds can be carried out with free enzymes and high *STY* in aqueous suspensions or emulsions.

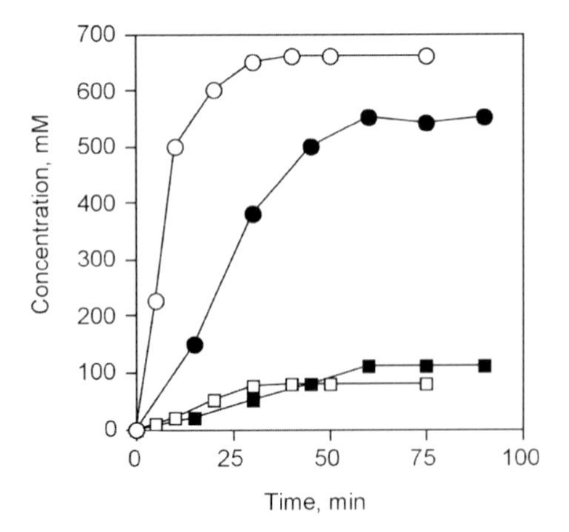

Figure 10.17 Kinetically controlled synthesis of *N*-acetyl-Tyr-Arg-NH_2 (circles) in aqueous suspensions with precipitated substrate *N*-acetyl-Tyr-ethyl ester (ATEE), with free (open symbols) and immobilized (closed symbols) α-chymotrypsin at 25 °C. Starting conditions: pH 9.0 (carbonate buffer, $I = 0.2$ M), 750 mM ATEE and 1000 mM Arg-NH_2; free enzyme 20 μg ml^{-1}; immobilized enzyme 200 μl ml^{-1} immobilized in Sepharose (20 μM in the support). The product *N*-acetyl-Tyr (squares) did not precipitate during the reaction. The pH was kept constant at pH 9.0 during the reaction. The free ATEE concentration was ≤20 mM (Kasche and Galunsky, 1995).

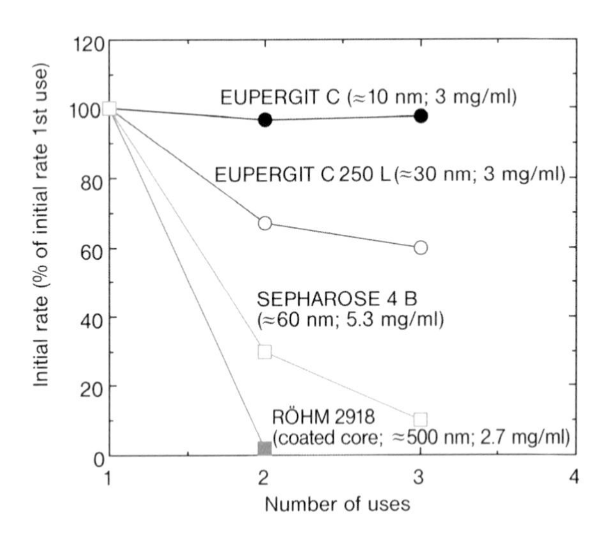

Figure 10.18 Influence of product precipitation on the initial rate as a percentage of rate for the first use for the hydrolysis of 400 mM (*R*)-phenylglycine amide at pH 7.5 and 25 °C catalyzed by penicillin amidase from *E. coli* immobilized in supports with different radii. The support pore radius and the immobilized enzyme concentration are given in the parentheses after the support (Kasche and Galunsky, 1995).

Figure 10.17 shows that such processes with suspended substrate can also be carried out with immobilized enzymes (Kasche and Galunsky, 1995; Youschko *et al.*, 2002). Indeed, such processes can even be carried out with immobilized enzymes when the product precipitates, although in this situation particles with narrow pores must be used to prevent precipitation of the product in the pores of the immobilized biocatalyst particle, as this would limit reuse of the immobilized biocatalyst (Figure 10.18, see also Figure 2.23).

10.9
Improving the Performance of Immobilized Biocatalysts

In Section 10.4, it was shown that concentration gradients of substrates and products are formed in immobilized biocatalyst particles when the effectiveness factors are smaller than 1. In processes where H^+ is formed or produced, a pH gradient is formed, even in immobilized biocatalysts where the effectiveness factors are ≈ 1 (Figure 10.14). The problems caused by such gradients or electric double layers (see Section 10.6.2) in different enzyme processes are summarized in Table 10.3. The reader should try to suggest how the stated problem can be resolved.

The concentration gradients can be reduced by decreasing the Thiele modulus. This can be achieved by using smaller particles or by reducing the diffusion distance between the flowing free solution and the immobilized enzymes. The latter has been achieved in perfusible particles mainly developed for chromatography (Afeyan *et al.*, 1990), though until now these have been used only minimally in enzyme technology.

Recently, however, interest has increased in the study of small, especially nonporous magnetic and nonmagnetic particles in the micrometer and nanometer ranges (Jia, Zhu, and Wang, 2003; Bozhinova *et al.*, 2004; see also Section 8.2.3). Such nonporous and perfusible particles can be used for the selective hydrolysis of biopolymers with hydrolases, as the biopolymers can easily diffuse to the enzymes (Xu *et al.*, 2007). Examples for their potential application are the proteolytic processing of recombinant pro-proteins to the biologically active protein, such as the processing of recombinant human pro-insulin to insulin.

Magnetic particles can be easily separated from the reaction mixture at the end of the reaction. Larger *STY* values can be obtained with smaller porous particles (see Figure 10.5), but these cannot be used in PB reactors, due to the increased pressure drop. Another limit of small particles for use in PB reactors is shown in Figure 10.19. The total surface area available for convective mass transfer is reduced with decreasing particle radius, due to the increasing fraction of the total outer surface occupied by the particle–particle contact areas, where only diffusive mass transfer is possible. In packed beds, there are more than 10 such direct particle–particle contact areas (Hinberg, Korus, and O'Driscoll, 1974; Andersson and Walters, 1986; Renken, 1993).

Table 10.3 Problems caused by concentration gradients in enzyme processes with immobilized biocatalysts, and how they can be solved.

Process	Problems compared to systems with free biocatalyst	Examples	How to minimize the problem
I. Equilibrium-controlled process			Fill out!
1. Hydrolysis involving uncharged substrates/products	Increased product inhibition Reduced rate Increased (equilibrium control) or decreased (kinetic control) formation of by-products Electrical double layer with charged supports	Di-, oligosaccharide hydrolysis	
2. Hydrolysis involving acids and bases	Formation of pH gradients Decreased rate, biocatalyst stability, and yield See 1	Hydrolysis of antibiotics (cephalosporin, penicillin G and V), peptides, proteins, lipids, nucleic acids, and so on	
3. Kinetic resolution of racemates	See 1 and 2 Decreased steric purity of product	Industrial racemate resolutions	
II. Kinetically controlled process			
1. Synthesis of condensation products (antibiotics, peptides, etc.)	Decreased yield See I: 1 and 2	Synthesis of β-lactam antibiotics (cephalosporin, penicillin), peptides	
2. Kinetic resolution of racemates	Decreased steric purity and yield See I: 1–3		

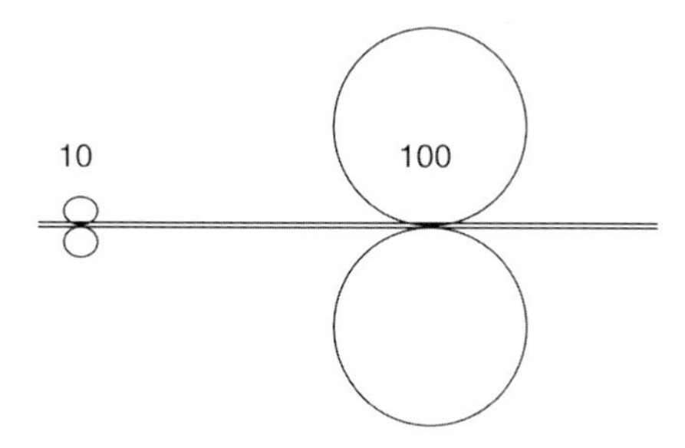

Figure 10.19 The fraction of a particle–particle contact area with a distance $<1\,\mu m$ (between the horizontal lines) of the total particle surface increases, when the particle diameter decreases. This affects the effective particle surface for convective mass transfer when the particle radius becomes less than $50\,\mu m$ (Renken, 1993).

Exercises

10.1 How can the substrate consumption rate in reactions catalyzed by an immobilized biocatalyst be measured in the free solution?

10.2 Define the stationary and operational effectiveness factors. How can they be determined?

10.3 Derive Eq. (10.12) for spherical particles.

10.4 What particle ranges are suitable for spherical immobilized biocatalysts (see Figure 10.5 and Section 10.9)? Why?

10.5 How can you increase the effectiveness factor of a given immobilized biocatalyst?

10.6 Explain the following results:

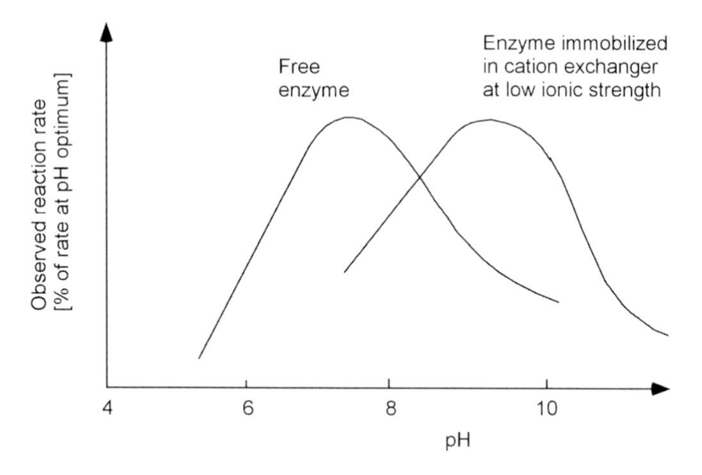

The pH is measured in the solution outside the particles. Why is $\eta > 1$ at some values of pH? What happens at higher ionic strengths?

10.7 Try to explain the influence of the effectiveness factors in Eq. (10.27).

10.8 How are Eqs. (10.20) and (10.24) influenced by competitive product inhibition? How can this be minimized in the case of the product H^+?

10.9 Why does the space–time yield of immobilized biocatalysts have an upper limit? (*Hint*: See Figure 10.1).

10.10 Explain the following graph for an immobilized biocatalyst:

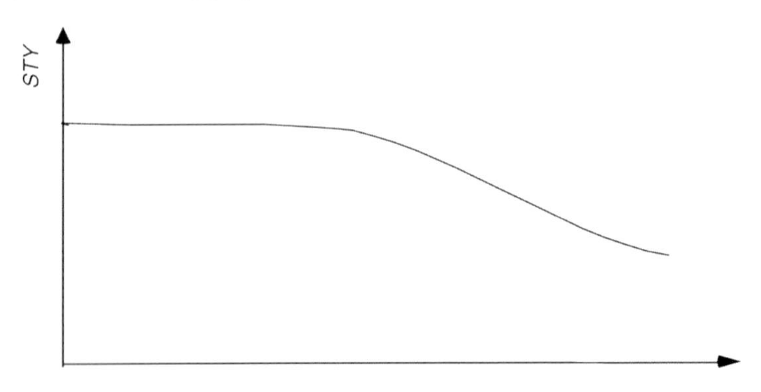

10.11 How can the *STY* be kept constant in an enzyme process with several reactors? The substrate content and the degree of substrate conversion cannot be changed.

10.12 Derive Eq. (10.31).

10.13 Determine the synthesis/hydrolysis selectivity for the free and immobilized enzyme in Figure 10.17. Why do they differ?

10.14 In the hydrolysis of maltose by glucoamylase, isomaltose is a by-product formed in the equilibrium-controlled reaction of the product:

$$\text{glucose} + \text{glucose} \rightleftharpoons \text{isomaltose}$$

In the hydrolysis of lactose by β-galactosidase, oligosaccharides formed in the kinetically controlled reaction between the intermediate product galactosyl-enzyme and the substrate lactose (Figure 1.9) are by-products. The selectivity of this process is assumed not to be changed upon immobilization of the enzyme.

Discuss how the by-product formation is influenced by the immobilization of the enzyme for the above enzyme processes. The same enzyme activity is used in the reactors with free and immobilized biocatalysts.

10.15 Precursors for the neuropeptide kyotorphin Tyr-Arg can be synthesized in the kinetically controlled process

$$\begin{aligned}&\text{X-Tyr-OEt} + \text{Arg-NH}_2 \text{ (or Arg-OEt)}\\&\quad\rightarrow \text{X-Tyr-Arg-NH}_2 \text{ (or X-Tyr-Arg-OEt)} + \text{EtOH}\end{aligned}$$

in aqueous suspension (X = acetyl, X-Tyr-OEt is suspended (Figures 2.23, 10.17 and 10.18) or in free aqueous solution (X = maleyl that increases the solubility in water, Section 4.2.3.3) catalyzed by the free or immobilized endopeptidase α-chymotrypsin.

a. Compare and discuss the advantages and disadvantages of these enzyme processes. (*Hints*: How can the products and by-products be isolated from the remaining substrates and the enzyme? Esters are unstable in water at high pH. EtOH destabilizes enzymes.)
b. How can kyotorphin be produced from the precursors formed in (a)?
c. What properties must an enzyme have to be a suitable biocatalyst for the production of kyotorphin with minimal by-product formation from the unprotected amino acids? (*Hints*: Figures 2.3–2.5 and Table 2.4.) How can the product be isolated from the remaining substrates and the enzyme (*Hint*: Isoelectric point of kyotorphin, Tyr, and Arg (see Section 6.5)?

10.16 The cancerogenic compound ethyl carbamate ($H_2N\text{-}CO\text{-}O\text{-}C_2H_5$) is formed in the following reaction between ethanol and urea

$$C_2H_5OH + H_2N\text{-}CO\text{-}NH_2 \rightarrow H_2N\text{-}CO\text{-}O\text{-}C_2H_5 + NH_3$$

Many countries have therefore set upper limits (voluntary as 15 ppb in the United States, or compulsory as 30 ppb in Canada) of ethyl carbamate in alcoholic beverages. Several methods exist to reduce the ethyl carbamate content in these (see the home page of FDA (www.FDA.gov) and search for ethyl carbamate). One is to reduce the urea content in these beverages by hydrolysis with urease (EC 3.5.1.5) that catalyzes the following reaction $H_2N\text{-}CO\text{-}NH_2 + H_2O \Leftrightarrow 2NH_3 + CO_2$. In this process, no other compounds except urease may be added to the beverages, and the urease must be removed after its use. In the European Union, the use of urease for this purpose is regulated (see http://eur-lex.europa.eu/LexUriServ/LexUriServ.do?uri=OJ:L:2008:127:0013:0013:EN:PDF). Therefore, the use of immobilized urease is the first choice for this process in which the urea content should be reduced below 0.01 mM. The pH of wine is in the range of 3–4, and its chemical composition is as follows: $\approx$12% ethanol, 1% glucose, 1.5% organic acids (acetic, citric, malic, tartaric, etc.), 1% inorganic ions (metal ions, F^-, etc.), 1% phenols (tannins and other polyphenols), and up to 0.1 mM urea. Many of these inhibit the enzyme (in order of importance: F^-, malate, ethanol, and the mainly uncharged phenolic compounds; Andrich, Esti, and Mortesi, 2009). Which urease would you use for this process? Which immobilization support would you use for the covalent immobilization of the enzyme? Which charge on the support should be minimized to reduce the inhibition due to the inhibitors mentioned above? (*Hints*: Use BRENDA, and the information in Chapter 8 on covalent immobilization, and the inactivation of nonreacted activated groups on the support after the immobilization of the enzyme.)

References

Afeyan, N.B., Gordon, N.F., Mazsaroff, I., Varady, L., Fulton, S.P., Yang, Y.B., and Regnier, F.E. (1990) Flow-through particles for the HPLC separation of biomolecules: perfusion chromatography. *J. Chromatogr.*, **519**, 1–29.

Aguade, J., Romero, M.D., Rodríguez, L., and Calles, J.A. (1995) Thermal deactivation of free and immobilized β-glucosidase from *Penicillium funiculosum*. *Biotechnol. Prog.*, **11**, 104–106.

Andersson, D.J. and Walters, R.R. (1986) Equilibrium and rate constants of immobilized concanavalin A determined by high-performance affinity chromatography. *J. Chromatogr.*, **376**, 69–105.

Andrich, L., Esti, M., and Mortesi, M. (2009) Urea degradation in model wine solutions by free and immobilized acid urease in a stirred bioreactor. *J. Agric. Food Chem.*, **57**, 3533–3542.

Anonymous (1983) Guidelines for the characterization of immobilized biocatalysts: worked out by the Working Party on Immobilized Biocatalysts within the European Federation of Biotechnology. *Enzyme Microbiol. Technol.*, **5**, 3304–3307.

Antrim, R.L. and Auterinen, A.-L. (1986) A new regenerable immobilized glucose isomerase. *Stärke*, **38**, 132–137.

Bonielli, C., Mayr, T., Klimant, I., Koenig, B., Riethorst, W., and Nidetzky, B. (2010) Intraparticle concentration gradients for substrate and acidic product in immobilized cephalosporin C amidase and their dependencies on carrier characteristics and reaction parameters. *Biotechnol. Bioeng.*, **106**, 528–540.

Bozhinova, D., Galunsky, B., Yueping, G., Franzreb, M., Köster, R., and Kasche, V. (2004) Evaluation of magnetic polymer micro-beads as carriers for immobilized biocatalysts for selective and stereoselective transformations. *Biotechnol. Lett.*, **26**, 343–350.

Buchholz, K., Ehrenthal, E., Gloger, M., Hennrich, N., Jaworek, D., Kasche, V., Klein, J., Krämer, D., Kula, M.R., Manecke, G., Palm, D., Scchlünsen, J., and Wagnmer, F. (1979) *Characterization of Immobilized Biocatalysts*, DECHEMA Monograph, vol. **84** (ed. K. Buchholz), Wiley-VCH Verlag GmbH, Weinheim, pp. 1–48.

Carleysmith, S.W., Dunnill, P.D., and Lilly, M.D. (1980) Kinetic behaviour of immobilized penicillin amidase. *Biotechnol. Bioeng.*, **22**, 735–756.

Chikere, A.C., Galunsky, B., Schünemann, V., and Kasche, V. (2001) Stability of immobilised soybean lipoxygenases: influence of coupling conditions on the ionisation state of the active site Fe. *Enzyme Microb. Technol.*, **210**, 168–175.

Damköhler, G. (1937) Influence of diffusion, fluid flow, and heat transport on the yield in chemical reactors. *Chem. Ing.*, **3**, 359–485; translated in *Int. Chem. Eng.*, 1988, **28**, 132–198.

Deranleau, D.A. (1969) Theory of the measurement of weak molecular complexes. I. General considerations. *J. Am. Chem. Soc.*, **91**, 4044–4049.

Ferreira, L., Ramos, M.A., Dordick, J.S., and Gil, M.H. (2003) Influence of different silica derivatives in the immobilization and stabilization of a *Bacillus licheniformis* protease (subtilisin Carlsberg). *J. Mol. Catal. B*, **21**, 189–199.

Fujii, R., Utsunomiya, Y., Hiratake, J., Sogabe, A., and Sakata, K. (2003) Highly sensitive active-site titration of lipase in microscale culture media using fluorescent organophosphorus ester. *Biochim. Biophys. Acta*, **1631**, 197–205.

Gabel, D. (1974) Active site titration of immobilized chymotrypsin with a fluorogenic reagent. *FEBS Lett.*, **49**, 280–281.

Gardossi, L., Poulsen, P., Ballesteros, A., Hult, K., Švedas, V., Vasić-Rački, D., Carrea, G., Magnusson, A., Schmid, A., Wohlgemuth, R., and Halling, P. (2010) Guidelines for reporting of biocatalytic reactions. *Trends Biotechnol.*, **28**, 171–180.

Goldstein, L., Levin, Y., and Katchalski, E. (1964) A water-insoluble polyanionic derivative of trypsin. II. Effect of the polyelectrolyte carrier on the kinetic behavior of the bound trypsin. *Biochemistry*, **3**, 1913–1919.

González-Caballero, B. and Shilov, V.N. (2002) Electric double layer at a colloid particle, in *Encyclopedia of Surface and Colloid Science*, Marcel Dekker, New York.

Haagensen, P., Karlsen, L.G., Petersen, J., and Villadesen, J. (1983) The kinetics of penicillin-V deacylation on an immobilized enzyme. *Biotechnol. Bioeng.*, **25**, 1873–1895.

Heinemann, M., Wagner, T., Doumèche, B., Ansorge-Schumacher, M., and Büchs, J. (2002) A new approach for the spatially resolved qualitative analysis of the protein distribution in hydrogel beads based on confocal laser scanning microscopy. *Biotechnol. Lett.*, **24**, 845–850.

Hinberg, I., Korus, R., and O'Driscoll, K.F. (1974) Gel-entrapped enzymes: kinetic studies of immobilized β-galactosidase. *Biotechnol. Bioeng.*, **16**, 943–963.

Hunter, R.J. (1993) *Foundations of Colloid Science*, Oxford University Press, Oxford.

Janssen, M.H., van Langen., L.M., Pereira, S.R.M., van Rantwijk, R., and Sheldon, R.A. (2002) Evaluation of the performance of immobilized penicillin G acylase using active-site titration. *Biotechnol. Bioeng.*, **78**, 425–432.

Jia, H., Zhu, G., and Wang, P. (2003) Catalytic behaviors of enzymes attached to nanoparticles: the effect of particle mobility. *Biotechnol. Bioeng.*, **84**, 406–414.

Kallenberg, A.I., van Rantwijk, F., and Sheldon, R.A. (2005) Immobilization of penicillin G acylase: the key to optimum performance. *Adv. Synth. Catal.*, **347**, 905–926.

Kasche, V. (1983) Correlation of theoretical and experimental data for immobilized biocatalysts. *Enzyme Microb. Technol.*, **5**, 2–14.

Kasche, V. (1986) Mechanism and yields in enzyme catalyzed equilibrium and kinetically controlled synthesis of β-lactam antibiotics peptides and other condensation products. *Enzyme Microb. Technol.*, **10**, 4–16.

Kasche, V. and Bergwall, M. (1973) *Insolubilized Enzymes* (eds M. Salmona, C. Saronio, and S. Garattini), Raven Press, New York, pp. 77–86.

Kasche, V. and Galunsky, B. (1995) Enzyme-catalyzed biotransformations in aqueous two-phase systems with precipitated substrate and/or product. *Biotechnol. Bioeng.*, **45**, 261–267.

Kasche, V., Galunsky, B., and Michaelis, G. (1991) Binding of organic solvent molecules influences the $P'_1 - P'_2$ stereo- and sequence specificity of α-chymotrypsin in kinetically controlled peptide synthesis. *Biotechnol. Lett.*, **13**, 75–100.

Kasche, V., Gottschlich, N., Lindberg, Å., Niebuhr-Redder, C., and Schmieding, J. (1994) Perfusible and non-perfusible supports with monoclonal antibodies for biospecific purification of *E. coli* penicillin amidase within its pH-stability range. *J. Chromatogr.*, **660**, 137–145.

Kasche, V. and Kuhlmann, G. (1980) Direct measurements of the thickness of the unstirred diffusion layer outside immobilized biocatalysts. *Enzyme Microb. Technol.*, **2**, 309–312.

Katchalski-Katzir, E. and Krämer, D. (2000) Eupergit® C, a carrier for immobilization of enzymes of industrial potential. *J. Mol. Catal. B*, **10**, 157–176.

Lummer, K., Riecks, A., Galunsky, B., and Kasche, V. (1999) pH-dependence of penicillin amidase enantioselectivity for charged substrates. *Biochim. Biophys. Acta*, **1433**, 327–334.

Mateo, C., Palomo, J.M., Fernandez-Lafuente, G., Guisan, J.M., and Fernadez-Lafuente, R. (2007) Improvement of enzyme activity, stability and selectivity via immobilization techniques. *Enzyme Microbiol. Technol.*, **40**, 1451–1463.

McLaren, A.D. and Packer, L. (1970) Some aspects of enzyme reactions in heterogeneous systems. *Adv. Enzymol.*, **33**, 245–303.

Monti, D., Carrea, G., Riva, S., Baldaro, E., and Frare, G. (2000) Characterization of an industrial biocatalyst: immobilized glutaryl-7-ACA acylase. *Biotechnol. Bioeng.*, **70**, 239–244.

Palomo, J.M., Muñoz, G., Fernándes-Lorente, G., Mateo, C., Fuentes, M., Guisan, J.M., and Fernándes-Lafuente, R. (2003) Modulation of *Mucor miehei* lipase properties via directed immobilization on different hetero-functional epoxy resins. Hydrolytic resolution of (*R,S*)-2-butyroyl-2-phenylacetic acid. *J. Mol. Catal. B*, **21**, 201–210.

Park, S.W., Choi, S.Y., Chung, K.H., Hong, S.I., and Kim, S.W. (2002) Characteristics of GL-7-ACA acylase immobilized on silica gel through silanization. *Biochem. Eng. J.*, **11**, 87–93.

Rashevsky, N. (1940) *Mathematical Biophysics*, The University of Chicago Press, Chicago, IL.

Renken, E. (1993) *Signalentstehung und Signalentwicklung in fluorimetrischen dynamischen und Affinitäts-Durchfluß-*

Biosensoren. Dissertation, TU Hamburg-Harburg.

Schellenberger, V., Jakubke, H.D., and Kasche, V. (1991) Electrostatic effects in the alpha-chymotrypsin-catalyzed acyl transfer. II. Efficiency of nucleophiles bearing charged groups in various locations. *Biochim. Biophys. Acta*, **1078**, 8–11.

Spieß, A. and Kasche, V. (2001) Direct measurement of pH profiles in immobilized enzymes during kinetically controlled synthesis using CLSM. *Biotechnol. Prog*, **17**, 294–303.

Spieß, A., Schlothauer, R., Hinrichs, J., Scheidat, B., and Kasche, V. (1999) pH gradients in heterogeneous biocatalysts and their influence on rates and yields of the catalysed processes. *Biotechnol. Bioeng.*, **62**, 267–277.

Svedas, V.K., Margolin, A.L., Sherstiuk, S.F., Klyosov, A.A., and Berezin, I.V. (1977) Inactivation of soluble and immobilized penicillin amidase from *E. coli* by phenylmethylsulphonylfluoride: kinetic analysis and titration of the active sites. *Bioorg. Khim.*, **3**, 546–553.

Thiele, E.W. (1939) Relations between catalytic activity and size of particle. *Ind. Eng. Chem.*, **31**, 916–920.

Tischer, W., Giesecke, U., Lang, G., Röder, A., and Wedekind, F. (1992) Biocatalytic 7-aminocephalosporanic acid production. *Ann. N. Y. Acad. Sci.*, **613**, 502–509.

Tischer, W. and Kasche, V. (1999) Immobilized enzymes: crystals or carriers? *Trends Biotechnol.*, **17**, 326–335.

Ulbrich, R., Golbik, R., and Schellenberger, A. (1991) Protein adsorption and leakage in carrier–enzyme systems. *Biotechnol. Bioeng.*, **37**, 280–287.

Villadsen, J. and Michelsen, M.L. (1978) *Solution of Differential Equation Models by Polynomial Approximation*, Prentice-Hall, Englewood Cliffs, NJ.

Warburg, O. (1923) Experiments on surviving carcinoma tissue. *Biochem. Z.*, **142**, 317–333.

Weisz, P.B. (1972) Diffusion and chemical reaction. *Science*, **179**, 433–440.

Wiesemann, T. (1991) *Enzymmodifikationen für analytische und präparative Zwecke: natürliche und künstliche Penicillinamidase-Varianten*. Dissertation, TU Hamburg-Harburg.

Xu, X., Deng, C., Yang, P., and Zhang, X. (2007) Immobilization of trypsin on supermagnetic nanoparticles for rapid and effective proteolysis. *J. Proteom. Res.*, **9**, 3849–3855.

Youschko, M.I., van Langen, L.M., de Vroom, E., van Rantwijk, F., Sheldon, R.A., and Svedas, V.K. (2002) Penicillin acylase-catalyzed ampicillin synthesis using a pH gradient: a new approach to optimization. *Biotechnol. Bioeng.*, **78**, 589–593.

Zavrel, M., Michalik, C., Schwendt, T., Schmidt, T., Ansorge-Schumacher, M., Janzen, C., Marquardt, W., Büchs, J., and Spiess, A.C. (2010) Systematic determination of intrinsic reaction parameters in enzyme immobilizates. *Chem. Eng. Sci.*, **65**, 2491–2499.

Zeldovich, Ya.B. (1939) The theory of reactions on powders and porous substances. *Acta Physicochim. URSS*, **10**, 583–592.

11
Reactors and Process Technology

General aims	Rational process development; economic and sustainable process design
Reactor types	Selection according to mass balances for the reaction
Standard reactors	Stirred tank and tubular reactor
Special reactor types	Reactors with recirculation and fluidized bed reactors for wastewater treatment
Process fundamentals	Measurement and calculation of residence time distribution; mixing; pressure drop; mass transfer; and energy requirement
Process technology: upstream- (substrate-) and downstream operations (product isolation and purification)	Unit operations: mixing, heat exchange, separation techniques

11.1
General Aspects, Biochemical Engineering, and Process Sustainability

The application of biocatalysts in industrial processes aims at the synthesis of valuable products or at the conversion and/or degradation of compounds in environmental technology. This must consider both economic and ecological conditions applying rational reaction and process design. To achieve these goals, biochemical engineering aims at the quantitative investigation of biotransformations and modeling of the processes, the development of bioreactors and downstream operations, and the transformation of laboratory results and theoretical

Biocatalysts and Enzyme Technology, Second Edition. Klaus Buchholz, Volker Kasche, and Uwe T. Bornscheuer

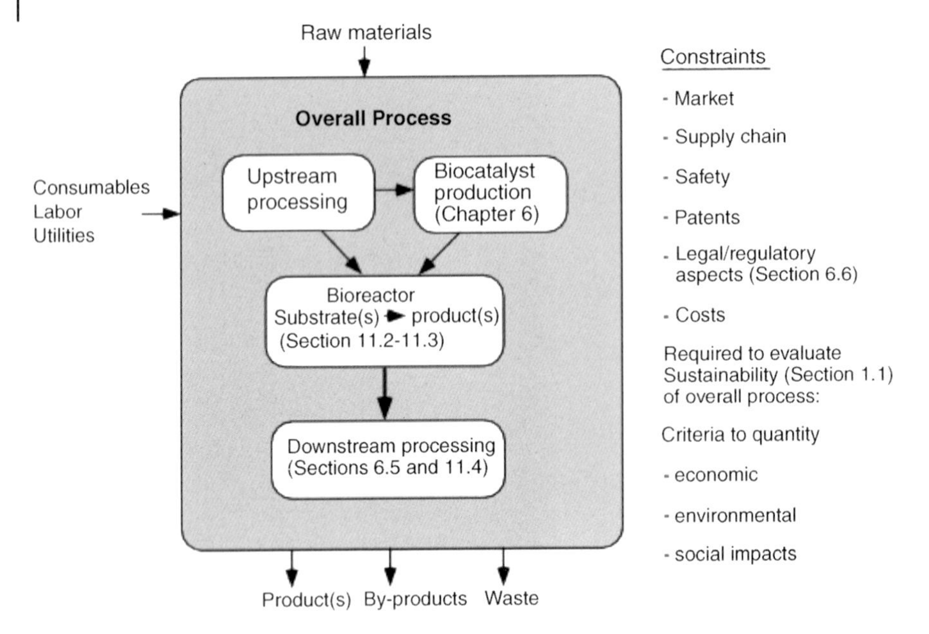

Figure 11.1 Integration of different process steps and external constraints that must be considered in the design of a new or improved process to produce a new or existing product. Consumables are materials or equipment parts that must be replaced from time to time (such as, filtration mermbranes, chromatographic resins, etc.). Energy consumed in the process by utilities is provided mainly by electricity, steam, and water.

approaches (models) into technical dimensions (scale-up) (Heinzle *et al.*, 2006; Hempel, 2007; Weuster-Botz and Hekmat *et al.*, 2007; Villadsen *et al.*, 2011).

Every enzyme process is integrated into an overall process as shown in Figure 11.1. From this follows that the design of an enzyme process, as shown in Figure 1.10 (Section 1.6.2), must be integrated at an early stage into the design of the overall process. This requires a close cooperation between research, development, and production departments. Figure 11.1 applies for both the design of a new production process and the improvement of an existing process to produce a new or existing product. The biochemical engineering and sustainability aspects for this are outlined in Section 11.1. The properties and selection of suitable reactors for an enzyme process are dealt with in Sections 11.2 and 11.3, and relevant aspects of process integration are covered in Section 11.4.

11.1.1
Biochemical Engineering Aspects

Tools for reaction and process development are stoichiometry, thermodynamics and kinetics, and reactor design. Stoichiometry is the basis for quantitative analysis of chemical and biochemical reactions (Nielsen *et al.*, 2003; Heinzle *et al.*, 2006). It involves relating the quantities of the reactants (substrates) to products that are formed. Thermodynamics provide data for equilibria of reactions, and exo- or endothermal process steps, and thus maximal product yields for given conditions (concentrations, temperature, and pH) (see Section 2.6) with minimal amounts of by-products formed. Kinetics define the time and/or biocatalyst amount required for a conversion as well as yield required for economic and ecological reasons (Chapter 2; Section 2.7). It provides the calculation basis for achieving short reaction time in an optimal reactor configuration, under conditions of sufficient biocatalyst stability, taking into account inactivation kinetics.

The bioreactor represents the central part of the plant as a whole, where the reaction takes place under controlled conditions. Reactor types and important properties of reactors are covered in Sections 11.2 and 11.3.

Despite the reactor occupying the central position, the equipment prior and subsequent to it for substrate preparation, product isolation, and purification up to confectioning contributes in terms of the overall cost of the process. Thus, up- and downstream processing provides for substrate(s) and product(s) of adequate quality and purity. Integrated processes take into account optimal conditions for reaction design and steps of up- and downstream operations. This notably aims at high yields at unfavorable equilibria by, for example, continuous product removal (Section 11.4.2). Likewise, fittings, pipes, and notably instrumentation for measurement and control, as well as data collection and processing, require high expenditure. Optimal integration of all these components forms the basis of process technology, and this is described briefly in Section 11.4.

Process sustainability takes into account both environmental and economic aspects, including minimal formation of by-products, recycling of by-products, and materials required, such as solvents, avoiding toxic substances, or adequate containment. Disposal of residues (e.g., wastewater, exhaust gas, etc.) must also be considered (Section 11.1.2). A detailed analysis of these topics is outside the scope of this book (for details, see Heinzle *et al.*, 2006).

Enzyme processes are integrated as one part of a process chain, which is of major importance for their design. To illustrate this, four case studies have been included in Chapter 12. These are as follows:

Integration of enzymatic steps into starch processing and the manufacture of high fructose corn syrup (HFCS) with up- and downstream operations, which are dealt with in Section 12.1.

Biofuel manufacture both from starch and from lignocellulosic biomass is discussed in Section 12.2.

As a newer enzyme process, the production of 7-aminocephalosporanic acid (7-ACA), a precursor for semisynthetic cephalosporin antibiotics, is covered in Section 12.3. The selection of the suitable reactor is essential to optimize the product yield and to minimize by-product formation. As enzyme processes are important in the production of β-lactam antibiotics (cephalosporins and penicillins), a short overview of this is also given.

Biocatalytic processes for the synthesis of the lipitor side chain are presented in Section 12.4.

11.1.2
Process Sustainability and Ecological Considerations

The now worldwide accepted definition of sustainable development has been outlined in the beginning of this section. It has been accepted by more than 100 nations. To evaluate the sustainability of a production process, its economic, environmental, and social impact must be quantified. Sustainability or sustainable development has been defined as the development that meets the needs of the present without compromising the ability of the future generations to meet their own needs. Others define it as the optimal growth path that maintains economic development while protecting the environment and optimizing the social conditions, relying on limited, exhaustible natural resources. Sustainable processes should consume less raw materials, be more energy efficient, and eco-friendly, compared to traditional ones, provide for high atom efficiency, thus produce low amounts of waste, and avoid toxic reagents and hazardous solvents. They should consider three dimensions: economic, environmental, and social development (Heinzle *et al.*, 2006, pp. 81–115; Wenda *et al.*, 2011). Example for improved processes under these aspects is 6-APA production (Section 1.4, Figure 1.7, and Section 12.3).

The *economic* impact *assessment* requires a cost and profitability analysis, including the estimation of capital investment and operating cost, that can be discussed only very briefly (for details, see Atkinson and Mavituna, 1991a; Heinzle *et al.*, 2006; Sheldon, 2011). The capital investment should be based on a process flow diagram with a list of the essential equipment for the process, including fermenter, media and product tanks, piping, downstream operation equipment, instrumentation, and control facilities. A multiplier covers further, for example, planning cost. The operating cost include raw materials, consumables, utilities, waste, and labor. Increasing the resource (energy, materials) efficiently by reducing the resource consumption per mass unit product reduces the operating costs. This is also required indirectly by current national legislation and international conventions (Kyoto and Biodiversity Convention, see Section 6.6) that require that the simultaneously produced waste must be reduced, for its negative influence on the environment. In many countries, current legislation also implies that a company must cover all costs of product safety and treatment of the waste it produces, to meet the limits set by government regulations. The main actions to reduce the waste treatment cost are to design a process where

- waste production is minimized,
- unavoidable waste is recycled as much as possible, and
- hazardous waste production is avoided.

As this is mainly required due to the environmental impact of waste, some quantitative criteria to evaluate this are given in the following paragraph.

The *environmental impact assessment* relies on two aspects, the process characteristics and the components properties, which result in environmental indices (EI).

The quantitative assessments start with the following mass indices: the *E*-factor and the mass, or intensity, index (MI), defined as (Sheldon, 2011):

$$E = \frac{\text{Total mass of waste}}{\text{Total mass of product}} \quad \text{and} \quad \text{MI} = \frac{\text{Total mass used in a process}}{\text{Total mass of product}}$$

The mass index can also be defined separately for the compounds entering and leaving the process (Heinzle *et al.*, 2006). The incoming mass index MI_{in} equals MI, and the outcoming $\text{MI}_{\text{out}} = E$. Both these mass indices should be used to evaluate environmental impacts as *E* evaluates only the influence of the compounds after the process. Different incoming substrates have different environmental impacts, as illustrated by the alternative substrates, cellulose and starch, for biofuel production. These mass indices are necessary but not sufficient for the quantitative evaluation of the environmental impact. Not all components have the same environmental impact. Their impact is given by environmental factors (EFs; Heinzle *et al.*, 2006) or environmental hazardous quotients *Q* (Sheldon, 2011; Van Aken *et al.*, 2006). Different properties (with high, medium, or low relevance) are used to estimate EF or *Q* of a component. These include raw material availability, biodegradability, acute and chronic toxicity, ecotoxicity, effect on global warming, acidification and eutrophication potential, competing uses of biomass (food–feed–fuel conflict), and so on. The total environmental impact (EI) is then obtained by multiplying the mass index, or *E*-factor contribution of each incoming or outcoming component of a process, with their EF or *Q*-values. Such calculations to compare the sustainability of different processes to produce a compound have been presented by Heinzle *et al.*, 2006 (pp. 81–115) and Sheldon, 2011. For the hydrolysis of penicillin, using data given in Figure 1.7, EF for the chemical process is 8.5 and for the biocatalytic process 0.34 (Heinzle *et al.*, 2006). The MI for both processes is almost equal when water is included. Without water, the MI for the biocatalytic process is an order of magnitude smaller. Thus, the biocatalytic process is more sustainable than the chemical process. A wide range of typical *E*-factors have been summarized with typical data for bulk chemicals (<1–5), fine chemicals (5–50), and pharmaceuticals (25–100) (Wenda *et al.*, 2011). The determination of the EF and *Q*-values is still under development. This development requires a close cooperation between biotechnical and chemical companies, professional societies, and regulatory bodies (see home pages of European Chemical Agency, a EU organization, EPA, FDA, American Chemical Society, and European Chemical Association, and Section 6.6). Finally, regulatory authorities will determine and update these impact factors.

The importance of evaluating the environmental impact of bio-based products with respect to their entire *life cycle* has been highlighted in a review, demonstrating that the choice of the raw material often turns out to be an important parameter influencing the life cycle performance (Hatti-Kaul *et al.*, 2007).

A range of these aspects have been introduced as aims in the management of development work: minimization of material and energy intensity of products and services, minimization of toxic effects, increase of recycling potential of products, maximal use of renewable resources, and increase of the life cycle of products. Yearly sustainability reports have been published by many biotechnical and chemical companies (see the respective home pages). BASF (Germany) even offers, as a service, the analysis of the eco-efficiency of products and processes (Brinkmann, 2009). A socio-eco-efficiency analysis, considering costs, environmental impact, and social effects can be performed with SEEbalance® developed by BASF. As a comparative life cycle assessment tool, it includes the raw material production as well as the production and use of products, and, in addition, pathways for recycling and disposal (Sustainable development at BASF; www.basf.com/group/sustainability_en/index). Further balance software is available (see (Heinzle *et al.*, 2006); www.chembiotec.de/files/cbt_review_nol_online.pdf).

As an example, the selective oxyfunctionalization by introduction of molecular oxygen into hydrocarbons has been investigated with respect to economic and ecological assessment (Kuhn *et al.*, 2010).

Indicators also play a key role in the *social assessment* of evolving technologies; however, they lack a general consensus. Such indicators have been developed by the Wuppertal Institute (justus.geibler@wupperinst.org) for the biotechnology sector. The political relevance refers to political initiatives such as the sustainability strategies of a government. The entrepreneurial and product relevance has been considered through a survey of biotech companies, including information from rating agencies. Such a survey identifies relevant social aspects, as well as the possible contribution of BT products to the satisfaction of human needs, and challenges and chances in the social field. Finally, eight significant aspects have been identified: health and safety, quality of working conditions, impact on employment policy, education and advanced training, knowledge management, innovative potential, customer acceptance, and societal dialogue (Heinzle *et al.*, 2006).

Manifold interactions exist between the three aspects of sustainability. Almost all environmental categories affect the economic and social sustainability. The raw material availability considers the depletion of natural resources that may lead to a strong increase in input material prices in the medium or long term. Thus, one should consider all three dimensions of sustainability early in process development and be aware of their possible interactions.

11.2 Types of Reactors

The bioreactor represents the central part of the plant as a whole, where the reaction takes place under controlled conditions. Based on chemical reaction engineering with few basic reactor types, a range of special reactor modifications and configurations may be derived that serve the aim to adapt the conversion to optimal conditions. In order to minimize the costs, the choice of the reactor must be based on the following aspects (see also Wenda *et al.*, 2011; Villadsen *et al.*, 2011):

High yield of product (in general >90% of the theoretical yield), at high concentration and high purity – that is, a low concentration of by-products, which are difficult to separate, and additives, such as buffer.

High catalyst productivity and STY (space–time yield) – that is, high effectiveness and good operational stability; STY should be in the range of 100 $gl^{-1}h^{-1}$ for bulk, or >0.1 $gl^{-1}h^{-1}$ for fine chemicals.
Low manufacturing and maintenance costs, as well as low space requirement, which means preferentially one or a few reactors with small volumes and minimal equipment (e.g., fittings, valves, and instrumentation).

As in most cases in technology development, these requirements tend to be contradictory: high conversion and catalyst productivity are more readily realized in reactor cascades compared to a single reactor; a constant and stable – as well as, with reference to reaction conditions, optimal – operation requires robust and multiple analytical instrumentation and control units. Although an economically viable decision can be made only by the manufacturer, the chemist, engineer, or biochemical engineer must elaborate appropriate basic information in terms of calculated and experimentally verified protocols and options.

11.2.1 Basic Types and Mass Balances

Although there are only two basic types of enzyme reactors, namely, the stirred tank and the tubular reactor (TR), combinations to form cascades of reactors play a major role in these processes. In particular, they are applied to experiments and mathematical modeling for the identification of optimal reactor configurations. In industrial practice, the stirred tank reactor (STR) is used in both the nonstationary (batch operation) and the stationary mode (continuous operation). The characteristic differences between the two types result from the correlations of concentrations (of educts and products) and space, as well as (residence) times. This relationship is shown schematically in Figure 11.2. It should be pointed out that while ideal reactors are discussed here, "real" reactors in which behavior deviates from the ideal situation are dealt with in Section 11.3.

Mass balances allow for the calculation of correlations of reactor size, or reaction (residence) time, and the amount of catalyst and required conversion. The more simple cases are dealt with subsequently (Reuss and Bajpai, 1991; Cabral and Tramper, 2000; Prenosil, *et al.*, 2009; for scale-up, see Kossen and Oosterhuis, 1985). Thus, the mass balances of the two limiting cases – the continuous stirred tank reactor (CSTR) and the continuous tubular reactor (see Figure 11.2) – are discussed at this point. The discontinuous stirred tank reactor (batch process) is included insofar as the mathematical treatment of its mass balance is equivalent to that of the continuous tubular reactor. For simplicity, it is assumed that parameters such as pH, temperature, and catalyst activity are constant.

The substrate converted ($\Delta[S]$) results from the initial $[S]_0$ and final $[S]_E$ substrate concentrations, ($\Delta[S] = [S]_0 - [S]_E$), and depends on

- the amount and activity of the catalyst (which are given by the maximal reaction rate V_{max});

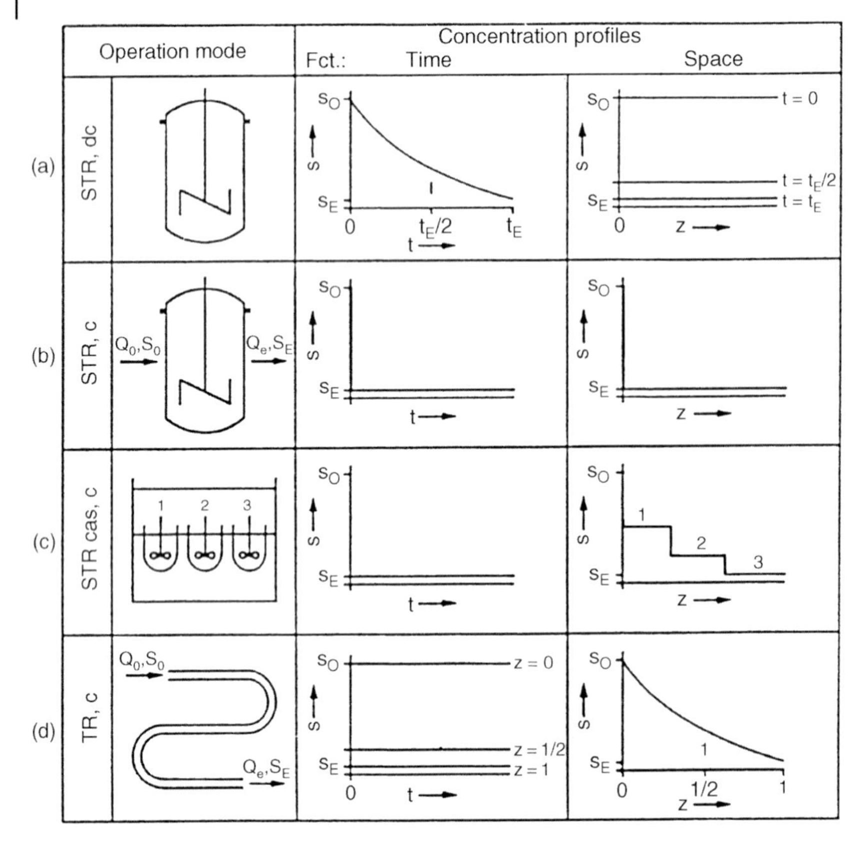

Figure 11.2 Basic (ideal) reactor types and operational modes as well as the respective concentration profiles for a substrate S as a function of (residence) time (t). (a) STR, dc: stirred tank reactor, discontinuous; (b) STR, c: continuous; (c) STR, cas, c: cascade, continuous; (d) TR, c: tubular reactor, continuous (Q = volumetric flow; z = length. Indices: entry = 0; exit = E).

- the reaction time (t) or residence time (τ) in the reactor; and
- the substrate concentration [S], which depends on the reaction time (t) or residence time (τ) (or the length in a tubular reactor).

Together with the substrate concentration [S], both the reaction rate v and the effectiveness factor η will vary, the latter referring to the ratio of the reaction rates of free and immobilized enzyme (see Section 10.3).

For the *ideal continuous stirred tank reactor* (Figure 11.2b), the substrate concentration is constant throughout space and time and is equal to the final concentration. Hence, this type of reactor is less favorable in general, as the reaction rate depends on the final substrate concentration, which should be low. Notably, for reactions with product inhibition it is inappropriate, as the reactor should operate at high product concentration. It may be favorable, however, for reactions with substrate inhibition when operating at the substrate concentration, which results in the maximal reaction rate.

To establish the mass balance of the CSTR in the steady state, the difference in the amounts of substrate introduced to and removed from the reactor (the products of $[S]_0$ and $[S]_E$, respectively, and the volumetric flow rate Q (ls^{-1})) are taken as equal to the substrate converted in the reactor volume V. For the simple case of Michaelis–Menten kinetics, one obtains (Eq. (10.17), Section 10.5.1):

$$Q[S]_0 - Q[S]_E = v([S]) \cdot V = \eta \cdot \frac{V_{max} \cdot [S]_E}{K_m + [S]_E} V. \quad (11.1)$$

In the case of the *discontinuous stirred tank reactor* and the continuous tubular reactor, the substrate concentrations vary with the reaction time (t) or residence time ($\tau = V/Q$), or length (z) in the reactor (see Figures 11.2a and d). The equation for the reaction rate (Eq. (11.2)), therefore, must be integrated in order to obtain the conversion $X = \Delta[S][S]_0^{-1}$ as a function of the residence time (τ) (or reaction time t) (Eqs. (11.3) and (11.4)). If mass transfer plays a role, the operational effectiveness factor η_0 must be introduced (see Section 10.3, Eq. (10.5), and Section 10.5).

$$v([S]) = -\frac{d[S]}{dt} = \eta_0 \frac{V_{max} \cdot [S]}{K_m + [S]}. \quad (11.2)$$

$$-\int_{[S]_0}^{[S]_E} (K_m + [S]) \cdot \frac{d[S]}{[S]} = \eta_0 \cdot V_{max} \cdot \int_0^t dt. \quad (11.3)$$

$$[S]_0 - [S]_E + K_m \cdot \ln([S]_0/[S]_E) = \eta_0 \cdot V_{max} \cdot t. \quad (11.4)$$

The *mass* balance is the basis for the calculation of the reactor volume (or the residence time τ) or the amount of biocatalyst (or the activity V_{max}) required for a certain conversion of a given amount of substrate and reaction time (see Section 10.5). Thus, it is possible to calculate from Eq. (11.4), for a given amount or activity of enzyme (with $V_{max} = k_{cat} \times [E]_0$), the reaction time required for the conversion of a certain amount of substrate (given by $[S]_0$, $[S]_E$). For a given reaction time t, it can also be used to calculate the enzyme activity per volume required for the conversion of a certain amount of substrate. For immobilized systems, the operational effectiveness factor η_0 for the range of concentration considered and the conversion ($X = ([S]_0 - [S]_E)/[S]_0$) must be taken into account.

The general equations for the product ($V_{max} \cdot \tau$) are summarized in Table 11.1, both for the continuous stirred tank and for the tubular reactors, for three different kinetics (simple Michaelis–Menten kinetics, as well as practically relevant cases with inhibition by product or substrate). In this table and in the subsequent equations, the effectiveness factor η is taken as 1. For immobilized biocatalyst, however, the reaction rate v must be multiplied by the effectiveness factor η or η_0 (Section 10.5).

The residence time τ required for the conversion of a component S at the reaction rate v can be calculated from the mass balance of the reactor under consideration.

Table 11.1 Equations for $V_{max}\,\tau$ for the continuous stirred tank and tubular reactors, and for three different kinetics (substrate conversion $X = \Delta\,[S]\,[S]_0^{-1}$; $P = \text{Product} = [S]_0 - [S]_E$; K_I = inhibitor association constant).

Reaction rate ν	Correlation for : $\frac{V_{max}\cdot V}{Q} = V_{max}\cdot\tau$	
	Stirred tank reactor	**Tubular reactor**
Michaelis–Menten		
$V_{max}\frac{[S]}{K_m+[S]}$	$X\cdot\left(\frac{K_m}{1-X}+[S]_0\right)$	$[S]_0\cdot X + K_m\cdot\ln(1-X)$
Competitive product inhibition		
$V_{max}\frac{[S]}{K_m\cdot(1+[P]/K_I)+[S]}$	$[S]_0\cdot X + K_m\frac{X}{(1-X)} + [S]_0\frac{K_m}{K_I}\cdot\frac{X^2}{(1-X)}$	$K_m\cdot(1+[S]_0/K_I)\cdot\ln(1-X)$ $+(1-K_m/K_I)\cdot[S]_0\cdot X$
Substrate inhibition		
$V_{max}\frac{[S]}{K_m+[S]+[S]^2/K_I}$	$[S]_0\cdot X\left[1+\frac{K_m}{[S]_0\cdot(1-X)}+\frac{(1-X)\cdot[S]_0}{K_I}\right]$	$[S]_0\cdot X - K_m\cdot\ln(1-X)+\frac{[S]_0^2}{2K_I}(2X-X^2)$

For the CSTR, when using the mass balance (Eq. (11.1)), the residence time is obtained from

$$\tau = V/Q = \frac{[S]_0 - [S]_E}{v([S])}. \tag{11.5}$$

For the continuous tubular reactor, one obtains for the residence time, with

$$\tau = \int_0^t dt = V/Q, \tag{11.6a}$$

the following correlation:

$$\tau = V/Q = -\int_{[S]_0}^{[S]_E} d[S](\nu([S]))^{-1}. \tag{11.6b}$$

In Figure 11.3, the two correlations for the residence time τ are interpreted by graphical means (according to Reuss, 1991). Figure 11.3a shows the case for a reaction of an order >0. The residence time required, or the reaction volume necessary (with equal amount of catalyst), for a given substrate amount and conversion is shown for the CSTR in terms of a rectangular plane with the coordinates ν^{-1} and [S], with the boundary conditions $[S]_0$ and the outlet concentration $[S]_E$ required (two cases with different outlet concentrations $[S]_1$ and $[S]_2$ are shown). The corresponding residence time for the tubular reactor results from the integral below the correlation given for ν^{-1} and [S], again with the boundary conditions $[S]_0$ and the outlet concentration $[S]_E$. It is obvious that the volume required for the tubular reactor is significantly smaller compared to that of the CSTR, the difference becoming larger with increasing conversion.

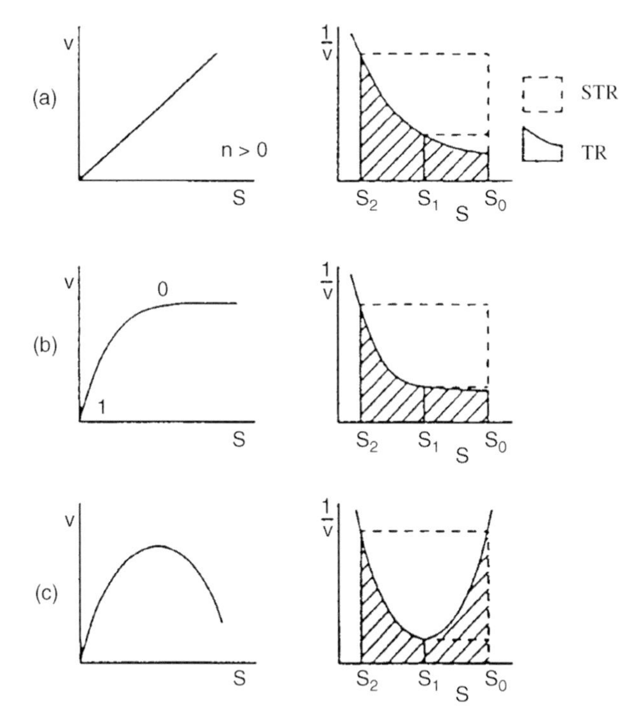

Figure 11.3 Graphical analysis of residence time required for a continuous stirred tank reactor and a tubular reactor, for different types of reactions. For explanation, see text.

The graphs in Figure 11.3b deal with Michaelis–Menten kinetics, which may often be applicable in biocatalysis. Here, it is clear that in the range of zero-order kinetics no difference exists as to the volume required with either a tubular or stirred tank reactor (both initial $[S]_0$ and final $[S]_E$ concentrations $> K_m$). If the final conversion should be higher (final concentration $[S]_2$), the reaction order is in a range of $n > 0$ and the conditions are the same as discussed before.

The graphs in Figure 11.3c show the case of substrate inhibition. The correlation of ν^{-1} and [S] point out that the optimal choice of the reactor depends again on the conversion required. At low conversion (final concentration $[S]_1$), the CSTR is superior, whereas at high conversion (final concentration $[S]_2$), the tubular reactor turns out to be the better choice. The optimal reactor configuration is a combination of a stirred tank as the first and a tubular reactor as the second one.

The following rule can be given for a general comparison of the efficiency of tubular and stirred tank reactors (both operating continuously): reactions with kinetics where the reaction rate decreases with increasing conversion (e.g., Michaelis–Menten kinetics), the tubular reactor provides for higher conversion compared to the stirred tank reactor, when both operate with the same volume and catalyst activity. These conclusions are similar to those elaborated for chemical reaction engineering.

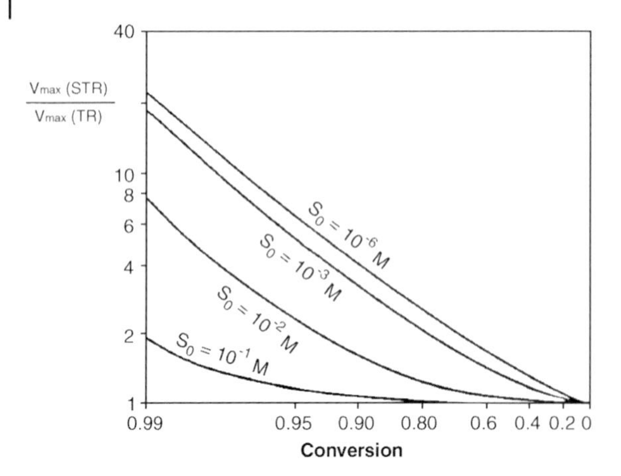

Figure 11.4 Ratio of enzyme activities required for a continuous stirred tank reactor and a tubular reactor, for different initial substrate concentrations $[S]_0$, given by simple Michaelis–Menten kinetics and a Michaelis constant (K_m) of 1 mM (Lilly, 1978).

In order to compare the reactor volumes required for a different conversion and for different initial substrate concentrations, Figure 11.4 shows the enzyme activities required for a CSTR and a tubular reactor (see Section 10.5).

Further advantages and disadvantages of different reactor types to be mentioned are as follows: batch stirred tank reactors offer simple equipment, while they need time-consuming downtime for loading and cleaning, and reaction conditions vary with time. Continuous stirred tank reactors offer constant reaction conditions, good external mass transfer, and easy control of environmental conditions, while they require flow-regulating equipment (pumps, valves). Continuous tubular, in most cases packed bed, reactors, with simple construction and high biocatalyst density, provide higher conversion in general, while external mass transfer may be insufficient, environmental control (notably pH) may be difficult, and high pressure drop requires mechanical strength of the biocatalyst. Fluidized bed reactors provide good external mass transfer, however, at rather low biocatalyst density. Membrane reactors offer straight separation of the catalyst from substrates and products; however, mass transfer limitations may occur, and membrane fouling and high pressure drop can be problems (Prenosil, *et al.*, 2009).

11.2.2
Other Reactor Types and Configurations: Application Examples

Simple standard reactors offer the advantage of flexibility – that is, the possibility to serve different applications. Often, older stirred tank reactors are used for newly developed processes in order to save or reduce investment cost. In contrast, special modifications of basic reactor types offer the advantage of being adjusted to create an optimal design to meet the requirements of a specific process.

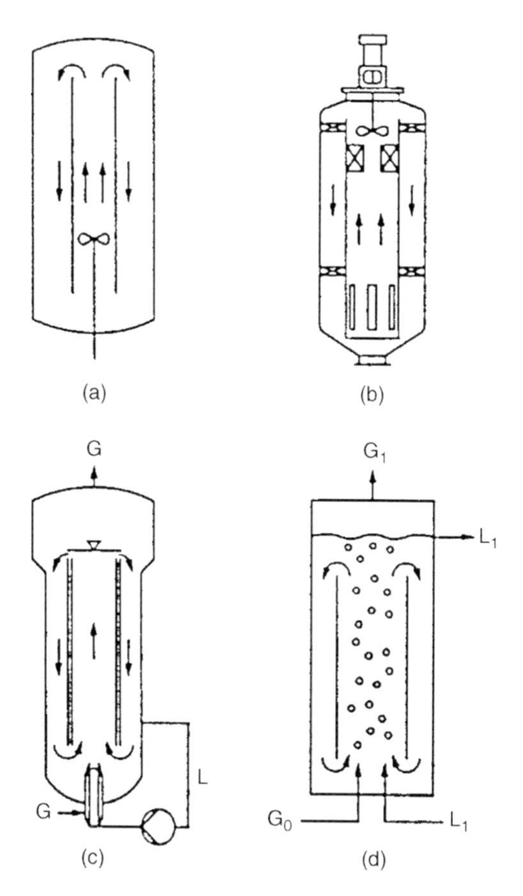

Figure 11.5 Loop reactors (G = gas phase; L = liquid phase). (a) Simple principle with internal tube and stirrer; (b) slim type with baffles/breakers; (c) reactor with *enlarged* head (for three-phase systems and degassing); (d) bubble column loop reactor.

First, some modifications and combinations of stirred tank reactors will be presented (for reviews, see Lilly, 1978; Cabral and Tramper, 2000; Prenosil, *et al.* 2009; Hempel, 2007).

Stirred tank reactors with internal loop – the so-called "loop reactors" (Figure 11.5) – offer short circulation times and thus provide for intense mixing of the reaction solution. For continuous operation, the circulated mass (or volume) should be at least 5- to 10-fold that of the feed in order to provide for short mixing time. With scale-up over several orders of magnitude, the mixing time – and thus the control of essential parameters such as pH – can be problematic; this is notably the case with solutions of high viscosity. In environmental technology, for example, in anaerobic wastewater treatment with immobilized biomass, scale-up must proceed from the laboratory scale with reactors of 1–10 l up to a reactor volume of several hundreds of m^3 – an increase by four to five orders of magnitude. The control of the mixing behavior, which is essential, in general requires special devices, for

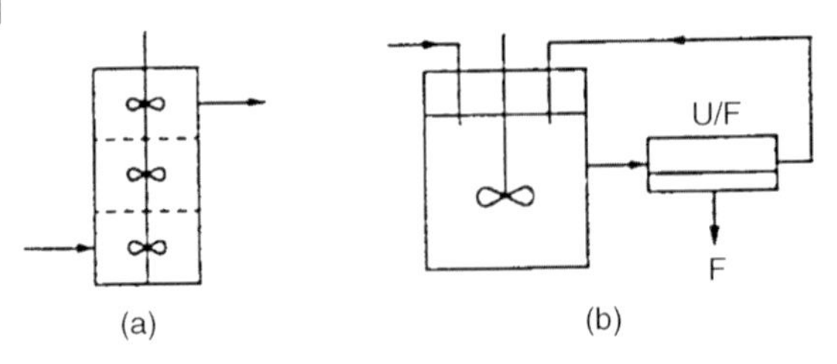

Figure 11.6 Reactor configurations with continuous stirred tank reactors. (a) Cascade of stirred tank reactors in the form of a tube subdivided into chambers. (b) Stirred tank reactor with subsequent ultrafiltration unit.

example, an internal tube (Figure 11.5a). The control of flow inside the reactor by an appropriate circulation rate ensures the control of the concentration distribution, which in turn depends on substrate inflow and reaction (Blenke, 1985).

A *cascade of stirred tank reactors* represents the most convenient method for continuous processes where the advantages of stirred tank reactors – simplicity and flexibility – can be utilized and the disadvantages – application of the biocatalyst under stationary conditions of low substrate and high product concentration – are avoided. The concentration levels of the cascade tend to approximate the profile of a tubular reactor (see Figure 11.2c versus d). The construction of the cascade can be rather simple in terms of a multistage tubular reactor equipped with sieve or perforated plates. A central shaft drive carries the stirrers in each chamber (Figure 11.6a).

The mass balance for the cascade is based on Eq. (11.1), by taking as the inlet concentration $[S]_i$ of a reactor the outlet concentration of the preceding reactor. For Michaelis–Menten kinetics, it follows that

$$\nu = \eta_{s,i} \frac{k_{cat} \cdot [E]_i \cdot [S]_i}{K_m + [S]_i}. \tag{11.7}$$

For the inlet concentration related to the exit concentration in each reactor, one may write (where X is the conversion $([S]_0 - [S]_E) \times [S]_0^{-1}$ and V the volume of a reactor; the residence time is $\tau = V \times Q^{-1}$):

$$[S]_i/[S]_{i-1} = 1 - V/Q(\nu_i/[S]_{i-1}) = 1 - X_i \tag{11.8a}$$

and

$$X_i = (V/Q) \cdot (\nu_i/[S]_{i-1}) = \tau \cdot \nu_i/[S]_{i-1}. \tag{11.8b}$$

Multiplying the singular equations gives the final outlet concentration of the cascade with i reactors related to the initial inlet concentration. For the final conversion X_i one obtains:

$$X_i = 1 - [S]_i/[S]_0 \tag{11.9}$$

(for the mass balances of other reactor modifications, for example, of reactors with recycle, see Moser, 1985).

Continuous stirred tank reactors are used for starch hydrolysis, both for liquefying with α-amylase and for further hydrolysis with glucoamylase. For this

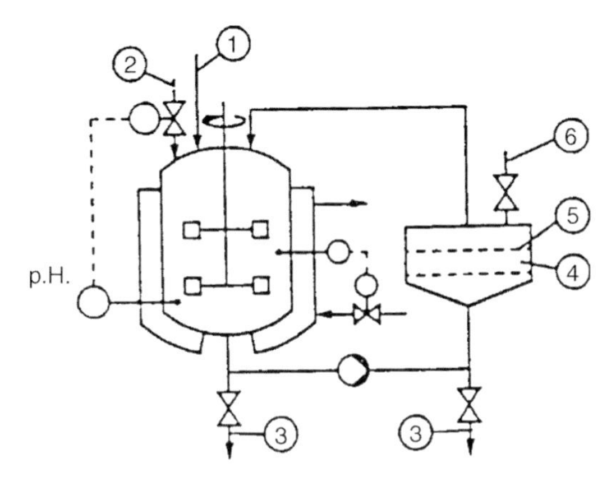

Figure 11.7 Reactor for penicillin hydrolysis, with differential fixed bed with immobilized penicillin amidase and a stirred tank reactor for neutralization of 6-aminopenicillanic acid formed. (1) Addition of penicillin, (2) addition of ammonia, (3) exit to extraction of 6-aminopenicillanic acid, (4) immobilized enzyme, (5) sieve plate, and (6) degassing.

technically important process, cascades of CSTRs are also used (see Section 12.1). For glucose manufacture, mainly cascades with 6–12 reactors are applied in order to avoid losses in yield of the final product (Uhlig, 1991; Olsen, 1995).

In continuous processes, a soluble biocatalyst is lost that, nevertheless, is economically feasible when cheap enzymes (as in the case of starch processing) are used. If the cost of the biocatalyst is higher (about >2% of the product price), reactors with recycle are required – that is, a separation unit for the biocatalyst must be installed downstream of the reactor. For soluble enzymes, ultrafiltration membrane systems are appropriate (see Section 7.4 and Figure 7.5).

Another reactor configuration to be used is the combined stirred tank reactor and differential fixed bed reactor (Figure 11.7). This is used in penicillin hydrolysis with immobilized penicillin amidase in a discontinuous process (see Section 12.3) (Liese *et al.*, 2000, pp. 286, 287; Liese *et al.*, 2006). Following a differential conversion of penicillin in the fixed bed of biocatalyst, mixing is required for pH control. The biocatalyst located in the fixed bed is neither subject to variations in pH occurring in large reactors nor is it subject to abrasion due to shear by the impeller (see Section 11.3.3).

Tubular reactors are applied mainly as *fixed bed reactors* with immobilized biocatalyst (spheres or granules; Figure 11.8a), as for glucose isomerization and for kinetic resolution of racemic amino acids (see Section 8.4.1, Figure 8.11, and Sections 8.4.3 and 12.1). Specific requirements (such as a sediment-free substrate solution) require appropriate upstream operations, and these are detailed in Section 11.4. The essential advantage of fixed bed reactors is simple continuous operation. When compared to CSTRs, a significantly higher catalyst productivity is obtained due to the *profiles* of substrate and product inside the tubular reactor, which are higher by a factor of about 1.7 in the cases of glucose isomerase and

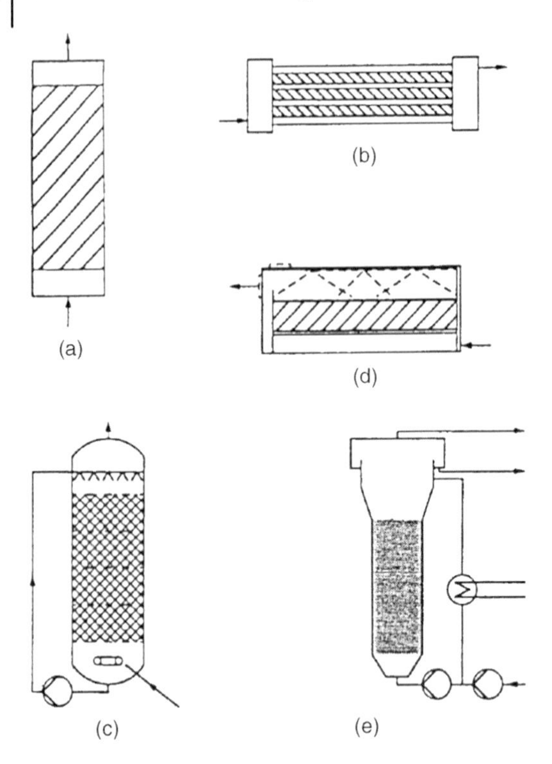

Figure 11.8 Different configurations of tubular reactors. (a) Column with fixed bed of biocatalyst, (b) hollow fiber membrane reactor, (c) trickle bed reactor, (d) biofilter, and (e) fluidized bed reactor.

amino acylase (Chibata, *et al.* 1987; Jensen and Rugh, 1987). Reactors with immobilized glucose isomerase have bed heights of 5–8 m, and the substrate feeding rate is 1–2 tons per m^3 a day.

Soluble enzymes can, in principle, be applied in simple tubular reactors. When scaling up to tubes of large diameter, however, back mixing may occur (see Section 11.3), though this can be controlled by applying internals. Enzyme recycling can be accomplished with a subsequent ultrafiltration unit (see Section 7.4; Figure 7.5d). Hollow fiber membrane reactors comprise bundles of hollow fibers in which the membrane serves for the retention of the enzyme(s) (Figure 11.8b).

Trickle bed reactors are applied in many varieties in biological exhaust gas purification (Figure 11.8c). These are three-phase reactors containing a carrier with an adhering mixed bacterial culture degrading organic pollutants; the aqueous phase flows from the top down through the reactor, absorbing the organic components from the exhaust air, which flows countercurrent from the lower part of the reactor to the top. The biofilter represent another system for exhaust gas treatment (Figure 11.8d; see also Section 9.6.2.4 for details of both systems).

Fluidized bed reactors (Figure 11.8e) can be applied as either two- or three-phase systems. The liquid (substrate-) phase enters the reactor from the downside, while the upflow keeps the biocatalyst suspended inside the reactor tube. The regime of flow rates is limited on the one hand by the minimum rate at which the particles are suspended (minimum fluidization) and on the other hand by the maximum rate where the particles tend to leave the reactor. The flow rate required is generally controlled by an external loop. The advantages of such systems are the tolerance (no clogging) to fluid media with suspended fine particles, or precipitates (occurring in many cases in wastewater treatment), and the ease of introducing or removing a gas phase (e.g., biogas). A drawback is mixing of the fluid phase, the residence time approaching that of a stirred tank reactor, if not prevented by internals or sieve plate installations.

Fluidized bed reactors are applied in large scale (up to $500\,m^3$) in anaerobic wastewater treatment (Figure 11.9). Anaerobic mixed cultures (including methanogenic bacteria) grow slowly, thus requiring retention inside, or recycle to the

Figure 11.9 Industrial fluidized bed reactor for anaerobic wastewater treatment; working volume $500\,m^3$, height 30 m. (Reproduced from Jördening *et al.*, 1996, with permission of Nordzucker AG, Braunschweig.)

bioreactor, in order to achieve good performance with high reaction rates. When immobilized by adhesion (e.g., on sand, pumice; see Section 9.6.2.2), they can be applied notably in fluidized bed reactors where the large amounts of biogas are formed separately from the liquid phase and leave the reactor at the top. Mixing by recirculation via an external loop is necessary in order to control the pH (Figure 11.8e). The mixed culture is able to adapt to low stationary substrate concentrations and to operate efficiently under these conditions.

The examples of industrial applications mentioned show that in the food and pharmaceutical industries, large reactors as well as sophisticated reaction engineering are important and standard techniques. The largest systems have been installed in environmental technology (ranging from 100 m^3 to several 1000 m^3). Here, reaction engineering plays an essential role in order to achieve process intensification and increase the efficiency of these systems.

Microstructured flow reactors are emerging tools for biocatalytic process development. They can serve for high-speed catalyst screening, and catalyst design, process development, monitoring and optimization, notably for the manufacture of pharmaceutical intermediates, also enabling reduced cost for substrates, and so on. Process development may also include green chemistry requirements. The reaction volume may be below 10 μl, and sensors for online monitoring of relevant parameters, such as O_2, can be integrated. A compelling design is that of the coated wall reactor where an immobilized enzyme is present as a surface layer attached to microchannel walls with multiple linear flow channels (Thomsen and Nidetzky, 2009; Karande *et al.*, 2011; Pollard and Woodley, 2007; Edlich *et al.*, 2009, 2010).

11.3 Residence Time Distribution, Mixing, Pressure Drop, and Mass Transfer in Reactors

11.3.1 Scale-Up, Dimensionless Numbers

Scale-up is a difficult task with conflicting requirements, such as constant power input, or constant stirrer tip speed, both of which have been proposed with reference to solid particle suspension, resulting in evident disparity. Transport processes are very dependent on scale, and this is the main reason that scale-up problems exist. While a process is determined by kinetic phenomena at small scale, it is determined by transport phenomena at large scale in general. The main problems are the mixing, suspending, and abrasion of particles in stirred tank reactors and pressure drop in tubular fixed bed reactors. Optimal conditions for a biocatalytic process often ask for conflicting process conditions. Furthermore, the accuracy of various scale-up correlations is questionable (Kossen and Oosterhuis, 1985; Middleton and Carpenter, 1992; Benz, 2011; Hempel and Dziallas, 1999).

In the laboratory – and in part also on the pilot scale – reactors may be constructed and operated at near-ideal conditions so that concentration profiles as shown initially (see Figure 11.2) are correct in good approximation. This includes

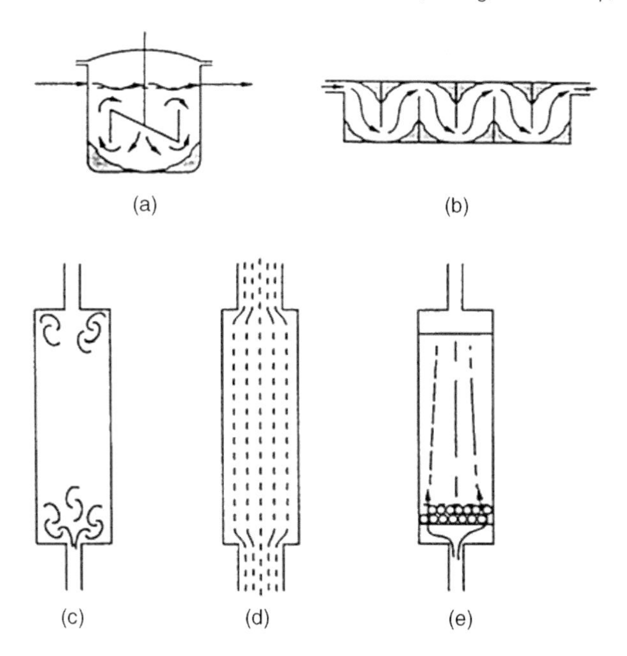

Figure 11.10 Schematic illustration of deviations from ideal flow in stirred vessels (a) and tubular reactors (b–e). (a) Stagnant zones and short-circuit flow/channeling, (b) stagnant zones, (c) back mixing in local turbulent field, (d) insufficient axial mixing, and (e) channeling in fixed bed with catalyst.

ideal mixing – that is, identical concentrations at any position in ideal stirred tank reactors for a given time or independent of time at continuous operation under steady state. In large reactors, however, significant deviations occur due to the problems of scale (Figure 11.10; Reuss and Bajpai, 1991; Schmalzried *et al.*, 2003).

In addition, several parameters do not play any role at the laboratory scale, or they may be neglected; examples include the pressure drop in packed bed reactors or the abrasion of particles in stirred vessels. A considerable number of parameters must be taken into account that require intercorrelation. In order to reduce these, groups of parameters are summarized in dimensionless numbers in physically meaningful ways, which in turn can be correlated, and this considerably reduces the number of correlations necessary to describe functions. These functions have been derived, for example, for external mass and heat transfer.

Dimensionless numbers, such as *Re*, *Ne*, *Sh*, and *Sc* are commonly used in order to correlate power consumption, and mass transfer, with hydrodynamics – that is, fluid flow at different flow regimes. Their definitions are as follows:

- For characterizing fluid flow: Reynolds number
  $Re = (\rho\, n \cdot d_i^{\,2}/\eta)$ (inertia forces/viscous forces) for stirred tank reactors
  $Re = (\rho\, u \cdot d_p/\eta)$ for tubular reactors with fixed bed of particles
- For characterizing power consumption: Newton number
  $Ne = (P_R/(\rho\, n^3 \cdot d_i^{\,5}))$

- For characterizing external mass transfer: Sherwood number and Schmidt number

  $Sh = (k_L \cdot d_p / D_s)$ (total mass transfer/mass transfer by diffusion)

  $Sc = (\nu / D_s)$ (hydrodynamics boundary layer/mass transfer boundary layer)

  ν = kinematic viscosity
- Another most relevant parameter (not dimensionless) for scale-up is the power input:

  $P_R = 6\,\rho\, n^3 \cdot d_i^5.$

11.3.2
Residence Time Distribution

Three parameters are important for describing fluid flow in reactors: (1) the (mean) residence time, (2) residence time distribution, and (3) mixing time. The residence time provides information about the time for which a volume element stays inside a reactor. In real reactors – and notably in CSTRs – different volume elements may have greatly differing residence times; this is described by the residence time distribution $E(t)$. This function gives the probability for a volume element (or a tracer pulse with the concentration c) that has entered the reactor at time $t = 0$ and leaves the reactor at time t (this also means the molar flow n' referring to the amount n_0 at time $t = 0$; Figure 11.11):

$$E(t) = \frac{\dot{n}}{n_0} = \frac{Q \cdot c(t)}{\int_0^\infty Q \cdot c(t)\mathrm{d}t}, \quad \text{where} \quad \int_0^\infty E(t)\mathrm{d}t = 1. \tag{11.10}$$

The mean hydraulic residence time τ_h can be calculated from the ratio of reactor volume and volumetric flow. It is often advantageous to relate the real-time distribution to this residence time, and to introduce a dimensionless time according to the following definition:

$$\tau_h = V \times Q^{-1} \text{ and } \theta = t \times \tau_h^{-1}. \tag{11.11}$$

The *integral* of the residence time distribution (Eq. (11.12)) gives the total amount $n(t)$ of a substance that has left the reactor at time t (see Eq. (11.10)). It may be described by the ratio of the actual concentration of a tracer $c(t)$ and the constant concentration at the inlet c_0.

$$F(\Theta) = \int_0^\Theta E(\Theta)\mathrm{d}\Theta = \frac{c(\Theta)}{c_0}. \tag{11.12}$$

For the experimental determination of the residence time function, a tracer is introduced to the reactor inlet and its concentration at the outlet is measured as a function of time. The tracer substance should not influence the reactor content physically (by density, viscosity) and it should be chemically inert; depending on the tracer, the electrical (addition of a salt solution) or thermal conductivity, or the

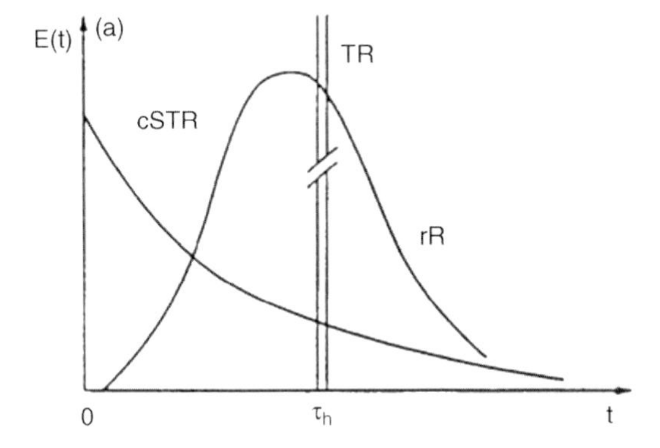

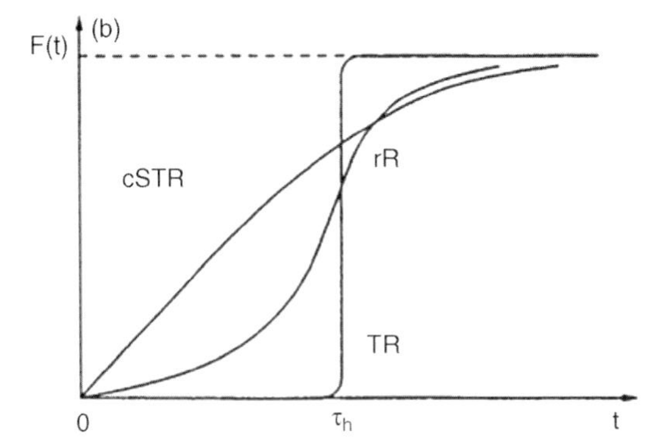

Figure 11.11 (a) Residence time distribution $E(t)$ for ideal continuous stirred tank (cSTR) and tubular (TR) and real reactors (rR); (b) *Integral* of the residence time distribution $F(\Theta)$ for ideal and real reactors.

absorption spectrum (e.g., a *lithium* salt with specific atomic absorption) can be measured. For the tracer addition, two different methods may be applied:

- In the first a pulse, a certain amount of tracer is added over a time interval, which is as short as possible. Continuous measurement of the tracer at the reactor outlet gives the residence time distribution as a result of the pulse. When comparing the different reactor types (Figure 11.11a), the ideal CSTR exhibits the broadest residence time distribution: when adding the tracer at time $t = 0$, it is immediately distributed equally throughout the reactor and leaves it with concentration c_0 ($t = 0$) at the outlet; at the hydraulic residence time $t = \tau_h$ about 25% of the tracer remains inside the reactor. In the ideal tubular reactor, the whole amount of the tracer leaves the reactor at the hydraulic residence time τ_h at the outlet.
- In the second method, a sudden change in the tracer solution is applied at the inlet of the reactor, for example, with a sudden increase in tracer concentration. The signal at the outlet corresponds to the integral of the residence time distribution $F(t)$ (Figure 11.11b).

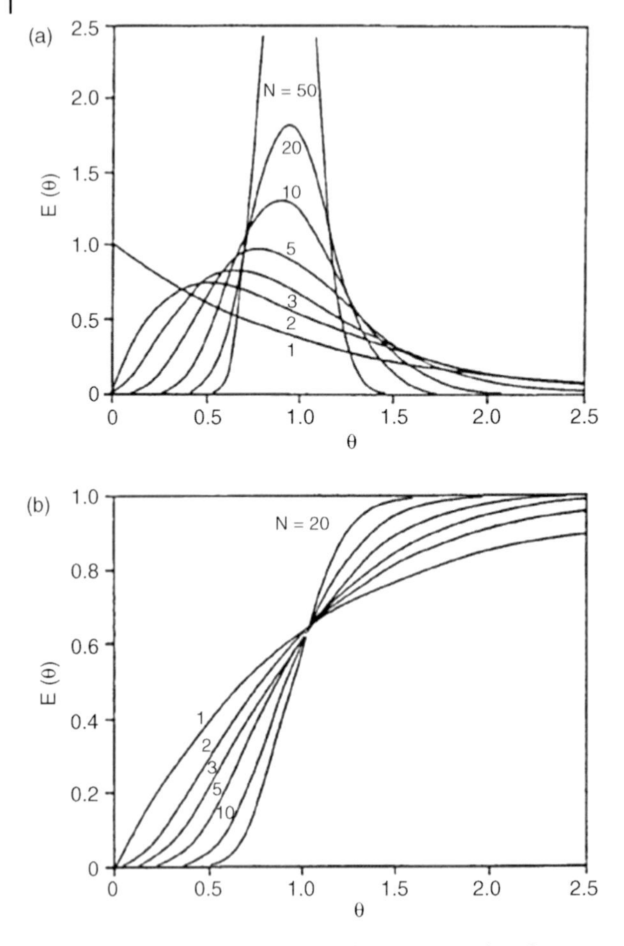

Figure 11.12 (a) Residence time distribution in cascades of continuous stirred tank reactors; parameters: number of vessels. (b) Integral of the residence time distribution for cascades; *N*: number of vessels.

The residence time functions for cascades of CSTRs are intermediate between the extremes of the ideal stirred tank and the tubular reactors. For practical purposes, three–five vessels are sufficient in order to provide for a favorable residence time distribution (e.g., with reference to the catalyst efficiency). The residence time distributions and their integrals for cascades of different numbers of vessels are presented in Figure 11.12.

The residence time distribution of a reactor is the most important criterion for its efficiency, and notably for the efficiency of the biocatalyst. Deviations are common in practice, and in general they have unfavorable consequences, especially for high conversion and/or high selectivity. The reasons for deviations of the residence time distribution from ideal behavior are shown schematically in Figure 11.10. For a stirred vessel, deviations may be due to unfavorable installation of inlet and outlet

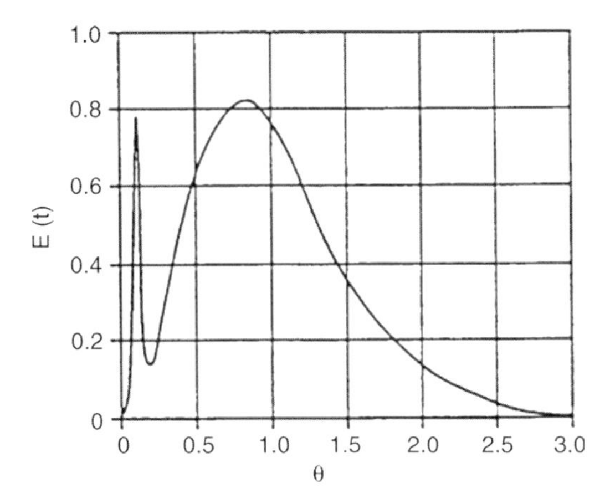

Figure 11.13 Residence time distribution with short circuit (simulated).

(e.g., at opposite positions) so that bypass or short-circuit flow may occur, and part of the substrate introduced flows directly toward the outlet, essentially without being mixed with the solution inside the reactor (see Figure 11.10a). Poorly mixed zones (dead zones) are likely to occur in viscous media (e.g., in starch hydrolysis, and generally at high substrate concentrations; Figure 11.10a and b). Figure 11.13 shows a model calculation for a residence time distribution exhibiting a large deviation from the behavior of an ideal reactor, which was simulated with the aid of a model with bypass. In practical cases, secondary maxima – both before and after the main maximum of the residence time distribution curve – may be observed for reactor cascades with insufficient mixing.

11.3.3
Mixing in Stirred Tank Reactors

In the laboratory, where energy demand can be neglected, stirred tank reactors are used experimentally and in model calculations as ideally mixed reactors. This situation is different in technical reactors, where electrical energy demand (stirrer motor) must be considered for economic reasons. Furthermore, the impeller dimension and speed must be limited according to the aspects of shear force and mechanical abrasion of the (immobilized) biocatalyst. On the other hand, intense mixing with short mixing time is essential for the transport of substrates with low solubility (e.g., oxygen) to the catalyst, or for the quick and efficient mixing of reactants (e.g., for neutralization as in penicillin hydrolysis).

The stirred tank reactor has been treated extensively for these reasons as it represents the dominant type of reactor in practice. The mixing time, which signifies a certain extent of distribution of a tracer pulse in the reaction volume, is taken as a characteristic parameter for the quality of mixing. Figure 11.14 shows schematically the flow patterns for fluid elements in reactors with one and two impellers.

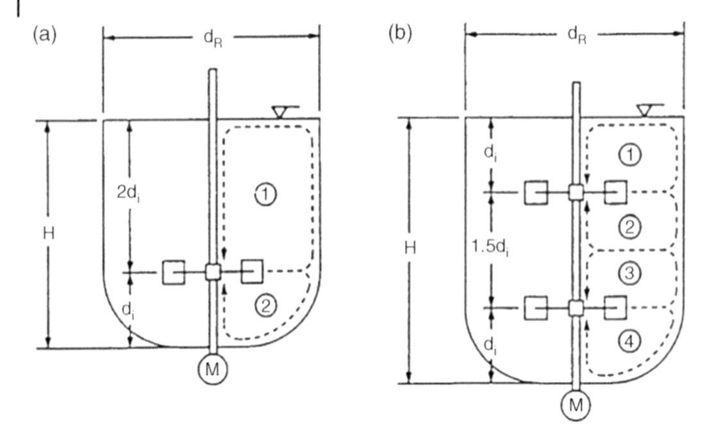

Figure 11.14 Flow patterns with circulation flow in stirred tank reactors. (a) Reactor with impeller in unsymmetrical position; (b) reactor with two impellers (H = height; d_R = reactor diameter; d_i = impeller diameter; (1–4) = circulation pathways) (Reuss and Bajpai, 1991).

For the measurement of mixing time, the conductivity of the solution may serve as a signal, for example, when salt is added as a tracer. Different models and their mathematical treatment have been presented by Reuss and Bajpai (1991) and Schmalzried *et al.* (2003).

The example illustrates the relevance of the mixing time t_m as a key parameter for stirred tank reactors. It gives the definition for the time at which a certain degree of mixing is obtained. It refers to an amplitude A_m of a signal (corresponding to the concentration of a tracer) that is taken as acceptable for a specific problem (Figure 11.15) (Reuss and Bajpai, 1991). It can be calculated from the following correlation (with the amplitude decay rate constant K_A):

$$t_m = K_A^{-1} \ln(2 \cdot A_m^{-1}). \tag{11.13}$$

It must be considered that both the position of the impeller (either in the middle or in the lower part of the reactor) and the impeller speed n (via the Reynolds number, Re) have an influence on this correlation; only the mixing time can, therefore, be estimated. K_A can be calculated approximately from the following equation:

$$\frac{n}{K_A}\left(\frac{d_i}{d_R}\right)^{2.3} \cong 0.5 \quad \text{for} \quad Re = \frac{nd_i^2}{\nu} > 2 \times 10^3. \tag{11.14}$$

The mean circulation time τ' (c.f. Figure 11.14) is correlated with the systems parameters by the correlation (Eq. (11.15)) given subsequently; here, asymmetries of the stirrer position inside the vessel are also neglected (Reuss and Bajpai, 1991) (mean circulation times are in the range of several seconds up to several minutes for small or large reactors):

$$n\,\tau' = 0.76\left(\frac{H}{d_R}\right)^{0.6} \cdot \left(\frac{d_R}{d_i}\right)^{2.7}. \tag{11.15}$$

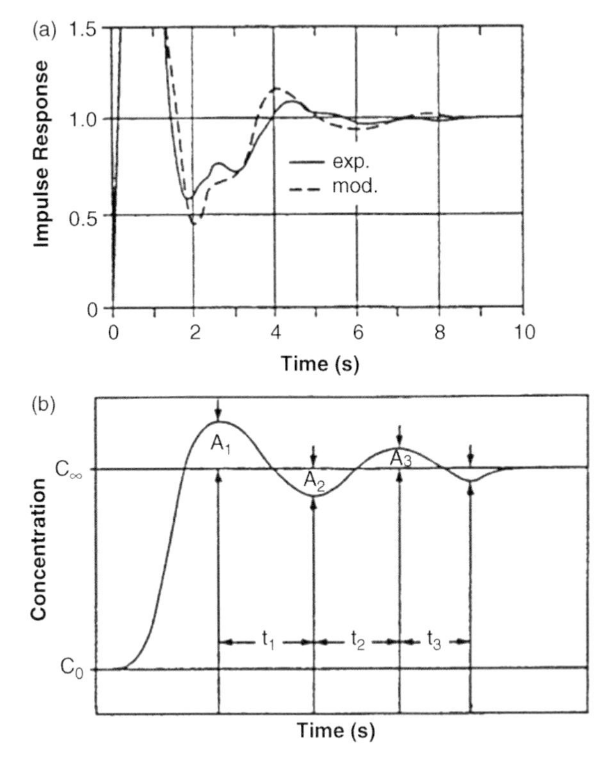

Figure 11.15 Experimental and simulated signal from mixing after a tracer pulse.

For an advanced insight, mixing and mixing time in a stirred tank reactor may be illustrated by figures taken from simulations based on computational fluid dynamics (CFD) (Schmalzried *et al.*, 2003). Such simulations can help and accelerate the process of scale-up. The Rushton turbine (see Figures 11.14 and 11.17(i) 1) used in the experiments and simulation is well established for mixing liquids with low viscosity. It generates a flow that leaves the impeller in radial and tangential directions and recirculates back from the wall into the impeller region, this being the main reason for the mixing capability. Figure 11.16 shows the simulated dynamics of the tracer distribution at different times after a pulse on the liquid surface in a stirred tank reactor. It is clear from this example that mixing requires an extended timescale (over 8 s in this example), which might cause significant side reactions in large reactors. An example of practical relevance is the enzymatic hydrolysis of penicillin and cephalosporin C (Section 12.3), where decomposition of the substrate and inactivation of the biocatalyst in a stirred tank can occur in fields of high pH near the point of addition of the base, which must be added at high concentration in order to avoid dilution of the product solution (Carleysmith *et al.*, 1980). Similarly, the substrate distribution in large stirred vessels depends strongly on the feeding position (Schmalzried *et al.*, 2003; Figure 11.14). Therefore, distributed

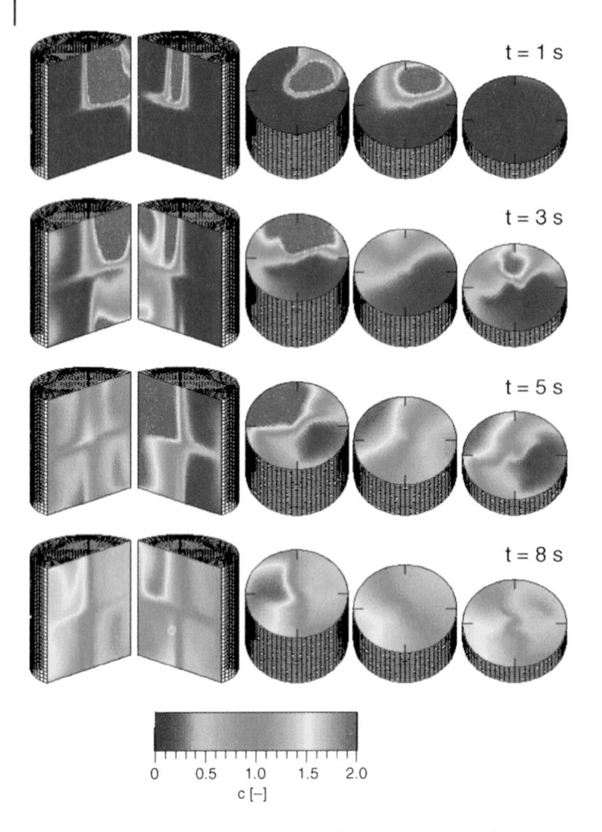

Figure 11.16 Simulated dynamics of the tracer distribution at different times after a pulse onto the liquid surface in a stirred tank reactor (height/tank diameter ratio = 1.0, Rushton turbine, impeller/tank diameter ratio = 0.3125, impeller clearance/height of liquid = 0.31). (Reproduced from Schmalzried *et al.*, 2003, with kind permission from Springer Science + Business Media.)

points for the addition of reactants near the stirrer tip (field of highest turbulence) should be used.

Several types of impeller are shown in Figure 11.17. The mixing time for a certain impeller can be taken from empirical correlations, as shown in Figure 11.17b for several types of impeller. The product of stirrer speed multiplied by the mixing time t_m is given as a function of the Reynolds number (*Re*).

The power consumption of a stirrer can also be taken from empirical correlations; therefore, the power number (Newton number, *Ne*) is correlated with the Reynolds number (*Re*) for different types of impeller (Brauer, 1985).

The Newton number (*Ne*) is, as is the mixing time, for a given type of stirrer and a certain geometrical arrangement a function of *Re* only:

$$Ne = \frac{P_R}{\rho \cdot n^3 \cdot d_i^5} = f(Re). \tag{11.16}$$

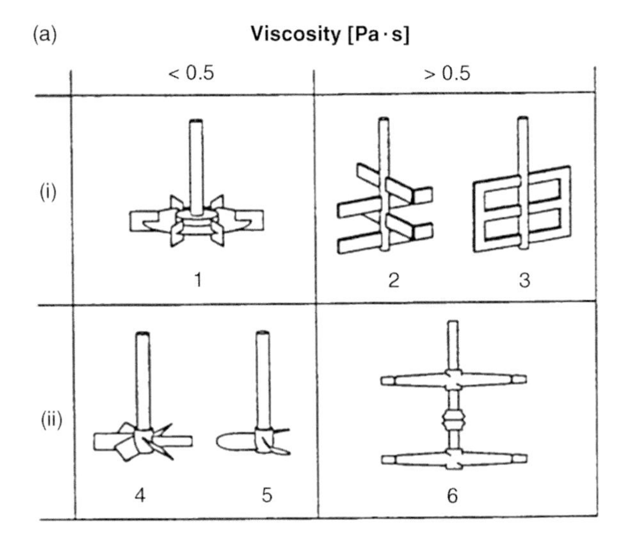

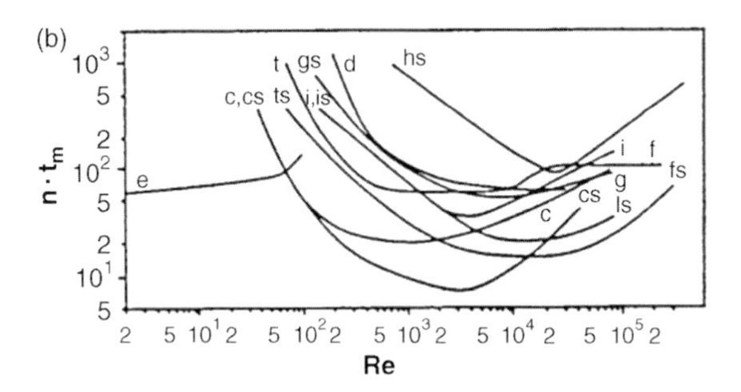

Figure 11.17 (a) Types of common impellers, in the sequence of application for media of different viscosity and the type of fluid flow; (i) = tangential to radial; (ii) axial fluid flow; (1) Rushton turbine; (2), (3) special types for high viscosity; (4) pitched blade turbine; (5) propeller; (6) MIG impeller (Ekato). (b) Correlations for estimating the mixing time for different types of impellers: s = with baffles; c = type 2; d = anchor; e = helical ribbon; f, fs = MIG impeller; gs = turbine; i = impeller; hs = propeller (Zlokarnik, 1972).

Three different ranges of flow pattern are to be considered (Brauer, 1985):

1) The laminar region, where the following correlation holds:

$$Ne \approx Re^{-1}, \text{ or } Ne \times Re = \text{const.} \tag{11.17}$$

2) The region of transition from laminar to turbulent fluid flow:

$$10^1 < Re < . \tag{11.18}$$

3) The turbulent flow region, where the influence of the baffles (which are essential) results in a 10-fold increase in the impeller power consumption, where,

depending upon the references,

$$Ne = \text{const.}; P_R = \text{const.} \cdot n^3 \times d_1^5 \times \rho \tag{11.19}$$

(Zlokarnik, 1972), or

$$Ne \approx Re^{-0.28} \tag{11.20}$$

(Brauer, 1985).

The *shear stress* increases in large-volume reactors, depending on the type of impeller, with $(N \times d_1^2)$, or with $(N \times d_i)^2$ (given by different sources) (Kossen and Oosterhuis, 1985); with the scale-up from a laboratory reactor of volume about 10 l (with $n = 5\,s^{-1}$, $d_i = 0.03$ m) to a technical reactor of 10 m^3 size (with $n = 2\,s^{-1}$, $d_i = 0.5$ m) and similar power input (proportional to $n^3 \times d_i^5$), the shear stress increases by a factor of 50, which may cause severe problems of abrasion for biocatalyst particles.

For *suspending particles* the general scale-up rule is to keep $P/V = \text{const.}$ (Kossen and Oosterhuis, 1985).

11.3.4
Mass Transfer in Reactors

External mass transfer depends on hydrodynamics in reactors; it is in general not rate limiting in stirred tank reactors under turbulent conditions, whereas it may be rather slow and eventually rate limiting in fixed bed reactors, depending on the flow rate (for overviews, see Prenosil, *et al.* 2009; Willaert, 2009). The phenomena involved are conveniently discussed with the aid of the simplified film model, assuming a quasi-stagnant layer of fluid thickness (Δ) that adheres to the outer catalyst surface (Buchholz, 1982). Substrates and products must penetrate this film by diffusion according to their concentration profile (see Figure 10.6). The rate of this mass transfer step is described by

$$k_L(c_S(\infty) - c_S(R)) = (D_S/\Delta)(c_S(\infty) - c_S(R)). \tag{11.21}$$

In the steady state, it is equal to the overall reaction rate:

$$v = k_L A_p (c_S(\infty) - c_S(R)) V_p^{-1}. \tag{11.22}$$

The importance of the external particle surface A_p is obvious. k_L can be determined experimentally (whereas Δ does not have a real physical significance for most experimental conditions). It is strongly correlated with the hydrodynamics of the reactor system, notably the flow rate in fixed bed and fluidized bed reactors, and the viscosity of the fluid.

The correlations for mass transfer and experimental conditions are commonly expressed in terms of dimensionless groups (Section 11.3.1):

$$Sh = c Re^a Sc^b. \tag{11.23}$$

Data compiled from the literature for packed bed and trickle bed reactors are as follows (Atkinson and Mavituna, 1991a):

$Sh = 0.99\ Re^{1/3}\ Sc^{1/3}$; $Re < 1$
$Sh = 0.95\ Re^{1/2}\ Sc^{1/3}$; $10 < Re < 10^4$
$Sh = 2 + 0.73\ Re^{1/2}\ Sc^{1/3}$; overall approximation of various correlations.

A correlation reported for isotropic turbulence is

$$Sh = 0.13\ Re^{3/4}\ Sc^{1/3}. \tag{11.24}$$

11.3.5
Pressure Drop and Fluidization in Tubular Reactors

In *fixed bed reactors*, the pressure drop is of major importance for the application of biocatalysts. These may be several meters in height, with pressure drops of several bar. A general correlation for the pressure drop Δp as a function of bed height H, flow rate u, voidage ε, and particle diameter d_p is given by Eq. (11.25) for laminar flow conditions (when $Re < 10$):

$$\Delta p\ H^{-1} = 150(1-\varepsilon)^2 \eta u\ (\varepsilon^3 d_p^2)^{-1}. \tag{11.25}$$

For turbulent flow, which is not likely to occur in large packed bed reactors, this equation must be extended (Buchholz, 1982; Eigenberger, 1992).

The correlation given before makes the high relevance of both bed porosity ε and particle diameter d_p obvious. In the choice of particle size of immobilized biocatalysts for technical applications, two opposite requirements must be considered: (1) high efficiency at high enzyme loading in order to provide for high space–time yield, with optimal size in the range of 50–100 μm, and (2) technical handling and performance in filtration, or sieve retention in batch, or fixed bed applications. Filtration time and pressure drop limit the particle size to a minimum range of 200–300 μm (see also Section 10.4.1). To limit the pressure drop in large, fixed bed reactors, the range of particle size applied is 400–1000 μm (Antrim and Auterinen, 1986; Pedersen and Christensen, 2000; Jensen and Rugh, 1987; Busse, 2004). The particle size distribution, which determines the void volume, is also essential, and should be narrow.

In *fluidized bed reactors*, the terminal settling velocity for a single particle is the limiting case where the particles tend to leave the reactor. It may be estimated from classical correlations (Jördening and Buchholz, 1999; Werther, 1992). One important point here is the fluidized bed expansion behavior, which may be calculated using the correlation of Richardson and Zaki (1954), most often as

$$u = u_T\ \varepsilon^n, \tag{11.26}$$

where u_T is the terminal velocity of isolated single particles. The exponent n is given as follows:

$n = 4.65$	$0 < Re < 0.2$
$n = 4.4\ Re^{-0.03}$	$0.2 < Re < 1$
$n = 4.4\ Re^{-0.1}$	$1 < Re < 500$
$n = 2.4$	$500 < Re$

where $Re = u_T\, d_p\, \nu^{-1}$.

Typical operation ranges for fluidization are for particles of low density, such as pumice, 5–10 $\mathrm{m h^{-1}}$, and for sand about 10–30 $\mathrm{m h^{-1}}$ (particle diameter 0.5 mm). The fluid bed pressure loss for low-density media is significantly lower than, for example, sand (Jördening and Buchholz, 1999).

11.4 Process Technology

Reaction engineering relevant to the technology of the bioconversion of educts (substrates) into products – that is, the reaction – has been dealt with in Sections 11.2 and 11.3. Here, the so-called upstream and the subsequent so-called downstream steps will be discussed. In practice, the reaction – as well as the final product quality – essentially depends on these operation steps, termed unit operations. First, a short survey will be provided, followed by actual trends that aim at the integration of bioconversion and downstream operations (Section 11.4.2). Aspects of reactor instrumentation will be treated in Section 11.4.3. Examples of industrial processes that include a range of generally relevant unit operations will be presented in Chapter 12 (see also Heinzle *et al.*, 2006).

A fascinating historical example of process development is that of penicillin manufacture in the early 1940s, when within 3 years screening, strain mutation, fermentation optimization, and several downstream operations, up to large-scale freeze drying, were accomplished (Buchholz and Collins, 2010) (Section 4.3.4).

11.4.1 Survey

Process technology cannot be treated here in depth, as this would require the treatment of all unit operations including the engineering fundamentals with mass transfer principles, as well as the design and construction of devices and equipment.

Practical work requires an in-depth knowledge of the fundamentals of these engineering aspects. Even if the solution of technical problems in many cases proceeds in empirical and experimental approaches, the selection of optional methods and devices would require many experimental investigations to be conducted, or the rational selection of approaches would not be possible. This requires the knowledge of unit operations and reaction engineering (for relevant reports on this topic, see Hempel, 2004; Arpe, 1988, 1992; www.mrw.interscience.wiley.com/emrw/). Biochemical engineering and key unit operations in biotechnology have been compiled in comprehensive special monographs (Moo-Young, 1985; Atkinson and Mavituna, 1991a; Heinzle *et al.*, 2006; Brauer, 1985; Kossen and Oosterhuis, 1985), as well as in special reviews (see the series of *Advances in Biochemical Engineering*, for example, Schmalzried *et al.*, 2003).

The conversion represents the central step of a biochemical or biotechnological process. However, in general it corresponds to less than half the equipment, operational cost, and investment required. The preparation of substrates and additives, the isolation and purification of products, and the utilization or recycling of by-products, and treatment of wastes mostly require more equipment, expenditure of work, and investment cost than the bioconversion itself. All individual steps of the overall process must be coordinated carefully in order to guarantee economic processing. The process as a whole must be controlled by appropriate measurement and instrumentation. The results from laboratory experiments – including separation operations such as chromatography – must be translated to the technical scale by careful scale-up.

A survey of the most important steps of a process, including the corresponding unit operations and equipment, is provided in Table 11.2. All individual steps must be coordinated corresponding to mass and heat flow and submitted to measurement and control. This comprises the volume of tanks, the energy requirement of pumps and stirrers, the cross-section of pipes and columns (e.g., of ion exchangers), and the capacity of filters and heat exchangers (for the calculation of dimensions and capacity of pumps and pipes, see Dunlap, 1985). Planning must also include the aspects of substrates and additives quality.

The aim of process technology and biochemical process design is to establish a process of

- high product quality and
- high productivity under economic conditions.

This is possible only as a compromise, since special aspects and individual steps often exhibit conflicting requirements. Scale-up imposes specific problems and challenges that have been discussed in brief in the preceding section. The general treatment is outlined in the specialized literature (Atkinson and Mavituna, 1991; Kossen and Oosterhuis, 1985; Heinzle *et al.*, 2006).

11.4.2 Process Integration

Process *integrated downstream operations* – ISPR, *in situ* product removal techniques – have found increased attention in recent years. They may offer considerable benefits for processes limited by reactions that are thermodynamically

Table 11.2 Elements of process technology.

Process step			Unit operation	Equipment
1.	Pretreatment of educts (substrates, additives)	Separation/ purification	Filtration	Tanks for storage
			Sterile filtration	Filter
			Adsorption and ion exchange chromatography	Columns
		Pumping		Pumps
		Concentrating	Evaporation	Evaporators
			Ultrafiltration	Membrane systems
			Reverse osmosis	
		Dilution (pH adjustment)	Mixing	Stirred tanks
		Heating		Heat exchangers
		Cooling		Heat exchangers
2.	Biochemical conversion	Reaction engineering[a)]		
3.	Product isolation and purification	Separation/ purification	See above, and crystallization; precipitation	
		Stabilization	Sterilization	
			Drying	Dryer
		Concentrating	See above	Spray dryer
4.	Recycling of auxiliary material (e.g., solvents)			
5.	Treatment of residual material and wastes (e.g., regeneration water from ion exchangers)			
6.	Confectioning (packaging, etc.) Concentrating			

a) See Sections 11.2 and 11.3.

unfavorable, include inhibitory or toxic products, unstable products or products that undergo further reactions. Systems most used include membrane separation, extraction in biphasic, including organic phases, aqueous two-phase systems and ionic liquids, dual-particle liquid–solid adsorption, including expanded bed systems, magnetic adsorbent particles, chromatography, crystallization, stripping, and foam fractionation. Industrial applications include pervaporation and *in situ* crystallization (Liese *et al.*, 2000,2006, pp. 313, 481–484). Furthermore, the production of unsaturated fatty acids from low-cost saturated substrates has been performed by a desaturase from a *Rhodococcus* spp. *in* an oil/water emulsion. The product (and residual substrate) is recovered by a hydrophobic membrane-based filtration (Liese *et al.*, 2000, pp. 165–168).[1)]

Several approaches are given in short versions subsequently, though much more complex. They need investigation and optimization of many parameters, as mentioned below for the case of solid particle adsorption in a fluidized bed. Thus, for evaluation and understanding of the examples mentioned here, the literature cited must be consulted.

Systematic selection methods for separation operations could help reduce the considerable laboratory effort required (Woodley *et al.*, 2008). In order to analyze the perspectives of ISPR in biotransformations and identify the most promising developments, simple model considerations have been performed (Bechtold and Panke, 2009; Asenjo and Andrews, 2008). (For an earlier review, see Panke *et al.*, 2004; Schugerl and Hubbuch, 2005). In biphasic systems, a reactive aqueous phase provides a natural enzyme environment, and an immiscible nonreactive phase serves for delivery of dissolved substrates at high concentrations and for extraction of products. An *in silico* screening (via a model estimation) of solvents for maximal conversion may reduce the considerable laboratory efforts required (Spieß *et al.*, 2008).

Approaches have been developed using membrane separation, including biphasic systems. A membrane system in a biofilm tube reactor used styrene as both substrate and organic phase for extraction of the product (*S*)-styrene oxide (see Section 9.7.1; Figure 9.12). In another example, an amino acid racemase with broad substrate specificity was used in a model-based experimental analysis for an integrated multistep process (e.g., dynamic kinetic resolution) in order to aim at nearly 100% chemical yield and 100% enantiomeric excess of L-methionine. Aqueous-organic mixtures (up to 60% methanol and 40% acetonitrile) were chosen for an integrated process scheme in an enzyme membrane reactor (EMR) setting. Problems of product concentration and biocatalyst stability were analyzed; acetonitrile or methanol drastically reduced the half-life of the racemase (Bechtold *et al.*, 2007). The kinetic resolution of racemic naproxen thioesters in isooctane, coupled with a continuous *in situ* racemization of the remaining (*R*)-thioester substrate, was performed with immobilized lipase. A hollow fiber membrane was integrated into the reactor system to reactively extract the desired (*S*)-naproxen out of the reaction medium (Lu *et al.*, 2002). Electrodialysis has proven to be efficient for the *in situ* removal of inhibitory products and pH control (Wong *et al.*, 2010).

Two-phase systems with ionic liquids have been used in a number of examples with enzymes and whole cells. Advantages, such as better substrate solubility, and disadvantages, for example, the price and additional efforts for downstream processing, are discussed by Wenda *et al.*, 2011.

1) For application in the pharmaceutical industry, the development of integrated downstream operations had to be included at an early stage of process development since all process steps will have to be fixed, and later changes will not be possible (Berensmeier, personal communication, January 2010).

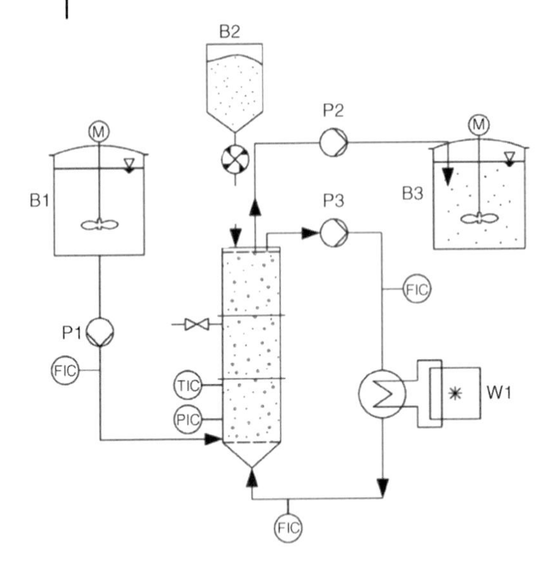

B1: Feed in (sucrose, glucose)

B2: Zeolite in

B3: Zeolite suspension with product

FIC, TIC, PIC: flow, temperature, and pressure indicator control

P1: Substrate pump

P2: Product pump

P3: Circulation pump

W1: Plate heat exchanger

Figure 11.18 Process flow sheet of fluidized multiphase bed reactor. Product recovery from the zeolite suspension was performed by centrifugation, recycling of the substrate solution, desorption of isomaltose from the zeolite with ethanol, distillation and ethanol recycle (Reproduced with permission of Elsevier © 2004, from Berensmeier *et al.*, 2004).

Product adsorption to solid particles in a specific reactor setup and biocatalyst optimization with the design of a genetically engineered enzyme were applied to improve selectivity and reduce downstream processing steps. A multiphase fluidized bed reactor with the suspended biocatalyst and a second solid phase (a zeolite) was conceived and operated continuously, with a special focus on the separation of the two solid phases by design of the fluidization properties of the two solid phases. The product, isomaltose, was synthesized by immobilized dextransucrase from sucrose and glucose; it was selectively adsorbed *in situ* on the zeolite and removed from the reaction phase in order to avoid subsequent glycosylation leading to a sequence of higher oligosaccharides (Figure 11.18). The fluidized bed operation required the appropriate design of the biocatalyst with respect to fluidization properties (particle diameter, density) and the zeolite that was fed continuously to the reactor and withdrawn subsequently from the reactor with the fluid phase for product recovery. This operation principle resulted in considerably higher overall isomaltose yields compared to a stirred tank reactor. Further design of the biocatalyst by engineering the dextransucrase significantly reduced the rate of subsequent glycosylation steps of the product and proved to be the most efficient step in optimization (Berensmeier *et al.*, 2004; Ergezinger *et al.*, 2006; Seibel *et al.*, 2010).

The approach presented here as a short version is much more complex: it requires biocatalyst optimization with respect to activity, particle dimensions and density, kinetics, adsorbent selection with respect to specificity, capacity, and dimensions, adsorption isotherms, residence time, and fluidization properties of biocatalyst and adsorbent in the fluidized bed, optimization of amounts or concentrations of all components, pH, and temperature.

A continuously recirculating bed of ion exchange resin was integrated into a bioreactor system that has been constructed and investigated for the simultaneous production and extraction of lactic acid using immobilized *Lactobacillus bulgaricus* (Patel *et al.*, 2008). In another example, a thermoplastic polymer was used, absorbing the product 3-methylcatechol into the polymer beads, which compared favorably to the single-phase operation (Prpich and Daugulis, 2007).

The potential of *magnetic adsorbent particles* has been examined in a review, analyzing where magnetic bead-based separations can fit into a downstream process, including an overview of magnetic separator technology. Examples of the separation and purification of proteins applying magnetic beads to a biosuspension have been given (Peuker *et al.*, 2010). Cells of *Pseudomonas aeruginosa* were immobilized in alginate together with magnetic particles and included by suspension polymerization to give spherical particles of polyurethane, aiming at the production of rhamnolipids. Continuous formation and separation of rhamnolipids was achieved in a stirred reactor, where the product was removed by introduction of gas and foam fractionation in a fractionation column, the rhamnolipids being concentrated in the foam phase, which was collected as collapsed foam in a separate receptacle. Particles carried with the foam were efficiently held in a high-gradient magnetic field separator and recycled (Figure 11.19). Complete separation of rhamnolipids from the production medium with an average enrichment of 15 in the collapsed foam was demonstrated, yielding a final rhamnolipid amount of 70 g after four production cycles (Heyd *et al.*, 2010; Heyd *et al.*, 2011). Lipases were immobilized on

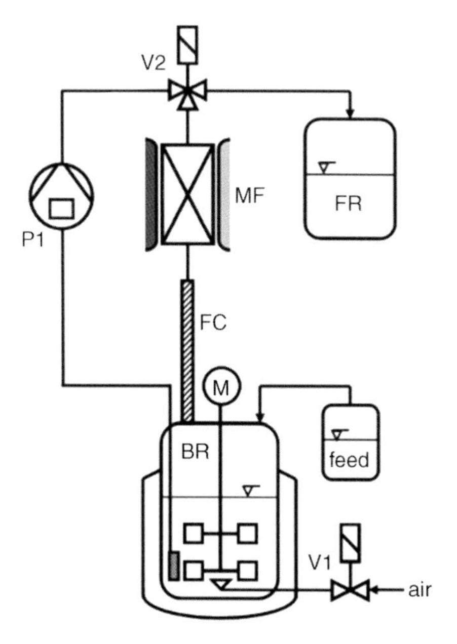

Figure 11.19 Scheme of integrated rhamnolipid production process management. BR: bioreactor; FC: fractionation column; FR: foam collection receptacle; MF: magnetic filter; P1 pump; V1/V2 valves (Heyd *et al.*, 2010).

acetylated magnetic nanoparticles. In an application for synthesis, the separation of the product is facilitated by a magnetic field where the immobilized lipase is held in place whereas the reaction solution is removed (Dyal *et al.*, 2003). Nucleic acids could be isolated directly from crude sample materials such as blood, tissue homogenates, cultivation media, and crude cell extracts by specifically functionalized magnetic particles. They could be removed easily and selectively even from viscous sample suspensions using small particles (diameter approximately 0.05–1 μm) in the presence of biological debris; the efficiency of magnetic separation is especially suited for large-scale purification (Berensmeier, 2006).

Combining *continuous chromatography* (see Section 6.5.3) and enzymatic reactions should allow for a new *in situ* or online product recovery process to achieve high reaction productivity and yield and make the biocatalysts economically more attractive. The integration imposes a series of constraints on the chromatographic separation, mainly on the applicable solvent, which is at the same time the reaction medium for the enzymatic reaction. For a model process, the integrated biocatalytic production of L-*allo*-threonine from glycine and acetaldehyde was investigated. Of crucial importance for this process is the separation of the two physicochemically similar amino acids, glycine and threonine, in particular in the presence of additional compounds, including enzyme cofactors. This separation was first investigated on a lab-scale simulated moving bed (SMB) using a triangle theory-based identification of SMB operating points. The two amino acids could be efficiently separated, applying aqueous eluents with minor content of organic cosolvent at neutral pH on a weak cation exchanger resin. Coupling the SMB to a continuously operated enzyme membrane reactor was achieved (Makart *et al.*, 2008). Another concept integrates chromatography and racemization for the preparation of L-methionine starting with a racemic mixture of DL-methionine. For the chromatographic unit, the impact of the mobile phase (buffer and methanol) on the stationary phase was evaluated in terms of resolution and selectivity as a function of temperature, varying content of methanol in the mobile phase, and pH. The experiments for determination of the racemization kinetics were performed for a compromised parameter set using crude lyophilizate (Petruševska-Seebach *et al.*, 2009). For the integration of a racemization reaction into a continuous chromatographic process, optimal process setups were determined indicating that a simple fully integrated SMB system with three zones is a particularly promising option (Palacios *et al.*, 2011).

The potential of *crystallization* as an ISPR technique for the biocatalytic production of crystalline compounds has been discussed in a review, including examples of relevant biocatalytic conversions for fine chemicals' production. These may include amino and carboxylic acids, chiral alcohols, and antibiotics. Crystallization using an external crystallizer may be favorable (Figure 11.20), inducing crystallization can be achieved as soon as supersaturation is reached, for example, by cooling, followed by seeding; finally, the product can be isolated by filtration (Buque-Taboada *et al.*, 2006).

In a case study, the production of 6*R*-dihydro-oxoisophorone (DOIP) has been investigated. The synthesis involves the asymmetric reduction of 4-oxoisophorone using *Saccharomyces cerevisiae* or *S. rouxii*. The product undergoes degradation by

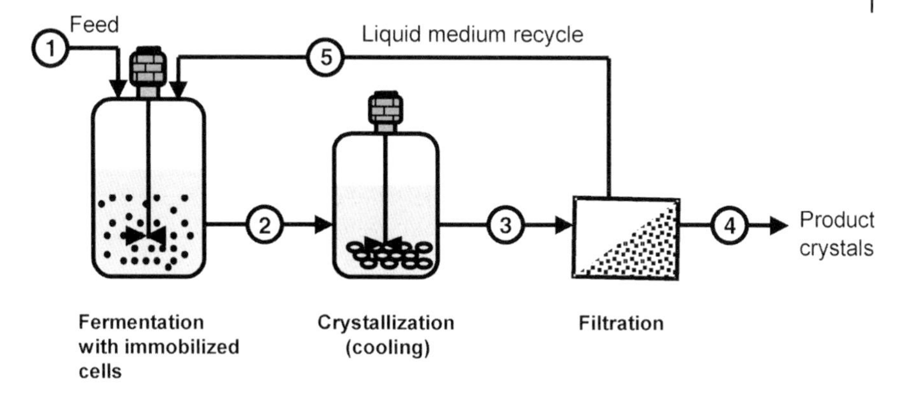

Figure 11.20 External configurations for ISPR with cooling crystallization. Cell retention is achieved by cell immobilization. The product solution is directly fed into the external crystallization loop to crystallize the product. The product-depleted mother liquor is recycled to the bioreactor (Buque-Taboada *et al.*, 2006, reproduced with permission from Springer Science + Business Media).

the yeast to an unwanted by-product. DOIP is used as a key intermediate in the production of some carotenoids. In the case study, immobilized *S. rouxii* was employed. For downstream processing, extraction of DOIP with ethyl acetate and subsequent crystallization was chosen. An ultrafiltration unit was applied to separate the liquid from the cells and to recycle the cells to the fermenter. Crystallization was perfomed by cooling to 5 °C; the kinetics of the biotransformation as well as of the crystallization had been investigated prior to the experiments. ISPR led to significant advantages over the nonintegrated case in terms of increased productivity and yield with a corresponding decrease both in the number of downstream processing steps and in the quantity of waste streams (Buque-Taboada *et al.*, 2006).

A recent work addressed the design and investigation of an enzymatic racemization with an integrated crystallization step in order to enhance the overall performance. A database was developed including the characterization of an amino acid racemase, the kinetics of the racemization, and the solubility ternary phase diagrams of the system. The proposed process concept couples preferential crystallization and racemization using a purified lyophilizate for the production of L-asparagine (L-Asn) using a racemic mixture of DL-asparagine (conglomerate-forming system) as a starting material. The concept was experimentally proven and the kinetics of the racemization was examined for D- and L-Asn in water (Petruševska-Seebach *et al.*, 2009). The synthesis of enantiopure *S*-mandelic acid derivatives was achieved using a hydroxynitrile lyase combined with enantioselective crystallization (Seidel-Morgenstern *et al.*, 2010).

Another approach is *product precipitation*. This not only can shift the equilibrium toward product formation but can also avoid product inhibition. Examples are the synthesis of Aspartame, in which the Z-protected sweetener precipitates from the reaction mixture (see Section 8.4.3.1), and the synthesis of the β-lactam antibiotic amoxicillin (Section 12.3) (Spieß and Kasche, 2001).

To overcome the limitation of product yield formation by the thermodynamic equilibrium, ISPR using a *stripping process* was applied, with an airflow removing the product, a volatile secondary alcohol from the reaction mixture. Whole lyophilized cells of an *Escherichia coli* overexpressing the alcohol dehydrogenase from *Rhodococcus ruber* were used for the asymmetric reduction of ketones to produce secondary alcohols. The conversion of selected biotransformations could thus be increased up to completeness (Goldberg *et al.*, 2006).

11.4.3
Reactor Instrumentation

Reactor equipment is shown in Figure 11.21 for another example, namely, the discontinuous hydrolysis of penicillin, together with details of measurement and control units for automated processing during 24 h in batch mode. The supplementation of concentrated and thermostated substrate solution, as well as deionized water, is controlled by computer, as is the temperature of the reactor. The most important factor is pH control, which is achieved by neutralization with ammonia (see Section 12.3). The rate of the amount of ammonia consumed indicates the progress of the reaction – that is, its reaction rate. These data are compiled continuously by a computer. A high standard of process control is required in order to ensure a stable processing and a high standard of quality.

Reactors as well as unit operations require further peripheral equipment and, in particular, instrumentation for measurement and control. The optimal and common solution is a central process control unit that provides all the services required: acquisition of measurements from online data analysis, computerized data analysis and data reduction, representation of results, including display in tabular and graphical formats, storage of results, and documentation, thereby providing a process database. Detailed information with regard to instrumentation, equipment for measurement, reliability, and control of failure has been reported in specialized publications (Atkinson and Mavituna, 1991b; Zabriskie, 1985). The economics of specific singular process steps, and also of whole processes, has been described by Atkinson and Mavituna (1991). For bulk products, it should be pointed out that enzyme cost must be kept very low in order to ensure that the process is economically feasible; for example, typical enzyme costs should be $<1\%$ of the product price (cf. glucose–fructose syrup).

Exercises

11.1 Exercise relating to process sustainability
For what type of products are the *E*-factor and the mass index most important, and why?

11.2 Exercise relating to reactor configuration
What reactor configuration should be chosen in case of relevant substrate inhibition?

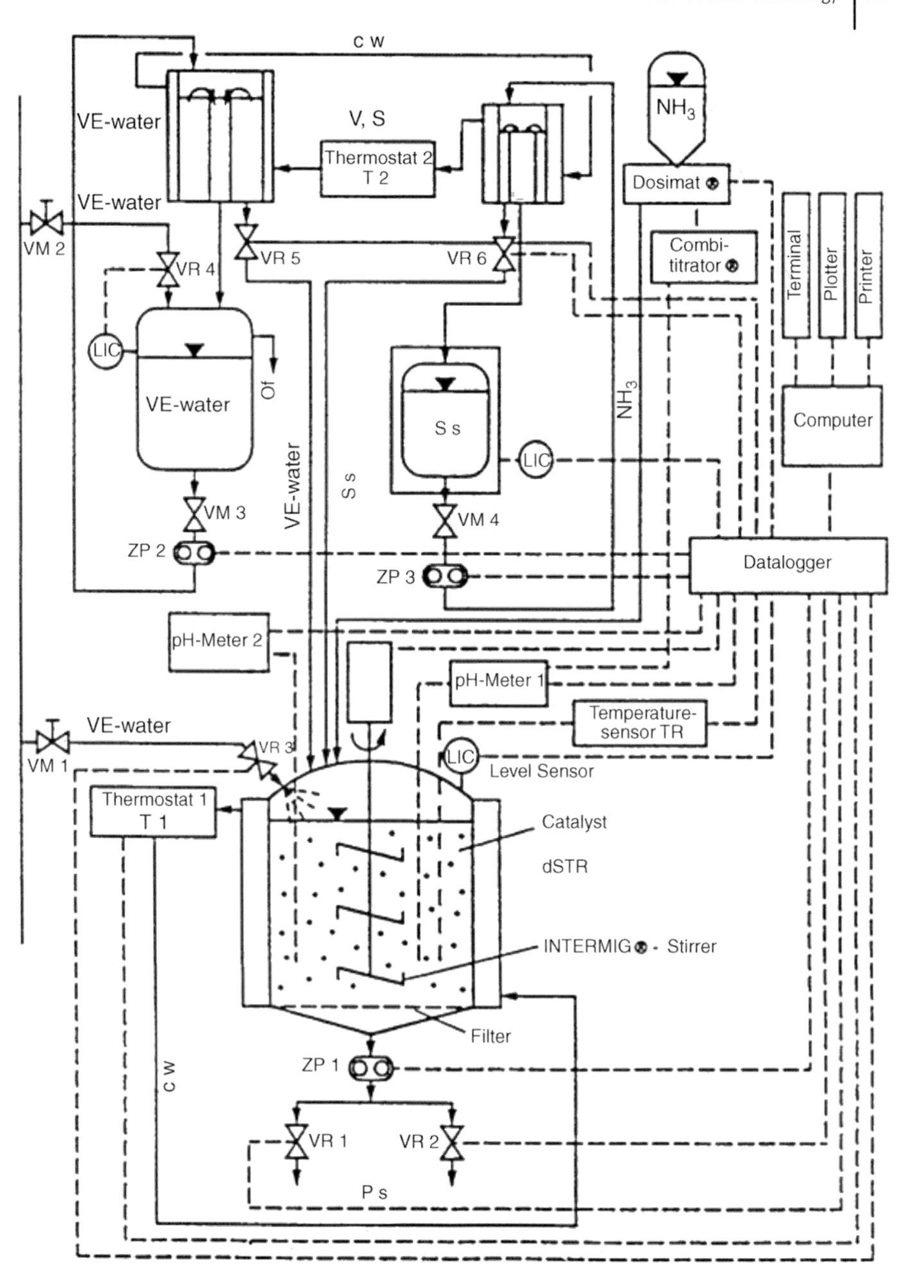

Figure 11.21 Measurement and control units of a laboratory reactor for automated discontinuous penicillin hydrolysis. VM, VR: mechanically or electrically controlled valves; VE: storage tank; ZP; pump (Tischer, 1990).

11.3 Exercises relating to mixing and shear

a. Calculate the mean circulation times for the following reactors (consider dimensions! Eq. (11.15)).
   i. $n = 5\,s^{-1}$; $H = 0.5$ m; $d_R = 0.3$ m; $d_i = 10$ cm
   ii. $n = 5\,s^{-1}$; $H = 2.0$ m; $d_R = 1.0$ m; $d_i = 0.3$ m.
   For what stirrer rates would equal mixing times be obtained? Compare the result for the value that can be taken from Figure 11.17b for a turbine and propeller in aqueous solution.
b. Calculate the increasing shear in stirred tank reactors with scale-up of a 10 l reactor to
   i. pilot scale with $0.5\,m^3$
   ii. production scale with $15\,m^3$
   in each case with constant energy input $P_R = 6\,\rho\,n^3 \cdot d_i^5$, with the stirrer diameter being one-third of the reactor diameter, and the dimensions of the stirred tank being $H: d_R = 3: 1$. The shear is (under turbulent conditions) proportional to $\rho(n \cdot d_i^2)$.

11.4 Exercises relating to process integration

a. What are obvious reasons to apply process integrated downstream processing?
b. What downstream operations are common to the isolation and purification of products?
c. What physicochemical properties of product(s) must be known in order to select optimal downstream processes?

Literature

General

Arpe, Hans-Jürgen (1992) *Ullmann's Encyclopedia of Industrial Chemistry: Principles of Chemical Reaction Engineering and Plant Design*, 5th edn, vol. **B4**, Wiley-VCH Verlag GmbH, Weinheim.

Arpe, Hans-Jürgen (1988) *Ullmann's Encyclopedia of Industrial Chemistry: Unit Operations I and II*, 5th edn, vol. **B2**, Wiley-VCH Verlag GmbH, Weinheim.

Atkinson, B. and Mavituna, F. (1991) *Biochemical Engineering and Biotechnology Handbook*, 2nd edn., MacMillan Publishers Ltd., New York.

Brauer, H. (1985) Stirred vessel reactors, in *Biotechnology*, vol. **2** (ed. H. Brauer), Wiley-VCH Verlag GmbH, Weinheim, pp. 395–444.

Kossen, N.W.F. and Oosterhuis, N.M.G. (1985) Modeling and scaling-up of bioreactors, in *Biotechnology*, vol. **2** (ed. H. Brauer), pp. Wiley-VCH Verlag GmbH, Weinheim, pp. 571–605.

Moo-Young, M. (1985) *Comprehensive Biotechnology*, vols **1 and 2**, Pergamon Press, Oxford.

Ullmann, F. and Gerhartz, W. (1988) *Ullmann's Encyclopedia of Industrial Chemistry: Unit Operations II*, 5th edn, vol. **B3**, Wiley-VCH Verlag GmbH, Weinheim.

References

Adrio, J.L., Hintermann, G.A., Demain, A.L., and Piret, J.M. (2002) Construction of hybrid bacterial deacetoxycephalosporin C synthases (expandases) by *in vivo* homologous recombination. *Enzyme Microb. Technol.*, **31**, 932–940.

Antrim, R.L. and Auterinen, A.-L. (1986) A new regenerable immobilized glucose isomerase. *Stärke*, **38**, 132–137.

Asenjo, J.A. and Andrews, B.A. (2008) Challenges and trends in bioseparations. *J. Chem. Technol. Biotechnol.*, **83** (2), 117–120.

Atkinson, B. and Mavituna, F. (1991a) *Biochemical Engineering and Biotechnology Handbook*, 2nd edn., MacMillan Publisher, New York, pp. 706, 707.

Atkinson, B. and Mavituna, F. (1991b) *Biochemical Engineering and Biotechnology Handbook*, 2nd edn., MacMillan Publisher, New York, pp. 1023–1057.

Atkinson, B. and Mavituna, F. (1991c) *Biochemical Engineering and Biotechnology Handbook*, 2nd edn., MacMillan Publisher, New York, pp. 1059–1109.

Bechtold, M., Makart, S. *et al.* (2007) Model-based characterization of an amino acid racemase from *Pseudomonas putida* DSM 3263 for application in medium-constrained continuous processes. *Biotechnol. Bioeng.*, **98** (4), 812–824.

Bechtold, M. and Panke, S. (2009) *In situ* product recovery integrated with biotransformations. *Chimia Int. J. Chem.*, **63** (6), 345–348.

Benz, G.T. (2011) Bioreactor design for chemical engineers. *Chem. Eng. Progress*, **107** (8), 21–26.

Berensmeier, S. (2006) Magnetic particles for the separation and purification of nucleic acids. *Appl. Microbiol. Biotechnol.*, **73** (3), 495–504.

Berensmeier, S., Ergezinger, M. *et al.* (2004) Design of immobilised dextransucrase for fluidised bed application. *J. Biotechnol.*, **114** (3), 255–267.

Blenke, H. (1985) Biochemical loop reactors, in *Biotechnology*, vol. **2** (eds H.-J. Rehm, G. Reed, and H. Brauer), Wiley-VCH Verlag GmbH, Weinheim, pp. 465–517.

Brinkmann, T. (2009) Nachhaltigkeit von Biotechnologischen Produkten. *CHEManager*, **9**, 23.

Buchholz, K. (1982) Reaction engineering parameters for immobilized biocatalysts. *Adv. Biochem. Eng.*, **24**, 39–71.

Buchholz, K. and Collins, J. (2010) *Concepts in Biotechnology: History, Science and Business*, Wiley-VCH Verlag GmbH, Weinheim, Germany.

Buque-Taboada, E., Straathof, A. *et al.* (2006) *In situ* product recovery (ISPR) by crystallization: basic principles, design, and potential applications in whole-cell biocatalysis. *Appl. Microbiol. Biotechnol.*, **71** (1), 1–12.

Cabral, J. and Tramper, J. (2000) Bioreactor design, in *Applied Biocatalysis* (eds A.J.J. Straathof and P. Adlercreutz), Harwood Academic Publishers, Amsterdam, pp. 339–378.

Carleysmith, S.W., Dunnil, P., and Lilly, M.D. (1980) Kinetic behavior of immobilized penicillin acylase. *Biotechnol. Bioeng.*, **22**, 735–756.

Chibata, I., Tosa, T., and Sato, T. (1987) Application of immobilized biocatalysts in pharmaceutical and chemical industries, in *Biotechnology*, vol. **7a** (eds H.-J. Rehm and G. Reed), Wiley-VCH Verlag GmbH, Weinheim, pp. 653–684.

Dunlap, C.E. (1985) Solids and liquids handling, in *Comprehensive Biotechnology*, vol. **2** (ed. M. Moo-Young), Pergamon Press, Oxford, pp. 237–271.

Dyal, A., Loos, K. *et al.* (2003) Activity of *Candida rugosa* lipase immobilized on gamma-Fe_2O_3 magnetic nanoparticles. *J. Am. Chem. Soc.*, **125** (7), 1684–1685.

Edlich, A., Sasse, V. *et al.* (2009) Mikroverfahrenstechnischer screening-reaktor zum monitoring biotechnologischer prozesse. *Chem. Ing. Tech.*, **81** (8), 1247.

Edlich, A., Magdanz, V. *et al.* (2010) Microfluidic reactor for continuous cultivation of *Saccharomyces cerevisiae*. *Biotechnol. Prog.*, **26** (5), 1259–1270.

Eigenberger, G. (1992) Fixed bed reactors, in *Ullmann's Encyclopedia of Industrial Chemistry: Principles of Chemical Reaction Engineering and Plant Design*, 5th edn., vol. **B4** Wiley-VCH Verlag GmbH, Weinheim, pp. 199–238.

Ergezinger, M., Bohnet, M. *et al.* (2006) Integrated enzymatic synthesis and adsorption of isomaltose in a multiphase fluidized bed reactor. *Eng. Life Sci.*, **6** (5), 481–487.

Goldberg, K., Edegger, K. *et al.* (2006) Overcoming the thermodynamic limitation in asymmetric hydrogen transfer reactions catalyzed by whole cells. *Biotechnol. Bioeng.*, **95** (1), 192–198.

Hatti-Kaul, R., Törnvall, U. *et al.* (2007) Industrial biotechnology for the production of bio-based chemicals: a cradle-to-grave perspective. *Trends Biotechnol.*, **25** (3), 119–124.

Heinzle, E., Biwer, A.P. *et al.* (2006) *Development of Sustainable Bioprocesses: Modeling and Assessment*, John Wiley & Sons, Ltd.

Hempel, D.C. (2007) Bioverfahrenstechnik, in *Dubbel, Taschenbuch für den Maschinenbau* (eds K.-H. Grote and J. Feldhusen), 21st edn, Springer-Verlag, Berlin, Heidelberg, New York (English version in preparation).

Hempel, D. and Dziallas, H. (1999) Scale-up, stirred-tank reactors. In: *Encyclopedia of Bioprocess Technology* (eds M. Flickinger and S. Drew), John Wiley & Sons, Inc.

Heyd, M., Franzreb, M. *et al.* (2010) Integrierte produktion und separation von biotensiden im mehrphasenreaktor. *Chem. Ing. Tech.*, **82** (1–2), 111–115.

Heyd, M., Franzreb, M. *et al.* (2011) Continuous rhamnolipid production with integrated product removal by foam fractionation and magnetic separation of immobilized *Pseudomonas aeruginosa*. *Biotechnol. Prog.*, **27** (3), 597–895.

Jensen, V.J. and Rugh, S. (1987) Industrial scale production and application of immobilized glucose isomerase, in *Methods in Enzymology*, vol. **136** (ed. K. Mosbach), Academic Press, Orlando, pp. 356–370.

Jördening, H.-J. and Buchholz, K. (1999) Fixed film stationary-bed and fluidized-bed reactors, in *Biotechnology*, vol. **11a** (ed. J. Winter), Wiley-VCH Verlag GmbH, pp. 493–515.

Jördening, H.-J., Mösche, M., and Küster, W. (1996) Entwicklung und BETRIEB VON Fließbettreaktoren für die anaerobe Abwasserreinigung. *Chem. Ing. Tech.*, **68**, 1152–1153; *Zuckerindustrie*, (1996), **121**, 847–854.

Karande, R., Schmid, A. *et al.* (2011) Miniaturizing biocatalysis: enzyme-catalyzed reactions in an aqueous/organic segmented flow capillary microreactor. *Adv. Synth. Catal*, **353** (13), 2511–2521.

Kuhn, D., Kholiq, M.A. *et al.* (2010) Intensification and economic and ecological assessment of a biocatalytic oxyfunctionalization process. *Green Chem.*, **12** (5), 815–827.

Liese, A., Seelbach, K., and Wandrey, C. (2000) *Industrial Biotransformations*, Wiley-VCH Verlag GmbH, Weinheim.

Liese, A., Seelbach, K. *et al.* (2000/2006) *Industrial Biotransformations*, Wiley-VCH Verlag GmbH, Weinheim, Germany.

Lilly, M.D. (1978) Immobilized enzyme reactors, in *Biotechnology, Dechema-Monographs*, vol. **82**, Verlag Chemie, Weinheim, pp. 165–180.

Lu, C.-H., Cheng, Y.-C., and Tsai, S.-W. (2002) Integration of membrane extraction. *Biotechnol. Bioeng.*, **79**, 31–34.

Makart, S., Bechtold, M. *et al.* (2008) Separation of amino acids by simulated moving bed under solvent constrained conditions for the integration of continuous chromatography and biotransformation. *Chem. Eng. Sci.*, **63** (21), 5347–5355.

Middleton, J.C. and Carpenter, K.J. (1992) Stirred-tank and loop reactors, in *Ullmann's Encyclopedia of Industrial Chemistry*, vol. **B4**, Wiley-VCH Verlag GmbH, Weinheim, pp. 167–180.

Moser, A. (1985) Imperfectly mixed bioreactor systems, in *Comprehensive Biotechnology*, vol. **2** (ed. M. Moo-Young), Pergamon Press, Oxford, pp. 77–98.

Nielsen, J.H., Villadsen, J. *et al.* (2003) *Bioreaction Engineering Principles*, Springer.

Olsen, H.S. (1995) Use of enzymes in food processing, in *Biotechnology*, vol. **9** (eds G. Reed and T.W. Nagodawithana), Wiley-VCH Verlag GmbH, Weinheim, pp. 663–736.

Palacios, J.G., Kaspereit, M. *et al.* (2011) Integrated simulated moving bed processes for production of single enantiomers. *Chem. Eng. Technol.*, **34** (5), 688–698.

Panke, S., Held, M. *et al.* (2004) Trends and innovations in industrial biocatalysis for the production of fine chemicals. *Curr. Opin. Biotechnol.*, **15** (4), 272–279.

Patel, M., Bassi, A.S. *et al.* (2008) Investigation of a dual-particle liquid–solid circulating fluidized bed bioreactor for extractive fermentation of lactic acid. *Biotechnol. Prog.*, **24** (4), 821–831.

Pedersen, S. and Christensen, M.W. (2000) Immobilized biocatalysts, in *Applied Biocatalysis* (eds A.J.J. Straathof and P. Adlercreutz), Harwood Academic Publishers, Amsterdam, pp. 213–228.

Petruševska-Seebach, K., Würges, K. *et al.* (2009) Enzyme-assisted physicochemical enantioseparation processes. Part II: Solid–liquid equilibria, preferential crystallization, chromatography and racemization reaction. *Chem. Eng. Sci.*, **64** (10), 2473–2482.

Peuker, U.A., Thomas, O. *et al.* (2010) Bioseparation, magnetic particle adsorbents, in *Encyclopedia of Industrial Biotechnology: Bioprocess, Bioseparation, and Cell Technology*, Wiley-Blackwell.

Pollard, D.J. and Woodley, J.M. (2007) Biocatalysis for pharmaceutical intermediates: the future is now. *Trends Biotechnol.*, **25** (2), 66–73.

Prenosil, J.E., Kut, Ö.M., Dunn, I.J., and Heinzle, E. (2009) Immobilized biocatalysts, in *Ullmann's Encyclopedia of Industrial Chemistry*, Wiley-VCH Verlag GmbH., Weinheim, Germany.

Prpich, G.P. and Daugulis, A.J. (2007) A novel solid–liquid two-phase partitioning bioreactor for the enhanced bioproduction of 3-methylcatechol. *Biotechnol. Bioeng.*, **98** (5), 1008–1016.

Reuss, M. and Bajpai, R. (1991) Stirred tank models, in *Biotechnology*, vol. **4** (eds H.-J. Rehm and G. Reed), Wiley-VCH Verlag GmbH, Weinheim, pp. 299–348.

Richardson, J.F. and Zaki, W.N. (1954) Sedimentation and fluidization: Part 1. *Trans. Inst. Chem. Eng.*, **32**, 35–53.

Schmalzried, S., Jenne, M., Mauch, K., and Reuss, M. (2003) Integration of physiology and fluid dynamics. *Adv. Biochem. Eng.*, **80**, 19–68.

Schugerl, K. and Hubbuch, J. (2005) Integrated bioprocesses. *Curr. Opin. Microbiol.*, **8** (3), 294–300.

Seibel, J., Jördening, H.J. *et al.* (2010) Extending synthetic routes for oligosaccharides by enzyme, substrate and reaction engineering. *Adv. Biochem. Eng. Biotechnol.*, **120**, 1–31.

Seidel-Morgenstern, A., von Langermann, J. *et al.* (2010) Kombination von biokatalyse und kristallisation zur darstellung enantiomerenreiner Mandelsäurederivate. *Chem. Ing. Tech.*, **82** (1–2), 93–100.

Sheldon, R.A. (2011) Reaction efficiencies and green chemistry metrics of biotransformations, in *Biocatalysis for Green Chemistry and Chemical Process Development* (eds J.A. Tao, J. Tao, and R.J. Kazlauskas), Wiley & Sons, New York, pp. 67–88.

Spieß, A. and Kasche, V. (2001) Enzymatic synthesis of amoxicillin: process integration using multiphase systems, in *Novel Frontiers in the Production of Compounds for Biomedical Use* (eds A. van Broekhoven *et al.*), Kluwer Academic Publishers, Amsterdam, pp. 169–192.

Spieß, A.C., Eberhard, W. *et al.* (2008) Prediction of partition coefficients using COSMO-RS: Solvent screening for maximum conversion in biocatalytic two-phase reaction systems. *Chem. Eng. Process. Process Intensif.*, **47** (6), 1034–1041.

Thomsen, M.S. and Nidetzky, B. (2009) Coated-wall microreactor for continuous biocatalytic transformations using immobilized enzymes. *Biotechnol. J.*, **4** (1), 98–107.

Tischer, W. (1990) Immobilisierte enzyme in der anwendung, in *Jahrbuch Biotechnologie* (eds P. Präve, M. Schlingmann, W. Crueger, K. Esser, R. Thauer, and J. Wagner), Carl Hanser Verlag, München, pp. 251–275.

Uhlig, H. (1991) *Enzyme Arbeiten für Uns*, Hanser-Verlag, München, pp. 214–241.

Van Aken, K., Strekowski, L. *et al.* (2006) EcoScale, a semi-quantitative tool to select an organic preparation based on economical and ecological parameters. *Beilstein J. Org. Chem.*, **2** (1), 3.

Villadsen, J., Nielsen, J., and Lidén, G. (2011) *Bioreaction Engineeering Principles*, 3rd edn, Springer, Berlin.

Wenda, S., Illner, S. *et al.* (2011) Industrial biotechnology: the future of green chemistry? *Green Chem.*, **13** (11), 3007–3047.

Werther, J. (1992) Fluidized bed reactors, in *Ullmann's Encyclopedia of Industrial Chemistry: Principles of Chemical Reaction Engineering and Plant Design*, 5th edn., vol. **B4**, Wiley-VCH Verlag GmbH, Weinheim, pp. 239–274.

Weuster-Botz, D., Hekmat, D. *et al.* (2007) Enabling technologies: fermentation and downstream processing. *Adv. Biochem. Eng. Biotechnol.*, **105**, 205–247.

Willaert, R. (2009) Cell immobilization: engineering aspects, *Encyclopedia of Industrial Biotechnology: Bioprocess, Bioseparation, and Cell Technology*, Wiley-Blackwell

Wong, M., Woodley, J.M. *et al.* (2010) Application of bipolar electrodialysis to *E. coli* fermentation for simultaneous acetate removal and pH control. *Biotechnol. Lett.*, **32** (8), 1053–1057.

Woodley, J.M., Bisschops, M. *et al.* (2008) Future directions for *in-situ* product removal ISPR. *J. Chem. Technol. Biotechnol.*, **83** (2), 121–123.

Zabriskie, D.W. (1985) Data analysis, in *Comprehensive Biotechnology*, vol. **2** (ed. M. Moo-Young), Pergamon Press, Oxford, pp. 175–190.

Zlokarnik, M. Rührtechnik (1972) in *Ullmann's Encyclopädie der Technischen Chemie*, vol. **2**, Verlag Chemie, Weinheim.

12
Case Studies

12.1
Starch Processing and Glucose Isomerization

12.1.1
Starch Processing

Enzymatic starch hydrolysis represents the first industrial enzyme process, established since the 1840s in France (see Section 1.3.1, Figure 1.3). Actually, starch processing is performed at a very large scale, some 40 million t/a. Major products are dextrins, glucose, and glucose–fructose syrup (Section 12.1.2), which are used mostly in food and ethanol manufacture (Section 12.2.1). Since starch is a polysaccharide, a macromolecular substrate, soluble enzymes must be applied for hydrolysis, and these must (necessarily) be available at low cost, in the range of €10/l (or €10/kg) or lower. Enzymes applied for starch processing comprise α-amylase, glucoamylase, pullulanase, and isoamylase, the latter two exclusively hydrolyzing α- (1 → 6) bonds. (For glucoamylase and α-amylase mechanisms, see Section 2.4, Figures 2.4 and 2.5.) Three-step processes are applied in general, except for dextrin production (Olsen, 1995) (for recent reviews, see Schäfer *et al.*, 2007; Bentley, 2010). A recent approach for modeling has been presented by Murthy and Johnston (Murthy *et al.*, 2011).

Starch is composed of two polysaccharides, with 15–30% amylose, a linear polyglucan with α-D-(1 → 4)-linked glucopyranosyl units, and 70–85% amylopectin, a polyglucan with main chains as in amylose, but with side chains linked by α-D-(1 → 6)-glycosidic bonds to the main chain (Section 7.3.2, Figure 7.3a and b). Native starch granules with partially crystalline amylose and amylopectin are essentially inaccessible to enzymes.

In an initial step, gelatinization of suspended ground starch is performed by thermal treatment (105–110 °C) in the so-called jet cooker, with steam injected through a jet, in order to make the polysaccharide particles accessible to α-amylase; this treatment, results in the loss of the crystalline structure of amylose and

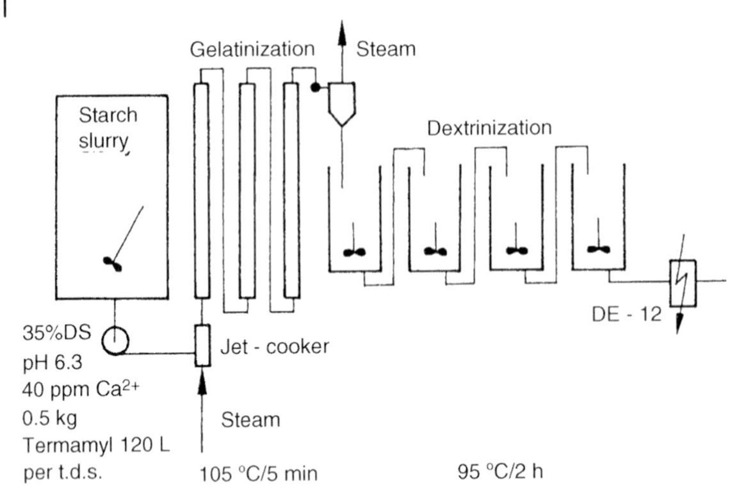

Figure 12.1 Reactor configuration for the starch liquefaction process. (Termamyl is a bacterial thermophilic α-amylase) (Olsen, 1995).

amylopectin, both due to thermal and due to shear effects, the resulting amorphous, hydrated structure being readily accessible to enzyme attack.[1] A process variant utilizes bacterial hyperthermophilic amylases (one part of the thermostable α-amylase is added to the gelatinization step at $T \geq 105\,^\circ C$) in order to utilize the synergism of thermal gelatinization and partial hydrolysis. The steam–enzyme–starch slurry proceeds to a tubular reactor with a residence time of 5–10 min.

The subsequent hydrolysis step – liquefaction, or dextrinization, partial hydrolysis by an endoglucanase (α-amylase) – follows at 85–95 °C for 1–2 h residence time in cascades of stirred tank reactors (Figure 12.1). The degree to which dextrinization is allowed to continue is determined by the type of product manufactured. Further technological conditions are pH 5.8–7 and 30–40% (dry solids) starch concentration with 0.4–1.2 l/t (dry matter) α-amylase present (Olsen, 1995). More detailed conditions for optimal enzyme activity are listed in Tables 12.1 and 12.2. (For different process variants, see Bentley and Williams, 1996; Bentley, 2010.) For wild-type α-amylases, appropriate concentrations of Ca^{2+} ions are essential for stabilizing the enzyme. Most enzymes used are produced by recombinant microorganisms that allow for lower prices. Some users require nonrecombinant enzymes that also are available commercially. This process step gives (malto-) dextrin ("soluble starch") of DE 10–15 as a commercial product (93% with a degree of polymerization (DP) > 4; 4% maltotriose, 3% maltose).[2] Such dextrins,

1) Recent developments, however, aim at enzymatic degradation of native starch starch (Tester *et al.*, 2006) (see also Section 12.2.1: *Future developments*).

2) The degree of hydrolysis is given by the DE (dextrose equivalent) value, which is determined by the reducing groups (equivalent to the amount of hydrolyzed bonds) taken as glucose equivalents per dry mass, DE 100 being pure glucose and 0 DE being unhydrolyzed long-chain starch.

Table 12.1 Enzymes used in industrial starch hydrolysis (Olsen, 1995; Bentley and Williams, 1996) (Bentley, 2010).

Enzyme	Source	Type[a]	Min. Dp substrate	Min. DP[b] product	Amount applied[c]	pH_{opt} (Ca)[e]	T_{opt} (°C)[d]
α-Amylase	*Aspergillus oryzae*	Endo-α-(1 → 4)	PS[f]	OS[f]		4–5 (50)	55–70
	A. niger					55–65	
	Bacillus subtilis				0.8–1.2	6–7 (150)	60–85
α-Amylase (thermostable)	*B. licheniformis* or *B. stearothermophilus*				0.6–1	5.8–7 (20)	105–110
ß-Amylase		Exo-α-(1 → 4)	3	2	0.3–0.7	4.7–5.0	55–60
Isoamylase[g]	*Pseudomonas* spp.,	Endo-α-(1 → 6)					
	Bacillus spp.						
Pullulanase[g]	*Pullularia*	Endo-	>3	3		3.5–5	55–65
	Streptococcus	α- (1 → 6)					
Glucoamylase (amyloglucosidase)	*A. oryzae*	Exo-α-(1 → 4)	2	1	0.5–1.1	4.2–4.6	55–65
	A. niger,	and α-(1 → 6)			(300 U ml^{-1})		
	Rhizopus oryzae	(Slow)					

a) Type of glycosidic bond hydrolyzed.
b) Minimal degree of polymerization (DP).
c) Preparation of mean activity (l or kg enzyme per ton starch dry substance).
d) Optimal application temperature.
e) Ca^{2+} required, ppm.
f) PS: polysaccharide; OS: oligosaccharides.
g) Applied in blends with glucoamylase.

Table 12.2 Process data for starch hydrolysis (Olsen, 1995; Bentley and Williams, 1996; Bentley, 2010).

Substrate and enzyme applied	Process step	Equipment/unit operation[b]	Conditions
Starch suspension 30–40% d.s.[a] α-Amylase (thermostable)	Heating, gelatinization	Storage tanks Heat exchanger Tubular reactor: Jet cooker	ca. 105–110 °C, pH 5.8–6,5, 5–10 min
	Flash cooling	Heat exchanger	
(DE 8–15)[2] 25–40% d.s. α-Amylase	Partial hydrolysis (solubilisation, dextrinization), cooling	Cascade of stirred tank reactors Heat exchanger	85–95 ° C, pH 6,5, 1–2 h Ca-ions (20–150 ppm), 35% d.s.[a]
Optional: Starch/dextrins (DE 10–15), ß-amylase	Hydrolysis to maltose syrup[c]		55–60 °C, pH 4.5–5, 24–40 h
Partial hydrolysate (DE 8–15), Glucoamylase[d]	Hydrolysis to glucose syrup	Cascade of 6–12 stirred tank reactors	55–60 °C, pH 4,5, 24–72 h
Hydrolysate (DE 42) or glucose syrup (DE 95)	Decoloring (active carbon) Deionisation (ion exchange) Concentration	Filtration or centrifugation Adsorption (activated carbon) Cation-, anion exchange Evaporators	
Product: Crystalline glucose	Crystallisation, centrifugation, drying	Crystallizers	

a) Dry substance.
b) For simplicity pumps and mixing operations (pH adjustment) are omitted.
c) Maltose content of 50, 60, or up to 80%, with higher maltooligosaccharides.
d) The enzymes used are mostly mixtures of enzymes with different activities and specificities.

for example, with oligosaccharide fractions of 4–20 glucosyl units, are used in large amounts in food industries.

Commercial products from extensive α-amylase hydrolysis typically have a DE of 30, or a DE close to 42, with about 19% glucose, 14% maltose, 12% maltotriose, and 55% higher oligosaccharides. The degree of hydrolysis may be controlled by rapid enzyme inactivation via acid and/or thermal treatment, for example, by another jet-cooking step at temperatures of $\geq$120 °C with a pH between 3.8 and 4.5 (Bentley, 2010). New recombinant α-amylases do not require Ca^{2+} and are active at pH down to about 5, so that little shift in pH is required for further hydrolysis by glucoamylase (Pedersen, 2002, personal communication). This substantially reduces upstream processing by an ion exchange operation. Additional use of isoamylase and/or pullulanase as debranching enzymes makes the starch conversion more efficient (see Section 7.3.2). Maltose syrups may be obtained by application of ß-amylases (Tables 12.1 and 12.2).

Further hydrolysis to produce a glucose syrup (DE 96–97%[2)]) (saccharification) requires another process step with glucoamylase at different conditions (Tables 12.1 and 12.2). The fungal enzyme releases α-(1 → 4)-bound glucose from the nonreducing end of amylose or amylopectin. It also hydrolyzes α-(1 → 6) bonds at a much lower rate. This step proceeds considerably faster when pullulanase or isoamylase are added, which are used in industrial processes. The reaction is carried out with substrate solutions of 25–40% (dry solids) at pH 4.3–4.5, 55–60 °C, and reaction times of 24–72 h, depending on the degree of hydrolysis required. Yields for glucose syrups are in the range of 96–97%, and the by-products are 2–3% disaccharides (maltose and isomaltose) and 1–2% higher oligosaccharides.

After completion of the reaction, the enzymes are inactivated at 120 °C and pH >4.5, or removed by ion exchange, in order to minimize the formation of reversion products. The conversion takes place in large stirred tanks at a high residence time.

Decoloring and ion exchange (elimination of metal ions) are essential subsequent steps. Elevated temperatures must be applied to avoid crystal formation while storing the product syrup. Syrups with high degree of DE up to those with essentially pure glucose, which can be crystallized to yield crystalline glucose or isomerized to a glucose–fructose syrup, can thus be produced. However, these require additional downstream processing, since besides low amounts of oligosaccharides from starch (i.e., maltose), other oligosaccharides are formed by condensation, that is, the reverse reaction catalyzed by the enzyme, but with different regioselectivity. The formation of such "reversion" (condensation) products is favored at high sugar concentrations, with preferential formation of an α-(1 → 6)-glycosidic bond, leading to isomaltose and higher isomaltooligosaccharides (see Section 2.4).

12.1.2
The Manufacture of Glucose–Fructose Syrup

Fructose exhibits a significantly (about twofold) higher sweetening power compared to glucose, which is produced in large amounts from corn (maize) starch. Consequently, glucose is isomerized enzymatically in part to fructose. (For the structure, mechanism, and immobilization of glucose isomerase, and reaction equilibrium, see Section 8.4.1; commercial biocatalysts are listed in Table 8.9.) This is the largest process performed with immobilized enzymes, using approximately 1000 tons of immobilized biocatalyst and producing a glucose–fructose syrup (high-fructose corn syrup, HFCS) with more than 12 million tons of product (dry matter) per year (Antrim and Auterinen, 1986; Misset, 2003; Pedersen and Christensen, 2000; Reilly and Antrim, 2003; Swaisgood, 2003; Antrim *et al.*, 1989).

Table 12.3 Process data and unit operations for the isomerization of glucose (Olsen, 1995; Misset, 2003; Bentley, 2010).

Process step	Unit operation/equipment	Reaction parameters
Substrate: glucose solution		40–50% d.s.[a], ≥ 95% glucose, peptides, organic substances
1. Substrate preparation	Adsorption, anion exchange	Elimination of Browning products, NO_3^-; (amino-) acids
2. Addition of (solution) $Na_2S_2O_8$, or SO_2, $MgSO_4$		SO_2 50–125 ppm, Mg^{2+} 50 ppm, (Ca^{2+} < 1ppm)
3. Adjustment of pH (NaOH)	Mixing vessel, storage tank	pH 7.5–8.2
4. Sterilization	Heat exchangers: steam, cooling water	
5. Isomerization	Tubular fixed bed reactors	55–60 °C, pH 7.8–8.2, 40–50% d.s.[a]
6. Product purification, finsihing	Filtration, adsorption (activated carbon), cation exchange, anion exchange	Elimination of browning products, ionic compounds
Evaporation, cooling	Heat exchangers	
Product: glucose–fructose syrup		≈ 70% d.s.[a], fructose: ≥ 42% (ref. to d.s.), or fructose: 55% [b]

a) d.s.: dry substance.
b) By addition of fructose to obtain higher sweetness.

At least three major companies provide and offer the process as a whole, this being a condition of successful marketing. The enzyme-catalyzed conversion is the central step of the process, but a considerable number of further operations must be carried out in a correct technical manner in order to ensure an economically feasible overall process.

The most relevant steps and unit operations for the overall process are detailed in Table 12.3. Further variations with modified process conditions are applied practically (Uhlig, 1991, 1998). In the preceding step, starch hydrolysis yields a glucose syrup of dextrose equivalents (DE) 95–97%. Glucose is isomerized subsequently to fructose by immobilized glucose isomerase. The endothermal reaction can only approach the chemical equilibrium with about 50% glucose and 50% fructose (depending on the temperature employed, see Table 2.7 and Section 8.4.1).

Several unit operations in downstream processing of starch hydrolysis must ensure the optimal conditions for the isomerization step. Thus, Ca^{2+} ions required for the stability of amylases must be removed by ion exchange as they inhibit glucose isomerase, which in turn requires Mg^{2+}. Furthermore, organic substances such as browning products, acids, peptides, and amino acids must be eliminated

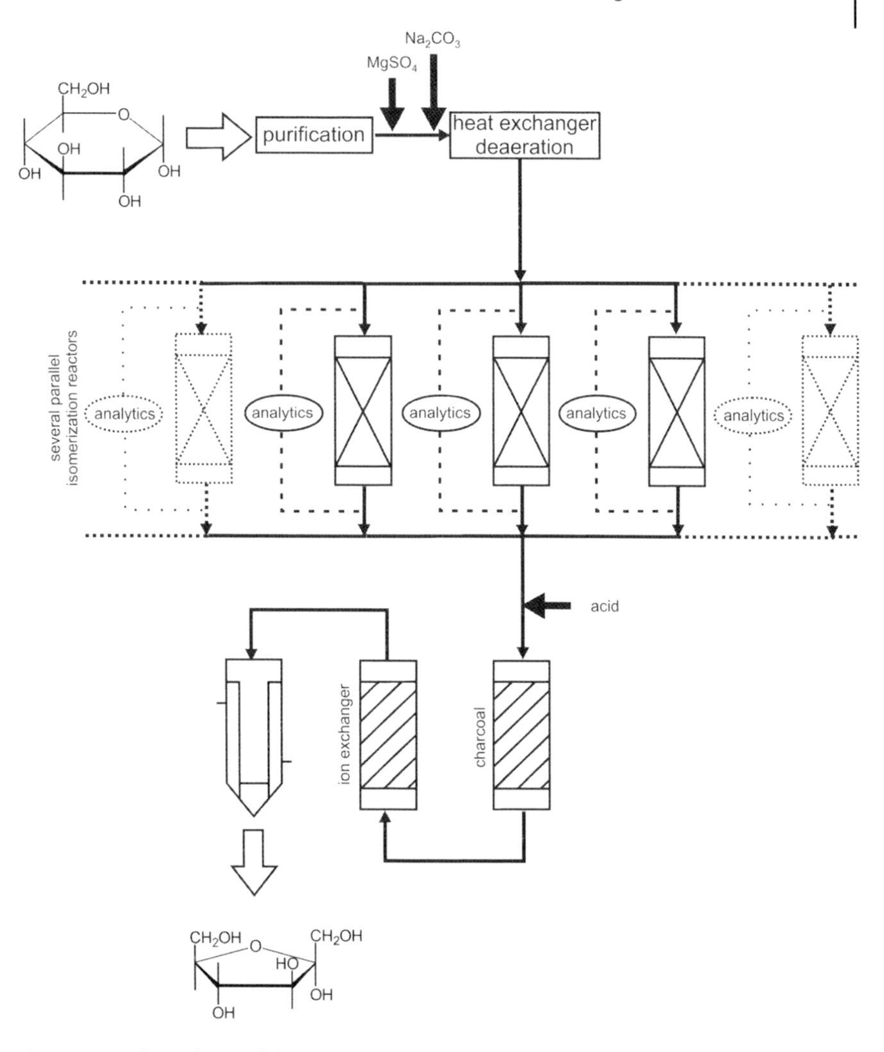

Figure 12.2 Flow scheme of glucose isomerization process with parallel fixed bed reactors, charcoal absorber, ion exchange column, and evaporator (with kind permission of Liese *et al.* 2000, 2006).

by adsorption to activated carbon. The dry matter content is adjusted by evaporation. The isomerization of glucose takes place in a series of tubular reactors with fixed beds of the immobilized glucose isomerase run in parallel (Figure 12.2).

The fixed bed reactors for isomerization have diameters in the range of 0.6 to 1.5 m, and heights between 2 and 5 m; thus, volumes in the range of 0.5–9 m^3. They are operated at 55–60 °C, the temperature range being limited by the thermal stability of glucose and fructose. The pressure drop may be up to 3 bar. Biocatalyst particle sizes are in the range of 0.3–1.0 mm, with an activity of about 300 U g^{-1}. A reactor with 1.5 m diameter and 5 m height contains about 4000 kg of the biocatalyst. In a plant with 20 reactors, about 1000 tons of product (with 42% fructose, d.m.) can be produced each day (Olsen, 1995; Misset, 2003).

Most commercial immobilized glucose isomerases show an exponential activity decay as a function of residence time. Typically, the catalyst is replaced when the activity falls to 12.5% of the initial value. Half-lives are in the range of 80–150 days, with the most stable commercial glucose isomerases exhibiting a half-life of around 200 days. Biocatalyst productivities are in the range of 12–15 tons (product) per kg (biocatalyst). In order to maintain a constant fructose concentration, the feed flow rate is adjusted according to the enzyme's actual activity. The production rate can be kept nearly constant while operating several reactors with different age of biocatalyst (Olsen, 1995; Misset, 2003; Swaisgood, 2003; Bentley, 2010).

The costs of the individual steps have been discussed by Olsen (1995). As in most processes with a low product price (often the case in the food industry), the raw material (starch) dominates the overall cost at 37% (about €8.8 per 100 kg HFCS), followed by milling and feed preparation at 29% (not included in the scheme discussed here). As suggested previously, the enzyme cost must be kept very low, and consequently liquefaction enzymes (α-amylases) comprise 0.7%, saccharification enzymes (glucoamylases and pullulanases) 1.5%, and glucose isomerase 1.6% of the overall cost. Total production costs have been reported as about €24 per 100 kg HFCS.

The product of the enzymatic isomerization must meet established quality standards, determining the value and price of the product. The most important of these are listed in Table 12.4. These standards are ensured by the application of several downstream unit operations (Table 12.3). They include its color, which depends on the formation of by-products, the so-called browning products. Not only the process parameters (e.g., pH, temperature, etc.) but also the nonsugar compounds in the

Table 12.4 Quality requirements for glucose–fructose syrups (HFCS, three specifications) (Aschengreen, 1984; Bonse and Schindler, 1986; Misset, 2003).

Parameter	A	B	C[a]
Fructose content (%)	42	42	55
Glucose content (%)	52–53	52–53	41
Oligosaccharides (%)	4–6		
Dry matter (%)	71	95–96	
Density ($kg l^{-1}$, 38 °C)	1.34		
pH	3.5–4.5		
Viscosity (Centipoise, 27 °C)	160		
Concentration of Ca^{2+} (ppm)		< 1	
Ash content (%)	0.05		
Turbidity	No	No	No
Extinction (UV and visible)	0.5[b] (UV 280 nm)	Low	Low
Relative sweetness (relating to sucrose = 100)	100	100	110

a) By blending with fructose syrup.
b) Value given for substrate.

substrate solution have significant influence on the formation of color. A question remains as to whether a purification step preceeding the enzymatic conversion is required and is economically feasible, or if appropriate process conditions can sufficiently avoid color formation. This decision must be made in the context of process technology.

12.2 Biofuels from Biomass

The need for biofuels from renewable biomass must not be discussed in depth – it is well known that fossil sources have been, and must be to a great extent, substituted by renewable sources. Not only high economic relevance but also heavy controversies, political, ecological, and social, characterize the production of biofuels. The problems of competition with food production as well as environmental considerations requiring massive reduction of CO_2 emissions underline the need to utilize lignocellulosic raw materials, as will be outlined in Section 12.2.2. Second-generation or advanced biofuels and recombinant technologies evoke expectations to reduce the dependence on fossil resources and to reduce significantly greenhouse gas emissions – a calculation of emissions gave 94 for gasoline, 77 for current bioethanol, and 11 for cellulosics-based ethanol (all kg of CO_2 equivalent per MJ of fuel production and burning) (C&EN, 2007). Advanced biofuels must achieve at least 50% reduction of emissions and include cellulosic ethanol and biomass-based diesel (from RSF, 2011, cited from: Renewable Fuel Standard, USA (http://www.nap.edu/catalog.php?record_id=13105). A large number of assessment studies, including life cycle assessment, have been conducted in recent years to assess the environmental merit of biofuels, with contrasting results. General lessons that emerge are that expansion of agricultural land usage for sugar and starch crops should proceed with extreme caution. A wider range of environmental impacts, including resource depletion, global warming, ozone depletion, acidification, eutrophication, human and ecological health, smog formation, and so on were considered, but came up with critical and divergent conclusions, including questionable social benefit (Solomon *et al.*, 2007; Von Blottnitz and Curran, 2007).

Biofuels from lignocellulosics represent a major challenge: they offer a huge potential for biofuel manufacture, providing new raw material sources, without competing with food production, and the option for considerably reduced greenhouse gas emissions. Cellulosic ethanol is projected to be more environmentally beneficial, and has a greater energy output to input ratio than grain ethanol (Balat, 2009; Solomon *et al.*, 2007).

Ethanol, and notably cellulose-based ethanol, are examples showing how research, technology development, market, and politics are intimately linked. Government support, funding of investment, and, more important, subsidies (direct and indirect via taxes), and legislation (fixing the ethanol content of fuels) are most important issues. Subsidies for biofuels were, from 1980 to 2009, over $1 billion (average) annually (Johnson, 2011b). Furthermore, market prices of oil and conventional fuel, as well as of feed materials, which compete in agricultural production, play the most important role (Coyle, 2010; Westcott and Usda, 2009). At the political level in the United States, ambitious targets were set by an energy law of 2007– the Energy Independence and Security Act of 2007 – that requires the production of 378 000 m^3/ per year of cellulosic biofuels by 2010, with a longer term goal of 60 million m^3/year, and further 19 million m^3/year of advanced biofuels and biodiesel by 2022 (Johnson, 2009) (RSF, 2011) (http://www.nap.edu/catalog.php?record_id=13105).

Supports at very high levels were offered: To spur commercial development, the US Department of Energy (DOE) announced grants of $564 million for 19 biorefinery projects at the pilot, demonstration, and commercial scale in December 2009, and the federal government of the United States committed more than $2 billion in support including research and private sector support. US DOE and Department of Agriculture state that there is enough biomass feedstock to displace approximately 40% of current US gasoline consumption, reclaiming that the use of marginal farmlands, as well as genetic manipulation of plants, to increase productivity would increase the potential for fuel production and diminish conflicts with food production (Iogen, 2010; Sticklen, 2008). However, a recent report by the US National Research Council says that cellulosic biofuel production lags behind, output is likely to be 25 000 m^3, far below the current RFS target of 945 000 m^3, and no commercial cellulosic biofuel plants exist. Key barriers are the high cost of producing cellulosic biofuels and uncertainties in future biofuel markets (Johnson, 2011a; RSF, 2011). An optimistic view sees commercialization of second-generation biofuels 5 years away from 2010 (Bacovsky, 2010). Large fuel, chemical, and car corporations, for example, BP, Exxon, Shell, General Motors, Novozymes, Dow, and DuPont collaborate with biofuel companies (Coyle, 2010; Westcott and Usda, 2009; Greer, 2008) in order to develop technical processes.

12.2.1 Starch-Based Ethanol Production

Ethanol currently, and traditionally, represents one of the most important BT products since the 19th century with up and downs depending on competing fuels and wartime requirements (Buchholz and Collins, 2010; Sections 3.3.1 and 4.3.3). At present, biofuels based on starch and sugar are produced on a very large scale. Ethanol production capacities have been 46 million m^3/a worldwide in 2006; in 2009, they were estimated at 47 and 28 (or even 40) million m^3/a in Brazil and the United States, respectively (Kosaric *et al.*, 2002; Johnson, 2007; Ritter, 2007; Coyle, 2010; Ingledew, 2010). However, production of corn-based ethanol in the United States will level off at 56–68 million m^3 per year. That volume would take around half the current US corn crop (Johnson, 2007). Other major ethanol producing countries are France, Germany, and Spain. Actually, ethanol can be blended with standard fuel up to 10% in the United States and up to 5% in Europe, whereas in Brazil all gasoline contains 20–25% ethanol, but also rather pure ethanol for use in modified motors. Most widespread *sources* for ethanol production by fermentation are sucrose, from sugar cane (Brazil) or sugar beet (EU), and glucose from corn (maize) (the United States), wheat and barley (EU), and sorghum, cassava, and sugarcane in China.[3)]

However, as has been mentioned, heavy controversies, political, ecological, and social, characterize the production of biofuels, notably from starch and sugar. A major *crisis* occured in 2007 and notably mid-2008, when food prices increased dramatically. Growing application of cereals for biofuels was thought to be part of this rise (estimated to about one-third by an OECD expert). Mainly poor people in third world countries suffered and protests turned into riots in two dozen countries, for example, in Haiti and Bangladesh (Solomon *et al.*, 2007; Sz, 2008; New York Times, 2008; von Braun, 2008). Even when in late 2008 food prices declined, critics argued that the food crisis was not over; ethical priority is required for nutrition (von Braun, 2008; von Braun, 2010). A life cycle analysis of starch- or grain-based bioethanol technology has been carried out, but advantages and disadvantages with respect to environmental benefits were difficult to balance (Schäfer *et al.*, 2007).

It is essential to note that economies of scale are crucial for plants, ethanol being a very cheap product (normally depending on governmental funding, and/or tax reduction, or legislation on blending of gasoline with bioethanol). Today, most fuel ethanol is produced by either the wet mill or the dry grind process of corn or wheat. The majority of the production in the United States and the growth in the industry worldwide is from dry grind plant construction (Bothast and Schlicher, 2005; Schäfer *et al.*, 2007). The first is an offshoot of the production of starch and sugar syrups (glucose and fructose) where virtually pure starch is used that is obtained from grain by a series of upstream processes including milling, hydrocyclone, centrifuge, filter separation, and multiple washing steps to separate starch from other grain components. Starch hydrolysis proceeds as described in Section 12.1.1. A recent approach for modeling has been presented by Murthy *et al.* (2011), a critical review of ethanol fermentation technologies including kinetic models is given by Bai *et al.* (2008). Most wet mills use continuous fermentation train technology, as outlined below (Ingledew, 2010).

The whole-grain *dry grind process* is becoming the working horse for bioethanol production both in the United States and in Europe. In this process, corn kernels or small grains such as wheat, barley, or triticale containing all starch and nonstarch components such as pentosans and other hemicelluloses, ß-glucans, proteins (soluble and insoluble), and fats are introduced into the process and thus influence the fermentation and subsequent steps. The three main stages are upstream processing, simultaneous saccharification and fermentation, and downstream processing (Figure 12.3) (Bothast and Schlicher, 2005; Kunz, 2007; Ingledew, 2010).

3) On March 20, 2008, China announced its latest 11th Five-Year Program on Renewable Energy Development with a target of increasing the production of sorghum-, cassava-, and sugarcane-based ethanol to 7000 m^3 per day by 2010 (O'Kray, 2010).

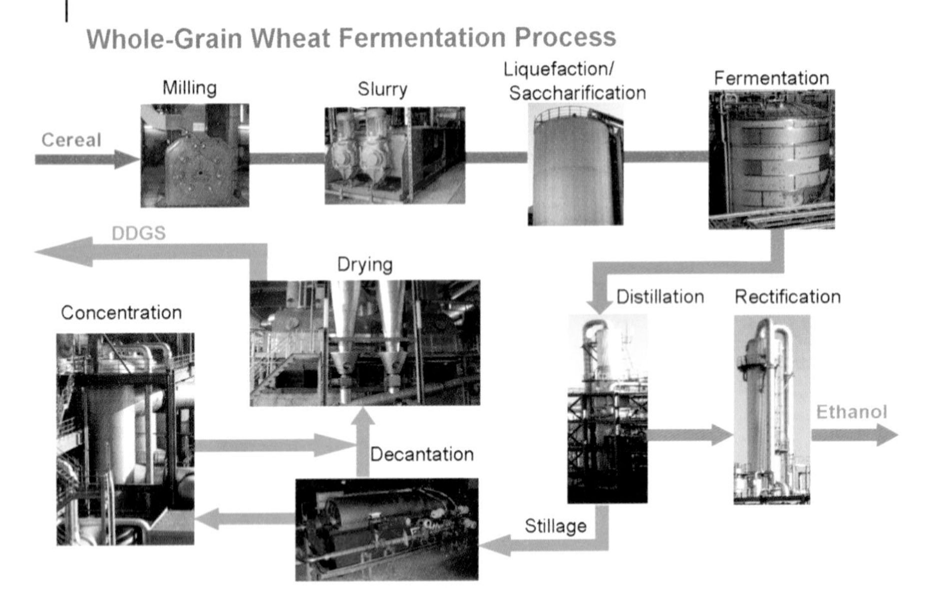

Figure 12.3 Whole dry grind process for bioethanol manufacture (DDGS: Dried Distillers Grains with Solubles) (Kunz, 2007).

In *upstream processing*, the major steps are hammer milling of the whole grain to a defined screen size range, starch prehydrolysis and enzymatic degradation of hemicelluloses, and sterilization. The ground grain is slurried in warm water (back-set from later process stages) and injected into a "jet cooker" where starch is converted by synergistic action of thermal denaturation (gelatinization) and hydrolysis with thermostable α-amylases first at about 90–120 °C, then at temperatures less than 90 °C, adding more α-amylases, to yield dextrins (oligoglucosides with 4–20 glucose units) (1–2 h).This mash is added directly to the fermenter after cooling. Enzymes applied additionally are proteases, ß-glucanases to convert ß-glucans to glucose, arabinoxylanases, xylanases, and other hemicellulases and pectinases in order to partially hydrolyze the corresponding polysaccharides and thus reduce the viscosity of the solution. Table 12.5 gives a summary of the enzymes used (Kunz, 2007; Schäfer *et al.*, 2007; Ingledew, 2010).

Simultaneous saccharification and fermentation (SSF) is subsequently carried out at 30–35 °C where, after yeast addition, three events ensue, dextrin conversion to glucose by glucoamylase (also called amyloglucosidase), yeast growth, and glucose fermentation to ethanol and carbon dioxide, which all take place simultaneously. The objective of SSF is to add glucoamylase in an appropriate dose so that glucose generation and consumption by yeast are balanced, which is rarely achieved in practice. More enzymes are added to hydrolyze soluble ß-glucans, xylans, and protein (Table 12.5). While most fermentation processes have worked in batch or fed batch mode, continuous systems with cell recycle are currently preferred, since productivity (in the range of 30–50 g/lh) is higher by an order of magnitude over batch fermentation (in the range of 2–2,5 g/lh). Trains of four or five fermenters

Table 12.5 Summary of the enzymes used in the dry mill ethanol process (Schäfer *et al.*, 2007).

Enzymes	Step	Purpose
α-Amylases	Liquefaction	Hydrolyze starch to dextrins and reduce solution viscosity
ß-Glucanases, cellulases, pectinases, arabinoxylanases, xylanases	Preliquefaction	Reduce viscosity
Glucoamylase	Simultaneous saccharification and fermentation	Provide glucose for yeast
Proteases	Fermentation	Provide nutrients for yeast and improve starch accessibility

operating in sequence are used.[4] Yeast in parts of the train will be constantly in the logarithmic phase where ethanol productivity is highest (up to 30-fold higher than in the stationary phase).[5] The biomass from the fermentation broth is separated by centrifugation and returned to the reactor(s). As substrate for fermentation, an equivalent of about 22% glucose solution is fed to yield 12–14% (v/v) ethanol, when over 90% of starch is converted to ethanol (approximately 48 h residence time). Fermenter capacities are up to 1000 m^3. Yields of products from 25.4 kg standardized grain (21 kg dry weight) are approximately 9.8–10.6 l of (190 US proof) ethanol, 8.2 kg of distillers dried grain, and 7.25–8.1 kg of CO_2 (usually not recovered). Recycling of the thin stillage and the reduction of freshwater consumption are crucial to reduce energy consumption. Continuous fermentation at a commercial plant is run for 2–3 weeks, with inoculation in the first fermenter. Virtually, all sugars or starches should be converted during fermentation (Kosaric *et al.*, 2002; Kunz, 2007; Schäfer *et al.*, 2007).

In *downstream operations*, ethanol is concentrated to about 94% v/v by distillation in a two-step procedure, or multistage high-pressure evaporation cascades, to reduce energy consumption. The aqueous alcohol is further dried commonly using molecular sieves. By-product utilization plays a crucial role both due to their high volume and due to their economic significance; per ton of ethanol about 1.2 ton of

4) Large plants (in the United States) produce more than 400 m^3/day; self-flocculating yeast, retained by baffles within the fermenter, is used in a plant with a capacity of 680 m^3/day, with a system of six fermenters with volumes of 1000 m^3 each arranged in a cascade, using a corn meal hydrolysate with a sugar concentration of 200–220 g/l. The final ethanol concentration was 11–12% (v/v) (Brethauer and Wyman, 2010). Production cost decreases with the size of the plant to a level around 0.30 at a capacity of about 750 kt/year of raw material (Galbe and Zacchi, 2007).

5) *Saccharomyces cervisiae* has been, and still is the workhorse for industrial fermentation. The ethanol yield and productivity of *Zymomonas mobilis* are higher; however, due to the undesirability of its biomass to be used as animal feed, this species cannot readily replace *S. cerevisiae* in ethanol production. The immobilization of yeast cells, despite research and development over several decades, has not proven successful for ethanol production, because in such systems the growth of the yeast cells is restrained and the slowly growing yeast cells are difficult to remove from the systems, which furthermore were not sufficiently stable (Bai *et al.*, 2008).

Figure 12.4 Production facility for bioethanol manufacture. The figure shows ethanol tanks (left), milling, saccharification (behind), fermentation station (background, left), and grain silos (background, middle). with kind permission of Suedzucker AG, Germany.

solid by-products (bran, husk, etc., largely sold as feed material), plus 1.1 ton CO_2, is to be marketed or disposed of. (For downstream processing including by-product processing, see Kosaric *et al.*, 2002; Ingledew, 2010.) Large-scale processes offer significant economic advantages; a capacity of at least 200 000 m^3 per year has been considered as advantageous; a large production facility is shown in Figure 12.4.

Future developments and challenges for research and improvements of existing technology are as follows (Kunz, 2007; Schäfer *et al.*, 2007; Ingledew, 2010):

- Improved hydrolytic enzymes for raw material hydrolysis at low temperature, and to reduce the influence of ß-glucans, pentosans, and proteins, which all impact viscosity of the thin stillage, in order to minimize the overall energy consumption in ethanol processing. Raw starch hydrolysis (RSH), conversion of starch below gelatinization temperature, offers advantages because of potential energy savings. Significant advances in enzyme efficiency for conversion of raw starch recently enabled commercial dry mill ethanol facilities of Broin and Associates (the United States) to utilize this technology. Hydrolysis and saccharification of starch is performed at approximately 48 °C, the biggest fear being the growth of thermotolerant lactobacilli and losses due to the production of lactic acid. It is believed that savings of 10–20% of the fuel value of ethanol can be realized. (Schäfer *et al.*, 2007; Ingledew, 2010).
- Optimized *Saccharomyces cervisiae,* modified by metabolic engineering, for extended substrate range to include, for example, pentoses, and having higher

osmotic, temperature, and alcohol tolerance and sufficiently high fermentation rates (Kunz, 2007).

- New advanced bioengineering and downstream engineering concepts, for example, ultrafiltration systems for high cell density processing, resulting in an ethanol productivity of up to 41 $gl^{-1}h^{-1}$ (Ben Chaabane *et al.*, 2006), and development of very high-gravity ethanol fermentation, where levels of ethanol of 20–23% v/v can be obtained. The concept of very high-gravity (VGH) fermentation technology has successfully been implemented in plants of Broin and Associates who recently announced the production of over 20% v/v of alcohol on average. New yeast strains exhibiting increased temperature stability and ethanol tolerance are now seen in some commercial plants (Devantier *et al.*, 2005; Ingledew, 2010; for more approaches, see Ingledew, 2010).
- Significant areas of research are in the production of plant hybrids with a higher starch content or a higher extractable starch content in the conversion of the corn kernel fiber fraction to ethanol (Bothast and Schlicher, 2005; Schäfer *et al.*, 2007).

12.2.2 Lignocellulose-Based Biofuels

12.2.2.1 General

Biofuels from lignocellulosics represent a major challenge: they offer a huge potential for biofuel manufacture, providing new raw material sources, without competing with food production, and the option for considerably reduced greenhouse gas emissions. A DOE life cycle analysis states that ethanol from cellulose reduces greenhouse gases by 90% compared to gasoline (Iogen, 2010). The "energy bill" of 2007 of the United States requires that in 2022, 15% of the annual fuel consumption must be substituted by biofuels (137 million m^3), of this 60% from lignocellulosics (Koltermann, 2009). But severe open problems remain to be solved, with respect to infrastructure, logistics, and transportation, and technical as well as economic problems, competition with other fuels, including biofuels. A range of recent *reviews* deal with the different aspects, feedstocks, pretreatment, fermentation, modeling, and technoeconomical issues (Olsson, 2007; Banerjee *et al.*, 2010; Pandey, 2010). One should keep in mind that this field has been a topic approached much earlier, subsequent to the first oil crisis during the early 1970s, as well as later initiatives, without success with respect to application.[6]

6) At the Sixth International Fermentation Symposium held in London, Ontario, (Canada) in 1980, 15 contributions were devoted to lignocellulosics utilization, and, with the decreasing attention due to declining oil prices, at the Third European Congress on Biotechnology, held in Munich in 1984, the number of such contributions was still 5 (Proceedings, 1981; Preprints, 1984), (Moo-Young, 1981; Kosaric *et al.*, 1981). After a promising (new) restart, about 2005, "much optimism has faded" currently – "after it became clear that the industry wouldn't be building commercial-scale plants as quickly as once thought." (Service, 2010).

Beyond political boundary conditions, a myriad of logistical and technological issues are to be considered and addressed in order to make lignocellulosic ethanol production feasible: raw material supply with sufficient volume and acceptable price, logistics ensuring transportation of large amounts of bulky low-value biomass, pretreatment, efficient and cheap enzyme and fermentation systems, economical downstream operation to isolate ethanol appropriate for use in fuel, as well as for isolating by-products, energy generation, and efficiency, minimal water requirements, and wastewater recycling.[7)] Some companies will exploit already existing streams of forestry waste and agricultural biomass (Iogen, 2010; Coyle, 2010; Westcott and Usda, 2009). With respect to economics, the topic is confronted with three main obstacles: (1) a narrow margin between feedstock and product prices, (2) high processing costs, and (3) huge capital investment (~\$15–40/l of annual ethanol production capacity, estimated at three–four times those of current plants, in the range from \$650 million to \$900 million for a 400 000 m^3 capacity plant. Effective coutilization of lignin and hemicellulose will be vital to economics; lignocellulose cocomponents may have higher selling prices (>\$1.0/kg) than ethanol (Coyle, 2010; Westcott and Usda, 2009; Zhang, 2008). Estimates suggest that the cost of producing cellulosic ethanol is about \$0.5/l, almost twice as high compared to starch-based ethanol.[8)] Competitive technologies must be considered, which include gasification or pyrolysis, to produce syngas, hydrothermal conversion, or electrochemical upgrading of biomass, and notably burning of lignocellulosics in efficient systems producing both electrical and thermal energy.[9)] Furthermore, electromobility is an important trend, so that only part of current liquid fuel consumption will be required in the future (Lens, 2005; Kersten *et al.*, 2006; Merino and Cherry, 2007; Greer, 2008; Gnansounou and Dauriat, 2010; Neubauer, 2011).

12.2.2.2 **Raw Materials**

A rather broad range of substrates, wood or parts of it, residues from agricultural raw materials processing in food industries, crop residues, for example, corn stover, corncobs, sugar cane bagasse, and energy crops, notably switchgrass and

7) Richard (2010) points out the logistical challenges for lignocellulosic biofuels that will require major changes in the supply chain infrastructure. Decentralized conversion and integrated systems may meet part of the logistic and of scale-up problems. The logistics of harvest, storage, transport and processing weave a complex web of interactions that will require massive investment in densification technology, new highway, rail, and pipe infrastructure, and they must meet economic, environmental, and community goals. The transportation fraction of energy required for lignocellulosics is two–three times that of grains and oilseeds. Competing at community scale is conversion of biomass into energy in combined heat and power systems, with minimum infrastructure cost, and at efficiencies of over 80% (Richard, 2010).

8) Several reviews conclude about the high spread of current and projected production costs of lignocellulosic ethanol due to the significant differences in assumptions concerning the following factors: composition and cost of feedstock, process design, conversion efficiency, valorization of coproducts, and energy conservation (Merino and Cherry, 2007; Gnansounou and Dauriat, 2010).

9) Next-generation biofuels, besides ethanol, include advanced biodiesel, hydrogenated vegetable oils, Fischer–Tropsch (FT) diesel, methyl-*tert*-butyl-ether (MTBE), biogas, hydrogen, and others. Some core processes such as FT synthesis are very efficient, whereas others like biogas are based on simple technology (Zinoviev *et al.*, 2010).

hybrid poplar, are available and have been taken into account. The global bioenergy potential from dedicated biomass plantations in the twenty-first century has been estimated under a range of sustainability requirements to safeguard food production, biodiversity, and terrestrial carbon storage. The combination of all biomass sources might provide 15–25% equivalents of the world's future energy demand in 2050[10] (Beringer *et al.*, 2011; Henry, 2010).

Agricultural residues in general contain approximately 30–40% cellulose and 20–30% hemicelluloses and in the range of 15–30% lignin, while softwood and hardwood content of cellulose is in the range from 40 to 42%, hemicellulose roughly from 20 to 30%, and lignin from 25 to 32%. By-product utilization, notably from lignin, is considered essential for a viable process. Pretreatment of the materials and the enzymatic hydrolysis of polysaccharides are key process steps. Hydrolysis can result in yields of between 0.5 and 0.75 g of C6 sugars per gram of cellulose for agricultural residues (corn stover), hard- and softwood (poplar and black spruce, respectively). Thus, it can be expected that bioethanol yields from agricultural residues have a potential range between 0.11 and 0.27 m^3t^{-1}, while wood residues could deliver bioethanol yields between 0.12 and 0.30 m^3t^{-1}. Prices for feedstocks are assumed to be in the range of $40–$60 per dry ton (corn stover may offer lower cost of $5/t) (Mabee and Saddler, 2009; Coyle, 2010; Westcott and Usda, 2009; Merino and Cherry, 2007).

Typically, a five-step process for the conversion of cellulosic biomass involves physically reducing the size by milling or chopping, pretreatment, enzymatic hydrolysis, fermentation, ethanol recovery by distillation or other separation technology, and by-product utilization (Figure 12.5); integration of part of the steps is considered as preferential (Merino and Cherry, 2007).

12.2.2.3 Pretreatment

Pretreatment is essential in order to improve the digestibility of the lignocellulosic biomass, the main aims being to make cellulosic solids digestible with yields of enzymatic hydrolysis higher than 90% in less than 3 days, to give high sugar concentration (above 10%, for glucose, additional arabinose, and xylose) for appropriate fermentation systems after enzymatic hydrolysis. The key factors are to dissolve hemicellulose and alter the lignin structure, providing an improved accessibility of cellulose for hydrolytic enzymes by a deprotected, disrupted, and hydrated structure with increased pore volume. The substrate characteristics that impact the rate of enzymatic hydrolysis

10) More data have been presented on estimated productivity (ethanol yield per ha per year) of current and potential bioenergy crops, including corn with grain and stover, sugarcane with sugar and bagasse, *Miscanthus* as a grass species, and poplar (Somerville *et al.*, 2010). According to the authors, evidence suggests that perennial grasses and trees can be produced sustainably, water efficiency being a key topic, favoring water-efficient grasses, or *Agave* spp. The authors discuss issues on a broad basis, such as yields, fertilizer needs, erosion, transportation, and drought-resistant plants for use in regions not used for food production. An essential step would be the use of all polysaccharide components, including pentosans, of plants for fuel production. The strategies should go beyond current crops and land used for food and feed, and avoiding the clearing of natural ecosystems (Somerville *et al.*, 2010).

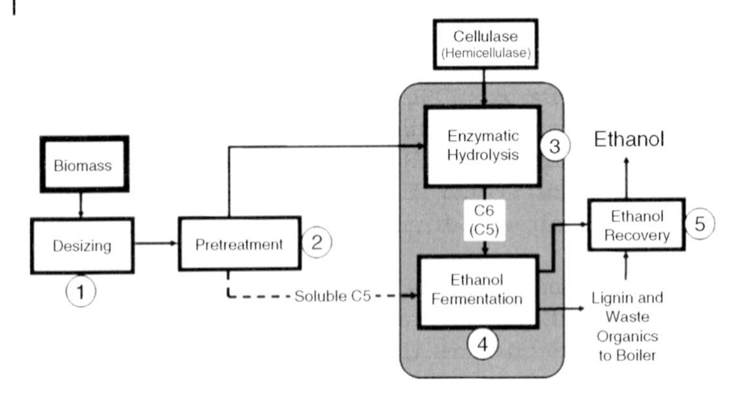

Figure 12.5 Five-step process for the conversion of biomass to ethanol (Merino and Cherry, 2007, reproduced with permission from Springer Science + Business Media).

include accessibility, degree of cellulose crystallinity, with regions of high crystallinity separated by amorphous regions, surrounded by hemicelluloses, and protected by lignin. The formation of fermentation inhibitors (formic, acetic, and levulinic acids, furan derivatives, and phenolic compounds) must be minimized. Minimum heat and power requirements and moderate costs are essential. It will be discussed here only shortly, the emphasis being on enzymatic hydrolysis (Hu *et al.*, 2008; Chandra *et al.*, 2007; Hendriks and Zeeman, 2009; Yang and Wyman, 2008; Alvira *et al.*, 2010).

The elementary fibrils of cellulose and their aggregations are determined by nature in native fibers such as cotton or wood pulp fibers and are laid down in various cell wall layers in a typical manner. Figure 12.6 shows the structural

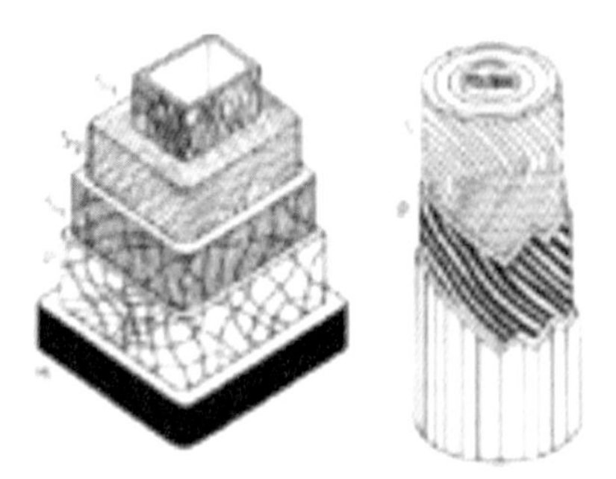

Figure 12.6 Positioning of the *cellulose* fibrils in wood (left) and cotton fibers (right) (from outer to inner position). Wood fibers: (M) middle lamella (lignin and hemicelluloses); (P) primary wall (fibril position unarranged); (S_1) secondary wall I (two or more fibrillar layers crossing one another and positioned spirally along the fiber axis); (S_2) secondary wall II (fibrils wound spirally around the fiber axis); (S_3) secondary wall III (fibrils tightly interlaced). Cotton fibers: (P) primary wall (interlaced fibrils); (S) secondary wall (fibrils wound spirally around the fiber axis; at distinct distances along the fiber axis the spiral reverses direction) (Krässig *et al.*, 2004).

organization of wood pulp and cotton fibers. The three different types of polymers, cellulose, hemicelluloses, and lignin, are intimately associated with each other, inhibiting enzymatic degradation. Cellulose exists in D-glucose subunits, linked by β-1,4 glycosidic bonds. Hemicelluloses represent complex carbohydrate structures that consist of different polymers like pentoses (e.g., xylan and araban), hexoses (mainly mannan, glucan, glucomanan, and galactan), and pectin. These are more easily hydrolyzable polysaccharides. Lignin is an amorphous heteropolymer consisting of three different phenylpropane units, with ester linkages to hemicelluloses, that provide the plant structural support, impermeability, and resistance against microbial (and enzymatic) attack (Alvira *et al.*, 2010; Krässig *et al.*, 2004; Hendriks and Zeeman, 2009).

Steam pretreatment (steam explosion), liquid hot water pretreatment (both developed since the 1980s), lime pretreatment, and ammonia-based pretreatments are concluded to provide high potential, and therefore the first two will be treated in some detail.[11] *Steam explosion* is the most widely employed physicochemical pretreatment for lignocellulosic biomass, which is subjected to pressurized steam at high temperature (up to 240 °C) for up to several minutes, and then suddenly depressurized. This pretreatment combines mechanical forces due to explosive decompression, and chemical effects due to the hydrolysis by the protons of water and the acetyl groups present in hemicelluloses (formation of acetic acid). Lignin is redistributed and to some extent removed from the material, and removal of hemicelluloses exposes the cellulose surface and increases enzyme accessibility to the cellulose. The pretreatment gives good yields in enzymatic hydrolysis with high sugar recovery, avoiding the formation of inhibitors. It has been proven efficient for a wide range of raw materials such as poplar hardwoods and agricultural residues such as corn stover and wheat straw. It is feasible for industrial-scale development, requires lower capital investment, and offers more potential for energy efficiency compared to other technologies (Hendriks and Zeeman, 2009; Alvira *et al.*, 2010). A concept of three-stage steam explosion of corn stover could reduce

11) Lime pretreatment with added $Ca(OH)_2$ (0.1 g per g substrate, 85–150 °C, 3–13 h), increases the digestibility of low-lignin containing biomass, that is, switchgrass, corn stover and poplar wood, by removal of lignin and thus increasing cellulose accessibility. Ammonia fiber explosion pretreatment (AFEX process, application of liquid ammonia at 60–100 °C, high pressure) leads to considerable (up to fivefold) increased digestibility of agricultural residues; however, the complexity of ammonia recovery may lead to significant associated cost (Alvira *et al.*, 2010; Hendriks and Zeeman, 2009). A promising approach may also be hydrothermal pre-treatment with carbon dioxide which considerably influenced enzymatic hydrolysis, and which avoids aggressive and/or toxic acids or solvents. The yield of glucose on the basis of cellulose reached about 80% (Matsushita *et al.*, 2010; Alvira *et al.*, 2010; Hendriks and Zeeman, 2009).Other methods, like biological (fungal), mechanical, acid, oxidative, thermal combined with chemical (e.g., acid), solvent or ionic liquid pretreatment, will not be discussed, either because of environmental impact, requiring full containment of chemicals, or side effects, for example, formation of toxic products, or high energy requirement, and/or high cost involved. It must be kept in mind that each material has different reactivity to pretreatment conditions (Hendriks and Zeeman, 2009; Alvira *et al.*, 2010).

the time required for pretreatment with a yield of 63% increased to a yield of over 70.2% (Yang *et al.*, 2010).

For *liquid hot water pretreatment* (LHW), water is applied in the liquid state under pressure at elevated temperatures (2–25 bar; 160–240 °C) to provoke alterations in the structure of the lignocelluloses induced by the change in pH value of liquid water under these conditions.[12] The objective is to solubilize mainly the hemicellulose, to make the cellulose more accessible, and to avoid the formation of inhibitors. Two fractions are obtained, one solid cellulose rich, and a liquid fraction rich in hemicellulose oligomers. pH control (in the range between 4 and 7) is essential to avoid the formation of inhibitors. Raw materials investigated are, for example, corn stover, sugarcane bagasse, and wheat straw.[13] LHW is attractive for its cost-saving potential, no catalyst or chemical requirement, and low-cost reactor construction (low-corrosion potential) (Yang and Wyman, 2008; Alvira *et al.*, 2010; Hendriks and Zeeman, 2009; Zetzl *et al.*, 2011). The firm Renmatix, with investment from BASF, has developed technology to extract C5 and C6 sugars to be fermented to biofuels, or basic chemicals, from cellulosic biomass using supercritical water (Bomgardner, 2012).

12.2.2.4 **Enzymes**

The complex structure of lignocelluloses (see before) requires multienzyme complexes for efficient hydrolysis (called cellulosomes at the genetic level) consisting of multiple subunits and acting synergistically; they include cellulases, with endo-β-glucanases (EG) and exo-β-glucanases, cellobiohydrolases (CBH), hemicellulases, with β-glucosidases (BG), xylanases, arabinofuranosidases, pectinases, pectin methyl esterase, pectate lyases, and so on. Most active systems comprise a large excess of cellobiohydrolases, cellobiose being an inhibitor for preceeding depolymerization steps. Much recent work has been devoted to enzyme research and recombinant production. Enzyme systems from the filamentous fungi, especially *Trichoderma reesei*, comprise three classes of cellulolytic enzymes, CBH (EC 3.2.1.91), EG (EC 3.2.1.4), and BG (EC 3.2.1.21). Eight major cellulase genes

12) Optimal conditions for LHW were identified, to be 200 °C, 10 min, which resulted in the highest fermentable sugar yield with minimal formation of sugar decomposition products. The LHW pretreatment solubilized 62% of hemicellulose as soluble oligomers and resulted in 54% glucose yield by 15 FPU cellulase per gram glucan after 120 h; the hydrolysate contained 56 g/l glucose and 12 g/l xylose – results still significantly too low for economical processing. Flow-through systems removed more hemicellulose and lignin from corn stover than batch systems (Kim, 2009).

13) In one example, dried distillers' grains with solubles (DDGS), a coproduct of corn ethanol production, and sugar cane bagasse were pretreated with liquid hot water (LHW) and ammonia, or ammonia fiber explosion (AFEX, 2 NH_3: 1 biomass, 100 °C, 30 min.) processes. Cellulose was readily converted to glucose from both LHW- and AFEX-treated DDGS using a mixture of commercial cellulase and β-glucosidase; adding pectinase preparations was necessary for releasing arabinose and xylose. Yields were 278 and 261 g sugars (i.e., total of arabinose, xylose, and glucose) per kg of DDGS (dry basis) for AFEX and LHW-pretreated DDGS, respectively (Dien *et al.*, 2008; Prior and Day, 2008).

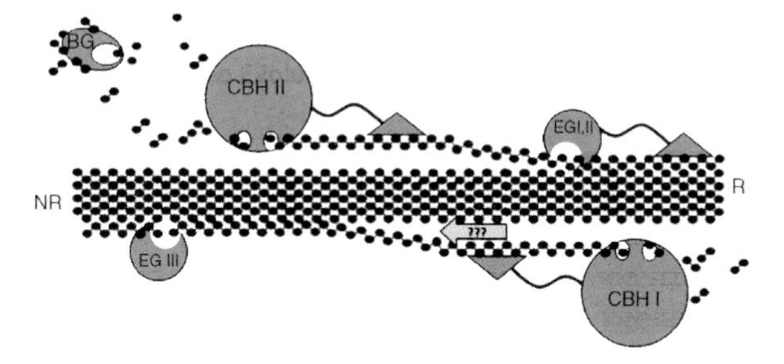

Figure 12.7 Schematic view of the primary *T. reesei* enzymes involved in hydrolysis of cellulose. Cellulose is represented as stacked chains of *black circles* with reducing (R) and nonreducing (NR) ends indicated. There are two major cellobiohydrolases that attack the cellulose chain ends processively from the reducing (CBH I) and nonreducing (CBH II) ends of the chain, releasing the glucose disaccharide cellobiose. In addition, there are three major endoglucanases depicted (EGI, II, and III) that attack the cellulose chain randomly, and two ß-glucosidases (BG) that hydrolyze cellobiose released by the CBHs to glucose. *Triangles* represent cellulose binding motifs, and the *arrow* represents an additional hypothetical protein component that may assist in cellulase action by disrupting the cellulose crystal structure (Merino and Cherry, 2007, reproduced with permission from Springer Science + Business Media).

have been identified from the *T. reesei* genome, two CBH, six EG, and two BG. All, with one exception, have a two-modular structure, consisting of a catalytic module and a carbohydrate binding module (CBM) connected with a linker region[14) (Figure 12.7). These enzymes act synergistically to catalyze the hydrolysis of cellulose. Two types of CBH constitute roughly 60 and 20% of the secreted protein mix and are critical to efficient hydrolysis. They hydrolyze the cellulose chain processively from the reducing and nonreducing chain ends, respectively, releasing the glucose disaccharide cellobiose. Endoglucanases constitute roughly 15% and hydrolyze ß-1,4 linkages within the cellulose chains, creating new reducing and nonreducing ends that can then be attacked by the CBH. ß-glucosidases constitute about 0.5% and hydrolyze cellobiose and some other short-chain cellodextrins into glucose (Merino and Cherry, 2007; Kumar *et al.*, 2008; Viikari *et al.*, 2007; Martinez *et al.*, 2008).

The *mechanisms* of cellulases are not completely understood yet, and differ in part from other known glucanase mechanisms. Thus, a number of glycoside hydrolases do not follow the classical catalytic mechanisms, as they lack a typical catalytic

14) In binding domains aromatic residues, tyrosine and tryptophane, and polar residues, proline, glutamine, and asparagine are responsible for binding to the cellulose surface (Mosier *et al.*, 1999). The accessibility of cellulase and xylanase enzymes to glucan and xylan, respectively, and its change with conversion were measured for different pretreated substrates. Xylanase supplementation of cellulases and ß-glucosidases resulted in an (incremental) increase in glucose release (Kumar and Wyman, 2009). Activities obtained in fermentation, for example, with *T. reesei*, exhibit clear correlation, for example, between the secreted amount of xylanase and mannanase enzymes induced by the presence of xylan in the carbon source (pretreated corn stover) (Kumar *et al.*, 2009; Viikari *et al.*, 2007).

base/nucleophile; a variety of mechanisms are used to replace this function[15] (Vuong and Wilson, 2010; Harris *et al.*, 2010). A recent review provides an overview of the structure–function relationships of the enzymes involved in plant cell wall deconstruction (Gilbert, 2010).

No simple, straight *kinetics* can be applied due to a range of complex reaction events and sequences, with different enzymes, a heterogeneous system of two phases, adsorption to different substrate surfaces, rates of hydrolysis and synergism of different enzymes with different substrates, and product inhibition patterns (Vlasenko *et al.*, 2010). A recent review by Bansal *et al.* (2009) provides an overview of published models and critically summarizes results and models of cellulase kinetics, to which the subsequent short summary refers (Bansal *et al.*, 2009). Furthermore, a detailed mechanistic model of enzymatic cellulose hydrolysis has been developed, including product inhibition, thermal deactivation, and surface heterogeneities of the substrate (Levine *et al.*, 2010). The major steps involved in hydrolysis are adsorption onto the substrate via the binding domain (CBM), formation of enzyme–substrate complex, hydrolysis of the ß-glycosidic bond, and simultaneous forward sliding of the enzyme along the cellulose chain (Divne *et al.*, 1998; Mulakala and Reilly, 2005), hydrolysis of cellobiose to glucose by ß-glucosidase; product inhibition must be considered, which must be taken into account in reaction engineering (see below)[16] (Andric *et al.*, 2010; Mosier *et al.*, 1999). The function of endoglucanases is primarily to increase the number of free insoluble cellulose chain ends, both reducing and nonreducing, and changing the specific surface area, increasing initially and leveling off subsequently. Rates of hydrolysis depend much on type of substrate, due to origin and pretreatment, and reaction progress, as well as synergism of different enzymes, notably endoglucanases and cellobiohydrolases, and xylosidases. Amorphous cellulose regions are hydrolyzed fast. Most important is a dramatic rate slowdown at high degrees of conversion. No consistent model that includes all effects and findings is available, due to the complexity of the problem. For application in cellulosics conversion, experimental investigations remain mandatory, with respect to (pretreated) substrates and their concentration, different enzyme mixtures and concentration, reaction time, pH, and temperature, in order to obtain optimal results with respect to minimized enzyme amount, high conversion and yield, and high end product concentration (Merino and Cherry, 2007). As an example, experiments using Celluclast 1.5, derived from *T. reesei* (Novozymes A/S, Bagsværd, Denmark) showed the feasibility of designing minimal enzyme mixtures for pretreated lignocellulosic biomass, different from the original enzyme mixture, employing statistically designed

15) Vlasenko *et al.* (2010) characterized 30 endoglucanases from 6 glycoside hydrolase families with respect to substrate specificity (Vlasenko *et al.*, 2010) and Juhasz *et al.* (2005) characterized cellulases and hemicellulases produced by *T. reesei* on various carbon sources (Juhasz *et al.*, 2005).

16) Endoglucanases and cellobiohydrolases are inhibited by their products, cellobiose and glucose, while ß-glucosidases are inhibited by glucose (Mosier *et al.*, 1999; Bansal *et al.*, 2009). These complex kinetics can neither be treated here in detail nor do they enable modeling of optimal reaction conditions.

combinations of the four main activities (two CBHs and two EGs) (Rosgaard *et al.*, 2007; Meyer *et al.*, 2009).

A large number of empirical models have been published that help in quantifying various effects observed. South *et al.* expressed the reaction rate constant in terms of conversion[17] (South *et al.*, 1995):

$$k(x) = k(1 - x)^n + c,$$

where k is the reaction rate constant for hydrolysis, x is conversion, $k(x)$ is the reaction rate constant at conversion x, n is the exponent of declining rate constant, and c is a constant. The parameters n and c were estimated by approximating $k(x)$ by fitting it to conversion (x). This expression was later used in modeling SSF with staged reactors and intermediate feeding of enzyme and substrate (Shao *et al.*, 2009a,b). Adsorption of enzymes from solution to the substrate surfaces determines the effective enzyme concentration at the cellulose surfaces, with CBM providing for highly efficient adsorption to cellulose. Adsorption equations, for example, the Langmuir isotherm, have been incorporated into hydrolysis models, taking into account the conversion range and four different cellulosic substrates (Nidetzky and Steiner, 1993; Bansal *et al.*, 2009).

Most of the experimental studies showed that the rate of hydrolysis drops by two to three orders of magnitude at high degrees of conversions (Figure 12.8). The contributing factors to decreasing rates (other than product inhibition) accounted for in the existing models include (a) enzyme deactivation, (b) biphasic composition of cellulose (amorphous and crystalline), (c) decrease in substrate reactivity, (d) decrease in substrate accessibility, in part due to lignin, and (e) adsorption of cellulases to lignin. The change in substrate reactivity and substrate accessability has been included in a number of models to explain the reduced digestibility of hydrolyzed cellulose, for both lignocellulosic and pure cellulosic substrates[18] (Bansal *et al.*, 2009).

Synergism of cellulase components has been obvious from a number of investigations, that is, a mixture of cellulase components, cellobiohydrolases, and

17) An empirical expression developed by Ohmine *et al.* (1983) describes the rate of hydrolysis that decreases continuously over time, where P is the product concentration, S_0 is the initial substrate concentration, v_0 is the initial rate, k is the rate retardation constant, and t is time (Ohmine *et al.*, 1983): $P = (S_0/k)\ \ln(1 + v_0kt/S_0)$.

18) Empirical models do not explain mechanistic aspects, of course, and they are not applicable outside the conditions under which they have been developed. Michaelis–Menten-based models fit experimental data under the conditions they were developed, and may consider product inhibition, but they overpredict the hydrolysis data since they do not take into account substrate heterogeneity changing with conversion.Also fractal kinetics has been applied to cellulase kinetics. Fractal kinetics is said to occur when reactions take place in spatially constrained media; such reaction conditions give rise to nonuniformly mixed reaction species, apparent rate orders, and time-dependent rate constants. Although the inclusion of the rate constant or substrate reactivity as a function of conversion may fit the data well, a physical interpretation of the constants in these equations is not possible (Bansal *et al.*, 2009).

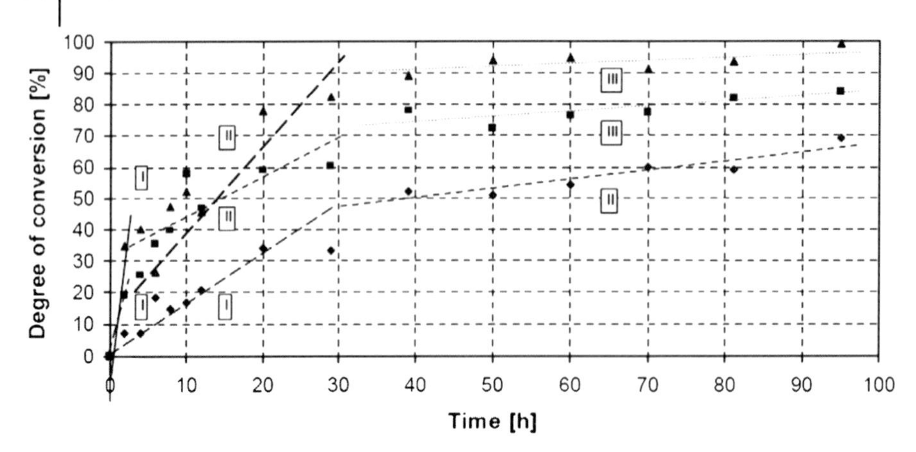

Figure 12.8 Conversion-time behavior of nonpretreated Avicel at optimal ratio of activities of β-glucosidase/cellulase 1 : 20 (Reproduced with permission of Elsevier © 2008, from Bommarius *et al.*, 2008); $T = 50\,^\circ C$, pH 5.0; $V = 8$ mL. (Filled circle) 1.5 U cellulase; (filled square) 15 U cellulase; (filled triangle) 30 U cellulase. I, II, and III denote the three kinetic phases identified.

endoglucanases has higher activity than the individual components alone. For example, a Michaelis–Menten-based model for synergism was proposed by Nidetzky *et al.*, where an additional term for synergism was added to the equation (Nidetzky *et al.*, 1994):

$$\nu(E_1, E_2) = \nu(E_1) + \nu(E_2) + \nu_{syn}(E_1, E_2),$$

where $\nu(E_1,E_2)$ is the hydrolysis rate in the presence of two enzymes E_1 and E_2, $\nu(E_1)$ and $\nu(E_2)$ are the individual hydrolysis rates, and $\nu_{syn}(E_1,E_2)$ is the synergistic hydrolysis rate. However, such models based on the Michaelis–Menten scheme have limitations, as discussed before. A deeper understanding of enzyme synergism is needed to optimize the mixtures of endoglucanases and cellobiohydrolases. Significant synergism was observed with ß-glucosidase (BG) enzymes from *Aspergillus oryzae* that were either supplemented to standard *T. reesei* cellulases or expressed by engineered *T. reesei*, which could reduce the required cellulase loading by as much as twofold[19] (Bansal *et al.*, 2009).

Industrial strains of *T. reesei* produce enzyme systems at very high levels of activity and low cost, by focussing on improving the efficiency of known enzymes, identification of new, more active enzymes, creating enzyme mixes optimized for selected pretreated substrates, and minimization of production cost. The most

19) Further synergism was observed with glycosyl hydrolase family 61 (GH61) enzymes; synergistic action of hemicellulases that break down arabinoxylans, the major hemicellulose component in wheat polysaccharides, is also a relevant factor, with α-L-arabinofuranosidases and endo-1,4-ß-xylanases as most relevant activities (Bansal *et al.*, 2009). Addition of new fungal cellulases to a benchmark blend of a commercial cellulase product increased the glucose yields significantly (Rosgaard *et al.*, 2007).

common commercial enzymes are produced by submerged fermentation of *T. reesei*, of which numerous mutants have been isolated and engineered to increase the productivity by over 20-fold.[20] The enzymes exhibit high efficiency and are typically secreted into the growth medium, allowing cost-efficient separation in a liquid form for application, or application of the whole fermentation broth. For isolation, enzymes are typically fermented, filtered (usually rotating vacuum filters), ultrafiltered for concentration of the enzymes, filtered again (blank filtration), and finally formulated (Pedersen, personal communication, 2010). Appropriate systems have been developed in recent years by Novozymes and Genencor (now with Danisco, part of DuPont) with large 4-year R&D projects funded by US DOE. Novozymes unveiled a cheaper enzyme system making it possible to produce cellulosic ethanol for less than $2 per gal[21] (C&EN, 2010b; www.biotimes.com; www.genencor.com). Realistically, enzyme cost targets in the range of $0.08/l should be achievable in the near future by avoiding formulation, transportation, and storage costs, with on-site enzyme production[22] (Tolan and Foody, 1999; Merino and Cherry, 2007). Both performance and enzyme cost could be reduced by operating at elevated temperature; several reviews deal with the use of thermostable enzymes in lignocellulose and starch conversion, as well as expressing recombinant enzymes in thermophilic hosts[23] (Viikari *et al.*, 2007; Turner *et al.*, 2007; Anbar *et al.*, 2010).

12.2.2.5 **Processing and Reaction Engineering**

Reaction engineering must provide reactor concepts for residence time required to obtain both high conversion and product concentration levels. Most important is to minimize feedback inhibition in order to minimize the amount of enzymes required. This means to operate at low concentrations of cellobiose and glucose at, however, acceptable reaction rate and high cellulosic substrate concentration (avoiding dilution). The process steps of pretreatment, hydrolysis, and fermentation need to be viewed holistically to maximize ethanol yield and overall process cost. The enzymatic hydrolysis can either be performed separately from fermentation

20) During World War II, the US Army was concerned about the deterioration of cotton tents and clothing in the South Pacific. By 1953, Reese and coworkers found extracellular cellulases produced by *Trichoderma* spp., and, together with Mandels, mutants were selected that produced levels of 3 IU $l^{-1}h^{-1}$. The research interest dramatically increased in the 1970s, and further selection led to cellulose productivity of over 100 IU $l^{-1}h^{-1}$. Further improvement by recombinant technologies in the late 1990s succeeded in establishing productivities of 400 IU $l^{-1}h^{-1}$ in commercial fermentations (Tolan and Foody, 1999).

21) Some years before enzyme cost for corn stover hydrolysis was estimated at $1.4/l, development work aimed at a 10-fold reduction, which seems to have been realized (Merino and Cherry, 2007).

22) The production of a broad variety of fungal enzymes that degrade the lignocellulosic material and their recombinant production has recently been claimed, as well as novel combinations of enzymes, including those that provide a synergistic release of sugars from plant biomass. Claims of the patent include cellobiohydrolase, endoglucanase, xylanase, β-glucosidase, hemicellulase, glucoamylase, pectate lyase, acetylxylan esterase, ferulic acid esterase, arabinofuranosidase, pectin methyl esterase, arabinase, and β-xylosidase activities (Gusakov *et al.*, 2009).

23) The limiting activities of cellobiohydrolases from for example, *Humicola grisea* and *Chaetomium thermophilum*, have been engineered for improved thermostability and expressed in *Trichoderma reesei* (Aehle, 2006; Voutilainen *et al.*, 2007).

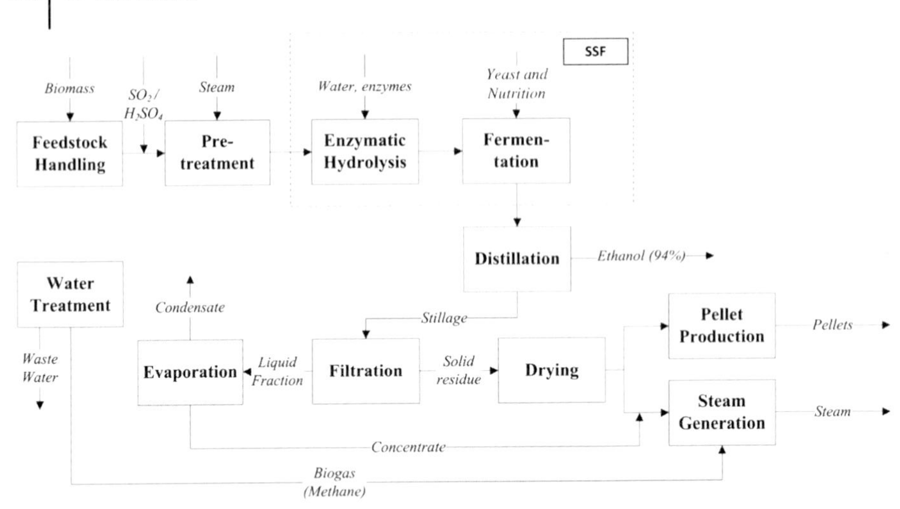

Figure 12.9 Schematic flowsheet for the production of ethanol using acid-catalyzed steam pretreatment followed by enzymatic hydrolysis/SSF of cellulose-containing materials (Galbe and Zacchi, 2007, reproduced with permission from Springer Science + Business Media).

or in combination with fermentation (SSF, simultaneous saccharification and fermentation), which is one of the most promising process configurations to convert pretreated lignocellulosic biomass (Figure 12.9). This avoids the primary drawback of separate hydrolysis caused by product inhibition of the enzymes by the released mono- and disaccharides. The SSF process is capable of improved hydrolysis rates, yields, and product concentrations because of the continuous removal of the sugars by yeast, preventing competitive inhibition of some of the enzymes used. Compromises in reaction temperature and pH must be made in the process design; hydrolysis by *T. reesei* enzymes is optimally performed at pH 5 and 50 °C, but adjustment to fermentation conditions at pH 5.5–7 and 30–40 °C is required. Yeast may be recycled by filtration to increase productivity.[24] SSF has also been shown to be less capital intensive and to result in higher overall ethanol yields (Merino and Cherry, 2007; Brethauer and Wyman, 2010; Galbe and Zacchi, 2007; Galbe *et al.*, 2007).

A model for a SSF process comprised all the major process steps, based on SO_2-catalyzed steam pretreatment, implemented in the commercial flow-sheeting program (Aspen Plus); the model input was based on data recently obtained on lab scale or in a process development unit (Galbe and Zacchi, 2007). The SSF takes place in 12 agitated nonsterile fermenters with a total volume of 920 m^3 each. An SSF cycle including filling, fermentation, draining and cleaning lasts for 60 h. The number of the fermenters was calculated from the cycle time, which was assumed to be 15 h for all stages. The coproduct revenue had a considerable effect on the

24) Today, almost no industrial process employs immobilized yeast cells due to additional cost and worries about long-term stability (Brethauer and Wyman, 2009).

process economics, and utilization of the excess solid residue for heat and power production was economically highly favorable. Theoretically, 426 l of ethanol can be produced from hexose sugars per dry ton of raw material. An additional 59 l can be produced if pentose sugars are also converted to ethanol (see also Li and Kim, 2012). The proposed ethanol plant is assumed to have a capacity to process 200 000 tons of dry raw material per year. Yeast is cultivated in aerated propagation tanks on sugars present in a liquid stream that is separated from the pretreated slurry. SSF is performed at a water-insoluble solid (WIS) concentration of 10% with 2 g/l yeast and an enzyme dosage corresponding to 15 filter paper units (FPU) per gram WIS. Most of the hexose sugars (glucose, mannose, and galactose) are fermented to ethanol, resulting in an SSF broth with 4.0% (w/w) ethanol. The overall ethanol yield, including ethanol losses in the process, and taking into consideration the fact that part of the sugars are used for yeast production, is 296 l per ton dry feedstock. This corresponds to 69% of the theoretical value, based on the hexosan content in the raw material. Most of the evaporation condensates are recycled to the process to reduce the use of freshwater. A wastewater treatment facility was also included in the model, with anaerobic digestion, followed by an aerobic step. It was assumed that 50% of the carbon–oxygen demand (COD) is converted to biogas, which is burnt to generate steam (Sassner and Zacchi, 2008). Integration of anaerobic digestion may result in an enhanced yield of biofuels by 30% (Uellendahl and Ahring, 2010). Life cycle analysis as well has shown that utilization of by-products, including protein, is essential for benefits with respect to renewability and fossil fuel reduction potential (Brehmer and Sanders, 2009). There is obviously still a huge scale-up problem in going from batch pretreatment reactors and fermenters on the liter scale to vessels of several cubic meters and continuous operation, for example, considering the very low ethanol concentration obtained in experiments (Galbe and Zacchi, 2007).

12.2.2.6 **Pilot Studies**

A range of pilot studies and/or commercial production facilities have been announced in recent years; however, only a few have been in operation on a relevant scale: by POET (USA) and Iogen Corporation (Canada).[25] Earlier, in 1995, a pilot plant conceptualized by the US National Renewal Energy Laboratory had been put into operation, with a cascade of three 9 m^3 fermenters and 36 h mean residence time in each[26] (Brethauer and Wyman, 2010).

POET (USA), one of the world's largest bioethanol producers, making more than 6 million m^3/a of ethanol, will extend an existing 190 000 m^3/a corn ethanol facility in Emmetsburg, Iowa, which will include a commercial 95 000 m^3/year

25) Some five companies announced significant (in the range of 40 000 m^3/year or more) ethanol production units by 2012 (Greer, 2008). A complete list of companies active with pilot studies or intending to produce cellulosic ethanol has been compiled (Coyle, 2010).

26) It was designed to process 900 kg dry biomass per day, using pretreated corn fiber, a corn wet milling by-product. An ethanol concentration of about 40 g/l was obtained early in the runs, before bacterial contamination with *Lactobacilli* was reported, illustrating one of the key problems of scale-up (Brethauer and Wyman, 2009).

cellulosic ethanol plant, called Project Liberty, to start production in 2013. By 2022, POET plans to be responsible for 3.5 billion gallons (13 million m^3) of cellulosic ethanol production by extending existing facilities, licensing to other producers, and using other forms of biomass such as wheat straw, switchgrass, and municipal waste (Broin, 2010). POET is receiving a total of $100 million in funding from different authorities in the context of the National Environmental Policy Act (NEPA). The cellulosic ethanol facility will be based on corncobs and corn fiber as raw materials and take advantage of the existing biorefinery infrastructure, including roads, railroads, utilities, and land. POET recently started a pilot-scale cellulosic ethanol plant making 75 m^3/a of ethanol (Broin, 2010; Greer, 2008; see also http://www.poet.com).

Iogen Corporation, based in Ottawa, Canada, has become one of Canada's leading biotech firms, with Royal DutchShell, Petro-Canada, and Goldman Sachs as major investors. It operates the first demonstration facility in Ottawa (Canada) since 2004, using a combination of thermal, chemical, and biochemical techniques to convert biomass, wheat, oat, and barley straw, into ethanol. Pretreatment is achieved by a modified steam explosion process. For enzymatic hydrolysis, Iogen developed a multistage system followed by fermentation. Efficient cellulase enzyme systems are produced in-house by *T. reesei*. Fermentation is performed by microorganisms that convert both C6 and C5 sugars into ethanol. The yield is 340 l per ton of fiber. The current ethanol process is a precommercial business that produced 580 m^3 ethanol from wheat straw in 2009; a demonstration plant is designed to process about 20–30 tons/day of feedstock to produce approximately 5–6 m^3/day (some 1400 m^3/a) of ethanol. Large-scale process design includes efficient heat integration, water recycling, and coproduct utilization for overall process economy (Iogen, 2010). Unfortunately, no detailed information was available in 2011, neither for the POET nor for the Iogen pilot processes.

Recently, another small pilot plant has been opened by Genera Energy of the University of Tennessee using a process developed by DuPont Danisco cellulosic ethanol, which should produce 950 m^3 ethanol from switchgrass (C&EN, 2010a).

Inbicon announced a demonstration plant (at a cost of 54 mio. Euro) to be devoted for processing 30 000 t/a of straw, and to produce 5400 m^3/a of bioethanol, 13000 t/a of lignin pellets, and 11 100 t/a of C5 molasses, using enzymes supplied by Danisco Genencor and Novozymes (www.inbicon.com; www.dongenergy.de). Only few commercial-scale cellulosic ethanol plants are in planning stages: an Abengoa plant and an expansion of POET's conventional ethanol plant (both in the United States) (Johnson, 2011a). The Gruppo Mossi&Ghisolfi began to build a 40 000 t/a plant in Italy. A key part is the firm's Proesa process that produces sugar in two steps, a steam and pressure step, breaking down the biomass, and enzyme hydrolysis (www.gruppomg.com/index.php).

12.2.2.7 Alternative Biocatalyst-Based Biofuels

A range of different approaches have been developed in recent years, which cannot be treated here; only a few shall be mentioned in order to illustrate the diverse options.

Enzymatic biodiesel transesterification has been discussed in Section 8.4.3.2. Examples of engineered cells to establish novel routes to biofuels have been presented in Chapter 5, including an engineered pathway for production of fatty acid derivatives (Figure 5.4) and cellulose hydrolysis and fermentation in one step for the conversion of lignocellulosic biomass into fuels (Figure 5.5).

A variety of biosynthesis pathways for the production of other potential fuel molecules, for example, by pumping the microbial well, making use of advances in microbial engineering, have been reviewed recently (Fortman *et al.*, 2008). A Japanese patent describes a new concept for the transformation of sugars, carbon dioxide, and light into ethanol via three steps: (a) fermenting a sugar with yeast cells to produce ethanol and carbon dioxide, (b) converting carbon dioxide produced in the fermentation step into sugar using a plant chloroplast under irradiation, and (c) fermenting the sugar produced from carbon dioxide into ethanol (Kuzuu, 2008). An ambitious target is hydrogen: a synthetic enzymatic pathway consisting of 13 enzymes for producing hydrogen from starch and water was proposed by Zhang *et al.* (2007). The stoichiometric reaction is $C_6H_{10}O_5 + 7\ H_2O \rightarrow 12\ H_2 + 6\ CO_2$. The overall process is spontaneous and unidirectional, because of a negative Gibbs free energy and separation of the gaseous products with the aqueous reactants, and occurs at 30 °C, and high H_2 yields have been claimed. In a combination of silicon photovoltaic cells, biocatalysts (engineered microbes) may be used synthesizing triglycerides from CO_2 to be processed to biodiesel (Ritter, 2011). For reviews of algal biofuels, see Pandey (2010). For example, genetically engineered photosynthetic algae should produce a hydrocarbon to be processed into fuels. Highlighting the economic relevance of the topic, Exxon will invest, together with Synthetic Genomics Inc., founded by Craig Venter, as much as $600 million to develop biofuels (C&EN, 2009; C&EN, 2010c).

Plant biotechnology provides tools to modify plant characteristics to improve and to increase the efficiency of conversion to biofuels and biomaterials. Options include the construction of a single-plant transformation, including molecular-level modifications in constituents such as lignocelluloses, the construction of vectors carrying several genes that encode a complete set of lignocellulosic hydrolase activities in plants as processing aids, and the production of high-value coproducts (Davies, Campbell, *et al.*, 2010; Fan and Yuan, 2010; Henry, 2010).

12.3 Case Study: the One-Step Enzymatic Process to Produce 7-ACA from Cephalosporin C

12.3.1 Enzyme Processes for the Production of β-Lactam Antibiotics

Penicillins and cephalosporins belong to the class of β-lactam antibiotics that are formed from the common precursor tripeptide isopenicillin N (Figure 12.10). The β-lactam structure is formed by ring-closure reactions between Cys and Val, where

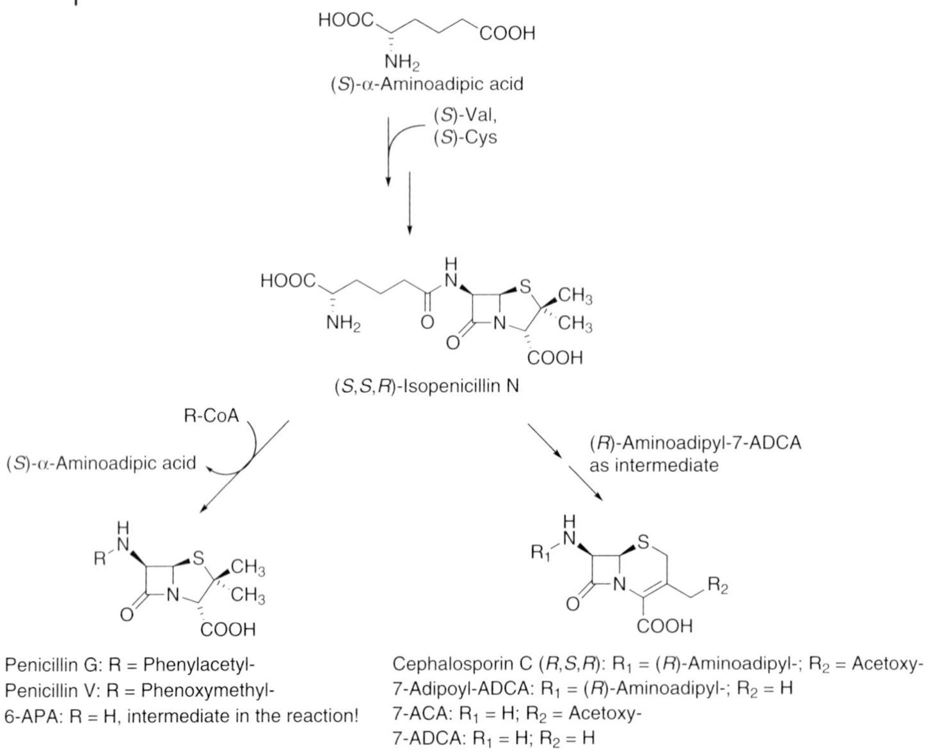

(S)-α-Aminoadipic acid

(S)-Val, (S)-Cys

(S,S,R)-Isopenicillin N

R-CoA

(S)-α-Aminoadipic acid

(R)-Aminoadipyl-7-ADCA as intermediate

Penicillin G: R = Phenylacetyl-
Penicillin V: R = Phenoxymethyl-
6-APA: R = H, intermediate in the reaction!

Cephalosporin C (R,S,R): R_1 = (R)-Aminoadipyl-; R_2 = Acetoxy-
7-Adipoyl-ADCA: R_1 = (R)-Aminoadipyl-; R_2 = H
7-ACA: R_1 = H; R_2 = Acetoxy-
7-ADCA: R_1 = H; R_2 = H

Figure 12.10 Biosynthesis of penicillins and cephalosporins used for the production of semisynthetic β-lactam antibiotics. Note the change in isomer structure of the amino acids given in brackets (Ingolia and Queener, 1989; Crawford *et al.*, 1995; Weil *et al.*, 1995).

(*S*)-Val is isomerized to (*R*)-Val. Penicillins and cephalosporins are the main antibiotics for human use,[27] with a market share of more than 50% (Hamad, 2010). A short history of these antibiotics is given in Box 12.1. The β-lactam precursors of all penicillins and cephalosporins are produced by fermentation in fermenters of up to about 1000 m^3. The concentration of the products in the medium at the end of the fermentation that takes between 5 and 7 days is now more than 100 gl^{-1} (penicillin G or V) and 20 gl^{-1} (cephalosporin C). These figures are extrapolations from previously published figures, as the producers publish only exact data that they achieved some years ago. The lower yields for the cephalosporin C explains why antibiotics derived from this compound are more expensive than those derived from penicillins. The yields have been and are continuously increased by improvement of the β-lactam-producing strains, the fermentation process, and the downstream processing, and this further reduces the processing costs.

27) In order to avoid the unnecessary selection and transmission (via food) of resistant bacterial strains, the same antibiotics should not be used for humans and animals. Antibiotics should therefore also not be used to prevent infections in animal feed. See also Box 12.2.

Box 12.1 : A short history of penicillin and cephalosporin antibiotics and two enzymes that are important for their successful use and production.

That fungi, especially molds on plants, and foods derived from them such as bread, tortillas, and similar foods could be used to treat infections in wounds, had been observed by many cultures in Asia, America, and Europe (Selwyn, 1980). For this aim, they have been used for more than 2000 years. The oldest written evidence for this is in the Old Testament. In the Psalm 51 verse 7, it is stated, "purge me with hyssop, and I shall be clean." From this plant, the mold *Penicillium notatum* can be isolated (Selwyn, 1980, p. 2). This indicated that molds can produce antibacterial substances. The first scientific report for this appeared in 1929 (Fleming, 1929). He observed that *S. aureus* bacteria growing near the mold *Penicillium notatum* were lyzed. This shows that an antibacterial substance was produced in the mold that diffused to the bacteria and destroyed them. He called the substance penicillin.

In the end of the 1930s, studies on penicillin producing molds started in several countries in America, Asia, and Europe. The development of the production of penicillin, and determination of its chemical structure, was considerably accelerated by a close and secret cooperation between government and university laboratories, and mainly chemical companies in United Kingdom and United States during the second world war. Its first aim was to produce penicillin that could be used to treat and prevent infections in wounded soldiers. The large-scale production by submerged fermentation in large reactors was first developed in the United States. This contributed much to the development of biochemical engineering. However, most of the chemical companies that produced penicillin thought that this production process would eventually be replaced by the chemical synthesis of penicillin. They invested much to develop the chemical synthesis, but failed to develop a process that was competitive with the biotechnical synthesis of penicillin (Sheehan, 1982; Bud, 2007). Later Alfred Elder, the US coordinator of the penicillin program, remarked, " I was ridiculed by some of my closest friends for allowing myself to become associated with what obviously was to be a flop – namely, the commercial production of penicillin by a fermentation process" (Elder, 1970). Chemical structures, one of them the β-lactam structure for penicillin, were first proposed in 1943 by scientists at Oxford University in a secret report (Abraham *et al.*, 1943). The β-lactam structure was confirmed by X-ray crystallography, first only documented in secret reports, and published after the second world war (Crowfoot (later Hodgkin), 1948).

Due to this cooperation, large amounts of penicillin (mainly G) were produced that after the war also became available for the general public in an increasing number of countries. Studies on penicillin were also performed in other countries (among others China, Germany, Japan, and the Netherlands) during the second world war. They even realized its production on a much smaller scale than in the United States and United Kingdom (Bud, 2007).

In 1940 and 1950, the first observations of two enzymes were reported. The first has had a large influence on the use of penicillin, the second later become important in the production of new penicillins and cephalosporins.

The first, penicillinase or ß-lactamase, produced in staphylococcal bacteria, was observed in 1940 (Abraham and Chain, 1940) and found to destroy the antibiotic function of penicillin, due to the hydrolysis of the C—N bond in the ß-lactam structure. In this study, the authors indicated – before the production and use of penicillin – that this could become a problem and limit the use of penicillin as an antibiotic. This was also observed some years after the introduction of penicillin to treat infections. An increasing number of penicillin-resistant *Staphylococcus aureus* appeared in patients treated with penicillin. These bacteria produced penicillinases. Such resistant bacteria survive, that is, selected, when penicillin is used to treat infections. Since then a large number of penicillinases (or ß-lactamases) have been found. The genes encoding for these enzymes can be transferred to penicillin-sensitive strains by plasmids transforming them to become penicillin resistant. This general and negative side effect of antibiotics requires a rational use of antibiotics to minimize the selection of resistant bacterial strains (Davies and Davies, 2010; Jovetic *et al.*, 2010). To achieve this, an antibiotic should be used to treat only infections caused by bacteria that are sensitive to the antibiotic. An antibiotic must never be used to prevent infections. This was, however, neglected for a long time (see Box 12.2). One reason for this was that cephalosporin C, detected in 1948 by Brotzu in Italy, was found to be resistant to the then known penicillinases (Selwyn, 1980).

The other enzyme, penicillin amidase, was found to hydrolyze the amide bond between the side chain and 6-APA in penicillin G (Sakaguchi and Murao, 1950). Due to its low acid stability, penicillin G could not be given orally. This and other properties of penicillins were found to be very dependent on the side chain. In 1953, Penicillin V was found to be so acid stable that it could be given orally (Brandl *et al.*, 1953). Thus, the penicillin-producing companies focused their research on the development of methods to produce 6-APA, 7-ACA, and later 7-ADCA that could be condensed with different side chains. New semisynthetic penicillins and cephalosporins with suitable antibiotic properties could be protected by patents. In the 1950s, huge efforts were made by – the mainly chemical – penicillin-producing companies to develop the chemical synthesis of 6-APA and 7-ACA. This failed, and chemical methods were developed to produce them from penicillin G or V, and cephalosporin C (Bud, 2007).

Based on the detection of penicillin amidase, and later cephalosporinases that could hydrolyze the amide bond between the side chain and 6-APA in penicillin G, and 7-ACA in Cephalosporin C, these compounds could also be produced in enzyme-catalyzed reactions. In the 1950s, the development of such processes started in two companies. One, Beecham (UK), was at this time not involved in the production of penicillin. The other, Hoechst (Germany), produced penicillin. Hoechst, however, evaluated the possible use of such a process negatively and stopped research on penicillin amidase (Nesemann, microbiologist at Hoechst, personal communication). They sold their know-how to Bayer (Germany). Bayer and Beecham applied for patents in 1959, and published their work on the hydrolysis of penicillin G using penicillin amidase from *E. coli* in 1960 (Kaufmann and Bauer, 1960; Rolinson *et al.*, 1960). At that time all the main producers of penicillin had started to develop enzymatic methods to produce 6-APA from penicillin G and V, using penicillin amidase from microorganisms other than *E. coli* (Huang *et al.*, 1963). The chemical methods to produce 6-APA were replaced by the enzymatic process from the beginning of the 1970s (see Figure 1.7). With these much less nonrecyclable waste was produced compared to the chemical process. This reduced the cost of waste treatment. Pioneers here were Beecham and Bayer in cooperation with Gist Brocades (now DSM, the Netherlands).

The enzymatic and chemical production processes for semisynthetic penicillins and cephalosporins derived from the fermentation products in Figure 12.10 are summarized in Figure 12.11. The market value of the products is approximately €25 × 10^9 (Hamad, 2010). A considerable proportion of the penicillins and cephalosporin C produced by fermentation is hydrolyzed to obtain 6-aminopenicillanic acid (6-APA) and 7-aminocephalosporanic acid (7-ACA), both of which are used in the production of semisynthetic β-lactam antibiotics such as ampicillin, amoxicillin, cephalexin, and cephaloglycin (Figure 12.11). This also applies for the precursor of 7-ACA 7-amino-deacetoxy-cephalosporinic acid (7-ADCA) that until 2000 was prepared chemically from penicillin G (Bruggink, 2001). Alternatives to produce 7-ADCA by fermentation and enzyme processes are shown in Figure 12.12.

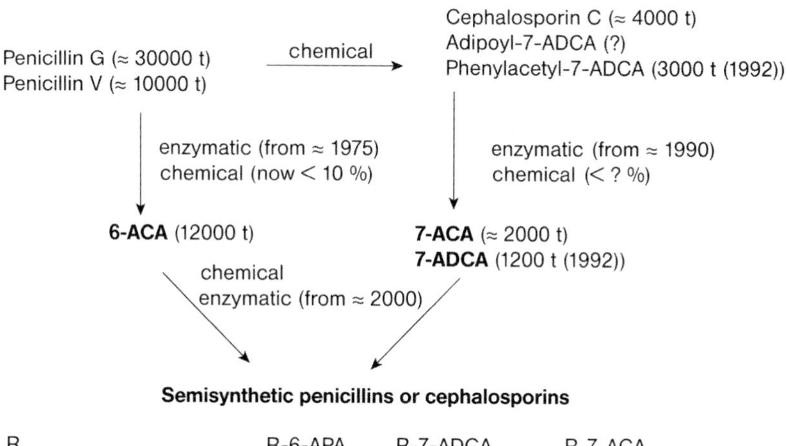

R	R-6-APA	R-7-ADCA	R-7-ACA
(R)-Phenylglycyl	Ampicillin	Cephalexin	Cephaloglycin
(R)-Hydroxy-phenylglycyl	Amoxicillin	Cephadroxil	

Figure 12.11 Enzymatic and chemical production of semisynthetic penicillins and cephalosporins from 6-APA, 7-ACA, and 7-ADCA, by hydrolysis of the fermentation products given in Figure 12.10. The by-products phenylacetate and adipate can be recycled in the fermentation. The amounts produced are estimated from literature data (Bruggink, 2001; Elander, 2003).

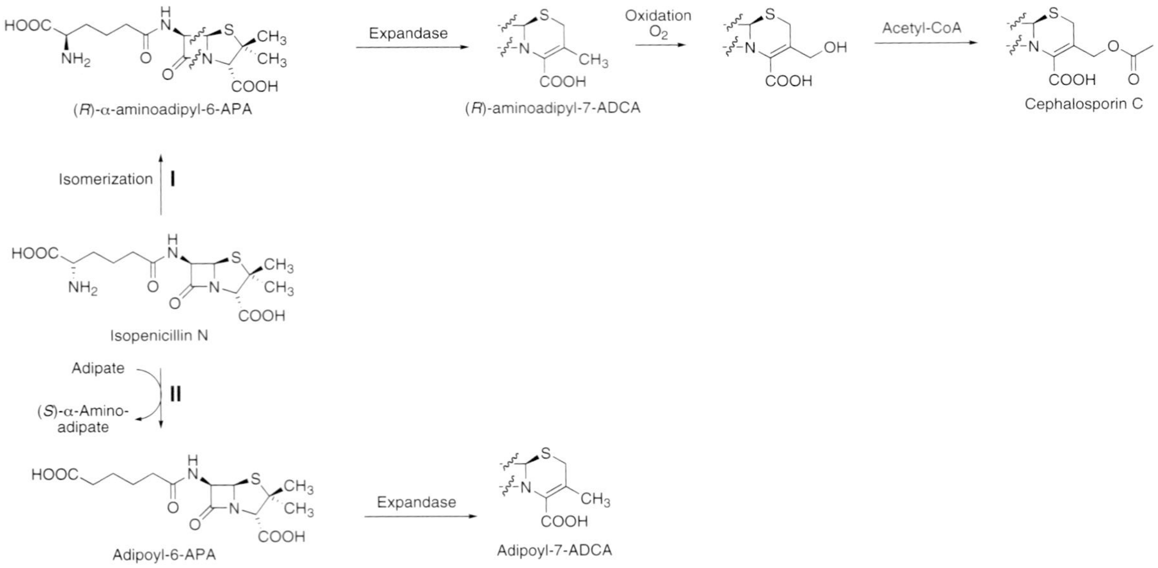

Figure 12.12 Alternative processes for the production of (*R*)-7-ADCA from isopenicillin N in metabolically engineered cells. (a) In cells where the oxidation of (*R*)-aminoadipyl-7-ADCA has been blocked, only the part of the ring that is expanded is shown (Adrio *et al.*, 2002). (b) In cells with expandase but without the racemase that catalyzes isomerization of the aminoadipyl-group in isopenicillin N (Crawford *et al.*, 1995; Bruggink, 2001).

In the biosynthesis of cephalosporin C, (*R*)-aminoadipyl-7-ADCA is an intermediate. When the enzyme that catalyzes the oxidation of this compound is deleted, 7-ADCA can be obtained from (*R*)-aminoadipyl-7-ADCA by using the same enzyme-catalyzed process as for the production of 7-ACA from cephalosporin C. This two-step procedure could be replaced by a one-step hydrolysis procedure.

Another one-step process is the following (Figure 12.12a). An adipoyl group is bound to 6-APA instead of a phenylacetyl group in the conversion of isopenicillin N to penicillin G. The adipoyl-6-APA produced can be converted to adipoyl-7-ADCA in the cells with the enzyme that catalyzes the ring expansion in Figure 12.12 (Baldwin *et al.*, 1987). The adipoyl-7-ADCA can then be hydrolyzed by the glutaryl amidase already used for the hydrolysis of Glu-7-ACA. This elegant example for a successful metabolic engineering is now used on an industrial scale (Bruggink, 2001). In this process as for the biocatalytic hydrolysis of penicillin G (Figure 1.7), much less nonrecycled or hazardous waste is produced than in the formerly used chemical process.

The β-lactam ring is essential for the biological function as an antibiotic. β-Lactam antibiotics act as a practically irreversible inhibitor of a transpeptidase (EC 3.4.16.4) that catalyzes the cross-linking of peptidoglycan chains in the cell wall of Gram-positive bacteria. This destabilizes the call wall and leads to lysis of the bacterial cell, that is, cell death. Similar enzymes do not exist in mammals, and this explains the organism-specific function and fewer adverse side effects of β-lactams compared to other antibiotics. It is possible that β-lactams may acylate proteins, leading to an allergic reaction when they act as haptens and cause antibody formation. By increasing the purity of β-lactam antibiotics, this effect has been reduced. When given in large doses for a long time (e.g., to treat bone infections), β-lactam antibiotics have been shown to influence the biosynthesis of nucleic acids (Do *et al.*, 1987). Beside these side effects of antibiotics, the inevitable development of antibiotic-resistant bacterial strains described in Box 12.2 must also be considered.

Box 12.2: Evolution of antibiotics and antibiotic resistance. The latter cannot be avoided, but can be delayed or reduced.

During the evolution, bacteria and molds had to survive in the same environment. This favored the selection of molds that could produce antibiotics. Concerning bacteria this favored the selection of those that produced enzymes that could metabolize the antibiotics, such as ß-lactamases for penicillin antibiotics. This also follows from the first observation of the properties of penicillin and penicillin-resistant bacterial strains (Fleming, 1929; Abraham and Chain, 1940).

With the introduction of penicillins and other antibiotics to treat infections in humans and animals, the rate of such evolution by selection of antibiotic resistance in bacteria was increased (Palumbi, 2001; Davies and Davies, 2010). Initailly, this increase was largest in hospitals. Now, this causes about 1 hospital-aquired infection caused by antibiotic-resistant bacteria per 1000 inhabitants in Europe and the United States (Taubes, 2008; Jovetic *et al.*, 2010). In the past decades, the occurrence of antibiotic-resistant bacteria outside hospitals that can cause community-acquired infections has continuously increased (Klevens *et al.*, 2007).

The treatment of bacterial infections is thus inevitably connected with the selection of antibiotic-resistant bacteria. This was observed already in the 1940s, but was, however, neglected in the first decades of the treatment of such infections with antibiotics. This was mainly due to the developments of new antibiotics. In recent decades, the rate of development of new antibiotics has, unfortunately, continuously decreased (Taubes, 2008).

From this follows that the continuing successful use of antibiotics to treat bacterial infections requires that the rate of evolution and selection of new antibiotic-resistant bacterial strains must be delayed, and that the appearance of such strains in and outside hospitals must be reduced. To achieve this the following is required:

- Antibiotics must be used only to treat bacterial infections and never used to prevent these.
- Different antibiotics must be used to treat bacterial infections in humans and animals.
- Development of rapid and cheap tests to identify the bacterial strain that causes the infection. Then, the suitable antibiotic to treat the infection can be selected. Such tests are also required to test patients that enter a hospital. When they carry antibiotic-resistant bacterial strains, they must be isolated to prevent their distribution in the hospital.
- Further research to improve hygienic practices in hospitals to reduce the frequence of antibiotic-resistant bacteria in hospitals and hospital-acquired bacterial infections.
- Development of new antibiotics.

The hydrolysis of penicillins to produce 6-APA or 7-ADCA, or of cephalosporin C to produce 7-ACA, can be carried out as either a chemical or a biotechnical (enzyme) process. The latter is now the dominating process due to lower costs and a reduced production of nonrecycled waste. In the equilibrium-controlled hydrolysis of natural penicillins, phenylacetyl-7-ADCA, adipoyl-7-ADCA, and cephalosporin C, the enzyme used as a biocatalyst cannot influence the end point of the reaction with the maximal yield, that is, the equilibrium (see Section 2.6) that is a function of pH, temperature, and ionic strength (see Section 2.8). To obtain a high yield, it is essential to utilize conditions where the hydrolysis yield is near 100%. Under these conditions, the product or enzyme may be unstable, and the enzyme kinetic properties not optimal. To design an optimal enzyme process, it is necessary to consider the steps shown in Figure 1.10. Finally, other side chains are added to the amino group of 6-APA, 7-ACA, and 7-ADCA to produce semisynthetic β-lactam antibiotics with better properties than penicillin G or V and cephalosporin C (acid stable for oral use, a wider range of antibiotic activity, etc.) (Figure 12.11). Although chemical synthesis remains the main procedure, synthesis is also possible using the same enzymes as used for the hydrolysis of penicillin G as biocatalysts (Cole, 1969). Equilibrium-controlled procedures are not possible here as the possible products yields are too low, and much higher yields can be obtained in kinetically controlled processes (see Sections 2.6 and 2.8, and Figure 2.20). Enzyme-catalyzed kinetically controlled processes for the synthesis of β-lactam antibiotics have now been introduced on an industrial scale (Bruggink, 2001). In the following section, the design of a recently introduced process for the one-step enzymatic hydrolysis of cephalosporin C to produce 7-ACA will be discussed in detail based on the scheme given in Figure 1.10. This has replaced the two-step enzymatic process Fig. 12.14 that was discussed in the previous edition of this book. This was achieved by a rational protein engineering to improve the cephalosporin C amidase (or glutaryl amidase, EC 3.5.1.93) that previously was used to hydrolyze glutaryl-7-ACA (Oh *et al.*, 2003; Boniello *et al.*, 2010).

A short history of the development from the first scientific reports of and patents on these enzymatic processes to their realization on an industrial scale is given in Box 12.3.

Box 12.3: From the first scientific report, the resulting patent(s) to the realization of an industrial process.

The first patents on the enzyme-catalyzed production of 6-APA by hydrolysis of penicillin G using penicillin amidase as biocatalyst by Bayer (Germany) and Beecham (UK) were filed in 1959, almost 10 years after the first paper on penicillin amidase was published (see Box 12.1). It then took about 15 years until this enzyme process started to replace the chemical process first used to produce 6-APA from penicillin G.

The first results on the one-step enzymatic hydrolysis of CephC to produce 7-ACA appeared in a patent in 1966 (Walton, 1966). This and a later patent in 1973 claimed a large number of microorganisms (bacteria, fungi, and yeasts) as producers of the enzyme that catalyzed the hydrolysis (Takahashi, 1973). The companies that owned these patents did not develop an industrial one-step enzyme process, probably due to the high enzyme costs. These patents probably hindered other companies to develop such a process, until the patents expired. The same probably applies for a patent on the first step in the two-step enzymatic process using D-amino-oxidase as a biocatalyst shown in Figure 12.14 (Arnold *et al.*, 1972). That glutaryl amidases (or cephalosporin amidases) from *Pseudomonas* strains are suitable biocatalysts for the hydrolysis of glutaryl-7-ACA in the two-step enzyme process to produce 7-ACA was first published in a patent in 1976 and a scientific journal in 1981 (Matsuda *et al.*, 1976; Ishikawa *et al.*, 1981). Based on these results, several companies started to develop the two-step enzymatic process during the 1980s. These could be realized once the general patents from 1966 and 1973 expired. Therefore, the two-step enzymatic process started to replace the chemical process in the beginning of the 1990s. As for the enzymatic process to hydrolyze PenG or PenV, it was an environmentally friendlier process, due to reduction of nonrecycled wastes and costs to treat them. In these processes, mainly recombinant wild-type enzymes were used.

Table 12.6 Time lags between the first publication or patent on an enzyme technological process and the production of semisynthetic cephalosporins and penicillins.

Process	First publication	First patents	Time lag: publication–industrial process (years)	Time lag: patent–industrial process (years)
Penicillin G hydrolysis	1950	1959	≈ 25	≈ 15
Cephalosporin C hydrolysis				
Two step	1972	1972	≈ 20	≈ 20
One step	1966	1966	≈ 40	≈ 40[a]
Synthesis and hydrolysis of adipoyl-7-ADCA	1987	1994	< 15	< 10
Kinetically controlled synthesis of semisynthetic penicillins and cephalosporins	1969	1967	≈ 30	≈ 30

a) The time lag was reduced once methods and information required for a rational protein engineering became available.

The glutaryl amidases from *Pseudomonas* spp. were modified by site-directed mutagenesis in the 1990s. The goal was to increase their (k_{cat}/K_m) and K_i-values for the hydrolysis of CephC (see Section 2.11). However, the engineered enzymes were still too expensive for a one-step enzyme process on an industrial scale. First, after the improvement of the enzyme by a rational protein engineering, based on its 3D-structure, an industrial process on a 1000 ton/year scale could be realized (Kim *et al.*, 2001; Fritz-Wolf *et al.*, 2002; Oh *et al.*, 2003; Shin *et al.*, 2005; Boniello *et al.*, 2010). This process is described in Sections 12.3.2–12.3.7.

The first report that demonstrated that the expandase in Figure 12.12a can expand the ring of penicillin G or adipoyl-6-APA *in vitro* appeared in 1987 (Baldwin *et al.*, 1987). That this can also be achieved *in vivo* in metabolically engineered cells with the expandase that produced penicillin G or adipoyl-6-APA was shown in a patent application filed by Merck (USA) in 1994 (Conder *et al.*, 1996). The patent was later bought by DSM (the Netherlands). They now use this process to produce adipoyl-7-ADCA on an industrial scale. From this they produce 7-ADCA, for the production of semisynthetic cephalosporins, by a one-step enzyme-catalyzed hydrolysis (Bruggink, 2001).

As a catalyst, penicillin amidase can also catalyze the reverse reaction, that is, the condensation of a side chain and 6-APA, 7-ACA, and 7-ADCA. The pH optimum for this is at 4–5, where PA has a low pH stability, and the product yield is low (Figures 2.9b, 2.20, and 2.28). Higher yields of semisynthetic penicillins and cephalosporins can be obtained when PA acts as a transferase in kinetically controlled processes using an activated side chain (see Section 2.6). This was demonstrated in a patent and a publication in the 1960s, where the kinetically controlled process was verified only in the publication (Ankerfarm, 1968; Cole,

1969). Of the wild-type penicillin amidase, the one from *E. coli* has been found to be the best transferase (Table 2.11). The enzymatic process to synthesize semi-synthetic penicillins and cephalosporins started to replace the previously used chemical processes after 2000 (Bruggink, 2001). Even here this resulted in reduced production of nonrecycled waste and costs for waste treatment.

The above is summarized in Table 12.6 that gives the time lag between the first publication or patent on a new enzyme technological process and the realization of an industrial process where this knowledge is applied. It also shows, as discussed above, that patents can delay this.

12.3.2
Overall Process for the Production of 7-ACA

The process is given as a flow sheet in Figure 12.13. Here, the design of the last three process steps will be mainly discussed. For this aim, it is essential to know the concentration and purity of the cephalosporin C solution after the chromatographic isolation procedure. The cephalosporin C (CephC) and by-products, mainly deacetoxy-cephalosporin C (d-CephC) (Figures 12.10 and 12.12) in the filtrate from the fermentation medium, are adsorbed to a hydrophobic adsorbent at a pH $\approx$ 2.5 and separated from all ions and uncharged polar molecules in this filtrate. CephC and its byproducts are then desorbed with an isopropanol–water mixture. The pH

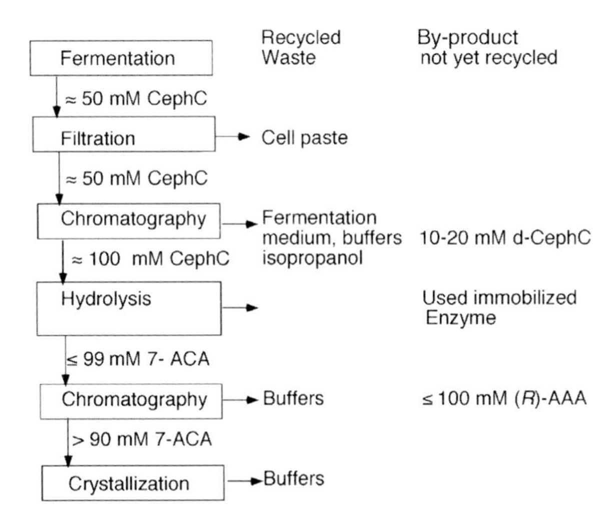

Figure 12.13 Flowsheet for the production of 7-ACA, with concentrations of the precursors after each processing step. The increase in concentration after the chromatographic step is due to the use of displacement chromatography or nanofiltration. The recycled and nonrecycled waste is shown; d-CephC is deacetoxy-cephalosporin C and (*R*)-AAA is (*R*)-amino-adipic acid.

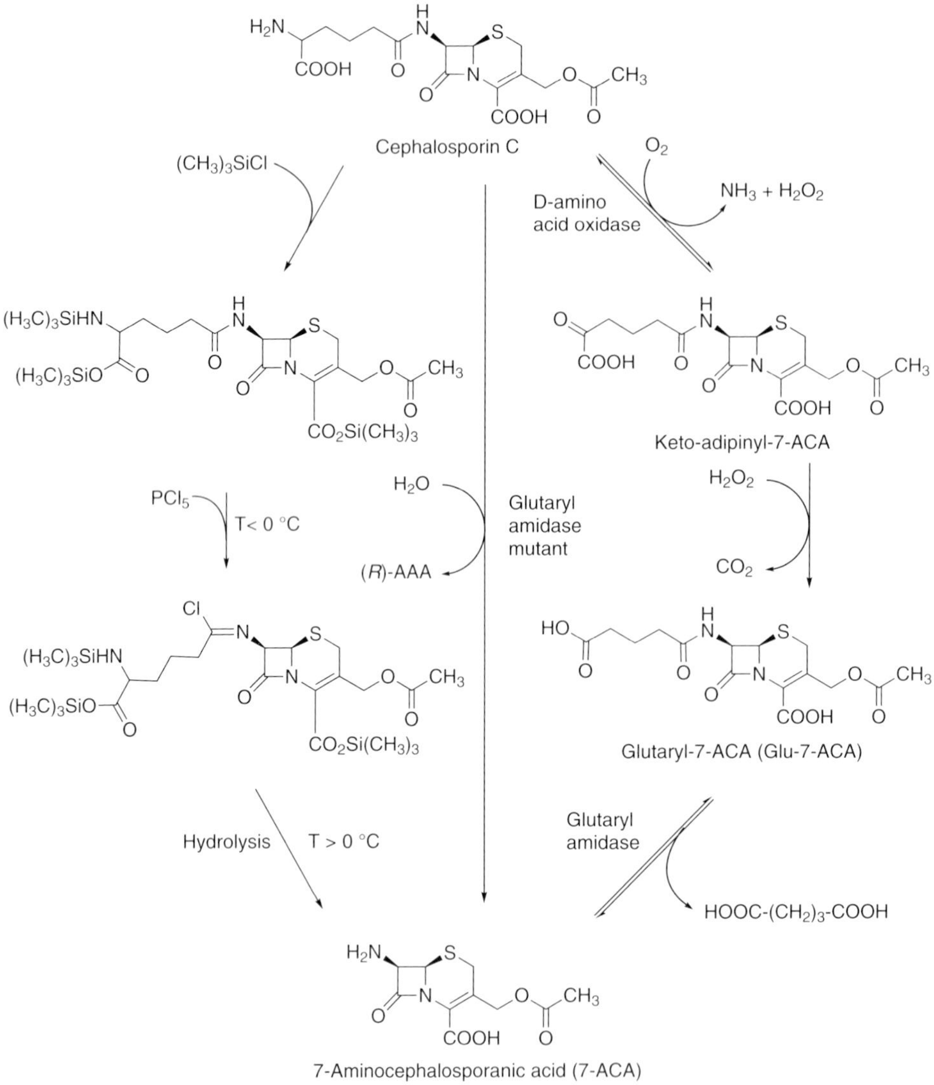

Figure 12.14 Chemical (left) and enzyme technological new one-step (center) and older (right) two-step processes to produce 7-aminocephalosporanic acid (7-ACA) from cephalosporin C. (Note that the charges in the acidic and basic functional groups of the compounds at the process pH are not given.) The one-step enzyme process has now replaced the two-step enzyme process.

in the desorbed solution is increased to $\approx$ 5 and CephC and its byproducts are adsorbed to an anion exchange column. In the desorption step with an acetic acid containing eluent, CephC can be isolated with a purity of >90–95% (concentration $\approx$ 0.1 M). This chromatographic separation procedure was originally developed during the 1970s for the subsequent chemical hydrolysis reaction (Figure 12.14)

(Voser, 1973). The byproduct d-CephC is on a weight basis about 10–20% of the produced CephC, and is still a waste product. The cell biomass after the filtration can be recycled as a soil additive, the solutions after the chromatographic steps can be partly recycled directly or after the separation of isopropanol from water. The aqueous solutions that cannot be directly recycled are recycled after a wastewater treatment.

The complete process shown in Figure 12.13 in a simplified manner has not yet been disclosed by the company that has developed this one-step hydrolysis process (Sandoz, a subsidiary of Novartis, Switzerland). The following is based on published data, and bears similarities with the previously used two-step process that was analyzed in the previous edition of this book.

12.3.3
Conversion of Cephalosporin C to 7-ACA

The possible chemical and enzyme processes for this conversion are shown in Figure 12.14. The main reason for developing the enzyme processes was to reduce the amount of nonrecycled or hazardous waste produced in the chemical process. This was successful, and waste production was reduced typically from 31 to ~ 0.3 tons per ton 7-ACA synthesized. Consequently, enzyme processes have now replaced the chemical process.

The two-step enzyme process could be improved considerably if a single enzyme (a hydrolase) would be available that would catalyze a one-step hydrolysis of cephalosporin C to 7-ACA. This has now been achieved.

12.3.4
Reaction Characterization and Identification of Constrainsts: Hydrolysis of Cephalosporin C

The pH and temperature dependence of the apparent equilibrium constant of the hydrolysis of glutaryl-7-ACA has been determined (Spiess *et al.*, 1999). For more than 95% yield, the reaction must be carried out at pH $\geq$8 (Figure 12.15). The yield increases with the ionic strength.

The apparent dissociation constant K_{app} for the hydrolysis of CephC is (see Section 2.8.2, Eq. (2.40))

$$K_{\text{app}} = \frac{[R\text{-AA}][7\text{-ACA}]}{[\text{CephC}]} = K_{\text{ref.D}}[\text{H}_2\text{O}]$$
$$= \frac{\left(1 + ([\text{H}^+]/K_1') + (K_2'/[\text{H}^+]) + (K_2'K_3'/[\text{H}^+]^2)\right)\left(1 + ([\text{H}^+]/K_1'') + (K_2''/[\text{H}^+])\right)}{\left(1 + ([\text{H}^+]/K_1''') + (K_2'''/[\text{H}^+]) + (K_2'''K_3'''/[\text{H}^+]^2)\right)}, \tag{12.1}$$

where $K_{\text{ref,D}}$ is the pH-independent dissociation constant, the reference dissociation states are those with zero net charge; *R*-AA is *R*-α-amino-adipic acid with the dissociation constants $pK_1' = 2.5$, $pK_2' \approx 3$, and $pK_3' \approx 10$, for 7-ACA they are

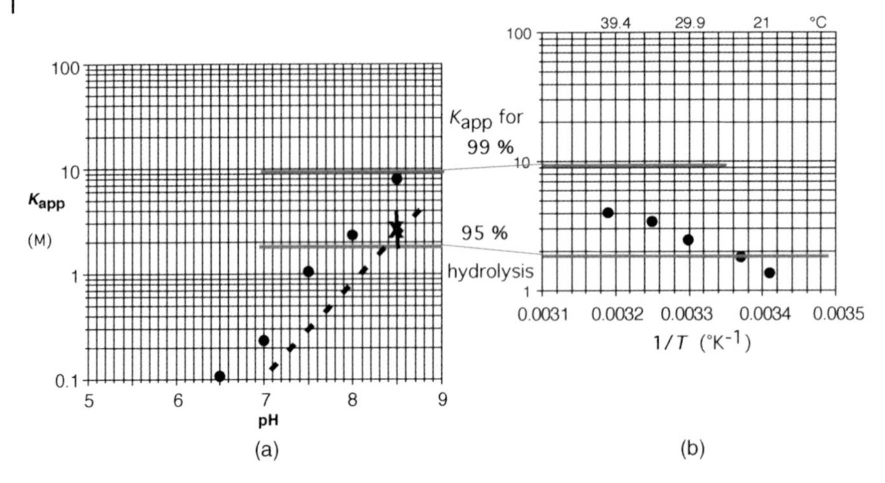

Figure 12.15 Temperature and pH dependence of the apparent equilibrium constant (K_{app} = [glutarate][7-ACA]/[Glu-7-ACA]) for the hydrolysis of 100 mM Glu-7-ACA. (a) pH dependence at 30 °C with phosphate buffers with NaCl (I = 1.5 M); (b): temperature dependence at pH 8.0 in phosphate buffer (I = 0.2 M) (Spieß *et al.*, 1999). The estimated K_{app} for the hydrolysis of CephC at pH 8.5 and 25 °C is given by X, and its pH dependence by the dotted line (Shin *et al.*, 2005).

$pK_1'' = 2.5$ and $pK_2' = 5$, and for CephC they are $pK_1'' = 2.5$, $pK_2'' \approx$, and $pK_3'' \approx 9.5$ (https://www.ebi.ac.uk/chembldb/compound). For pH values above 6, the pH dependence of K_{app} depends only on the last bracket in the nominator, that is, on the dissociation state of 7-ACA. The same applies for the hydrolysis of glutaryl-7-ACA. Thus, the pH dependence of K_{app} for the hydrolysis of CephC for pH >6 is the same as for glutaryl-7-ACA. One value for K_{app} at 25 °C and pH 8.5 for the hydrolysis of CephC can be estimated from figures given in a patent application from 2005 (Shin *et al.*, 2005). It is included in Figure 12.15 with error bars. No data exist on whether the hydrolysis of CephC is exo- or endothermal. The published value at 25 °C shows that the yield of 7-ACA is more than 95% at pH 8.5 and up to 99% at pH ≥9. To achieve this, the hydrolysis time must be short as 7-ACA is an unstable product.

During the hydrolysis, the 7-ACA yield is reduced due to the following monomolecular reactions:

1) 7-ACA → desacetyl-7-ACA + HAc
2) 7-ACA → 7-ACA-lactone
   with a total rate constant k_1, and a bimolecular reaction (Tischer *et al.*, 1992): k_2
3) 7-ACA + 7-ACA → Dimers.

The rate of reaction (2) is increased in the presence of compounds with free amino groups (Yamana and Tsuji, 1976; V. Kasche *et al.*, unpublished results). The pH and temperature dependence of reaction rates in the ranges where the hydrolysis is carried out are given in Figure 12.16. From these data, it follows that the loss

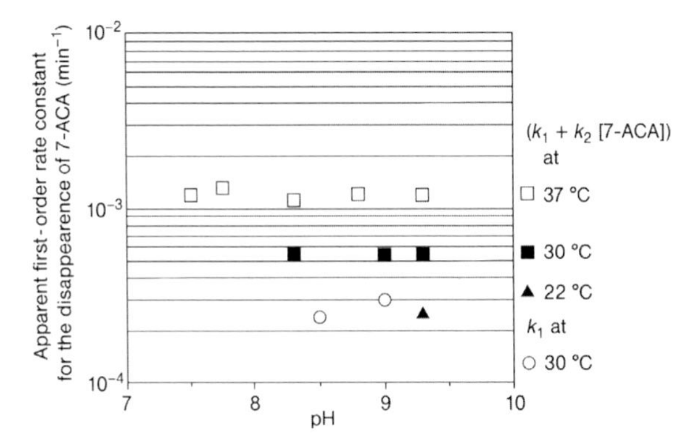

Figure 12.16 pH and temperature dependence of the apparent first-order rate constant ($k_1 + k_2$[7-ACA]) with [7-ACA]) = 100 mM for the reactions (1)–(3) that reduce the yield of 7-ACA.

of 7-ACA due to the reactions (1)–(3) in a solution of 100 mM at pH 8–9.5 will be ≈ 6, 3, and 1.5% after 1 h at 37, 30, and 22 °C, respectively. During the hydrolysis of 100 mM CephC, the 7-ACA concentration is less than 100 mM. One can estimate that the loss of 7-ACA in a hydrolysis cycle of 1 h is about half the above values. Thus, the temperature during the hydrolysis reaction must be selected to keep the loss of 7-ACA below an acceptable value. The loss of 7-ACA during a hydrolysis cycle is also shown in Figure 12.13 of the patent application for this process with the mutant CA (Shin *et al.*, 2005). The conversion of CephC is larger than the concentration of the formed 7-ACA.

12.3.5
Enzyme Characterization and Identification of Constraints: Cephalosporin Acylase (or Glutaryl Acylase or Amidase)

The first molecular biology studies on a cephalosporin acylase (CA, also called glutaryl amidase or acylase) produced by mutant *Pseudomonas* strains were published in 1981 (Ishikawa *et al.*, 1981). They indicated similarities with penicillin amidase (PA, EC 3.5.1.11). The active CA consisted of two polypeptide chains, was periplasmic, and its production had a similar temperature dependence (see Section 6.4.1). This was further stressed by the similarity of the proteolytic processing of the proenzyme to the active enzyme, starting with an intramolecular proteolytic reaction (Kim and Kim, 2001). The CA has, however, been given a different EC number (EC 3.5.1.93). Different studied wild-type CAs could be used to hydrolyze glutaryl-7-ACA in the two-step process, but their activity for the direct hydrolysis of CephC was too low for a competitive one-step process.

Based on the 3D structure of CA from *Pseudomonas diminuta* (N176) with and without glutaryl-7-ACA and glutarate, several groups applied protein engineering to improve its activity for the hydrolysis of CephC (Kim and Kim, 2001; Oh *et al.*, 2002; Pollegioni *et al.*, 2005). The best results were obtained by a rational protein design (site-directed mutagenesis) based on the structure of the active site of CA

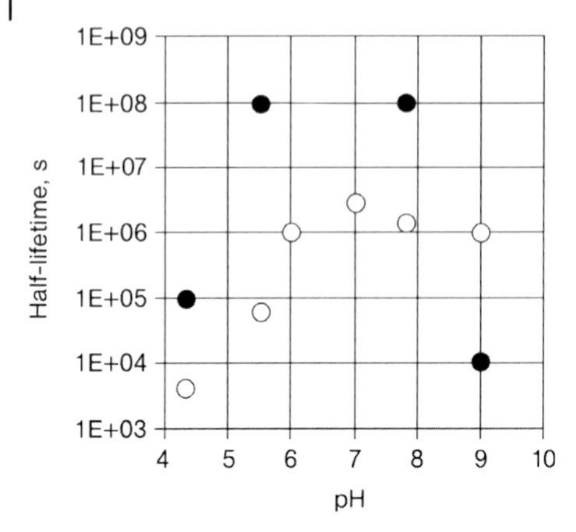

Figure 12.17 The half-life of wild-type glutaryl amidase (blank circle) and penicillin amidase (filled circle) produced in *E. coli* as function of pH (25 °C, buffer with $I = 0.2$ M).

with and without bound substrate and random mutagenesis of the best site-directed mutagenesis variants (see Chapter 3). The enzyme kinetic constants, at pH 8.5 and 25 °C, for the hydrolysis of CephC were up to an order of magnitude better for the best mutant (six amino acid changes) compared to the wild-type CA (data in brackets): $k_{cat} = 2.4$ (0.16) s^{-1}; $K_m = 8$ (50) mM; $K_i = 2$ (0.4) mM (Shin *et al.*, 2005). All these changes improve its use for the hydrolysis of CephC (see Section 2.11).

Based on these results, and probably further, yet nondisclosed, protein engineering to improve these properties and the stability of CA, Sandoz (Switzerland) has developed the one-step process on an industrial scale. They now produce more than 1000 tons of 7-ACA per year using the immobilized mutant CA produced in *Escherichia coli* as biocatalyst (Boniello *et al.*, 2010).

No data on the pH stability of the mutant CA have been published. It is expected that it has a similar pH dependence as determined for a wild-type recombinant CA or glutaryl amidase (Figure 12.17). The lower stability than for PA in the pH range 6–8 is probably due to lack of tightly bound Ca^{2+} in CA (see Figure 2.6 and Section 6.4.1). When immobilized, its half-lifetime under process conditions has been found to be about 20 days (Monti *et al.*, 2000). The same should apply for the yield using a mutant CA in the process. For the recombinant wild-type CA, the yield was up to 5 g/l (Koller *et al.*, 1998).

12.3.6
Evaluation of Process Options

12.3.6.1 Process Window

The pH-T process window for >95 and 99% hydrolysis of 100 mM CephC is derived from Figures 12.15–12.17. The upper temperature limit is determined by the acceptable loss of 7-ACA. The lower temperature limit is given by requirements on the minimal reaction rate. This process window is shown in Figure 12.18.

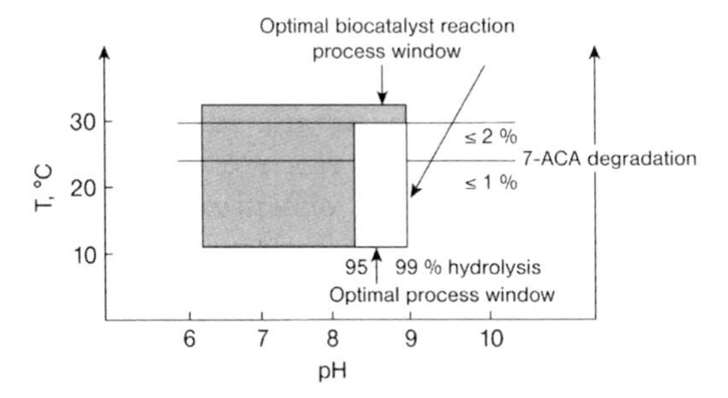

Figure 12.18 The pH-T process window for the hydrolysis of 100 mM CephC in 1 h is the overlapping part of the optimal biocatalyst and reaction windows (≥ 95% hydrolysis). The upper temperature is limited by the inactivation of 7-ACA and not by the biocatalyst properties. The process temperature can be increased by reducing the hydrolysis time, to keep the 7-ACA loss constant.

12.3.6.2 **Suitable Reactors and pH-Controlling Buffers**

A 7-ACA plant as discussed here produces about 1000 tons per year. Each day, a certain amount of CephC must be converted to 7-ACA, and this can be carried out as either a continuous or a discontinuous process. Due to product inhibition, a CSTR is unfavorable, as much more enzyme is required for a given space–time yield than for a packed bed reactor. The latter is, however, not suitable due to the pH change in the bed caused by the hydrolysis reaction (see Section 10.6.2.1). It can be replaced by several shorter fixed bed reactors in series with pH adjustments between the reactors. Such a reactor system has been shown to carry out the continuous hydrolysis of penicillin G on a laboratory scale (Spiess *et al.*, 1999). For such a reactor system, more control equipment is required than for reactors for the discontinuous process with one or several parallel reactors (Figure 12.19, when the reactors in this figure are coupled in series with a pH adjustment between two consecutive reactors, it could be used for a continuous hydrolysis process).

The hydrolysis of CephC must be carried out rapidly in order to minimize losses in 7-ACA. This requires large quantities of enzyme in the immobilized enzyme reactor, and consequently a batch reactor is not suitable, due to abrasion caused by particle–particle collisions at immobilized enzyme contents >10% (v/v). As a fixed bed reactor cannot be used, a recirculation reactor consisting of a fixed bed and a mixing vessel where the pH is adjusted is a suitable reactor configuration (Figure 11.7 with one reactor or Figure 12.19 with more reactors). The substrate is recirculated through the fixed beds until the end point of the reaction has been reached. In such a reactor, 20–30% of the total reactor volume (fixed bed + mixing vessel) can be occupied by immobilized enzyme.

The reactors must be designed to ensure a minimal space–time yield given by the amount of cephalosporin C that must be converted to 7-ACA per day. The enzyme will undergo inactivation and must be replaced by new immobilized enzyme. One possibility is to use a reactor as shown in Figure 11.7 that initially is

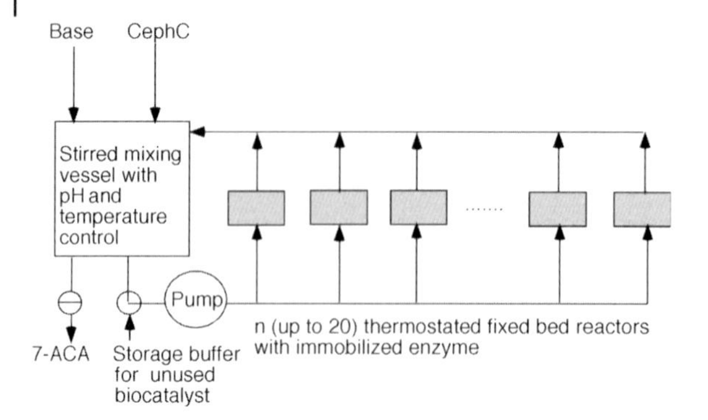

Figure 12.19 A differential reactor with a high content of immobilized enzyme and minimal abrasion. The system consists of a mixing vessel for pH control and one or more parallel fixed bed reactors through which the substrate solution is pumped until the end point of the reaction is reached. The temperature in both the mixing vessel and the packed beds must be controlled. With this reactor system, short hydrolysis times, to reduce the loss of 7-ACA, can be kept within a desired time interval that cannot be achieved with a single reactor. The productivity (kg product/used kg immobilized biocatalyst) is also increased. This illustrated for 20 reactors each filled with ($V_{max,0}/10$), where $V_{max,0}$ is the biocatalyst amount required to hydrolyze a given amount of CephC solution in a given time. At first only 10 reactors are used. When they have lost 10% of their activity, the 11th reactor is added for the hydrolysis, when these 11 reactors have lost 10% of their activity, the 12th reactor is added, and so on. When after the addition of the 20th reactor, the reactors have lost 10% of their activity, the first of the first 10 reactors used is replaced by a reactor with unused biocatalyst. When this reactor system has lost 10% of its activity, the second of the first 10 reactors is replaced, and so on. With this reactor system, the hydrolysis times for each batch differ maximally by 10%. When the first used 10 reactors have been replaced by reactors with unused biocatalyst, the immobilized biocatalyst is used for more than two half-lifetimes. The interval within which the hydrolysis time varies is increased and the productivity reduced when the number of reactors is decreased.

filled with $2V'_{max,0}$ immobilized enzyme activity, where $V'_{max,0}$ is the activity required to hydrolyze 100 mM CephC in 1 h with the acceptable loss of 7-ACA. This reactor could be used until its activity is $V'_{max,0}/2$, that is, for two half-lifetimes of the immobilized enzyme. In the second half-lifetime, the loss of 7-ACA could be reduced by increasing the temperature in the reactor that reduces the hydrolysis time and reducing the temperature in the mixing vessel. Alternatively, the CephC hydrolysis can be performed using the fixed bed reactor consisting of several parallel fixed beds where the inactivated biocatalyst is replaced by a new one in order to maintain the time of one hydrolysis cycle almost constant, and the 7-ACA loss below the acceptable value (Figure 12.19).

The pH in the mixing vessel must be kept constant in the range pH 8.5–9. The most suitable bases to be used here are those with pK-values in this range, for example, NH_3 and CO_3^{2-} (see Section 10.6.2.1). The latter is the better choice as it reduces costs for the denitrification of the wastewater produced. NaOH is not a suitable buffer as the local pH in the mixing chamber may increase to >10, thereby decreasing the stability of the biocatalyst, substrates, and products.

12.3.6.3 Reaction End Point and Immobilized Enzyme Requirement for Minimum Space–Time Yield

The hydrolysis of 100 mM CephC must be carried out within 60 min in order to reduce losses due to degradation of 7-ACA to less than 1%. As shown above (Figures 12.15 and 12.18), 99% hydrolysis is possible. This is a suitable end point. With $[S]_0 = 100$ mM and $t = 60$ min, the required space–time yield is ≈ 1.6 mM min^{-1}.

The amount of biocatalyst per unit volume $V'_{max.0}$ required to obtain this result can be calculated as shown in Sections 10.5 and 10.7. For a 60 min hydrolysis, the deactivation of the enzyme can be neglected. For 99% hydrolysis, the degradation of 7-ACA and the reverse reaction can be neglected. Then, $V'_{max.0}$ can be calculated using Eq. (10.31), written to include biocatalyst inactivation.

$$V'_{max.0} = \frac{K'_m(1 + [S]_0/K_i) \cdot \ln([S]_0/[S]_t) + (1 - K'_m/K_i)([S]_0 - [S]_t)}{\eta_0 \times (1 - e^{-k_i t})/k_i}. \quad (12.2)$$

This equation was derived for a similar process, namely, the hydrolysis of penicillin G. Without biocatalyst inactivation, the denominator equals $\eta_0 t$. With $K_m = 8$ mM in the pH interval 8–9, both (*R*)-amino-adipic acid and 7-ACA are competitive inhibitors. Only the K_i-value for 7-ACA has been determined, it is 2 mM in this pH interval (Shin *et al.*, 2005). With these values in Eq. (12.2), $\eta_0 V'_{max.0}$ is 25 000 U l^{-1}. Immobilized CA preparations have activities of up to 200 000 U/(l wet support) (Boniello *et al.*, 2010). With the data above and the diffusion coefficient for CephC $\approx 5\ 10^{-6}$ cm^2/s and particle radius 100 μm, η_0 can be estimated to be close to 1 (see Section 10.4.2. and Figure 10.8). From this it follows that ≈10% of the enzyme reactor volume must be taken up by immobilized biocatalyst. As pH gradients in the particles cannot be completely avoided (see Figure 10.14, Spiess and Kasche, 2001; Boniello *et al.*, 2010), the enzyme is inactivated, and as the values used for this estimation were determined at different temperatures, a higher value (20–30%) must be used in practice. Alternatively, the required $V'_{max.0}$ can be determined by pilot-scale experiemts.

For the size of reactors used for the industrial enzyme process, see Exercise 12.15.

12.3.6.4 Product Isolation

7-ACA has a low solubility (<5 mM) in the pH range 3–4. This property can, however, not be used to isolate it from (*R*)-amino-adipic acid (*R*)-AAA), as the latter also has a low solubility in this pH range. How 7-ACA is purified after the hydrolysis reaction has not yet been disclosed by the company that developed the one-step hydrolysis of CephC. It must be isolated rapidly due to its instability. As the solution after the hydrolysis practically is a two-component system 7-ACA and (*R*)-AAA, a possible method is a continuous chromatography, using an adsorbent that binds 7-ACA better than (*R*)-AAA or vice versa (see Exercise 12.16). The isolated 7-ACA can then be crystallized at pH ≈ 4, as is done in the previously used two-step process. The (*R*)-AAA is still a waste product, about 600 g per kg of 7-ACA produced.

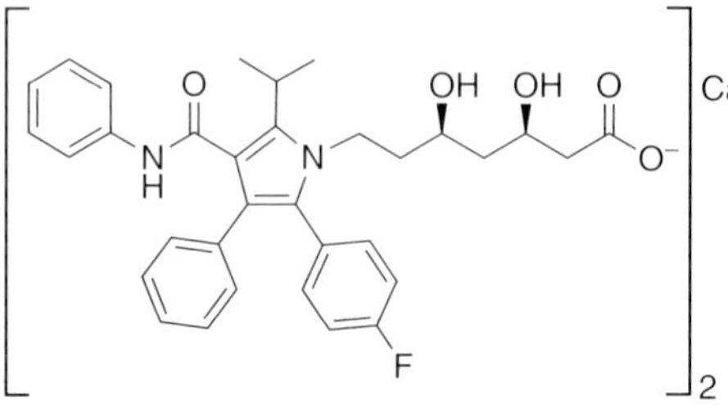

Figure 12.20 Structure of atorvastatin calcium salt.

12.4
Case Study: Biocatalytic Process for the Synthesis of the Lipitor Side Chain

Atorvastatin is the active ingredient of the cholesterol-lowering drug Lipitor®, a drug with annual sales exceeding US$10 billion per year. This pharmaceutical belongs to the statin family of HMG-CoA reductase inhibitors that block the synthesis of cholesterol in the liver.

Similar to other statins, atorvastatin contains a chiral 3,5-dihydroxy carboxylate side chain attached to an achiral nucleus (Figure 12.20). In principle, these two chiral centers could be both introduced by enzymatic means; however, most routes focus on the synthesis of ethyl (*R*)-4-cyano-3-hydroxybutyrate as the second chiral center can then be introduced in a subsequent chemical reaction by diastereomeric induction.

Due to the huge market for atorvastatin (the demand has been estimated to more than 100 metric tons per year) and the requirement of high chemical and optical purity, a range of enzymatic routes have been developed for this compound and related statins (for reviews, see Huisman *et al.*, 2010; Müller, 2005; Panke and Wubbolts, 2005; Tao and Xu, 2009). Early approaches used a microbial resolution or lipases (Hoff and Anthonsen, 1999). A very interesting concept was the use of a nitrilase discovered by the company Diversa Inc. (now Verenium Inc., USA) in metagenome libraries followed by optimization using protein engineering (DeSantis *et al.*, 2003b; DeSantis *et al.*, 2002). The best variant allowed the production of the (*R*)-acid at 3 M substrate concentration with 96% yield at 98.5% ee (see also Figure 4.58, Section 4.2.3.6). This process nicely fitted into a chemical route developed by Dow (USA) starting from epichlorohydrin and chemical synthesis of the prochiral substrate 3-hydroxy glutaronitrile (Figure 12.21). The optimized

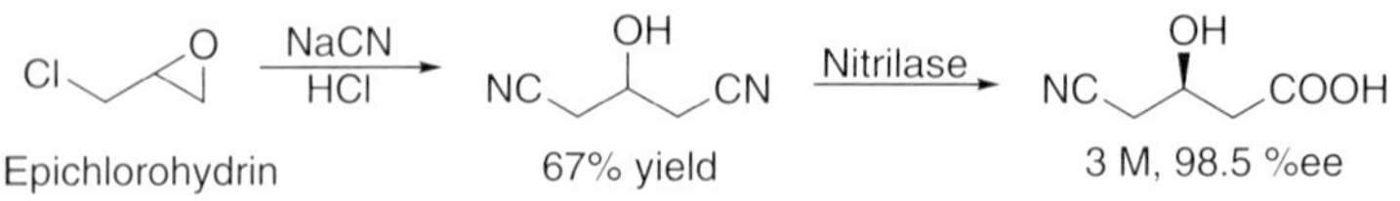

Figure 12.21 Route to optically pure (*R*)-4-cyano-3-hydroxy-butyric acid starting from readily available epichlorohydrin using a nitrilase.

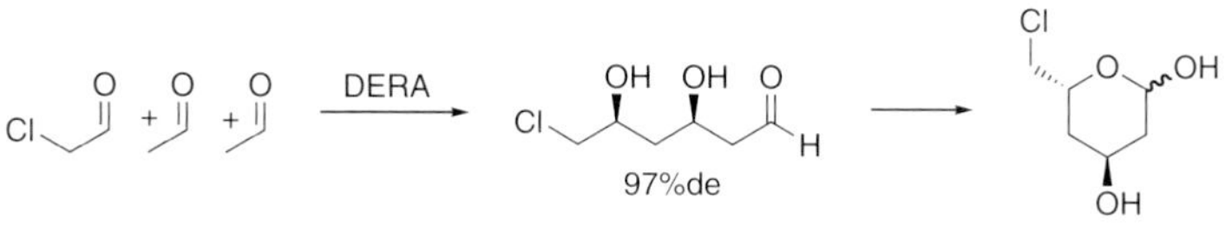

Figure 12.22 An aldolase (DERA) catalyzes the direct formation of two stereocenters from acetaldehyde and chloroacetaldehyde with subsequent formation of a hemiacetal.

nitrilase could be produced recombinantly in *P. fluorescens* in an efficient way (OD 400, 100 g/l cell dry weight, 25 g nitrilase per liter culture, 50% of total protein). In the biocatalytic process, 6 wt% enzyme enabled complete conversion within 16 h leading to the chiral building block in 82% isolated yield (Bergeron *et al.*, 2006).

Another biocatalytic route uses an aldolase (deoxyribose-5-phosphate aldolase, DERA), yielding the (3*R*,5*S*)-dihydroxy-hexanoate intermediate starting from the simple precursors acetaldehyde and chloroacetaldehyde (Figure 12.22). Interestingly – as initially discovered by the Wong group – the enzyme first combines two molecules of acetaldehyde followed by aldol condensation with one equivalent of chloroacetaldehyde in a highly selective manner. The resulting (3*R*,5*S*)-6-chloro-3,5-dihydroxyhexanal is stabilized as a hemiacetal. This process was optimized independently by DSM (up to 100 g/l product) and Diversa (>99.5% ee, 96.6% de, 93 g/l, 2 wt% enzyme loading (DeSantis *et al.*, 2003a; Greenberg *et al.*, 2004)). Later, a DERA variant (S238D) was described (Liu *et al.*, 2004), which allows to use azido-propionaldehyde alternative to chloroacetaldehyde yielding an advanced intermediate of atorvastatin.

An elegant and commercialized approach uses the combination of three enzymes in a two-step process (Figure 12.13). Starting from a prochiral ketone readily available from diketene, first a ketoreductase (KRED) in combination with glucose dehydrogenase (GDH) for cofactor regeneration yields the chiral ethyl (*R*)-4-chloro-3-hydroxybutyrate. The key step is the halohydrin dehalogenase (HHDH) catalyzed formation of a chiral epoxide followed by ring-opening with HCN at neutral pH to afford the ethyl (*R*)-4-cyano-3-hydroxybutyrate building block in excellent yield and optical purity (Ma *et al.*, 2010).

In order to develop an economic route, the KRED and the HHDH needed to be adapted substantially by protein engineering. Table 12.7 lists the parameters of the final process conditions; details about the optimization can be found in a publication by Ma *et al.*, 2010, in which the economical and environmental aspects of the biocatalytic route are also addressed. One issue in the reduction was the formation of an emulsion, which hampered the phase separation and hence reduced the isolated yield. This was overcome by improving the activity of the KRED (13-fold) and the GDH (7-fold) by DNA shuffling as then the enzyme loading could be lowered 9.5-fold.

The wild-type HHDH exhibited very low activity in the cyanation and poor stability in the presence of substrate and product. In addition, filtration and phase separation were difficult and the enzyme was strongly inhibted by the product. Protein engineering solved these issues and resulted in a variant with >2500-fold improved activity compared to the starting enzyme (Fox *et al.*, 2007).

Table 12.7 Conditions of the optimized process using engineered ketoreductase and halohydrin dehalogenase (Ma *et al.*, 2010).

Process parameters	**Conditions**	
	KRED[a]	**HHDH**[b]
Substrate concentration (g/l)	160	140
Reaction time (h)	8	5
Biocatalyst loading (g/l)	0.9	1.2
Isolated product yield (%)	96	92
Chemical purity (%)	>98	>98
Enantiomeric excess (% ee)	>99.9	>99.5
Space–time yield (g/l*d)	480	672
Catalyst performance ($g_{product}/g_{catalyst}$)	178	117

See also Figure 12.23 for details.
a) For the reduction of the ketone to the chiral alcohol.
b) For the formation of the cyanide (Figure 12.23).

In conclusion, several efficient enzymatic processes for the synthesis of the optically pure key building block of atorvastatin were developed. Key parameters were the integration into chemical routes based on the availability of cheap precursors, excellent enantioselectivity, and high activity of the biocatalyst to reduce enzyme loading and to achieve high yields of product. In all cases, the enzymes were successfully optimized by protein engineering to meet the process requirements. This success highlights the importance of an integrated approach where recombinant methods of biocatalyst optimization helped to solve downstream processing problems.

Figure 12.23 Route to optically pure (*R*)-4-cyano-3-hydroxy-butyric acid starting from a prochiral ketone using a combination of a ketoreductase (KRED), a glucose dehydrogenase (GDH), and a halohydrin dehalogenase (HHDH). The reduction was performed with 240 g ketone at pH 7 in a buffer (570 ml) containing 100 mM triethanolamine and butyl acetate (370 ml) for 8 h at 25 °C. Both enzymes (854 mg KRED and 578 mg GDH) were added as lyophilized powders. After addition of Celite and filtration, the product was extracted with butyl acetate yielding 96% alcohol. The subsequent cyanation step was performed in an airtight reactor (due to the presence of gaseous HCN) at pH 7.3 and 40 °C with 1.03 g HHDH (supplied as aqueous solution) with 70 g alcohol in a ~400 ml volume. After complete conversion and removal of residual HCN, the mixture was treated with Celite and bleach followed by filtration and product extraction with ethyl acetate yielding 93% final product (Ma *et al.*, 2010).

Exercises

12.1 Enzymatic conversion of starch is part of several technical steps – what are these and what is their function; what steps can or should be integrated? What nontechnical conditions are to be taken into account?

12.2 Enumerate bottlenecks to lignocellulose-based biofuels.

12.3 Can you imagine existing solutions as models for logistics and transport?

12.4 Enzymatic conversion lignocellulose is part of the whole process; what other steps are essential?

12.5 What steps should or could be integrated?

12.6 What may be the most simple, low-investment technology, to be operated locally, competing with lignocellulosic ethanol?

12.7 What are boundary conditions of nontechnical nature that are essential for cellulosic fuel production?

12.8 What technical operations, besides efficient enzyme and fermentation systems, are essential for economical cellulosic fuel production?

12.9 How can *R*-α-amino-adipic acid (or adipic acid) formed in the hydrolysis of cephalosporin C or adipoyl-7-ADCA (Figure 12.12) be recycled?

12.10 Discuss how the frequency in the selection of ß-lactam-resistant bacteria can be reduced.

12.11 How can the results in Figure 12.15 be determined? Is the hydrolysis of Glu-7-ACA endo- or exothermal? Determine the free energy change at 25 °C, and the enthalpy and entropy change of the hydrolysis reaction in the studied temperature interval.

12.12 Under what conditions can the loss of 7-ACA be kept at $< 1\%$ in the hydrolysis of CephC (Figure 12.16)?

12.13 Suggest methods to recycle up to 200 tons of d-CephC and 600 tons of (*R*)-AAA that are produced per year in the plant producing 1000 ton 7-ACA per year. They are now waste products. Can the formation of d-CephC in the fermentation and medium filtration and chromatography step in Figure 12.13 be reduced?

12.14 Could the improved cephalosporin acylase be used for kinetic resolution of racemates or to deprotect peptides?

12.15 Plants exist for the production of 1000 tons 7-ACA per year. Assuming a continuous production of CephC in such a plant, 3.8 tons CephC must be hydrolyzed per day. This is contained in 100 m^3 with 100 mM CephC. Each hydrolysis cycle is 60 min, with 99% substrate conversion, and the enzyme content per liter immobilized biocatalyst is 200 000 U. Other conditions are given in Section 12.3.6.3:

a. What is the minimal total packed bed reactor size (Figure 11.6) to hydrolyze this with 10 hydrolysis cycles per day without biocatalyst inactivation?
b. Propose a method for determining the end of each hydrolysis cycle.
c. Discuss the pros and cons of the discontinuous hydrolysis of CephC with the reactor system shown in Figure 12.19 with one or several

packed bed reactors, and the reactor system for the continuous hydrolysis (Figure 12.19 with the reactors coupled in series with pH adjustment between the reactors). Enzyme inactivation must be considered.

d. (Advanced) Use a graphical representation of V'_{max} (logarithmic scale) at function of time in use (linear scale) with $V'_{max,0}$ at zero time to verify the conclusions in the legend of Figure 12.19. The immobilized enzyme has a half-lifetime of 20 days under hydrolysis conditions. After which time is the first of the initially used 10 reactors replaced by a new reactor with unused immobilized enzyme? Note that the hydrolysis time required to reach the given end-point (determined as given in the solution to 12.15.b) of the reaction will increase before a new reactor is added, or a used reactor is replaced.

12.16 How can 7-ACA and (*R*)-AAA be separated by chromatography after the hydrolysis of CephC?

12.17 Analyze based on Figure 12.10 whether it is possible to produce 7-ACA or 6-APA directly by fermentation.

12.18 Discuss the pros and cons of the different routes shown in Figures 12.21–12.23.

12.19 How can the different chiral products from the different routes be transformed into the building blocks for the desired side chain of the final atorvastatin product (hint: read the cited literature!)

References

Abraham, E.P. and Chain, E. (1940) An enzyme from bacteria able to destroy penicillin. *Nature*, **146**, 837.

Abraham, E.P., Chain, E., Baker, W., and Robinson, R. (1943) Further studies on the degradation of penicillin. VII. Confidential PEN-Report 103 (cited in Sheehan, 1982) 114.

Adrio, J.L., Hintermann, G.A., Demain, A.L., and Piret, J.M. (2002) Construction of hybrid bacterial deacetoxycephalosporin C synthases (expandases) by *in vivo* homologous recombination. *Enzyme Microb. Technol.*, **31**, 932–940.

Aehle, W., Caldwell, R.M., Dankmeyer, L. *et al.* (2006) Variant hyprocrea Jecorina CBH2 cellulases. WO/2006/074005. I. Genencor International.

Alvira, P., Tomás-Pejó, E. *et al.* (2010) Pretreatment technologies for an efficient bioethanol production process based on enzymatic hydrolysis: a review. *Bioresour. Technol.*, **101**, 4851–4861.

Anbar, M., Lamed, R. *et al.* (2010) Thermostability enhancement of *Clostridium thermocellum* cellulosomal endoglucanase Cel8A by a single glycine substitution. *ChemCatChem*, **2**, 997–1003.

Andric, P., Meyer, A.S. *et al.* (2010) Reactor design for minimizing product inhibition during enzymatic lignocellulose hydrolysis: I. Significance and mechanism of cellobiose and glucose inhibition on cellulolytic enzymes. *Biotechnol. Adv.*, **28**, 308–324.

Ankerfarm, S.p.A. (1968) Penicillins, British Patent GB1213492.

Antrim, R.L. and Auterinen, A.-L. (1986) A new regenerable immobilized glucose isomerase. *Stärke*, **38**, 132–137.

Antrim, R.L., Lloyd, N.E., and Auterinen, A.L. (1989) New isomerization technology for high fructose syrup production. *Starch/Stärke*, **41**, 155–159.

Arnold, B.H., Fildes, R.A., and Gilbert, D.A. (1972) Cephalosporin Derivatives. US Patent 3,658,649.

Aschengreen, N.H. (1984) Technical isomerization of glucose, in *Immobilisierte Biokatalysatoren*, DECHEMA-Kurs, Braunschweig.

Bacovsky, D. (2010) How close are second-generation biofuels? *Biofuels Bioprod. Biorefin.*, **4**, 249–252.

Bai, F., Anderson, W. *et al.* (2008) Ethanol fermentation technologies from sugar and starch feedstocks. *Biotechnol. Adv.*, **26**, 89–105.

Balat, M. (2009) Global status of biomass energy use. *Energy Sources A*, **31**, 1160–1173.

Baldwin, J.E., Adlington, R.M., Crabbe, M.J., Knight, G., Nomoto, T., Schofield, C.J., and Ting, H.-H. (1987) The enzymatic ring expansion of penicillins to cephalosporins: side chain specificity. *Tetrahedron*, **43**, 3009–3014.

Banerjee, S., Mudliar, S. *et al.* (2010) Commercializing lignocellulosic bioethanol: technology bottlenecks and possible remedies. *Biofuels Bioprod. Biorefin.*, **4**, 77–93.

Bansal, P., Hall, M. *et al.* (2009) Modeling cellulase kinetics on lignocellulosic substrates. *Biotechnol. Adv.*, **27**, 833–848.

Ben Chaabane, F., Aldiguier, A. *et al.* (2006) Very high ethanol productivity in an innovative continuous two-stage bioreactor with cell recycle. *Bioproc. Biosyst. Eng.*, **29**, 49–57.

Bentley, I.S., and Williams, E.C (1996) Starch conversion. In: Industrial Enzymology (Godfrey, T., West, S.I. Eds.) pp. 339–357, Macmillan Press, London.

Bentley, I.S. (2010) *Enzymes, Starch Conversion. Encyclopedia of Industrial Biotechnology*, John Wiley & Sons, Inc, New York, http://tinyurl.com/wileyEIB.

Bergeron, S., Chaplin, D.A., Edwards, J.H., Ellis, B.S.W., Hill, C.L., Holt-Tiffin, K., Knight, J.R., Mahoney, T., Osborne, A.P., and Ruecroft, G. (2006) Nitrilase-catalysed desymmetrisation of 3-hydroxyglutaronitrile: preparation of a statin side-chain intermediate. *Org. Proc. Res. Dev.*, **10**, 661–665.

Beringer, T., Lucht, W. *et al.* (2011) Bioenergy production potential of global biomass plantations under environmental and agricultural constraints. *GCB Bioenergy*, **3**, 299–312.

Bomgardner, M. (2012) BASF, DuPont back biofeedstocks. *Chem. Eng. News*, January **9**, 8.

Bommarius, A.S., Katona, A. *et al.* (2008) Cellulase kinetics as a function of cellulose pretreatment. *Metab. Eng.*, **10**, 370–381.

Boniello, C., Mayr, T., Klimant, I., Koenig, B., Riethorst, W., and Nidetsky, B. (2010) Intraparticle concentration gradients for substrate and acidic product in immobilized cephalosporin C amidase and their dependencies on carrier characteristics and reaction parameters. *Biotechnol. Bioeng.*, **106**, 528–540.

Bonse, D. and Schindler, H. (1986) Immobilisierte glucoseisomerase, in *Immobilisierte Biokatalysatoren*, DECHEMA-Kurs, Braunschweig.

Bothast, R. and Schlicher, M. (2005) Biotechnological processes for conversion of corn into ethanol. *Appl. Microbiol. Biotechnol.*, **67**, 19–25.

Bothast, R.J. and Schlicher, M.A. (2005) Biotechnological processes for conversion of corn into ethanol. *Appl. Microbiol. Biotechnol.*, **67**, 19–25.

Brandl, E., Giovannini, M., and Margreiter, H. (1953) Studies on the acid stable, orally efficacious phenoxymethylpenivillin (penicillinV). *Wien. Med. Wochenschr.*, **103**, 602–607.

Brehmer, B. and Sanders, J. (2009) Implementing an energetic life cycle analysis to prove the benefits of lignocellulosic feedstocks with protein separation for the chemical industry from the existing bioethanol industry. *Biotechnol. Bioeng.*, **102**, 767–777.

Brethauer, S. and Wyman, C.E. (2010) Review: continuous hydrolysis and fermentation for cellulosic ethanol production. *Bioresour. Technol.*, **101**, 4862–4874.

Broin, J. (2010) Speech to the National Press Club in Washington D.C., April 21. Washington D.C.

Bruggink, A. (2001) *Synthesis of ß-Lactam Antibiotics*, Kluwer Academic Publishers, Dordrecht.

Buchholz, K. and Collins, J. (2010) *Concepts in Biotechnology: History, Science and Business*, Wiley-VCH Verlag GmbH, Weinheim, Germany.

Bud, R. (2007) *Penicillin: Triumph and Tragedy*, Oxford University Press, Oxford.

C&EN (2/2010b) *Chem. Eng. News*, Feb. 22, 20.

C&EN (2007) The costs of biofuels (emissions adapted from *Science*, 2006, 311, 506). *Chem. Eng. News* (December 17), 12–16.

C&EN (2009) *Chem. Eng. News*, July 20, 15.

C&EN (3/2010c) *Chem. Eng. News*, July 19, 6.

C&EN (2010a) *Chem. Eng. News*, Feb. 8, 28.

Chandra, R.P., Bura, R. *et al.* (2007) Substrate pretreatment: the key to effective enzymatic hydrolysis of lignocellulosics? *Biofuels*, **108**, 67–93.

Cole, M. (1969) Factors affecting the synthesis of amoxicillin and hydroxypenicillins by the cell-bound penicillin acylase from *E. coli*. *Biochem. J.*, **115**, 757–764.

Conder, M.J., McAda, P.C., and Rambosek, J.A. (1994) Recombinant expandase bioprocess for preparing 7-ADCA. US Patent 5,318,896.

Coyle, W. (2010) Next-Generation biofuels: near term challenges and implications for agriculture. BIO-01-01, USDA Economic Research Service.

Crawford, L., Stepan, A.M., McAda, P.C., Ramsbosek, J.A., Conder, M.J., Vinci, V.A., and Reeves, C.D. (1995) Production of cephalosporin intermediates by feeding adipic acid to recombinant *Penicillium chrysogenum* strains expressing ring expansion activity. *Nat. Biotechnol.*, **13**, 58–62.

Crowfoot, D. (1948) (later Hodgkin D.C.), X-ray crystallographic studies of compounds of biochemical interest. *Ann. Rev. Biochem.*, **17**, 115–146.

Davies, J. and Davies, D. (2010) Origins and evolution of antibiotic resistance. *Microbiol. Mol. Biol. Rev.*, **74**, 417–433.

Davies, M., Campbell, M. *et al.* (2010) The role of plant biotechnology in bio-energy production. *Plant Biotechnol. J.*, **8**, 243–1243.

DeSantis, G., Zhu, Z., Greenberg, W.A., Wong, K., Chaplin, J., Hanson, S.R., Farwell, B., Nicholson, L.W., Rand, C.L., Weiner, D.P., Robertson, D.E., and Burk, M.J. (2002) An enzyme library approach to biocatalysis: development of nitrilases for enantioselective production of carboxylic acid derivatives. *J. Am. Chem. Soc.*, **124**, 9024–9025.

DeSantis, G., Liu, J., Clark, D.P., Heine, A., Wilson, I.A., and Wong, C.H. (2003a) Structure-based mutagenesis approaches toward expanding the substrate specificity of D-2-deoxyribose-5-phosphate aldolase. *Bioorg. Med. Chem.*, **11**, 43–52.

DeSantis, G., Wong, K., Farwell, B., Chatman, K., Zhu, Z., Tomlinson, G., Huang, H., Tan, X., Bibbs, L., Chen, P., Kretz, K., and Burk, M.J. (2003b) Creation of a productive, highly enantioselective nitrilase through gene site saturation mutagenesis (GSSM). *J. Am. Chem. Soc.*, **125**, 11476–11477.

Devantier, R., Scheithauer, B. *et al.* (2005) Metabolite profiling for analysis of yeast stress response during very high gravity ethanol fermentations. *Biotechnol. Bioeng.*, **90**, 703–714.

Dien, B.S., Ximenes, E.A. *et al.* (2008) Enzyme characterization for hydrolysis of AFEX and liquid hot-water pretreated distillers' grains and their conversion to ethanol. *Bioresour. Technol.*, **99**, 5216–5225.

Divne, C., Ståhlberg, J. *et al.* (1998) High-resolution crystal structures reveal how a cellulose chain is bound in the 50 Å long tunnel of cellobiohydrolase I from *Trichoderma reesei 1. J. Mol. Biol.*, **275**, 309–325.

Do, U.H., Neftel, K.A., Spadari, S., and Hübscher, U. (1987) Beta-lactam antibiotics interfere with eucaryotic DNA-replication by inhibiting DNA polymerase. *Nucleic Acid. Res.*, **15**, 10945.

Drauz, K. and Waldmann, H. (eds.) (2002) *Enzyme Catalysis in Organic Synthesis*, 2nd edn, Vols **1–3**, Wiley-VCH Verlag GmbH, Weinheim

Elander, R.P. (2003) Industrial production of β-lactam antibiotics. *Appl. Microbiol. Biotechnol.*, **61**, 385–392.

Elder, A. (1970) The role of the government in the penicillin program. Chemical Engineering Progress Symposium Series No. 100, Chem. Eng. Prog., 66, 1–11.

European Federation of Biotechnology (1984) *Third European Congress on Biotechnology*, vol. II, Verlag Chemie (Preprints, Topics C1 and C5).

Fan, Z. and Yuan, L. (2010) Production of multifunctional chimaeric enzymes in plants: a promising approach for degrading plant cell wall from within. *Plant Biotechnol. J.*, **8**, 308–315.

Fildes, R.A., Potts, J.R., and Farthing, J.E. (1974) Process for preparing cephalosporin derivatives. US Patent 3,801,458.

Fleming, A. (1929) On the antibacterial action of cultures of a *Penicillium*, with special reference to their use in the isolation of *B. influencae*. *Brit. J. Exp. Pathol.*, **10**, 226–236.

Fortman, J., Chhabra, S. *et al.* (2008) Biofuel alternatives to ethanol: pumping the microbial well. *Trends Biotechnol.*, **26**, 375–381.

Fox, R.J., Davis, S.C., Mundorff, E.C., Newman, L.M., Gavrilovic, V., Ma, S.K., Chung, L.M., Ching, C., Tam, S., Muley, S., Grate, J., Gruber, J., Whitman, J.C., Sheldon, R.A., and Huisman, G.W. (2007) Improving catalytic function by ProSAR-driven enzyme evolution. *Nat. Biotechnol.*, **25**, 338–344.

Fritz-Wolf, K., Koller, K.P., Lange, G., Liesum, A., Sauber, K., Schreuder, H., Aretz, W., and Kabsch, W. (2002) Structure based prediction of modifications in glutarylamidase to allow single-step enzymatic production of 7-aminocephalosporanic acid from cephalosporin C. *Protein Sci.*, **11**, 92–103.

Galbe, M. and Zacchi, G. (2007) Pretreatment of lignocellulosic materials for efficient bioethanol production. *Biofuels*, **108**, 41–65.

Galbe, M. and Zacchi, G. (2007) Pretreatment of lignocellulosic materials for efficient bioethanol production. *Adv. Biochem. Eng. Biotechnol. Biofuels*, **108**, 41–65.

Galbe, M., Sassner, P. *et al.* (2007) Process engineering economics of bioethanol production. *Adv. Biochem. Eng. Biotechnol. Biofuels*, **108**, 303–327.

Gilbert, H.J. (2010) The biochemistry and structural biology of plant cell wall deconstruction. *Plant Physiol.*, **153**, 444–455.

Gnansounou, E. and Dauriat, A. (2010) Techno-economic analysis of lignocellulosic ethanol: A review. *Bioresour. Technol.*, **101**, 4980–4991.

Greenberg, W.A., Varvak, A., Hanson, S.R., Wong, K., Huang, H., Chen, P., and Burk, M.J. (2004) Development of an efficient, scalable, aldolase-catalyzed process for enantioselective synthesis of statin intermediates. *Proc. Natl. Acad. Sci. USA*, **101**, 5788–5793.

Greer, D. (2008) Commercializing cellulosic ethanol. *Biocycle*, **49**, 47.

Gusakov, A.V., Punt, P.J. *et al.* (2009) Novel fungal enzymes. Patent Application 20090280105.

Hamad, B. (2010) The antibiotics market. *Nature Rev. Drug Discov.*, **9**, 675–676.

Harris, P.V., Welner, D. *et al.* (2010) Stimulation of lignocellulosic biomass hydrolysis by proteins of glycoside hydrolase family 61: structure and function of a large, enigmatic family. *Biochemistry*, **49**, 3305–3316.

Hendriks, A. and Zeeman, G. (2009) Pretreatments to enhance the digestibility of lignocellulosic biomass. *Bioresour. Technol.*, **100**, 10–18.

Henry, R.J. (2010) Evaluation of plant biomass resources available for replacement of fossil oil. *Plant Biotechnol. J.*, **8**, 288–293.

Hoff, F.H. and Anthonsen, T. (1999) Lipase-catalyzed resolution of esters of 4-chloro-3-hydroxy-butanoic acid: effects of the alkoxy group and solvent on the enantiomeric ratio. *Tetrahed. Asym.*, **10**, 1401–1412.

Hu, G., Heitmann, J.A. *et al.* (2008) Feedstock pretreatment strategies for producing ethanol from wood, bark, and forest residues. *BioResources*, **3**, 270–294.

Huang, H.T., Seto, T.A., and Shull, G.M. (1963) Distribution and substrate specificity of benzylpenicillin acylase. *Appl. Microbiol.*, **11**, 1–6.

Huisman, G.W., Liang, J., and Krebber, A. (2010) Practical chiral alcohol manufacture using ketoreductases. *Curr. Opin. Chem. Biol.*, **14**, 122–129.

Ingledew, W.M. (2010) Ethanol fuel production: yeast processes, in *Encyclopedia Industrial Biotechnol*, John Wiley & Sons, Inc., New York.

Ingolia, T. and Queener, S.W. (1989) Beta-lactam biosynthetic genes. *Med. Res. Rev.*, **9**, 245–264.

Iogen (2010) www.iogen.ca/company/demo_plant/index.html.

Ishikawa, S.S., Shibuya, Y., Fujii, T., Komatsu, K., and Kodaira, R. (1981) Purification and properties of 7β-(4-carboxybutanamido)-cephalosporinic acid acylase produced by mutants from *Pseudomonas*. *Agri. Biol. Chem.*, **45**, 2231–2236.

Isogai *et al.* (1991) Construction of a 7-aminocephalosporanic acid (7ACA) biosynthetic operon and direct production of 7ACA in *Acremonium chrysogenum*. *Nat. Biotechnol.*, **9**, 188–191.

Johnson, J. (2007) Corn based ethanol. *Chem. Eng. News*, May 7: 50–52.

Johnson, J. (2009) Supporting biofuels. *Chem. Eng. News*, May 7.

Johnson, J. (2011a) Cellulosic ethanol production lags. *Chem. Eng. News*, October 10, 12.

Johnson, J. (2011b) Long history of U.S. energy subsidies. *Chem. Eng. News*, December 19, 30–31.

Jovetic, S., Zhu, Y., Marcone, G.L., Marinelli, F., and Trampere, J. (2010) β-Lactam and glycopeptide antibiotics: first and last line of defense? *Trends Btechnol.*, **28**, 596–604.

Juhasz, T., Szengyel, Z. *et al.* (2005) Characterization of cellulases and hemicellulases produced by *Trichoderma reesei* on various carbon sources. *Process Biochem.*, **40**, 3519–3525.

Kaufmann, W. and Bauer, K. (1960) Enzymatische spüaltung und resynthese von penicillin. *Naturwiss.*, **47**, 474–475.

Kersten, S.R.A. Potic, B. *et al.* (2006) Gasification of model compounds and wood in hot compressed water. *Ind. Eng. Chem. Res*, **45**, 4169–4177.

Kim, S. and Kim, Y. (2001) Active site residues of cephalosporin acylase are critical not only for enzymatic catalysis but also for post-tranlational modification. *J. Biol. Chem.*, **276**, 48376–48381.

Kim, Y., Mosier, N.S., and Ladisch, M.R. (2009) Enzymatic digestion of liquid hot water pretreated hybrid poplar. *Biotechnol. Prog.*, **25**, 340–348.

Klevens, R.M., Morrison, M.A., Nadel, J., Petit, S., Gershman, K., Ray, S., Harrison, L. H., Dumyati, G., Townes, J.M., Craig, A.S., Zell, E.R., Fosheim, G.E., McDougal, L.K., Cary, R.B., and Fridkin, S.K. (2007) Invasive methicillin-resistant *Staphylococcus aureus* infections in the United States. *JAMA*, **298**, 1763–1771.

Koller, K.P., Riess, G.J., and Aretz, W. (1998) Process for the preparation of glutarylacylase in large quantities. US Patent 5,830,743.

Koltermann, A. (2009) Von der petrochemie zur biorafinerie. *CHEManager*, **9**, 8.

Kosaric, N., Duvnjak, Z. et al. (1981) *Fuel Ethanol from Biomass: Production, Economics and Energy*, Springer.

Kosaric, N., Duvnjak, Z., Farkas, A., Sahm, H., Bringer-Meyer, S., Goebel, O., and Mayer, D. (2002) Ethanol, in *In Ullmann's Encyclopedia of Industrial Chemistry: Electronic Edition*, Wiley-VCH Verlag GmbH, Weinheim, Germany.

Krässig, H., Schurz, J. *et al.* (2004) Cellulose, in *Ullmann's Encyclopedia of Industrial Chemistry*, Wiley-VCH Verlag GmbH, Weinheim, Germany.

Kumar, R. and Wyman, C.E. (2009) Does change in accessibility with conversion depend on both the substrate and pretreatment technology? *Bioresour. Technol.*, **100**, 4193–4202.

Kumar, R., Singh, S. *et al.* (2008) Bioconversion of lignocellulosic biomass: biochemical and molecular perspectives. *J. Ind. Microbiol. Biotechnol.*, **35**, 377–391.

Kumar, R., Mago, G. *et al.* (2009) Physical and chemical characterizations of corn stover and poplar solids resulting from leading pretreatment technologies. *Bioresour. Technol.*, **100**, 3948–3962.

Kunz, M. (2007) Bioethanol: experiences from running plants, optimization and prospects. *Biocat. Biotrans.*, **26**, 128–132.

Kuzuu, M. (2008) Ethanol production process. WO/2009/093367.

Lens, P. (2005) Biofuels for fuel cells: renewable energy from biomass fermentation. *International Water Association.*

Levine, S.E., Fox, J.M. *et al.* (2010) A mechanistic model of the enzymatic hydrolysis of cellulose. *Biotechnol. Bioeng.*, **107**, 37–51.

Li, X. and Kim, T.H. (2012) Bioconversion of corn stover derived pentose and hexose to ethanol using cascade simultaneous saccharification and fermentation. *Bioproc. Biosyst. Eng.*, **35**, 99–104.

Liese, A., Seelbach, K. *et al.* (2000/2006) *Industrial Biotransformations*, Wiley-VCH Verlag GmbH, Weinheim, Germany.

Liu, J., Hsu, C.C., and Wong, C.H. (2004) Sequential aldol condensation catalyzed by DERA mutant Ser238Asp and a formal total synthesis of atorvastatin. *Tetrahedron*, **45**, 2439–2441.

Ma, S.K., Gruber, J., Davis, C., Newman, L., Gray, D., Wang, A., Grate, J., Huisman, G.W., and Sheldon, R.A. (2010) A green-by-design biocatalytic process for atorvastatin intermediate. *Green Chem.*, **12**, 81–86.

Mabee, W.E. and Saddler, J.N. (2009) Bioethanol from lignocellulosics: status and perspectives in Canada. *Bioresour. Technol.*, **101**, 4806–4813.

Martinez, D., Berka, R.M. *et al.* (2008) Genome sequencing and analysis of the biomass-degrading fungus *Trichoderma reesei* (syn. *Hypocrea jecorina*). *Nat. Biotechnol.*, **26**, 553–560.

Matsuda, T., Yamaguchi, T., Fujii, T., Matsumoto, K., Morishita, M., Fukushima, M., and Shibuya, Y. (1976) Process for the production of 7-amino-cephem compounds. US Patent 3,960,662.

Matsushita, Y., Yamauchi, K. *et al.* (2010) Enzymatic saccharification of eucalyptus bark using hydrothermal pre-treatment with carbon dioxide. *Bioresour. Technol.*, **101**, 4936–4939.

Merino, S. and Cherry, J. (2007) Progress and challenges in enzyme development for biomass utilization. *Biofuels. Adv. Biochem. Eng. Biotechnol.*, **108**, 95–120.

Meyer, A.S., Rosgaard, L. *et al.* (2009) The minimal enzyme cocktail concept for biomass processing. *J. Cereal Sci.*, **50**, 337–344.

Misset, O. (2003) Xylose (glucose) isomerase, in *Handbook of Food Enzymology* (eds J.R. Whitaker, A.G.J. Voragen, and D.W.S. Wong), Marcel Dekker, New York, pp. 1057–1077.

Monti, D., Carrea, G., Riva, S., and Baldaro, E., Frare, G. (2000) Characterization of an industrial biocatalyst: immobilized glutaryl-7-ACA acylase. *Biotechnol. Bioeng*, **70**, 239–244.

Moo-Young, M. (ed.) (1981) *Advances in Biotechnology*, vol. **II**, Pergamon Press, Toronto.

Mosier, N., Hall, P. *et al.* (1999) Reaction kinetics, molecular action, and mechanisms of cellulolytic proteins. In *Recent Progress in Bioconversion of Lignocellulosics* (ed. G.T. Tsao), pp. 23–40.

Mulakala, C. and Reilly, P.J. (2005) *Hypocrea jecorina* (*Trichoderma reesei*) Cel7A as a molecular machine: a docking study. *Proteins Struct. Funct. Bioinf.*, **60**, 598–605.

Müller, M. (2005) Chemoenzymatic synthesis of building blocks for statin side chains. *Angew. Chem. Int. Ed.*, **44**, 362–365.

Murthy, G.S., Johnston, D.B. *et al.* (2011) Starch hydrolysis modeling: application to fuel ethanol production. *Bioproc. Biosyst. Eng.*, **34**, 879–890.

Neubauer, Y. (2011) Nutzung von biomasse zur energiegewinnung. *Chem. Ing. Tech.*, **83** (11) 1753–2063.

Nidetzky, B. and Steiner, W. (1993) A new approach for modeling cellulase–cellulose adsorption and the kinetics of the enzymatic hydrolysis of microcrystalline cellulose. *Biotechnol. Bioeng.*, **42**, 469–479.

Nidetzky, B., Steiner, W. *et al.* (1994) Cellulose hydrolysis by the cellulases from *Trichoderma reesei*: a new model for synergistic interaction. *Biochem. J.*, **298** (Pt 3), 705.

O'Kray, C. and Wu, K. (2010) U. D. o. Agriculture, East-West-Center Honolulu, Hawai.

Oh, B., Kim, M., Yoon, J., Chung, K., Shin, Y., Lee, D., and Kim, Y. (2003) Deacylation activity of cephalosporin acylase C is improved by changing the side-chain conformation of active-site residues. *Biochem. Biophys. Res. Commun.*, **310**, 19–27.

Ohmine, K., Ooshima, H. *et al.* (1983) Kinetic study on enzymatic hydrolysis of cellulose by cellulose from *Trichoderma viride*. *Biotechnol. Bioeng.*, **25**, 2041–2053.

Olsen, H.S. (1995) Use of enzymes in food processing, in *Biotechnology*, vol. 9 (eds G. Reed and T.W. Nagodawithana), Wiley-VCH Verlag GmbH, Weinheim, pp. 663–736.

Olsson, L., Otero, J.N., and Panagiotou G, (2007) Biofuels. *Adv. Biochem. Eng. Biotechnol.*, **108**, 1–40.

Palumbi, S.R. (2001) Humans as the world's greatest evolutionary force. *Science*, **293**, 1786–1790.

Pandey, A. (2010) Lignocellulosic bioethanol: current status and perspectives. *Bioresour. Technol.*, **101**, 4743–5042.

Panke, S. and Wubbolts, M. (2005) Advances in biocatalytic synthesis of pharmaceutical intermediates. *Curr. Opin. Chem. Biol.*, **9**, 188–194.

Pedersen, S. and Christensen, M.W. (2000) Immobilized biocatalysts, in *Applied Biocatalysis* (eds A.J.J. Straathof and P. Adlercreutz), Harwood Academic Publishers, Amsterdam, pp. 213–228.

Pollegioni, L., Lorenzi, S., Rosini, E., Marcone, G.L., Molla, G., Verga, R., Cabri, W., and Pilone, M.S. (2005) Evolution of an acylase active on cephalosporin C, *Protein Science*, **14**, 3064–3076.

Prior, B.A. and Day, D.F. (2008) Hydrolysis of ammonia-pretreated sugar cane bagasse with cellulase, β-glucosidase, and hemicellulase preparations. Appl. *Biotechnol. Fuels Chem.*, 271–284.

Reilly, P. and Antrim, R.L. (2003) Enzymes in grain wet milling, in *Ullmann's Encyclopedia of Industrial Chemistry*, Wiley-VCH Verlag GmbH, Weinheim.

Richard, T.L. (2010) Challenges in scaling up biofuels infrastructure. *Science*, **329**, 793.

Ritter, S. (2007) Biofuel nations. *Chem. Eng. News*, June 4, **85**, 9.

Ritter, S.K. (2011) Electrofuels bump up solar efficiency. *Chem. Eng. News*, December 19, **30**, 31.

Rolinson, G.N., Batchelor, F.R., Butterworth, D., Cameron-Wood, J., Cole, M., Eustace, G.C., Hart, M.V., Richards, M., and Chain, E.B. (1960) Formation of 6-aminopenicillanic acid from penicillin by enzymatic hydrolysis. *Nature*, **187**, 236–237.

Rosgaard, L., Pedersen, S. *et al.* (2007) Evaluation of minimal *Trichoderma reesei* cellulase mixtures on differently pretreated barley straw substrates. *Biotechnol. Prog.*, **23**, 1270–1276.

Sakaguchi, K. and Murao, S. (1950) A new enzyme, penicillin-amidase. *Nippon Nogei Kagaku Kaishi*, **23**, 411.

Sassner, P. and Zacchi, G. (2008) Integration options for high energy efficiency and improved economics in a wood-to-ethanol process. *Biotechnol. Biofuels*, **1**, 1754–6834.

Schäfer, T., Borchert, T.W. *et al.* (2007) Industrial enzymes. *Adv. Biochem. Eng. Biotechnol.*, **105**, 59–131.

Selwyn, S. (1980) *The Beta-Lactam Antibiotics: Penicillins and Cephalosporins in Perspective*, Hodder and Stoughton, London.

Service, R. (2010) Is there a road ahead for cellulosic ethanol? *Science*, **329**, 784.

Shao, X., Lynd, L. *et al.* (2009a) Kinetic modeling of cellulosic biomass to ethanol via simultaneous saccharification and fermentation: Part II. Experimental validation using waste paper sludge and anticipation of CFD analysis. *Biotechnol. Bioeng.*, **102**, 66–72.

Shao, X., Lynd, L. *et al.* (2009b) Kinetic modeling of cellulosic biomass to ethanol via simultaneous saccharification and fermentation: Part I. Accommodation of intermittent feeding and analysis of staged reactors. *Biotechnol. Bioeng.*, **102**, 59–65.

Sheehan, J.C. (1982) *The Enchanted Ring: The Untold Story of Penicillin*, MIT Press, Cambridge.

Shin, Y.C., Jeon, J.Y.J., Jung, K.H., Park, M.R., and Kim, Y. (2005) Cephalosporin C acylase mutant and method for preparing 7-ACA using same WO 2005/014821 A1.

Solomon, B., Barnes, J. *et al.* (2007) Grain and cellulosic ethanol: history, economics, and energy policy. *Biomass Bioenerg.*, **31**, 416–425.

Somerville, C., Youngs, H. *et al.* (2010) Feedstocks for lignocellulosic biofuels. *Science*, **329**, 790.

South, C.R., Hogsett, D.A.L. *et al.* (1995) Modeling simultaneous saccharification and fermentation of lignocellulose to ethanol in batch and continuous reactors. *Enzyme Microb. Technol.*, **17**, 797–803.

Spiess, A. and Kasche, V. (2001) Direct measurement of pH profiles in immobilized enzyme carriers during kinetically controlled synthesis using CLSM. *Biotechnol. Prog.*, **17**, 294–303.

Sticklen, M.B. (2008) Plant genetic engineering for biofuel production: towards affordable cellulosic ethanol. *Nat. Rev. Genet.*, **9**, 433–443.

Swaisgood, H.J. (2003) Use of immobilized enzymes in the food industry, in *Handbook of Food Enzymology* (eds J.R. Whitaker, A.G. J. Voragen, and D.W.S. Wong), Marcel Dekker, New York, pp. 359–366.

Sz, A. (2008) *Sueddeutsche Zeitung*, May 30, 19.

Tao, J. and Xu, J.H. (2009) Biocatalysis in development of green pharmaceutical processes. *Curr. Opin. Chem. Biol.*, **13**, 43–50.

Taubes, G. (2008) The bacteria fight back. *Science*, **321**, 356–361.

Tester, R., Qi, X. *et al.* (2006) Hydrolysis of native starches with amylases. *Anim. Feed Sci. Technol.*, **130** (1–2), 39–54.

The New York Times (2008) *The New York Times*, **1**, 4.

Tischer, W., Giesecke, U., Lang, G., Röder, A., and Wedekind, F. (1992) Biocatalytic 7-aminocephalosporanic acid production. *Ann. N.Y. Acad. Sci.*, **612**, 502–509.

Tolan, J. and Foody, B. (1999) *Cellulase from submerged fermentation.* In *Recent Progress in Bioconversion of Lignocellulosics* (ed. G.T. Tsao), pp. 41–67.

Turner, P., Mamo, G. *et al.* (2007) Potential and utilization of thermophiles and thermostable enzymes in biorefining. *Microb. Cell Fact.*, **6**, 9.

Uellendahl, H. and Ahring, B.K. (2010) Anaerobic digestion as final step of a cellulosic ethanol biorefinery: biogas production from fermentation effluent in a UASB reactor – pilot-scale results. *Biotechnol. Bioeng.*, **107**, 59–64.

Uhlig, H. (1991) *Enzyme Arbeiten für uns*, Hanser-Verlag, München, pp. 214–241.

Uhlig, H. (1998) *Industrial Enzymes and their Application*, John Wiley & Sons, Inc., New York.

Viikari, L., Alapuranen, M. *et al.* (2007) Thermostable enzymes in lignocellulose hydrolysis. *Adv. Biochem. Eng. Biotechnol. Biofuels*, **108**, 121–145.

Viikari, L., Alapuranen, M. *et al.* (2007) Thermostable enzymes in lignocellulose hydrolysis. *Biofuels*, 121–145.

Vlasenko, E. Schülein, M. *et al.* (2010) Substrate specificity of family 5, 6, 7, 9, 12, and 45 endoglucanases. *Bioresour. Technol.*, **101**, 2405–2411.

Von Blottnitz, H. and Curran, M. (2007) A review of assessments conducted on bio-ethanol as a transportation fuel from a net energy, greenhouse gas, and environmental life cycle perspective. *J. Cleaner Prod.*, **15**, 607–619.

von Braun, J. (2008) The food crisis isn't over. *Nature*, **456**, 701.

von Braun, J. (2010) Food insecurity, hunger and malnutrition: necessary policy and technology changes. *New Biotechnol.*, **27**, 449–452.

Voser, W. (1973) Process for the recovery of hydrophobic antibiotics. US Patent 3,725,400.

Voutilainen, S.P. Boer, H. *et al.* (2007) Heterologous expression of *Melanocarpus albomyces* cellobiohydrolase Cel7B, and random mutagenesis to improve its thermostability. *Enzyme Microb. Technol.*, **41**, 234–243.

Vuong, T.V. and Wilson, D.B. (2010) Glycoside hydrolases: catalytic base/nucleophile diversity. *Biotechnol. Bioeng.*, **107**, 195–205.

Walton, R.B. (1966) Process for producing 7-amino-cephalosporanic acid. US Patent 3,239,394.

Westcott, P.C. and Usda, E.R.S. (2009) Full throttle US ethanol expansion faces challenges down the road. *Amber Waves*, **7**

Yamana, T. and Tsuji, A. (1976) Comparative stability of cephalosporins in aqueous solution: kinetics and mechanism of degradation. *J. Pharm. Sci.*, **65**, 1563–1574.

Yang, B. and Wyman, C.E. (2008) Pretreatment: the key to unlocking low-cost cellulosic ethanol. *Biofuels Bioprod. Biorefin.*, **2**, 26–40.

Yang, J., Zhang, X. *et al.* (2010) Three-stage hydrolysis to enhance enzymatic saccharification of steam-exploded corn stover. *Bioresour. Technol.*, **101**, 4930–4935.

Zetzl, C., Gairola, K. *et al.* (2011) High pressure processes in biorefineries. *Chem. Ing. Tech.*, **83**, 1016–1025.

Zhang, Y.H.P. (2008) Reviving the carbohydrate economy via multi-product lignocellulose biorefineries. *J. Ind. Microbiol. Biotechnol.*, **35**, 367–375.

Zhang, Y., Evans, B. *et al.* (2007) High-yield hydrogen production from starch and water by a synthetic enzymatic pathway. *PLoS One*, **2**

Zinoviev, S., Müller Langer, F. *et al.* (2010) Next generation biofuels: survey of emerging technologies and sustainability issues. *ChemSusChem*, **3**, 1106–1133.

Appendix A: The World of Biotechnology Information: Seven Points for Reflecting on Your Information Behavior

Prepared with the assistance of Thomas Hapke

> It is crucial that biotechnology students are able to access the relevant information for their studies and can critically evaluate information and its sources. Information literacy is part of lifelong learning and prepares biotechnology graduates for their careers.
>
> (H. Ward and J. Hockey, 2007, p. 374)

A.1
Thinking about Your Information Behavior

Information literacy is a crucial key skill for self-directed learning in scholarly and professional everyday life. In addition to efficient retrieval and navigation strategies, it includes – above all – the creativity to organize and shape one's own information process in a conscious and demand-oriented way. For the searcher, it is no longer questionable to find some information, but rather to filter reliable information from many similar offers. In a time, "where information and data are cheap, proliferating through digital environments and always at the end of a search engine query," thinking critically about information includes "understanding the process through which truth become authenticated, and the underlying assumptions, values, biases, presuppositions and belief systems which inform that process" (Tredinnick, 2008, p. 114).

Like every subject, biotechnology has its own special information media, in addition to particular retrieval strategies, to meet the subject-related information needs. Which of the available databases match your specific needs and are reliable? It is generally important to consider and focus on information under the aspect of reliability. There is a considerable range of publications, especially peer-reviewed journals, available that underwent a critical review process that should provide information of adequate quality and relevance. The so-called "invisible web" or "deep web" contains information sources that are not collected by most search

Biocatalysts and Enzyme Technology, Second Edition. Klaus Buchholz, Volker Kasche, and Uwe T. Bornscheuer

engines such as Google™ – that is, it includes the content of special databases, websites secured by password access or available only in an intranet, and script-based websites.

There is a whole range of reasons for reading and information for research: To provide you with ideas and enhance your creativity; to understand and be able to effectively criticize what other researchers have done in your subject; to broaden your perspective and view your work in context to others (direct personal experience is never enough); to legitimize your arguments; to avoid double efforts in research; "to learn more about research methods and their application in practice"; and to find new areas for research (Blaxter *et al.*, 2010, p. 100). Before beginning to search for information, first reflect on your topic and specific information needs, gather background information, and then focus on your research.

A.2
Subject Gateways, Tutorials, and Literature Guides

Subject gateways on the Internet, tutorials, and literature guides help to inform you about searching information. The so-called "subject gateways" are good starting points for relevant web sites containing collections of subject-specific links. Two examples are the Engineering Subject Gateway of the TIB, the German National Library of Science and Technology, at http://vifatec.tib.uni-hannover.de/index.php3?L=e, and the web site of the "U.S. National Center for Biotechnology Information" (http://www.ncbi.nlm.nih.gov). Special link collections in biotechnology can be found in web catalogs such as the Open Directory Projekt (http://www.dmoz.org/Science/Biology/Biotechnology/).

DISCUS (Developing Information Skills and Competence for University Students) is an example of a web-based bilingual (German, English) learning tutorial for information literacy in engineering that can be used independent of time and space. It also contains a module on biotechnology. DISCUS was developed at the University Library of the Hamburg University of Science and Technology and is available at http://discus.tu-harburg.de. Two more examples for online tutorials are provided by the Dutch Delft University of Technology (http://ocw.tudelft.nl/courses/information-skills/tulib/) and the Danish Aalborg University (http://content.aub.aau.dk/swim/).

Literature guides provide a comprehensive overview about all forms of primary sources like journal articles, reports, dissertations, patents or preprints, and secondary literature (textbooks, monographs, encyclopedias, reviews, abstracting services, etc.) of the treated subject. Guides like MacLeod and Corlett's *Information Sources in Engineering* (4th edition, 2005) and Osif's *Using the Engineering Literature* (Osif, 2012) contain chapters of specific value for the process and biochemical engineer. A guide for the life sciences comes from Schmidt *et al.* (2002).

A.3
Orientation through Using Encyclopedias and Your Local Library

A number of encyclopedias in chemical and process engineering have been published, including volumes dedicated to biotechnology. These are listed in this book at the end of each chapter. Encyclopedias – such as the *Encyclopedia of Industrial Biotechnology, Bioprocess, Bioseparation, and Cell Technology* (ed. by Michael C. Flickinger. Hoboken, NJ: Wiley, 2010) or *Ullmann's Encyclopedia of Industrial Chemistry* (7th ed. Weinheim: Wiley-VCH, 2011) – contain a detailed view of evaluated knowledge, in addition to references for further reading. Libraries offer a selected range of such reference works in printed form in their reading rooms. Electronic versions may be available in the local intranet.

The easily accessible Internet source Wikipedia covers a broad range of articles related to biotechnology and biocatalysis. The users should be aware that the information does not come from validated scientific sources. Hence, extensive coverage of a specific field may be misleading or may include even wrong information. The reliability of the information found in Wikipedia can be evaluated by the literature it cites. The reliability should increase with the number of citations from peer-reviewed journals.

Even in the Internet age, a visit to the local university library can help information retrieval. If they do not possess the item you are interested in, library union catalogs offer a wide range of library materials that can be ordered through inter-library loan or document delivery. In many countries, special libraries function as National Library for Science and Technology (e.g., in Germany the TIB in Hanover; http://www.tib.uni-hannover.de). Databases available in the local intranet also provide references to further information (e.g., journal articles) not necessarily housed by the library itself. In addition, subject librarians can provide information consultation.

A.4
Playing with Databases and Search Terms

When searching a database, it is important to use appropriate key words that allow retrieval of the desired information. Too general key words lead to too many hits from which often only a fraction is useful; when using too specific key words, important information might not be found. It is also recommended to use logical, the so-called Boolean operators (AND, OR, NOT), to link search terms and to use wildcard (joker) symbols ("*" or "?" or "$", which one depends on the search interface). For example, searching with "biodegr*" retrieves documents containing "biodegradation or biodegradable or biodegraded or biodegradability or . . ." Today's user interfaces allow the so-called "facetted search" or "drilldowns" to reduce the number of results after searching. In addition, search results are arranged as default setting – in the past often ordered by descending date – according relevance. Here, you are faced with the difficulty that it is unknown how relevance is determined by the search engine or database.

A search term work sheet can help to structure your query and to find additional search terms like synonyms. For this, the topic must be divided into components and key words chosen for every component. For searching, terms in each of the work sheet's columns have to be combined with the "OR" and the resulting sets with "AND":

Topic: Microbial degradation of aromatic compounds in soil

Component 1	Component 2	Component 3
microbi? degrad?	aromat?	soil?
biodegrad?	polyaromat?	clay?
bioremed?	benzene	compost?
microbi? decompos?	PAH	sediment?

A.5
Searching Journal Articles, Patents, and Data

Today, most recent research results are published in scientific journals and subject-specific text books. More and more of these are integrated with databases and other digital media that become more accessible, sociable, and personalized (Hull, *et al.*, 2008). The difference between primary and secondary sources vanishes. However, it should be emphasized that sources that appeared in peer-reviewed journals should be preferred. A literature search (see above) provides rapid identification of specific journal articles, reviews, and recent books. The original publications either can be downloaded through the university library homepage as gateway or are available as printed versions in the library (see also, "German Electronic Journals Library" at http://www.bibliothek.uni-regensburg.de/ezeit/ or the "DOAJ Directory of Open Access Journals" at http://www.doaj.org). In addition, patents are an important and often less frequently used source by academia.

For searching specific papers, the Internet offers a huge diversity of databases. Free databases like Google™ Scholar often lead you to full texts. In case you are asked for a login or for your credit card number, remember that perhaps the library offers the print version of the article or a further e-version that is available only from another source through the local intranet.

Publishers' portals like ScienceDirect, SpringerLink, or the Wiley Online Library offer full text searching for their own e-books and e-journals, a feature which reference databases like Web of Science or INSPEC don't offer. The searchability of information depends on choices made by authors, publishers, and database providers (Falciola, 2009). Be prepared for change: information sources on the Internet and their user interfaces are updated and enhanced constantly.

A range of databases for information retrieval in chemistry, biotechnology, and related fields available for free either only within the intranet of universities or companies is listed in Table A.1.

Table A.1 Internet databases useful for biocatalysis (selection).

Searching for	Database name and web site	Comments
Articles in journals	Chemical Abstracts Service at http://www.cas.org	For all areas of chemistry and related sciences like the materials sciences and the environmental sciences (with user interface SciFinder perhaps in your local intranet)
	PubMed at http://www.ncbi.nlm.nih.gov/pubmed	Interdisciplinary for medicine; also of great importance for biotechnology as it allows search of nucleotide or protein sequences, genome data and is linked to enzyme structure databases. Free access
	"Science Citation Index" in the "Web of Knowledge" from Thomson Reuters (formerly Institute of Scientific Information, ISI) at http://wokinfo.com/	Interdisciplinary citation database where you can search with documents as "search terms" and answer questions such as Who have cited a specific document? How often is a document cited?
	SciVerse Scopus of the publisher Elsevier at http://www.scopus.com	Interdisciplinary citation database; compared to Web of Knowledge, it also contains proceedings and its contents as well as more publications in other languages than in English
	CEABA (Chemical Engineering and Biotechnology Abstracts – Verfahrenstechnische Berichte) at http://www.wti-frankfurt.de/index.php/datenbanken, see also http://www.wti-frankfurt.de/images/stories/download/en-ceab.pdf	Produced in Germany by the Dechema until May 2011, now by WTI-Frankfurt
	COMPENDEX (COMPuterized ENgineering InDEX) at http://www.ei.org/compendex	Most important and comprehensive database for general engineering

(continued)

Table A.1 (*Continued*)

Searching for	Database name and web site	Comments
	INSPEC (Information Service in Physics, Electrotechnology, Computer and Control) at http://www.theiet.org/publishing/inspec/index.cfm	Of importance because information technology plays a considerable role in all areas of engineering today
	WTI-Frankfurt (Wissenschaftlich Technische Information, formerly Fachinformationszentrum, Specialized Information Center of Technology) at http://www.wti-frankfurt.de	Databases also contain German resources
	Agricola at http://agricola.nal.usda.gov Ulidat at http://doku.uba.de	For agricultural sciences. Free access German database for the environmental sciences. Free access
Patents	DEPATISnet at http://depatisnet.dpma.de	The German patent information system contains the full text of every German and American patent in pdf format, also patents from other countries. You have to know the exact patent number. Searching in other database fields – for example, title, patent inventor, or abstract field – is possible. Available data in a respective search field may vary. Free access
	Esp@cenet, at http://worldwide.espacenet.com/ US Patent and Trademark Office at http://patft.uspto.gov	European patents (European Patent Office). Free Access Example for full text access to national patents. Free Access
Chemicals	ChemBioFinder at http://chemfinder.camsoft.com ChemSpider at http://www.chemspider.com/	Meta-search engine for chemical substances information. Free access Meta-search engine for chemical substances information. Free access

Table A.1 (*Continued*)

	NIST Webbook at http://webbook.nist.gov/chemistry/	Detailed data for many common substances. Free access
	InfoTherm by FIZ Chemie at http://www.infotherm.de	Thermophysical experimental data for the daily use of process engineers, tables, and charts for about 34 000 mixtures and 9000 pure substances
	Physical Properties Sources Index (PPSI) at http://www.eqi.ethz.ch/en/	Lists recommended databases, handbooks, and web sites (data, definition, and measurement) for physicochemical and other material properties. Free access
	PubChem at http://pubchem.ncbi.nlm.nih.gov/	Provides information about chemicals especially relevant for medical sciences. Free access
	ChEMBL at https://www.ebi.ac.uk/chembldb/	A database on properties of chemicals where also p*K*-values are given. Free access
Hazardous substances	GESTIS at http://www.dguv.de/ifa/en/gestis/stoffdb/index.jsp	Information system of the German institutions for statutory accident insurance and prevention (English version available). Free access
	TOXNET, Toxicological Data Network, of the US National Library of Medicine at http://toxnet.nlm.nih.gov	Toxicology and hazardous substances. Free access
	International Chemical Safety Cards (ICSCs): International Programme on Chemical Safety at http://www.cdc.gov/niosh/ipcs/icstart.html	Available in a lot of languages. Free Access
Enzyme manufacturers (see also Table 7.4)	Association of Manufacturers and Formulators of Enzyme	Contains information on enzymes, safety rules for their use, and links to similar

(*continued*)

Table A.1 (*Continued*)

Searching for	Database name and web site	Comments
	Products at http://www.amfep.org	organizations, the companies, and organizations of importance for the regulation of enzymes (EU, FAO, FDA, WHO, etc.). Free Access
Enzyme classification and structure	Enzyme nomenclature by the International Union of Biochemistry and Molecular Biology at http://www.chem.qmul.ac.uk/iubmb/enzyme	Enzyme classification. Free access
	ExPASy, Bioinformatics Resource Portal, of the Swiss Institute of Bioinformatics at http://www.expasy.ch	Extensive information on all aspects of proteins/enzymes. Includes enzyme nomenclature database and links to more specialized databases on enzymes. Free access
	RSCB Protein Data Bank at http://www.rcsb.org/pdb	A database for 3D structures of proteins/enzymes and cofactors important for structure and function. Free access
	CATH protein Structure Classification at http://www.cathdb.info	A database on enzyme classification based on 3-D structures. Free access
Enzyme properties	BRENDA, The Comprehensive Enzyme Information System at http://www.brenda-enzymes.org	A comprehensive database on enzyme properties (k_{cat}, K_m for different substrate, cofactors, inhibitors, stability, etc.) Free access
Enzyme catalyzed reactions	Thermodynamics of Enzyme-Catalyzed Reactions at http://xpdb.nist.gov/enzyme_thermodynamics	NIST Standard Reference Database 74. Free access
	KEGG LIGAND Database at http://www.genome.jp/kegg/ligand.html	A database to find an enzyme that can catalyze the biotransformation of a compound (ligand). Links to metabolic charts that show the enzyme that catalyzes the metabolic reactions. Free access

Table A.1 (*Continued*)

Enzymes, specific	MEROPS – The peptidase database at http://merops.sanger.ac.uk	A database for peptides. Free access
	The Lipase Engineering Database: http://www.led.uni-stuttgart.de	A database for lipases. Free access
	CAZy Carbohydrate-Active enZYmes Database at http://www.cazy.org	A database on enzymes catalyzing reactions with carbohydrates. Free access
Bioinformatics	EMBL-EBI from the European Bioinformatics Institute at http://www.ebi.ac.uk	A database from the European Bioinformatics Institute on the bioinformatics of proteins/enzymes with links to the main databases and tools for sequence and structure analysis (gene or protein), alignment of enzyme sequences, and so on. Free access
Biocatalysis	UM-BBD University of Minnesota Biocatalysis/Biodegradation Database at http://umbbd.msi.umn.edu	Biocatalysis/biodegradation database. Free access
National laws and regulations and international conventions influencing the production and use of enzymes and cells as biocatalysts	Laws and regulations based on the Cartagena protocol on biosafety for most countries can be found in the homepage of the Biosafety Clearing House http://bch.cbd.int/about Additional Internet resources for laws and regulations of importance can be found in Section 6.6	

A.6
After Searching: Evaluating and Processing Information

After searching successfully, you have to evaluate your search findings with respect to relevance. How to be sure that all the potentially important documents are included in your resulting set? How to modify your query to reach this goal? But it is also important to evaluate critically the quality of the documents you have found. In case the document is published in a scholarly peer-reviewed journal, the article has been evaluated by independent experts before acceptance/publication. Who is the author and what is his or her background? Why is the document being provided? How current is it?

The process of information retrieval interweaves more and more – especially in the digital age – with writing and communication processes (Cottrell, 2008; Divan, 2009; Hofmann, 2010; Johnson and Scott, 2009; Lebrun, 2009). Networking and collaboration opportunities like weblogs, wikis, and other tools of the "social web" are central themes today and enhance data sharing and new ways to stay current (Cann *et al.*, 2011; Oliver, 2009). Subject-specific tools of the "social web" for the bioengineer include resources like http://openwetware.org, a wiki "for researchers and groups who are working in biology and biological engineering," http://www.proteopedia.org, a 3D-encyclopedia of proteins and other molecules, or http://www.cazypedia.org, an encyclopedia of carbohydrate-active enzymes. The need to keep yourself up-to-date by getting a free table of contents via e-mail from publishers of journals can now be met by using RSS feeds that can be collected in social feed readers like Netvibes, see for a personal example, http://www.netvibes.com/thapke. Subscribing to subject-specific mailing lists is substituted through reading subject-specific weblogs or following a researcher via Twitter.

Note-taking strategies today occur by using reference management software like Zotero, Jabref, Mendeley, Endnote, or Citavi. Such software allows organizing references, quotations, and full text files. Formatting texts in specific citation styles supports the publishing process.

In addition to impact factors that allow the evaluation and ranking of journals (Journal Citation Reports,[1] http://www.eigenfactor.org), there exist further quantitative measures like citation rates and personal impact factors as the h-index to evaluate research through citation analysis (De Bellis, 2009), for example, within Web of Science or Scopus. Software using Google™ Scholar as source for evaluation is also available (Harzing, 2011). But "no matter how considerate and extensive a [bibliometric] evaluation is, it will be implemented only to the extent that it is in consonance with the prevailing power structure" at the commissioning higher education institution (Seglen, 2003, p. 151).

A.7
Information and the World

What is publication – what is an author, a document, a journal, a collection, or a library? In the electronic world of the Internet all of these terms have changed their meaning and use. Thinking about information is particularly of interest in

1) Look at http://thomsonreuters.com/products_services/science/science_products/a-z/journal_citation_reports/.

biotechnology (Braman, 2004). At a time when historians of science describe ". . . biology's metamorphosis in an information science" (Lenoir, 1999, p.43), it is necessary to reflect about information and its communication and use (Feather and Sturges, 2003). Even new uses of the word biotechnology arise as the following citation shows: "I also would treat as biotechnology those affective technologies including so-called new media technologies that have permitted us to rethink the body in terms of digitization" (Clough, 2007, p. 312).

In spite of information overload, only a limited part of information is freely available on the Internet. Access to commercial information sources for scholarly research such as reference databases and the full text of a specific journal is usually subject to a license fee and controlled by password. However, they are often offered within the intranet of universities or companies. Open access activities try to free access to scholarly publications at least for research and educational purposes. Examples are the journal *PLoS Biology* at http://www.plosbiology.org and the journals of BioMed Central at http://www.biomedcentral.com.

Issues of intellectual property and copyright increase in a "cut-and-paste" environment. They are especially part of biotechnology research (Castle, 2009). Why is it important to cite sources of information? What is the right way to cite? Questions of information ethics (plagiarism) as well as information policy (ownership, access, privacy) become important. Does there exist a digital divide? Even think of the preservation and long-term stability of information. What will be happening with electronic records or data in 30 or 50 years' time?

References

Blaxter, L., Hughes, C., and Tight, M. (2010) *How to Research*, 4th edn, Open University Press, Maidenhead.

Braman, S. (ed.) (2004) *Biotechnology and Communication: The Meta-Technologies of Information*, Lawrence Erlbaum, Mahwal, NJ.

Cann, A., Dimitriou, K., and Hooley, T. (2011) *Social Media: A Guide for Researchers*, Research Information Network, London, http://www.rin.ac.uk/social-media-guide. Access date June 26, 2012.

Castle, D. (ed.) (2009) *The Role of Intellectual Property Rights in Biotechnology Innovation*, Elgar, Cheltenham.

Clough, P.D. (2007) Biotechnology and digital information. *Theor. Cult. Soc.*, **24**, 312–314.

Cottrell, S. (2008) *The Study Skills Handbook*, 3rd edn, Palgrave Macmillan, Basingstoke.

De Bellis, N. (2009) *Bibliometrics and Citation Analysis: From the Science Citation Index to Cybermetrics*, Scarecrow Press, Lanham, Md.

Divan, A. (2009) *Communication Skills for the Biosciences: A Graduate Guide*, Oxford University Press, Oxford.

Falciola, L. (2009) Searching biotechnology information: a case study. *World Pat. Info.*, **31** (1), 36–47.

Feather, J. and Sturges, P. (eds.) (2003) *International Encyclopedia of Information and Library Science*, 2nd edn, Routledge, London.

Harzing, A.W. (2011) *Publish or Perish*, version 3.1.4097, available at http://www.harzing.com/pop.htm. Access date June 26, 2012.

Hull, D., Pettifer, S.R., and Kell, D.B. (2008) Defrosting the digital library: bibliographic tools for the next generation web. *PLoS Comput. Biol.*, **4** (10), e1000204.

Hofmann, A.H. (2010) *Scientific Writing and Communication: Papers, Proposals, and*

Presentations, Oxford University Press, New York.

Johnson, S. and Scott, J. (2009) *Study and Communication Skills for the Biosciences*, Oxford University Press, Oxford.

Lebrun, J.-L. (2009) *Scientific Writing: A Reader and Writer's Guide*, World Scientific, Singapore.

Lenoir, T. (1999) Shaping biomedicine as an information science, in *Proceedings of the 1998 Conference on the History and Heritage of Science Information Systems* (eds M.E. Bowden, T.B. Hahn, and R.V. Williams), Information Today, Medford, NJ, pp. 27–45.

MacLeod, R.A. and Corlett, J. (eds) (2005) *Information Sources in Engineering*, 4th edn, Saur, Munich.

Oliver, A.L. (2009) *Networks for Learning and Knowledge-Creation in Biotechnology*, Cambridge University Press, Cambridge.

Osif, B.A. (ed.) (2012) *Using the Engineering Literature*, 2nd edn, CRC Press, Boca Raton, FL.

Schmidt, D., Davis, E.B., and Jacobs, P.F. (2002) *Using the Biological Literature: A Practical Guide*, 3rd edn, Dekker, New York.

Seglen, P.O. (2003) Bibliometric analysis: what can it tell us? in *Science Between Evaluation and Innovation: A Conference on Peer Review. Ringberg-Symposium April 2002*, Max-Planck-Gesellschaft, Munich.

Tredinnick, L. (2008) *Digital Information Culture: The Individual and Society in the Digital Age*, Chandos Publ., Oxford.

Ward, H. and Hockey, J. (2007) Engaging the learner: embedding information literacy skills into a biotechnology degree. *Biochem. Mol. Biol. Educ.*, **35**, 374–380.

Appendix B: Solutions to Exercises

Chapter 1

1.1 See Section 1.3.1.

1.2 Reduce the number of processing steps. This can be achieved when all enzyme processes are carried out at the same pH, without cofactors (Mg^{2+} or Ca^{2+}) or with the same cofactor, and the same temperature (except for the first process). This requires other enzymes as biocatalysts that can be found by screening or developed by gene technological methods (see Chapter 3 and Case Study 1 in Chapter 12).

1.3 Especially the properties of the substrates and products (solubility, stability) and of the process (equilibrium or kinetically controlled), and how they are influenced by pH or temperature.

1.4 The by-products are the monosaccharides formed in an equilibrium-controlled process. Their formation is determined by the ratio of the synthesis rate/hydrolysis rate (selectivity) of the enzyme used as biocatalyst. The higher this selectivity, the lower the by-product formation. Thus, the selectivity of the enzymes and its temperature and pH dependence must be known. (For a possible pH dependence, see Sections 2.7 and 2.8.)

1.5 List your consumption and compare.

1.6 Inhalation and ingestion of enzymes must be avoided. Thus, the processes must be carried out in closed containers, from which aerosols or dust cannot escape. In the case of dry enzyme production (used in detergents), the enzyme must be produced in particle sizes that are too large to be inhaled deep into the lungs ($\geq$100 μm).

Biocatalysts and Enzyme Technology, Second Edition. Klaus Buchholz, Volker Kasche, and Uwe T. Bornscheuer

Chapter 2

2.1 In phosphate buffer as in this case the pH hardly changes with temperature. Tris should be avoided in enzyme kinetic studies as it can act as a nucleophile (Kasche and Zöllner,). An alternative to phosphate buffer here is HEPES.

2.2 They catalyze oxidations or reductions of substrates; without a cosubstrate the enzyme would be reduced or oxidized in the process and could not further act as a catalyst.

2.3 Consider the influence of charge–charge interactions. The difference between the exopeptidase carboxypeptidase A and the endopeptidase α-chymotrypsin is that only the former enzyme can bind negatively charged P_1' residues in their positively charged S_1' binding site. This difference is reduced with increasing ionic strength as the binding of P_1' to the S_1' binding sites is reduced for the exopeptidase and increased for the endopeptidase. The enzyme function is divided in substrate binding, catalytic reaction(s), and the dissociation of the products. Product inhibition cannot be avoided.

2.4 Ca^{2+} binding site, the conserved residues, and the 3D structure far from the active site (helices, turns, etc.).

2.5 See Section 2.7.1.1.

2.6 Use the Arrhenius equation (see Eq. (2.22), note that $\ln x = 2.3 \log x$) $\log V_{max} = \log k_{cat}[E] = \text{constant} - (\Delta G^{\#}_{ES})/2.3RT = \text{constant} - (\Delta H^{\#}_{ES} - T\Delta S^{\#}_{ES})/\ 2.3RT$, where $\Delta G^{\#}_{ES}$, $\Delta H^{\#}_{ES}$, and $\Delta S^{\#}_{ES}$ are the activation free energy, activation energy (enthalpy), and activation entropy that are considered to be temperature independent in the studied temperature interval. Plot $\log V_{max}$ as a function of $1/T$ (Kelvin!). The slope of the line is $-(\Delta H^{\#}_{ES})/2.3R$, where R is the gas constant (8.3 J/mol^{-1} K^{-1}). This gives the activation energy $\Delta H^{\#}_{ES} = 33\ \text{kJ mol}^{-1}$ for the enzyme-catalyzed reaction. The activation free energy can be determined when $\Delta S^{\#}_{ES}$ is known. How can this activation entropy be determined? (*Hint*: What is the difference of the Arrhenius equation for two temperatures?)

2.7 See Section 2.7.1.1.

2.8 Trypsin and α-chymotrypsin are specific for positively charged or hydrophobic amino acid residues in P_1, respectively. Both are endopeptidases (P_1' residue uncharged) and are much better esterases than amidases. Penicillin amidase has a (*R*)-specific and α-chymotrypsin a (*S*)-specific S_1 binding site. As expected, the properties of the same enzyme from different sources can differ (penicillin amidase). For the hydrolysis of penicillin, the specificity constant of this enzyme cannot be considerably increased by evolution as its size indicates that the reaction is almost diffusion controlled (see the text around the tables).

2.9 They are esters that can be hydrolyzed or synthesized by esterases (EC 3.1, such as lipase) or glycosidases (EC 3.2).

2.10 Use the Arrhenius equation given in Exercise 2.6. The relation between the half-life time and the rate constant for the monomolecular inactivation of the enzyme k_i (Figure 2.28) is (half-life time) $= 0.70/k_i$. From the slopes, the following activation energies are obtained: for the thermal denaturation, 470 kJ mol^{-1}; for the hydrolysis of the (*S*)-substrate, 48 kJ mol^{-1}; for the hydrolysis of the (*R*)-substrate, 18 kJ mol^{-1}. For the latter two, the activation energy for the acylation of the enzyme is determined. Note that *E* can be increased by ≈ 1 order of magnitude by decreasing the temperature. This may be an alternative to increase *E* by molecular evolution (see Section 2.11 and Chapter 3). With increasing temperature, the enzyme becomes (*S*)-specific. (The error in the determination of the slopes here and in Exercise 2.6 does not allow to give more than two significant numbers.)

2.11 Search for this enzyme in the BRENDA database or other Internet resources for information on the active site given in Appendix A. You will then also find that this enzyme is important in many physiological processes.

The catalytically active residue is Cys. This gives the following mechanism:

- acylation

$$\text{E-SH} + \text{R-(Gln)-(CO)-NH}_2 \leftrightarrow \text{ES-(CO)-R} + \text{NH}_3$$

- deacylation by a Lys-NH_2 on the protein acylated to the enzyme

$$\text{E-S-(CO)-R} + \text{R}'\text{-(Lys)-NH}_2 \leftrightarrow \text{E-SH} + \text{R-(CO)-(NH)-(Lys)-R}'$$

that is formation of an intramolecular cross-link between Gln and Lys residues on the surface of the protein located near each other.

The nucleophile R-NH_2 used for the deacylation can also be

- another protein. Then proteins are cross-linked that can change the viscosity (texture) of proteins in food. This is a possible biotechnological application of this enzyme.
- located on the surface of porous particles. This can be used to apply this enzyme to immobilize proteins (is this process reversible?)
- a fluorescent molecule. Then this can be used to label proteins with fluorochromes. Such labeled proteins can be used for analytical purposes (see http://probes.invitrogen.com/).

The reaction is equilibrium controlled, as the enzyme is a transferase where hydrolysis of the acyl-enzyme can be neglected. The pH dependence K_{app} can be determined as shown in Section 2.8.2, Eq. (2.40) (see also Figure 2.20).

The following relation is derived:

$$K_{app} = K_{ref}([H^+] + K_{NH_3})/([H^+] + K_{Lys\text{-}NH_2}),$$

where $K_{NH_3} = 10^{-9.3}$ and $K_{Lys\text{-}NH_2} = 10^{-10.5}$ (the latter value can, however, be dependent on the localization of the Lys residue on the protein surface). Plot log K_{app} as a function of pH. From this, or the above equation, follows that the equilibrium product yield is almost independent of pH for pH $< \approx 8$, and that K_{app} increases by almost an order of magnitude with pH in the range 8–11.

The rate of the reaction, however, depends on the pH–activity profile of the enzyme.

2.12 As large product and substrate concentrations (up to more than 1 M) are obtained and used in enzyme technology.

2.13
- Be a substrate (see also Exercise 2.21).
- Environment-friendly.
- Be easily contained, separated from products and substrates, and recycled.

2.14 The K_{app} values for 95 and 99% yields are 5.4 and 29.4 M, respectively. From the above plots follows that 95% yield can be obtained for pH ≥ 8.0 and $T \geq 30\,°C$; 99% yield can be obtained for pH ≥ 8.5 and $T \geq 40\,°C$ (or a higher pH and lower temperature). From Figure 2.28 follows that the enzyme from *A. faecalis* should be used as a biocatalyst here as it has a better stability at the pH values given above. Due to the thermal denaturation (Exercise 2.10), the temperature during the hydrolysis should be kept as low as possible to reach the above yields (i.e., around 30 °C). The best buffer here is $NaHCO_3$ or $KHCO_3$ (pK = 10.3) that is better than NH_3 (pK = 9.3) (Why?).

2.15 The specific heat of water is 4.2 J $(g^{-1}\,°C^{-1}$ at the temperatures used in these processes. For the hydrolysis of penicillin, the temperature will decrease 1.4 °C. For the hydrolysis of saccharose, the temperature will increase 3.5 °C.

2.16 (See Figure 2.22 and Eq. (2.42)) The reaction takes place in the interfacial diffusion layer, where the mass transfer occurs only by diffusion and in the bulk liquid phase in which the buffer can be rapidly mixed by stirring whose magnitude is expressed by the Sherwood number (see Eq. (10.7)). The stirring also influences the thickness of the diffusion layer where the pH adjustment takes longer time than in the bulk phase. Assume that the solid particles are spherical. At $Sh = 10$ (a value that is normally reached), the thickness of the diffusion layer is $R/5$, where R is the radius of the solid particles. The ratios between the bulk liquid phase and the volume of the diffusion layer at different solid contents (vol%) and at $Sh = 10$ are as follows:

Solid content (vol%)	Bulk liquid volume/diffusion layer volume
10	11
20	4.3
30	2.2
40	1
50	0.4

The rate of the reaction is higher in the bulk phase than in the diffusion layer, where the pH adjustment is slow. Thus, in this case the solid content should not be too high ($\leq$30%). It is also essential to use small particles to increase the mass transfer rate from the solid to the liquid phase (Eq. (2.42)), and to use a buffer with high buffer capacity.

2.17 The binding of the negatively charged 6-APA to its binding site on the enzyme depends on the ionic strength when the binding site on the enzyme is positively charged that has now been shown to be the case.

2.18 The solution is given in the text.

2.19 By increasing the temperature stability and changing the pH optimum to that of the preceding reaction of the enzyme. At higher temperatures, glucose can react with amino groups (of Lys) on the enzyme. For this critical Lys residues must be exchanged. That this can be achieved is shown in US Patents 5,290,690 (1994) and 5,384,257 (1995). See also Section 2.11.

2.20 a. At equilibrium, the maltose and isomaltose concentrations are x and y M, respectively. The ratio of the equilibrium constants shows that $y = 33.3x$, that is, $y >> x$. Then y can be calculated by solving the second grade equation in y derived from the equilibrium constant with $x = 0$. This gives $y = 0.1$ M and $x = 0.003$ M.

b. Reducing the temperature and/or changing the specificity constant of the enzyme for the formation of isomaltose and maltose and increasing it for the hydrolysis of maltose.

2.21 The concentration of dissolved CO_2 in water does not depend on pH (the total amount of dissolved CO_2, its hydration product, and their dissociation forms do, however, increase with pH). The reaction is equilibrium controlled, and the pH dependence of K_{app} for the reaction in aqueous solutions can be determined as shown in Eq. (2.40). This gives

$$K_{app} = K_{ref}([H^+] + K_{Pyr})/[H^+],$$

thus K_{app} and the equilibrium product yield increase with pH above pK_{Pyr} ($\approx$3). Plot log K_{app} as a function of pH. Thus, the reaction must be carried out at pH >5 for high yields. (Try to find the pH optimum of the enzyme for this reaction using the Internet resources given in Chapter 9.)

An alternative solvent for this reaction is supercritical CO_2 where the substrate content is much larger than that in aqueous solutions. Then the solubility of acetaldehyde is a limiting factor. This may not be possible, as the statement of the authors to try this has not resulted in a publication yet. Acetaldehyde is known to inactivate biocatalysts (see Exercise 3.5), and this may also be a problem.

Is this a realistic process to reduce the CO_2 content in the atmosphere due to the combustion of fossil fuels? (Pyruvate may be converted to lactate that can be used as monomer to produce biodegradable polymers.)

2.22 a. *Equilibrium-controlled processes*: Amidate the peptides with a suitable amidase or an exopeptidase such as carboxypeptidase A. *Kinetically controlled processes*: Then only enzymes that form acyl-enzymes can be used. At high ionic strength, trypsin can be used with Gly-NH_2 as nucleophile for vasopressin synthesis. Some exopeptidases such as carboxypeptidase Y could also be used (Kasche, 2001).
b. Do this.

2.23 On the MEROPS home page start SEARCH and search for which peptidase can hydrolyze this peptide bond.

For the mini-proinsulin you will find that trypsin is a suitable peptidase for both peptide bonds. When it hydrolyzes mini-proinsulin in the presence of Thr butyl ester, the B-chain amino acid sequence is complete, and the ester bond is hydrolyzed at a basic pH that shifts the equilibrium toward hydrolysis (Figure 2.20a). Alternatively, this can also be achieved by adding an excess of an organic acid as trifluoroacetic acid.

For proinsulin, trypsin can hydrolyze the Arg–Gly and the Arg–Arg peptide bond to the right of the Thr–Arg bond that must also be hydrolyzed. When the first two bonds are hydrolyzed, the last bond can be hydrolyzed by an exopeptidase (carboxypeptidase B).

All reactions are equilibrium controlled except the reactions in the presence of the Thr ester that is kinetically controlled.

The enzymes used in the patents for these processes were free, but it would be better to use them immobilized in wide-pore adsorbents. This would reduce the enzyme impurities in the finally purified insulin.

Chapter 3

3.1 Directed evolution is essentially a random mutagenesis combined with screening or selection to identify desired enzyme variants in a mutant library.

3.2 In a recombining (sexual) method, several related genes – that is, homologous enzymes from different microorganisms or best variants from an initial round of directed evolution – are subjected to a recombination method such as gene (DNA) shuffling. In a nonrecombining (asexual) method, a single protein encoding gene is subjected to random mutagenesis, for example, by error-prone PCR.

3.3 In contrast to a standard PCR method, in an error-prone PCR the conditions are varied to increase the error rate. In typical protocols, one dNTP is supplied at nonstoichiometric (usually 10-fold lower) concentration relative to the three other dNTPs, Mn^{2+} ions are added (the polymerase used is Mg^{2+} dependent), the concentration of Mg^{2+} ions can be changed, a nonproofreading polymerase (i.e., the *Taq* polymerase from *Thermus aquaticus* instead of a proofreading polymerase such as the enzyme from *Pyrococcus furiosus*) must be used, and finally the numbers of cycles and the temperature program can be varied.

3.4 A selection can take place if a key metabolite cannot be produced any longer by the host microorganism. Insertion of a mutant library expressing an enzyme to complement this missing pathway allows survival and hence growth of the host microorganism. Consequently, only those clones expressing this enzyme variant with the missing and necessary activity can grow on an agar plate. Advantages are that rather large mutant libraries can be easily subjected to selection as only a few clones are expected to contain a hit enzyme and grow. Disadvantages are that in most cases only enzymes involved in metabolic pathways can be targeted. Moreover, it often turns out that the host microorganism finds its way around this evolutionary pressure and can grow even without the missing enzyme activity. For instance, deletion of an L-alanine dehydrogenase (catalyzing the formation of L-Ala from pyruvic acid) gene must not mean that the growth assay in the absence of L-Ala will work as L-Ala can also be generated by a transaminase or by a racemase from D-Ala.

3.5 In an iterative saturation mutagenesis, distinct positions in a protein are subjected to saturation mutagenesis. First one position (A) is randomized and the best hit is identified, for example, after biocatalysis and analysis for the desired property. Then this best hit serves as a template for the next position B, and so on. If four positions (A, B, C, D) are mutated, there are $4! = 24$ ($4 \times 3 \times 2 \times 1$) possible combination paths. Note that in a *simultaneous* saturation mutagenesis of four positions, $20^4 = 160\,000$ ($20 \times 20 \times 20 \times 20$) combinations are possible.

Chapter 4

4.1 In contrast to a kinetic resolution, a dynamic kinetic resolution can yield (theoretically) 100% optically pure product. This requires that the unwanted enantiomer racemizes faster than the enzymatic resolution step takes place. Also, racemization and enzymatic resolution have to take place in a one-pot reaction.

4.2 *Pros with hydrolase*: no need for cofactor recycling. Good chances to get access to both enantiomers using *one* hydrolase (using either hydrolysis or acylation yields opposite enantiomers); higher solvent stability. *Cons for hydrolase*: dynamic kinetic resolution often difficult to achieve. *Pros for dehydrogenase*: usually high yield and optical purity possible without problem. *Cons for dehydrogenase*: cofactor recycling needed (or in a whole-cell system: risk of side reactions/presence of interfering alcohol dehydrogenase); ketones less available than alcohols; often low solvent stability; requires two different dehydrogenases (Prelog and anti-Prelog) to get access to both enantiomers.

4.3 *Pros for whole-cell system*: *in situ* cofactor recycling, higher enzyme stability. *Cons for whole-cell system*: risk of side reactions/presence of interfering enzymes, high dilution, more difficult product isolation. *Determinants*: activity/stability of P450 enzyme, NADH or NADPH dependent?, cost of product versus cost of substrate, and so on.

4.4 a. amino acid dehydrogenase,
b. transaminase,
c. acylase/amidase/esterase/lipase,
d. hydantoinase and carbamoylase,
e. amino acid oxidase.

4.5 Use isopropenyl acetate (yields acetone) or non-enol ester acyl donors (i.e., diketene, acid anhydride); perform protein engineering to replace sensitive lysine residues in the protein.

4.6 *Easiest way*: growth on D-amino acid as sole carbon and nitrogen source. (*Note*: There could be, for example, a D-amino acid oxidase or transaminase acting on this enantiomer.)

4.7 A straightforward access to a single enantiomer of mandelic acid in theoretically 100% yield is possible by the use of a ketoreductase starting from the corresponding ketone (Scheme 4.1, route I). However, this reaction requires the cofactor NAD(P)H and hence efficient recycling is needed, which adds to the overall costs of the process. The alternative C—C bond formation from benzaldehyde and HCN by a hydroxynitrile lyase (route II) has the advantages of readily available starting materials and various enzymes to choose from, but

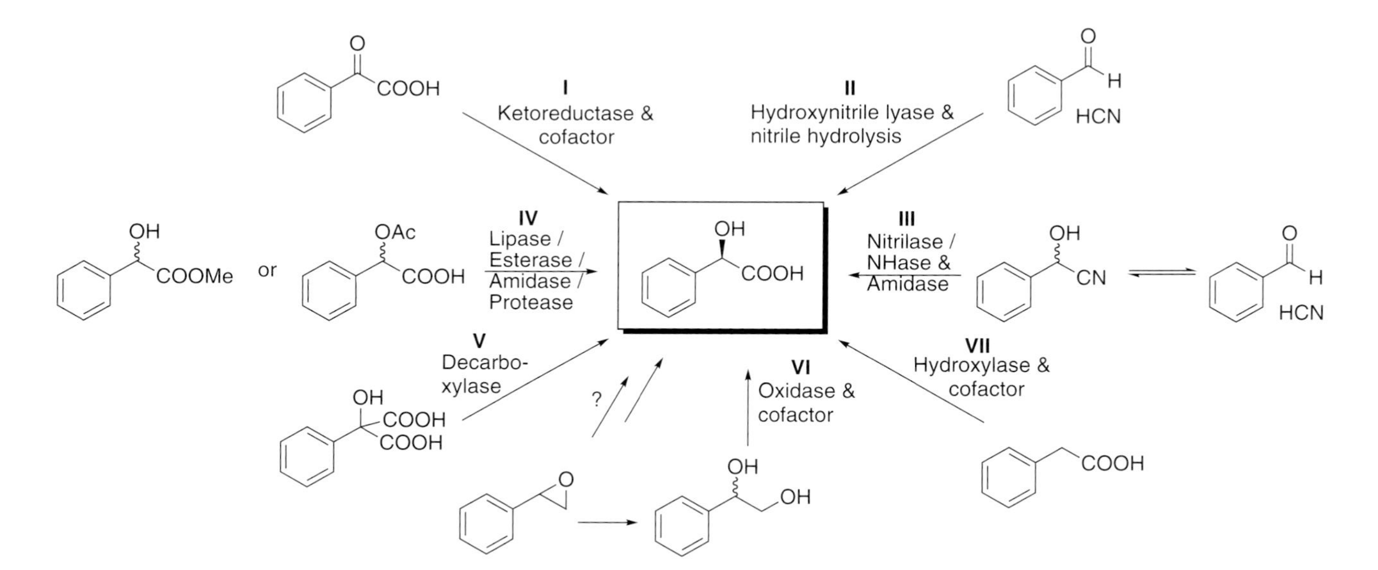

Scheme 4.1 Possible enzymatic routes toward chiral mandelic acid. Some of the routes have been indeed described in the literature; others are theoretical. General criteria are availability of the enzyme with desired properties and of the starting material. Furthermore, problems encountered with product isolation (i.e., substrate and product cannot be separated efficiently or cost effective) can determine that an alternative enzymatic route needs to be chosen.

the disadvantages that the chemical background formation of racemic mandelic acid at certain pH values must be suppressed and that safe handling of the highly toxic HCN must be ensured in the factory. In addition, chemical hydrolysis of the α-hydroxynitrile needs to be performed. A good atom efficiency is also feasible by using either a nitrilase or a combination of nitrile hydratase with an amidase starting from racemic α-hydroxynitrile (route III). Although this is at first glance a kinetic resolution with only 50% yield of the desired enantiomer, the disproportion of α-hydroxynitrile to benzaldehyde and HCN – the same chemical background reaction that affects the hydroxynitrile lyase-catalyzed route negatively – enables here a dynamic kinetic resolution. The use of hydrolases such as lipases, proteases, or amidases is a reasonably good route despite that it is a kinetic resolution (route IV). The strength of this approach lies in the low prices and high stabilities and activities of the enzymes (especially lipases), which reduces the costs for the biocatalysts and hence can make such a process economical although one enantiomer needs to be recycled or even discarded. The rather elegant application of decarboxylases (route V) is hampered by the lack of availability of the dicarboxylic acid starting material. The route starting from styrene oxide or the corresponding diol suffers from the lack of availability of suitable enzymes and the cofactor requirement (route VI). Finally, the direct and stereoselective hydroxylation of phenylacetic acid (route VII) would be very useful, but the P450 monooxygenases required for this approach show very little turnover numbers and stereoselectivity. Furthermore, their stability is a major problem and cofactor recycling and oxygen supply have to be ensured as well. This simple example compound demonstrates that various enzymatic routes can be imagined, but a careful evaluation of the entire process is needed, before a decision for a specific route can be made.

Chapter 5

5.1 As the metabolite concentration is in the range of K_m, the flux J_2 will increase when the concentration of M_1 increases.

5.2 Yes.

Chapter 6

6.1 No it is a class II (2) organism.

6.2 To avoid the selection of resistant microorganisms in the production of these enzymes. Look at the home page of the enzyme-producing companies (main producers Genencor and Novozymes) and AMFEP.

6.3 (The % given are weight %.) Theoretically they can be produced until the whole fermenter is filled with cells. The cylindrical cells can be packed as a long cylinder of cells where each cell has direct contact with six (or eight) cells. The former case is easier to calculate. Then the cells occupy $\pi\sqrt{3}/6$ or $\approx$90% of the total fermenter volume. The amount of protein is then $0.15 \times 0.9 \times \rho$ kg $l^{-1} = 0.145 \times \rho$ kg l^{-1}, where ρ is the cell density in kg l^{-1} ($\approx$1.2). This is larger than that for real cells.

6.4 See the last paragraph in Section 6.4.1. For PA from *A. faecalis*, this would give $\approx 24\rho$ g l^{-1}. Such values have been obtained for other enzymes and give upper limits. Cells can, however, not be grown to the densities in Exercise 6.3, as it is difficult to supply sufficient substrates and oxygen, and to mix the fermenter contents at such high cell densities.

6.5 The solution is given in the text.

6.6 To be able to bind these ions, whose concentrations are below 1 μM in the cell.

6.7 No, as additional processing steps and more materials are used for the solubilization and renaturation of the inclusion bodies.

6.8 Heat-stable enzymes are not precipitated when the homogenates are heated. Thus, they can be enriched when the homogenate is heated, so that the *E. coli* proteins are precipitated. This simplifies their separation in comparison with the enzyme with the same function from *E. coli*. The allergenic potential of the enzyme must be studied. If it is too high, it cannot be used.

6.9 It comes from cells that have been damaged or lyzed. Its concentration can be up to 0.1 g l^{-1}. It can be separated using the adsorbents in Exercise 6.9 or by cation-exchange adsorbents. (They should, owever, not be used for enzymes that bind metal ions (Why?).) In this case, anion-exchange adsorbents can be used (see Figure 4.10). DNA can bind to basic proteins, and this may be a problem in the purification of basic enzymes. This can be avoided when DNA is hydrolyzed by DNases (Benzonase®) before the chromatographic separation step(s). To avoid competition for the adsorbent sites by small molecules, ultrafiltration should be performed before this to reduce their content in the protein solution. The increased enzyme concentration after this step also increases the rate of adsorption to the chromatographic adsorbent.

6.10 a. Less proteolytic processing of the A-chain.
b. They should be immobilized rapidly to reduce intermolecular autoproteolysis that has been observed for PA.
c. Be reduced as small proteolytic peptides hydrolyzed from the C-end of the A-chain that activate the enzyme will be filtered off.

6.11 a. Use Eqs. (6.5) and (6.6).
b. It is better to change the column diameter, as the same flow rate can be used. When changing the column length, a higher flow rate must be used that increases the back pressure.

6.12 The answer is given in the text.

6.13 The rate of membrane transport is higher through the Sec than the Tat transport system, either due to a higher number of translocation systems or due to a higher rate through a Sec translocation system (Ignatova *et al.*, 2002). The *E. coli* PA with a Sec signal peptide should give higher yields. Alternatively, cells with a larger number of Tat translocation systems should be constructed.

6.14 Look in the home page of the Biosafety Clearing-House (Appendix A) or Internet resources in the literature list to this chapter.

Chapter 7

Exercise 1: Production of Invert Sugar by Invertase

A. Reaction temperature: 20 °C (no consideration of enzyme inactivation)
Result with the data given:

$$(1.75 - 0.0877) \times 10^3 + 10 \ln 20 = V_{max} \times 3.6 \times 10^4$$

(obviously the term '10 ln 20 = 30' can be neglected; that means the reaction is zero order in the range considered here)

$$1.66 \times 103 = V_{max} \times 3.6 \times 10^4$$
$$V_{max} = 4.6 \times 10^{-2} (\text{kat m}^{-3}) = 2.76 \times 10^6 \text{ U m}^{-3}$$

2.76×10^6 U (1 kat = 6×10^7 U) corresponds to the enzyme amount required per m^3 of reaction volume.
Inactivation can be neglected at 20 °C.

B. Reaction temperature: 50 °C (consideration of inactivation)
Result for the rate constant of inactivation: k_i (50 °C) = 8.29×10^{-4} (min^{-1}).
Residual activity (E/E_0) after 10 h: ln (E_0/E) = k_i t; E_0/E = 1.64, which corresponds to a residual activity of 60.8 %.
For the example discussed (50 °C) the effective mean activity is

$$V_{max,m} = 0.788\, V_{max,0}$$

The rate constant k_2 at 50 °C is higher as compared to k_1 at 20 °C, corresponding to:

$$k_2/k_1 = (\exp - E_A/RT_2)/(\exp - E_A/RT_1)$$

$$\ln(k_2/k_1) = (E_A/R)(1/T_1 - 1/T_2)$$
$$\ln(k_2/k_1) = (39.1/8.315 \times 10^{-3})(1/293.15 - 1/323.15) = 1.489$$
$$k_2/k_1 = 4.43$$

It follows that only 1/4 of the enzyme amount or activity, respectively, is required at 50 °C, as compared to that at 20 °C, in order to obtain the same reaction rate. However, due to inactivation the effective amount required is:

$$V_{max,0} = (1/0.788)\,V_{max,m} = 1.27\,V_{max,m}$$

The total amount required therefore is:

$$V_{max,0} = (1/4.43)1.27 \cdot 2.76 \cdot 10^6\ \mathrm{U\,m^{-3}}$$
$$V_{max,m} = 0.79 \cdot 10^6\ \mathrm{U\,m^{-3}}$$

That corresponds to 29 % of the amount required at 20 °C.

C. Calculate the productivity of the enzyme catalyst in terms of kg (products) per kg (enzyme) (enzyme activity assumed: 300 $\mathrm{Umg^{-1}}$).
For the productivity, it follows that:

$$P(t) = 600 \times 0.95\ \mathrm{kg\,(invert\ sugar/m^3)}/13.17 \times 10^{-3}\ (\mathrm{kat/m^3})$$
$$\approx 43.3\ t/\mathrm{kat},\ \text{or}\ 0.72\ \mathrm{g}\ U^{-1}$$
$$(\text{for}\ t = 10\ \mathrm{h}).$$

For an enzyme activity of 300 $\mathrm{U\,mg^{-1}}$ this corresponds to ca. 216000 kg invert sugar per kg enzyme. For a technical enzyme with an activity of 10 $\mathrm{U\,mg^{-1}}$, this would correspond to ca. 7200 kg product per kg enzyme.

Exercise 2: Calculation of Enzyme Inactivation under Different Conditions

1. Inactivation time for *Thermoanaerobacterium* GA: 374 min, or 6.2 h; for *A. niger* GA: 18.3 min.
2. no sorbitol: half-life of 11.2 min
   2 M sorbitol: half-life of 86.6 min
   Stabilization factor: 7.7
3. k_i at 60 °C: 0.0105 $\mathrm{min^{-1}}$; at 70 °C: 0.153 $\mathrm{min^{-1}}$.
   Activation energy $E_A = 248\ \mathrm{kJ\,K\,mol^{-1}}$; time for 90 % inactivation at 80 °C is 1.23 min.

Exercise 3: Hydrolysis of Penicillin with Penicillin Amidase

The amount of enzyme required for a given conversion of substrate in a given time can be calculated from Eqs. (2.24) and (2.30) and Table 2.6 for different cases. The integration of the Michalis–Menten equation for competitive product inhibition gives (per L reaction volume)

$$\begin{aligned} V_{max} &= (([S]_0 - [S])(1 - K_m/K_i) + (K_m + K_m[S]_0/K_i)\ln([S]_0/[S]))\,t^{-1} \\ &= ((0.3 - 0.015)\,(1 - 0.5) + (10^{-5} + 0.15)\ln(20))/120 \\ &= 0.6 \times 10^6/120\ [\mu\mathrm{M/L}\ min] \\ &= 5 \times 10^3\ [\mathrm{U/L}] \end{aligned}$$

Amount of enzyme required for 1 kg of 6-APA:

$$5 \times 10^3 \, (0.285 \times 0.216)^{-1} = 5000/0.062 = 80600 \text{ U}$$

Enzyme cost per kg 6-APA: ~2000 €.

Consequences: For a cost range of 5 € kg^{-1}, the enzyme must be re-used more than 400 times in order to make the process economically competitive. The enzyme must therefore be applied as an immobilized bioctalyst.

Chapter 8 and 9

The exercises are laboratory experiments described in the text

Chapter 10

10.1 By measuring the concentration of substrate(s) and product(s) in the solution outside the particles as a function of time. HPLC is a suitable method for this purpose.

10.2 See Section 10.3.

10.3 The solution is given in the text.

10.4 100–200 μm.

10.5 Reduce its particle size.

10.6 In the cation exchanger, the local pH on the surface of the pores in the particle is lower than that in the bulk solution as the concentration of H^+ is larger on the surface. Thus, the apparent pH optimum is shifted to larger pH values (pH in bulk solution). This difference is reduced at higher ionic strengths, as other positively charged ions reduce the H^+ content on the surfaces of the particle. The observation that $\eta > 1$ is an artifact as the rates for the free and immobilized enzyme are determined at different pH values around the enzyme (see definitions in Section 10.3).

10.7 Use Figures 8.7 and 8.8. From these follows that for the same process the operational effectiveness factor is larger than the stationary effectiveness factor for the reaction in the CST reactor. This difference can be reduced when instead of one CST reactor a cascade of CST reactors are used (see Section 11.2.2). This does, however, not apply when product inhibition is considered (see the next exercise).

10.8 The influence of product inhibition is more marked in the CST reactor than in the PB reactor, as the product concentration is larger in the CST reactor. In the case of inhibition by H^+, this can be reduced by controlling the pH by addition of buffer in this reactor. This, however, is more difficult in the PB reactor (where it can be reduced by dividing it into several PB reactors in series with pH adjustment between the reactors).

10.9 Because the maximum STY is limited by the maximal mass transfer rate to the particles.

10.10 The reactor is mass transfer controlled. Its STY is first reduced when the active biocatalyst amount is reduced so that the reactor becomes reaction controlled.

10.11 (a) For uncharged substrates as in the case of glucose isomerization. The enzyme is immobilized by adsorption to an ion-exchange adsorbent in packed beds. At least three packed beds are used in series. At the beginning, enzyme is only adsorbed to the first packed bed. When the STY is reduced due to biocatalyst inactivation, new enzyme is added to the packed beds to keep it constant. When the third bed is filled with enzyme, the first column is replaced with a column with new enzyme. The replaced column is then regenerated by desorbing the inactivated enzyme (see Chapter 8, Antrim and Auterinen, 1986). (b) For charged substrates, the enzymes used as biocatalysts must be covalently adsorbed. Then several reactors are used either in series or in parallel. In the beginning, only some are used to reach the necessary STY. When it is reduced, a new reactor is added that replaces the oldest reactor to maintain the STY at a given level, and so on.

10.12 The solution is given in the text.

10.13 The selectivity is larger for the free enzyme. The difference is probably due to the pH gradient formed in the particles with immobilized enzyme. The pH is lower than that in the bulk solution and this reduces the concentration of uncharged α-amino groups on Arg-NH_2 that determines the rate of formation of the dipeptide, and this reduces the selectivity.

10.14 In equilibrium-controlled processes involving products, by-product formation is larger when immobilized enzymes are used as biocatalysts, as the product concentration is larger in the biocatalyst particles than in solution. For kinetically controlled processes involving substrates, the by-product formation is reduced when the enzyme is immobilized as the substrate concentration is lower in the biocatalyst particle than in the free solution.

10.15 a. In both cases, it is favorable to use immobilized enzyme, to be able to reuse it. It can also be easily separated from the product by filtration. Arg-NH_2 should be better as a nucleophile than Arg-OEt to avoid hydrolysis of the latter (unprotected Arg is a poor nucleophile here). At equal total concentrations of X-Tyr-OEt, the ionic strength in the solution is larger with X = maleyl- (charged) than with X = acetyl-. This reduces the binding of the positively charged nucleophile to the negatively charged S'_1 binding site on the enzyme, and the product yield decreases (Schellenberger, Jakubke, and Kasche, 1991). At the end of the reactions, all substrates and products are in solution. After filtration of the enzyme, they can be isolated by HPLC using reverse-phase adsorbents or by ion-exchange chromatography.

b. The acyl group can be removed by an aminoacylase (EC 3.5.1.14) and the amide group by an amidase (EC 3.5.1.4) with suitable substrate specificity.

c. This would be an equilibrium-controlled process, where Try-Try, Arg-Try, and Arg-Arg are possible by-products. A dipeptidase (EC 3.4.13.-) with a high P_1 and P'_1 specificity for Tyr and Arg, respectively, would be a suitable biocatalyst. Alternatively, an exopeptidase with the same substrate specificity can possibly be used. Use the hint to discuss the separation procedure.

10.16 This exercise should train the reader to select possible alternatives that must be evaluated further to find the most suitable solution to this problem.

Chapter 11

11.1 For bulk products with scale of, or over, 10 000 t/a, because for unfavorable *E*-factor and mass index, thousands t/a of by-products to be used and/or waste to be disposed of would be produced. If many steps are involved in the synthesis of a product, an *E*-factor 100 may result that also means high amounts of by-products and/or waste that should be avoided.

11.2 A combination of continuous stirred tank reactor and subsequent tubular reactor(s), where the first reactor should be run at a conversion, and thus substrate concentration where inhibition is considerably reduced.

11.4 a. Processes limited by reactions that are thermodynamically unfavorable; ready extraction of inhibitory or toxic products from the reaction medium; to quickly extract unstable products for stabilization; to quickly extract products that undergo further reactions.

b. Membrane separation; extraction in biphasic systems, including organic phases, or aqueous two-phase systems, or ionic liquids; solid particle adsorption to magnetic adsorbent particles; chromatography; crystallization.

c. Molecular weight; surface properties such as hydrophilic, hydrophobic, ionic; solubility in phases to be applied; adsorption properties, and isotherms on solid phases to be applied; stability under conditions of pH, temperature.

Chapter 12

12.1 Grain milling to obtain appropriate particle size for enzymatic hydrolysis; thermal denaturation (gelatinization) to provide access to enzymes; enzymatic hydrolysis with different enzymes, α-amylases, ß-glucanases, cellulases, pectinases, arabino-xylanases, xylanases glucoamylase; fermentation; and alcohol distillation. Integration of thermal treatment and α-amylase action is favorable, due to a synergistic effect; enzymatic hydrolysis and fermentation should be integrated; and fermentation and alcohol removal (e.g., by membrane systems) can be favorable. Economic factors are by-product utilization; governmental subsidies for fuel alcohol.

12.2 Logistics and transport; pretreartment; and utilization of residual material.

12.3 Existing systems for coordinated, organized transport of agricultural products to factories: sugar cane, sugar beet, and by-products, for example, sugar beet pulp.

12.4 Logistics or raw material transport; pretreatment, physicochemical (thermal, for example, steam explosion), fermentation, and alcohol separation, by product isolation and utilization.

12.5 Integration of enzymatic hydrolysis and fermentation are highly favorable, in order to minimize product inhibition, notably by glucose that should be converted by yeast to maintain a low steady-state concentration. Integration of fermentation and alcohol removal is as well considered to be favorable.

12.6 Burning to produce electricity and steam.

12.7 Political subsidies or funding, raw materials supply with sufficient volume and acceptable price, and logistics ensuring transportation.

12.8 Energy-efficient heat integration, for example, by lignin utilization, water recycling, and coproduct utilization.

12.9 Adipate can be recycled in the production of adipoyl-7-ADCA. (*R*)-amino-adipic acid cannot be recycled this way as (*S*)-amino adipic acid is the precursor for the synthesis of cephalosporin C.

12.10 They must not be used to prevent infections. They must not be used as antibiotics for animal care, where other antibiotics should (better must) be used.

12.11 The results can be determined by measuring the concentrations at equilibrium using HPLC. The hydrolysis of glutaryl-7-ACA is endothermal. The free energy, enthalpy, and entropy changes can be determined from the following relations:

$$\Delta G = (\Delta H - T\Delta S) = -2.3\ RT \log K_{app} \tag{12.1}$$

that can be rewritten as

$$\log K_{app} = -\Delta H/2.3\ RT + \Delta S/2.3\ R \tag{12.2}$$

ΔG at 25 °C is determined from K_{app} at this temperature using Eq. (12.1)
ΔH is determined from the slope of the plot of log K_{app} as function of $1/T$ (°K !) using Eq. (12.2), ΔS is then calculated using Eq. (12.1).

12.12 Low temperature ($<$ 22 °C) and short hydrolysis times ($<$ 1 h) (see Section 12.3.4).

12.13 d-Ceph: acylate with acetate (Figure 2.20a) at a low pH that shifts the equilibrium toward CephC with or without a suitable esterase.
(*R*)-AAA can be converted to (*S*)-AAA with a suitable recemase, and recycled for the production of CephC.

12.14 Yes, the similar enzyme penicillin amidase is already used for these purposes (Drauz and Waldmann, 2002).

12.15 a. 1.25 m^3
b. When the rate of addition of base (volume/time unit) to the mixing vessel is below a value near zero.
c. The system with only one reactor is the simplest with minimal control units that can fail. (For safety reasons, at least one extra reactor must be in use.)
d. Calculate the time in which the enzyme has lost 10 % of its activity and multiply this with 20.

12.16 The best to do is to separate them rapidly in the solution under the conditions after the hydrolysis. They can be separated by size exclusion or anion exchange chromatography. As the pH of the solution is $\approx$ 9, (*R*)-AAA has a negative net charge that is larger than that of 7-ACA and would be better bound to the anion exchanger. In both cases, continuous chromatography (see Chapter 6) can be used. The isolated 7-ACA in the raffinate (see Figure 6.14) can then be purified by precipitation at about pH 3 where it has a low solubility.

12.17 6-APA is formed by the hydrolysis of isopenicillin N with a hydrolase, then a transferase catalyzes the addition of the side chain R. When the latter enzyme is deleted, 6-APA could accumulate and be transported out of the cells.
7-ACA is not an intermediate; it could, however, be formed when the cephalosporin acylase is expressed in the cells producing CephC (Isogai

et al., 1991). The pH during the growth of these cells was at pH ≤ 7, where the hydrolysis equilibrium is unfavorable (Figure 12.15). The yield was less than 10 % (of the formed CephC).

12.18 The nitrilase route in Figure 12.21 has the advantage that readily available and cheap starting materials can be used, that the biocatalytic reaction is an asymmetric synthesis with up to 100% yield of the chiral product, and that the engineered nitrilases work at very high substrate concentration. A possible disadvantage is that additional steps are needed to convert the chiral product into the desired building block for the coupling step to yield the final drug. The aldolase route (Figure 12.22) has the advantage that again readily available and cheap starting materials can be used and that the biocatalytic reaction is an asymmetric synthesis. The major advantage is that the required two chiral centers are created simultaneously. A possible disadvantage could be the inactivation of the enzyme by the high concentration of acetaldehyde. The route using a three-enzyme system depicted in Figure 12.23 again is an asymmetric synthesis and advantageously the cofactor recycling by GDH could be efficiently integrated into the overall process. Although only one chiral center is created, this route is used to manufacture the Lipitor side chain. A potential disadvantage is that all three enzymes need to be produced and their activity and stability under process conditions must be balanced.

12.19 This is given in detail in Ma *et al.* (2010), Figure 2.

Appendix C: Symbols and Abreviations

a	Surface area per unit volume	(m^{-1})
A_p	Particle outer surface area	(m^2)
A_m	Amplitude in concentration during mixing	$(mol\ l^{-1})$
a	annum: year	
AA	Amino acid(s)	
7-ACA	7-aminocephalosporanic acid	
7-ADCA	7-amino-desacetoxy-cephalosporinic acid	
ADH	Alcohol dehydrogenase	
ADM	Archer Daniels Midland (USA)	
6-APA	6-aminopenicillanic acid	
ATA	Amine transaminase	
BDW	Biomass dry weight	
B-FIT	B-factor iterative test	
BG	ß-glucosidases	
BLAST	Basic local alignment search tool	
BSA	Bovine serum albumin	
BVMO	Baeyer–Villiger monooxygenase	
c	Extent of a reaction in racemate resolutions/Conversion	
c	Concentration	$(mol\ l^{-1})$
$c_S(r)$	Substrate concentration at distance r from the particle center	$(mol\ l^{-1})$
$c_S(\infty)$	Bulk substrate concentration	$(mol\ l^{-1})$
$c_S(z)$	Dimensionless substrate concentration	
CAST	Combinatorial active-site saturation test	
CBH	Cellobiohydrolases	
CBM	Carbohydrate binding module	
CDW	Cell dry weight	
CHMO	Cyclohexanone monooxygenase	
d_i	Stirrer diameter	(m)
d_p	Particle diameter	(m)

Biocatalysts and Enzyme Technology, Second Edition. Klaus Buchholz, Volker Kasche, and Uwe T. Bornscheuer

dNTP	Deoxynucleotide triphosphate	
d_R	Reactor diameter	(m)
D	Diameter (e.g., of a chromatographic column)	(m)
D_p	Pore diffusion coefficient of a molecule in the pores of an adsorbent particle	($cm^2 s^{-1}$)
$D_{p,eff}$	Effective pore diffusion coefficient of a molecule in the pores of an adsorbent particle	($cm^2 s^{-1}$)
D_S	Diffusion coefficient of a substrate molecule in solution	($cm^2 s^{-1}$)
D'_S	Diffusion coefficient of a substrate molecule in the immobilized biocatalyst particle	($cm^2 s^{-1}$)
3D	Three dimensional	
DS	Dry substance content (wt%)	
DE	Directed evolution/Dextrose equivalent	
DMF	Dimethyl formamide	
DNA	Desoxyribonucleic acid	
DOE	Department of Energy (USA)	
E	Enantioselectivity	
EA	Activation energy	
ee	Enantiomeric excess	
E	Enantio- or stereoselectivity/Enantiomeric ratio	
E_{eq}	Enantio- or stereoselectivity for an equilibrium controlled process	
EG	Endo-ß-1,4-glucanases	
EGSB	Expanded granular sludge bed reactor	
E_{kin}	Enantio- or stereoselectivity for a kinetically controlled process	
epPCR	Error-prone polymerase chain reaction	
ERED	Ene reductase	
$E(t)$	Residence time distribution	(s^{-1})
EWG	Electron withdrawing group	
FACS	Fluorescence-activated cell sorting	
FAD	Flavine adenine dinucleotide	
FMN	Flavine adenine mononucleotide	
FOS	Fructooligosaccharides	
FPU	Filter paper units (cellulase activity)	
FTF	Fructosyltransferase, fructansucrase	
Gal	Gallon, 3,78 L	
GC	Gas chromatography	
GLP	Good laboratory practice	
GMP	Good manufacturing practice	
h	Hour	
H	Height	(m)
HETP	Height equivalent to a theoretical plate	
HPLC	High-performance liquid chromatography	

I	Ionic strength	(mol l^{-1})
ISM	Iterative saturation mutagenesis	
(I)SSM	(iterative) site-directed saturation mutagenesis	
k_{cat}, k'_{cat}	Turnover number of a free and immobilized enzyme	(s^{-1})
k_H	Apparent hydrolysis rate constant in a kinetically controlled process	$(\text{l mol}^{-1}\text{ s}^{-1})$
k_i	Apparent first-order rate constant for the inactivation of enzymes	(s^{-1})
K_i	Dissociation constant for the binding of an inhibitor to an enzyme	(mol l^{-1})
k_L	Mass transfer coefficient	(m s^{-1})
K_m, K'_m	Michaelis–Menten constant for a free or immobilized enzyme	(mol l^{-1})
K_{ref}	pH-independent reference equilibrium constant for a hydrolysis reaction	(mol l^{-1})
k_T	Apparent transferase rate constant in a kinetically controlled process	$(\text{l mol}^{-1}\text{ s}^{-1})$
K	Equilibrium constant (always as dissociation constants)	(mol l^{-1})
K_A	Amplitude	(s^{-1})
K_{app}	Apparent dissociation constant for a hydrolysis reaction	(mol l^{-1})
lb	Pound, 0.454 kg	
LHW	Liquid hot water pretreatment	
M	Concentration	(mol l^{-1})
MS	Mass spectrometry	
MTEA	Tris-(2-hydroxyethyl)-methylammonium methylsulfate	
MTP	Microtiter plate	
MW	Molecular weight	(Da)
n	Stirrer speed	(s^{-1})
n	Static capacity of a chromatographic adsorbent (Chapter 6) or number of particles per unit volume (Chapter 10)	(mol l^{-1}), (l^{-1})
n_E	Concentration of active immobilized enzyme	(mol l^{-1})
n_0	Molar amount of a compound at $t = 0$	(mol)
$n\cdot$	Molar flow rate	(mol s^{-1})
Ne	Newton number	
NMR	Nuclear magnetic resonance	
odm	Organic dry matter	
OS	Oligosaccharide(s)	
OYE	Old yellow enzyme	
p	Partition coefficient	
p	Pressure	(Pa)
P	Productivity (kg product/kg biocatalyst)	
P	Permeability	(m s^{-1})
P_R	Power input	(W s^{-1})
[P]	Product concentration	(mol l^{-1})

PCR	Polymerase chain reaction	
PDC	Pyruvate decarboxylase	
PFE	*Pseudomonas fluorescence* esterase	
PLP	Pyridoxal-5′-phosphate	
PMP	Pyridoxamine phosphate	
Pound	0.454 kg	
ProSAR	Protein sequence activity relationships	
Q	Volumetric flow rate	($m^3\ s^{-1}$)
QM	Quantum mechanical	
r	Distance from particle center	(m)
Re	Reynolds number	
R	Particle or cell radius	(m)
RD	Rational design	
RM	Random mutagenesis	
RSF	Renewable Fuel Standard, USA	
Sh	Sherwood number	
Sc	Schmidt number	
$[S]_E$	Substrate concentration at the end point of a reaction	($mol\ l^{-1}$)
$[S]_0$	Substrate concentration at $t = 0$	($mol\ l^{-1}$)
$[S]_i$	Substrate concentration after the reactor i	($mol\ l^{-1}$)
$[S]_t$	Substrate concentration at t	($mol\ l^{-1}$)
SMB	Simulated moving bed	
(S)RD	(semi-) rational design	
SSF	Simultaneous saccharification and fermentation	
SSM	Site-directed saturation mutagenesis	
StEP	Staggered extension process	
STY	Space–time yield	(mol/(l s)) or (g/ (l s))
t	Time	(s)
t_{12}	Time required for the substrate conversion from $[S]_1$ to $[S]_2$	(s)
$t_{1/2}$	Half-lifetime	(s)
t_m	Mixing time	(s)
T	Temperature	(K, °C)
T_M	Melting temperature	
TA	Transaminase	
TAG	Structured triglycerides	
t/d	Tons per day	
TTN	Total turnover number	
u	Linear flow rate	($m\ s^{-1}$)
U	Unit for enzyme activity	($\mu mol\ min^{-1}$)
UASB	Upflow anaerobic sludge blanket reactor	
USDA	US Department of Agriculture	

v, V_{max}, V'_{max}	Reaction rate, maximal reaction rate of an enzyme catalyzed reaction with free or immobilized enzyme	(mol/(l s))
v_{obs}	Observed rate of a reaction catalyzed by a biocatalyst	(mol/(l s))
V	Reactor volume	(m^3)
V_m	Mobile phase volume of a chromatographic column	(m^3)
V_p	Particle (cell) volume	(m^3)
V_{CST}, V_{PB}	Volume of a continuous stirred tank (CST) or packed bed (PB) reactor	(m^3)
V_s	Stationary phase volume of a chromatographic column	(m^3)
v/v	Volume per volume	
WIS	Water-insoluble solids	
WT	Wild type (organism)	
wt %	Weight %	
x	Cells per volume unit	(l^{-1})
X	Degree of substrate conversion $\Delta[[S]/[S]_0 = ([S]_0 - [S]_t)/[S]_0)$	
y	Year	
z	Dimensionless distance from particle center (r/R_p)	
α	Volumetric fraction (immobilized biocatalyst, solid) in a reactor	
γ	Relative substrate concentration $= [S]/K_m$	
Δ	Thickness of diffusion layer	(m)
φ	Thiele modulus	
ε	Porosity of a packed bed	
ε_p	Particle porosity	
η	Stationary effectiveness factor	
η_o	Operational effectiveness factor	
η	Viscosity	(kg m^{-1} s^{-1})
ρ	Density	(kg m^{-3})
σ^2	Variance of the circulation distribution	
τ	Residence time	(s)
τ'	Circulation time	(s)
τ_h	Hydrodynamic residence time	(s)
ν	Kinematic viscosity	(m^2 s^{-1})

Index

Biocatalysts and Enzyme Technology, Second Edition. Klaus Buchholz, Volker Kasche, and Uwe T. Bornscheuer

d

n

t